Pearson New International Edition

Systems Engineering and Analysis
Blanchard Fabrycky
Fifth Edition

Pearson Education Limited
Edinburgh Gate
Harlow
Essex CM20 2JE
England and Associated Companies throughout the world

Visit us on the World Wide Web at: www.pearsoned.co.uk

© Pearson Education Limited 2014

 ISBN 10: 1-292-02597-2
ISBN 13: 978-1-292-02597-1

British Library Cataloguing-in-Publication Data
A catalogue record for this book is available from the British Library

Printed and bound by CPI Group (UK) Ltd, Croydon, CR0 4YY

Table of Contents

Systems Science and Engineering

From Chapter 1 of *Systems Engineering and Analysis,* Fifth Edition, Benjamin S. Blanchard, Wolter J. Fabrycky. Copyright © 2011 by Pearson Education, Inc. Published by Pearson Prentice Hall. All rights reserved.

1

Systems Science and Engineering

Systems are as pervasive as the universe in which they exist. They are as grand as the universe itself or as infinitesimal as the atom. Systems appeared first in natural forms, but with the advent of human beings, a variety of human-made systems have come into existence. In recent decades, we have begun to understand the underlying structure and characteristics of natural and human-made systems in a scientific way.

In this chapter, some system definitions and systems science concepts are presented to provide a basis for the study of systems engineering and analysis. They include definitions of system characteristics, a classification of systems into various types, consideration of the current state of systems science, and a discussion of the transition to the Systems Age. Finally, the chapter presents technology and the nature and role of engineering in the Systems Age and ends with a number of commonly accepted definitions of systems engineering.

Upon completion of this chapter, the reader will have obtained essential insight into systems and systems thinking, with an orientation toward systems engineering and analysis. The system definitions, classifications, and concepts presented in this chapter are intended to impart a general understanding about the following:

- System classifications, similarities, and dissimilarities;
- The fundamental distinction between natural and human-made systems;
- The elements of a system and the position of the system in the hierarchy of systems;
- The domain of systems science, with consideration of cybernetics, general systems theory, and systemology;
- Technology as the progenitor for the creation of technical systems, recognizing its impact on the natural world;
- The transition from the machine or industrial age to the Systems Age, with recognition of its impact upon people and society;
- System complexity and scope and the demands these factors make on engineering in the Systems Age; and
- The range of contemporary definitions of systems engineering used within the profession.

The final section of this chapter provides a summary of the key concepts and ideas pertaining to systems science and engineering. It is augmented with selected references and website addresses recommended for further inquiry.

1 SYSTEM DEFINITIONS AND ELEMENTS

A *system* is an assemblage or combination of functionally related elements or parts forming a unitary whole, such as a river system or a transportation system. Not every set of items, facts, methods, or procedures is a system. A random group of items in a room would constitute a set with definite relationships between the items, but it would not qualify as a system because of the absence of functional relationships. This text deals primarily with systems that include physical elements and have useful purposes, including systems associated with all kinds of products, structures, and services, as well as those that consist of a coordinated body of methods or a complex scheme or plan of procedure.[1]

1.1 The Elements of a System

Systems are composed of components, attributes, and relationships. These are described as follows:

1. *Components* are the parts of a system.
2. *Attributes* are the properties (characteristics, configuration, qualities, powers, constraints, and state) of the components and of the system as a whole.
3. *Relationships* between pairs of linked components are the result of engineering the attributes of both components so that the pair operates together effectively in contributing to the system's purpose(s).

The state is the situation (condition and location) at a point in time of the system, or of a system component, with regard to its attributes and relationships. The situation of a system may change over time in only certain ways, as in the *on* or *off* state of an electrical switching system. A connected series of changes in the state over time comprise a *behavior*. The set of all behaviors with their relative sequence and timing comprise the *process*. The process of a component may control the process of another component.

A system is a set of interrelated components functioning together toward some common objective(s) or purpose(s). The set of components meets the following requirements:

1. The properties and behavior of each component of the set have an effect on the properties and behavior of the set as a whole.
2. The properties and behavior of each component of the set depend on the properties and behavior of at least one other component in the set.
3. Each possible subset of components meets the two requirements listed above; the components cannot be divided into independent subsets.

[1]This definition-like description was influenced by the *Random House Webster's Unabridged Dictionary*, 2nd ed. (New York: Random House, Inc., 2001).

The previous requirements ensure that the set of components constituting a system always has some property, or behavior pattern, that cannot be exhibited by any of its subsets acting alone. A system is more than the sum of its component parts. However, the components of a system may themselves be systems, and every system may be part of a larger system in a hierarchy.

When designing a system, the objective(s) or purpose(s) of the system must be explicitly defined and understood so that system components may be engineered to provide the desired function(s), such as a desired output for each given set of inputs. Once defined, the objective(s) or purpose(s) make possible the derivation of measures of effectiveness indicating how well the system performs. Achieving the intended purpose(s) of a human-made system and defining its measures of effectiveness are usually challenging tasks.

The purposeful action performed by a system is its *function*. A common system function is that of altering material, energy, or information. This alteration embraces input, process, and output. Some examples are the materials processing in a manufacturing system or a digestive system, the conversion of coal to electricity in a power plant system, and the information processing in a computer system or a customer service system.

Systems that alter material, energy, or information are composed of structural components, operating components, and flow components. *Structural components* are the static parts; *operating components* are the parts that perform the processing; and *flow components* are the material, energy, or information being altered. A motive force must be present to provide the alteration within the restrictions set by structural and operating components.

System components have *attributes* that determine the component's contribution to the system's function. Examples of component attributes include the color of an automobile (a characteristic), the strength of a steel beam (a quality), the number and arrangement of bridge piers (a configuration), the capacitance of an electrical circuit (a power), the maximum speed permitted by the governor of a turbine (a constraint), and whether or not a person is talking on the telephone (a state). An example of a system-level attribute is the length of runway required by an aircraft for takeoff and landing. The runway length requirement is determined by the attributes and relationships of the aircraft as a component and by the configuration attributes of the air transportation system.

A single relationship exists between two and only two components based on their attributes. The two components are directly connected in some way, though they are not necessarily physically adjacent. In a system with more than two components, at least one of the components in the relationship also has at least one relationship with some other component. Each component in a relationship provides something that the other component needs so that it can contribute to the system's function. In order to form a relationship of maximum effectiveness, the attributes of each component must be engineered so that the collaborative functioning of the two components is optimized.

Relationships that are functionally necessary to both components may be characterized as *first order*. An example is symbiosis, the association of two unlike organisms for the benefit of each other. *Second-order* relationships, called *synergistic*, are those that are complementary and add to system performance. *Redundancy* in a system exists when duplicate components are present for the purpose of assuring continuation of the system function in case of component failure.

1.2 Systems and Subsystems

The definition of a system is not complete without consideration of its position in the hierarchy of systems. Every system is made up of *components*, and many components can be broken down into smaller components. If two hierarchical levels are involved in a given system, the lower is conveniently called a *subsystem*. For example, in an air transportation system, the aircraft, control tower, and terminals are subsystems. Equipment, people, and software are components. The designation of system, subsystem, and component are relative, because the system at one level in the hierarchy is the subsystem or component at another.

In any particular situation, it is important to define the system under consideration by specifying its limits, boundaries, or scope. Everything that remains outside the boundaries of the system is considered to be the *environment*. However, no system is completely isolated from its environment. Material, energy, and/or information must often pass through the boundaries as *inputs* to the system. In reverse, material, energy, and/or information that pass from the system to the environment are called *outputs*. That which enters the system in one form and leaves the system in another form is usually called *throughput*.

The total system, at whatever level in the hierarchy, consists of all components, attributes, and relationships needed to accomplish one or more objectives. Each system has objective(s) providing purpose(s) for which all system components, attributes, and relationships have been organized. Constraints placed on the system limit its operation and define the boundary within which it is intended to operate. Similarly, the system places boundaries and constraints on its subsystems.

An example of a total system is a fire department. The subsystems of this "fire control system" are the building, the fire engines, the firefighters with personal equipment, the communication equipment, and maintenance facilities. Each of these subsystems has several contributing components. At each level in the hierarchy, the description must include all components, all attributes, and all relationships.

Systems thinking and the systems viewpoint looks at a system from the top down rather than from the bottom up. Attention is first directed to the system as a black box that interacts with its environment. Next, attention is focused on how the smaller black boxes (subsystems) combine to achieve the system objective(s). The lowest level of concern is then with individual components.

The process of bringing systems into being and of improving systems already in existence, in a holistic way, is receiving increasing attention. By bounding the total system for study purposes, the systems engineer or analyst will be more likely to obtain a satisfactory result. The focus on systems, subsystems, and components in a hierarchy forces consideration of the pertinent functional relationships. Components and attributes are important, but only to the end that the purpose of the whole system is achieved through the functional relationships linking them.

2 A CLASSIFICATION OF SYSTEMS

Systems may be classified for convenience and to provide insight into their wide range. In this section, classification will be accomplished by several dichotomies conceptually contrasting system similarities and dissimilarities. Descriptions are given of natural and human-made systems, physical and conceptual systems, static and dynamic systems, and closed and open systems.[2]

---

[2]The classifications in this section are only some of those that could be presented. All system types have embedded information flow components and, therefore, information systems are not included as a separate classification.

2.1 Natural and Human-Made Systems

The origin of systems gives the most important classification opportunity. *Natural systems* are those that came into being by natural processes. *Human-made systems* are those in which human beings have intervened through components, attributes, and relationships. A *human-modified system* is a natural system into which a human-made system has been integrated as a subsystem.

All human-made systems, when brought into being, are embedded into the natural world. Important interfaces often exist between human-made systems and natural systems. Each affects the other in some way. The effect of human-made systems on the natural world has only recently become a keen subject for study by concerned people, especially in those instances where the effect is undesirable. In some cases, this study is facilitated by analyzing the natural system as a human-modified system.

When designing a human-made system, undesirable effects can be minimized—and the natural system can sometimes be improved—by engineering the larger human-modified system instead of engineering only the human-made system. If analysis, evaluation, and validation of the human-modified system are appropriate, then the boundary of the environmental system—drawn to include the human-made system—should be considered the boundary of the human-modified system.

Natural systems exhibit a high degree of order and equilibrium. This is evidenced in the seasons, the food chain, the water cycle, and so on. Organisms and plant life adapt themselves to maintain equilibrium with the environment. Every event in nature is accompanied by an appropriate adaptation, one of the most important being that material flows are cyclic. In the natural environment there are no dead ends, no wastes, only continual recirculation and regeneration.

Only recently have significant human-made systems appeared. These systems make up the human-made world, their chief engineer being human. The rapid evolution of human beings is not adequately understood, but their arrival has significantly affected the natural world, often in undesirable ways. Primitive beings had little impact on the natural world, for they had not yet developed potent and pervasive technologies.

An example of the impact of human-made systems on natural systems is the set of problems that arose from building the Aswan Dam on the Nile River. Construction of this massive dam ensures that the Nile will never flood again, solving an age-old problem. However, several new problems arose. The food chain was broken in the eastern Mediterranean, thereby reducing the fishing industry. Rapid erosion of the Nile Delta took place, introducing soil salinity into Upper Egypt. No longer limited by periodic dryness, the population of bilharzia (a waterborne snail parasite) has produced an epidemic of disease along the Nile. These side effects were not adequately anticipated by those responsible for the project. A systems view encompassing both natural and human-made elements, as a human-modified system, might have led to a better solution to the problem of flooding.

2.2 Physical and Conceptual Systems

Physical systems are those that manifest themselves in physical form. They are composed of real components and may be contrasted with *conceptual systems*, where symbols represent the attributes of components. Ideas, plans, concepts, and hypotheses are examples of conceptual systems.

A physical system consumes physical space, whereas conceptual systems are organizations of ideas. One type of conceptual system is the set of plans and specifications for a physical system before it is actually brought into being. A proposed physical system may be simulated in the abstract by a mathematical or other conceptual model. Conceptual systems often play an essential role in the operation of physical systems in the real world.

The totality of elements encompassed by all components, attributes, and relationships focused on a given result employ a process that determines the state changes (behaviors) of a system. A process may be mental (thinking, planning, and learning), mental-motor (writing, drawing, and testing), or mechanical (operating, functioning, and producing). Processes exist in both physical and conceptual systems.

Process occurs at many different levels within systems. The subordinate process essential for the operation of a total system is provided by the subsystem. The subsystem may, in turn, depend on more detailed subsystems. System complexity is the feature that defines the number of subsystems present and, consequently, the number of processes involved. A system may be bounded for the purpose of study at any process or subsystem level. Also, related systems that are normally analyzed individually may be studied as a group, and the group is often called a *system-of-systems* (SOS).

2.3 Static and Dynamic Systems

Another system dichotomy is the distinction of static and dynamic types. A *static system* is one whose states do not change because it has structural components but no operating or flow components, as exemplified by a bridge. A *dynamic system* exhibits behaviors because it combines structural components with operating and/or flow components. An example is a school, combining a building, students, teachers, books, curricula, and knowledge.

A dynamic conception of the universe has become a necessity. Yet, a general definition of a system as an ongoing process is incomplete. Many systems would not be included under this broad definition because they lack operating and flow components. A highway system is static, yet it contains the system elements of components, attributes, and functional relationships.

A system is static only in a limited frame of reference. A bridge system is constructed, maintained, and altered over a period of time. This is a dynamic process conducted by a construction subsystem operating on a flow of construction materials. A structural engineer must view the bridge's members as operating components that expand and contract as they experience temperature changes.

A static system serves a useful purpose only as a component or subsystem of a dynamic system. For example, a static bridge is part of a dynamic system with an overpass operating component processing a traffic flow component and with an underpass component handling water or traffic flow.

Systems may be characterized as having random properties. In almost all systems in both the natural and human-made categories, the inputs, process, and output can only be described in statistical terms. Uncertainty often occurs in both the number of inputs and the distribution of these inputs over time. For example, it is difficult to predict exactly the number of passengers who will check in for a flight, or the exact time each will arrive at the airport. However, because these factors can be described in terms of probability distributions, system operation may be considered *probabilistic* in its behavior.

For centuries, humans viewed the universe of phenomena as immutable and unchanging. People habitually thought in terms of certainties and constants. The substitution of a process-oriented description for the static description of the universe, and the idea that almost anything can be improved, distinguishes modern science and engineering from earlier thinking.

2.4 Closed and Open Systems

A *closed system* is one that does not interact significantly with its environment. The environment provides only a context for the system. Closed systems usually exhibit the characteristic of equilibrium resulting from internal rigidity that maintains the system in spite of influences from the environment. An example is the chemical equilibrium eventually reached in a closed vessel when various reactants are mixed together, provided that the reaction does not increase pressure to the point that the vessel explodes. The reaction and pressure can be predicted from a set of initial conditions. Closed systems involve deterministic interactions, with a one-to-one correspondence between initial and final states. There are relatively few closed systems in the natural and the human-made world.

An *open system* allows information, energy, and matter to cross its boundaries. Open systems interact with their environment, examples being plants, ecological systems, and business organizations. They exhibit the characteristics of *steady state*, wherein a dynamic interaction of system elements adjusts to changes in the environment. Because of this steady state, open systems are self-regulatory and often self-adaptive.

It is not always easy to classify a system as either open or closed. Open systems are typical of those that have come into being by natural processes. Human-made systems have both open and closed characteristics. They may reproduce natural conditions not manageable in the natural world. They are closed when designed for invariant input and statistically predictable output, as in the case of a spacecraft in flight.

Both closed and open systems exhibit the property of entropy. *Entropy* is defined here as the degree of disorganization in a system and is analogous to the use of the term in thermodynamics. In the thermodynamic usage, entropy is the energy unavailable for work resulting from energy transformation from one form to another.

In systems, increased entropy means increased disorganization. A decrease in entropy occurs as order occurs. Life represents a transition from disorder to order. Atoms of carbon, hydrogen, oxygen, and other elements become arranged in a complex and orderly fashion to produce a living organism. A conscious decrease in entropy must occur to create a human-made system. All human-made systems, from the most primitive to the most complex, consume entropy because they involve the creation of more orderly states from less orderly states.

3 SCIENCE AND SYSTEMS SCIENCE

The significant accumulation of scientific knowledge, which began in the eighteenth century and rapidly expanded in the twentieth, made it necessary to classify what was discovered into scientific disciplines. Science began its separation from philosophy almost two centuries ago. It then proliferated into more than 100 distinct disciplines. A relatively recent unifying development is the idea that systems have general characteristics, independent of the area of science to which they belong. In this section, the evolution of a science of systems is presented through an examination of cybernetics, general systems theory, and systemology.

3.1 Cybernetics

The word *cybernetics* was first used in 1947 by Norbert Wiener, but it is not explicitly defined in his classical book.[3] Cybernetics comes from the Greek word meaning "steersman" and is a cognate of "governor." In its narrow view, cybernetics is equivalent to servo theory in

---

[3]N. Wiener, *Cybernetics* (New York: John Wiley & Sons, Inc., 1948). Also see H. S. Tsien, *Engineering Cybernetics* (New York: McGrew Hill Book Co., 1954).

engineering. In its broad view, it may encompass much of natural science. Cybernetics has to do with self-regulation, whether mechanical, electromechanical, electrical, or biological.

The concept of feedback is central to cybernetic theory. All goal-seeking behavior is controlled by the feedback of corrective information about deviation from a desired state. The best known and most easily explained illustration of feedback is the action of a thermostat. The thermometer component of a thermostat senses temperature. When the actual temperature falls below that set into the thermostat, an internal contact is made, activating the heating system. When the temperature rises above that set into the thermostat, the contact is broken, shutting off the heating system.

Biological organisms are endowed with the capacity for self-regulation, called *homeostasis*. The biological organism and the physical world are both very complex. And, analogies exist between them and human-made systems. Through these analogies humans have learned some things about their properties that might have not been learned from the study of natural systems alone. As people develop even more complex systems, we will gain a better understanding of how to control them and our environment.

The science of cybernetics has made three important contributions to the area of regulation and control. First, it stresses the concept of information flow as a distinct system component and clarifies the distinction between the activating power and the information signal. Second, it recognizes that similarities in the action of control mechanisms involve principles that are fundamentally identical. Third, the basic principles of feedback control are subject to mathematical treatment.

A practical application of cybernetics has been the remarkable development of automatic equipment and automated processes, most controlled by microprocessors. However, its significance is greater than this technological contribution. The science of cybernetics is important not only for the control engineer but also for the purest of scientists. Cybernetics is the science of purposeful and optimal control found in complex natural processes and applicable in society as well as commercial enterprises.

3.2 General Systems Theory

An even broader unifying concept than cybernetics evolved during the late 1940s. It was the idea that basic principles common to all systems could be found that go beyond the concept of control and self-regulation. A unifying principle for science and a common ground for interdisciplinary relationships needed in the study of complex systems were being sought. Ludwig von Bertalanffy used the phrase *general systems theory* around 1950 to describe this endeavor.[4] A related contribution was made by Kenneth Boulding.[5]

General systems theory is concerned with developing a systematic framework for describing general relationships in the natural and the human-made world. The need for a general theory of systems arises out of the problem of communication among various disciplines. Although the scientific method brings similarity between the methods of approach, the results are often difficult to communicate across disciplinary boundaries. Concepts and hypotheses formulated in one area seldom carry over to another, where they could lead to significant forward progress.

One approach to an orderly framework is the structuring of a hierarchy of levels of complexity for individual systems studied in the various fields of inquiry. A *hierarchy of levels*

---

[4]L. von Bertalanffy, "General System Theory: A New Approach to Unity of Science," *Human Biology*, December 1951.

[5]K. Boulding, "General Systems Theory: The Skeleton of Science," *Management Science*, April 1956.

can lead to a systematic approach to systems that has broad application. Boulding suggested such a hierarchy. It begins with the simplest level and proceeds to increasingly complex levels that usually incorporate the capabilities of all the previous levels, summarized approximately as follows:

1. The level of static structure or *frameworks*, ranging from the pattern of the atom to the anatomy of an animal to a map of the earth to the geography of the universe.
2. The level of the simple dynamic system, or *clockworks*, adding predetermined, necessary motions, such as the pulley, the steam engine, and the solar system.
3. The level of the *thermostat* or cybernetic system, adding the transmission and interpretation of information.
4. The level of the *cell*, the open system where life begins to be evident, adding self-maintenance of structure in the midst of a throughput of material.
5. The level of the *plant*, adding a genetic-societal structure with differentiated and mutually dependent parts, "blueprinted" growth, and primitive information receptors.
6. The level of the *animal*, adding mobility, teleological behavior, and self-awareness using specialized information receptors, a nervous system, and a brain with a knowledge structure.
7. The level of the *human*, adding self-consciousness; the ability to produce, absorb, and interpret symbols; and understanding of time, relationship, and history.
8. The level of *social organization*, adding roles, communication channels, the content and meaning of messages, value systems, transcription of images into historical record, art, music, poetry, and complex human emotion.
9. The level of the *transcendental system*, adding the ultimates and absolutes and unknowables.

The first level in Boulding's hierarchy is the most pervasive. Static systems are everywhere, and this category provides a basis for analysis and synthesis of systems at higher levels. Dynamic systems with predetermined outcomes are predominant in the natural sciences. At higher levels, cybernetic models are available, mostly in closed-loop form. Open systems are currently receiving scientific attention, but modeling difficulties arise regarding their self-regulating properties. Beyond this level, there is little systematic knowledge available. However, general systems theory provides science with a useful framework within which each specialized discipline may contribute. It allows scientists to compare concepts and similar findings, with its greatest benefit being that of communication across disciplines.

3.3 Systemology and Synthesis

The science of systems or their formation is called *systemology*. Problems and problem complexes faced by humankind are not organized along disciplinary lines. A new organization of scientific and professional effort based on the common attributes and characteristics of problems will likely accelerate beneficial progress. As systems science is promulgated by the formation and acceptance of interdisciplines, humankind will benefit from systemology and systems thinking.

Disciplines in science and the humanities developed largely by what society permitted scientists and humanists to investigate. Areas that provided the least challenge to cultural, social, and moral beliefs were given priority. The survival of science was also of concern in the progress of certain disciplines. However, recent developments have added to the acceptance of a scientific approach in most areas. Much credit for this can be given to the recent respectability of interdisciplinary inquiry. One of the most important contributions of systemology is that it offers a single vocabulary and a unified set of concepts applicable to many types of systems.

During the 1940s, scientists of established reputation in their respective fields accepted the challenge of attempting to understand a number of common processes in military operations. Their team effort was called *operations research*, and the focus of their attention was the optimization of operational military systems. After the war, this interdisciplinary area began to take on the attributes of a discipline and a profession. Today a body of systematic knowledge exists for both military and commercial operations. But operations research is not the only science of systems available today. Cybernetics, general systems research, organizational and policy sciences, management science, and the information sciences are others.

Formation of interdisciplines began in the middle of the last century and has brought about an evolutionary synthesis of knowledge. This has occurred not only within science but also between science and technology and between science and the humanities. The forward progress of systemology in the study of large-scale complex systems requires a synthesis of science and the humanities as well as a synthesis of science and technology. Synthesis, sometimes referred to as an interdisciplinary discipline, is the central activity of people often considered to be *synthesists*.

4 TECHNOLOGY AND TECHNICAL SYSTEMS

Technology is broadly defined as the branch of knowledge that deals with the mechanical and industrial arts, applied science, and engineering, or the sum of the ways in which social groups provide themselves with the material objects and the services of their civilization.[6] A key attribute of a civilization is the inherent ability and the knowledge to maintain its store of technology. Modern civilizations possess pervasive and potent technology that makes possible needed systems manifested as products, which include structures and services.

4.1 Technology and Society

Human society is characterized by its culture. Each human culture manifests itself through the media of technology. In turn, the manifestation of culture is an important indicator of the degree to which a society is technologically advanced.

The entire history of humankind is closely related to the progress of technology. But, technological progress is often stressful on people and their cultures alike. This need not be. The challenge should be to find ways for humans to live better lives as a result of new technological capability and social organizational structure.

In general, the complexity of systems desired by societies is increasing. As new technologies become available, the pull of "want" is augmented by a push to incorporate these new capabilities into both new and existing systems. The desire for better systems produces an ever-changing set of requirements. The identification of the "true" need in answer to a problem and the elicitation of "real" requirements is, in itself, a technological challenge.

Transition from the past to present and future technological states is not a one-step process. Continuing technical advances become available to society as time unfolds. Societal

---

[6]This definition was adapted from the *Random House Webster's Unabridged Dictionary*, 2nd ed. (New York: Random House, Inc., 2001).

response is often to make one transition and then to adopt a static pattern of behavior. A better response would be to seek new and well-thought-out possibilities for continuous advancement.

Improvement in technological literacy should increase the population of individuals capable of participating in this desirable activity. One key to imparting this literacy is the communication technologies now expanding at a rapid pace. Thus, technology in this sphere may act favorably to aid the understanding and subsequent evaluation by society of technologies in other spheres.

4.2 Technical Systems[7]

Science and technology is a phrase used often. Translated into its systems counterpart, this phrase prompts consideration of the link between systems science and technical systems. Technical or engineered systems have their foundation in both the natural and the systems sciences. They are a prominent and pervasive sector of the human-made world.

The phrase *technical system* may be used to represent all types of human-made artifacts, including technical products and processes. Accordingly, the technical system is the subject of the collection of activities that are performed by engineers within the processes of engineering design, including generating, retrieving, processing, and transmitting product information. It is also the subject of production processes, including work preparation and planning. It is also the subject of many economic considerations, both within enterprises and in society.

In museums, thousands of technical objects are on display, and they are recognized as products of technology. Their variety of functions, form, size, and so forth tends to obscure common properties and features. But vast variety also exists in nature, and in those circumstances clearly defined kingdoms of natural objects have been defined for study in the natural sciences. Likewise, attempts have been made to define terms that conceptually describe classes of technical objects.

Technical objects can be referred to simply as objects, entities, things, machines, implements, products, documents, or technical works. The results of a manufacturing activity, as the conceptual content of technology, can be termed *artifacts* or *instrumentum*. Such definitions are meant to include all manner of machines, appliances, implements, structures, weapons, and vessels that represent the technical means by which humans achieve their ends. But, to be complete, this definition must recognize the hierarchical nature of systems and the interactions that occur between levels in the hierarchy. For example, the "system" of interest may be a transportation system, an airline system within the transportation system, or an aircraft system contained within the airline system.

Little difficulty exists in the classification of systems as either natural, technical (human-made), or human-modified. But it is difficult to classify technical systems accurately. One approach is to classify in accordance with the well-established subdivisions of technology in industry, for example, civil engineering, electrical engineering, and mechanical engineering. However, from a practical and organizational viewpoint, this does not permit a precise definition of a mechanical system or electrical system because no firm boundary can be established by describing these systems as outcomes of mechanical or electrical engineering.

Modern developments of technical systems have generally blurred the boundaries. Electronic and computer products, especially software, are increasingly used together with mechanical and human interfaces. Each acts as a subsystem to a system of greater complexity and purpose. Most systems in use today are hybrids of the simple systems of the past.

---

[7]This section was adapted from V. Hubka and W. E. Eder, *Theory of Technical Systems* (Berlin: Springer-Verlag, 1988).

4.3 Technological Growth and Change

Technological growth and change is occurring continuously and is initiated by attempts to respond to unmet deficiencies and by attempts to perform ongoing activities in a more effective and efficient manner. In addition, changes are being stimulated by political objectives, social factors, and ecological concerns.

In general, people are not satisfied with the impact of the human-made or technical systems on the natural world and on humankind. Because engineering and the applied sciences are largely responsible for bringing technical systems into being, it is not surprising that there is some dissatisfaction with these fields of endeavor. Accordingly, technical and economic feasibility can no longer be the sole determinants of what engineers do. Ecological, political, social, cultural, and even psychological influences are equally important considerations. The number of factors in any given engineering project has multiplied. Because of the shifts in social attitudes toward moral responsibility, the ethics of personal decisions are becoming a major professional concern. Engineering is not alone in facing up to these newer considerations.

As examples, environmental concerns have resulted in recent legislation and regulations requiring new methods for crop protection from insects, new means for the disposal of medical waste, and new methods for treating solid waste. Concern for shortages of fossil fuel sources as well as ecological impacts brought about a great focus on energy conservation and alternative energy sources. These and other comparable situations were created through both properly planned programs and as a result of panic situations. A common outcome is that all have stimulated beneficial technological innovation.

The world is increasing in complexity because of human intervention. Through the advent of advanced technologies, transportation times have been greatly reduced, and vastly more efficient means of communication have been introduced. Every aspect of human existence has become more intimate and interactive. The need for integration of ideas and conflict resolution becomes more important. At the same time, increasing populations and the desire for larger and better systems is leading to the accelerated exploitation of resources and increased environmental impact. A variety of technically literate specialists, if properly organized and incentivized, can meet most needs that arise from technological advancement and change.

5 TRANSITION TO THE SYSTEMS AGE[8]

Evidence suggests that the advanced nations of the world are leaving one technological age and entering another. It appears that this transition is bringing about a change in the conception of the world in which we live. This conception is both a realization of the complexity of natural and human-made systems and a basis for improvement in humankind's management of these systems. Long-term sustainability of both human-made systems and the natural world is becoming a common desideratum.

5.1 The Machine Age

Two ideas have been dominant in the way people seek to understand the world in which we live. The first is called *reductionism*. It consists of the belief that everything can be reduced, decomposed, or disassembled to simple indivisible parts. These were taken to be atoms in

---

[8]The first part of this section was adapted from R. L. Ackoff, *Redesigning the Future* (New York: John Wiley & Sons, Inc., 1974).

physics; simple substances in chemistry; cells in biology; and monads, instincts, drives, motives, and needs in psychology.

Reductionism gives rise to an analytical way of thinking about the world, a way of seeking explanations and understanding. Analysis consists, first, of taking apart what is to be explained, disassembling it, if possible, down to the independent and indivisible parts of which it is composed; second, of explaining the behavior of these parts; and, finally, of aggregating these partial explanations into an explanation of the whole. For example, the analysis of a problem consists of breaking it down into a set of as simple problems as possible, solving each, and assembling their solutions into a solution of the whole. If the analyst succeeds in decomposing a problem into simpler problems that are independent of each other, aggregating the partial solutions is not required because the solution to the whole is simply the sum of the solutions to its independent parts. In the industrial or *Machine Age*, understanding the world was taken to be the sum, or result, of an understanding of its parts, which were conceptualized as independently of each other as was possible.

The second basic idea was that of *mechanism*. All phenomena were believed to be explainable by using only one ultimately simple relation, cause and effect. One thing or event was taken to be the *cause* of another (its *effect*) if it was both necessary and sufficient for the other. Because a cause was taken to be sufficient for its effect, nothing was required to explain the effect other than the cause. Consequently, the search for causes was environment free. It employed what is now called "closed-system" thinking. Laws such as that of freely falling bodies were formulated so as to exclude environmental effects. Specially designed environments, called *laboratories*, were used so as to exclude environmental effects on phenomena under study.

Causal-based laws permit no exceptions. Effects are completely determined by causes. Hence, the prevailing view of the world was deterministic. It was also mechanistic because science found no need for teleological concepts (such as functions, goals, purposes, choice, and free will) in explaining any natural phenomenon. They considered such concepts to be unnecessary, illusory, or meaningless. The commitment to causal thinking yielded a conception of the world as a machine; it was taken to be like a hermetically sealed clock—a self-contained mechanism whose behavior was completely determined by its own structure.

The *Industrial Revolution* brought about *mechanization*, the substitution of machines for people as a source of physical work. This process affected the nature of work left for people to do. They no longer did all the things necessary to make a product; they repeatedly performed a simple operation in the production process. Consequently, the more machines were used as a substitute for people at work, the more workers were made to behave like machines. The dehumanization of work was an irony of the Industrial Revolution and the Machine Age.

5.2 The Systems Age

Although eras do not have precise beginnings and endings, the 1940s can be said to have contained the beginning of the end of the Machine Age and the beginning of the *Systems Age*. This new age is the result of a new intellectual framework in which the doctrines of reductionism and mechanism and the analytical mode of thought are being supplemented by the doctrines of expansionism, teleology, and a new synthetic (or systems) mode of thought.

Expansionism is a doctrine that considers all objects and events, and all experiences of them, as parts of larger wholes. It does not deny that they have parts, but it focuses on the wholes of which they are part. It provides another way of viewing things, a way that is different from, but compatible with, reductionism. It turns attention from ultimate elements to a whole with interrelated parts—to systems.

Preoccupation with systems brings with it the synthetic mode of thought. In the *analytic* mode, an explanation of the whole was derived from explanations of its parts. In *synthetic* thinking, something to be explained is viewed as part of a larger system and is explained in terms of its role in that larger system. The Systems Age is more interested in putting things together than in taking them apart.

Analytic thinking is outside-in thinking; synthetic thinking is inside-out thinking. Neither negates the value of the other, but by synthetic thinking one can gain understanding that cannot be obtained through analysis, particularly of collective phenomena.

The synthetic mode of thought, when applied to systems problems, is called *systems thinking* or the *systems approach*. This approach is based on the observation that when each part of a system performs as well as possible, the system as a whole may not perform as well as possible. This follows from the fact that the sum of the functioning of the parts is seldom equal to the functioning of the whole. Accordingly, the synthetic mode seeks to overcome the often observed predisposition to perfect details and ignore system outcomes.

Because the Systems Age is *teleologically oriented*, it is preoccupied with systems that are goal seeking or purposeful; that is, systems that offer the choice of either means or ends, or both. It is interested in purely mechanical systems only insofar as they can be used as enablers for purposeful systems. Furthermore, the Systems Age is largely concerned with purposeful systems, some of whose parts are purposeful; in the human domain, these are called *social groups*. The most important class of social groups is the one containing systems whose parts perform different functions that have a division of functional labor; these are called *organizations*.

In the Systems Age, attention is focused on groups and on organizations as parts of larger purposeful societal systems. Participative management, collaboration, group decision making, and total quality management are new working arrangements within the organization. Among organizations is now found a keen concern for social and environmental factors, with economic competition continuing to increase worldwide.

5.3 Engineering in the Systems Age

Engineering activities of analysis and design are not ends in themselves, but are a means for satisfying human wants. Thus, modern engineering has two aspects. One aspect concerns itself with the materials and forces of nature; the other is concerned with the needs of people.

In the Systems Age, successful accomplishment of engineering objectives usually requires a combination of technical specialties and expertise. Engineering in the Systems Age must be a team activity where various individuals involved are cognizant of the important relationships between specialties and between economic factors, ecological factors, political factors, and societal factors. The engineering decisions of today require consideration of these factors in the early stage of system design and development, and the results of such decisions have a definite impact on these factors. Conversely, these factors usually impose constraints on the design process. Thus, technical expertise must include not only the basic knowledge of individual specialty fields of engineering but also knowledge of the context of the system being brought into being.

Although relatively small products, such as a wireless communication device, an electrical household appliance, or even an automobile, may employ a limited number of direct engineering personnel and supporting resources, there are many large-scale systems that require the combined input of specialists representing a wide variety of engineering and related disciplines. An example is that of a ground-based mass-transit system.

Civil engineers are required for the layout and/or design of the railway, tunnels, bridges, and facilities. Electrical engineers are involved in the design and provision of automatic controls, traction power, substations for power distribution, automatic fare collection, digital data systems, and so on. Mechanical engineers are necessary in the design of passenger vehicles and related mechanical equipment. Architectural engineers provide design support for the construction of passenger terminals. Reliability and maintainability engineers are involved in the design for system availability and the incorporation of supportability characteristics. Industrial engineers deal with the production and utilization aspects of passenger vehicles and human components. Test engineers evaluate the system to ensure that all performance, effectiveness, and system support requirements are met. Engineers in the planning and marketing areas are required to keep the public informed, to explain the technical aspects of the system, and to gather and incorporate public input. General systems engineers are required to ensure that all aspects of the system are properly integrated and function together as a single entity.

Although the preceding example is not all-inclusive, it is evident that many different disciplines are needed. In fact, there are some large projects, such as the development of a new aircraft, where the number of engineers needed to perform engineering functions is in the thousands. In addition, the different engineering types often range in the hundreds. These engineers, forming a part of a large organization, must not only be able to communicate with each other but must also be conversant with such interface areas as purchasing, accounting, personnel, and legal.

Another major factor associated with large projects is that considerable system development, production, evaluation, and support are often accomplished at supplier (sometimes known as *subcontractor*) facilities located throughout the world. Often there is a prime producer or contractor who is ultimately responsible for the development and production of the total system as an entity, and there are numerous suppliers providing different system components. Thus, much of the project work and many of the associated engineering functions may be accomplished at dispersed locations, often worldwide.

5.4 Engineering Education for the Systems Age

Engineering education has been subjected to in-depth study every decade or so, beginning with the Mann Report in 1918.[9] The most recent and authoritative study was conducted by the National Academy of Engineering (NAE) and published in 2005 under the title *Educating the Engineer of 2020.*[10]

Although acknowledging that certain basics of engineering will not change, this NAE report concluded that the explosion of knowledge, the global economy, and the way engineers will work will reflect an ongoing evolution that began to gain momentum at the end of the twentieth century. The report gives three overarching trends to be reckoned with by engineering educators, interacting with engineering leaders in government and industry:

1. The economy in which we work will be strongly influenced by the global marketplace for engineering services, evidenced by the outsourcing of engineering jobs, a growing need for interdisciplinary and system-based approaches, demands for new paradigms of customization, and an increasingly international talent pool.

---

[9]C. R. Mann, "A Study of Engineering Education," Bulletin No. 11, The Carnegie Foundation for the Advancement of Teaching, 1918.
[10]National Academy of Engineering, *Educating the Engineer of 2020* (Washington, DC: National Academies Press, 2005).

2. The steady integration of technology in our public infrastructures and lives will call for more involvement by engineers in the setting of public policy and for participation in the civic arena.

3. The external forces in society, the economy, and the professional environment will all challenge the stability of the engineering workforce and affect our ability to attract the most talented individuals to an engineering career.

Continuing technological advances have created an increasing demand for engineers in most fields. But certain engineering and technical specialties will be merged or become obsolete with time. There will always be a demand for engineers who can synthesize and adapt to changes. The astute engineer should be able to detect trends and plan for satisfactory transitions by acquiring knowledge to broaden his or her capability.

6 SYSTEMS ENGINEERING

To this day, there is no commonly accepted definition of systems engineering in the literature. Almost a half-century ago, Hendrick W. Bode, writing on "The Systems Approach" in *Applied Science-Technological Progress*, said that "It seems natural to begin the discussion with an immediate formal definition of systems engineering. However, systems engineering is an amorphous, slippery subject that does not lend itself to such formal, didactic treatment. One does much better with a broader, more loose-jointed approach. Some writers have, in fact, sidestepped the issue entirely by simply saying that 'systems engineering is what systems engineers do.'"[11]

The definition of systems engineering and the systems approach is usually based on the background and experience of the individual or the performing organization. The variations are evident from the following five published definitions:

1. "An interdisciplinary approach and means to enable the realization of successful systems."[12]

2. "An interdisciplinary approach encompassing the entire technical effort to evolve into and verify an integrated and life-cycle balanced set of system people, product, and process solutions that satisfy customer needs. Systems engineering encompasses (a) the technical efforts related to the development, manufacturing, verification, deployment, operations, support, disposal of, and user training for, system products and processes; (b) the definition and management of the system configuration; (c) the translation of the system definition into work breakdown structures; and (d) development of information for management decision making."[13]

3. "The application of scientific and engineering efforts to (a) transform an operational need into a description of system performance parameters and a system configuration through the use of an iterative process of definition, synthesis, analysis, design, test, and evaluation; (b) integrate related technical parameters and ensure compatibility of

---

[11]H. Bode, Report to the Committee on Science and Astronautics, U.S. House of Representatives, Washington, DC, 1967.

[12]International Council on Systems Engineering (INCOSE), 7670 Opportunity Road, Suite 220, San Diego, CA, USA.

[13]EIA/IS 632, "Systems Engineering," Electronic Industries Alliance, 2500 Wilson Boulevard, Arlington, VA, 1994.

all physical, functional, and program interfaces in a manner that optimizes the total system definition and design; and (c) integrate reliability, maintainability, safety, survivability, human engineering, and other such factors into the total engineering effort to meet cost, schedule, supportability, and technical performance objectives."[14]

4. "An interdisciplinary collaborative approach to derive, evolve, and verify a life-cycle balanced system solution which satisfies customer expectations and meets public acceptability."[15]

5. "An approach to translate operational needs and requirements into operationally suitable blocks of systems. The approach shall consist of a top-down, iterative process of requirements analysis, functional analysis and allocation, design synthesis and verification, and system analysis and control. Systems engineering shall permeate design, manufacturing, test and evaluation, and support of the product. Systems engineering principles shall influence the balance between performance, risk, cost, and schedule."[16]

Although the definitions vary, there are some common threads. Basically, systems engineering is good engineering with special areas of emphasis. Some of these are the following:

1. A *top-down* approach that views the system as a whole. Although engineering activities in the past have adequately covered the design of various system components (representing a bottom-up approach), the necessary overview and understanding of how these components effectively perform together is frequently overlooked.

2. A *life-cycle* orientation that addresses all phases to include system design and development, production and/or construction, distribution, operation, maintenance and support, retirement, phase-out, and disposal. Emphasis in the past has been placed primarily on design and system acquisition activities, with little (if any) consideration given to their impact on production, operations, maintenance, support, and disposal. If one is to adequately identify risks associated with the up-front decision-making process, then such decisions must be based on life-cycle considerations.

3. A better and more complete effort is required regarding the initial *definition of system requirements*, relating these requirements to specific design criteria, and the follow-on analysis effort to ensure the effectiveness of early decision making in the design process. The true system requirements need to be well defined and specified and the traceability of these requirements from the system level downward needs to be visible. In the past, the early "front-end" analysis as applied to many new systems has been minimal. The lack of defining an early "baseline" has resulted in greater individual design efforts downstream.

4. An *interdisciplinary* or team approach throughout the system design and development process to ensure that all design objectives are addressed in an effective and efficient manner. This requires a complete understanding of many different design disciplines and their interrelationships, together with the methods, techniques, and tools that can be applied to facilitate implementation of the system engineering process.

[14]DSMC, *Systems Engineering Management Guide*, Defense Systems Management College, Superintendent of Documents, U.S. Government Printing Office, Washington, DC, 1990.
[15]IEEE P1220, "Standard for Application and Management of the Systems Engineering Process," Institute of Electrical and Electronics Engineers, 345 East 47th Street, New York, NY, 1994.
[16]Enclosure 12 on *Systems Engineering* in DOD Instruction 5000.02, "Operation of the Defense Acquisition System," December 8, 2008.

Systems engineering is not a traditional engineering discipline in the same sense as civil engineering, electrical engineering, industrial engineering, mechanical engineering, reliability engineering, or any of the other engineering specialties. It should not be organized in a similar manner, nor does the implementation of systems engineering (or its methods) require extensive resources. However, a well-planned and highly disciplined approach must be followed. The systems engineering process involves the use of appropriate technologies and management principles in a synergetic manner. Its application requires synthesis and a focus on process, along with a new "thought process" that is compatible with the needs of the Systems Age.

7 SUMMARY AND EXTENSIONS

This chapter is devoted to helping the reader gain essential insight into systems in general, and *systems thinking* in particular, with orientation toward the engineering and analysis of technical systems.

System definitions, a discussion of system elements, and a high-level classification of systems provide an opening panorama. It is here that a consideration of the origin of systems provides an orientation to natural and human-made domains as an overarching dichotomy. The importance of this dichotomy cannot be overemphasized in the study and application of systems engineering and analysis. It is the suggested frame of reference for considering and understanding the interface and impact of the human-made world on the natural world and on humans.

Individuals interested in obtaining an in-depth appreciation for this interface and the mitigation of environmental impacts are encouraged to read T. E. Graedel and B. R. Allenby, *Industrial Ecology*, 2nd ed., Prentice Hall, 2003. Also of contemporary interest is the issue of sustainability treated as part of an integrated approach to sustainable engineering by P. Stasinopoulos, M. H. Smith, K. Hargroves, and C. Desha, *Whole System Design*, Earthscan Publishing, 2009. These works are recommended as an extension to this chapter, because they illuminate and address the sensitive interface between the natural and the human-made.

This chapter is also anchored by the domains of *systems science* and *systems engineering*, beginning with the former and ending with the latter. Accordingly, it is important to recognize that at least one professional organization exists for each domain. For systems science, there is the International Society for the System Sciences (ISSS), originally named the "Society for General Systems Research." ISSS was established at the 1956 meeting of the American Association for the Advancement of Science under the leadership of biologist Ludwig von Bertalanffy, economist Kenneth Boulding, mathematician–biologist Anatol Rapoport, neurophysiologist Ralph Gerard, psychologist James Miller, and anthropologist Margaret Mead.

The founders of the International Society for the System Sciences felt strongly that the *systematic* (*holistic*) aspect of reality was being overlooked or downgraded by the conventional disciplines, which emphasize specialization and *reductionist* approaches to science. They stressed the need for more general principles and theories and sought to create a professional organization that would transcend the tendency toward fragmentation in the scientific enterprise. The reader interested in exploring the field of systems science and learning more about the work of the International Society for System Sciences, may visit the ISSS website at *http://www.isss.org*.

Technology, human-made, and human-modified systems comprise the core of this chapter, with science and systems science as the foundation. Accordingly, an in-depth understanding of the engineered system (through a focused definition and description) is not part of this chapter. The purpose is to clarify the distinction between systems that are engineered and systems that exist naturally.

Transition to the Systems Age spawned the systems sciences and, driven by potent technologies, established a compelling need for systems engineering. Accordingly, selected definitions of systems engineering are given at the end of this chapter. Many others exist, but are not included herein. However, systems engineering may be described as a *technologically-based interdisciplinary process* for bringing systems into being. Systems engineering is an *engineering interdiscipline* in its own right, with important engineering domain manifestations.

The most prominent professional organization for systems engineering is the International Council on Systems Engineering (INCOSE). Originally named the "National Council on Systems Engineering," NCOSE was chartered in 1991 in the United States; it has now expanded worldwide to become the leading society to develop, nurture, and enhance the interdisciplinary approach and means to enable the realization of successful systems. INCOSE has strong and enduring ties with industry, academia, and government to achieve the following goals: (1) to provide a focal point for the dissemination of systems engineering knowledge; (2) to promote collaboration in systems engineering education and research; (3) to assure the establishment of professional standards for integrity in the practice of systems engineering; (4) to improve the professional status of persons engaged in the practice of systems engineering; and (5) to encourage governmental and industrial support for research and educational programs that will improve the systems engineering process and its practice. An expanded view of systems engineering, as promulgated through the International Council on Systems Engineering, as well as a window into a wealth of information about this relatively new engineering interdiscipline may be obtained at *http://www.incose.org*.

Most scientific and professional societies in the United States interact and collaborate with cognizant but independent honor societies. The cognizant honor society for systems engineering is the Omega Alpha Association (OAA), emerging under the motto "Think About the End Before the Beginning." Chartered in 2006 as an international honor association, OAA has the overarching objective of advancing the systems engineering process and its professional practice in service to humankind. Among subordinate objectives are opportunities to (1) inculcate a greater appreciation within the engineering profession that every human design decision shapes the human-made world and determines its impact upon the natural world and upon people; (2) advance system design and development morphology through a better comprehension and adaptation of the da Vinci philosophy of thinking about the end before the beginning; that is, determining what designed entities are intended to do before specifying what the entities are and concentrating on the provision of functionality, capability, or a solution before designing the entities per se; and (3) encourage excellence in systems engineering education and research through collaboration with academic institutions and professional societies to evolve robust policies and procedures for recognizing superb academic programs and students. The OAA website, *http://www.omegalpha.org*, provides information about OAA goals and objectives, as well as the OAA vision for recognizing and advancing excellence in systems engineering, particularly in academia.

QUESTIONS AND PROBLEMS

1. Pick a system with which you are familiar and verify that it is indeed a system as per the system definition given at the beginning of Section 1.

2. Name and identify the components, attributes, and relationships in the system you picked in Question 1.

3. Pick a system that alters material and identify its structural components, operating components, and flow components.

4. Select a complex system and discuss it in terms of the hierarchy of systems.

5. Select a complex system and identify some different ways of establishing its boundaries.

6. Identify and contrast a physical and conceptual system.

7. Identify and contrast a static and a dynamic system.

8. Identify and contrast a closed and an open system.

9. Pick a natural system and describe it in terms of components, attributes, and relationships; repeat for a human-made system and for a human-modified system.

10. Identify the purpose(s) of the above human-made system and name some possible measures of worth.

11. For the above human-made system, describe its state at some point in time, describe one of its behaviors, and summarize its process.

12. For the above human-made system, name two components that have a relationship, identify what need each component fills for the other component, and describe how the attributes of these two components must be engineered so that the pair functions together effectively in contributing to the system's purpose(s).

13. Give an example of a first-order relationship, a second-order relationship, and redundance.

14. For a human-modified system, identify some of the ways in which the modified natural system could be degraded and some of the ways in which it could be improved.

15. Give an example of a random dynamic system property and of a steady-state dynamic system property.

16. Give an example of a system that reaches equilibrium and of a system that disintegrates over time.

17. Is a government with executive, legislative, and judicial branches three systems or a single system? Why?

18. Identify a system-of-systems whose analysis could yield insights not available by separately analyzing the individual systems of which it is composed.

19. Explain cybernetics by using an example of your choice.

20. Give a system example at any five of the levels in Boulding's hierarchy.

21. Select a system at one of the higher levels in Boulding's hierarchy and describe if it does or does not incorporate the lower levels.

22. Identify a societal need, define the requirements of a system that would fill that need, and define the objective(s) of that system.

23. What are the similarities between systemology and synthesis?

24. What difficulty is encountered in attempting to classify technical systems?

25. Name some of the factors driving technological advancement and change.

26. What benefits could result from improving systems thinking in society?

27. Identify the attributes of the Machine or Industrial Age and the Systems Age.

28. Explain the difference between analytic and synthetic thinking.

29. What are the special engineering requirements and challenges in the Systems Age?

30. What are the differences (and similarities) between systems engineering and the traditional engineering disciplines?

31. Given the recommendations in *Educating the Engineer of 2020*, what should be added to the curriculum with which you are familiar?

32. Give an example of a problem requiring an interdisciplinary approach and identify the needed disciplines.

33. Name an interdiscipline and identify the disciplines from which it was drawn.

34. Write your own (preferred) definition of systems engineering.

35. Go to the website of ISSS given in Section 7 and summarize the goal of the society.

36. Go to the website of INCOSE given in Section 7 and summarize the goals of the council.

37. Contrast the goals of ISSS and INCOSE as given in Section 7 or on the Web.

38. Go to the website of OAA and compare this honor society with one that you are familiar with.

Bringing Systems Into Being

The world in which we live may be divided into the natural world and the human-made world. Included in the former are all elements of the world that came into being by natural processes. The human-made world is made up of all human-originated systems, their product subsystems (including structures and services), and their other subsystems (such as those for production and support).

Systems engineering and analysis reveals unexpected ways of using technology to bring new and improved systems and products into being that will be more competitive in the global economy. New and emerging technologies are expanding physically realizable design options and enhancing capabilities for developing more cost-effective systems. And, unprecedented improvement possibilities arise from the proper application of the concepts and principles of systems engineering and analysis to legacy systems.

This chapter introduces a technologically based interdisciplinary process encompassing an extension of engineering through all phases of the system life cycle; that is, design and development, production or construction, utilization and support, and phase-out and disposal. Upon completion, this chapter should provide the reader with an in-depth understanding of the following:

- A detailed definition and description for the category of systems that are human-made, in contrast with a definition of general systems;
- The product as part of the engineered system, yet distinguishable from it, with emphasis on the system as the overarching entity to be brought into being;
- Product and system categories with life-cycle engineering and design as a generic paradigm for the realization of competitive products and systems;
- Engineering the relationships among systems to achieve sustainability of the product and the environment, synergy among human-made systems, and continuous improvement of human existence;
- System design evaluation and the multiple-criteria domain within which it is best pursued;
- Integration and iteration in system design, invoking the major activities of synthesis, analysis, and evaluation;

- A morphology for synthesis, analysis, and evaluation and its effective utilization within the systems engineering process;
- The importance of investing in systems thinking and engineering early in the life cycle and the importance thereto of systems engineering management; and
- Potential benefits to be obtained from the proper and timely implementation of systems engineering and analysis.

The final section of this chapter provides a summary and extension of the key concepts and ideas pertaining to the process of bringing systems into being. It is augmented with some annotated references and website addresses identifying opportunities for further inquiry.

1 THE ENGINEERED SYSTEM[1]

The tangible outcome of systems engineering and analysis is an engineered or technical system, whether human-made or human-modified. This section and the material that follows pertain to the organized technological activities for bringing engineered systems into being. To begin on solid ground, it is necessary to define the engineered system in terms of its characteristics.

1.1 Characteristics of the Engineered System

An engineered or technical system is a human-made or human-modified system designed to meet functional purposes or objectives. Systems can be engineered well or poorly. The phrase "engineered system" in this book implies a well-engineered system. A well-engineered system has the following characteristics:

1. Engineered systems have *functional purposes* in response to identified *needs* and have the ability to achieve stated *operational objectives*.
2. Engineered systems are *brought into being* and *operate* over a life cycle, beginning with identification of needs and ending with phase-out and disposal.
3. Engineered systems have design momentum that steadily increases throughout design, production, and deployment, and then decreases throughout phase-out, retirement, and disposal.
4. Engineered systems are composed of a harmonized *combination of resources*, such as facilities, equipment, materials, people, information, software, and money.
5. Engineered systems are composed of *subsystems* and related *components* that *interact* with each other to produce a desired system response or behavior.
6. Engineered systems are part of a *hierarchy* and are influenced by external factors from larger systems of which they are a part and from sibling systems from which they are composed.

---

[1]The phrase "engineering system" is occasionally used. However, it will be used in this text only to refer to the organized activity of technical and supporting people, together with design and evaluation tools and facilities utilized in the process of bringing the engineered system into being.

7. Engineered systems are *embedded* into the natural world and *interact* with it in desirable as well as undesirable ways.

Systems engineering is defined in several ways. Basically, systems engineering is a functionally-oriented, technologically-based interdisciplinary process for bringing systems and products (human-made entities) into being as well as for improving existing systems. The outcome of systems engineering is the engineered system as previously described. Its overarching purpose is to make the world better, primarily for people. Accordingly, human-made entities should be designed to satisfy human needs and/or objectives effectively while minimizing system life-cycle cost, as well as the intangible costs of societal and ecological impacts.

Organization, humankind's most important innovation, is the time-tested means for bringing human-made entities into being. While the main focus is nominally on the entities themselves, systems engineering embraces a better strategy. Systems engineering concentrates on *what the entities are intended to do* before determining *what the entities are composed of*. As simply stated within the profession of architecture, *form follows function*. Thus, instead of offering systems or system elements and products per se, the organizational focus should shift to designing, delivering, and sustaining functionality, a capability, or a solution.

1.2 Product and System Categories

It is interesting and useful to note that systems are often known by their products. They are identified in terms of the products they propose, produce, deliver, or in other ways bring into being. Examples are manufacturing systems that produce products, construction systems that erect structures, transportation systems that move people or goods, traffic control systems that manage vehicle or aircraft flow, maintenance systems that repair or restore, and service systems that meet the need of a consumer or patient. What the system does is manifested through the product it provides. The product and its companion system are inexorably linked.

As frameworks for study, or baselines, two generic product/system categories are presented and characterized in this section. Consideration of these categories is intended to serve two purposes. First, it will help explain and clarify the topics and steps in the process of bringing engineered or technical systems into being. Then, in subsequent chapters, these categories and examples will provide opportunities for look-back reference to generic situations. Although there are other less generic examples in those chapters, greater understanding of them may be imparted by reference to the categories established in this section.

Single-Entity Product Systems. A single-entity product system, for example, may manifest itself as a bridge, a custom-designed home entertainment center, a custom software system, or a unique consulting session. The product may be a consumable (a nonstandard banquet or a counseling session) or a repairable (a highway or a supercomputer). Another useful classification is to distinguish consumer goods from producer goods, the latter being employed to produce the former. A product, as considered in this textbook, is not an engineered system no matter how complex it might be.

The product standing alone is not an engineered system. Consider a bridge constructed to meet the need for crossing an obstacle (a river, a water body, or another roadway). The engineered system is composed of the bridge structure plus a construction subsystem, a maintenance subsystem, an operating and support subsystem, and a phase-out and demolition process. Likewise, an item of equipment for producer or consumer use is not a system within the definition and description of an engineered system given in Section 1.1.

Manufacturing plants that produce repairable or consumable products, warehouses or shopping centers that distribute products, hospital facilities that provide health care services, and air traffic control systems that produce orderly traffic flow are also single-entity systems when the plant, shopping center, or hospital is the product being brought into being. These entities, in combination with appended and companion subsystems, may rightfully be considered engineered systems.

The preceding recognizes the engineered system as more than just the consumable or repairable product, be it a single entity, a population of homogenous entities, or a flow of entities. The product must be treated as part of a system to be engineered, deployed, and operated. Although the product subsystem (including structure or service) directly meets the customer's need, this need must be functionally decomposed and allocated to the subsystems and components comprising the overall system.

Multiple-Entity Population Systems. Multiple-entity populations, often homogenous in nature, are quite common. Thinking of these populations as being aggregated generic products permits them to be studied probabilistically. However, the engineered system is more than just a single entity in the population, or even the entities as a population. It is composed of the population together with the subsystems of production, maintenance and support, regeneration, and phase-out and disposal.

A set of needs provides justification for bringing the population into being. This set of needs drives the life-cycle phases of acquisition and utilization, made up of design, construction or production, maintenance and support, renovation, and eventually ending with phase-out and demolition/disposal. As with single-entity product systems, the product is the subsystem that directly meets the customer's need.

Examples of repairable-entity populations include the following: The airlines and the military acquire, operate, and maintain aircraft with population characteristics. In ground transit, vehicles (such as taxicabs, rental automobiles, and commercial trucks) constitute repairable equipment populations. Production equipment types (such as machine tools, weaving looms, and autoclaves) are populations of equipment classified as producer goods.

Also consider repairable (renovatable) populations of structures, often homogenous in nature. In multi-family housing, a population of structures is composed of individual dwelling units constructed to meet the need for shelter at a certain location. In multi-tenant office buildings, the population of individual offices constitutes a population of renovatable entities. And in urban or suburban areas, public clinics constitute a distributed population of structures to provide health care.

The simplest multi-entity populations are called inventories. These inventories may be made up of consumables or repairables. Examples of consumables are small appliances, batteries, foodstuffs, toiletries, publications, and many other entities that are a part of everyday life. Repairable-entity inventories are often subsystems or components for prime equipment. For example, aircraft hydraulic pumps, automobile starters and alternators, and automation controllers are repairable entities that are components of higher-level systems.

Homogenous populations lend themselves to designs that are targeted to the end product or prime equipment, as well as to the population as a whole. Economies of scale, production and maintenance learning, mortality considerations, operational analyses based on probability and statistics, and so on, all apply to the repairable-entity population to a greater or lesser degree. But the system to be brought into being must be larger in scope than the population itself, if the end result is to be satisfactory to the producer and customer.

1.3 Engineering the Product and the System

People often acquire diverse products to meet specific needs without companion contributing systems to ensure the best overall results, and without adequately considering the effects of the products on the natural world, on humans, and on other human-made systems. Proper application of systems engineering and analysis ensures timely and balanced evaluation of all issues to harmonize overall results from human investments, minimizing problems and maximizing satisfaction.

Engineering has always been concerned with the economical use of limited resources to achieve objectives. The purpose of the engineering activities of design and analysis is to determine how physical and conceptual factors may be altered to create the most utility for the least cost, in terms of product cost, product service cost, social cost, and environmental cost. Viewed in this context, engineering should be practiced in an expanded way, with engineering of the system placed ahead of concern for product components thereof.[2]

Classical engineering focuses on physical factors such as the selection and design of physical components and their behaviors and interfaces. Achieving the best overall results requires focusing initially on conceptual factors, such as needs, requirements, and functions. The ultimate system, however, is manifested physically where the main objective is usually considered to be product performance, rather than the design and development of the overall system of which the product is a part. A product cannot come into being and be sustained without a production or construction capability, without support and maintenance, and so on. Engineering the system and product usually requires an interdisciplinary approach embracing both the product and associated capabilities for production or construction, product and production system maintenance, support and regeneration, logistics, connected system relationships, and phase-out and disposal.

A product may also be a service such as health care, learning modules, entertainment packets, financial services and controls, and orderly traffic flow. In these service examples, the engineered system is a health care system, an educational system, an entertainment system, a financial system, and a traffic control system.

Systems and their associated products are designed, developed, deployed, renewed, and phased out in accordance with processes that are not as well understood as they might be. The cost-effectiveness of the resulting technical entities can be enhanced by placing emphasis on the following:

1. Improving methods for determining the scope of needs to be met by the system. All aspects of a systems engineering project are profoundly affected by the scope of needs, so this determination should be accomplished first. Initial consideration of a broad set of needs often yields a consolidated solution that addresses multiple needs in a more cost-effective manner than a separate solution for each need.

2. Improving methods for defining product and system requirements as they relate to verified customer needs and external mandates. This should be done early in the design phase, along with a determination of performance, effectiveness, and specification of essential system characteristics.

---

[2]According to the definition of engineering adopted by the Accreditation Board for Engineering and Technology (ABET), "Engineering is the profession in which a knowledge of the mathematical and natural sciences gained by study, experience, and practice is applied with judgment to develop ways to utilize economically the materials and forces of nature for the benefit of mankind." This definition is understood herein to encompass both systems and products, with the product often being a structure or service.

3. Addressing the total system with all of its elements from a life-cycle perspective, and from the product to its elements of support and renewal. This means defining the system in functional terms before identifying hardware, software, people, facilities, information, or combinations thereof.

4. Considering interactions in the overall system hierarchy. This includes relationships between pairs of system components, between higher and lower levels within the system hierarchy, and between sibling systems or subsystems.

5. Organizing and integrating the necessary engineering and related disciplines into a top-down systems-engineering effort in a concurrent and timely manner.

6. Establishing a disciplined approach with appropriate review, evaluation, and feedback provisions to ensure orderly and efficient progress from the initial identification of needs through phase-out and disposal.

Any useful system must respond to identified *functional needs*. Accordingly, the elements of a system must not only include those items that relate directly to the accomplishment of a given operational scenario or mission profile but must also include those elements of logistics and maintenance support for use should failure of a prime element(s) occur. To ensure the successful completion of a mission, all necessary supporting elements must be available, in place, and ready to respond. System sustainability can help insure that the system continues meeting the functional needs in a competitive manner as the needs and the competition evolves. And, system sustainability contributes to overall sustainability of the environment.

1.4 Engineering for Product Competitiveness

Product competitiveness is desired by both commercial and public-sector producers worldwide. Thus, the systems engineering challenge is to bring products and systems into being that meet customer expectations cost-effectively.

Because of intensifying international competition, producers are seeking ways to gain a sustainable competitive advantage in the marketplace. Acquisitions, mergers, and advertising campaigns seem unable to create the intrinsic wealth and goodwill necessary for the long-term health of the organization. Economic competitiveness is essential. Engineering with an emphasis on economic competitiveness must become coequal with concerns for advertising, production distribution, finance, and the like.

Available human and physical resources are dwindling. The industrial base is expanding worldwide, and international competition is increasing rapidly. Many organizations are downsizing, seeking to improve their operations, and considering international partners. Competition has reduced the number of suppliers and subcontractors able to respond. This is occurring at a time when the number of qualified team members required for complex system development is increasing. As a consequence, needed new systems are being deferred in favor of extending the life of existing systems.

All other factors being equal, people will meet their needs by purchasing products and services that offer the highest value–cost ratio, subjectively evaluated. This ratio can be increased by giving more attention to the resource-constrained world within which engineering is practiced. To ensure economic competitiveness of the product and enabling system, engineering must become more closely associated with economics and economic feasibility. This is best accomplished through a system life-cycle approach to engineering.

2 SYSTEM LIFE-CYCLE ENGINEERING

Experience over many decades indicates that a properly functioning system that is effective and economically competitive cannot be achieved through efforts applied largely after it comes into being. Accordingly, it is essential that anticipated outcomes during, as well as after, system utilization be considered during the early stages of design and development. Responsibility for *life-cycle engineering,* largely neglected in the past, must become the central engineering focus.

2.1 The Product and System Life Cycles

Fundamental to the application of systems engineering is an understanding of the life-cycle process, illustrated for the product in Figure 1. The product life cycle begins with the identification of a need and extends through conceptual and preliminary design, detail design and development, production or construction, distribution, utilization, support, phase-out, and disposal. The life-cycle phases are classified as *acquisition* and *utilization* to recognize producer and customer activities.[3]

System life-cycle engineering goes beyond the product life cycle viewed in isolation. It must simultaneously embrace the life cycle of the production or construction subsystem, the life cycle of the maintenance and support subsystem, and the life cycle for retirement, phase-out, reuse, and disposal as another subsystem. The overall system is made up of four concurrent life cycles progressing in parallel, as is illustrated in Figure 2. This conceptualization is the basis for *concurrent engineering.*[4]

The need for the product comes into focus first. This recognition initiates conceptual design to meet the need. Then, during conceptual design of the product, consideration should simultaneously be given to its production. This gives rise to a parallel life cycle for a production and/or construction capability. Many producer-related activities are needed to prepare for the production of a product, whether the production capability is a manufacturing plant, construction contractors, or a service activity.

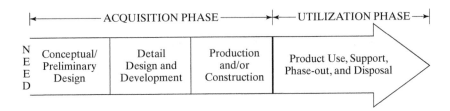

Figure 1   The product life cycle.

---

[3]This classification represents a generic approach. Sometimes the *acquisition* process may involve both the customer (or procuring agency) and the producer (or contractor), whereas utilization may include a combination of contractor and customer (or ultimate user) activities. In some instances, the customer may not be the ultimate user (as is the case in the defense sector) but represents the user's interests during the acquisition process.

[4]*Concurrent engineering* is defined as a systematic approach to creating a system design that simultaneously considers all phases of the life cycle, from conception through disposal, to include consideration of production, distribution, maintenance, phase-out, and so on.

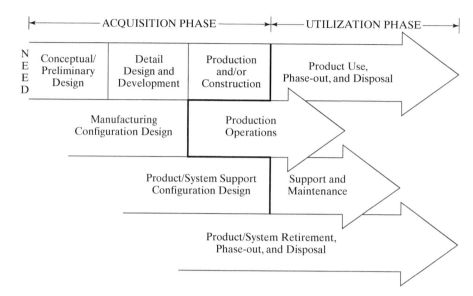

Figure 2 Life cycles of the system.

Also shown in Figure 2 is another life cycle of considerable importance that is often neglected until product and production design is completed. This is the life cycle for support activities, including the maintenance, logistic support, and technical skills needed to service the product during use, to support the production capability during its duty cycle, and to maintain the viability of the entire system. Logistic, maintenance, technical support, and regeneration requirements planning should begin during product conceptual design in a coordinated manner.

As each of the life cycles is considered, design features should be integrated to facilitate phase-out, regeneration, or retirement having minimal impact on interrelated systems. For example, attention to end-of-life recyclability, reusability, and disposability will contribute to environmental sustainability. Also, the system should be made ready for regeneration by anticipating and addressing changes in requirements, such as increases in complexity, incorporation of planned new technology, likely new regulations, market expansion, and others.

In addition, the interactions between the product and system and any related systems should begin receiving compatibility attention during conceptual design to minimize the need for product and system redesign. Whether the interrelated system is a companion product sold by the same company, an environmental system that may be degraded, or a computer system on which a software product runs, the relationship with the product and system under development must be engineered concurrently.

2.2 Designing for the Life Cycle

Design within the system life-cycle context differs from design in the ordinary sense. Life-cycle-guided design is simultaneously responsive to customer needs (i.e., to requirements expressed in functional terms) and to life-cycle outcomes. Design should not only transform a need into a system configuration but should also ensure the design's compatibility with related physical and functional requirements. Further, it should consider operational outcomes expressed as producibility, reliability, maintainability, usability, supportability, serviceability, disposability, sustainability, and others, in addition to performance, effectiveness, and affordability.

A detailed presentation of the elaborate technological activities and interactions that must be integrated over the system life-cycle process is given in Figure 3. The progression is iterative from left to right and not serial in nature, as might be inferred.

Although the level of activity and detail may vary, the life-cycle functions described and illustrated are generic. They are applicable whenever a new need or changed requirement is identified, with the process being common to large as well as small-scale systems. It is essential that this process be implemented completely—at an appropriate level of detail—not only in the engineering of new systems but also in the re-engineering of existing or legacy systems.

Major technical activities performed during the design, production or construction, utilization, support, and phase-out phases of the life cycle are highlighted in Figure 3. These are initiated when a new need is identified. A planning function is followed by conceptual, preliminary, and detail design activities. Producing and/or constructing the system are the function that completes the acquisition phase. System operation and support functions occur during the utilization phase of the life cycle. Phase-out and disposal are important final functions of utilization to be considered as part of design for the life cycle.

The numbered blocks in Figure 3 "map" and elaborate on the phases of the life cycles depicted in Figure 2 as follows:

1. The acquisition phase—Figure 3, Blocks 1–4.
2. The utilization phase—Figure 3, Blocks 5 and 6.
3. The design phase—Figure 3, Blocks 1–3.
4. The startup phase—Figure 3, Block 4.
5. The operation phase—Figure 3, Block 5.
6. The retirement phase—Figure 3, Block 6.

The communication and coordination needed to design and develop the product, the production capability, the system support capability, and the relationships with interrelated systems—so that they traverse the life cycle together seamlessly—is not easy to accomplish. Progress in this area is facilitated by technologies that make more timely acquisition and use of design information possible. Computer-Aided Design (CAD) technology with internet/intranet connectivity enables a geographically dispersed multidiscipline team to collaborate effectively on complex physical designs.

For certain products, the addition of Computer-Aided Manufacturing (CAM) software can automatically translate approved three-dimensional CAD drawings into manufacturing instructions for numerically controlled equipment. Generic or custom parametric CAD software can facilitate exploration of alternative design solutions. Once a design has been created in CAD/CAM, iterative improvements to the design are relatively easy to make. The CAD drawings also facilitate maintenance, technical support, regeneration (re-engineering), and disposal. A broad range of other electronic communication and collaboration tools can help integrate relevant geographically dispersed design and development activities over the life cycle of the system.

Concern for the entire life cycle is particularly strong within the U.S. Department of Defense (DOD) and its non-U.S. counterparts. This may be attributed to the fact that acquired defense systems are owned, operated, and maintained by the DOD. This is unlike the situation most often encountered in the private sector, where the consumer or user is usually not the producer. Those private firms serving as defense contractors are obliged to

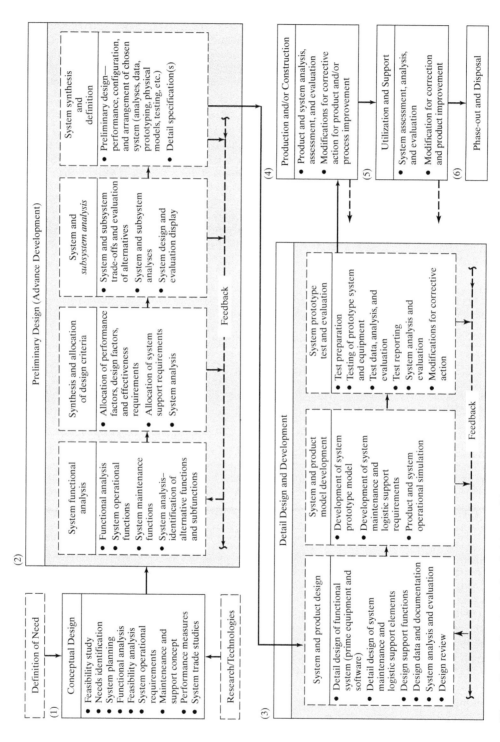

Figure 3 Technological activities and interactions within the system life-cycle process.

32

design and develop in accordance with DOD directives, specifications, and standards. Because the DOD is the customer and also the user of the resulting system, considerable DOD intervention occurs during the acquisition phase.[5]

Many firms that produce for private-sector markets have chosen to design with the life cycle in mind. For example, design for energy efficiency is now common in appliances such as water heaters and air conditioners. Fuel efficiency is a required design characteristic for automobiles. Some truck manufacturers promise that life-cycle maintenance costs will be within stated limits. These developments are commendable, but they do not go far enough. When the producer is not the consumer, it is less likely that potential operational problems will be addressed during development. Undesirable outcomes too often end up as problems for the user of the product instead of the producer.

3 THE SYSTEMS ENGINEERING PROCESS

Although there is general agreement regarding the principles and objectives of systems engineering, its actual implementation will vary from one system and engineering team to the next. The process approach and steps used will depend on the nature of the system application and the backgrounds and experiences of the individuals on the team. To establish a common frame of reference for improving communication and understanding, it is important that a "baseline" be defined that describes the systems engineering process, along with the essential life-cycle phases and steps within that process. Augmenting this common frame of reference are top-down and bottom-up approaches. And, there are other process models that have attracted various degrees of attention. Each of these topics is presented in this section.

3.1 Life-Cycle Process Phases and Steps

Figure 4 illustrates the major life-cycle process phases and selected milestones for a generic system. This is the "model" that will serve as a frame of reference for material presented in subsequent chapters. Included are the basic steps in the systems engineering process (i.e., requirements analysis, functional analysis and allocation, synthesis, trade-off studies, design evaluation, and so on).[6]

A newly identified need, or an evolving need, reveals a new system requirement. If a decision is made to seek a solution for the need, then a decision is needed whether to consider other needs in designing the solution. Based on an initial determination regarding the scope of needs, the basic phases of conceptual design and onward through system retirement and phase-out are then applicable, as described in the paragraphs that follow. The scope of needs may contract or expand, but the scope should be stabilized as early as possible during conceptual design, preferably based on an evaluation of value and cost by the customer.

Program phases described in Figure 4 are not intended to convey specific tasks, or time periods, or levels of funding, or numbers of iterations. Individual program requirements will vary from one application to the next. The figure exhibits an overall *process* that needs to be

---

[5]This intervention is widely supported by a variety of both past and current Department of Defense Directives and Instructions. Among the most recent and notable is DOD Instruction 5000.02, "Operation of the Defense Acquisition System," December 8, 2008.

[6]Figure 4 exhibits more information and detail than can be adequately explained in this chapter. Its purpose is to provide context by consolidating terms and notation on a single page.

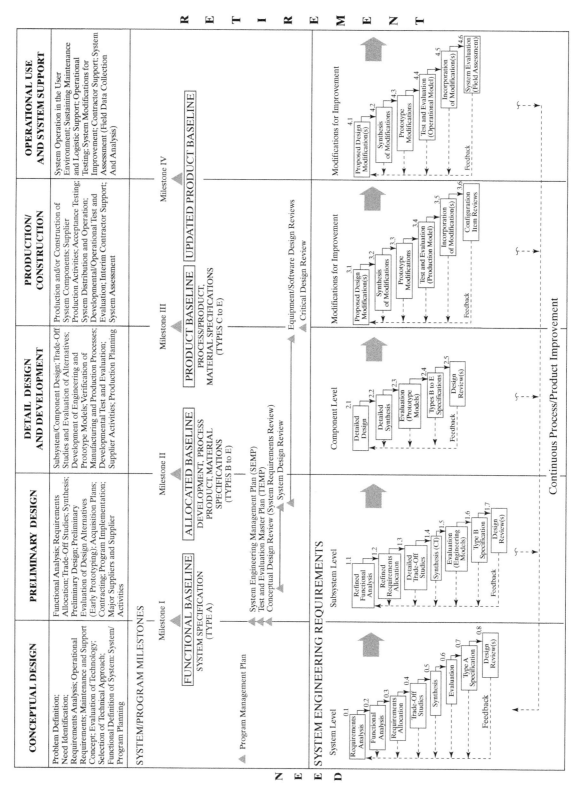

Figure 4 System process activities and interactions over the life cycle.

followed during system acquisition and deployment. Regardless of the type, size, and complexity of the system, there is a conceptual design requirement (i.e., to include requirements analysis), a preliminary design requirement, and so on. Also, to ensure maximum effectiveness, the concepts presented in Figure 4 must be properly "tailored" to the particular system application being addressed.

Figure 4 (Blocks 0.1–0.8, Blocks 1.1–1.7, and Blocks 2.1–2.5) shows the basic steps in the systems engineering process to be iterative in nature, providing a top-down definition of the system, and then proceeding down to the subsystem level (and below as necessary). Focused on the needs, and beginning with conceptual design, the completion of Block 0.2 defines the system in *functional* terms (having identified the "whats" from a requirements perspective). These "whats" are translated into an applicable set of "hows" through the iterative process of functional partitioning and requirements allocation, together with conceptual design synthesis, analysis, and evaluation. This conceptual design phase is where the initial configuration of the system (or system architecture) is defined.

During preliminary design, completion of Block 1.1 defines the system in *refined functional* terms providing a top-down definition of subsystems with preparation for moving down to the component level. Here the "whats" are extracted from (provided by) the conceptual design phase. These "whats" are translated into an applicable set of "hows" through the iterative process of functional partitioning and requirements allocation, together with preliminary design synthesis, analysis, and evaluation. This preliminary design phase is where the initial configuration of subsystems (or subsystem architecture) is defined.

Blocks 1.1–1.7 are an evolution from Blocks 0.1–0.8, Blocks 2.1–2.5 are an evolution from Blocks 1.1–1.7, and Blocks 3.1–3.6 are an evolution from Blocks 2.1–2.5. The overall process reflected in the figure constitutes an evolutionary design and development process. With appropriate feedback and design refinement provisions incorporated, the process should eventually converge to a successful design. The functional definition of the system, its subsystems, and its components serves as the baseline for the identification of resource requirements for production and then operational use (i.e., hardware, software, people, facilities, data, elements of support, or a combination thereof).

3.2 Other Systems Engineering Process Models

The overarching objective is to describe a *process* (as a frame of reference) that should be "tailored" to the specific program need.

The illustration in Figure 4 is not intended to emphasize any particular model, such as the "waterfall" model, the "spiral" model, the "vee" model, or equivalent. These well-known process models are illustrated and briefly described in Figure 5.

4 SYSTEM DESIGN CONSIDERATIONS

The systems engineering process is suggested as a preferred approach for bringing systems and their products into being that will be cost-effective and globally competitive. An essential technical activity within the process is that of system design evaluation. Evaluation must be inherent within the systems engineering process and be invoked regularly as the system design activity progresses. However, system evaluation should not proceed without either accepting guidance from customer requirements and applicable system design criteria, or direct involvement of the customer. When conducted with full

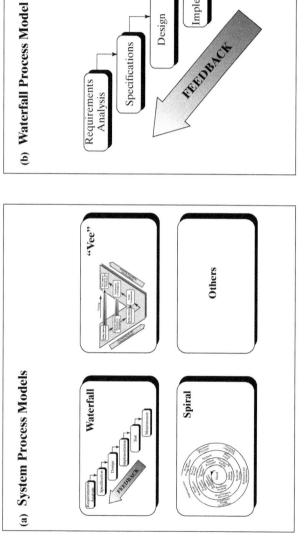

(a) System Process Models

Waterfall

"Vee"

Spiral

Others

• It is observed that the preference expressed by individuals and groups for one of the system models is subjective.

• A study of the literature and current practice is needed to identify which model fits a specific situation best.

(b) Waterfall Process Model

Requirements Analysis → Specifications → Design → Implementation → Test → Maintenance

FEEDBACK

• The waterfall model, introduced by Royce in 1970, initially was used for software development. This model usually consists of five to seven series of steps or phases for systems engineering or software development. Boehm expanded this into an eight-step series of activities in 1981.

• A similar model splits the hardware and software into two distinct efforts. Ideally, each phase is carried out to completion in sequence until the product is delivered. However, this rarely is the case. When deficiencies are found, phases must be repeated until the product is correct.

Figure 5 Some systems engineering process models (sheet 1).

(d) "Vee" Process Model

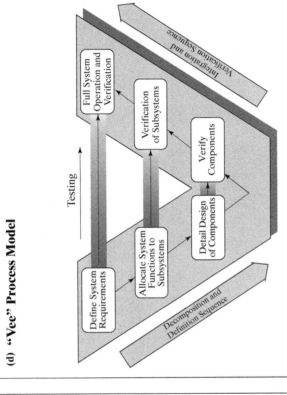

(c) Spiral Process Model

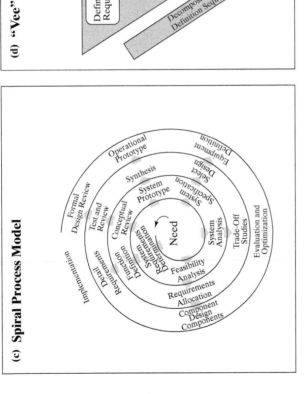

- The spiral process model of the development life cycle (developed by Boehm in 1986 using Hall's work in systems engineering from 1969) is intended to introduce a risk-driven approach for the development of products or systems.
- This model is an adaptation of the waterfall model, which does not mandate the use of prototypes. The spiral model incorporates features from other models, such as feedback, etc.
- Application of the spiral model is iterative and proceeds through the several phases each time a different type of prototype is developed. It allows for an evaluation of risk before proceeding to a subsequent phase.

- Forsberg and Mooz describe what they call "the technical aspect of the project cycle" by the "Vee" process model. This model starts with user needs on the upper left and ends with a user-validated system on the upper right.
- On the left side, decomposition and definition activities resolve the system architecture, creating details of the design. Integration and verification flows upward to the right as successively higher levels of subsystems are verified, culminating at the system level.
- Verification and validation progress from the component level to the validation of the operational system. At each level of testing, the originating specifications and requirements documents are consulted to ensure that component/subsystems/system meet the specifications.

Figure 5   Some systems engineering process models (sheet 2).

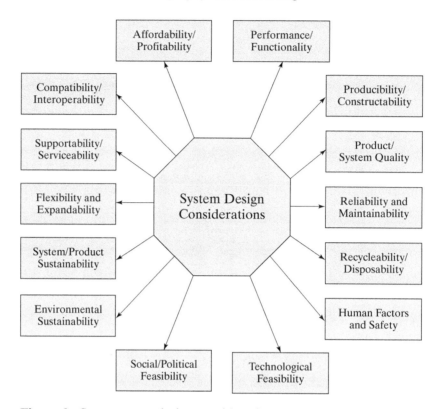

Figure 6   Some system design considerations.

recognition of customer requirements and design criteria, evaluation enhances assurance of continuous design improvement.

There are numerous system design considerations that should be identified and studied when developing design criteria. These are shown in Figure 6. Design considerations provide a broad range of possibilities from which the derivation of design criteria may evolve. A general discussion of this important topic is the focus of this section. Formal analytical and modeling approaches for performing systems analysis and design evaluation, incorporating multiple criteria, are not presented here.

4.1 Development of Design Criteria[7]

As depicted in Figure 7, the definition of needs at the system level is the starting point for determining customer requirements and developing design criteria. The requirements for the system as an entity are established by describing the functions that must be performed. The operational and support functions (i.e., those required to accomplish a specified mission scenario, or series of missions, and those required to ensure that the system is able to perform the needed functions when required) must be described at the top level. Also,

---

[7]Design *criteria* constitute a set of "design-to" requirements, which can be expressed in both qualitative and quantitative terms. These requirements represent bounds on the "design-space" within which the designer must "negotiate" when engaged in the iterative process of synthesis, analysis, and evaluation.

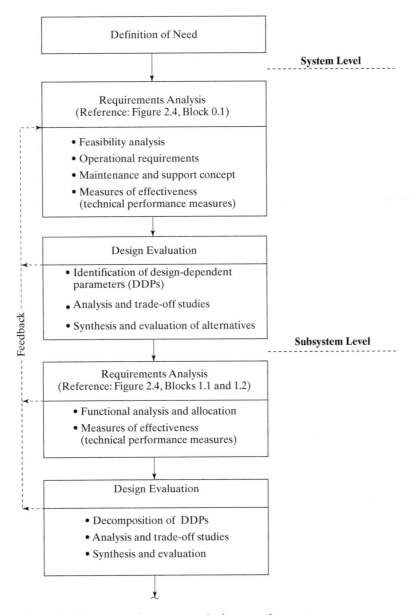

Figure 7 Decomposing system design requirements.

the general concepts and nonnegotiable requirements for production, systems integration, and retirement must be described.

In design evaluation, an early step that fully recognizes design criteria is to establish a *baseline* against which a given alternative or design configuration may be evaluated (refer to Figure 4, Block 0.1). This baseline is determined through the iterative process of requirements analysis (i.e., identification of needs, analysis of feasibility, definition of system operational and support requirements, and selection of concepts for production, systems integration, and retirement). A more specific baseline is developed for each *system level* in Figure 7. The functions that the system must perform to satisfy a specific scope of customer needs should be described, along with expectations for cycle time, frequency, speed, cost, effectiveness, and other relevant factors. Functional requirements must be met by incorporating design characteristics within the system at the appropriate level.

As part of this process, it is necessary to establish some system "metrics" related to performance, effectiveness, cost, and similar quantitative factors as required to meet customer expectations. For instance, what functions must the system perform, where are these to be accomplished, at what frequency, with what degree of reliability, and at what cost? Some of these factors may be considered to be more important than others by the customer which will, in turn, influence the design process by placing different levels of emphasis on meeting criteria. Candidate systems result from design synthesis and become the appropriate targets for design analysis and evaluation.

Evaluation is invoked to determine the degree to which each candidate system satisfies design criteria. Applicable criteria regarding the system should be expressed in terms of *technical performance measures* (TPMs) and exhibited at the system level. TPMs are measures for characteristics that are, or derive from, attributes inherent in the design itself. Attributes that depend directly on design characteristics are called *design-dependent parameters* (DDPs), with specific measures thereof being the TPMs. In contrast, relevant factors external to the design are called *design-independent parameters* (DIPs).

It is essential that the development of *design criteria* be based on an appropriate set of *design considerations*, considerations that lead to the identification of both *design-dependent* and *design-independent parameters* and that support the derivation of *technical performance measures*. More precise definitions for these terms are as follows:

1. Design considerations—the full range of attributes and characteristics that could be exhibited by an engineered system, product, or service. These are of interest to both the producer and the customer (see Figure 6).
2. Design-dependent parameters (DDPs)—attributes and/or characteristics inherent in the design for which predicted or estimated measures are required or desired (e.g., design life, weight, reliability, producibility, maintainability, pollutability, and others).
3. Design-independent parameters (DIPs)—factors external to the design that must be estimated and/or forecasted for use during design evaluation (e.g., fuel cost per pound, labor rates, material cost per pound, interest rates, and others). These depend upon the production and operating environment for the system.
4. Technical performance measures (TPMs)—predicted and/or estimated values for DDPs. They also include values for higher level (derived) performance considerations (e.g., availability, cost, flexibility, and supportability).
5. Design criteria—customer specified or negotiated target values for technical performance measures. Also, desired values for TPMs as specified by the customer as requirements.

The issue and impact of multiple criteria will be presented in the paragraphs that follow. Then, the next section will direct attention to design criteria as an important part of a morphology for synthesis, analysis, and evaluation. In so doing, the terms defined above will be better related to each other and to the system realization process.

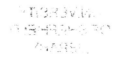

4.2 Considering Multiple Criteria

In Figure 7, the prioritized TPMs at the top level reflect the overall performance characteristics of the system as it accomplishes its mission objectives in response to customer needs. There may be numerous factors, such as system size and weight, range and accuracy, speed of performance, capacity, operational availability, reliability and maintainability, supportability, cost, and so on. These *measures of effectiveness* (MOEs) must be specified in terms of a level of importance, as determined by the customer based on the criticality of the functions to be performed.

For example, there may be certain mission scenarios where system availability is critical, with reliability being less important as long as there are maintainability considerations built into the system that facilitate ease of repair. Conversely, for missions where the accomplishment of maintenance is not feasible, reliability becomes more important. Thus, the nature and criticality of the mission(s) to be accomplished will lead to the identification of specific requirements and the relative levels of importance of the applicable TPMs.

Given the requirements at the top level, it may be appropriate to develop a "design-objectives" tree similar to that presented in Figure 8. First-, second-, and third-order (and lower-level) considerations are noted. Based on the established MOEs for the system, a top-down breakout of requirements will lead to the identification of characteristics that should be included and made inherent within the design; for example, a first-order consideration may be *system value*, which, in turn, may be subdivided into *economic* factors and *technical* factors.

Technical factors may be expressed in terms of *system effectiveness*, which is a function of performance, operational availability, dependability, and so on. This leads to the consideration of such features as speed of performance, reliability and maintainability, size and weight, and flexibility. Assuming that maintainability represents a high priority in design, then such features as packaging, accessibility, diagnostics, mounting, and interchangeability should be stressed in the design. Thus, the *criteria* for design and the associated DDPs should be established early, during conceptual design, and then carried through the entire design cycle. The DDPs establish the extent and scope of the design space within which *trade-off* decisions may be made. During the process of making these trade-offs, requirements must be related to the appropriate hierarchical level in the system structure (i.e., system, subsystem, and configuration item) as in Figure 7.

5 SYSTEM SYNTHESIS, ANALYSIS, AND EVALUATION

System design is the prime mover of systems engineering, with system design evaluation being its compass. System design requires both integration and iteration, invoking a process that coordinates synthesis, analysis, and evaluation, as is shown conceptually in Figure 9. It is essential that the technological activities of synthesis, analysis, and evaluation be integrated and applied iteratively and continuously over the system life cycle. The benefits of continuous improvement in system design are thereby more likely to be obtained.

5.1 A Morphology for Synthesis, Analysis, and Evaluation

Figure 10 presents a high-level schematic of the systems engineering process from a product realization perspective. It is a morphology for linking applied research and technologies (Block 0) to customer needs (Block 1). It also provides a structure for visualizing the technological activities of synthesis, analysis, and evaluation. Each of these activities is summarized in the paragraphs that follow, with reference to relevant blocks within the morphology.

Bringing Systems Into Being

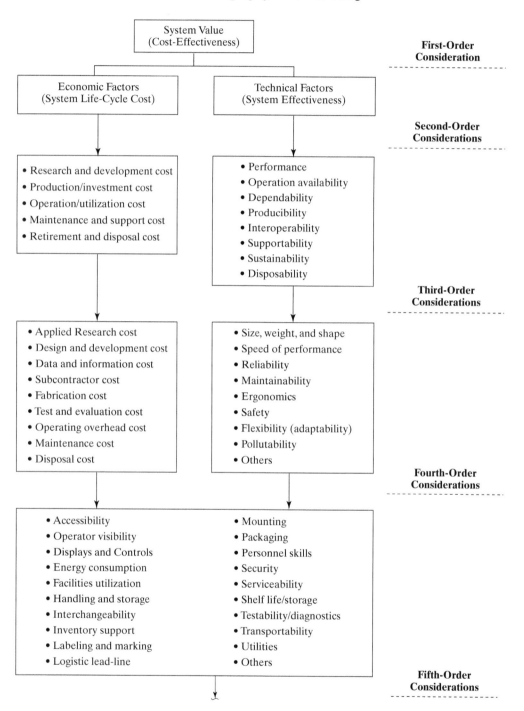

Figure 8 A hierarchy of system design considerations.

Bringing Systems Into Being

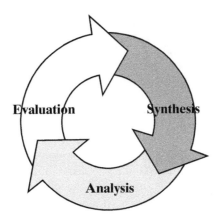

Figure 9 The relationship of synthesis, analysis, and evaluation.

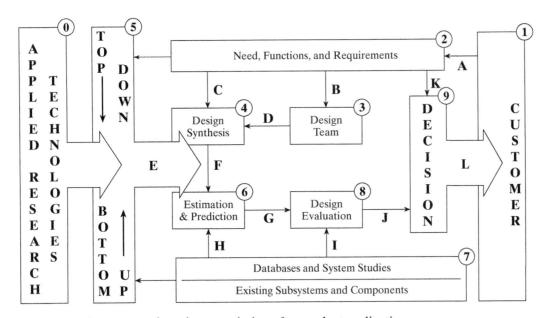

Figure 10 Systems engineering morphology for product realization.

Synthesis. To design is to synthesize, project, and propose what might be for a specific set of customer needs and requirements, normally expressed in functional terms (Block 2). Synthesis is the creative process of putting known things together into new and more useful combinations. Meeting a need in compliance with customer requirements is the objective of design synthesis.

The primary elements enabling design synthesis are the design team (Block 3), supported by traditional and computer-based tools for design synthesis (Block 4). Design synthesis is best accomplished by combining top-down and bottom-up activities (Block 5). Existing and newly developed components, parts, and subsystems are then integrated to generate candidate system designs in a form ready for analysis and evaluation.

Analysis. Analysis of candidate system and product designs is a necessary but not sufficient ingredient in system design evaluation. It involves the functions of estimation and prediction of DDP values (TPMs) (Block 6) and the determining or forecasting of DIP values from information found in physical and economic databases (Block 7).

Systems analysis and operations research provides a step on the way to system design evaluation, but adaptation of those models and methods to the domain of design is necessary. The adaptation explicitly recognizes DDPs.

Evaluation. Each candidate design (or design alternative) should be evaluated against other candidates and checked for compliance with customer requirements. Evaluation of each candidate (Block 8) is accomplished after receiving DDP values for the candidate from Block 6. It is the specific values for DDPs (the TPMs) that differentiate (or instance) candidate designs.

DIP values determined in Block 7 are externalities. They apply to and across all candidate designs being presented for evaluation. Each candidate is optimized in Block 8 before being presented for design decision (Block 9). It is in Block 9 that the best candidate is sought. Since the preferred choice is subjective, it should ultimately be made by the customer.

5.2 Discussion of the 10-Block Morphology

This section presents and discusses the functions accomplished by each block in the system design morphology that is exhibited in Figure 10. The discussion will be at a greater level of detail than the general description of synthesis, analysis, and evaluation given above.

The Technologies (Block 0). Technologies, exhibited in Block 0, are the product of applied research. They evolve from the activities of engineering research and development and are available to be considered for incorporation into candidate system designs. As a driving force for innovation, technologies are the most potent ingredient for advancing the capabilities of human-made entities.

It is the responsibility of the designer/producer to propose and help the customer understand what might be for each technological choice. Those producers able to articulate and deliver better technological solutions, on time and within budget, will attain and retain a competitive edge in the global marketplace.

The Customer (Block 1). The purpose of system design is to satisfy customer (and stakeholder) needs and expectations. This must be with the full realization that the perceived success of a particular design is ultimately determined by the customer, identified in Block 1; the customary being Number 1.

During the design process, all functions to be provided and all requirements to be satisfied should be determined from the perspective of the customer or the customer's representative. Stakeholder and any other special interests should also be included in the "voice of the customer" in a way that reflects all needs and concerns. Included among these must be ecological and human impacts. Arrow A represents the elicitation of customer needs, desired functionality, and requirements.

Need, Functions, and Requirements (Block 2). The purpose of this block is to identify and specify the desired behavior of the system or product in functional terms. A market study identifies a need, an opportunity, or a deficiency. From the need comes a definition of the basic requirements, often stated in functional terms. Requirements are the input for design and operational criteria, and criteria are the basis for the evaluation of candidate system configurations.

At this point, the system and its product should be defined by its function, not its form. Arrow A indicates customer inputs that define need, functionality, and operational requirements. Arrows B and C depict the translation and transfer of this information to the design process.

The Design Team (Block 3).   The design team should be organized to incorporate in-depth technical expertise, as well as a broader systems view. Included must be expertise in each of the product life-cycle phases and elements contained within the set of system requirements.

Balanced consideration should be present for each phase of the design. Included should be the satisfaction of intended purpose, followed by producibility, reliability, maintainability, disposability, sustainability, and others. Arrow B depicts requirements and design criteria being made available to the design team and Arrow D indicates the team's contributed synthesis effort wherein need, functions, and requirements are the overarching consideration (Arrow C).

Design Synthesis (Block 4).   To design is to project and propose what might be. Design synthesis is a creative activity that relies on the knowledge of experts about the state of the art as well as the state of technology. From this knowledge, a number of feasible design alternatives are fashioned and presented for analysis. Depending upon the phase of the product life cycle, the synthesis can be in conceptual, preliminary, or detailed form.

The candidate design is driven by both a top-down functional decomposition from Block 2 and a bottom-up combinatorial approach utilizing available system elements from Block 7. Arrow E represents a blending of these approaches. Adequate definition of each design alternative must be obtained to allow for life-cycle analysis in view of the requirements. Arrow F highlights this definition process as it pertains to the passing of candidate design alternatives to design analysis in Block 6.

Top-Down and Bottom-Up (Block 5).   Traditional engineering design methodology is based largely on a bottom-up approach. Starting with a set of defined elements, designers synthesize the system/product by finding the most appropriate combination of elements. The bottom-up process is iterative with the number of iterations determined by the creativity and skill of the design team, as well as by the complexity of the system design.

A top-down approach to design is inherent within systems engineering. Starting with requirements for the external behavior of any component of the system (in terms of the function provided by that component), that behavior is then decomposed. These decomposed functional behaviors are then described in more detail and made specific through an analysis process. Then, the appropriateness of the choice of functional components is verified by synthesizing the original entity. Most systems and products are realized through an intelligent combination of the top-down and bottom-up approaches, with the best mix being largely a matter of judgment and experience.

Design Analysis (Block 6).   Design analysis is focused largely on determining values for cost and effectiveness measures generated during estimation and prediction activities. Models, database information, and simulation are employed to obtain DDP values (or TPMs) for each synthesized design alternative from Block 4. Output Arrow G passes the analysis results to design evaluation (Block 8).

The TPM values provide the basis for comparing system designs against input criteria to determine the relative merit of each candidate. Arrow H represents input from the available databases and from relevant studies.

Physical and Economic Databases (Block 7). Block 7 provides a resource for the design process, rather than being an actual step in the process flow. There exists a body of knowledge and information that engineers, technologists, economists, and others rely on to perform the tasks of analysis and evaluation. This knowledge consists of physical laws, empirical data, price information, economic forecasts, and numerous other studies and models.

Block 7 also includes descriptions of existing system components, parts, and subsystems, often "commercial off-the-shelf." It is important to use existing databases in doing analysis and synthesis to avoid duplication of effort. This body of knowledge and experience can be utilized both formally and informally in performing needed studies, as well as in supporting the decisions to follow.

At this point, and as represented by Arrow I, DIP values are estimated or forecasted and provided to the activity of design evaluation in Block 8.

Design Evaluation (Block 8). Design evaluation is an essential activity within system and product design and the systems engineering process. It should be embedded appropriately within the process and then pursued continuously as design and development progresses.

Life-cycle cost is one basis for comparing alternative designs that otherwise meet minimum requirements under performance criteria. The life-cycle cost of each alternative is determined based on the activity of estimation and prediction just completed. Arrow J indicates the passing of the evaluated candidates to the decision process. The selection of preferred alternative(s) can only be made after the life-cycle cost analysis is completed and after effectiveness measures are defined and applied.

Design Decision (Block 9). Given the variety of customer needs and perceptions as collected in Block 2, choosing a preferred alternative is not just the simple task of picking the least expensive design. Input criteria, derived from customer and product requirements, are represented by Arrow K and by the DDP values (TPMs) and life-cycle costs indicated by Arrow J. The customer or decision maker must now trade off life-cycle cost against effectiveness criteria subjectively. The result is the identification of one or more preferred alternatives that can be used to take the design process to the next level of detail.

Alternatives must ultimately be judged by the customer. Accordingly, arrow L depicts the passing of evaluated candidate designs to the customer for review and decision. Alternatives that are found to be unacceptable in performance can be either discarded or reworked and new alternatives created. Alternatives that meet all, or the most important, performance criteria can then be evaluated based on estimations and predictions of TPM values, along with an assessment of risk.

6 IMPLEMENTING SYSTEMS ENGINEERING

Within the context of synthesis, analysis, and evaluation is the opportunity to implement systems engineering over the system life cycle in measured ways that can help ensure its effectiveness. These measured implementations are necessary because the complexity of technical systems continues to increase, and many systems in use today are not meeting the needs of the customer in terms of performance, effectiveness, and overall cost. New technologies are being introduced on a continuing basis, while the life cycles for many systems are being extended. The length of time that it takes to develop and acquire a new system needs to be reduced, the costs of modifying existing systems are increasing, and available resources are dwindling. At the same time, there is a greater degree of international cooperation, and competition is increasing worldwide.

6.1 Application Domains for Systems Engineering

There are many categories of human-made systems, and there are several application domains where the concepts and principles of systems engineering can be effectively implemented. Every time that there is a newly identified need to accomplish some function, a new *system* requirement is established. In each instance, there is a new design and development effort that must be accomplished at the *system* level. This, in turn, may lead to a variety of approaches at the subsystem level and below (i.e., the design and development of new equipment and software, the selection and integration of new commercial off-the-shelf items, the modification of existing items already in use, or combinations thereof).

Accordingly, for every new customer requirement, there is a needed design effort for the system overall, to which the steps described in Section 4 are applicable. Although the extent and depth of effort will vary, the concepts and principles for bringing a system into being are basically the same. Some specific application areas are highlighted in Figure 11, and application domains include the following:

1. Large-scale systems with many components, such as a space-based system, an urban transportation system, a hydroelectric power-generating system, or a health-care delivery system.
2. Small-scale systems with relatively few components, such as a local area communications network, a computer system, a hydraulic system, or a mechanical braking system, or a cash receipt system.

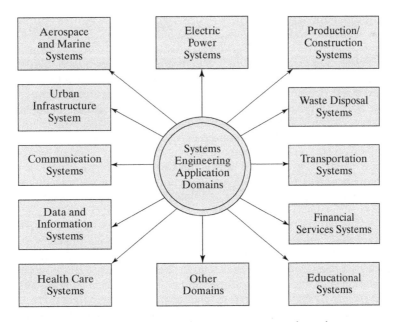

Figure 11 Application domains for systems engineering.

3. Manufacturing or production systems where there are input–output relationships, processes, processors, control software, facilities, and people.

4. Systems where a great deal of new design and development effort is required (e.g., in the introduction of advanced technologies).

5. Systems where the design is based largely on the use of existing equipment, commercial software, or existing facilities.

6. Systems that are highly equipment, software, facilities, or data intensive.

7. Systems where there are several suppliers involved in the design and development process at the local, and possibly international, level.

8. Systems being designed and developed for use in the defense, civilian, commercial, or private sectors separately or jointly.

9. Human-modified systems wherein a natural system is altered or augmented to make it serve human needs more completely, while being retained/sustained largely in its natural state.

6.2 Recognizing and Managing Life-Cycle Impacts

In evaluating past experiences regarding the development of technical systems, it is discovered that most of the problems experienced have been the direct result of not applying a *disciplined* top-down "systems approach." The overall requirements for the system were not defined well from the beginning; the perspective in terms of meeting a need has been relatively "short term" in nature; and, in many instances, the approach followed has been to "deliver it now and fix it later," using a bottom-up approach to design. In essence, the systems design and development process has suffered from the lack of good early planning and the subsequent definition and allocation of requirements in a complete and methodical manner. Yet, it is at this early stage in the life cycle when decisions are made that have a large impact on the overall effectiveness and cost of the system. This is illustrated conceptually in Figure 12.

Referring to Figure 12, experience indicates that there can be a large commitment in terms of technology applications, the establishment of a system configuration and its performance characteristics, the obligation of resources, and potential life-cycle cost at the early stages of a program. It is at this point when system-specific knowledge is limited, but when major decisions are made pertaining to the selection of technologies, the selection of materials and potential sources of supply, equipment packaging schemes and levels of diagnostics, the selection of a manufacturing process, the establishment of a maintenance approach, and so on. It is estimated that from 50% to 75% of the projected life-cycle cost for a given system can be committed (i.e., "locked in") based on engineering design and management decisions made during the early stages of conceptual and preliminary design. Thus, it is at this stage where the implementation of systems engineering concepts and principles is critical. It is essential that one start off with a good understanding of the customer need and a definition of system requirements.

The systems engineering process is applicable over all phases of the life cycle, with the greatest benefit being derived from its emphasis on the early stages, as illustrated in Figure 12. The objective is to influence design early, in an effective and efficient manner, through a comprehensive needs analysis, requirements definition, functional analysis and allocation, and then to address the follow-on activities in a logical and progressive manner with the provision of appropriate feedback. As conveyed in Figure 13, the overall objective is to influence design in the early phases of system acquisition, leading to the identification of individual discipline-based design needs. These should be applied in a timely manner as

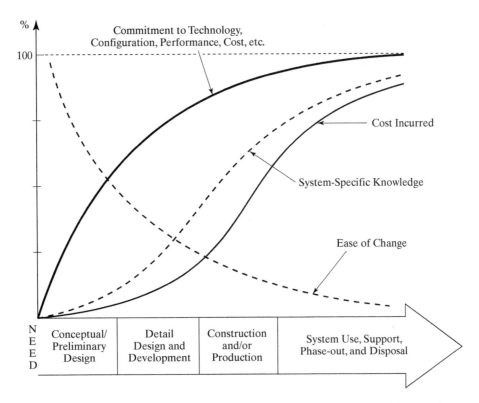

Figure 12 Life-cycle commitment, system-specific knowledge, and incurred cost.

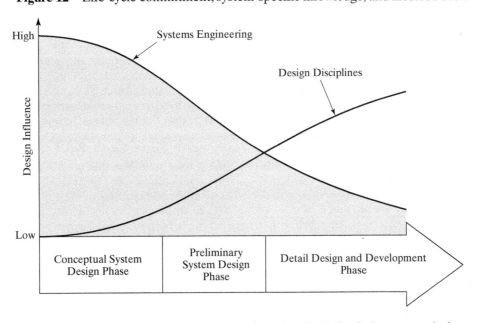

Figure 13 Systems engineering versus engineering discipline influence on design.

one evolves from system-level requirements to the design of various subsystems and components thereof.[8]

6.3 Potential Benefits from Systems Engineering

An understanding of the interrelationships among the factors identified in Figures 12 and 13 is essential if the full benefit of systems engineering is to be realized. There is a need to ensure that the applicable engineering disciplines responsible for the design of individual system elements are properly integrated. This need extends to the proper implementation of concurrent engineering to address the life cycles for the product and for the supporting capabilities of production, support, and phase-out, as is illustrated in Figure 2. A good communication network, with local-area and wide-area capability, must also be in place and available to all critical project personnel. This is a particular challenge when essential project personnel are located remotely, often worldwide.

Successful implementation of the systems engineering process depends not only on the availability and application of the appropriate technologies and tools but also on the planning and management of the activities required to accomplish the overall objective. Although the steps described in Section 3 may be specified for a given program, successful implementation (and the benefits to be derived) will not be realized unless the proper organizational environment is established that will encourage it to happen. There have been numerous instances where a project organization included a "systems engineering" function but where the impact on design has been almost nonexistent, resulting in objectives not being met.

Although some of the benefits associated with application of the concepts and principles of systems engineering have been provided throughout this chapter, it may be helpful to provide a compact summary for reference. Accordingly, application of the systems engineering process can lead to the following benefits:

1. Reduction in the cost of system design and development, production and/or construction, system operation and support, system retirement and material disposal; hence, a reduction in life-cycle cost should occur. Often it is perceived that the implementation of systems engineering will increase the cost of system acquisition. Although there may be a few more steps to perform during the early (conceptual and preliminary) system design phases, this investment could significantly reduce the requirements in the integration, test, and evaluation efforts accomplished late in the detail design and development phase. The bottom-up approach involved in making the system work can be simplified if a holistic engineering effort is initiated from the beginning. In addition experience indicates that the early emphasis on systems engineering can result in a cost savings later on in the production, operations and support, and retirement phases of the life cycle.

2. Reduction in system acquisition time (or the time from the initial identification of a customer need to the delivery of a system to the customer). Evaluation of all feasible alternative approaches to design early in the life cycle (with the support of available design aids such as the use of CAD technology) should help to promote greater design maturity earlier. Changes can be incorporated at an early stage before the

---

[8]In Figure 13, the intent is to convey the degree of "design influence" imparted by application of the systems engineering process, and not to imply levels of human effort or cost. A single individual with the appropriate experience and technical expertise can exert a great deal of influence on design, whereas the establishment of a new organization and the assignment of many people to a project may have little beneficial effect.

design is "fixed" and more costly to modify. Further, the results should enable a reduction in the time that it takes for final system integration, tests, and evaluation.

3. More visibility and a reduction in the risks associated with the design decision-making process. Increased visibility is provided through viewing the system from a long-term and life-cycle perspective. The long-term impacts as a result of early design decisions and "cause-and-effect" relationships can be assessed at an early stage. This should cause a reduction in potential risks, resulting in greater customer satisfaction.

Without the proper organizational emphasis from top management, the establishment of an environment that will allow for creativity and innovation, a leadership style that will promote a "team" approach to design, and so on, implementation of the concepts and methodologies described herein may not occur. Thus, systems engineering must be implemented in terms of both *technology* and *management*. This joint implementation of systems engineering is the responsibility of systems engineering management.

7 SUMMARY AND EXTENSIONS

The overarching goal of systems engineering is embodied in the title of this chapter. This goal is to bring successful systems and their products into being. Accordingly, it is appropriate to devote this chapter to a high-level presentation of the essentials involved in the engineering of systems.

The engineered or technical system is to be brought into being; it is a system destined to become part of the human-made world. Therefore, the definition and description of the engineered system is given early in this chapter. In most cases, there is a product coexistent with or within the system, and in others the system is the product. But in either case, there must exist a human need to be met.

Since systems are often known by their products, product and system categories are identified as frameworks for study in this and subsequent chapters. Major categories are single-entity product systems and multiple-entity population systems. Availability of these example categories is intended to help underpin and clarify the topics and steps in the process of bringing engineered or technical systems into being.

The product and system life cycle is the *enduring paradigm* used throughout this book. It is argued that the defense origin of this life-cycle paradigm has profitable applications in the private sector. The life cycle is first introduced in Section 2 with two simple diagrams; the first provides the product and the second gives an expanded concurrent life-cycle view. Then, designing for the life cycle is addressed with the aid of more elaborate life-cycle diagrams, showing many more activities and interactions. Other systems engineering process models are then exhibited to conclude an overview of the popular process structures for bringing systems and products into being.

Since design is the fundamental technical activity for both the product and the system, it is important to proceed with full knowledge of all system design considerations. The identification of DDPs and their counterparts, DIPs, follows. Emanating from DDPs are technical performance measures to be predicted and/or estimated. The deviation or difference between predicted TPMs and customer-specified criteria provides the basis for design improvement through iteration, with the expectation of convergence to a preferred design. During this design activity, criteria or requirements must be given center stage. Accordingly, the largest section of this chapter is devoted to an explanation of design evaluation based on customer-specified criteria. The explanation is enhanced by the development and presentation of a 10-block morphology for synthesis, analysis, and evaluation.

This chapter closes with some challenges and opportunities that will surely arise during the implementation of systems engineering. The available application domains are numerous. A general notion is that systems engineering is an engineering interdiscipline in its own right, with important engineering domain manifestations. It is hereby conjectured that the systems engineering body of systematic knowledge will not advance significantly without engineering domain opportunities for application. However, it is clear that significant improvements in domain-specific projects do occur when resources are allocated to systems engineering activities early in the life cycle. Two views of this observed benefit are illustrated in this chapter.

It is recognized that some readers may need and desire to probe beyond the content of this textbook. If so, we would recommend two edited works: The *Handbook of Systems Engineering and Management*, A. P. Sage and W. B. Rouse (Eds), John Wiley & Sons, Inc., 2009, augments the technical and managerial topics encompassed by systems engineering. *Design and Systems*, A. Collen and W. W. Gasparski (Eds), Transaction Publishers, 1995, makes visible the pervasive nature of design in the many arenas of human endeavor from a philosophical and praxiological perspective.

Regarding the body of systems engineering knowledge, there is a timely project being pursued within the INCOSE—the SEBoK (Systems Engineering Body of Knowledge) activity involving hundreds of members. Interested individuals may review the current state of development and/or make contributions to it by visiting *http://www.incose.org*. An earlier effort along this line was to engage the intellectual leaders of INCOSE (including the authors) in the writing of 16 seminal articles. These were published in the inaugural issue of *Systems Engineering*, Vol. 1, No. 1, July–September 1994. Copies of this special issue and subsequent issues of the journal may be obtained through the INCOSE website.

QUESTIONS AND PROBLEMS

1. What are some of the characteristics of a human-made or engineered system that distinguish it from a natural system?

2. Describe some of the interfaces between the natural world and the human-made world as they pertain to the process of bringing systems/products into being.

3. Identify and describe a natural system of your choice that has been human-modified and identify what distinguishes it from a human-made system.

4. Describe the product or prime equipment as a component of the system; provide an explanation of the functions provided by each entity.

5. Put a face on the generic single-entity product system of Section 1.2 by picking a real structure or service with which you are familiar. Then, rewrite the textbook description based on the characteristics of the entity you picked.

6. Put a face on the generic multiple-entity population system of Section 1.2 by picking real equipment with which you are familiar. Then, rewrite the textbook description based on the characteristics of the equipment you picked.

7. Pick a consumer good and name the producer good(s) that need to be employed to bring this consumer good to market.

8. Pick a product, describe the enabling system that is required to bring it into being, and explain the importance of engineering the system and product together.

9. What are some of the essential factors in engineering for product competitiveness? Why is product competitiveness important?

10. What does system life-cycle thinking add to engineering as currently practiced? What are the expected benefits to be gained from this thinking?

11. Various phases of the product life cycle are shown in Figure 1 and expanded in Figure 2. Describe some of the interfaces and interactions between the life cycles of the system and the product life cycle.

12. What is the full meaning of the phrase "designing for the life cycle"?

13. Select a system of your choice and describe the applicable life-cycle phases and activities, tailoring your description to that system.

14. As best you can, identify life-cycle activities that occur in the waterfall model, the spiral model, and the "vee" model. Of these models, pick the one you prefer and explain why.

15. Design considerations are the first step on the way to deriving technical performance measures. Outline all of the steps, emphasizing the design-dependent parameter concept.

16. How are requirements related to technical performance measures? What is the remedy when requirements and TPMs are not in agreement?

17. Pick a design situation of your choice and itemize the multiple criteria that should be addressed.

18. Pick a top-level requirement and decompose it in accordance with the structure shown in Figure 7.

19. Candidate systems result from design synthesis and become the object of design analysis and evaluation. Explain.

20. Pick up a design consideration at the lowest level in Figure 8. Discuss its position and impact on each of the next higher levels.

21. Take synthesis, analysis, and evaluation as depicted in Figure 9 and then classify each activity exhibited by application of elements in the 10-block morphology of Figure 10.

22. Identify some of the engineering domain manifestations of systems engineering.

23. What are some of the impediments to the implementation of systems engineering?

24. What are some of the benefits that may result from the utilization of systems thinking and engineering?

25. Go to the INCOSE website and find the page about the journal *Systems Engineering*. Pick an article that touches upon a topic in this chapter and relate it thereto in no more than one paragraph.

26. Go to the INCOSE website and identify one individual from the Fellows group who most closely matches your own interest in systems engineering. Say why you would like to meet this person.

Conceptual System Design

From Chapter 3 of *Systems Engineering and Analysis,* Fifth Edition, Benjamin S. Blanchard, Wolter J. Fabrycky. Copyright © 2011 by Pearson Education, Inc. Published by Pearson Prentice Hall. All rights reserved.

Conceptual System Design

Conceptual design is the first and most important phase of the system design and development process. It is an early and high-level life-cycle activity with the potential to establish, commit, and otherwise predetermine the function, form, cost, and development schedule of the desired system and its product(s). The identification of a problem and an associated definition of need provide a valid and appropriate starting point for conceptual system design.

Selection of a path forward for the design and development of a preferred system architecture that will ultimately be responsive to the identified customer need is a major purpose of conceptual design. Establishing this foundation early, as well as initiating the early planning and evaluation of alternative technological approaches, is a critical initial step in the implementation of the systems engineering process. Systems engineering, from an organizational perspective, should take the lead in the solicitation of system *requirements* from the beginning and then address them in an integrated life-cycle manner.

This chapter addresses certain steps in the systems engineering process and, in doing so, provides basic insight and knowledge about the following:

- Identifying and translating a problem or deficiency into a definition of need for a system that will provide a preferred solution;
- Accomplishing advanced system planning and architecting in response to the identified need;
- Developing system operational requirements describing the functions that the system must perform to accomplish its intended purpose(s) or mission(s);
- Conducting exploratory studies leading to the definition of a technical approach for system design;
- Proposing a maintenance concept for the sustaining support of the system throughout its planned life cycle;
- Identifying and prioritizing technical performance measures (TPMs) and related criteria for design;
- Accomplishing a system-level functional analysis and allocating requirements to various subsystems and components;
- Performing systems analysis and producing trade-off studies;

- Developing a system specification; and
- Conducting a conceptual design review.

The completion of the steps above constitutes the *system definition* process at the conceptual level. Although the depth, effort, and cost of accomplishing these steps may vary, the process is applicable to any type or category of system, complex or simple, large or small. It is important that these steps, which encompass the front end of the systems engineering process, be thoroughly understood. Collectively, they serve as a learning objective with the goal being to provide a comprehensive step-by-step approach for addressing this critical early phase of the systems engineering process.

1 PROBLEM DEFINITION AND NEED IDENTIFICATION

The systems engineering process generally commences with the identification of a "want" or "desire" for something based on some "real" deficiency. For instance, suppose that a current system capability is not adequate in terms of meeting certain required performance goals, is not available when needed, cannot be properly supported, or is too costly to operate. Or, there is a lack of capability to communicate between point A and point B, at a desired rate X, with reliability of Y, and within a specified cost of Z. Or, a regional transportation authority is faced with the problem of providing for increased two-way traffic flow across a river that divides a growing municipality (to illustrate the overall process, this particular example is developed further in Sections 3, 4.

It is important to commence by first defining the "problem" and then defining the need for a specific system capability that (hopefully) is responsive. It is not uncommon to first identify some "perceived" need which, in the end, doesn't really solve the problem at hand. In other words, *why is this particular system capability needed?* Given the problem definition, a new system requirement is defined along with the priority for introduction, the date when the new system capability is required for customer use, and an estimate of the resources necessary for its acquisition. To ensure a good start, a comprehensive *statement of the problem* should be presented in specific qualitative and quantitative terms and in enough detail to justify progressing to the next step. It is essential that the process begin by defining a "real" problem and its importance.

The necessity for identifying the need may seem to be basic or self-evident; however, a design effort is often initiated as a result of a personal interest or a political whim, without the requirements first having been adequately defined. In the software and information technology area, in particular, there is a tendency to accomplish considerable coding and software development at the detailed level before adequately identifying the real need. In addition, there are instances when engineers sincerely believe that they know what the customer needs, without having involved the customer in the "discovery" process. The "design-it-now-fix-it-later" philosophy often prevails, which, in turn, leads to unnecessary cost and delivery delay.

Defining the problem is often the most difficult part of the process, particularly if there is a rush to get underway. The number of false starts and the resulting cost commitment can be significant unless a good foundation is laid from the beginning. A complete description of the need, expressed in quantitatively related criteria whenever possible, is essential. It is important that the problem definition reflects true customer requirements, particularly in an environment of limited resources.

Having defined the problem completely and thoroughly, a *needs analysis* should be performed with the objective of translating a broadly defined "want" into a more specific system-level requirement. The questions are as follows: *What is required of the system in "functional" terms? What functions must the system perform? What are the "primary" functions? What are the "secondary" functions? What must be accomplished to alleviate the stated deficiency? When must this be accomplished? Where is it to be accomplished? How many times or at what frequency must this be accomplished?*

There are many basic questions of this nature, making it important to describe the customer requirements in a *functional* manner to avoid a premature commitment to a specific design concept or configuration. Unless form follows function, there is likely to be an unnecessary expenditure of valuable resources. The ultimate objective is to define the *WHATs* first, deferring the *HOWs* until later.

Identifying the problem and accomplishing a needs analysis in a satisfactory manner can best be realized through a team approach involving the customer, the ultimate consumer or user (if different from the customer), the prime contractor or producer, and major suppliers, as appropriate. The objective is to ensure that proper and effective communications exist between all parties involved in the process. Above all, the "voice of the customer" must be heard, providing the system developer(s) an opportunity to respond in a timely and appropriate manner.

2 ADVANCED SYSTEM PLANNING AND ARCHITECTING

Given an identified "need" for a new or improved system, the advanced stages of system planning and architecting can be initiated. Planning and architecting are essential and coequal activities for bringing a new or improved capability into being. The overall "program requirements" for bringing the capability into being initiate an advanced system planning activity and the development of a *program management plan* (PMP), shown as the second block in Figure 1. While the specific nomenclature for this top-level plan may differ with each program, the objective is to prepare a "management-related" plan providing the necessary guidance for all subsequent managerial and technical activities.

Referring to Figure 1, the PMP guides the development of requirements for implementation of a systems engineering program and the preparation of a *systems engineering management plan* (SEMP), or *system engineering plan* (SEP). The "technical requirements" for the system are simultaneously determined. This involves development of a system-level architecture (functional first and physical later) to include development of system operational requirements, determination of a functional architecture, proposing alternative technical concepts, performing feasibility analysis of proposed concepts, selecting a maintenance and support approach, and so on, as is illustrated by Figure 2. The results lead to the preparation of the *system specification* (Type *A*). The preparation of the SEMP and the system specification should be accomplished concurrently in a coordinated manner. The two documents must "talk" to each other and be mutually supportive.

It can be observed from Figures 1 and 2 that the identified requirements are directly aligned and supportive of the activities and milestones shown in Figure A.1. The *system specification* (Type *A*) contains the highest-level architecture and forms the basis for the preparation of all lower-level specifications in a top-down manner. These lower-level specifications include *development* (Type *B*), *product* (Type *C*), *process* (Type *D*), and *material* (Type *E*) *specifications*, and are described further in Section 9. The *systems engineering management plan* is not detailed here. The systems engineering process and the steps illustrated in Figure 2 are described in the remaining sections of this chapter.

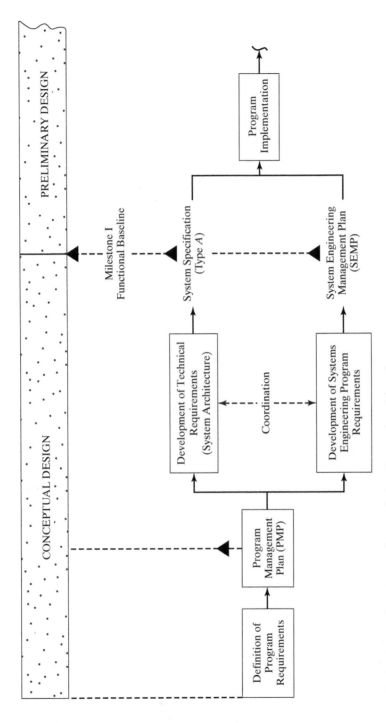

Figure 1 Early system advanced planning and architecting.

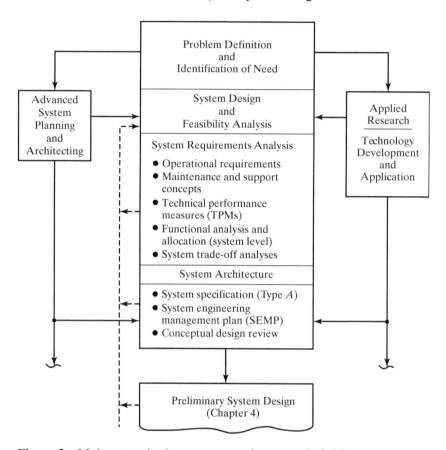

Figure 2 Major steps in the system requirements definition process.

3 SYSTEM DESIGN AND FEASIBILITY ANALYSIS

Having justified the need for a new system, it is necessary to (1) identify various system-level design approaches or alternatives that could be pursued in response to the need; (2) evaluate the feasible approaches to find the most desirable in terms of performance, effectiveness, maintenance and sustaining support, and life-cycle economic criteria; and (3) recommend a preferred course of action. There may be many possible alternatives; however, the number of these must be narrowed down to those that are physically feasible and realizable within schedule requirements and available resources.

In considering alternative system design approaches, different technology applications are investigated. For instance, in response to the river crossing problem (identified in Section 1), alternative design concepts may include a tunnel under the river, a bridge spanning the river, an airlift capability over the river, the use of barges and ferries on the river, or possibly re-routing the river itself. Then a feasibility study would be accomplished to determine a preferred approach. In performing such a study, one must address limiting factors such as geological and geotechnical, atmospheric and weather, hydrology and water flow, as well as the projected capability of each alternative to meet life-cycle cost objectives. In this case, the feasibility results might tentatively indicate that some type of bridge structure spanning the river appears to be best.

At a more detailed level, in the design of a communications system, is a fiber optics technology or the conventional twisted-wire approach preferred? In aircraft design, to what extent should the use of composite materials be considered? In automobile design, should high-speed electronic circuitry in a certain control application be incorporated or should an electromechanical approach be utilized? In the design of a data transmission capability, should a digital or an analogue format be used? In the design of a process, to what extent should embedded computer capabilities be incorporated? Included in the evaluation process are considerations pertaining to the type and maturity of the technology, its stability and growth potential, the anticipated lifetime of the technology, the number of supplier sources, and so on.

It is at this early stage of the system life cycle that major decisions are made relative to adopting a specific design approach and related technology application. Accordingly, it is at this stage that the results of such design decisions can have a great impact on the ultimate behavioral characteristics and life-cycle cost of a system. Technology applications are evaluated and, in some instances where there is not enough information available (or a good solution is not readily evident), research may be initiated with the objective of developing new knowledge to enable other approaches. Finally, it must be agreed that the "need" should dictate and drive the "technology," and not vice versa.

The identification of alternatives and feasibility considerations will significantly impact the operational characteristics of the system and its design for constructability, producibility, supportability, sustainability, disposability, and other design characteristics. The selection and application of a given technology has reliability and maintainability implications, may impact human performance, may affect construction or manufacturing and assembly operations in terms of the processes required, and may significantly impact the need for system maintenance and support. Each will certainly affect life-cycle cost differently. Thus, it is essential that life-cycle considerations be an inherent part of the process of determining the feasible set of system design alternatives.

4 SYSTEM OPERATIONAL REQUIREMENTS

Once the need and technical approach have been defined, it is necessary to translate this into some form of an "operational scenario," or a set of operational requirements. At this point, the following questions may be asked: *What are the anticipated types and quantities of equipment, software, personnel, facilities, information, and so on, required, and where are they to be located? How is the system to be utilized, and for how long? What is the anticipated environment at each operational site (user location)? What are the expected interoperability requirements (i.e., interfaces with other "operating" systems in the area)? How is the system to be supported, by whom, and for how long?* The answer to these and comparable questions leads to the definition of system operational requirements, the follow-on maintenance and support concept, and the identification of specific design-to criteria, and related guidelines.

4.1 Defining System Operational Requirements

System operational requirements should be identified and defined early, carefully, and as completely as possible, based on an established need and selected technical approach. The operational concept and scenario as defined herein is identified in Figure 2. It should include the following:

1. *Mission definition:* Identification of the prime and alternate or secondary missions of the system. What is the system to accomplish? How will the system accomplish

its objectives? The mission may be defined through one or a set of scenarios or operational profiles. It is important that the *dynamics* of system operating conditions be identified to the extent possible.

2. *Performance and physical parameters:* Definition of the operating characteristics or functions of the system (e.g., size, weight, speed, range, accuracy, flow rate, capacity, transmit, receive, throughput, etc.). What are the critical system performance parameters? How are they related to the mission scenario(s)?

3. *Operational deployment or distribution:* Identification of the quantity of equipment, software, personnel, facilities, and so on and the expected geographical location to include transportation and mobility requirements. How much equipment and associated software is to be distributed, and where is it to be located and for how long? When does the system become fully operational?

4. *Operational life cycle (horizon):* Anticipated time that the system will be in operational use (expected period of sustainment). What is the total inventory profile throughout the system life cycle? Who will be operating the system and for what period of time?

5. *Utilization requirements:* Anticipated usage of the system and its elements (e.g., hours of operation per day, percentage of total capacity, operational cycles per month, facility loading). How is the system to be used by the customer, operator, or operating authority in the field?

6. *Effectiveness factors:* System requirements specified as figures-of-merit (FOMs) such as cost/system effectiveness, operational availability (A_o), readiness rate, dependability, logistic support effectiveness, mean time between maintenance (MTBM), failure rate (λ), maintenance downtime (MDT), facility utilization (in percent), operator skill levels and task accomplishment requirements, and personnel efficiency. Given that the system will perform, how effective or efficient is it? How are these factors related to the mission scenario(s)?

7. *Environmental factors:* Definition of the environment in which the system is expected to operate (e.g., temperature, humidity, arctic or tropics, mountainous or flat terrain, airborne, ground, or shipboard). This should include a range of values as applicable and should cover all transportation, handling, and storage modes. How will the system be handled in transit? To what will the system be subjected during its operational use, and for how long? A complete environmental profile should be developed.

In addition to defining operational requirements that are system specific, the system being developed may be imbedded within an overall higher-level structure making it necessary to give consideration to *interoperability requirements*. For example, an aircraft system may be contained within a higher-level airline transportation system, which is part of a regional transportation capability, and so on. There may be both ground and marine transportation systems within the same overall structure, where major interface requirements must be addressed when system operational requirements are being defined.

In some instances, there may be both vertical and horizontal impacts when addressing the system in question, within the context of some larger overall configuration. There are two important questions to be addressed: *What is the potential impact of this new system on the other systems in the same SOS configuration? What are the external impacts from the other systems within the same SOS structure on this new system?*

4.2 Illustrating System Operational Requirements

Further consideration of system operational requirements (as presented in Section 4.1) is provided through five sample illustrations, each covering different degrees or levels of detail. The first illustration is an extension of the river crossing problem introduced in Section 1. The second illustration, covering operational requirements in more depth, is an aircraft system with worldwide deployment. The third illustration is a communication system with ground and airborne applications. The fourth illustration deals with commercial airline capability for a metropolitan area. The fifth illustration considers a hospital as part of a community healthcare system. Finally, it is noted that there may exist many other applications and situations to which the illustrated methodology applies.

The intent of these examples is to encourage consideration of operational requirements at a greater depth than before and to do so early in the system life cycle when the specification of such requirements will have the greatest impact on design. While it may be easier to delay such considerations until later in the system design process, the consequences of such are likely to be very costly in the long term. The objective in systems engineering is to "force" considerations of operational requirements as early as practicable in the design process.

Illustration 1: River Crossing Problem. Returning to the regional public transportation authority facing the problem of providing for capability that will allow for a significant increase in the two-way traffic flow across a river dividing a growing municipality (a *what*). Further study of the problem (i.e., the current deficiency) revealed requirements for the two-way flow of private vehicles, taxicabs, buses, rail and rapid transit cars, commercial vehicles, large trucks, people on motor cycles and bicycles, and pedestrians across the river. Through advanced system planning and consideration of possible architectures, various river crossing concepts were proposed and evaluated for physical and economic feasibility. These included going under the river, on the river, spanning the river, over the river, or possibly re-routing the river itself. Feasibility considerations determined that the river is not a good candidate for rerouting, both physically and due to its role in providing navigable traffic flow upstream and downstream.

Results from the study indicated that the most attractive approach is the construction of some type of a bridge structure spanning the river (a *how*). From this point on, it is necessary to delve further into the operational requirements leading to the selection and evaluation of a bridge type (suspension, pier and superstructure, causeway, etc.) by considering some detailed "design-to" factors as below:

1. *Mission definition and performance parameters:* The peak traffic flow rate(s) in units/people per hour during any 24-hour period shall be 4,000 for passenger cars, 120 for taxicabs, 20 for buses and trolley cars, 320 for commercial vehicles, 120 for large trucks, 100 people on bicycles, and 180 pedestrians.

2. *Operational deployment and distribution:* The proposed bridge shall be located at point *ABC*, along a straight part of the river, connecting the two community urban centers (e.g., Main Street in Community *X* and Center Street in Community *Y*). The location shall be based on the results from the study dealing with geological, geotechnical, hydrology and water flow, and related factors (refer to Section 3).

3. *Operational life cycle (horizon):* The proposed bridge shall be constructed and fully operational 5 years from the date of initial contract award, and the operational life cycle for the bridge shall be 50 years.

4. *Effectiveness factors:* The operational availability (A_o) for the overall system (i.e., the bridge itself along with all of its operational infrastructure) shall be at least 99.5%, the MTBM shall be 5 years or greater, and the maintenance downtime (MDT) shall be 1 day or less.

5. *Environmental factors:* The proposed bridge shall be fully operational in an environment with temperatures ranging from +125°F to –50°F ambient; in 100% humidity; in rain, sleet, and/or snow; able to withstand any shock and vibration due to a fully loaded traffic flow; and able to withstand any earth tremors of up to 6.0 on the Richter Scale. In the event of inclement weather (i.e., snow, sleet, or ice), the bridge road conditions shall be returned to complete operational status within 2 hours or less.

6. *Environmental sustainment factors:* There shall be no degradation to the river flow, adjoining river embankment, air quality, acoustical emission, and/or view-scope aesthetics. The material resources utilized in the construction of the bridge and its infrastructure shall be completely replenished in 5 years, or less, from the point in time when the bridge initially becomes fully operational.

7. *Economic factors:* The bridge and its infrastructure shall be designed such that the projected life-cycle cost (LCC) shall not exceed $X, and the annual operational and maintenance cost shall not exceed 1% of the system acquisition cost (design, construction, and test and evaluation). All future design decisions (i.e., modifications or otherwise) must be justified and based on total system life-cycle cost.

While some of the specific design-to qualitative and quantitative factors introduced in this example may vary from one project to the next, this example is presented with the intention of illustrating those considerations that must be addressed early in conceptual design as it pertains to the river crossing problem.

Illustration 2: Aircraft System. Based on the results from feasibility analysis of conceptual design alternatives, it was determined that there is a need for a new aircraft. These aircraft are to be procured and deployed in multiple quantities throughout the world. The anticipated geographical locations, estimated quantities per location, and average aircraft utilization times are projected in Figure 3.

In terms of missions, each aircraft will be required to fly three different mission profiles, as illustrated in Figure 4 and described in *Specification 12345*. Basically, an aircraft will be prepared for flight, take off and complete a specific mission scenario, return to its base, undergo maintenance as required, and be returned to a "ready" status. For planning purposes, all aircraft will be required to fly at least one each of the three different mission profiles per week.

The aircraft shall meet performance requirements in accordance with *Specification 12345*; the operational availability (A_o) shall be at least 90%; the MDT shall not exceed 3 hours; the maintenance time ($\overline{M}ct$) shall be 30 minutes or less; M_{max} at the 90th percentile shall be 2 hours; the maintenance labor hours per operating hour (MLH/OH) for the aircraft shall be 10 or less; and the cost per maintenance action at the organizational level shall not exceed $10,000. The aircraft will incorporate a built-in test capability that will allow for fault isolation to the unit level with an 85% self-test thoroughness. No special external support equipment will be allowed. Relative to the support infrastructure, there will be an intermediate-level maintenance capability located at each operational base. In addition, there will be two depot-level maintenance facilities. The overall support concept is illustrated in Figure 5.

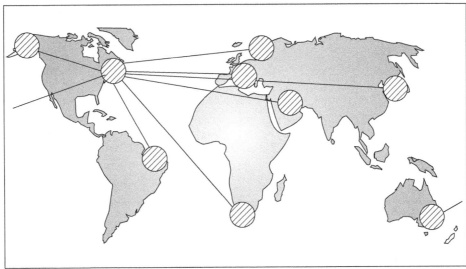

Number of Units in Operational Use per Year

Geographical Operational Areas	Year Number										Total Units
	1	2	3	4	5	6	7	8	9	10	
1. North and South America	–	–	10	20	40	60	60	60	35	25	310
2. Europe	–	–	12	24	24	24	24	24	24	24	180
3. Middle East	–	–	12	12	12	24	24	24	24	24	156
4. South Africa	–	–	12	24	24	24	24	24	24	24	180
5. Pacific Rim 1	–	–	12	12	12	24	24	24	12	12	132
6. Pacific Rim 2	–	–	12	12	12	12	12	12	12	12	96
Total	–	–	70	104	124	168	168	168	131	121	1,054

Average Utilization: 4 Hours per Day, 365 Days per Year

Figure 3   System operational requirements (distribution and utilization).

Although the description given here is rather cursory in nature considering the total spectrum of system operational requirements, it is necessary as an *input* to design to define (1) the geographical location and the anticipated environment in which the system is to be utilized; (2) one or more typical mission scenarios in order to identify operational sequences, potential stresses on the system, and system effectiveness requirements; and (3) the system operational life cycle in order to determine performance factors, reliability and maintainability requirements, human engineering requirements, and the anticipated length and magnitude of the required maintenance and sustaining support capability. In essence, the overall concept conveyed in Figures 3–5, supplemented with the appropriate quantitative measures, must be defined to establish a future baseline for system design and development. While many of the specific quantitative factors specified in this problem are not introduced, the objective here is to further emphasize that such factors must be addressed early in determining operational requirements as an input to the design process.

Illustration 3: Communication System.   A new communication system with an increased range capability and improved reliability is needed to replace several existing

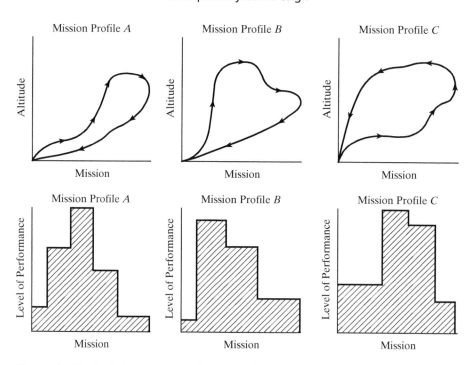

Figure 4 Typical aircraft operational profiles.

capabilities that are currently distributed in multiple quantities throughout the world. The system must accomplish three basic missions:

1. *Mission Scenario 1:* System elements are to be installed in low-flying light aircraft (10,000 feet altitude or less) in quantities of one per aircraft. The system shall enable communication with ground vehicles dispersed throughout mountainous and flat terrain and with a centralized area communication facility. It is anticipated that each aircraft will fly 15 missions per month with an average mission duration of 2 hours. A typical mission profile is illustrated in Figure 6. The communication system utilization requirement is 110% (1.1 hours of system operation for every hour of aircraft operation, which includes air time plus some ground time). The system must be operationally available 99.5% of the time and have a reliability mean time between failures (MTBF) of not less than 2,000 hours.

2. *Mission Scenario 2:* System elements are to be installed in ground vehicular equipment (e.g., car, light truck, or equivalent) in quantities of one per vehicle. The system shall enable communication with other vehicles at a range of 200 miles in relatively flat terrain, with overhead aircraft at an altitude of 10,000 feet or less, and with a centralized area communication facility. Sixty-five percent of the vehicles will be in operational use at any given point in time and the system shall be utilized 100% of the time for those vehicles that are operational. The system must have a reliability MTBF of at least 1,800 hours and a mean corrective maintenance time ($\overline{M}ct$) of 1 hour or less.

3. *Mission Scenario 3:* System elements are to be installed in 20 area communication facilities located throughout the world with 5 operational systems assigned to each

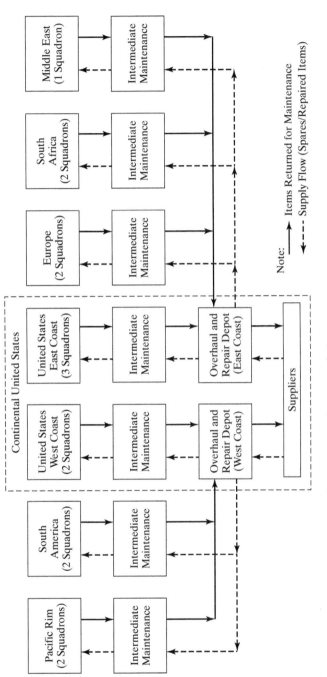

Figure 5 Top-level system maintenance and support infrastructure.

Conceptual System Design

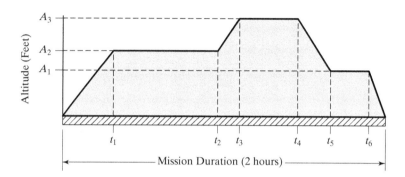

Figure 6 Mission profile.

facility. The system shall enable communication with aircraft flying at an altitude of 10,000 feet or less and within a radius of 500 miles from the facility and with ground vehicles at a range of 300 miles in relatively flat terrain. Four of the systems are utilized 24 hours a day, and the remaining system is a backup and used an average of 6 hours per day. Each operational system shall have a reliability MTBF of at least 2,500 hours and a $\overline{M}ct$ of 30 minutes or less. Each communication facility shall be located at an airport.

In the interest of minimizing the total cost of support (e.g., test and support equipment, spares, and personnel), the transmitter–receiver, which is a major element of the system, shall be a common design for the vehicular, airborne, and ground applications. The antenna configuration may be unique in each instance.

Operational prime equipment/software shall be introduced into the inventory commencing 4 years from this date, and a maximum complement is acquired by 8 years. The maximum complement must be sustained for 10 years, after which a gradual phase-out will occur through attrition. The last equipment is expected to be phased out of the inventory in 25 years. The program schedule is illustrated in Figure 7.

Specifically, the requirements dictate the need for (1) 20 centralized communication facilities; (2) 11 aircraft assigned to each communication facility; and (3) 55 vehicles

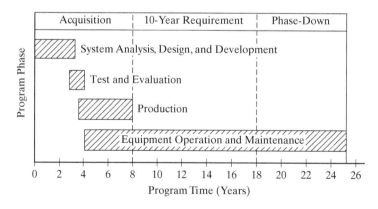

Figure 7 Program schedule.

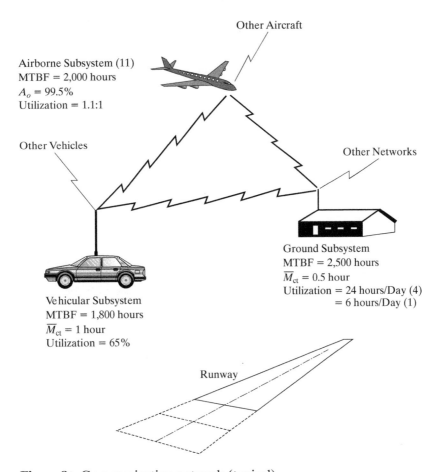

Airborne Subsystem (11)
MTBF = 2,000 hours
A_o = 99.5%
Utilization = 1.1:1

Other Aircraft

Other Vehicles

Other Networks

Ground Subsystem
MTBF = 2,500 hours
$\overline{M}_{ct}$ = 0.5 hour
Utilization = 24 hours/Day (4)
            = 6 hours/Day (1)

Vehicular Subsystem
MTBF = 1,800 hours
$\overline{M}_{ct}$ = 1 hour
Utilization = 65%

Runway

Figure 8   Communication network (typical).

assigned to each communication facility. Based on the three mission scenarios just defined, there is a total requirement for 1,420 prime equipments deployed in a series of communication networks (i.e., the total system), as illustrated in Figure 8.

In support of the program schedule and the basic need, it is necessary to develop an equipment inventory profile, as shown in Figure 9. This profile provides an indication of the total quantity of prime equipment in the user's inventory during any given year in the life cycle. The front end of the profile represents the production rate, which, of course, may vary considerably, depending on the type and complexity of equipment/software, the capacity of the production facility, and the cost of production. The total quantity of prime equipment produced is 1,491, which assumes (1) that 5% of the equipment will be condemned during the 10-year full-complement period due to loss or damage beyond economical repair, and (2) that production is accomplished on a one-time basis (to avoid production startup and shutdown costs). In other words, assuming that production is continuous, 1,491 pieces of equipment must be produced to cover attrition and yet maintain the operational requirements of 1,420 equipments through the 10-year period. After the 10-year period, the number of units is reduced by attrition and/or phase-out owing to obsolescence until the inventory is completely depleted.

Conceptual System Design

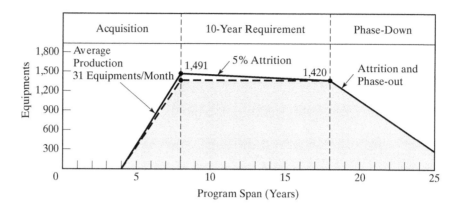

Figure 9 Basic inventory requirements over the system life cycle.

Definition of the operational requirements for the communication system (i.e., distribution, utilization, effectiveness factors, reliability and maintainability measures, etc.) provides the basis for determining the maintenance and support concept (refer to Section 5) and also the specific *design-to* requirements. As the system design and development process progresses, the operational requirements for the communication system are further refined on an iterative basis.

Illustration 4: Commercial Airline Upgrade.   Three commercial airline carriers are proposing to serve a large metropolitan region 8 years hence. As future growth is expected, additional airline companies may become involved at a later time. For planning purposes, the combined anticipated passenger handling requirement follows the projection in Figure 10. The combined airline requirements are as shown.

1. Anticipated flight arrivals/departures are evenly spaced in the time periods indicated. It is assumed that 100 passengers constitute an average flight load.
2. The aircraft A_0 is 95%. That is, 95% of all flights must be fully operational when scheduled (discounting aborts due to weather). Allowable factors for scheduled maintenance and passenger loading are as follows:

Function	Frequency	Downtime
Through service	Each through flight	30 minutes
Turnaround service	Each turnaround	1 hour
Termination check	Each terminal flight	6 hours
Service check	15 days	9 hours

Periodic and main base checkouts will be accomplished elsewhere.

3. Allowable unscheduled maintenance in the area shall be limited to the organizational level and will include the removal and replacement of line replaceable items, tire changes, and engine replacements as required. The specific MDT limits are as follows:
   - Engine change — 6 hours
   - Tire change — 1 hour
   - Any other item — 1 hour

Time Period	Anticipated Flights per Day			
	Point A	Point B	Point C	Point D
6:00 A.M. – 11:00 A.M.	33	65	95	105
11:00 A.M. – 4:00 P.M.	17	43	50	55
4:00 P.M. – 9:00 P.M.	33	60	90	100
9:00 P.M. – 6:00 A.M.	3	10	15	18
Total	86	178	250	278

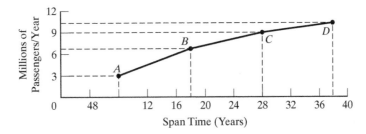

Figure 10   Projected passenger-handling requirement.

4. The metropolitan area must provide the necessary ground facilities to support the following types of aircraft (fully loaded): A-320, A-321, A-330, B-727, B-737, B-747, B-757, B-767, B-777, L-1011, and MD-80. In addition, provisions must be made for cargo handling and storage.

The airline carriers have identified a need to provide air transportation service but, from the airlines' standpoint, the metropolitan area must provide the necessary logistic resources (facilities, test and support equipment, ground handling equipment, people movers, operating and maintenance personnel, etc.) to support this service. This involves selecting a site for an air transportation facility; designing and constructing the facility; providing local transportation to and from the airline terminal; acquiring the test and support equipment, spare/repair parts, personnel, and data to support airline operations; and maintaining the total capability on a sustaining basis throughout the planned operational period.

Because of the size of the program and the growth characteristics projected in Figure 10, a three-phased construction program is planned. The schedule is presented in Figure 11.

The initial step is to select a location for the air transportation facility. The selection process considers available land, terrain, geology, wind effects, distance from the metropolitan area, access via highway and/or public transportation, noise and ecology requirements, and cost. Once a site has been established, the facility must include runways, holding area, flight control equipment, airline terminal, control tower, operations building, hangars, maintenance docks, fuel docks, cargo handling and storage capability, utilities, and all the required support directly associated with passenger needs and comfort.

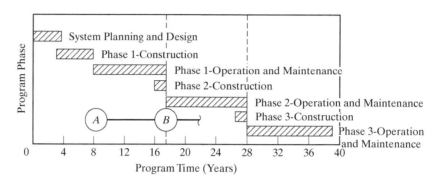

Conceptual System Design

Figure 11   Program schedule.

The ultimate design configuration of the air transportation facility is based directly on the operational requirements—anticipated airline flight arrivals/departures, passenger loading, aircraft turnaround times, and maintenance and servicing requirements. For example, at a point 18 years hence (Figure 10, point B), the anticipated average number of flights is 65 per day between the hours of 6:00 A.M. and 11:00 A.M. The number of through flights is 30, the gate time for each is 45 minutes (to accomplish servicing and passenger loading), and the number of turnarounds is 35 with a gate time of 1 hour each. The total gate time required (considering no delays and assuming one gate for each aircraft arrival/departure) is 50 hours during the 6:00 A.M.–11:00 A.M. time period; thus, at least 10 gates are required in the passenger terminal to satisfy the load. This requirement, in turn, influences the size of the passenger waiting lobby and the number of airline personnel agents required. Further, the servicing and ground handling of the aircraft require certain consumables (fuel, oil, lubricants, etc.), spare/repair parts, test and support equipment (towing vehicles, fuel trucks, etc.), and technical data (operating and maintenance instructions). The possibility of unscheduled maintenance dictates the need for trained maintenance personnel, spare/repair parts, data, and a backup maintenance dock or hangar.

One can go on identifying requirements to support the basic need of the metropolitan area and the commercial airline companies. It readily becomes obvious that an understanding of operational requirements and logistic support plays a major role. There are requirements associated with the aircraft, the air transportation facility, and the transportation mode between the air transportation facility and the metropolitan area. The commercial airline illustration presented here only touches on a small segment of the problem. The problem should be addressed from a total systems approach considering the functions associated with all facets of the operation.

Illustration 5: Community Hospital.   A different type of system was selected to illustrate that the same approach can be applied in the design of a wide variety of system configurations. More specifically, and based on the results of a feasibility study, the objective here is to design and construct a new hospital to meet the needs of a community with a population of approximately 150,000, including a 10% per year growth. This new hospital, to be considered as a system by itself, is to be part of a larger "Community Healthcare" capability, as illustrated in Figure 12. Note that we are addressing a *system-of- systems* (SOS) configuration.

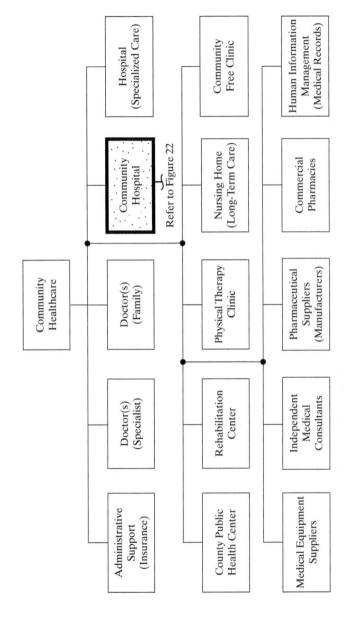

Figure 12 Community healthcare infrastructure.

The anticipated mission of the new hospital is to provide a variety of services for those in need, commencing 2 years from this date and with a planning horizon of 50 years. More specifically, the hospital shall be capable of the following:

1. Provide support for 120 patients who are admitted for surgical and related reasons. It is assumed that the average patient internship (stay) in the hopital shall be 3 days.

2. Process 40 patients per day through the "outpatient surgery" department (for minor surgical and same-day care procedures). The preregistration process shall be no longer than 30 minutes.

3. Process 100 patients per day for diagnostic imaging and/or laboratory testing. The preregistration process shall take 30 minutes or less.

4. Process 50 patients per day through its Emergency Room (ER), or Emergency Care, capability.

5. Transmit/receive the appropriate medical records within 15 minutes from the time of request.

6. Respond to the above requirements with an overall A_o of 95%, or greater.

7. The cost per patient per day shall not exceed X dollars ($).

Given these basic requirements, one can then proceed with the design of the hospital facility, its proposed location and layout, and so on. Further, it is essential that the design and construction of the community hospital be accomplished within the context of the overall "Community Healthcare" configuration shown in Figure 12, considering all of the interfaces that exist with the various other associated human-care capabilities shown. While this example is not presented to the same level of detail as the first four illustrations, it is hoped that one can understand the importance for defining operational requirements early in conceptual design, regardless of the type of system being addressed.

Additional Illustrations. The five illustrations presented, derived from the results of feasibility analyses conducted earlier, are representative of typical "needs" for which system operational requirements must be defined at the inception of a program, and must serve as the basis for all subsequent program activities. These requirements must not only be implemented within the bounds of the specific system configuration in question but must also consider all possible external interfaces that may exist.

For example, as illustrated in Figure 13, one may be dealing with a number of different systems, all of which are closely related and may have a direct impact on each other. Also, there may be some "sharing" of capabilities across the board; for example, the air and ground transportation systems utilizing some of the same components which operate as part of the communication systems. Thus, the design of any new system must consider all possible impacts that it will have on other systems within the same SOS configuration (as shown in the figure), as well as those possible impacts that the other systems may have on the new system.

The methodology employed is basically the same for any system, whether the subject is a relatively small item as part of the river crossing bridge, installed in an aircraft or on a ship, a factory, or a large one-of-a-kind project such as the community hospital involving design and construction. In any case, the system must be defined in terms of its projected mission, performance, operational deployment, life cycle, utilization, effectiveness factors, and the anticipated environment.

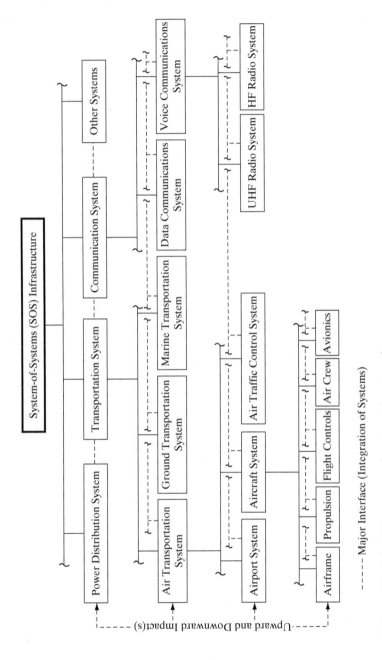

Figure 13 Multiple systems (system-of-systems).

75

5 SYSTEM MAINTENANCE AND SUPPORT

In addressing system requirements, the normal tendency is to deal primarily with those elements of the system that relate directly to the "performance of the mission," that is, prime equipment, operator personnel, operational software, operating facilities, and associated operational data and information. At the same time, too little attention is given to system maintenance and support and the sustainment of the system throughout its planned life cycle. In general, the emphasis has been directed toward only *part* of the system and not the entire system as an entity, which has led to costly results in the past.

To realize the overall benefits of systems engineering, it is essential that *all* elements of the system be considered on an integrated basis from the beginning. This includes not only the prime mission-related elements of the system but the maintenance and support infrastructure as well. The prime system elements must be designed in such a way that they can be effectively and efficiently supported through the entire system life cycle, and the maintenance and support infrastructure must be responsive to this requirement. This, in turn, means that one should also address the design characteristics as they pertain to transportation and handling equipment, test and support equipment, maintenance facilities, the supply chain process, and other applicable elements of logistic support.

The maintenance and support *concept* developed during the conceptual design phase evolves from the definition of system operational requirements described in Section 4. It constitutes a before-the-fact series of illustrations and statements leading to the definition of reliability, maintainability, human factors and safety, constructability and producibility, supportability, sustainability, disposability, and related requirements for design. It constitutes an "input" to the design process, whereas the maintenance *plan* (developed later) defines the follow-on requirements for system support based on a known design configuration and the results of supportability analysis.

The maintenance and support concept is reflected by the network and the activities and their interrelationships, illustrated in Figure 14. The network exists whenever there are requirements for corrective and/or preventive maintenance at any time and throughout the system life cycle. By reviewing these requirements, one should address such issues as the levels of maintenance, functions to be performed at each level, responsibilities for the accomplishment of these functions, design criteria pertaining to the various elements of support (e.g., type of spares and levels of inventory, reliability of the test equipment, personnel quantities and skill levels), and the effectiveness factors and "design-to" requirements for the overall maintenance and support infrastructure. Although the design of the prime elements of the system may appear to be adequate, the overall ability of the system to perform its intended mission objective highly depends on the effectiveness of the support infrastructure as well.

While there may be some variations that arise, depending on the type and nature of the system, the maintenance and support concept generally includes the following items:

1. *Levels of maintenance:* Corrective and preventive maintenance may be performed on the system itself (or an element thereof) at the site where the system is operating and used by the customer, in an intermediate shop near the customer's operational site, and/or at a depot or manufacturer's facility. Maintenance level pertains to the division of functions and tasks for each area where maintenance is performed. Anticipated frequency of maintenance, task complexity, personnel skill-level requirements, special facility needs, supply chain requirements, and so on, dictate to a great extent the specific functions to be accomplished at each level. Depending on the nature and mission of the system, there may be two, three, or four levels of maintenance; however, for the purposes of further discussion, maintenance may be classified as *organizational, intermediate*, and *manufacturer/ depot/supplier*. Figure 15 describes the basic criteria and differences between these levels.

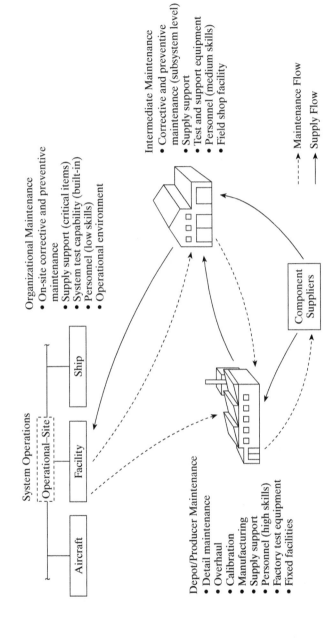

Figure 14 System operational and maintenance flow.

Criteria	Organizational Maintenance	Intermediate Maintenance		Supplier/Manufacturer/Depot Maintenance
		Mobile or semimobile units	Fixed units	
Done where?	At the operational site or wherever the prime elements of the system are located	Truck, van, portable shelter, or equivalent	Fixed field shop	Supplier/manufacturer/depot facility
				Specialized repair activity or manufacturer's plant
Done by whom?	System/equipment operating personnel (low-maintenance skills)	Personnel assigned to mobile, semimobile, or fixed units (intermediate-maintenance skills)		Depot facility personnel or manufacturer's production personnel (high-maintenance skills)
On whose equipment?	Using organization's equipment	Equipment owned by using organization		
Type of work accomplished?	Visual inspection Operational checkout Minor servicing External adjustments Removal and replacement of some components	Detailed inspection and system checkout Major servicing Major equipment repair and modifications Complicated adjustments Limited calibration Overload from organizational level of maintenance		Complicated factory adjustments Complex equipments repairs and modifications Overhaul and rebuild Detailed calibration Supply support Overload from intermediate level of maintenance

Figure 15 Major levels of maintenance.

2. *Repair policies:* Within the constraints illustrated in Figures 14 and 15, there are a number of possible policies specifying the extent to which the repair of an element or component of a system should be accomplished (if at all). A repair policy may dictate that an item should be designed such that, in the event of failure, it should be *nonrepairable, partially repairable,* or *fully repairable.* Stemming from the operational requirements described in Section 4 (refer to the five system illustrations), an initial "repair policy" for the system being developed should be established with the objective of providing some early guidelines for the design of the different components that make up the system. Referring to the example of the repair policy, illustrated in Figure 16, it can be seen that there are numerous quantitative factors, which were initially derived from the definition of system operational requirement, that provide "design-to" guidelines as an input to the overall design process; for example, the system shall be designed such that the MTBM shall be 175 hours or greater, the MDT shall be 2 hours or less, the MLH/OH shall not exceed 0.1, and so on. A repair policy should be initially developed and established during the conceptual design phase, and subsequently updated as the design progresses and the results of the level-of-repair and supportability analyses become available.

3. *Organizational responsibilities:* The accomplishment of maintenance may be the responsibility of the customer, the producer (or supplier), a third party, or a combination thereof. In addition, the responsibilities may vary, not only with different components of the system but also as one progresses in time through the system operational use and sustaining support phase. Decisions pertaining to organizational responsibilities may affect system design from a diagnostic and packaging standpoint, as well as dictate repair policies, product warranty provisions, and the like. Although conditions may change, some initial assumptions are required at this point in time.

4. *Maintenance support elements:* As part of the initial maintenance concept, criteria must be established relating to the various elements of maintenance support. These elements include supply support (spares and repair parts, associated inventories, and provisioning data), test and support equipment, personnel and training, transportation and handling equipment, facilities, data, and computer resources. Such criteria, as an input to design, may cover self-test provisions, built-in versus external test requirements, packaging and standardization factors, personnel quantities and skill levels, transportation and handling factors, constraints, and so on. The maintenance concept provides some initial system design criteria pertaining to the activities illustrated in Figure 14, and the final determination of specific logistic and maintenance support requirements will occur through the completion of a supportability analysis as design progresses.

5. *Effectiveness requirements:* These constitute the effectiveness factors associated with the support capability. In the supply support area, they may include a spare-part demand rate, the probability that a spare part will be available when required, the probability of mission success given a designated quantity of spares in the inventory, and the economic order quantity as related to inventory procurement. For test equipment, the length of the queue while waiting for test, the test station process time, and the test equipment reliability are key factors. In transportation, transportation rates,

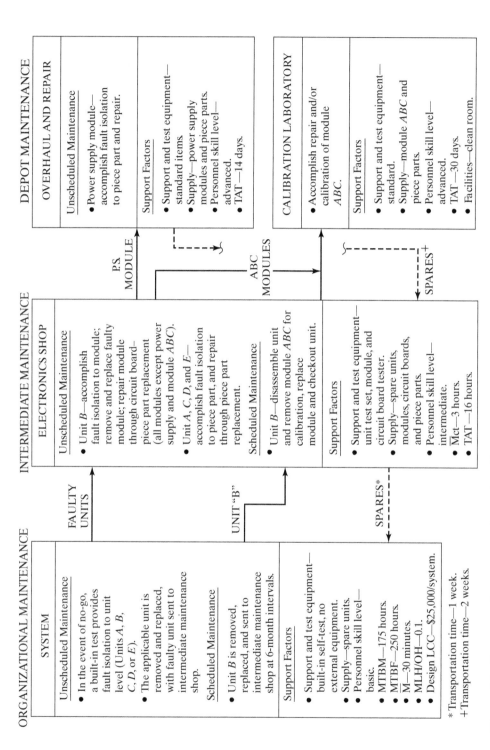

Figure 16 System maintenance and repair policy.

transportation times, the reliability of transportation, and transportation costs are of significance. For personnel and training, one should be interested in personnel quantities and skill levels, human error rates, training rates, training times, and training equipment reliability. In software, the number of errors per mission segment, per module of software, or per line of code may be important measures. For the supply chain overall, reliability of service, item processing time, and cost per item processed may be appropriate metrics to consider. These factors, as related to a specific system-level requirement, must be addressed. It is meaningless to specify a tight quantitative requirement applicable to the repair of a prime element of the system when it takes 6 months to acquire a needed spare part (for example). The effectiveness requirements applicable to the supply chain and support capability must complement the requirements for the system overall.

6. *Environment:* Definition of environmental requirements as they pertain to the maintenance and support infrastructure is equally important. This includes the impact of external factors such as temperature, shock and vibration, humidity, noise, arctic versus tropical applications, operating in mountainous versus flat terrain country, shipboard versus ground conditions, and so on, on the design of the maintenance and support infrastructure. In addition, it is also necessary to address possible "outward" environmental impact(s) of the maintenance and support infrastructure on other systems and on the environment in general (with the "design for sustainability" as being a major objective).

The maintenance concept provides the foundation that leads to the design and development of the maintenance and support infrastructure and defines the specific design-to requirements for the various elements of support (e.g., the supply support capability, transportation and handling equipment, test and support equipment, and facilities). These requirements, as they apply to system life-cycle support, can have a significant "feedback" effect (impact) on the prime elements of the system as well. Thus, the definition of system operational requirements and the development of the maintenance concept must be accomplished concurrently and early during the conceptual design phase. The combined result forms the basis for development, particularly with regard to the subject areas of design for reliability, design for maintainability, design for usability (human factors), design for supportability, and others.

In summary, when defining the maintenance concept, it is important that consideration be given to the interfaces that may exist between the support requirements and infrastructure for the new system being developed and those comparable requirements for other systems that may be contained within the same overall *system-of-systems* configuration (refer to Figure 13). The requirements for this new system must first be defined, the impacts on (and from) the other systems evaluated, major conflicting areas noted, and finally modifications be incorporated as required. Care must be taken to ensure that the requirements for this new system are not compromised in any way. Finally, the selection of the ultimate maintenance and support infrastructure configuration must be justified on the basis of the LCC. What may seem to be a least-cost approach in providing maintenance support at the local level may not be such when considering the costs associated with all of the supply chain activities for the system in question.

6 TECHNICAL PERFORMANCE MEASURES

Technical performance measures (TPMs) are quantitative values (estimated, predicted, and/or measured) that describe system performance. TPMs are measures of the attributes and/or characteristics that are inherent within the design. *Design-dependent parameters* (DDPs) are quantified and are the bases for the determination of TPMs. During conceptual design, these values are best determined by parametric methods.

6.1 TPM Identification and Evolution

TPMs may include such quantitative factors as reliability MTBF, maintainability MTBM and MDT, A_0, logistics response time, information processing time, facility utilization rate, a specific value of LCC as a design requirement, and so on. There may be any number, or combination, of TPM values specified for a particular system during the conceptual design phase (as "design-to" requirements). The objective is to influence the system design process to incorporate the right attributes/characteristics to produce a system that will ultimately meet customer requirements effectively and efficiently.

TPMs evolve primarily from the development of system operational requirements and the maintenance and support concept. Referring to the five system examples in Section 4 and the illustrated support network and repair policy in Section 5, there are many different quantitative parameters that constitute relevant performance measures (of one type or another) to which the ultimate system design configuration must comply. Some of these specified values may be contradictory when it comes to determining the specific characteristics that must be incorporated into the design. For example, in the design of a vehicle, is *speed* more important than *size*? For a manufacturing plant, is *production quantity* more important than *production quality*? In a communication system, is *range* more important than *clarity of message*? For a computer capability, is *capacity* more important than *speed*? Is *reliability* more important than *maintainability*? Are *human factors* more important than *life-cycle cost*? There may be any number of different design objectives, and the designer needs to understand which are more important than others and, if a trade-off has to be made, wherein the design compromises must be accepted in order to meet a higher-level requirement.

Figure 17 conveys the results from a TPM identification and prioritization effort involving a team of individuals representing the customer (user), the appropriate designer(s), the area of system maintenance and support, a major supplier(s), and key management personnel. The "team" met on several occasions and, with the customer providing the necessary guidance, decided the most important objectives in design. Note that critical TPMs are identified along with their relative degrees of importance. The performance factors of *velocity, availability*, and *size* are the most critical in this instance and where emphasis in design must be directed. The specific features or attributes to be incorporated into the design (e.g., use of standardized components, accurate and thorough diagnostic provisions, use of high-reliability components, and use of lightweight materials) must be responsive to these requirements, and there must be a *traceability* of requirements from the system level down to its various elements.

6.2 Quality Function Deployment

A useful tool that can be applied to aid in the establishment and prioritization of TPMs is the *quality function deployment* (QFD) model. QFD constitutes a team approach to help

Technical Performance Measure	Quantitative Requirement ("Metric")	Current "Benchmark" (Competing Systems)	Relative Importance (Customer Desires) (%)
Process time (days)	30 days (maximum)	45 days (system M)	10
Velocity (mph)	100 mph (minimum)	115 mph (system B)	32
Availability (operational)	98.5% (minimum)	98.9% (system H)	21
Size (feet)	10 feet long 6 feet wide 4 feet high (maximum)	9 feet long 8 feet wide 4 feet high (system M)	17
Human factors	Less than 1% error rate per year	2% per year (system B)	5
Weight (pounds)	600 pounds (maximum)	650 pounds (system H)	6
Maintainability (MTBM)	300 miles (minimum)	275 miles (system H)	9
			100

Figure 17 Prioritization of technical performance measures (TPMs).

ensure that the "voice of the customer" is reflected in the ultimate design. The purpose is to establish the necessary requirements and to translate those requirements into technical solutions. Customer requirements are defined and classified as *attributes*, which are then weighted based on the degree of importance. The QFD method provides the design team an understanding of customer desires, forces the customer to prioritize those desires, and enables a comparison of one design approach against another. Each customer attribute is then satisfied by a technical solution.[1]

The QFD process involves constructing one or more matrices, the first of which is often referred to as the *House of Quality* (HOQ). A modified version of the HOQ is presented in Figure 18. Starting on the left side of the structure is the identification of customer needs and the ranking of those needs in terms of priority, the levels of importance being specified quantitatively. This side reflects the *whats* that must be addressed. The top part of the HOQ identifies the designer's *technical* response relative to the attributes that must be incorporated into the design in order to respond to the needs (i.e., the voice of the

[1]The QFD method was first developed at the Kobe Shipyard of Mitsubishi Heavy Industries, Ltd., Japan, in the late 1960s and has evolved since. Some references are (1) Yoji Akao, *Quality Function Deployment: Integrating Customer Requirements into Product Design* (New York, NY: Productivity Press, 1990); (2) L. Cohen, *Quality Function Deployment: How to Make QFD Work for You* (Reading, MA: Addison-Wesley, 1995); and (3) J. B. Revelle, J. W. Moran, and C. Cox, *The QFD Handbook* (Hoboken, NJ: John Wiley & Sons, Inc., 1997).

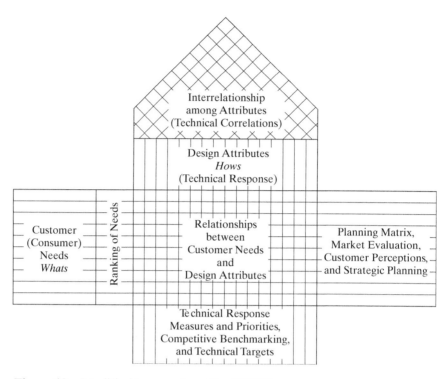

Interrelationship
among Attributes
(Technical Correlations)

Design Attributes
Hows
(Technical Response)

Customer
(Consumer)
Needs
Whats

Ranking of Needs

Relationships
between
Customer Needs
and
Design Attributes

Planning Matrix,
Market Evaluation,
Customer Perceptions,
and Strategic Planning

Technical Response
Measures and Priorities,
Competitive Benchmarking,
and Technical Targets

Figure 18 Modified house of quality (HOQ).

customer). This part constitutes the *hows*, and there should be at least one technical solution for each identified customer need. The interrelationships among attributes (or technical correlations) are identified, as well as possible areas of conflict. The center part of the HOQ conveys the strength or impact of the proposed technical response on the identified requirement. The bottom part allows a comparison between possible alternatives, and the right side of the HOQ is used for planning purposes.

The QFD method is used to facilitate the translation of a prioritized set of subjective customer requirements into a set of system-level requirements during conceptual design. A similar approach may be used to subsequently translate system-level requirements into a more detailed set of requirements at each stage in the design and development process. In Figure 19, the *hows* from one house become the *whats* for a succeeding house. Requirements may be developed for the system, subsystem, component, manufacturing process, support infrastructure, and so on. The objective is to ensure the required justification and traceability of requirements from the top down. Further, requirements should be stated in *functional* terms.

Although the QFD method may not be the only approach used in helping define the specific requirements for system design, it does constitute an excellent tool for creating the necessary visibility from the beginning. One of the largest contributors to *risk* is the lack of a good set of requirements and an adequate system specification. Inherent within the system specification should be the identification and prioritization of TPMs. The TPM, its associated metric, its relative importance, and benchmark objective in terms of what is currently available will provide designers with the necessary guidance for accomplishing their task. This is essential for establishing the appropriate levels of design emphasis, for

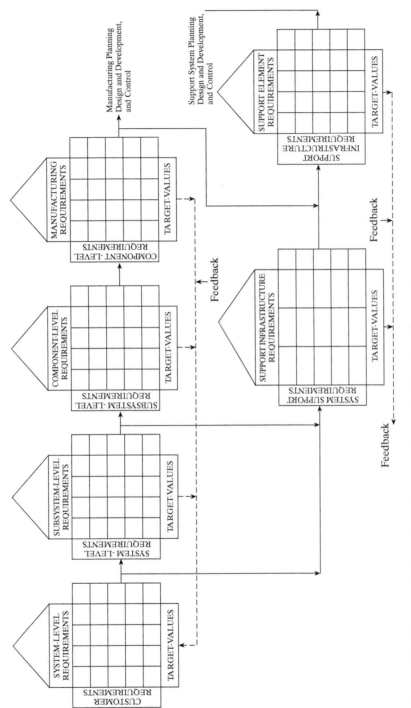

Figure 19 The traceability of requirements through a "family of houses."

defining the criteria as an input to the design, and for identifying the levels of possible risk should the requirements not be met. Again, care must be taken to ensure that these system-level requirements are not adversely impacted by external factors from other systems within the same overall SOS structure.

7 FUNCTIONAL ANALYSIS AND ALLOCATION

An essential activity in early conceptual and preliminary design is the development of a *functional* description of the system to serve as a basis for identification of the resources necessary for the system to accomplish its mission. A *function* refers to a specific or discrete action (or series of actions) that is necessary to achieve a given objective; that is, an operation that the system must perform, or a maintenance action that is necessary to restore a faulty system to operational use. Such actions may ultimately be accomplished through the use of equipment, software, people, facilities, data, or various combinations thereof. However, at this point in the life cycle, the objective is to specify the *whats* and not the *hows*; that is, *what* needs to be accomplished versus *how* it is to be done.

The *functional analysis* is an iterative process of translating system requirements into detailed design criteria and the subsequent identification of the resources required for system operation and support. It includes breaking requirements at the system level down to the subsystem, and as far down the hierarchical structure as necessary to identify input design criteria and/or constraints for the various elements of the system. The purpose is to develop the top-level *system architecture*, which deals with both "requirements" and "structure."

Referring to the iterative process, the functional analysis actually commences (in a broad context) in conceptual design as part of the problem definition and needs analysis task (refer to Section 1). Subsequently, "operational" and "maintenance" functions are identified, leading to the development of top-level systems requirements, as described in Sections 4–6. The purpose of the "functional analysis" is to present an overall integrated and composite description of the system's *functional architecture*, to establish a functional baseline for all subsequent design and support activities, and to provide a foundation from which all physical resource requirements are identified and justified; that is, the system's *physical architecture*.

7.1 Functional Flow Block Diagrams

Accomplishment of the functional analysis is facilitated through the use of *functional flow block diagrams* (FFBDs). The preparation of these diagrams may be accomplished through the application of any one of a number of graphical methods, including the Integrated DEFinition (IDEF) modeling method, the Behavioral Diagram method, the N-Squared Charting method, and so on. Although the graphical presentations are different, the ultimate objectives are similar. The approach assumed here is illustrated in Figures 20–22.

In Figure 20, a simplified flow diagram with some decomposition is shown. Top-level functions are broken down into second-level functions, second-level functions into third-level functions, and so on, down to the level necessary to adequately describe the system and its various elements in functional terms to show the various applicable functional interface relationships and to identify the resources needed for functional implementation. Block

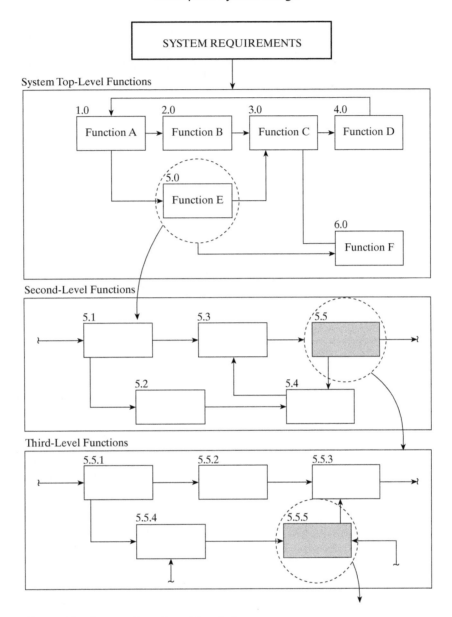

Figure 20 System functional breakdown.

numbers are used to show sequential and parallel relationships, initially for the purpose of providing top-down "traceability" of requirements, and later as a bottom-up "traceability" and justification of the physical resources necessary to accomplish these functions.

Figure 21 shows an expansion of a functional flow block diagram (FFBD), identifying a partial top level of activity, a breakdown of Function 4.0 into a top "operational" flow, and a breakdown of Function 4.2 into a top "maintenance" flow in the event that

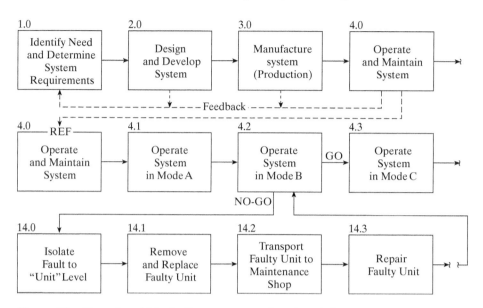

Figure 21 Functional block diagram expansion (partial).

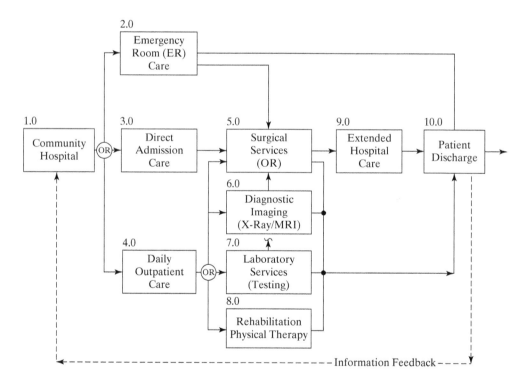

Figure 22 Health care functional flow diagram (extension of Figure 12)

Conceptual System Design

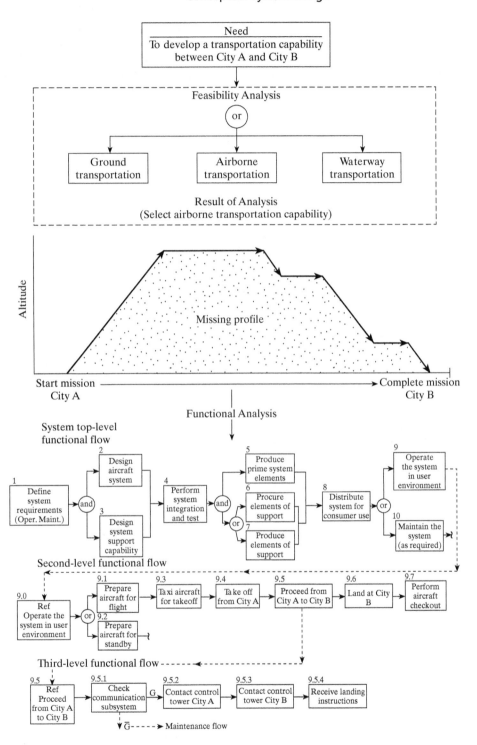

Figure 23 Progression from the "need" to the functional analysis.

this function does not "perform" as required. Note that the words in each block are "action-oriented." Each block represents some operational or maintenance support function that must be performed for the system to accomplish its designated mission, and there are performance measures (i.e., metrics) associated with each block that are allocated from the top. In addition, each block can be expanded (through further downward iteration) and then evaluated in terms of inputs, outputs, controls and/or constraints, and enabling mechanisms. Basically, the "mechanisms" lead to the identification of the physical resources necessary to accomplish the function, evolving from the *whats* to the *hows*. The identification of the appropriate resources in terms of equipment needs, software, people, facilities, information, data, and so on, is a result of one or more trade-off studies leading to a preferred approach as to *how best* to accomplish a given function.

Figure 22 shows an expansion of one of the functions identified in the overall basic community health care infrastructure illustrated in Figure 12. The objective here is to show traceability from the definition of operational requirements in Section 4 by selecting a specific functional block from one of the example illustrations; that is, the "Community Hospital" in Illustration 5.

The functional analysis evolves through a series of steps illustrated in Figure 23. Initially, there is a need to accomplish one or more functions (Sections 1 and 3). Through the definition and system operational requirements (Section 4) and the maintenance and support concept (Section 5), the required functions are further delineated, with the functional analysis providing an overall description of system requirements; that is, *functional architecture*. FFBDs are developed for the primary purpose of structuring these requirements by illustrating organizational and functional interfaces.

As one progresses through the functional analysis, and particularly when developing a new system that is within a higher-level SOS structure, there may be functions identified as "common" and shared with a different system. For example, a "transmitter" or "receiver" function may be common for two different and separate communication systems; a "power supply" function may provide the necessary power for more than one system; an "imaging center" may provide the necessary medical diagnostic services for more than one hospital and/or doctor's office complex (refer to Figure 12), and so on. Figure 24 illustrates the application of several "common functions" as part of the functional breakdown for three different systems; that is, Systems *A*, *B*, and *C*.

The formal functional analysis is fully initiated during the latter stages of conceptual design, and is intended to enable the completion of the system design and development process in a comprehensive and logical manner. More specifically, the functional approach helps to ensure the following:

1. All facets of system design and development, production, operation, support, and phase-out are considered (i.e., all significant activities within the system life cycle).
2. All elements of the system are fully recognized and defined (i.e., prime equipment, spare/repair parts, test and support equipment, facilities, personnel, data, and software).
3. A means is provided for relating system packaging concepts and support requirements to specific system functions (i.e., satisfying the requirements of good functional design).
4. The proper sequences of activity and design relationships are established along with critical design interfaces.

Finally, it should be emphasized that the functional analysis provides the baseline from which reliability requirements, maintainability requirements, human factors requirements,

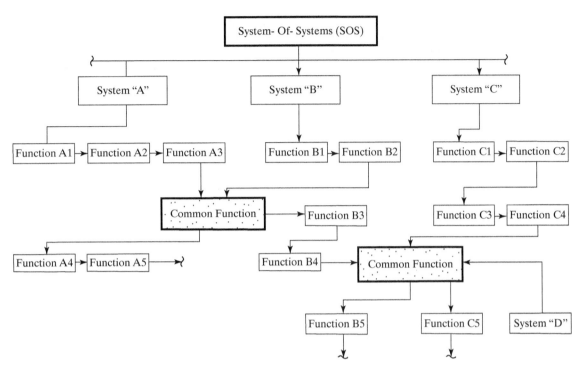

Figure 24 Functional interfaces in a system-of-systems (SOS) configuration.

and supportability requirements are determined. The *functional baseline* leads to the *allocated baseline*, which leads to the *product baseline*.

7.2 Functional Allocation

Given a top-level description of the system through the functional analysis, the next step is to break the system down into elements (or components) by *partitioning*. The challenge is to identify and group closely related functions into packages employing a common resource (e.g., an equipment item or a software package) and to accomplish multiple functions to the extent possible. Although it may be relatively easy to identify individual functional requirements and associated resources on an independent basis, this process may turn out to be rather costly when it comes to packaging system components, weight, size, and so on. The questions are as follows: *What hardware or software (or other) can be selected that will perform multiple functions reliably, effectively, and efficiently? How can new functional requirements be added without adding new physical elements to the system structure?*

The partitioning of the system into elements is evolutionary in nature. Common functions may be grouped or combined to provide a system packaging scheme, with the following objectives in mind:

1. System elements may be grouped by geographical location, a common environment, or by similar types of items (e.g., equipment, software, data packages, etc.) that have similar functions.

2. Individual system packages should be as independent as possible with a minimum of "interaction effects" with other packages. A design objective is to enable the removal and replacement of a given package without having to remove and replace other packages in the process, or requiring an extensive amount of alignment and adjustment as a result.

3. In breaking a system down into subsystems, select a configuration in which the "communications" between the various different subsystems is minimized. In other words, whereas the subsystem's *internal* complexity may be high, the *external* complexity should be low. Breaking the system down into packages requiring high rates of information exchange between these packages should be avoided.

4. An objective is to pursue an *open-architecture* approach in system design. This includes the application of common and standard modules with well-defined standard interfaces, grouped in such a way as to allow for system upgrade modifications without destroying the overall functionality of the system.

An overall design objective is to break the system down into elements such that only a few critical events can influence or change the inner workings of the various packages that make up the system architecture.

Through the process of partitioning and functional packaging, trade-off studies are conducted in evaluating the different design approaches that can be followed in responding to a given functional requirement. It may be appropriate to perform a designated function through the use of equipment, software, people, facilities, data, and/or various combinations thereof. The proper mix is established, and the result may take the form of a system structure similar to the example presented in Figure 25. With this structure representing the proposed system "make-up," the next step is to determine the design-to requirements for each of the system elements; that is, system operator, Equipment 123, Unit *B*, computer resources, and facilities.

Referring to Section 6 (and Figure 17), specific quantitative design-to requirements have been established at the top level for the *system*. These TPMs, which evolved from the definition of operational requirements and the maintenance support concept (Sections 4 and 5), must be *allocated* or *apportioned* down to the appropriate subsystems or the ele-

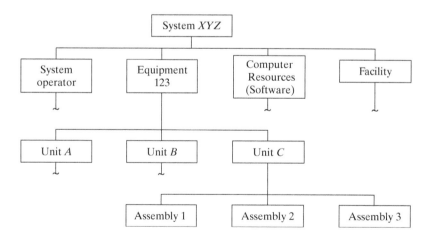

Figure 25 The functional breakdown of the system into components.

ments that make up the system. For instance, given an operational availability A_o requirement of 0.985 for the system (Figure 17), *what design requirements should be specified for the equipment, software, facility, and operator such that, when combined, they will meet the availability requirements for the overall system?* In other words, to guarantee an ultimate system design configuration that will meet all customer (user) requirements, there must be a top-down allocation of design criteria from the beginning; that is, during the latter stages of conceptual design. The allocation process starts at this point and continues with the appropriate subsystems and below as required.

7.3 The System Architecture

With the definition of system operational requirements, the maintenance and support concept, and the identification and prioritization of the TPMs, the basic *system architecture* has been established. The architecture deals with a top-level system structure (configuration), its operational interfaces, anticipated utilization profiles (mission scenarios), and the environment within which it is to operate.

Architecture describes how various requirements for the system interact. This, in turn, leads into a description of the *functional architecture*, which evolves from a functional analysis and is a description of the system in functional terms. From this analysis, and through the requirements allocation process and the definition of the various resource requirements necessary for the system to accomplish its mission, the *physical architecture* is defined. Through application of this process, one is able to evolve from the *whats* to the *hows*.

8 SYSTEM TRADE-OFF ANALYSES

Many different trade-offs are possible as the system design progresses. Decisions must be made regarding the evaluation and selection of appropriate technologies, the evaluation and selection of commercial off-the-shelf (COTS) components, subsystem and component packaging schemes, possible degrees of automation, alternative test and diagnostic routines, various maintenance and support policies, and so on. Later in the design cycle, there may be alternative manufacturing processes, alternative factory maintenance plans, alternative logistic support structures, and alternative methods of material phase-out, recycling, and/or disposal.

One must first define the problem and then identify the design criteria or measures against which the various alternatives will be evaluated (i.e., the applicable TPMs), select the appropriate evaluation techniques, select or develop a model to facilitate the evaluation process, acquire the necessary input data, evaluate each of the candidates under consideration, perform a sensitivity analysis to identify potential areas of risk, and finally recommend a preferred approach. This process is illustrated in Figure 26, and can be tailored and applied at any point in the life cycle. Only the depth of the analysis and evaluation effort will vary, depending on the nature of the problem.

Trade-off analysis involves *synthesis*. Synthesis refers to the combining and structuring of components to create a feasible system configuration. Synthesis is *design*. Initially, synthesis is used in the development of preliminary concepts and to establish relationships among various components of the system. Later, when sufficient functional definition and decomposition have occurred, synthesis is used to further define the *hows* at a lower level.

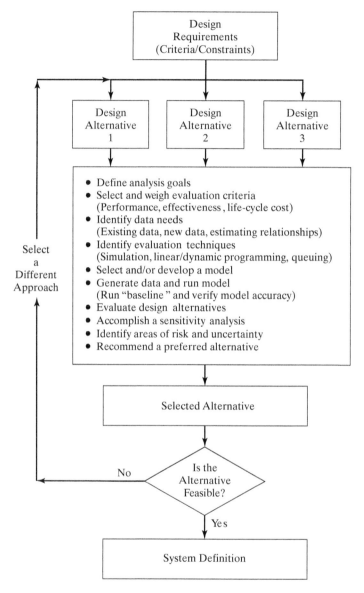

Conceptual System Design

Figure 26   Trade-off analysis process.

Synthesis involves the creation of a configuration that could be representative of the form that the system will ultimately take, although a final configuration should not be assumed at this early point in the design process.

   Given a synthesized configuration, its characteristics need to be evaluated in terms of the system requirements initially specified. Changes are incorporated as required, leading to a preferred design configuration. This iterative process of synthesis, analysis, evaluation, and design refinement leads initially to the establishment of the *functional baseline*, then the *allocated baseline*, and finally the *product baseline*. A good description of these configuration baselines, combined with a *disciplined* approach to baseline management, is essential for the successful implementation of the systems engineering process.

9 SYSTEM SPECIFICATION

Throughout the conceptual system design phase (commencing with the needs analysis), one of the major objectives is to develop and define the specific "design-to" requirements for the *system* as an entity. The results from the activities described in Sections 3–8 are combined, integrated, and included in a *system specification* (Type *A*). This specification constitutes the top "technical-requirements" document that provides overall guidance for system design from the beginning. Referring to Figure 1, this specification is usually prepared at the conclusion of conceptual design. Further, this top-level specification provides the baseline for the development of all lower-level specifications to include *development* (Type *B*), *product* (Type *C*), *process* (Type *D*), and *material* (Type *E*) *specifications*. While there may be a variety of formats used in the preparation of the system specification, an example of one approach is presented in Figure 27.

10 CONCEPTUAL DESIGN REVIEW

Design progresses from an abstract notion to something that has form and function, is firm, and can ultimately be reproduced in designated quantities to satisfy a need. Initially, a need is identified. From this point, design evolves through a series of stages (i.e., conceptual design, preliminary system design, and detail design and development). In each major stage of the design process, an evaluative function is accomplished to ensure that the design is correct at that point before proceeding with the next stage. The evaluative function includes both the informal day-to-day project coordination and data/documentation review and the formal design review. Design information is released and reviewed for compliance with the basic system-equipment requirements (i.e., performance, reliability, maintainability, usability, supportability, sustainability, etc., as defined by the system specification). If the requirements are satisfied, the design is approved as is. If not, recommendations for corrective action are initiated and discussed as part of the formal design review.

The formal design review constitutes a coordinated activity (including a meeting or series of meetings) directed to satisfy the interests of the design engineer and the technical discipline support areas (reliability, maintainability, human factors, logistics, manufacturing engineering, quality assurance, and program management). The purpose of the design review is to formally and logically cover the proposed design from the total system standpoint in the most effective and economical manner through a combined integrated review effort. The formal design review serves a number of purposes.

1. It provides a formalized check (audit) of the proposed system/subsystem design with respect to specification requirements. Major problem areas are discussed and corrective action is taken as required.
2. It provides a common baseline for all project personnel. The design engineer is provided the opportunity to explain and justify his or her design approach, and

System Specification
1.0   Scope
2.0   Applicable Documents
3.0   Requirements
3.1   System Definition
3.1.1   General Description
3.1.2   Operational Requirements (Need, Mission, Use Profile, Distribution, Life Cycle)
3.1.3   Maintenance Concept
3.1.4   Functional Analysis and System Definition
3.1.5   Allocation of Requirements
3.1.6   Functional Interfaces and Criteria
3.2   System Characteristics
3.2.1   Performance Characteristics
3.2.2   Physical Characteristics
3.2.3   Effectiveness Requirements
3.2.4   Reliability
3.2.5   Maintainability
3.2.6   Usability (Human Factors)
3.2.7   Supportability
3.2.8   Transportability/Mobility
3.2.9   Flexibility
3.2.10 Sustainability
3.2.11 Security
3.3   Design and Construction
3.3.1   CAD/CAM Requirements
3.3.2   Materials, Processes, and Parts
3.3.3   Mounting and Labeling
3.3.4   Electromagnetic Radiation
3.3.5   Safety
3.3.6   Interchangeability
3.3.7   Workmanship
3.3.8   Testability
3.3.9   Economic Feasibility
3.4   Documentation/Data
3.5   Logistics
3.5.1   Maintenance Requirements
3.5.2   Supply Support
3.5.3   Test and Support Equipment
3.5.4   Personnel and Training
3.5.5   Facilities and Equipment
3.5.6   Packaging, Handling, Storage, and Transportation
3.5.7   Computer Resources (Software)
3.5.8   Technical Data/Information
3.5.9   Customer Services
3.6   Producibility
3.7   Disposability
3.8   Affordability
4.0   Test and Evaluation
5.0   Quality Assurance Provisions
6.0   Distribution and Customer Service
7.0   Retirement and Material Recycling/Disposal

Figure 27   Type *A* system specification format (example).

representatives from the various supporting organizations (e.g., maintainability, logistic support, and marketing) are provided the opportunity to learn of the design engineer's problems. This serves as an excellent communication medium and creates a better understanding among design and support personnel.

3. It provides a means for solving interface problems and promotes the assurance that all system elements will be compatible, internally and externally.

4. It provides a formalized record of what design decisions were made and the reasons for making them. Analyses, predictions, and trade-off study reports are noted and are available to support design decisions. Compromises to performance, reliability, maintainability, human factors, cost, and/or logistic support are documented and included in the trade-off study reports.

5. It promotes a higher probability of mature design, as well as the incorporation of the latest techniques (where appropriate). Group review may identify new ideas, possibly resulting in simplified processes and ultimate cost savings.

The formal design review, when appropriately scheduled and conducted in an effective manner, leads to reduction in the producer's risk relative to meeting specification requirements and results in improvement of the producer's methods of operation. Also, the customer often benefits from the receipt of a better product.

Design reviews are generally scheduled before each major evolutionary step in the design process. In some instances, this may entail a single review toward the end of each stage (i.e., conceptual, preliminary system design, and detail design and development). For the other projects, where a large system is involved and the amount of new design is extensive, a series of formal reviews may be conducted on designated elements of the system. This may be desirable to allow for the early processing of some items while concentrating on the more complex, high-risk items.

Although the number and type of design reviews scheduled may vary from program to program, four basic types are readily identifiable and are common to most programs. They include the conceptual design review (i.e., system requirements review), the system design review, the equipment/software design review, and the critical design review. Of particular interest relative to the activities discussed in this chapter is the *conceptual design review*, which is dedicated to the review and validation of system operational requirements, maintenance and support concept, specified TPMs, and the functional analysis and allocation of requirements at the system level. This review is usually conducted at the end of the conceptual design phase and prior to the accomplishment of preliminary design.

11 SUMMARY AND EXTENSIONS

This chapter addresses the basic *requirements* for the design and development of a typical system and its interfaces. Whether a relatively large or small system is being acquired, the steps discussed herein are applicable and can be "tailored" as appropriate.

This chapter commences with the definition of a problem and the development of a needs analysis and continues through the accomplishment of a feasibility analysis, definition of system operational requirements and the maintenance and support concept, identification and prioritization of TPMs, functional analysis and allocation, system trade-off studies, and the development of a system specification (Type *A*). The objective of this chapter is to emphasize the steps involved and the importance of defining the requirements for bringing a system into being. This, in turn, leads to the implementation of systems engineering principles and concepts as the system design progresses through preliminary system design, detail design and development, and system test and evaluation.

Of particular significance is the importance of defining a good and complete set of *requirements* from the beginning and initiating a comprehensive *plan* in response. The specification of requirements must cover a *single* system, a SOS configuration, and all associated *interfaces*, as applicable. Defining the scope (i.e., bounds) and all of the system operational concepts for the entire life cycle is critical. It is not uncommon for a *system specification* to be written in rather vague terms, with the objective of keeping the requirements "loose" in the beginning to provide the maximum flexibility in design, allowing for "innovation" on the part of the designer up to the last minute. However, without a good foundation on which to build, the follow-on *B*, *C*, *D*, and *E* specifications, which are usually stated in rather specific terms, may not properly reflect what is ultimately desired by the customer in terms of system performance. Further, these lower-level specifications may not be mutually supportive or compatible if there is a poor description of the system upon which to base the requirements.

Thus, it is essential that a well-written *performance-based* specification be prepared in the beginning. This, of course, depends on the thoroughness and depth to which the earlier requirements have been identified; that is, the activities described in Sections 3–8. Given a good "specification," there needs to be a well-written comprehensive plan for the implementation of a program for developing a system to meet the stated requirements; that is, the *SEMP* (or *SEP*) described in Section 2. Both the *specification* and the *plan* must be prepared jointly and must "communicate with each other."

For a more detailed and in-depth coverage of the systems engineering process and the subject of requirements, the reader can review the following references: ANSI/GEIA EIA-632, IEEE 1220-1998, INCOSE's *Systems Engineering Handbook*, ISO/IEC 15288, and *System Engineering Management* (B. S. Blanchard). In addition, the reader should visit the website for the International Council on Systems Engineering (INCOSE) at *http://www.incose.org*, where a number of useful links may be found.

QUESTIONS AND PROBLEMS

1. In accomplishing a needs analysis in response to a given deficiency, what type of information would you include? Describe the process that you would use in developing the necessary information.

2. What is the purpose of the feasibility analysis? What considerations should be addressed in the completion of such an analysis?

3. Through a review of the literature, describe the QFD approach and how it could be applied in helping to define the requirements for a given system design.

4. Why is the definition of system operational requirements important? What type of information is included?

5. What specific challenges exist in defining the operational requirements for a *system-of-systems* (SOS) configuration? What is meant by *interoperability*? Provide an example.

6. Why is it important to define specific mission scenarios (or operational profiles) within the context of the system operational requirements?

7. What information should be included in the system maintenance concept? How is it developed (describe the steps), and at what point in the system life cycle should it be developed?

8. How do system operational requirements influence the maintenance concept (if at all)?

9. How does the maintenance concept affect system/product design? Give some specific examples.

10. Select a system of your choice and develop the operational requirements for that system. Based on the results, develop the maintenance concept for the system. Construct the necessary operational and maintenance flows, identify repair policies, and apply quantitative effectiveness factors as appropriate.

11. Refer to Figure 12 and assume that a similar infrastructure exists in your own community. Identify the various system capabilities (functions), illustrate (draw) an overall configuration structure (similar to that in the figure), and identify some of the critical metrics required as an input to the design of such.

12. Refer to Figure 13. For a SOS configuration of your choice, describe some of the critical requirements in the design of a SOS.

13. In evaluating whether or not a two- or three-level maintenance concept should be specified, what factors would you consider in the evaluation process?

14. In developing the maintenance concept, it is essential that all levels of maintenance be considered on an integrated basis. Why?

15. Why is the development of technical performance measures (TPMs) important?

16. Refer to Figure 17. Describe the steps that you would complete in developing the information included in the figure. Be specific.

17. Given the information provided in Figure 17, how would you apply this information in the system design process (if at all)?

18. What is meant by functional analysis? When in the system life cycle is it accomplished? What purpose does it serve? Identify some of the benefits derived. Can a functional analysis be accomplished for any system? Can a functional analysis be accomplished for a SOS configuration?

19. What is meant by a *common function* in the functional analysis? How are common functions determined?

20. How does the functional analysis lead into the definition of specific resource requirements in the form of hardware, software, people, data, facilities, and so on? Briefly describe the steps in the process, and include an example. What is the purpose of the block numbering shown in Figures 20–22?

21. What is the purpose of allocation? To what depth in the system hierarchical structure should allocation be accomplished? How does it impact system design (if at all)? How can allocation be applied for a SOS configuration (if at all)?

22. In conceptual design, there are a number of different requirements for predicting or estimating various system metrics. What approach (steps) would you apply in accomplishing such?

23. What is the purpose of the formal design review? What are some of the benefits derived from the conduct of design reviews? Describe some of the negative aspects.

24. What are the basic objectives in conducting a *conceptual design review*?

Preliminary System Design

From Chapter 4 of *Systems Engineering and Analysis,* Fifth Edition, Benjamin S. Blanchard, Wolter J. Fabrycky. Copyright © 2011 by Pearson Education, Inc. Published by Pearson Prentice Hall. All rights reserved.

Preliminary System Design

The conceptual design process leads to the selection of a tentatively preferred, conceptual system design architecture or configuration. Top-level requirements, as defined there, enable early design evolution that follows in the *preliminary system design phase*. This phase of the life cycle progresses by addressing the definition and development of the preferred system concept and the allocated requirements for subsystems and the major elements thereof.

An essential purpose of preliminary design is to demonstrate that the selected system concept will conform to performance and design specifications, and that it can be produced and/or constructed with available methods, and that established cost and schedule constraints can be met. Some products of preliminary design include the functional analysis and allocation of requirements at the subsystem level and below, the identification of design criteria as an "input" to the design process, the application of models and analytical methods in conducting design trade-offs, the conduct of formal design reviews throughout the system development process, and planning for the *detail design and development phase*.

This chapter addresses the following steps in the systems engineering process:

- Developing design requirements for subsystems and major system elements from system-level requirements;
- Preparing *development, product, process*, and *material* specifications applicable to subsystems;

- Accomplishing functional analysis and allocation to and below the subsystem level;
- Establishing detailed design requirements and developing plans for their handoff to engineering domain specialists;
- Identifying and utilizing appropriate engineering design tools and technologies;
- Conducting trade-off studies to achieve design and operational effectiveness; and
- Conducting design reviews at predetermined points in time.

Completion of these steps implements the process illustrated by Blocks 1.1–1.7 in Figure A.1. While the depth, level of effort, and costs of accomplishing this may vary from one application to the next, the process outlined is applicable to the development of any type and size system. As a learning objective, the goal is to provide the reader with a comprehensive and valid approach for addressing preliminary design.

1 PRELIMINARY DESIGN REQUIREMENTS

Preliminary design requirements evolve from "system" design requirements, which are determined through the definition of system operational requirements, the maintenance and support concept, and the identification and prioritization of TPMs. These requirements are documented through the preparation of the *system specification* (Type *A*), prepared in the conceptual design phase. These requirements become the criteria by which preliminary design alternatives are judged.

The *whats* initiating conceptual design produce *hows* from the conceptual design evaluation effort applied to feasible conceptual design concepts. Next, the *hows* are taken into preliminary design through the means of allocated requirements. There they become *whats* and drive preliminary design to address *hows* at this lower level. This is a cascading process following the pattern exhibited in Figure A.1. It emanates from a process giving attention to *what the system is intended to do before determining what the system is.*

Requirements for the design of subsystems and the major elements of the system are defined through an extension of the functional analysis and allocation, the conduct of design trade-off studies, and so on. This involves an iterative process of top-down/bottom-up design, which continues until the next lowest level of system components are identified and configured.

Consider a regional public transportation authority facing the need to increase the capacity for two-way traffic flow across a river that separates a growing municipality. From the results of the conceptual design phase, a bridge spanning the river is selected from among other mutually exclusive river crossing alternatives. Each preliminary bridge design alternative is evaluated through consideration and analysis of its subsystem components. For example, if the pier and superstructure alternative is under evaluation, it will be the abutments, piers, and superstructure that have to be synthesized, sized, and evaluated. Trade-off and optimization is accomplished to determine the pier spacing that will minimize the sum of the first cost of piers and of superstructure, estimated cost of maintenance and support, projected end-of-life cost, and total life-cycle cost. The optimized result provides the basis for comparing this preliminary design alternative on an equivalent basis with other bridge design alternatives.

Lower-level requirements then emanate from the allocated requirements for the tentatively chosen (preliminary) bridge design. These lower-level requirements become the design criteria for subsystems and components of the pier and superstructure bridge. In this particular illustration, the major subsystems identified for the river crossing bridge include the basic road and railway bed, passenger walkway and bicycle path, toll facilities, the maintenance and support infrastructure, and others. Subsystems are further broken down into their respective system elements, such as, superstructure, substructures, piles and footings, foundations, retaining walls, and construction materials. These lower-level requirements are then documented through *development, product, process,* and/or *material* specifications. Design and development of the specific system elements is accomplished in the *detail design and development phase.*

2 DEVELOPMENT, PRODUCT, PROCESS, AND MATERIAL SPECIFICATIONS

The *technical* requirements for the system and its elements are documented through a series of specifications, as indicated in Figure A.1. This series commences with the preparation of the *system specification* (Type *A*) prepared in the conceptual design phase. This, in turn, leads to one or more subordinate specifications and/or standards covering applicable subsystems, configuration items, equipment, software, and other components of the system. In addition, there may be any number of supplemental ANSI (American National Standards Institute), EIA (Electronic Industries Alliance), IEC (International Electrotechnical Commission), ISO (International Organization for Standardization), and related standards that are required in support of the basic program-related specifications.

Although the individual specifications for a given program may assume a different set of designations, a generic approach is used throughout this text. The categories assumed herein are described below and illustrated in Figure 1. Referring to the figure, the development of a *specification tree* is recommended for each program, showing a hierarchical relationship in terms of which specification has "preference" in the event of conflict. Further, it is critical that all specifications and standards be prepared in such a way as to ensure that there is a *traceability* of requirements from the top down. Preparation of the *development specification* (Type *B*), *product specification* (Type *C*), and so on, must include the appropriate TPM requirements that will support an overall system-level requirement; for example, operational availability (A_o) of 0.98. The traceability of requirements, through a specification tree, is particularly important in view of current trends pertaining to increasing globalization, greater outsourcing, and the increasing utilization of external suppliers, where variations often occur in implementing different practices and standards.

1. *System specification (Type A):* includes the technical, performance, operational, and support characteristics for the system as an entity; the results of a feasibility analysis, operational requirements, and the maintenance and support concept; the appropriate TPM requirements at the system level; a functional description of the system; design requirements for the system; and an allocation of design requirements to the subsystem level.

2. *Development specification (Type B):* includes the technical requirements (qualitative and quantitative) for any new item below the system level where research, design, and development are needed. This may cover an item of equipment, assembly, computer program, facility, critical item of support, data item, and so on. Each specification must include the performance, effectiveness, and support characteristics that are required in the evolving of design from the system level and down.

3. *Product specification (Type C):* includes the technical requirements (qualitative and quantitative) for any item below the system level that is currently in inventory and can be procured "off the shelf." This may cover any commercial off-the-shelf (COTS) equipment, software module, component, item of support, or equivalent.

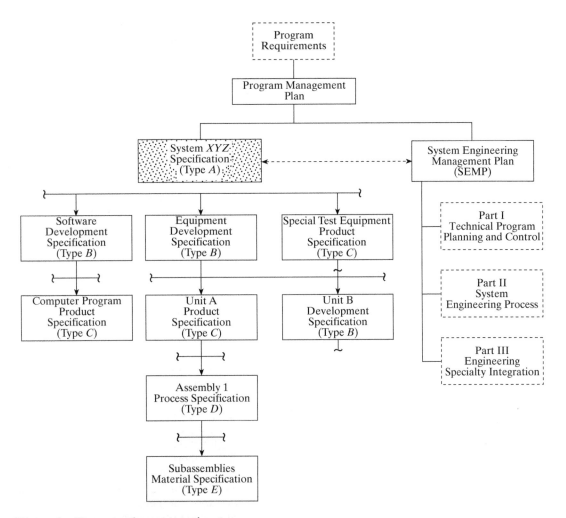

Figure 1 Program documentation tree.

4. *Process specification (Type D):* includes the technical requirements (qualitative and quantitative) associated with a process and/or a service performed on any element of a system or in the accomplishment of some functional requirement. This may include a manufacturing process (e.g., machining, bending, and welding), a logistics process (e.g., materials handling and transportation), an information handling process, and so on.

5. *Material specification (Type E):* includes the technical requirements that pertain to raw materials (e.g., metals, ore, and sand), liquids (e.g., paints and chemical compounds), semifabricated materials (e.g., electrical cable and piping), and so on.

Each applicable specification must be direct, complete, and written in *performance-related* terms and must describe the appropriate design requirements in terms of the *whats*; that is, the function(s) that the item in question must perform. Further, the specification must be properly tailored to its application, and care must be taken to ensure that it is not overspecifying or underspecifying. While individual programs may vary in applying a different set of designations, or specific content within each specification, it is important that a complete top-down approach be implemented encompassing the requirements for the entire system and all of its elements.

3 FUNCTIONAL ANALYSIS AND ALLOCATION (SUBSYSTEM)

With the basic objectives in accomplishing a *functional analysis* and the process for the development of *functional flow block diagrams* (FFBDs), the next step is to extend the functional analysis from the system level down to the subsystem and below as required. The depth of such an analysis (i.e., the breakdown in developing FFBDs from the system level to the second level, third level, and so on) will vary depending on the degree of visibility desired, whether new or existing design is anticipated, and/or to the level at which the designer wishes to establish some specific design-to requirements as an input. As mentioned earlier, it is important to establish the proper architecture describing structure, interrelationships, and related requirements.

3.1 The Functional Analysis Process

There are a variety of illustrations showing a breakdown of functions into subfunctions and ultimately describing major subsystems. Figure A.2 shows the general sequence of steps leading from the system level down to a communications subsystem (refer to Block 9.5.1). While this shows only one of the many subsystems required to meet an airborne transportation need, the same approach can be applied in defining other subsystems. The development of *operational* FFBDs can then lead to the development of maintenance FFBDs, as shown in Figure A.3. Given completion of the operational and maintenance FFBDs that reflect the *whats*, one must next determine the *hows*; that is, *how will each function be accomplished?* This is realized by evaluating each individual block of an FFBD, defining the necessary *inputs* and expected *outputs*, describing the external *controls* and *constraints*, and determining the *mechanisms* or the physical resources required for accomplishing the function; that is, equipment, software, people, facilities, data/information, or various combinations thereof. An example of the process is presented in Figure 2.

Referring to the figure, note that just one of the blocks in Figure 2 is addressed, where the resource requirements are identified as *mechanisms*. As there may be a number of different approaches for accomplishing a given function, trade-off studies are conducted with a preferred approach being selected. The result leads to the determination and compilation of the resource requirements for each function, and ultimately for all of the functions included in the functional analysis.

In performing the analysis process depicted in Figure 2, a documentation format, similar to that illustrated in Figure 3 (or something equivalent), should be used. In the figure, the functions pertaining to system design and development are identified along with required inputs, expected outputs, and anticipated resource requirements. While this particular example is "qualitative" by nature, there are many functions where specific metrics (i.e., TPMs) can be applied in the form of design-to "constraints." Referring to Figure A.2, for example, there is a functional requirement at the system level that states, "Operate the system in the user environment," represented by Block 9.0. Assuming that there is a system TPM requirement for *operational availability (A_o)* of 0.985, this measure of effectiveness will constitute a design requirement for the function in Block 9.0, and the appropriate resources must be applied accordingly. The same approach can be applied in relating all of the TPM requirements, which are specified for the system, to one or more functions. Accordingly, the purpose of including Figure 3 is to promote a disciplined approach in accomplishing a functional analysis.

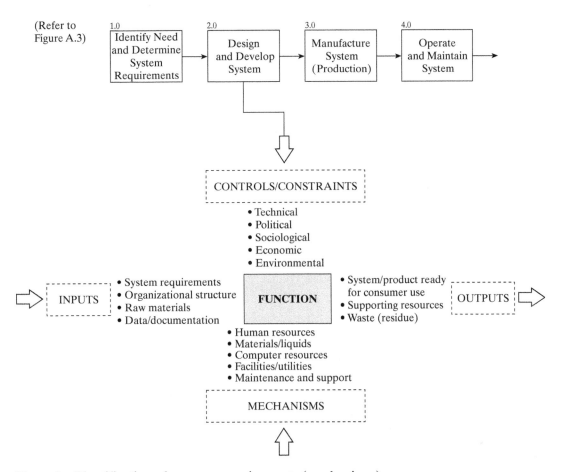

Figure 2 Identification of resource requirements (mechanisms).

In conducting trade-off studies pertaining to the best approach in responding to a functional requirement (i.e., the mechanisms), the results may point toward the selection of hardware, software, people, facilities, data, or various combinations thereof. Figure 4 gives an example where the requirements for hardware, software, and the human are identified. Stemming from the functional analysis (Block 0.2 in Figure A.1), the individual design and development steps for each is shown, and a plan is prepared for the acquisition of these system elements. From a systems engineering perspective, it is essential that these activities be coordinated and integrated, across the life cycles, from the beginning. In other words, an ongoing "communication(s)" must exist throughout the design and development of the hardware, software, and human elements of the system.

3.2 Requirements Allocation

Lower-level elements of the system are defined through the functional analysis and subsequently by partitioning (or grouping) similar functions into logical subdivisions, identifying major subsystems, configuration items, units, assemblies, modules, and so forth. Figure 5

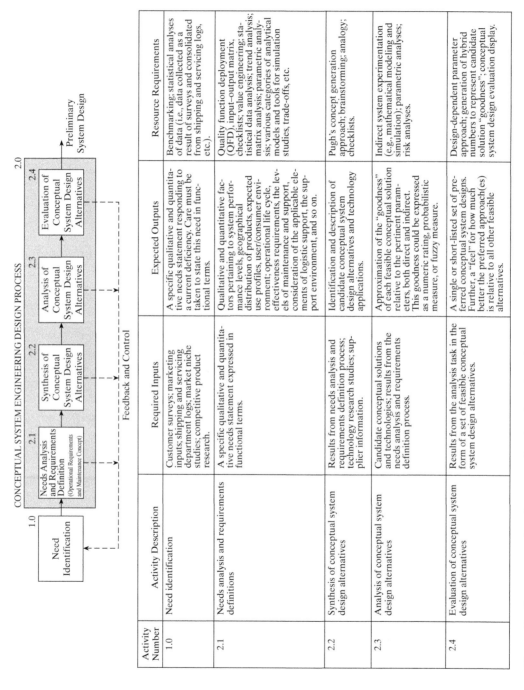

CONCEPTUAL SYSTEM ENGINEERING DESIGN PROCESS

Activity Number	Activity Description	Required Inputs	Expected Outputs	Resource Requirements
1.0	Need identification	Customer surveys; marketing inputs; shipping and servicing department logs; market niche studies; competitive product research.	A specific qualitative and quantitative needs statement responding to a current deficiency. Care must be taken to state this need in functional terms.	Benchmarking; statistical analyses of data (i.e., data collected as a result of surveys and consolidated from shipping and servicing logs, etc.).
2.1	Needs analysis and requirements definitions	A specific qualitative and quantitative needs statement expressed in functional terms.	Qualitative and quantitative factors pertaining to system performance levels, geographical distribution of products, expected use profiles, user/consumer environment; operational life cycle, effectiveness requirements, the levels of maintenance and support, consideration of the applicable elements of logistic support, the support environment, and so on.	Quality function deployment (QFD), input–output matrix, checklists; value engineering; statistical data analysis; trend analysis; matrix analysis; parametric analysis; various categories of analytical models and tools for simulation studies, trade-offs, etc.
2.2	Synthesis of conceptual system design alternatives	Results from needs analysis and requirements definition process; technology research studies; supplier information.	Identification and description of candidate conceptual system design alternatives and technology applications.	Pugh's concept generation approach; brainstorming; analogy; checklists.
2.3	Analysis of conceptual system design alternatives	Candidate conceptual solutions and technologies; results from the needs analysis and requirements definition process.	Approximation of the "goodness" of each feasible conceptual solution relative to the pertinent parameters, both direct and indirect. This goodness could be expressed as a numeric rating, probabilistic measure, or fuzzy measure.	Indirect system experimentation (e.g., mathematical modeling and simulation); parametric analyses; risk analyses.
2.4	Evaluation of conceptual system design alternatives	Results from the analysis task in the form of a set of feasible conceptual system design alternatives.	A single or short-listed set of preferred conceptual system designs. Further, a "feel" for how much better the preferred approach(es) is relative to all other feasible alternatives.	Design-dependent parameter approach; generation of hybrid numbers to represent candidate solution "goodness"; conceptual system design evaluation display.

Figure 3 Documentation format reflecting functions, input–output requirements, and resource requirements (partial).

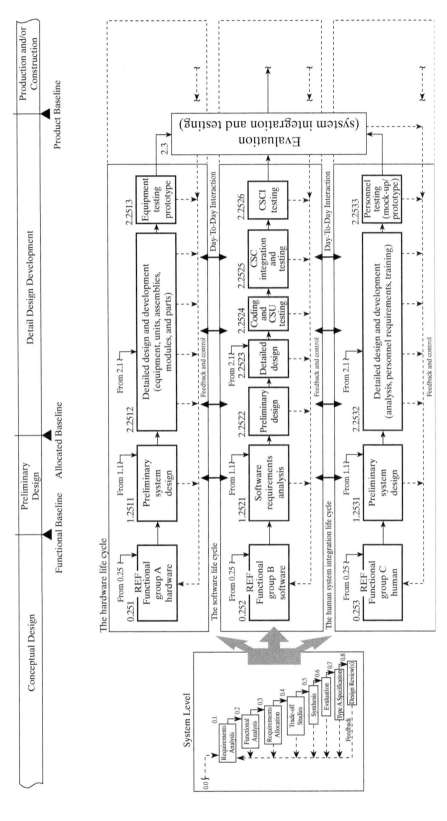

Figure 4 The evolution of hardware, software, and human requirements from the functional analysis (refer to Figure A.1).

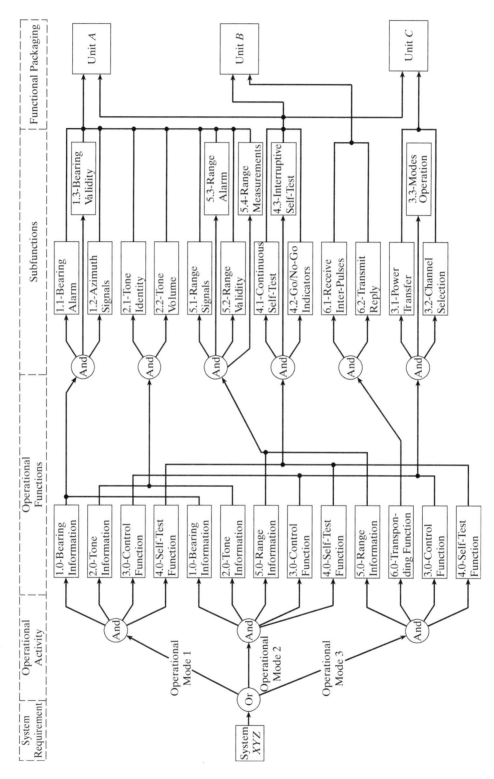

Figure 5 The functional packaging of the system into major elements.

presents an overview of this process, evolving from a functional definition of System *XYZ* to the packaging of the system into three units; that is, Units *A*, *B*, and *C*.

Given the packaging concept shown in Figure 5, it is now appropriate to determine the "design-to" requirements for each one of the three units. This is accomplished through the process of *allocation* (or *apportionment*). In the development of design goals at the unit level, priorities are established based on the TPMs, and both quantitative and qualitative design requirements are determined. Such requirements then lead to the incorporation of the appropriate design characteristics (attributes) in the design of Units *A*, *B*, and *C*. Such design characteristics should be "tailored" in response to the relative importance of each as it impacts the system-level requirements. These design characteristics are initially viewed from the top down and are compared. Trade-off studies are conducted to evaluate the interaction effects, and those characteristics most significant in meeting the overall system objectives are selected.

Figure 6 presents an example resulting from the allocation process described. System-level TPMs are identified along with the design-to metrics for each of the three units, as well as the requirements for Assemblies 1, 2, and 3 within Unit *B*. The requirements at the system level have been allocated downward. Further, the requirements established at the unit level, when combined, must be compatible with the higher-level requirements. Thus, there is

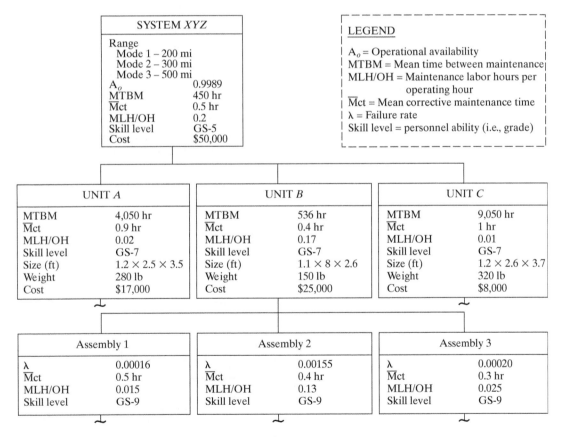

Figure 6 System *XYZ* requirements allocation.

a top-down/bottom-up relationship, and there may be trade-offs conducted comprehensively at the unit level in order to achieve the proper balance of requirements overall. Meeting these quantitative design-to requirements leads to the incorporation of the proper characteristics in the design of the item in question.

The objective of including Figure 6 with all of its metrics is to emphasize the *process* and its importance early in system design. Although the metrics shown are primarily related to reliability, maintainability, availability, and design-to-life-cycle-cost factors, it should be noted that in the allocation process, one needs to include all performance factors, human factors, physical features, producibility and supportability factors, sustainability and disposability factors, and so on.

In situations where there are a number of different systems in a system-of-systems (SOS) configuration, or where there are *common* functions, the allocation process becomes a little more complex. One needs to not only comply with the top-down/bottom-up traceability requirements for each system in the overall configuration but consider the compatibility (or operability) requirements among the systems as well. Referring to Figure 7, for example, there are two systems, System *ABC* and System *XYZ*, within a given SOS structure, with a "common" function being shared by each. Given such, several possibilities may exist:

1. One of the systems already exists, is operational, and the design is basically "fixed," while the other system is new and in the early stages of design and development. Assuming that System *ABC* is operational, the design characteristics for the unit identified as being "Common" in the figure and required as a functioning element of *ABC* are essentially "fixed." This, in turn, may have a significant impact on the overall effectiveness of System *XYZ*. In order to meet the overall requirements for *XYZ*, more stringent design input factors may have to be placed on the design of new Units *C* and *D* for that system. Or the common unit must be modified to be compatible with higher-level *XYZ* requirements, which (in turn) would likely impact the operational effectiveness of System *ABC*. Care must be taken to ensure that the overall requirements for both System *ABC* and System *XYZ* will be met. This can be accomplished by establishing a fixed requirement for the "common unit" and by modifying the design

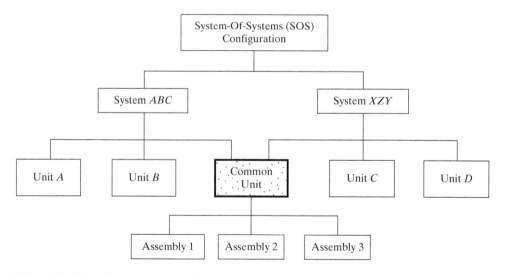

Figure 7 Requirements allocation for systems with a common function.

requirements for Assemblies *1, 2*, and *3* within that unit to meet allocated requirements from both *ABC* and *XYZ*.

2. Both System *ABC* and System *XYZ* are new and are being developed concurrently. The allocation process illustrated in Figure 6 is initially accomplished for each of the systems, ensuring that the overall requirements at the system level are established. The design-to characteristics for the "common unit" are compared and a single set of requirements is identified for the unit through the accomplishment of trade-off analyses, and the requirements for each of the various system *units* are then modified as necessary across-the-board while ensuring that the top system-level requirements are maintained. This may constitute an iterative process of analysis, feedback, and so on.

As a final point, those quantitative and qualitative requirements for the various elements of the system (i.e., subsystems, units, and assemblies) must be included in the appropriate specification, as identified in Section 2 (and Figure 1). Further, there must be a top-down/bottom-up "traceability" of requirements throughout the overall hierarchical structure for each of the systems in question.

3.3 Applications of the Functional Analysis

A major objective of systems engineering is to develop a complete set of requirements in order to define a single "baseline" from which all lower-level requirements may evolve; that is, to develop a *functional baseline* in conceptual design and later an *allocated baseline* in the preliminary system design phase (refer to Figure A.1). The results of the functional analysis constitute a required input for a number of design-related activities that occur subsequently. Most important is breaking the system (and its elements) down into *functional* entities through functional packaging and the development of an *open-architecture* configuration. A prime objective is to develop a configuration that can be easily upgraded as required (through new technology insertions) and easily supported throughout its life cycle (through a modularized approach in maintenance). In addition, the functional analysis provides a foundation upon which many of the subsequent analytical tasks and associated documentation are based, some of which are listed below:

1. Reliability analysis: reliability models and block diagrams; failure mode, effects, and criticality analysis (FMECA); fault-tree analysis (FTA); reliability prediction.
2. Maintainability analysis: maintainability models; reliability-centered maintenance (RCM); level-of-repair analysis (LORA); maintenance task analysis (MTA); total productive maintenance (TPM); maintainability prediction.
3. Human factors analysis: operator task analysis (OTA); operational sequence diagrams (OSDs); safety/hazard analysis; personnel training requirements.
4. Maintenance and logistic support: supply chain and supportability analysis leading to the definition of maintenance and support requirements—spares/repair parts and associated inventories, test and support equipment, transportation and handling equipment, maintenance personnel, facilities, technical data, information.
5. Producibility, disposability, and sustainability analysis.
6. Affordability analysis: life-cycle and total ownership cost.

4 PRELIMINARY DESIGN CRITERIA

The basic design objective(s) for the system and its elements must (1) be compatible with the system operational requirements, maintenance and support concept, and the prioritized TPMs; (2) comply with the allocated design-to criteria described in Section 3.2; and (3) meet all of the requirements in the various applicable specifications. The particular design characteristics to be incorporated will vary from one instance to the next, depending on the type and complexity of the system and the mission or purpose that it is intended to accomplish. In all cases, the design team activity must address the downstream life-cycle outcomes considering the phases of production and construction, system utilization and sustaining support, and retirement and material recycling/disposal. While *all* considerations in system design must be addressed, a few require some additional emphasis:

1. *Design for functional capability*—Functional capability derives from the characteristics of design that relate to the technical performance of the system; that is, the technical characteristics that are required for the system to accomplish its intended mission(s). This includes such factors as size, weight, volume, shape, accuracy, capacity, flow rate, throughput, units per time period, speed of travel, power output, and all of the technical and physical characteristics that the system (when operating) must exhibit to accomplish its objective(s). Functional capability is the main concern of aeronautical design, civil design, chemical design, electrical design, mechanical design, structural design, and others.

2. *Design for interoperability*—Interoperability pertains to the ability of the system to operate successfully in an environment where there are many other operating systems already in existence. An objective is to minimize the interaction effects between the new system and those other systems already in the operational inventory; that is, the impact of this new system on others as well as the external impact of other systems on this new system.

3. *Design for sustainability*—Sustainability, as defined herein, applies to the sustaining operation and support of systems throughout their respective life cycles without causing any degradation to the environment or to the earth's natural resources. The presumption is that whatever "resources" are initially consumed must be replaceable or are replaced. The objective is to design a system so as to eliminate wastes, greenhouse gases, toxic substances, air and water pollution, and any other factors that would cause degradation to the environment. Other terms used to describe this area of activity include *environmental engineering*, *green engineering*, and *sustainability engineering*.

4. *Design for reliability*—Reliability is that characteristic of design and installation concerned with the successful operation of the system throughout its planned mission and for the duration of its life cycle. Reliability (R) is often expressed as the probability of success, or is measured in terms of *mean time between failure* (MTBF), *mean time to failure* (MTTF), *failure rate* (λ), or a combination of these. An objective is to maximize operational reliability while minimizing system failure.

5. *Design for maintainability*—Maintainability is that characteristic of design and installation that reflects the ease, accuracy, safety, and economy of performing maintenance actions. Maintainability can be measured in terms of maintenance times (*mean corrective maintenance time* or $\overline{M}ct$, *mean preventive maintenance time* or $\overline{M}pt$, *active maintenance time* or $\overline{M}$, *maximum maintenance time* or Mmax, *logistics delay time* or LDT, *administrative delay time* or ADT, and *total maintenance downtime* or MDT); maintenance frequency (*mean time between maintenance* or MTBM); maintenance

labor (*maintenance labor hours* or MLH); and/or maintenance cost. The objective is to minimize maintenance times and labor hours while maximizing supportability characteristics in design (e.g., accessibility, diagnostic provisions, standardization, and interchangeability), minimize the logistic support resources required in the performance of maintenance (e.g., spares and supporting inventories, test equipment, maintenance personnel, and facilities), and minimize maintenance cost.

6. *Design for usability and safety*—Usability is that characteristic of design concerned with the interfaces between the human and hardware, the human and software, the human and facilities, the human and information/data, and so on; that is, ensuring the compatibility between, and safety of, system physical and functional design features and the human element in the operation, maintenance, and support of the system. Considerations must include anthropometric, human sensory, physiological, and psychological factors as they apply to system operability and supportability. An objective is to minimize the number of people and skill-level requirements, minimize training requirements, and minimize human errors while maximizing productivity and safety.

7. *Design for security*—Security, in this instance, pertains to those characteristics of design that will prevent (or at least deter) one or more individuals from intentionally inducing faults that will destroy the system, cause harm to personnel, and/or have an impact that will endanger society and the associated environment. Such characteristics should include an external alarm capability that will detect the presence of unauthorized personnel and prevent them from gaining access to the system, a condition-based monitoring capability that will enable one to check the status of the system and its elements on a continuing basis, and a built-in detection and diagnostic capability leading directly to the cause of any recurring problem(s) and to subsequent self-repair, or rapid repair, of the system.

8. *Design for supportability and serviceability*—Supportability and serviceability refer to the characteristics of design that ensure that the system can ultimately be serviced and supported effectively and efficiently throughout its planned life cycle. An objective is to consider both the internal characteristics of the prime mission-related elements of the system (equipment, software, facilities, and personnel used in system operation and in the accomplishment of a mission) and the design of the maintenance and logistic support infrastructure. Supportability includes the consideration and application of reliability, maintainability, and human factors in design.

9. *Design for producibility and disposability*—Producibility is that characteristic of design that pertains to the ease and economy with which a system or product can be produced. The objective is to design an entity that can be produced (in multiple quantities) easily and economically, using conventional and flexible manufacturing methods and processes without sacrificing function, performance, effectiveness, or quality. Considerations include the utilization of standard parts, the construction and packaging for the ease of assembly (and disassembly), the use of standard equipment and tools in manufacturing, and so on. Disposability is that characteristic of design that allows for the disassembly and disposal of elements or components of the system easily, rapidly, and economically without causing environmental degradation.

10. *Design for affordability*—Economic feasibility (or affordability) refers to the characteristics of design and installation that impact total system cost and overall budgetary

constraints. An objective is to justify design decisions based on total system *life-cycle* cost, and not just on system acquisition cost (or purchase price). Economic feasibility depends on the combined and balanced incorporation of reliability, maintainability, human factors, supportability, producibility, and other characteristics of design.

Accomplishing the overall system design objective requires an appropriate *balance* among these and many other related factors. There are performance requirements, reliability and maintainability requirements, human factors and supportability requirements, sustainability requirements, and so on, and acquiring the proper balance among these is often difficult to attain, since some stated objectives may appear to be in opposition to others. For example, incorporating high performance and highly sophisticated functional design may reduce the reliability and increase logistic support requirements and life-cycle cost; incorporating too much reliability may require the use of costly system components and thus increase both acquisition cost and life-cycle cost; incorporating too much accessibility for improving maintainability may cause a significant increase in equipment size and weight. The goal is to conduct the necessary trade-off studies and to incorporate only those features considered to be essential to meet the requirements—not too many or too few. One must be careful not to overdesign or underdesign, as measured in terms of the need.

Figure 8 is an extension of the concept in Figures A.1 and 4. Referring to Figure 4, the emphasis is on the three individual life cycles (i.e., hardware, software, and the human) with the evolution of design from the system level to the subsystem level illustrated. As one proceeds through the design process, the specific characteristics of design must be defined for each level. Reliability considerations are important at each level as are maintainability considerations, and so forth, and the various characteristics must be complementary as one proceeds downward. These characteristics must also support the allocated requirements identified in Figure 6. Finally, appropriate design characteristics identified at each level must be included in the applicable development, product, process, or material specification(s). Figure 8 also identifies some of the design tasks and methods utilized in helping to ultimately attain the overall system design objectives.

5 DESIGN ENGINEERING ACTIVITIES

The day-to-day design activities begin with the implementation of the appropriate planning that was initiated in the conceptual design phase (i.e., the program management and system engineering management plans). This includes the establishment of the design team and the initiation of specific design tasks, the ongoing liaison and working with various responsible designers throughout the project organization, the development of design data, the accomplishment of periodic design reviews, and the initiation of corrective action as necessary.

Success in systems engineering derives from the realization that the design activity requires a *team* approach. As one proceeds from conceptual design to the subsequent phases of the life cycle, the actual team "make-up" will vary in terms of the specific expertise required and the number of project personnel assigned. Early in the conceptual and preliminary design phases, there is a need for a *few* highly qualified individuals with broad technical knowledge who understand the customer's requirements and the user's operational environment, the major functional elements of the system and their interface relationships, and the general process for bringing a system into being. These key individuals are those who understand and believe in the *systems approach* and know when to call on the appropriate disciplinary expertise for assistance. The objective is to ensure the early consideration of all of the "design-for" requirements described in Section 4.

INTEGRATION OF THE HARDWARE, SOFTWARE, AND HUMAN LIFE CYCLES

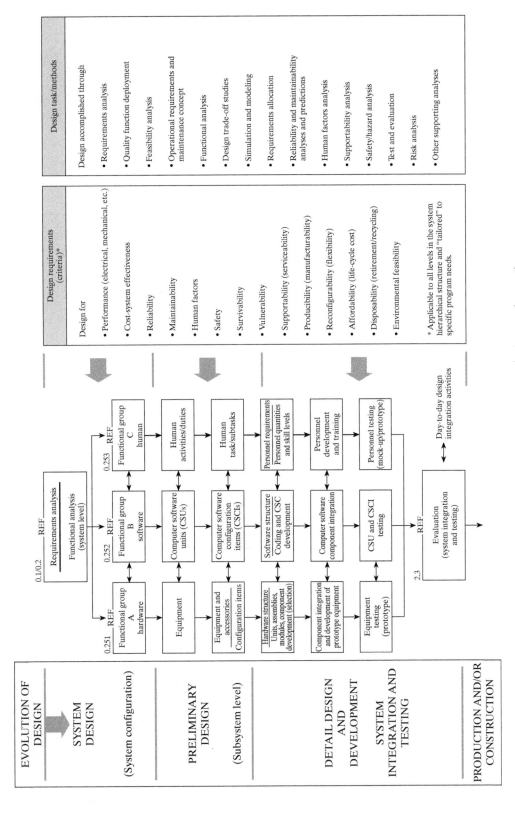

Figure 8 The breakout of design requirements by system indenture level (refer to Figure 4).

As the design progresses, the need for representation from the various design disciplines increases. Referring to Figure 9, a systems engineering implementation goal is to ensure the proper integration of the design disciplines as appropriate to the need. Depending on the project, there may be relatively few individuals assigned, or there may be hundreds involved. Further, some of the expertise desired may be located within the same physical facility, whereas other members of the project team may be remotely located in supplier organizations (both locally and internationally). In a system-of-systems (SOS) context, the project team may need to include individuals representing major "interfacing" systems.

A major goal in the implementation of the system engineering process is first to understand system requirements and the expectations of the customer and then to provide the *technical* guidance necessary to ensure that the ultimate system configuration will meet the need. Realization of this goal depends on providing the right personnel and material resources at the right location and in a timely manner. Such resources may include a combination of the following:

1. Engineering technical expertise (e.g., aeronautical engineers, civil engineers, electical engineers, mechanical engineers, software engineers, reliability engineers, logistics engineers, and environmental engineers).

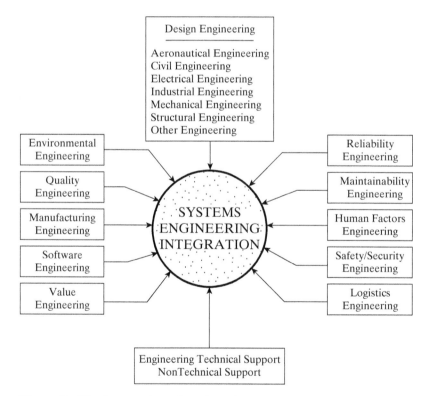

Figure 9   The integration of engineering disciplines.

2. Engineering technical support (e.g., technicians, component part specialists, computer programmers, model builders, drafting personnel, test technicians, and data analysts).

3. Nontechnical support (e.g., marketing, purchasing and procurement, contracts, budgeting and accounting, industrial relations, manufacturing personnel, and logistics supply chain specialists).

Whatever the case, the objective of systems engineering is to promote the "team" approach and to create the proper working environment for the necessary ongoing communications and exchange of information on a continuing day-to-day basis. Figure 9 illustrates this essential integration function, and the "system engineer" must be knowledgeable of the various disciplines, their respective objectives, and when to integrate these requirements into the overall design process.

6 ENGINEERING DESIGN TOOLS AND TECHNOLOGIES

The successful implementation of systems engineering highly depends on the selection and application of the proper analytical models and computer-based tools in accomplishing many of the essential program tasks. A few key points pertaining to computer-aided tools and analytical models are presented in this section.

6.1 Application of Computer-Aided Tools

The design team must solve a wide variety of problems and evaluate numerous alternatives in a limited amount of time, and still must design a system that will meet all requirements effectively and efficiently and in a highly competitive international environment. Meeting this challenge requires the utilization of a wide range of computer-based design aids. Generic categories of computer-aided methods may include *computer-aided engineering (CAE), computer-aided design (CAD), computer-aided design data (CADD), computer-aided manufacturing (CAM), computer-aided support (CAS), computer geometric modeling (feature-based parametric modeling),* and/or similar tools of an equivalent capability.

Computer usage in the design process is now quite mature. It is used to generate drawings and three-dimensional graphic displays to facilitate the accomplishment of many different analyses, to generate materials and parts lists, and to support many additional design-related functions, both of an administrative and of a technical nature. However, these various applications are not yet being accomplished in a fully integrated way. One computer program has been developed to produce three-dimensional graphic displays, another program to produce design drawings, another to accomplish a reliability analysis, another to generate a component parts list, and so on. Further, the language requirements, platforms, and program formats are different, and there is limited (if any) integration of these design methods. The expression relating to *islands of automation* addresses reality in many areas. A desirable objective is to implement the systems engineering process effectively, efficiently, and in a seamless manner. This, of course, depends not only on one's knowledge of the process and implementation requirements but also on the availability of the proper tools in a timely manner and the resources necessary to support their application. An integrated life-cycle approach to the application of computer-aided methods is required. One way of reflecting such an approach is through the concept of Macro-CAD, illustrated in Figure 10.

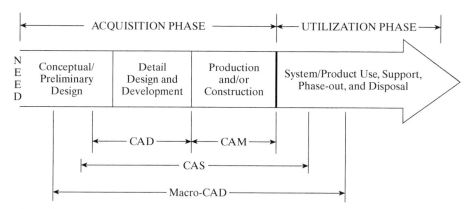

Figure 10 The relationship of CAD, CAM, CAS, and Macro-CAD.

The objective of Macro-CAD is to link early life-cycle design decisions with operational outcomes, using simulation methods and mathematical modeling. This approach is feasible because of the large body of knowledge available about operational modeling and simulation. Key to this is the prediction and/or estimation of design-dependent parameter values (that is, TPMs) for each design alternative, followed by timely optimization over design variables. For each set of design-dependent system parameter values, one can determine an optimal value for the chosen measure of evaluation. By comparing these optimal values for a set of alternatives in the face of multiple criteria, a preferred design configuration can be selected.

In summary, the use of CAD, CAM, CAS, and related methods, implemented on an integrated basis, offers many advantages. A few of these are noted below:

1. The designer can address many different alternatives in a relatively short interval of time. With the capability of evaluating a greater number of possible options, the risks associated with design decision-making are reduced.

2. The designer is able to simulate, and verify, design for a greater number of configurations by using three-dimensional projections. Employing an electronic database to facilitate this objective may eliminate the need to build a physical model later on, thus reducing cost.

3. The ability to incorporate design changes is enhanced, both in terms of the reduced time for accomplishing them and in the accuracy of data presentation. If the information flow is properly integrated, changes will be reflected throughout the database as applicable. Also, the vehicle for incorporating a change is greatly simplified.

4. The quality of design data is improved, both in terms of the methods for data presentation and in the reproduction of individual data elements.

5. The availability of an improved database earlier in the system life cycle (with better methods for presenting design configurations) facilitates the training of personnel assigned to the project. Not only is it possible to better describe the design through graphical means but also a common database will help to ensure that all design activities are tracking the same baseline.

6.2 Analytical Models and Modeling

The design evaluation process may be further facilitated through the use of various analytical models, methods, and tools in support of the Macro-CAD objective. A model, in this context, is a simplified representation of the real world that abstracts features of the situation relative to the problem being analyzed. It is a tool employed by an analyst to assess the likely consequences of various alternative courses of action being examined. The model must be adapted to the problem at hand and the output must be oriented to the selected evaluation criteria. The model, in itself, is not the decision maker but is a tool that provides the necessary data in a timely manner in support of the decision-making process.

The extensiveness of the *model* will depend on the nature of the problem, the number of variables, input parameter relationships, number of alternatives being evaluated, and the complexity of operation. The ultimate objective in the selection and development of a model is simplicity and usefulness. The model used should incorporate the following features:

1. The model should represent the dynamics of the system configuration being evaluated in a way that is simple enough to understand and manipulate, and yet close enough to the operating reality to yield successful results.

2. The model should highlight those factors that are most relevant to the problem at hand and suppress (with discretion) those that are not as important.

3. The model should be comprehensive, by including *all* relevant factors, and be reliable in terms of repeatability of results.

4. Model design should be simple enough to allow for timely implementation in problem solving. Unless the tool can be utilized in a timely and efficient manner by the analyst (or the manager), it is of little value. If the model is large and highly complex, it may be appropriate to develop a series of models where the output of one can be tied to the input of another. Also, it may be desirable to evaluate a specific element of the system independent of other elements.

5. Model design should incorporate provisions for ease of modification or expansion to permit the evaluation of additional factors as required. Successful model development often includes a series of trials before the overall objective is met. Initial attempts may suggest information gaps, which are not immediately apparent and consequently may suggest beneficial changes.

The use of mathematical models offers significant benefits. In terms of system application, several considerations exist—operational considerations, design considerations, product/construction considerations, testing considerations, logistic support considerations, and recycling and disposal considerations. There are many interrelated elements that must be integrated as a system and not treated on an individual basis. The mathematical model makes it possible to deal with the problem as an entity and allows consideration of all major variables of the problem on a simultaneous basis. More specifically:

1. The mathematical model will uncover relations between the various aspects of a problem that are not apparent in the verbal description.

2. The mathematical model enables a comparison of *many* possible solutions and aids in selecting the best among them rapidly and efficiently.

3. The mathematical model often explains situations that have been left unexplained in the past by indicating cause-and-effect relationships.

4. The mathematical model readily indicates the type of data that should be collected to deal with the problem in a quantitative manner.

5. The mathematical model facilitates the prediction of future events, such as effectiveness factors, reliability and maintainability parameters, logistics requirements, and so on.

6. The mathematical model aids in identifying areas of risk and uncertainty.

When analyzing a problem in terms of selecting a mathematical model for evaluation purposes, it is desirable to first investigate the tools that are currently available. If a model already exists and is proven, then it may be feasible to adopt that model. However, extreme care must be exercised to relate the right technique with the problem being addressed and to apply it to the depth necessary to provide the sensitivity required in arriving at a solution. Improper application may not provide the results desired, and the consequence may be costly.

Conversely, it might be necessary to construct a new model. In accomplishing this task, one should generate a comprehensive list of system parameters that will describe the situation being simulated. Next, it is necessary to develop a matrix showing parameter relationships, each parameter being analyzed with respect to every other parameter to determine the magnitude of relationship. Model input–output factors and parameter feedback relationships must be established. The model is constructed by combining the various factors and then testing it for validity. Testing is difficult to do because the problems addressed primarily deal with actions in the future that are impossible to verify. However, it may be possible to select a known system or equipment item that has been in existence for several years and exercise the model using established parameters. Data and relationships are known and can be compared with historical experience. In any event, the analyst might attempt to answer the following questions: *Can the model describe known facts and situations sufficiently well? When major input parameters are varied, do the results remain consistent and are they realistic? Relative to system application, is the model sensitive to changes in operational requirements, design, production/construction, and logistics and maintenance support? Can cause-and-effect relationships be established?*

Trade-off and system optimization relies on appropriate systems analysis methods and their application. A fundamental knowledge of probability and statistics, economic analysis methods, modeling and optimization, simulation, queuing theory, control techniques, and the other analytical techniques is essential to accomplishing effective systems analysis.

7 TRADE-OFF STUDIES AND DESIGN DEFINITION

As the design evolves, the system synthesis, analysis, and evaluation process continues. Proposed configurations for subsystems and major elements of the system are synthesized, trade-off studies are conducted, alternatives are evaluated, and a preferred design approach is selected. This process continues throughout the conceptual design, preliminary system design, and the detail design and development phases, leading to the definition of the system configuration down to the detailed component level (refer to Figure 8).

This iterative process of systems analysis is shown in a generic and simplified context in Figure 11. Referring to the figure, a key challenge is the application of the appropriate analytical techniques and models. As conveyed in the previous section, knowledge of analytical methods is required to accomplish the steps reflected by Blocks 4 and 5 (5a and 5b) of Figure 11. Further, when selecting and/or developing a specific analytical model/tool to facilitate the analysis effort, care must be exercised to ensure that the right computer-based tool is selected for the application intended.

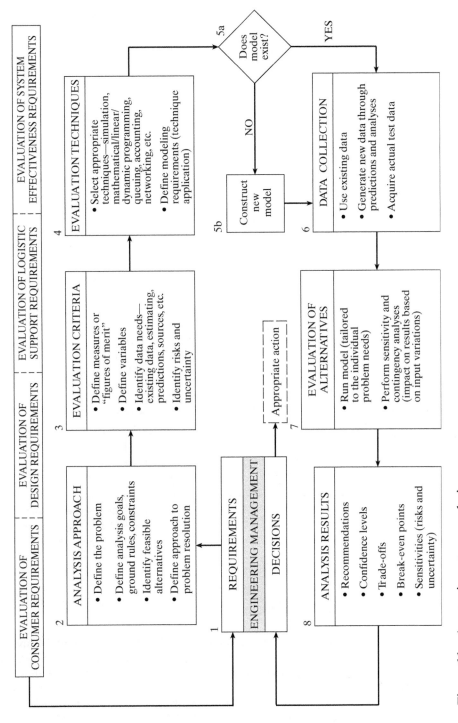

Figure 11   A generic systems analysis process.

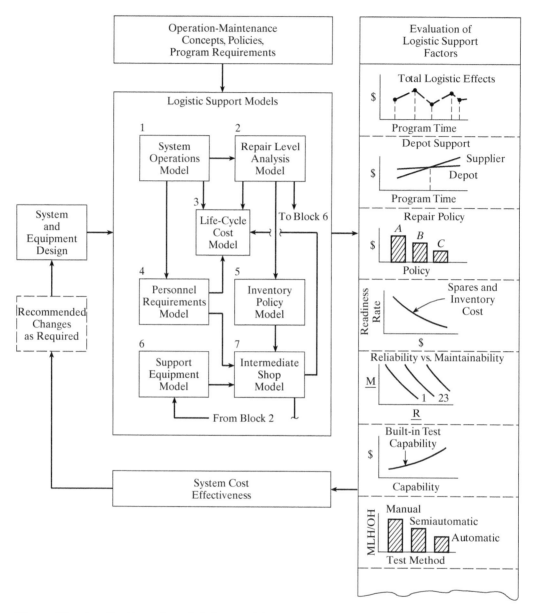

Figure 12 Application of models (example).

Figure 12 shows the application of a number of different analytical models used in the evaluation of alternative maintenance and logistic support policies. This illustration presents just one of a number of examples where there may be a multiple mix (or combination) of tools used to evaluate a specific design configuration. While the approach conveyed in the figure may not appear to be unique, it should be noted that there are many different computer-based models that are available in the commercial market and are advertised as solutions to a wide variety of problems. Most of these models were developed on a relatively "independent" or "isolated" basis in terms of selected platform, context or computer language, input data requirements, varying degrees of "user friendliness," and so on. In general, many of the models advertised today do not "talk to each other," are too complex, require too much input data, and can only be effectively used in the "downstream" portion of the life cycle during the detail design and development phase when there are a lot of data available.

From a systems engineering perspective, a good objective is to select or develop an *integrated design workstation* (incorporating a Macro-CAD approach) that can be utilized in all phases of the system life cycle and that can be adapted to the different levels of design definition as one progresses from the conceptual design to the detail design and development phase (refer to Figure 8). For example, this workstation should incorporate the right tools that can be applied at a high level in conceptual design and again at a more in-depth level in detail design and development.

8 DESIGN REVIEW, EVALUATION, AND FEEDBACK

The basic objectives and benefits of the design review, evaluation, and feedback process include two facets of the activity shown in Figure 13. First, there is an ongoing informal review and evaluation of the results of the design, accomplished on a day-to-day basis, where the responsible designer provides applicable technical data and information to all project personnel as the design progresses. Through subsequent review, discussion, and feedback, the proposed design is either approved or recommended changes are submitted for consideration. Second, there is a structured series of formal design reviews conducted at specific times in the overall system development process. While the specific types, titles, and scheduling of these formal reviews will vary from one program to the next, it is assumed herein that formal reviews will include the following:

1. The *conceptual design review* is usually scheduled toward the end of the conceptual design phase and prior to entering the preliminary system design phase of the program. The objective is to review and evaluate the requirements and the functional baseline for the system, and the material to be covered through this review should include the results from the feasibility analysis, system operational requirements, the maintenance and support concept, applicable prioritized TPMs, the functional analysis (top level for the system), system specification (Type A), a systems engineering management plan (SEMP), a test and evaluation master plan (TEMP), and supporting design criteria and data/documentation. Refer to Figure A.1.

2. *System design reviews* are generally scheduled during the preliminary system design phase when functional requirements and allocations are defined, preliminary design layouts and detailed specifications are prepared, system-level trade-off studies are conducted, and so on. These reviews are oriented to the overall system configuration (as subsystems and major system elements are defined), rather than to individual equipment items, software, and other lower-level components of the system. There may be one or more formal reviews scheduled, depending on the size of the system and the complexity of design. System design reviews may cover a variety of topics, including the following: functional analysis and the allocation of requirements; development, product, process, and material specifications (Types B, C, D, and E); applicable TPMs; significant design criteria for major system elements; trade-off study and analysis reports; predictions; and applicable design data (layouts, drawings, parts/material lists, supplier reports, and data).

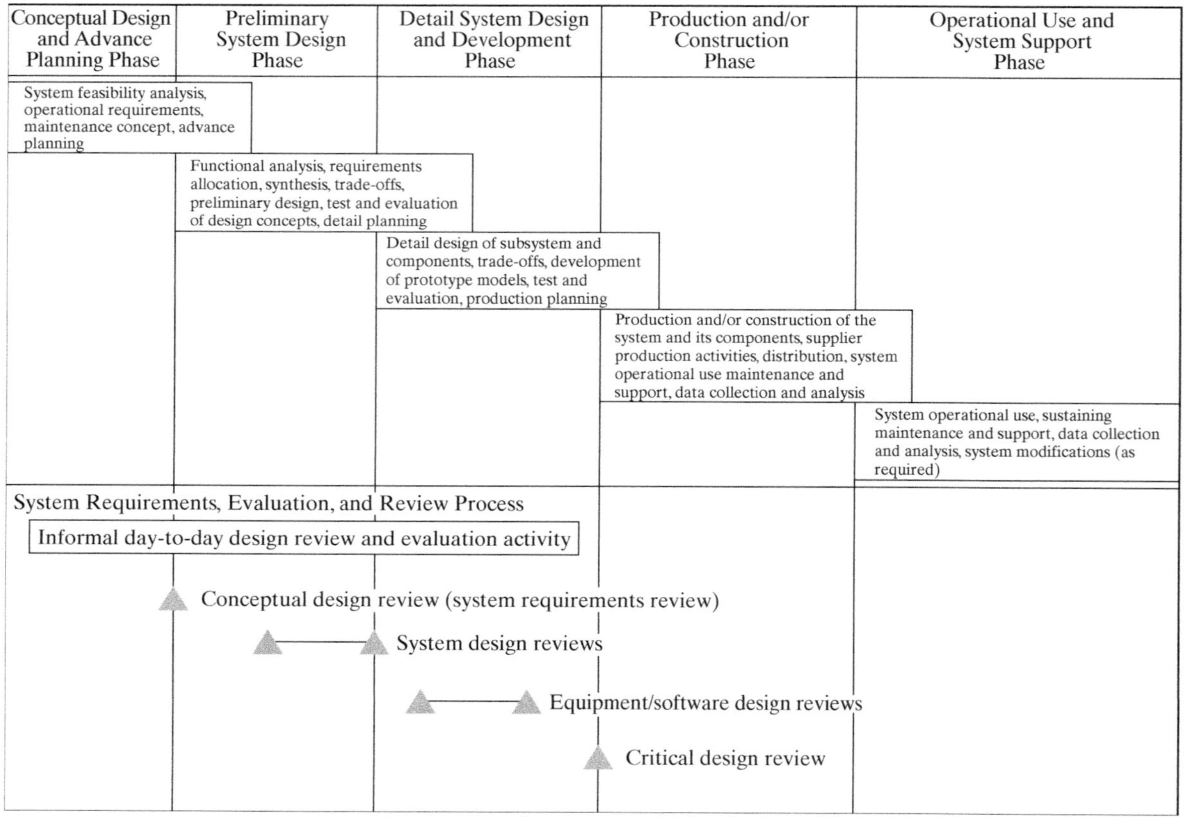

Conceptual Design and Advance Planning Phase	Preliminary System Design Phase	Detail System Design and Development Phase	Production and/or Construction Phase	Operational Use and System Support Phase
System feasibility analysis, operational requirements, maintenance concept, advance planning	Functional analysis, requirements allocation, synthesis, trade-offs, preliminary design, test and evaluation of design concepts, detail planning	Detail design of subsystem and components, trade-offs, development of prototype models, test and evaluation, production planning	Production and/or construction of the system and its components, supplier production activities, distribution, system operational use maintenance and support, data collection and analysis	System operational use, sustaining maintenance and support, data collection and analysis, system modifications (as required)

System Requirements, Evaluation, and Review Process

Informal day-to-day design review and evaluation activity

Conceptual design review (system requirements review)

System design reviews

Equipment/software design reviews

Critical design review

Figure 13   Formal design reviews (example).

3. *Equipment/software design reviews* are scheduled during the detail design and development phase and usually cover such topics as product, process, and material specifications (Types *C*, *D*, and *E*); design data defining major subsystems, equipment, software, and other elements of the system as applicable (assembly drawings, specification control drawings, construction drawings, installation drawings, logic drawings, schematics, materials/parts lists, and supplier data); analyses, predictions, trade-off study reports, and other related design documentation; and engineering models, laboratory models, mock-ups, and/or prototypes used to support a specific design configuration.

4. The *critical design review* is generally scheduled after the completion of detailed design, but prior to the release of firm design data for production and/or construction. Design is essentially "fixed" at this point, and the proposed configuration is evaluated in terms of adequacy, producibility, and/or constructability. The critical design review may include the following topics: a complete package of final design data and documentation; applicable analyses, trade-off study reports, predictions, and related design documentation; detailed production/construction plans; operational and sustainability plans; detailed maintenance plans; and a system retirement and material recycling/disposal plan. The results of the critical design review describe the final system configuration product baseline prior to entering into production and/or construction.

The review, evaluation, and feedback process is continuous throughout system design and development and encompasses conceptual, preliminary, and detail design.

9 SUMMARY AND EXTENSIONS

This chapter provides a continuation of the system design process and covers the transition from the "Conceptual Design Phase" to and throughout the "Preliminary Design Phase." The systems engineering process establishes the framework.

Specifically, this chapter includes a description of the design requirements for subsystems and major elements of the system, preparing detailed specifications, accomplishing functional analysis and the allocation of requirements to the subsystem level and below, developing detailed design requirements, initiating design engineering activities and establishing the design team, utilizing technologies and analytical models in design, conducting detailed trade-off studies, and describing the types and scheduling of design reviews.

Of particular significance is the proper integration of the various design engineering disciplines (illustrated in Figure 9) and establishing the "design team" in Section 5 and the selection and/or development of the appropriate technologies and analytical models/tools for application throughout the implementation of the systems engineering process described in Section 6. The successful implementation of this process requires that the system designer, or design team, be able to (1) accomplish all design objectives rapidly, accurately, and reliably throughout system acquisition; (2) transmit applicable design data and supporting documentation to many different locations, both nationally and internationally, rapidly and on a concurrent basis; and (3) implement design changes efficiently while at the same time maintaining effective configuration control. Included within this continuous process is the accomplishment of many different trade-off studies covering a wide variety of design-related issues. Responding to these objectives requires a comprehensive knowledge of the analytical techniques and the models/tools that are available to the designer (design team). Thus, a "qualified" systems engineer needs to be conversant and thoroughly familiar with the various analytical methods and their application in system design.

Challenges pertaining to shortening the system acquisition cycle for highly complex systems involving a wide mix of suppliers worldwide dictate the need for improvements in the use of automated computer-based design aids and global networks. There is a great need for improved methods for accomplishing design analyses, technical communications, and information/data processing. As an objective, the development and implementation of an *integrated design workstation*, described in Section 6, should be addressed. In addition, the "systems engineer" must have an in-depth knowledge of systems concepts, interface requirements, interdisciplinary relationships, and so on, and should assume a leadership position in attaining these and related objectives.

For more detailed knowledge of systems engineering design-related requirements, the reader should review in particular: ANSI/GEIA EIA-632, IEEE 1220-1998, INCOSE's *Systems Engineering Handbook, Systems Engineering Guidebook*, and *System Engineering Management* (B. S. Blanchard). In addition, one should visit the website for the International Council on Systems Engineering (INCOSE) at *http://www.incose.org*.

QUESTIONS AND PROBLEMS

1. Select a system of your choice and develop operational functional flow block diagrams (FFBDs) to the third level. Select one of the functional blocks and develop maintenance functional flows to the second level. Show how the maintenance functional flow diagrams evolve from the operational flows.

2. Describe how specific resource requirements (i.e., hardware, software, people, facilities, data, and elements of support) are derived from the functional analysis.

3. Describe the steps involved in transitioning from the functional analysis to a "packaging scheme" for the system. Provide an example.

4. Refer to the allocation in Figure 6. Explain how the quantitative factors (i.e., TPMs) at the unit and assembly levels were derived.

5. What steps would you take in accomplishing the allocation of requirements for a system-of-systems (SOS) configuration? Describe the overall process.

6. Select a system of your choice and assign some top-level TPMs. Allocate these requirements as appropriate to the second and third levels.

7. Refer to Figure 7. How would you define the design-to requirements for the *common unit*?

8. Describe what is meant by *interoperability*? Why is it important?

9. Describe what is meant by *environmental sustainability*? Identify some specific objectives in the design for such.

10. Why is the *design for security* important? Identify some specific design objectives.

11. What is meant by *design criteria*? How are they developed? How are they applied in the design process?

12. Refer to Figure 1. Why is the development of a *specification tree* important? What inherent characteristics should be included?

13. Refer to Figure 2. How are metrics established for the function shown? Give an example.

14. Refer to Figure 4. Describe some of the interfaces or interactions that must occur to ensure a completely integrated approach in the development of the hardware, software, and human system requirements. Be specific.

15. Define CAD, CAM, CAS, Macro-CAD, and their interrelationships. Include an illustration showing interfaces, information/data flow, etc.

16. Assume that you are a design engineer and are looking for some analytical models/tools to aid you in the synthesis, analysis, and evaluation process. Develop the criteria that you would apply in selecting the most appropriate tools for your application.

17. Assume that you have selected an analytical model for a specific application. Explain how you would validate that the model is adequate for the application in question.

19. List some of the benefits that can be derived from the use of computer-based models. Identify some of the concerns associated with the application of such.

20. Refer to Figure 8. Identify some of the objectives in selecting the appropriate methods/tools for accomplishment of the tasks in the right-hand columns.

21. Refer to Figure 12. Identify some of the objectives in designing a "tool set" as shown by the seven blocks in the figure.

22. Briefly identify some inputs and outputs for the *system design review*, the *equipment/software design review*, and the *critical design review*.

23. On what basis are formal design reviews scheduled? What are some of the benefits derived from conducting formal design reviews?

24. How are supplier requirements determined? How are these requirements passed on to the supplier?

25. Design is a *team* effort. Explain why! How should the design team be structured? How does systems engineering fit into the process?

26. What is the desired output of the Preliminary System Design Phase?

Appendix: Figures

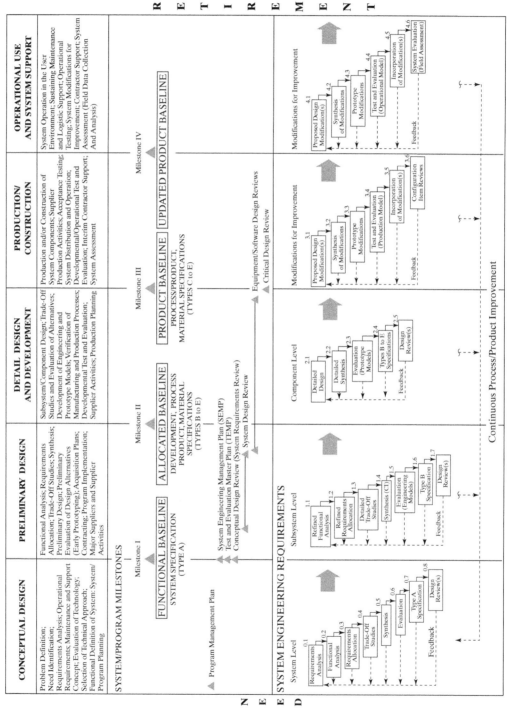

Figure A.1 System process activities and interactions over the life cycle.

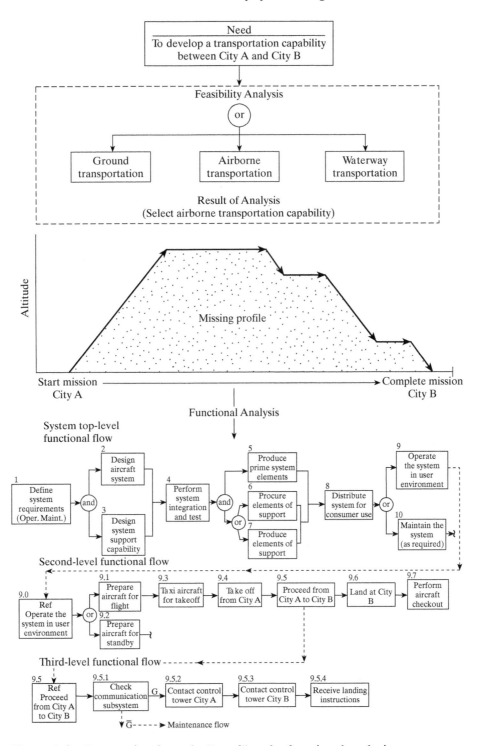

Figure A.2 Progression from the "need" to the functional analysis.

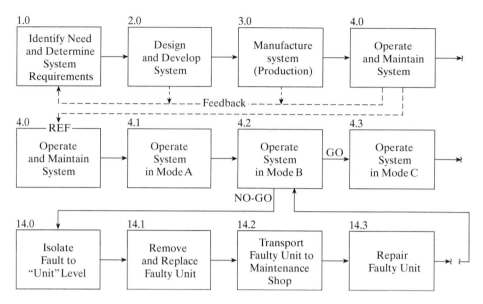

Figure A.3 Functional block diagram expansion (partial).

Detail Design and
Development

From Chapter 5 of *Systems Engineering and Analysis,* Fifth Edition, Benjamin S. Blanchard, Wolter J. Fabrycky. Copyright © 2011 by Pearson Education, Inc. Published by Pearson Prentice Hall. All rights reserved.

Detail Design and Development

The *detail design and development* phase of the system life cycle is a continuation of the iterative development process. The definition of system requirements, the establishment of a top-level functional baseline, and the preparation of a system specification (Type *A*) lead to the definition and development of subsystems and major elements of the system. Functional analysis, the allocation of design-to requirements below the system level, the accomplishment of synthesis and trade-off studies, the preparation of lower-level specifications (Types *B–E*), and the conduct of formal design reviews all provide a foundation upon which to base detailed design decisions that go down to the component level.

With the *functional baseline* developed as an output of the conceptual system design phase and with the *allocated baseline* derived during the preliminary design phase, the design team may now proceed with reasonable confidence in the realization of specific components as well as the "make-up" of the system configuration at the lowest level in the hierarchy. Realization includes the accomplishment of activities that (1) describe subsystems, units, assemblies, lower-level components, software modules, people, facilities, elements of maintenance and support, and so on, that make up the system and address their interrelationships; (2) prepare specifications and design data for all system components; and (3) acquire and integrate the selected components into a final system configuration.

This chapter addresses eight essential steps in the detail design and development process, while simultaneously providing an understanding of the intricacy of detail design and development within the systems engineering context. This includes

- Developing design requirements for all lower-level components of the system;
- Implementing the necessary technical activities to fulfill all design objectives;
- Integrating system elements and activities;
- Selecting and utilizing design tools and aids;

- Preparing design data and documentation;
- Developing engineering and prototype models;
- Implementing a design review, evaluation, and feedback capability; and
- Incorporating design changes as appropriate.

It is important for these steps to be thoroughly understood and reviewed, as they fall within the overall systems engineering process. As a learning objective, the intent is to provide a relatively comprehensive approach that addresses the detailed aspects of design. The chapter summary and extensions contribute insights by calling attention to applicable design standards and supporting documentation.

1 DETAIL DESIGN REQUIREMENTS

Specific requirements at this stage in the system design process are derived from the system specification (Type *A*) and evolve through applicable lower-level specifications (Types *B–E*). Included within these specifications are applicable design-dependent parameters (DDPs), technical performance measures (TPMs), and supporting design-to criteria leading to identification of the specific characteristics (attributes) that must be incorporated into the design configuration of elements and components. This is influenced through the requirements allocation process, where the appropriate built-in characteristics must be such that the allocated quantitative requirements in the figure will be met.

Given this top-down approach for establishing requirements at each level in the system hierarchical structure, the design process evolves through the iterative steps of synthesis, analysis, and evaluation, and to the definition of components leading to the establishment of a *product baseline*. At this point, the procurement and acquisition of system components begin, components are tested and integrated into a next higher entity (e.g., subassembly, assembly, and unit), and a physical model (or replica) of the system is constructed for test and evaluation. The integration, test, and evaluation steps constitute a bottom-up approach, and should result in a configuration that can be assessed for compliance with the initially specified customer requirements. This top-down/bottom-up approach is guided by the steps in the "vee" process model.

Progressing through the system design and development process in an expeditious manner is essential in today's competitive environment. Minimizing the time that it takes from the initial identification of a need to the ultimate delivery of the system to the customer is critical. This requires that certain design activities be accomplished on a *concurrent* basis. Figure 1 further illustrates the importance of "concurrency" in system design.

Referring to Figure 1, the designer(s) must think in terms of the four life cycles and their interrelationships, concurrently and in an integrated manner, in lieu of the sequential approach to design often followed in the past. The realization of this necessity became readily apparent in the late 1980s and resulted in concepts promulgated as *simultaneous engineering, concurrent engineering, integrated product development* (IPD), and others. These concepts must be inherent within the systems engineering process if the benefits of that process are to be realized.

THE SYSTEM/PRODUCT LIFE CYCLE—SERIAL APPROACH

System/product design and development	Production and/or construction	System/product utilization
		Maintenance and support

THE SYSTEM/PRODUCT LIFE CYCLE—CONCURRENT APPROACH

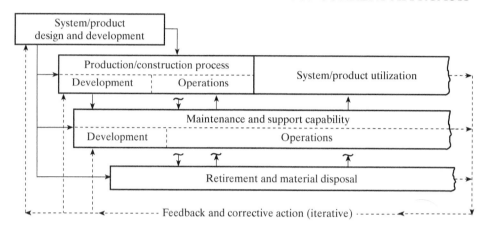

Figure 1 Sequential versus concurrent approaches in system design.

2 THE EVOLUTION OF DETAIL DESIGN

The evolution of detail design is based on the results from the requirements established during the conceptual and preliminary system design phases. As an example, for a river crossing problem addressed by a bridge system, the primary top-level requirements were identified in conceptual design. These requirements were decomposed further, expanded during the preliminary design phase through the functional analysis and allocation of requirements down to major subsystems to include the roadway and railbed, passenger walkway and bicycle path, toll collection facilities, and the maintenance and support infrastructure; and then the more detailed design-to requirements for the various lower-level elements of the system are defined (in this chapter) down to the bridge substructures, foundations, piles and footings, retaining walls, toll collectors, lighting, and so on. This top-down/bottom-up process is illustrated in Figure 2.

Referring to the figure, the design-to *requirements* are identified from the "top down," with the cross-hatched block defined in conceptual design along with the allocation and requirements for the major subsystems noted by the shaded blocks (superstructure, substructure, etc.). These requirements are further expanded, through functional analysis and allocation, during the preliminary design phase to define the specific requirements for the lower-level elements of the bridge represented by the white blocks in the figure (road and railway deck, piles/piers, footings, toll facilities, etc.). The basic "requirements" for the river crossing bridge are defined and allocated (i.e., "driven") from the "top down," while the detailed design and follow-on construction is accomplished from the "bottom up." This is why one often invokes a top-down/bottom-up process versus assuming only a "bottom-up" approach.

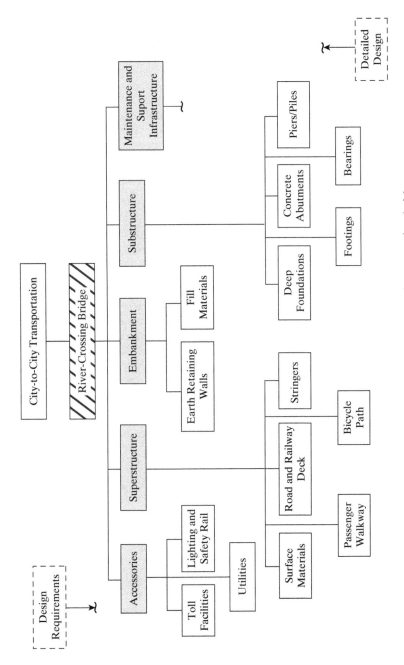

Figure 2 The top-down/bottom-up design approach for the river crossing bridge system.

Having established the basic top-level requirements for the overall system and the preliminary design requirements, the design process from this point forward is essentially *evolutionary* by nature as described earlier. The design team has been established with the overall objective of integrating the various system elements into a final system configuration. Such elements include not only mission-related hardware and software but also people, real estate and facilities, data/information, consumables, and the materials and resources necessary for the operation and sustaining support of the system throughout its planned life cycle. The integration of these elements is emphasized in Figure 3.

Detail design evolution follows the basic sequence of activities shown in Figure 4. The process is iterative, proceeding from system-level definition to a product configuration that can be constructed or produced in multiple quantities. There are "checks and balances" in the form of reviews at each stage of design progression and a feedback loop allows for corrective action as necessary. These reviews may be relatively informal and occur continuously, as compared with the formal design reviews scheduled at specific milestones. In this respect, the process is similar to the synthesis, analysis, evaluation, and product definition, accomplished in the preliminary design stage, except that the requirements are at a lower level (i.e., units, assemblies, subassemblies, etc.). As the level of detail increases, actual definition is accomplished through the development of data describing the item being designed. These data may be presented in the form of a digital description of the item(s) in an electronic format, design drawings in physical and electronic form, material and part lists, reports and analyses, computer programs, and so on.

The actual process of design iteration may occur through the use of the World Wide Web (WWW), a local area network (LAN), the use of telecommunications and compressed video conferences, or some equivalent form of communication. The design configuration selected may be the best possible in the eyes of the responsible designer. However, the results are practically useless unless properly documented, so that others can first understand what is

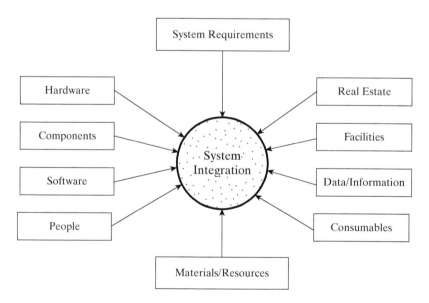

Figure 3 The integration of system elements.

Detail Design and Development

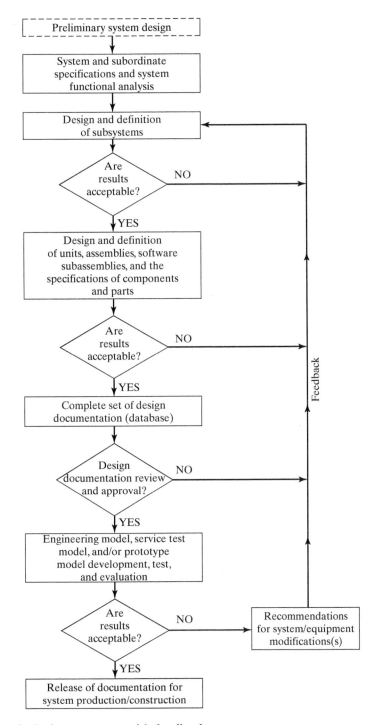

Figure 4 Basic design sequence with feedback.

being conveyed and then be able to translate and convert the output into an entity that can be constructed (as is the case for the river crossing bridge) or produced in multiple quantities.

3 INTEGRATING SYSTEM ELEMENTS AND ACTIVITIES

An important output from the functional analysis and allocation process is the identification of various elements of the system and the need for hardware (equipment), software, people, facilities, materials, data, or combinations thereof. The objective is to conduct the necessary trade-off analyses to determine the best way to respond to the *hows*.

Given the basic configuration of system elements, the designer must decide how best to meet the need in selecting a specific approach in responding to an equipment need, a software need, and so on. For example, there may be alternative approaches in selecting a specific resource, such as illustrated in Figure 5, with the following steps taken (in order of precedence) in arriving at a satisfactory result:

1. Select a standard component that is commercially available and for which there are a number of viable suppliers; for example, a commercial off-the-shelf (COTS) item, or equivalent. The objective is, of course, to gain the advantage of competition (at reduced cost) and to provide the assurance that the appropriate maintenance and support will be readily available in the future and throughout the system life cycle when required, or;

2. Modify an existing commercially available off-the-shelf item by providing a mounting for the purposes of installation, adding an adapter cable for the purposes of compatibility, providing a software interface module, and so on. Care must

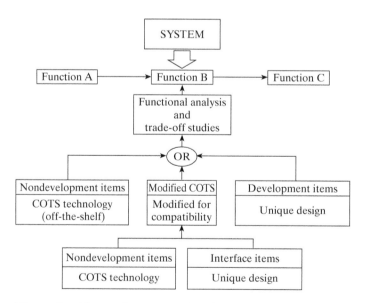

Figure 5 Alternative approaches in the selection of resources.

be taken to ensure that the proposed modification is relatively simple and inexpensive and doesn't result in the introduction of a lot of additional problems in the process, or;

3. Design and develop a new and unique component to meet a specific functional requirement. This approach will require that the component selected be properly integrated into the overall system design and development process in a timely and effective manner.

The most cost-effective solution seems to favor the utilization of commercial off-the-shelf (COTS) components, as the acquisition cost, item availability time, and risks associated with meeting a given system technical requirement are likely to be less. In any event, the decision-making process may occur at the subsystem level early in preliminary design, at the configuration-item level, and/or at the unit level. These decisions will be based on factors including the functions to be performed, availability and stability of current technology, number of sources of supply and supplier response times, reliability and maintainability requirements, supportability requirements, cost, and others. From the resulting decisions, the requirements for component acquisition (procurement) will be covered through the preparation of either a *product specification* (Type *C*) or *process specification* (Type *D*) for COTS components, or a *development specification* (Type *B*) for newly designed or modified components.

When the ultimate decision specifies the design and development of a new, or unique, system element, the follow-on design activity will include a series of steps, which evolve from the definition of need to the integration and test of the item as part of the system validation process. For example, the functional analysis may lead to the identification of *hardware*, *software*, and *human* requirements. The development of hardware usually leads to the identification of units, assemblies, modules, and down to the component-part level. The software acquisition process often involves the identification of computer software configuration items (CSCIs), computer software components (CSCs) and computer software units (CSUs), the development of code, and CSC integration and testing. The development of human requirements includes the identification of individual tasks, the combining of tasks into position descriptions, the determination of personnel quantities and skill levels, and personnel training requirements.

While the above example addresses only the three "mini" life cycles mentioned (i.e., hardware, software, and human), there may be other resource requirements evolving from the functional analysis to include facilities, data/documentation, elements of the maintenance and support infrastructure, and so on.. Associated with each will be a life cycle unique to the particular application.

There must be some form and degree of communication and integration throughout and across these component life cycles on a continuing basis. Quite often, design engineers become so engrossed in the development of hardware that they neglect the major interfaces and the impact that hardware design decisions have on the other elements of the system. Software engineers, like hardware engineers, often operate strictly within their own domain without considering the necessary interfaces with the hardware development process. On occasion, the hardware development process will lead to the need for software at a lower level. Further, consideration of the human being, as an element of the system, is often ignored in the design process altogether. Because of these practices in the past, the integration of the various system elements has not occurred until late in the detail design and development phase during systems integration and testing, often resulting in incompatibilities and the need for last-minute costly modifications in order to "make it work!"

Thus, a primary objective of systems engineering is to ensure the proper coordination and timely integration of all system elements (and the activities associated with each) from the beginning. This can be accomplished with a good initial definition of requirements through the preparation of well-written specifications, followed by a structured and disciplined approach to design. This includes the scheduling of an appropriate number of formal design reviews to ensure the proper communications across the project and to check that all elements of the system are compatible at the time of review. In the event of incompatibilities, the hardware design effort should not be allowed to proceed without first ensuring that the software is compatible; the software development effort should not be allowed to proceed without first ensuring compatibility with the hardware; and so on.

4 DESIGN TOOLS AND AIDS

The successful completion of the design process depends on the availability of the appropriate tools and design aids that will help the design team in accomplishing its objectives in an effective and efficient manner. The application of *computer-aided engineering* (CAE) and *computer-aided design* (CAD) tools enables the projection of many different design alternatives throughout the life cycle. At the early stages of design, it is often difficult to visualize a system configuration (or element thereof) in its true perspective, whereas simulating a three-dimensional view of an item is possible through the use of CAD. In some instances, the validation of system requirements can be accomplished through the use of *simulation* methods during the preliminary system design and detail design and development phases prior to the introduction of hardware, software, and so on. Through the proper use of these and related technologies (to include the appropriate application of selected analytical models), the design team is able to produce a robust design more quickly, while reducing the overall program technical risks.

As an additional aid to the designer, physical three-dimensional scale models or *mock-ups* are sometimes constructed to provide a realistic simulation of a proposed system configuration. Models, or mock-ups, can be developed to any desired scale and to varying depths of detail depending on the level of emphasis desired. Mock-ups can be developed for large as well as small systems and may be constructed of heavy cardboard, wood, metal, or a combination of different materials. Mock-ups can be developed on a relatively inexpensive basis and in a short period of time when employing the right materials and personnel services. Industrial design, human factors, or model-shop personnel are usually available in many organizations, are well-oriented to this area of activity, and should be utilized to the greatest extent possible. Some of the uses and values of a mock-up are that they:

1. Provide the design engineer with the opportunity of experimenting with different facility layouts, packaging schemes, panel displays, cable runs, and so on, before the preparation of final design data. A mock-up or engineering model of the proposed river crossing bridge can be developed to better visualize the overall structure, its location, interfaces with the communities on each side of the river, and so on.

2. Provide the reliability–maintainability–human factors engineer with the opportunity to accomplish a more effective review of a proposed design configuration for the incorporation of supportability characteristics. Problem areas readily become evident.

3. Provide the maintainability-human factors engineer with a tool for use in the accomplishment of predictions and detailed task analyses. It is often possible to simulate operator and maintenance tasks to acquire task sequence and time data.

4. Provide the design engineer with an excellent tool for conveying his or her final design approach during a formal design review.

5. Serve as an excellent marketing tool.

6. Can be employed to facilitate the training of system operator and maintenance personnel.

7. Can be utilized by production and industrial engineering personnel in developing fabrication and assembly processes and procedures and in the design of factory tooling and associated test fixtures.

8. Can serve as a tool at a later stage in the system life cycle for the verification of a modification kit design prior to the preparation of formal data and the development of kit hardware, software, and supporting materials.

In the domain of software development, in particular, designers are oriented toward the building of "one-of-a-kind" software packages. The issues in software development differ from those in other areas of engineering in that mass production is not the normal objective. Instead, the goal is to develop software that accurately portrays the features that are desired by the user (customer). For example, in the design of a complex workstation display, the user may not at first comprehend the implications of the proposed command routines and data format on the screen. When the system is ultimately delivered, problems occur, and the "user interface" is not acceptable for one reason or another. Changes are then recommended and implemented, and the costs of modification and rework are usually high.

The alternative is to develop a "protoype" early in the system design process, design the applicable software, involve the user in the operation of the prototype, identify areas that need improvement, incorporate the necessary changes, involve the user once again, and so on. This iterative and evolutionary process of software development, accomplished throughout the preliminary and detail design phases, is referred to as *rapid prototyping*. Rapid prototyping is a practice that is often implemented and is inherent within the systems engineering process, particularly in the development of large software-intensive systems.

5 DESIGN DATA, INFORMATION, AND INTEGRATION

As a result of advances in information systems technology, the methods for documenting design are changing rapidly. These advances have promoted the use of vast electronic databases for the purposes of information processing, storage, and retrieval. Through the use of CAD techniques, information can be stored in the form of three-dimensional representations, in regular two-dimensional line drawings, in digital format, or in combinations of these. By using computer graphics, word processing capability, e-mail communications, video-disc technology, and the like, design can be presented in more detail, in an easily modified format, and displayed faster.

Although many advances have been made in the application of computerized methods to data acquisition, storage, and retrieval, the need for some of the more conventional methods of design documentation remains. These include a combination of the following:

1. *Design drawings*—assembly drawings, control drawings, logic diagrams, structural layouts, installation drawings, schematics, and so on.

2. *Material and part lists*—part lists, material lists, long-lead-item lists, bulk-item lists, provisioning lists, and so on.

3. *Analyses and reports*—trade-off study reports supporting design decisions, reliability and maintainability analyses and predictions, human factors analyses, safety reports,

supportability analyses, configuration identification reports, computer documentation, installation and assembly procedures, and so on.

Design drawings, constituting a primary source of definition, may vary in form and function depending on the design objective; that is, the type of equipment being developed, the extent of development required, whether the design is to be subcontracted, and so on. Some typical types of drawings from the past are illustrated in Figure 6.

1. *Arrangement drawing*—shows in any projection or perspective, with or without controlling dimensions, the relationship of major units of the item covered.

2. *Assembly drawing*—depicts the assembled relationship of (a) two or more parts, (b) a combination of parts and subassemblies, or (c) a group of assemblies required to form the next higher indenture level of the equipment.

3. *Connection diagram*—shows the electrical connections of an installation or of its component devices or parts.

4. *Construction drawing*—delineates the design of buildings, structures, or related construction (including architectural and civil engineering operations).

5. *Control drawing*—an engineering drawing that discloses configuration and configuration limitations, performance and test requirements, weight and space limitations, access clearances, pipe and cable attachments, support requirements, etc., to the extent necessary that an item can be developed or procured on the commercial market to meet the stated requirements. Control drawings are identified as envelope control (i.e., configuration limitations), specification control, source control, interface control, and installation control.

6. *Detail drawing*—depicts complete end item requirements for the part(s) delineated on the drawing.

7. *Elevation drawing*—depicts vertical projections of buildings and structures or profiles of equipment.

8. *Engineering drawing*—an engineering document that discloses by means of pictorial or textual presentations, or a combination of both, the physical and functional end product requirements of an item.

9. *Installation drawing*—shows general configuration and complete information necessary to install an item relative to its supporting structure or to associated items.

10. *Logic diagram*—shows by means of graphic symbols the sequence and function of logic circuitry.

11. *Numerical control drawing*—depicts complete physical and functional engineering and product requirements of an item to facilitate production by tape control means.

12. *Piping diagram*—depicts the interconnection of components by piping, tubing, or hose, and when desired, the sequential flow of hydraulic fluids or pneumatic air in the system.

13. *Running (wire) list*—a book-form drawing consisting of tabular data and instructions required to establish wiring connections within or between items.

14. *Schematic diagram*—shows, by means of graphical symbols, the electrical connections and functions of a specific circuit arrangement.

15. *Software diagrams*—functional flow diagrams, process flows, and coding drawings.

16. *Wiring and cable harness drawing*—shows the path of a group of wires laced together in a specified configuration, so formed to simplify installation.

Figure 6 Typical engineering drawing categories.

During the process of detail design, engineering documentation is rather preliminary and then gradually progresses to the depth and extent of definition necessary to enable product manufacture. The responsible designer, using appropriate design aids, produces a functional diagram of the overall system. The system functions are analyzed and initial packaging concepts are assumed. With the aid of specialists representing various disciplines (e.g., civil, electrical, mechanical, structural, components, reliability, maintainability, and sustainability) and supplier data, detail design layouts are prepared for subsystems, units, assemblies, and subassemblies. The results are analyzed and evaluated in terms of functional capability, reliability, maintainability, human factors, safety, producibility, and other design parameters to assure compliance with the allocated requirements and the initially established design criteria. This review and evaluation occurs at each stage in the basic design sequence and generally follows the steps presented in Figure 7.

Engineering data are reviewed against design standards and checklist criteria. Throughout the industrial and governmental sectors are design standards manuals and handbooks developed to cover preferred component parts and supplier data, preferred design and manufacturing practices, designated levels of quality for specified products, requirements for safety, and the like. These standards (ANSI, EIA, ISO, etc.), as applicable, may serve as a basis for design review and evaluation.

Checklists may be developed and "tailored" for a specific system application. For example, the requirements definition process aids in the initial identification of technical performance measures (TPMs). These TPMs are then prioritized to reflect levels of importance. Top-level qualitative and quantitative factors are allocated to establish the appropriate design criteria at each level in the system hierarchical structure. A design parameter tree may be prepared to capture critical top-down relationships. Finally, questions are developed in support of each parameter and included in a comprehensive checklist. Each question should be traceable back to a given TPM. The resulting checklist is then used for the purpose of design evaluation. Through a review of the available design data, the degree of compliance with the checklist criteria is assessed and a recommendation is made as to whether to accept the item as is, to modify it, or to reject it.

Figure 8 constitutes an abbreviated checklist, which is generic in nature but which can serve as a baseline for the purposes of "tailoring." The questions, in turn, should be based on specific qualitative and quantitative criteria, with the appropriate weighting factors to indicate level of importance. Checklists may be developed and tailored to any level of detail desired. They serve as a reminder and aid to the design engineer in achieving a complete and robust design.

6 DEVELOPMENT OF ENGINEERING MODELS

As the system design and development effort progresses, the basic process evolves from a description of the design in the form of drawings, documentation, and databases to the construction of a physical model or mock-up, to the construction of an engineering or laboratory model, to the construction of a prototype, and ultimately to the production of a final product. The purpose in proceeding through these steps is to provide a solid basis for design evaluation and/or validation. The earlier in the design process that one can accomplish this purpose, the better, since the incorporation of any necessary changes for corrective action will be more costly and disruptive later when the design progresses toward the production/construction phase.

The first two steps in this development sequence are discussed in the previous sections: Section 4 discusses the development of a mock-up and Section 5 discusses the development of

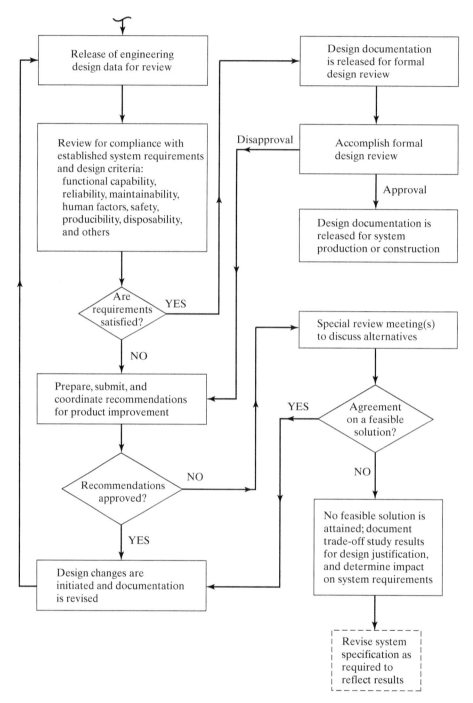

Figure 7 Design data review cycle (refer to Figure 4).

System Design Review Checklist

General

1. System operational requirements defined
2. Effectiveness factors established
3. System maintenance concept defined
4. Functional analysis and allocation accomplished
5. System trade-off studies documented
6. System specification and supporting specifications completed

7. System engineering management plan completed
8. Design documentation completed
9. Logistic support requirements defined
10. Ecological requirements met
11. Societal requirements met
12. Economic feasibility determined
13. Sustainability requirements met

Design Features—Does the design reflect adequate consideration of

1. Accessibility
2. Adjustments and alignments
3. Cables and connectors
4. Calibration
5. Disposability
6. Environment
7. Fasteners
8. Handling
9. Human factors
10. Interchangeability
11. Maintainability

12. Packaging and mounting
13. Panel displays and controls
14. Producibility
15. Reliability
16. Safety
17. Selection of parts/materials
18. Servicing and lubrication
19. Software
20. Standardization
21. Supportability
22. Testability

When reviewing design (layouts, drawings, parts lists, computer graphics, engineering reports, program plans), this checklist may prove beneficial in covering various program functions and design features. The response to each item listed should be YES.

Figure 8 Sample design review checklist.

data and documentation requirements. At this stage, the results have not produced an actual "working" model of the system.

At some point in the detail design and development phase, it may be appropriate to produce an *engineering model* (or a laboratory model). The objective is to demonstrate some (if not all) of the functions that the system is to ultimately perform for the customer by constructing an "operating" model and utilizing it in a research-oriented environment in an engineering shop, or equivalent. This model may be constructed using nonstandard and unqualified parts and will not necessarily reflect the design configuration that will ultimately be produced for the customer. The intent is to verify certain performance characteristics and to gain confidence that "all is well" at the time.

7 SYSTEM PROTOTYPE DEVELOPMENT

A *prototype model* represents the production/construction configuration of a system (and its elements) in all aspects of form, fit, and function except that it has not been fully "qualified" in terms of operational and environmental testing. It is constructed of common and approved component parts, using standard assembly and test processes, and is of the same design configuration that will ultimately be delivered and utilized by the customer in his/her operational environment. The prototype system is made up of the required prime equipment, operational software, operational facilities, technical data, and associated elements of the maintenance and support infrastructure. The objective is to accomplish a specified amount of testing for the purposes of design evaluation prior to entering a formal test and evaluation phase.

As part of the overall system design and development process, engineering evaluation functions include the utilization of analytical methods/tools, physical mock-ups, engineering models, prototype models, and/or various combinations of these. Depending on the type and nature of the system (and its functions), the selection and development of some form of system "replica" will vary. The purpose is to assist in the verification of technical concepts and various system design approaches. Areas of noncompliance with the specified requirements are identified and corrective action is initiated as required.

8 DESIGN REVIEW, EVALUATION, AND FEEDBACK

The design review and evaluation activity is continuous and includes both (1) an informal ongoing iterative day-to-day process of review and evaluation, and (2) the scheduling of formal design reviews at discrete points in design and throughout system acquisition. The purpose of conducting any type of a review is to assess if (and just how well) the design configuration, as envisioned at the time, is in compliance with the initially specified quantitative and qualitative requirements. The technical performance measures (TPMs), identified and prioritized in the conceptual design phase and allocated to the various elements of the system, must be measured to ensure compliance with these specified requirements. Further, these critical measures must be "tracked" from one review to the next.

8.1 Tracking and Controlling to TPMs

Five key TPMs have been selected in Figure 9 to illustrate the tracking of TPMs. It is assumed that the specific quantitative requirements have been established, along with an allowable "target" range for each of the values. For instance, the system operational availability (A_o) must be at least 0.90, but can be higher if feasible. As a result of a *system design review*, based on a prediction associated with the design configuration at the time, it was determined that A_o was only around 0.82. Given this, there is a need to develop a formal design plan indicating the steps that will be necessary to realize the required reliability growth in order to comply with the 0.90 requirement. Each of the specified TPMs must be monitored in a similar manner.

For relatively large systems, there may be several distinct TPM requirements that must be met and there should be priorities established to indicate relative degrees of importance. While all requirements may be important, there is a tendency on the part of some designers to favor one TPM requirement over another. Given that system design requires a "team" approach and that the different team members may represent different design disciplines, there may be a greater amount of effort expended to ensure that the design will meet one

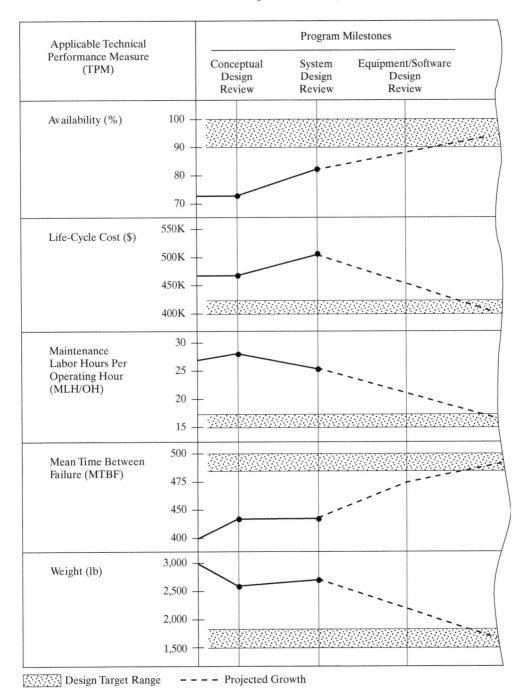

Figure 9 Parameter measurement and evaluation at design review ("tracking").

requirement at the expense of not complying with another. For instance, if the system being developed is an aerospace system, then weight, size, and speed are critical and the impact from an aeronautical engineering perspective is high. On the other hand, meeting an MLH/OH requirement may not be of concern to the aeronautical engineer assigned to the design team, but is highly important when considering the maintainability characteristics in design.

To ensure that *all* of the requirements are met, or at least seriously addressed, various design team members may be assigned to "track" specific TPMs throughout the design process. Figure 10 illustrates the relationships between TPMs and responsible design

Technical Performance Measures (TPMs) \ Engineering Design Functions	Aeronautical Engineering	Components Engineering	Cost Engineering	Electrical Engineering	Human Factors Engineering	Logistics Engineering	Maintainability Engineering	Manufacturing Engineering	Materials Engineering	Mechanical Engineering	Reliability Engineering	Structural Engineering	Systems Engineering
Availability (90%)	H	L	L	M	M	H	M	L	M	M	M	M	H
Diagnostics (95%)	L	M	L	H	L	M	H	M	M	H	M	L	M
Interchangeability (99%)	M	H	M	H	M	H	H	H	M	H	H	M	M
Life cycle cost ($350K, unit)	M	M	H	M	M	H	H	L	M	M	H	M	H
$\overline{M}$ct (30 min)	L	L	L	M	M	H	H	M	M	M	M	M	M
MDT (24 hr)	L	M	M	L	L	H	M	M	L	L	M	L	H
MLH/OH (15)	L	L	M	L	M	M	H	L	L	L	M	L	H
MTBF (300 hr)	L	H	L	M	L	L	M	H	H	M	H	M	M
MTBM (250 hr)	L	L	L	L	L	M	H	L	L	L	M	L	H
Personnel skill levels	M	L	M	M	H	M	H	L	L	L	L	L	H
Size (150 ft by 75 ft)	H	H	M	M	M	M	M	H	H	H	M	H	M
Speed (450 mph)	H	L	L	L	L	L	L	L	L	L	L	M	H
System effectiveness (80%)	M	L	L	M	L	M	M	L	L	M	M	M	H
Weight (150K lb)	H	H	M	M	M	M	M	H	H	H	L	H	M

Figure 10 The relationship between TPMs and responsible design disciplines (refer to Figure 3.17). H = high interest; M = medium interest; and L = low interest.

disciplines (in terms of likely interest) and indicates where the "tracking" assignments may be specified. In other words, a reliability engineer will likely "track" the requirements pertaining to the MTBF, a maintainability engineer will be interested in MLH/OH, MTBM, and LCC, and so on. From a systems engineering perspective, all of these factors must be "tracked" to the extent indicated by the distributed priorities.

8.2 Tracking of Other Design Considerations

As the designer evolves through the detail design and development process, including the conduct of the essential trade-off analyses leading to the selection of specific components, he/she must be cognizant of considerations other than the obvious TPMs. Two of these are presented here as examples.

The expected *life* of each of the components chosen for incorporation into the design is a consideration of importance. If the specified component "life" is less than the planned life cycle of the system (i.e., a "critical useful-life item"), then this leads to the requirement for a scheduled maintenance plan for component replacement. If, when combining and integrating the various components into a larger assembly, the expected life is less than the planned life cycle of the system, then once again there may be a need for a preplanned system maintenance cycle or a redesign for mitigation of this burden. In any case, the designer must track and address the issue of *obsolescence* in design "before the fact" instead of leaving it to chance later downstream in the system life cycle.

Another consideration to be tracked in today's environment pertains to implementing the overall system design process effectively, in a limited amount of time, and at reduced cost. Shortening the acquisition process (cycle) continues to be a desired objective. At the same time, there has been a tendency to continue the design process, on an evolutionary basis, for as long as possible and incorporate design improvements up to the last minute before delivering the system to the customer. As desirable as this might appear, a superb system delivered late and over budget may be considered a failure by the customer.

8.3 Conducting Design Reviews

Design reviews are scheduled periodically. Any review depends on the depth of planning, organization, and preparation prior to the review itself. An extensive amount of coordination is needed, involving the following factors:

1. Identification of the items to be reviewed.
2. A selected date for the review.
3. The location or facility where the review is to be conducted.
4. An agenda for the review (including a definition of the basic objectives).
5. A design review board representing the organizational elements and disciplines affected by the review. Basic design functions, reliability, maintainability, human factors, quality control, manufacturing, sustainability, and logistic support representation are included. Individual organizational responsibilities should be identified. Depending on the type of review, the customer and/or individual equipment suppliers may be included.
6. Equipment (hardware) and/or software requirements for the review. Engineering models, prototypes and/or mock-ups may be required to facilitate the review process.
7. Design data requirements for the review. This may include all applicable specifications, lists, drawings, predictions and analyses, logistic data, computer data, and special reports.
8. Funding requirements. Planning is necessary in identifying sources and a means for providing the funds for conducting the review.

9. Reporting requirements and the mechanism for accomplishing the necessary follow-up actions stemming from design review recommendations. Responsibilities and action-item time limits must be established.

The design review involves a number of different discipline areas and covers a wide variety of design data and in some instances the presence of hardware, software, and/or other selected elements of the system. In order to fulfill its objective expeditiously (i.e., review the design to ensure that all system requirements are met in an optimum manner), the design review must be well organized and firmly controlled by the design review board chairperson. Design review meetings should be brief and to the point and must not be allowed to drift away from the topics on the agenda. Attendance should be limited to those who have a direct interest and can contribute to the subject matter being presented. Specialists who participate should be authorized to speak and make decisions concerning their area of specialty. Finally, the design review must make provisions for the identification, recording, scheduling, and monitoring of any subsequent corrective action that is required. Specific responsibility for follow-up action must be designated by the chairperson of the design review board.

9 INCORPORATING DESIGN CHANGES

After a baseline has been established as a result of a formal design review, changes are frequently initiated for any one of a number of reasons such as to correct a design deficiency, improve a product, incorporate a new technology, improve the level of sustainability, respond to a change in operational requirements, compensate for an obsolete component, and so on. Changes may be initiated from within the project, or as a result of some new externally imposed requirement.

At first, it may appear that a change is relatively insignificant in nature and that it may constitute a change in the design of a prime equipment item, a software modification, a data revision, and/or a change in some process. However, what might initially appear to be minor often turns out to have a great impact on and throughout the system hierarchical structure. For instance, a change in the design configuration of prime equipment (e.g., a change in size, weight, repackaging, and added performance capability) will probably affect related software, design of test and support equipment, type and quantity of spares/repair parts, technical data, transportation and handling requirements, and so on. A change in software will likely have an impact on associated prime equipment, technical data, and test equipment. A change in any one item will likely have an impact on many other elements of the system. Further, if there are numerous changes being incorporated at the same time, the entire system configuration may be severely compromised in terms of maintaining some degree of requirements traceability.

Past experience with a variety of systems has indicated that many of the changes incorporated are introduced late in detail design and development, during production or construction, and/or early in the system utilization and sustaining support phase. The accomplishment of changes this far downstream in the life cycle can be very costly; for example, a small change in an equipment item can result in a subsequent change in software, technical data, facilities, the various elements of support, or a production process.

While the incorporation of changes (for one reason or another) is certainly inevitable, the process for doing this must be formalized and controlled to ensure traceability from one configuration baseline to another. Also, it is necessary to ensure that the incorporation of a change is consistent with the requirements in the system specification (Type *A*). Referring to Figure 11, a proposed change is initially submitted through the preparation of an *engineering*

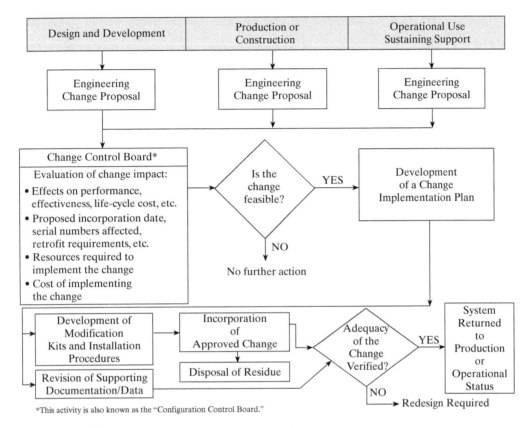

Detail Design and Development

Figure 11 Change control process.

change proposal (ECP), which, in turn, is reviewed by a Change Control Board (or Configuration Control Board). Each proposed change must be thoroughly evaluated in terms of its impact on other elements of the system, the specified TPMs, life-cycle cost, and the various considerations that have been addressed throughout the earlier stages of design (e.g., reliability, maintainability, human factors, producibility, sustainability, and disposability). Approved changes will then lead to the development of the required modification kit, installation instructions, and the ultimate incorporation of the change in the system configuration. Accordingly, there needs to be a highly disciplined *configuration management* (CM) process from the beginning and throughout the entire system life cycle. This is particularly important in the successful implementation of the systems engineering process.

10 SUMMARY AND EXTENSIONS

This chapter provides a continuation of the system design process and progresses to and through the "Detail Design and Development Phase."

Detailed design requirements (below the subsystem level), the design process and integration of design activities, utilization of design tools and aids, preparation of design data and documentation, development of engineering and prototype models, implementation of design review and evaluation, and the processing and incorporation of design changes are the concerns of this chapter.

In accomplishing engineering design activities, the designer (i.e., the design team) depends on a number of resources for guidance. Initially, there are system-level requirements; there are the specific qualitative and quantitative design criteria developed from the established requirements (e.g., TPM factors); and there is the preparation of the various categories of specifications (e.g., Types *A–E* specifications).

Providing additional guidance is a wide variety of ANSI, EIA, IEEE, ISO, and related *standards*, covering standard practices and the implementation of "standardization" in design. The systems engineer must not only be knowledgeable of the specifications that apply to a particular system design configuration but should also be conversant and knowledgeable of the appropriate standards that may also be applied throughout the overall design and development effort, such as ANSI/GEIA EIA-632, IEEE 1220-1998, ISO 9001, ISO/IEC 15288. To facilitate this objective, the following standards-oriented organizations are listed as additional sources recommended for further investigation:

1. American National Standards Institute (ANSI).
2. Electronic Industries Alliance (EIA).
3. Government Electronics and Information Technology Association (GEIA).
4. International Organization for Standardization (ISO).

The predisposition of many designers has been to incorporate the "latest and the greatest," even at the sacrifice of maintaining any form of baseline management and configuration control. Following this undesirable approach in the past has caused many problems downstream in the system life cycle. The costs of producing many different and unique products (some with certain changes incorporated and some without) have been high, and the costs of sustaining maintenance and support of these systems with slightly different configurations in the field have also been high. In other words, in the development of systems where the proper design controls have been loosely implemented (if not ignored altogether), the ultimate life-cycle costs have been high. Thus, while it is appropriate to adhere to the objectives of "modernizing" the system design and acquisition process, it is also necessary to maintain some form of *configuration control* (i.e., "change" control) in the process. The "baseline" management approach and the proper implementation of systems engineering objectives can aid in accomplishing this goal. A good reference for configuration management is ANSI/GEIA EIA-649-A, *National Consensus Standard for Configuration Management*, Electronic Industries Alliance (EIA), Arlington, VA. See *www.eia.org*

QUESTIONS AND PROBLEMS

1. What are the basic differences between *conceptual design, preliminary system design*, and *detailed design and development*? Are these stages of design applicable to the acquisition of all systems? Explain.

2. Design constitutes a *team* effort. Explain why. What constitutes the make-up of the design team? How can this be accomplished? How does systems engineering fit into the process?

3. Briefly describe the role of systems engineering in the overall design process.

4. Refer to Figure 1. What are some of the advantages of the concurrent approach in design? Identify some of the problems that could occur in its implementation.

5. As the systems engineering manager on a given program, what steps would you take to ensure that the proper integration of requirements occurs across the three life cycles (hardware, software, and human) from the beginning?

6. As a designer, one of your tasks is the selection of a component to fulfill a specific design objective. What priorities would you consider in the selection process (if any)?

7. Why are design standards (as applied to component parts and processes) important?

8. Why are engineering documentation and the establishment of a design database necessary?

9. Refer to Figure 5. When accomplishing the necessary trade-offs, there may be some confusion as to which of the three options to pursue. Describe what information is required as an input in order to evolve into a "clear-cut" approach.

10. Describe how the application of CAD, CAM, and CAS tools can facilitate the system design process. Identify some benefits. Address some of the problems that could occur in the event of misapplication.

11. How can CAD, CAM, and CAS tools be applied to *validate* the design? Provide an example of two of these.

12. What is the purpose of developing a physical model of the system, or an element thereof, early in the system design process?

13. What are some of the differences between a *mock-up*, an *engineering model*, and a *prototype*?

14. Select a system (or an element of a system) of your choice and develop a design review checklist that you can use for evaluation purposes. (Refer to Figure 8.)

15. What are some of the benefits that can be acquired through implementation of a formal design review process?

16. Refer to Figure 9. The predicted LCC value for the system, at the time of a *system design review*, is around $500K, which is well above the $420K design-to requirement. What steps would you take to ensure that the ultimate requirement will be met at (or before) the *critical design review*? Be specific.

17. Refer to Figure 10. As a systems engineering manager, how would you ensure that all of the TPM requirements are being properly "tracked"?

18. What determines whether or not a given design review has been successful?

19. In evaluating the feasibility of an ECP, what considerations need to be addressed?

20. Assume that an ECP has been approved by the CCB. What steps need to be taken in implementing the proposed change?

21. What is *configuration management*? When can it be implemented? Why is it important?

22. Why is *baseline management* so important in the implementation of the systems engineering process?

System Test, Evaluation, and Validation

From Chapter 6 of *Systems Engineering and Analysis,* Fifth Edition, Benjamin S. Blanchard, Wolter J. Fabrycky. Copyright © 2011 by Pearson Education, Inc. Published by Pearson Prentice Hall. All rights reserved.

System Test, Evaluation, and Validation

System test, evaluation, and validation activities should be established during the conceptual design phase of the life cycle, concurrently with the definition of the overall system design requirements. From that point on, the test and evaluation effort continues by the testing of individual components, the testing of various system elements and major subsystems, and then by the testing of the overall system as an integrated entity. The objective is to adopt a "progressive" approach that will lend itself to continuous implementation and improvement as the system design and development process evolves.

Test and evaluation activities discussed in this chapter can be aligned initially with concept to detailed design activities and then extended through the production/construction and the system utilization and support phases. These topics address an evolutionary treatment of the system design process and "look ahead" to downstream outcomes where the results will have defined a specific configuration, supported by a comprehensive design database and augmented by supplemental analyses.

The next step is that of *validation*. Validation, as defined herein, refers to the steps and the process needed to ensure that the system configuration, as designed, meets all requirements initially specified by the customer. The process of validation is somewhat evolutionary in nature. Referring to Figure A.1, the activities noted by Blocks 0.6, 1.5, 2.3, and 3.4 invoke an ongoing process of review, evaluation, feedback, and the ultimate verification of requirements. These activities have been primarily directed to address early design concepts and various system elements. However, up to this point, a total integrated approach for the validation of the system and its elements, as an integrated entity, has not been fully accomplished. Final system validation occurs when the system performs effectively and efficiently when operating within its associated higher-level system-of-systems (SOS) configuration (as applicable).

The purpose of this chapter is to present an integrated approach for system test and evaluation and to facilitate the necessary validation of the proposed system configuration to provide assurance that it will indeed meet customer requirements. This chapter addresses the following topics:

- Determining the requirements for system test, evaluation, and validation;
- Describing the categories of system test and evaluation;

- Planning for system test and evaluation;
- Preparing for system test and evaluation;
- Conducting the system test, collecting and analyzing the test data, comparing the results with the initially specified requirements, preparing a test report; and
- Incorporating system modifications as required.

Upon completion of this chapter, the reader should have acquired an understanding of how the evaluation and validation of system design should be accomplished. A section on summary and extensions closes the chapter, with suggested references for further study also provided. Also, several relevant website addresses are offered.

1 THE PROCESS OF SYSTEM TEST, EVALUATION, AND VALIDATION

As system-level requirements are identified through the definition of system operational requirements and the maintenance and support concept, a method must be established for measurement and evaluation and the subsequent determination as to whether or not these requirements are likely to be met. For each new requirement that is established, the question is, *How will we be able to determine if this requirement will be met, and what test and evaluation approach should be implemented to verify that it has or has not been met?* The specific needs for testing evolve from technical performance measures (TPMs), and the depth of testing will be influenced by the established levels of importance.

Figure 1 illustrates a simplified process that can be implemented from the beginning and used in iterative way. Initially, there are system-level requirements established, which, in turn, generates the need for test and evaluation later. This leads to the preparation of a *test and evaluation master plan* (TEMP), providing an overall integrated approach leading to system validation. The TEMP may be updated periodically as the system design and development process evolves. Referring to Figure A.1, the initial version of the TEMP is prepared toward the end of the conceptual design phase.

Given an initial definition of system-level requirements, the ongoing iterative process of evaluation may begin. Early in system design, various analytical techniques may be used to predict and evaluate the anticipated characteristics of a proposed system design configuration. CAD and related methods are often utilized to present an early three-dimensional view of a given design, showing major interfaces, accesses, size of components, human-equipment relationships, and so on. Potential interferences and design incompatibilities readily become visible.

As the design progresses to the point where mock-ups, engineering models, and prototypes are developed, the test and evaluation process becomes more meaningful because of the availability of physical models, actual hardware and software, and other elements of the system. Further, when preproduction and production models evolve, as the system configuration approaches its intended operational configuration, the effectiveness of the test and evaluation effort assumes an even greater degree of significance.

When addressing overall test requirements at the system level, it should be realized that a *true* test (i.e., that which is relevant from the standpoint of assessing total system performance and effectiveness) constitutes the evaluation of a system deployed and utilized in a user operational environment, subjected to all known operating conditions. For example, an aircraft should be tested while it is performing its intended mission during an actual flight in an operational environment; a communications network should be tested while actual field operations are in progress; a ground transportation vehicle should be tested in dry, dusty, and windy

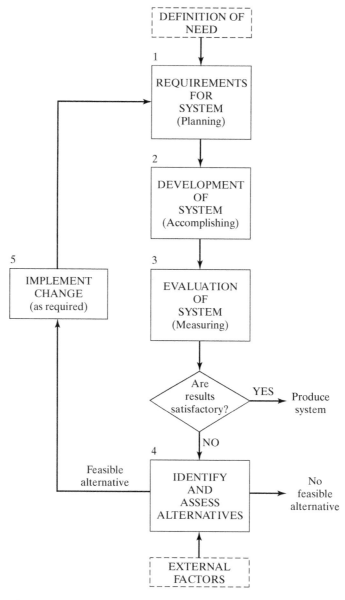

Figure 1 System requirements and evaluation relationships.

conditions if the system is to ultimately operate in this type of environment; and a river-crossing bridge should be tested when actually constructed and subjected to the desired traffic flow conditions. User personnel should accomplish operator and maintenance functions with the designated operating and maintenance procedures, support equipment, and so on. In these situations, actual experience in a realistic environment can be recorded, giving a true system validation result.

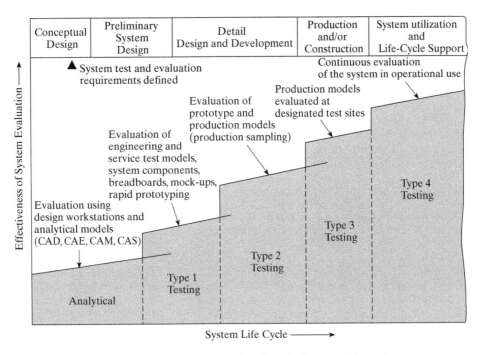

Figure 2 Stages of system test and evaluation during the life cycle.

While a true system validation effort cannot be fully completed until the system is fully operational in the user's environment, one must start early by implementing a "progressive" test and evaluation plan, as illustrated in Figure 2. The objective is to gain confidence, as early as possible, that all system requirements will be met. If it appears that certain requirements will not be met, then corrective action should be taken. Waiting until later in the life cycle before detecting a potential problem and incorporating a late change can be quite costly. Thus, the implementation of a progressive and evolving plan as shown in the figure is preferred. The objective is to commence with the validation effort as early as possible, as long as the results obtained are meaningful.

2 CATEGORIES OF SYSTEM TEST AND EVALUATION

An evolutionary approach to system test and evaluation is illustrated in Figure 2. Various categories of test and evaluation are identified by the program phase and the effectiveness of the evaluation effort increases when progressing through *Analytical, Type 1 Testing, Type 2 Testing*, and so on. The scope and depth of testing will depend on the type of system, whether new design and development are required, the complexity of the design, and the associated risks. Comprehensive testing may be required where a new design is underway, the external interfaces are numerous, unknowns exist, and the risks are high. On the other hand, commercial off-the-shelf (COTS) proven entities may not require as much testing, except to assure proper compatibility and integration with the other elements of the system. Not too much or too little testing must be accomplished, as the consequences in either direction may turn out to be quite costly.

2.1 Analytical and Simulation Evaluation

The utilization of analytical models, CAD methods, and various combinations of these can be effective in the verification of certain design relationships before the design is finalized. Analytical and simulation models are necessary in the initial phases of a program because of the lack of tangible entities to subject to testing. For example, the development of a three-dimensional spatial model of the system through computer graphics can illustrate the relationships among various system elements, the location of equipment, human-equipment interfaces, operational workstations, and accesses for maintenance and support. From this, potential interferences, space utilization conflicts, and so on, can be detected. Through the use of simulation methods, the designer can verify certain design concepts early in the life cycle. The objective is to gain as much visibility as possible at an early stage before the design configuration becomes relatively firm. For an early evaluation of the river crossing bridge system, the application of CAD and the presentation of a three-dimensional spatial model could be very productive.

2.2 Type 1 Testing

During the initial phases of detail design, breadboards, bench-test models, engineering models, engineering software, and service test models are built and applied with the intent of verifying certain physical design, performance, and operational characteristics. These models, representing either an entire system or a designated system component, usually operate functionally (electrically and mechanically) but do not by any means represent production equipment. In the development of software, the application of rapid prototyping is sometimes used to verify design adequacy.

Tests may involve operating and logistic support actions that are directly comparable to tasks performed in a real operational situation (e.g., measuring a performance parameter, accomplishing a remove-replace action, and meeting a servicing requirement). Although these tests are not formal demonstrations in a fully operational sense, information pertinent to actual system characteristics can be derived and used as an input in overall system evaluation and assessment. Such testing is usually performed in the producer/supplier's facility by engineering technicians using "jury-rigged" test equipment and engineering notes for test procedures. It is during this initial phase of testing that changes to the design can be incorporated at minimum cost. For the river crossing bridge design, various elements of the structure can be selected for early test and evaluation to gain confidence that they will perform as intended; for example, road- and railbed materials, walkway lighting, toll facilities, and segments of superstructure.

2.3 Type 2 Testing

Formal tests and demonstrations are accomplished during the latter part of detail design when preproduction prototype equipment, software, formal procedures, and the like are available. Prototype equipment/software is similar to a production configuration that will be utilized in the field but has not necessarily been fully *qualified* at this point in time. A "qualified" system is one that represents a full production configuration (as approved through a Critical Design Review) that has been tested and verified through the successful completion of performance tests, environmental qualification tests (e.g., temperature cycling, shock and vibration tests), reliability qualification, maintainability demonstration, and system compatibility testing. Type 2 testing primarily refers to that activity associated with the initial qualification of the system for operational use, and a given test program may include a series of individual tests tailored to the need. Such a program might include the following:

1. *Performance tests*—Tests are accomplished to verify individual system performance characteristics. For instance, tests are designed to determine whether the electric motor will provide the necessary output, whether the pipeline will withstand certain fluid pressures, whether the airplane will perform its intended mission successfully, whether a production process will provide x widgets per given period, whether a health care facility will be able to process y patients in a given time frame, whether a segment of the river crossing bridge will be able to handle the expected traffic flow conditions, and so on. Also, it is necessary to verify form, fit, interchangeability, product safety, and other comparable features.

2. *Environmental qualification*—Temperature cycling, shock and vibration, humidity, wind, salt spray, dust and sand, fungus, acoustic noise, pollution emission, explosion proofing, and electromagnetic interference tests are conducted. These factors are oriented to what the various system elements will be subjected to during operation, maintenance, and transportation and handling functions. In addition, the effects of the item(s) being tested on the overall external environment will be noted.

3. *Structural tests*—Tests are conducted to determine material characteristics relative to stress, strain, fatigue, bending, torsion, and general decomposition.

4. *Reliability qualification*—Tests are accomplished on one or more system elements to determine the MTBF and MTBM. Also, special tests are often designed to measure component life to evaluate degradation and to determine modes of failure.

5. *Maintainability demonstration*—Tests are conducted on one or more system elements to assess the values for mean active maintenance time $\overline{M}$ mean corrective maintenance time ($\overline{M}ct$), mean preventive maintenance time ($\overline{M}pt$), maintenance labor-hours per operating hour (MLH/OH), and so on. In addition, maintenance tasks, task times and sequences, prime equipment–test equipment interfaces, maintenance personnel quantities and skills, maintenance procedures, and maintenance facilities are verified to varying degrees. The elements of logistic support are initially evaluated on an individual basis.

6. *Support equipment compatibility tests*—Tests are often accomplished to verify compatibility among the prime equipment, test and support equipment, and transportation and handling equipment.

7. *Personnel test and evaluation*—Tests are often accomplished to verify the relationships between people and equipment, people and software, the personnel quantities and skill levels required, and training needs. Both operator and maintenance tasks are evaluated.

8. *Technical data verification*—Verification of operational and maintenance procedures is often accomplished.

9. *Software verification*—Verification of operational and maintenance software is accomplished. This includes computer software units (CSUs), computer software configuration items (CSCIs), hardware–software compatibility, software reliability and maintainability, and related testing.

10. *Supply chain element compatibility tests*—The verification (in terms of adequacy) of the various elements of the proposed supply chain with the applicable elements of the system is accomplished. This includes component packaging and handling methods,

transportation and distribution modes, warehousing and storage methods, and procurement approaches.

11. *Compatibility tests with other elements within the higher-level SOS structure (as applicable)*— Selected tests of a special and unique nature may be required at the subsystem (and below) level to ensure the compatibility between specific elements of the newly designed system and external elements of other "interfacing" systems within the same SOS structure. The objective is to identify, as early as possible, areas where incompatibilities may exist and where additional design activity may be required.

The ideal situation is to plan and schedule individual tests listed above so that they can be accomplished on an integrated basis as one overall test. Data output from one test may be beneficial as an input to another test. The intent is to provide the proper emphasis, consistent with the need, and to eliminate redundancy and excessive cost. Proper test planning is essential.

Another aspect of Type 2 testing involves production sampling tests when multiple quantities of an item are produced. The tests defined earlier basically "qualify" the item; that is, the equipment hardware configuration meets the requirements for production and operational use. However, once an item is initially qualified, some assurance must be provided that all subsequent replicas of that item are equally qualified. Thus, in a multiple-quantity situation, samples are selected from production and tested.

Production sampling tests may cover certain critical performance characteristics, reliability, or any other designated parameter that may significantly vary from one serial-numbered item to the next, or may vary as a result of a production process. Samples may be selected on the basis of a percentage of the total equipment produced or may tie in with a certain number of pieces of equipment in a given calendar time interval. This depends on the peculiarities of the system and the complexities of the production process. From production sampling tests, one can measure system growth (or degradation) throughout the production/construction phase.

Type 2 tests are generally performed in the producer or supplier's facility by people at that facility. Test and support equipment, designated for operational use, and preliminary technical manual procedures are employed where possible. User personnel often observe and/or participate in the testing activities. Design changes as a result of corrective action should be accommodated through a formalized engineering change procedure.

2.4 Type 3 Testing

Formal tests and demonstrations may be conducted following the initial system qualification testing and prior to completion of the production/construction phase. Such tests are generally accomplished by user personnel at some operational test site, which may constitute a ship, an aircraft, a space vehicle in flight, a health care facility, a facility in the arctic or in a desert, a highway/rail bridge, or a mobile land vehicle traveling between two points. Operational test and support equipment, operational software, operational spares, and formal system operating and maintenance procedures are used. Further, testing is accomplished with the system installed and operating within the applicable SOS configuration with all of the essential interfaces in place. Testing is generally continuous, accomplished over an extended period of time, and covers the evaluation of a number of system elements scheduled through a series of simulated operational exercises.

Type 3 testing provides the first time in the life cycle that all elements of the system can be operated and evaluated on a truly integrated basis. The compatibility between the prime equipment and software, personnel and equipment/software, the elements of support with the prime components of the system, the elements of support with each other, and the elements of other systems within the same SOS configuration is verified. TPMs related to operational and maintenance support functions are measured and verified for compliance with the initially specified requirements. Although Type 3 testing does not represent a complete operational situation, such testing can be planned and designed to provide a close approximation.

2.5 Type 4 Testing

During the system operational utilization and support phase, formal tests are sometimes conducted to gain further insight into a specific area. It may be desirable to vary a mission profile or system utilization rate to determine the impact on total system effectiveness and on life-cycle cost, or it might be feasible to evaluate alternative support policies to see whether system operational availability (A_o) can be improved, or it may be feasible to select a different supply chain approach to improve the overall effectiveness of the logistic support capability.

Even though the system is designed and operational in the customer environment, this is actually the first time that we can really assess and know its true capability. Hopefully, the system will be able to accomplish its objective in an effective and efficient manner; however, there is still the possibility that improvements can be realized by simply varying basic operational and maintenance support policies. In addition, formal testing may be accomplished in conjunction with the incorporation of technology enhancements and system upgrades to ensure that the appropriate measures of effectiveness (MOEs and/or TPMs) are being maintained as system operations continue.

3 PLANNING FOR SYSTEM TEST AND EVALUATION

Test planning should actually commence during the conceptual design phase when the technical requirements for the system are defined. If a system requirement is to be specified, then there must be a way to measure and evaluate the system later to ensure that the requirement has been met; hence, testing, evaluation, and validation requirements are intuitive at an early point in time. Test planning and supporting objectives are included in a *test and evaluation master plan* (TEMP), initially prepared during the latter stages of conceptual design (refer to Figure A.1). This plan must be closely integrated with the *systems engineering management plan* (SEMP).

Throughout the various stages of system development, a number of different and individual tests may be specified covering subsystems, equipment, software, or various combinations of system elements. Often, there is a tendency to specify a test to measure one system characteristic, another test to measure a different characteristic, and so on. Thus, the amount of testing specified may become overwhelming and prove to be quite costly.

Test requirements must be considered on an *integrated* basis. Where possible, individual test requirements should be reviewed in terms of the resources needed and expected output results, and should be scheduled in such a manner as to gain the maximum benefit possible. For instance, reliability data can be obtained from conducting system performance tests and, thus, can reduce some of the anticipated reliability testing needed. Maintainability data can be obtained from the accomplishment of reliability tests, resulting in a possible reduction in the amount of maintainability demonstration testing required. Support equipment compatibility data and personnel data can be acquired from both reliability and maintainability testing.

Thus, it may be feasible to schedule reliability qualification testing first, maintainability demonstration second, and so on. In some instances, the combining of testing requirements may be feasible as long as the proper characteristics (and TPMs) are measured, and the data output is compatible with the initial testing objectives. In any event, systems engineering requires a comprehensive knowledge and understanding of the overall test objectives, the objectives of each of the individual tests planned, expected input–output results, and so on.

For each system development effort, an integrated test plan is prepared, usually for implementation beginning in the preliminary system design phase. Although the specific content will vary depending on the nature, makeup, and complexity of the system, the plan will generally include the following:

1. An identification of all of the tests to be accomplished, the items to be evaluated, the schedule of each, required inputs and expected outputs, applicable test sequences, etc. In determining test requirements, some elements of the system may evolve through all of the categories shown in Figure 2, whereas other elements may undergo only a limited amount of testing. This is a function of the degree of design definition, whether it is a known entity with a good historical database or a new design with a high degree of complexity. In any event, a proposed test plan layout, such as shown in Figure 3, should be included.

2. An identification of the organization(s) responsible for the administration, those actually conducting the tests, and the personnel and facilities necessary to provide ongoing test support (i.e., test technicians, facilities, test and support equipment, special test jigs and fixtures, test procedures, test monitors, data collection and recording, and test reporting).

3. A description of the test location(s), local political/economic/social factors, logistics provisions, and the test environment (i.e., arctic, desert, mountainous or flat terrain, facilities, and safety/security provisions). Testing the system, within the context of a SOS structure involving many different and varied interfaces, will likely be accomplished at a combination of various different geographical locations. The objective herein is to convey an integrated approach illustrating the interrelationships among these different test sites (as applicable).

4. A description of the test preparation phase for each category (type) of testing as required. This includes the selection of a test method, preparation of specific test procedures, training of test and supporting personnel, acquisition of the required logistics and maintenance support infrastructure, and preparation of test site and supporting facilities.

5. A description of the formal test phase (i.e., personnel requirements, equipment and software requirements, facility requirements, data collection and analysis methods, and test reporting).

6. A plan and associated provisions for retesting (i.e., the approach and methods for conducting additional testing as required due to a "reject" situation, or equivalent).

7. A description of the final test report (i.e., format, distribution, feedback, and recommendations for corrective action as required).

While the emphasis thus far has been on test and evaluation of the system itself, as a separate and independent entity, a true validation of the system and its capabilities is based on just how well the system will perform when operating within its planned higher-level SOS configuration structure. For example, *Will the ground transportation system perform as required when operating within the higher-level overall transportation system structure?* Referring

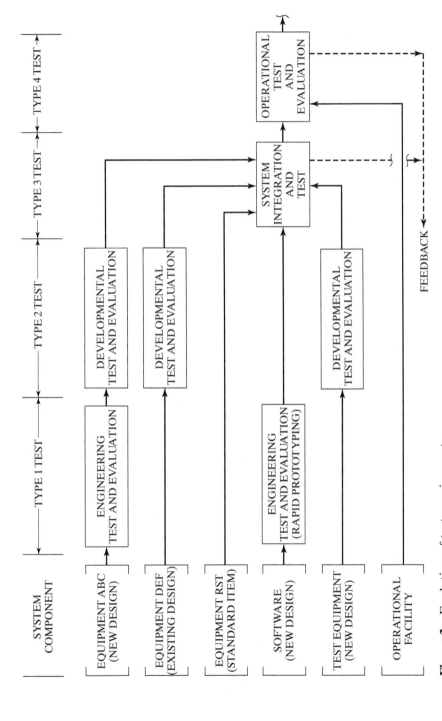

Figure 3 Evolution of test requirements.

to Figure A.2 (where there are "common" functions), *How well will System B perform when operating in conjunction with System A and System C?* System test and evaluation planning must address these critical interfaces, and the proper coordination with those external organizations who control and operate these other interface systems (i.e., Systems *A* and *C* in Figure A.2) must be initiated and maintained throughout the overall test program.

In general, the basic test plan serves as a valuable reference throughout system design, development, and production. It indicates what is to be accomplished, the requirements for testing, a schedule for the processing of equipment (and software) and material for test support, data collection and reporting methods, and so on. It serves as an integrating device for a number of individual tests in various categories, and a change in any single test requirement will likely affect the total plan.

4 PREPARATION FOR SYSTEM TEST AND EVALUATION

After initial planning and prior to the start of formal evaluation, a period of time is set aside for test *preparation*. During this period, the proper conditions must be established to ensure effective results. Although there is some variance depending on the type of evaluation, these conditions or prerequisites include the selection of the item(s) to be tested, establishment of test procedures, test site selection, selection and training of test personnel, preparation of test facilities and resources, the acquisition of support equipment, coordination with interfacing organizations when required for a SOS configuration, and the specified supply chain items necessary to ensure the successful accomplishment of the planned test program.

4.1 Selection of Test Item(s)

The equipment/software configuration used in the test must be representative of the operational item(s) to the maximum extent possible. For Type 1 tests, engineering models are used that are not often directly comparable with operational equipment; however, most subsequent testing will be accomplished at such a time that equipment (or software) representing the final configuration is available. A prerequisite task involves selecting the test model by serial number, defining the configuration in terms of incorporated versus unincorporated engineering changes (if any), and ensuring that it is available at the time needed. In addition, for the accomplishment of Types 3 and 4 testing where critical system interfaces exist in a SOS configuration, the selection process must include those external items that directly support the overall system being evaluated. Referring to Figure A.2 (for example), if System *B* is the system under test, then those common functions shared with Systems *A* and *C* must be available and operational as initially planned.

4.2 Test and Evaluation Procedures

Fulfillment of test objectives involves the accomplishment of both operator and maintenance tasks. Completion of these tasks should follow formal approved procedures, which are generally in the form of technical manuals developed during the latter phases of detail design. Following approved procedures is necessary to ensure that the system is operated and maintained in a proper manner. Deviation from these procedures may result in the introduction of personnel-induced failures and may distort the test data as recorded (e.g., the maintenance frequencies and task times). The identification of the procedures to be used in testing should be included in the evaluation plan. Such procedures covering the major "interface" items within a given SOS structure (i.e., the "common" functions in Figure A.2) must also be included for Types 3 and 4 testing as applicable.

4.3 Test-Site Selection

A single test site may be selected to evaluate the product in a designated environment; several test sites may be selected to evaluate the product under different environmental conditions (e.g., arctic versus tropical areas, mountainous versus flat terrain), or a variety of geographical locations may be selected to evaluate the product in terms of different markets and user conditions. This may also include a designated location where other elements within the same SOS configuration structure are operating. Any one or combination of these methods may be appropriate, depending on the product type and program requirements.

4.4 Test Personnel and Training

Test personnel will include (1) individuals who actually operate and maintain the system and equipment during the test, and (2) the supporting engineers, technicians, data recorders, analysts, and test administrators, as appropriate. Individuals assigned to the operation and maintenance of the system should possess the backgrounds and skill levels similar to consumer (i.e., user) personnel who will normally be assigned to operate and support the system throughout its life cycle. The required proficiency level(s) is attained through a combination of formal and on-the-job training.

4.5 Test Facilities and Resources

The necessary facilities, test chambers, capital equipment, environmental controls, special instrumentation, and associated resources (e.g., heat, water, air conditioning, gas, telephone, power, and lighting) must be identified and scheduled. In many instances, new design and construction are required, which directly affects the scheduling and duration of the test preparation period. A detailed description of the test facility and the facility layout should be included in test planning documentation and in subsequent test reports. Additional facilities and resources required to support any testing of the system and its external interfaces must also be identified (i.e., for a system operating within a SOS structure).

4.6 Test and Support Equipment

Test and support equipment requirements are initially considered in the maintenance concept and later defined through the maintenance and supportability analysis. During the latter phases of design, the necessary support items are procured and should be available for Types 2–4 testing. In the event that the proper type of support equipment is not available and alternative items are required, such items must be identified in the test and evaluation plan. The use of alternative items generally results in distorted test data (e.g., maintenance times), tends to cause personnel-induced failures in the system, and often forces a change in test procedures and facility requirements.

4.7 Test Supply Support

Supply support constitutes all materials, data, personnel, and related activities associated with the requirements, provisioning, and procurement of spare and repair parts and the sustaining maintenance of inventories for support of the system throughout the planned system testing cycle. This basically includes all of the required supply chain items that have been identified for the "operational" system and need to be available for evaluation along with the prime elements of the system. Specifically, this section addresses the following:

1. Initial and sustaining requirements for spares, repair parts, and consumables for the prime elements of the system. Spares are major replacement items and are repairable, while repair parts are nonrepairable, smaller components. Consumables refer to fuel, oil, lubricants, liquid oxygen, nitrogen, and so on.

2. Initial and sustaining requirements for spares, repair parts, material, and consumables for the various elements of logistic support (i.e., test and support equipment, transportation and handling equipment, training equipment, and facilities).

3. Facilities and warehousing required for the storage of spares, repair parts, and consumables. This involves consideration of space requirements, location, and environmental needs.

4. Personnel requirements for the accomplishment of supply support activities, such as provisioning, cataloging, receipt and issue, inventory management and control, shipment, and disposal of material. This includes the personnel requirements necessary to support those external components of other systems within the same SOS that are being evaluated on a concurrent basis.

5. Technical data requirements for supply support, which include initial and sustaining provisioning data, catalogs, material stock lists, receipt and issue reports, and material disposition reports.

The type of spare and repair parts needed at each level of maintenance depends on the maintenance concept and the supportability analysis. For Types 3 and 4 testing, spare and repair parts will generally be required for all levels since these tests primarily involve an evaluation of the system as an entity, to include its total maintenance and logistic support capability. The complete maintenance cycle, supply support provisions (the type and quantity of spares specified at each level), transportation times, distribution and warehousing requirements, turnaround times, and related factors are evaluated as appropriate. In certain instances, the producer's facility may provide depot-level support. Thus, it is important to establish a realistic supply chain infrastructure capability (or a close approximation thereto) for the system being tested. A complete description of this infrastructure should be included within the *test and evaluation master plan* (TEMP), and the data collection and results from the day-to-day system support throughout the test cycle should be included in the final test report subsequent to the completion of system testing.

5 SYSTEM TEST, DATA COLLECTION, REPORTING, AND FEEDBACK

As a prerequisite to system testing, supported by proper data handling, it is essential that the place of verification and validation in the technical aspect of the project cycle be understood and accepted. The objective is to start with user needs and end with a user-validated system. The systems engineering process model that best makes this objective visible is the "vee" model. Verification and validation progresses from the component level to validation of the operational system. At each level of testing, the originating specification and requirements documents (on the left side of the "vee") are consulted to ensure that components, higher-level system elements, and the system itself meet the specifications.

5.1 Formalizing Data Collection and Handling

With the necessary prerequisites established, the next step is to commence with the formal test and demonstration of the system. To accomplish a complete validation of the system as an entity (through the completion of Types 3 and 4 tests as identified in Figure 2—also, refer to Blocks 3.4 and 4.4 in Figure A.1) requires that the system be operated and maintained in a manner as prescribed by the applicable test plan, simulating a realistic user's operational environment to the maximum extent practicable. Throughout the testing process, data are collected and analyzed, which leads to an assessment of system performance and effectiveness against requirements. With the system in an "operational" status, the questions are, *What is the "true" performance and effectiveness of the system operating with all of its interfaces? What is the "true" performance and effectiveness of the logistics and maintenance support infrastructure? Are all of the initially specified TPM and related requirements being met? Is the system compatible with the other systems operating within the same SOS structure?*

Providing answers to these questions requires a comprehensive and formalized data collection, analysis, reporting, and feedback capability. The establishment of such a capability involves a two-step process: (1) the identification of the requirements and applications for such a cpability; and (2) the design, development, and implementation of a capability that will be responsive to these requirements.

The first step requires an understanding of the characteristics of the system that require evaluation and assessment. While there are many different data collection subsystems in existence and currently being applied for system assessment purposes, much of the data actually being collected does not provide the information needed for the purposes intended. Thus, the feedback as to what is really happening for many systems in the field today is lacking. Hence, it is important to understand what factors need to be measured and what information needs to be acquired. Figure 4 is presented (as an example) to show some of the objectives of a complete system validation effort. It should be noted that it is important to address not only the prime mission-related elements of the system but the logistics and maintenance support infrastructure as well.

With the data requirements defined, the second step is to develop the appropriate forms and the process for data collection, analysis, reporting, and feedback. These forms should be simple to understand and complete, as the tasks of recording the data in the field may be accomplished under adverse environmental conditions (in the rain, wind, cold, cramped conditions, etc.) and by a variety of personnel with different skill levels. If the forms are difficult, they may not be completed properly (if at all). At the same time, the personnel completing the data forms must first understand the need for the information and, second, be highly motivated to complete the forms correctly. Otherwise, the information desired will not be acquired, may not be useful if acquired, or may be misleading.

5.2 Feedback and System Design Modification

With the proper data recorded, the next step involves that of processing, accomplishing the appropriate analysis, evaluation, reporting, and feedback. It is important that the right information be reported and fed back to the responsible engineering and management personnel in an expeditious manner. They need to know exactly how the system is performing against specifications in the field quickly, so that design modifications can be initiated.

The process illustrated in Figure 5 includes steps that cover data collection, analysis, reporting, and feedback. Although the primary objective is to provide a good assessment of

1. *General System Operational and support Factors*
   (a) Evaluation of mission requirements (operational scenarios, times, frequencies).
   (b) Evaluation of system performance factors (capacity, output, size, weight, mobility, etc.).
   (c) Verification of cost and system effectiveness factors (TPMs—operational availability, reliability MTBF, maintainability MTBM/MDT, human factors, safety, life-cycle cost).
   (d) Verification of the logistics and maintenance support infrastructure (levels and locations of maintenance, repair policies, logistics and supply chain effectiveness, response times).
   (e) Evaluation of system security (protection against personnel-induced faults, terrorism).
   (f) Verification of system compatibility with other systems within the same SOS structure.
2. *Operational and Maintenance Software*
   (a) Verification of the compatibility of operational software with other system elements.
   (b) Verification of the compatibility of maintenances software with other system elements.
   (c) Verification of software reliability and maintainability characteristics.
3. *Operational and Maintenance Facilities*
   (a) Verification of operational facility adequacy, utilization, and maintenances support.
   (b) Verification of maintenance facility adequacy, utilization, and support.
   (c) Verification of warehousing facilities adequacy, utilization, and support.
   (d) Verification of training facility adequacy, utilization, and support.
4. *Transportation and Handling*
   (a) Verification of the transportation and handling capabilities for system operation and maintenance activities (adequacy, capacity, transportation times, response times).
   (b) Evaluation of the reliability, maintainability, human factors, safety, security, and related characteristics of transportation and handling equipment.
5. *Personnel and Training*
   (a) Verification of operational personnel quantities and skill levels by location.
   (b) verification of maintenance and support personnel quantities and skill levels by location.
   (c) Evaluation of personnel training policies and requirements (adequacy, throughput, etc.).
6. *Supply Support (Spares and Repair Parts)*
   (a) Verification of spare and repair part types and quantities by maintenance level/location.
   (b) Evaluation of supply responsiveness (spare part availability when required).
   (c) Evaluation of item replacement rates, condemnation rates, attrition rates, etc.
   (d) Evaluation of spare and repair part replacement and inventory policies.
7. *Test and Support Equipment*
   (a) Verification of support equipment type and quantity by operational/maintenance level.
   (b) Verification of support equipment availability, reliability, maintainability, safety, etc.
   (c) Evaluation of maintenance requirements for the support equipment (required resources).
8. *Technical Data and Information Handling*
   (a) Verification of technical data coverage (level, accuracy, availability, and method of information presentation for operating and maintenance manuals).
   (b) Verification of the adequacy of the management and technical information capability (accuracy, speed of processing, reliability, etc.).
   (c) Verification of adequacy of the field data collection, analysis, corrective-action, and reporting capability.
9. *Consumer (User) Response*
   (a) Evaluation of the degree of consumer (user) satisfaction.
   (b) Verification that the consumer (customer/user) needs are met.

Figure 4 System test, evaluation, and validation checklist (example).

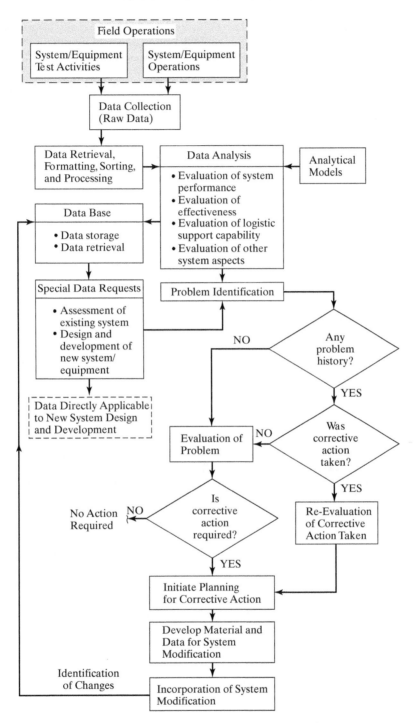

Figure 5 System evaluation and corrective action loop.

just how well the system is performing in the user's operational environment, a secondary objective is to identify any problems that may be detected and to initiate the required steps leading to corrective action and the incorporation of necessary design changes and system modifications as required.

Referring to Figure 5, the results of the data collection, analysis, reporting, and feedback process may indicate a problem pertaining to not meeting a given TPM (or equivalent) requirement. Through a subsequent analysis, the proper cause-and-effect relationships are established and the results may point to a deficiency in an element of prime equipment, a software module, an element of the maintenance support infrastructure, an interface problem with one or more other systems within the same SOS, and/or some procedure or process. As a result, a design change is initiated through the processing of an engineering change proposal (ECP).

6 SUMMARY AND EXTENSIONS

Test and evaluation activities actually take place throughout the life-cycle process illustrated in Figure A.1, commencing in conceptual design (Block 0.6) and extending through the system utilization and support phase as system modifications are incorporated for the purposes of technology upgrades (Block 4.4). Thus, these activities are continuous (to varying degrees) in all phases of the system life cycle, with the appropriate feedback loop incorporated for continuous product/process improvement.

The successful accomplishment of a good "validation" effort depends on (1) the development of a good integrated system test and evaluation plan from the beginning; (2) a complete and extensive knowledge of the methods/tools that can be applied to achieve an effective and efficient validation output; and (3) a comprehensive data collection, analysis, and reporting capability that provides an accurate account of the system's capability in the user's operational environment.

In the future, more emphasis needs to be applied on item (3); that is, the development and implementation of a robust and comprehensive data collection, analysis, reporting, and feedback capability. Traditionally, systems are designed and developed, tested to ensure compliance with the initially specified contractual requirements, delivered to the customer for operational use, and then often forgotten. While there are some data collection and reporting procedures/formats in existence for many systems in use today, the net output is that a great deal of data may be collected but with very little resulting "information" content provided. Further, there is limited (if any) feedback to the design community. Accordingly, there is a significant lack of field operational experience being fed back for application to follow-on design activities associated with newly developed systems and products and, as a result, producers tend to replicate some of the same mistakes over and over again.

In implementing the systems engineering process, it is critical that the general steps presented in Figure A.1 be properly adhered to (progressing from left to right). In addition, the various "feedback loops" (identified with the dashed lines) must be effectively utilized. The process involves an iterative forward flow of activities, with proper feedback incorporated in a timely manner.

For more detailed coverage of system validation requirements, it is recommended that the reader review, in particular: *Systems Engineering Handbook* (INCOSE), and *Systems Engineering Guidebook* (INCOSE). In addition, one should visit the following websites: INCOSE at *http://www.incose.org* and the International Test and Evaluation Association (ITEA) at *http://www.itea.org* as gateways to supporting information.

QUESTIONS AND PROBLEMS

1. How are the specific requirements for "system test, evaluation, and validation" determined?

2. Describe the "system test, evaluation, and validation" process. What are the objectives?

3. Describe the basic categories of test and their application. How do they fit into the total system evaluation process?

4. Refer to Figure A.1. How is "evaluation" accomplished in each of the phases shown (i.e., Blocks 0.6, 1.5, 2.3, 3.4, and 4.4)?

5. What special test requirements should be established for systems operating within a higher-level system-of-systems (SOS) structure? Identify the steps required to ensure compatibility with the other systems in the SOS configuration.

6. When should the planning for "system test, evaluation, and validation" commence? Why? What information is included?

7. Select a system of your choice and develop a system test and evaluation plan (i.e., a TEMP, or equivalent).

8. Why is it important to establish the proper level of logistics and maintenance support in conducting a system test? What would likely happen in the absence of adequate test procedures? What would likely happen in the event that the assigned test personnel were inadequately trained? What would likely happen in the absence of the proper test and support equipment? Be specific.

9. Why is it important to develop and implement a good data collection, analysis, reporting, and feedback capability? What would likely happen in its absence?

10. Describe some of the objectives in the development of a data collection, analysis, and reporting capability.

11. What data are required to measure the following: cost effectiveness, system effectiveness, operational availability, life-cycle cost, reliability, and maintainability?

12. How would you evaluate the adequacy of the system's supporting *supply chain*? What measures are required?

13. How would you measure and evaluate system *sustainability*?

14. In the event that the system evaluation process indicates a "noncompliance" with a specific system requirement, what steps should be taken to correct the situation?

15. In the event that a system design change is being recommended to correct an identified design deficiency, what should be done in response?

16. Why is configuration management so important in the implementation of the "system test, evaluation, and validation" process?

17. Why is the "feedback loop" important (assuming that it is important)?

18. Briefly describe the customer/producer/supplier activities (and interrelationships) in system evaluation.

19. What benefits can be derived from good test and evaluation reporting?

20. Describe the role of systems engineering in the "system test, evaluation, and validation" process.

Appendix: Figures

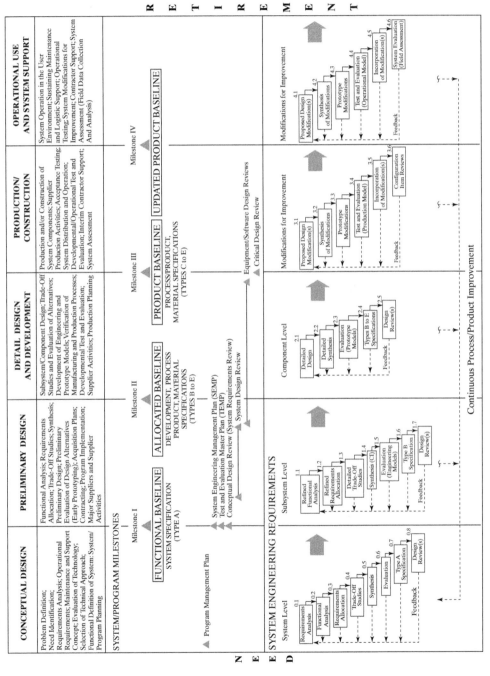

Figure A.1 System process activities and interactions over the life cycle.

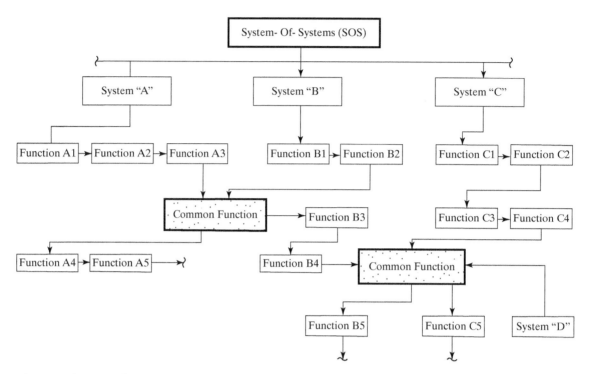

Figure A.2   Functional interfaces in a system-of-systems (SOS) configuration.

Alternatives and Models in Decision Making

From Chapter 7 of *Systems Engineering and Analysis,* Fifth Edition, Benjamin S. Blanchard, Wolter J. Fabrycky. Copyright © 2011 by Pearson Education, Inc. Published by Pearson Prentice Hall. All rights reserved.

Alternatives and Models in Decision Making

Most engineers, whether engaged in research, design, development, construction, production, operations, or a combination of these activities, are concerned with the efficient use of limited resources. When known opportunities fail to hold sufficient promise for the efficient employment of resources, more promising opportunities are sought. This view, accompanied by initiative, leads to exploratory activities aimed at finding better opportunities. In exploratory activities, steps are taken into the unknown to find new alternatives that may be evaluated to determine whether they could be superior to those now known.

An understanding of the decision-making process usually requires simplification of the complexity facing the decision maker. Conceptual simplifications of reality, or models, are a means to this end. In this chapter, a methodology for structuring alternatives and a general decision evaluation approach using models are presented to facilitate decision making in both design and operations.

Upon completion of this chapter, the reader will have obtained essential insight into alternatives from a rigorous perspective and as the foundation for decision model formulation. But decision models are only intended to take the decision maker part way to the point of decision. Decision evaluation is one of the steps and this is treated under the framework of decision evaluation theory. Then there is the issue of multiple criteria in decision making, as well as concern for risk and uncertainty.

The preceding topics are developed and presented in this chapter. Their study and consideration will impart a general understanding of the following concepts:

- Alternatives and their equivalent comparison as the basis for going beyond the known to explore opportunities for technological progress;
- A classification of models, models and indirect experimentation, and the formulation and validation of decision models;
- Two related paradigms for decision evaluation theory, money flow modeling and economic optimization modeling;

- Contrasting the role of models and modeling in design and operations;
- Multiple criteria considerations and some classical methods for choosing from among alternatives;
- The decision evaluation display as a means for presenting and considering alternatives under multiple criteria; and
- A number of classical approaches and rules for dealing with risk and uncertainty in decision making.

The final section of this chapter provides a summary and suggested extensions of the broad domain of decision making involving models and modeling. It provides insight into irreducible and intangible factors and emphasizes the importance of intuition and judgment in coming to a final decision.

1 ALTERNATIVES IN DECISION MAKING

A complete and all-inclusive alternative rarely emerges in its final state. It begins as a hazy but interesting idea. The attention of the individual or organization is then directed to analysis and synthesis, and the result is a definite proposal. In its final form, an alternative should consist of a complete description of its objectives and its requirements in terms of benefit and cost.

Both different ends and different methods are embraced by the term *alternative*. All proposed alternatives are not necessarily attainable. Some are proposed for analysis even though there seems to be little likelihood that they will prove feasible. The idea is that it is better to consider many alternatives than to overlook one that might be preferred. All alternatives should be presented for analysis even though there is little likelihood that all will prove to be feasible. Alternatives that are not considered cannot be adopted, no matter how desirable they may actually prove to be.

1.1 Limiting and Strategic Factors

Those factors that stand in the way of attaining objectives are known as *limiting factors*. An important element of the systems engineering process is the identification of the limiting factors restricting accomplishment of a desired objective. Once the limiting factors have been identified, they are examined to locate *strategic factors*, those factors that can be altered to make progress possible.

The identification of strategic factors is important, for it allows the decision maker to concentrate effort on those areas in which success is obtainable. This may require inventive ability, or the ability to put known things together in new combinations, and is distinctly creative in character. The means achieving the desired objective may consist of a procedure, a technical process, or an organizational or managerial change. Strategic factors limiting success may be circumvented by operating on engineering, human, and economic factors individually and jointly.

An important element of the process of defining alternatives is the identification of the limiting factors restricting the accomplishment of a desired objective. Once the limiting factors have been identified, they are examined to locate those strategic factors that can be altered in a cost-efficient way so that a selection from among the alternatives may be made.

1.2 Comparing Alternatives Equivalently

To compare alternatives equivalently, it is important that they be converted to a common measure. Conversion to a common measure permits comparison on the basis of equivalence. Money flow models and economic optimization models are central to the conversion process and permit alternatives to be compared equivalently.

On completion of the conversion step, quantitative and qualitative outputs and inputs for each alternative form the basis for comparison and decision. Quantitative measures should be obtained with the use of suitable models, and decisions between alternatives should be made on the basis of their differences. Thus, all identical factors can be canceled out for the comparison of any two or more alternatives at any step of a decision-making process.

After alternatives have been carefully analyzed and converted to equivalent bases, a decision may be made by the customer alone or by the customer and the producer jointly, but rarely by the producer alone. Whatever the case may be, decision evaluation is necessary and often facilitated by some common approaches and methods.

Several outcomes are evaluated for each alternative that is being examined. It is known that all the outcomes cannot occur, and that at any given time any outcome can occur. Thus, decision making or selection from among alternatives is often done under risk or uncertainty.

In addition to the alternatives formally set up for evaluation, another alternative is almost always present—that of making no decision. The decision not to decide may be a result of either active consideration or passive failure to act; it is usually motivated by the thought that there will be opportunities in the future that may prove more desirable than any known at present.

2 MODELS IN DECISION MAKING

Models and their manipulation (the process of simulation) are useful tools in systems analysis. A *model* may be used as a representation of a system to be brought into being, or to analyze a system already in being. Experimental investigation using a model yields design or operational decisions in less time and at less cost than direct manipulation of the system itself. This is particularly true when it is not possible to manipulate reality because the system is not yet in existence, or when manipulation is costly and disruptive as with complex human-made systems.

There is a fundamental difference between models used in science and engineering. Science is concerned with the natural world, whereas engineering is concerned primarily with the human-made world. Science uses models to gain an understanding of the way things are in the natural world. Engineering uses models of the human-made world in an attempt to achieve what ought to be. The validated models of science are used in engineering to establish bounds for engineering creations and to improve the products of such creations.

2.1 Classification of Models

When used as a noun, the word "model" implies representation. An aeronautical engineer may construct a wooden model of a possible configuration for a proposed aircraft type. An architect might represent a proposed building with a scale model of the building. An industrial engineer may use templates to represent a proposed layout of equipment in a factory. The word "model" may also be used as an adjective, carrying with it the implication of ideal. Thus, a man may be referred to as a model husband or a child praised as a model student. Finally, the word "model" may be used as a verb, as is the case where a person models clothes.

Models are designed to represent a system under study, by an idealized example of reality, to explain the essential relationships involved. They can be classified by distinguishing physical, analogue, schematic, and mathematical types. Physical models look like what they represent, analogue models behave like the original, schematic models graphically describe a situation or process, and mathematical models symbolically represent the principles of a situation being studied. These model types are used successfully in systems engineering and analysis.

Physical Models. Physical models are geometric equivalents, either as miniatures, enlargements, or duplicates made to the same scale. Globes are one example. They are used to demonstrate the shape and orientation of continents, water bodies, and other geographic features of the earth. A model of the solar system is used to demonstrate the orientation of the sun and planets in space. A model of an atomic structure would be similar in appearance but at the other extreme in dimensional reproduction. Each of these models represents reality and is used for demonstration.

Some physical models are used in the simulation process. An aeronautical engineer may test a specific tail assembly design with a model airplane in a wind tunnel. A pilot plant might be built by a chemical engineer to test a new chemical process for the purpose of locating operational difficulties before full-scale production. An environmental chamber is often used to create conditions anticipated for a component under test.

The use of templates in plant layout is an example of experimentation with a physical model. Templates are either two- or three-dimensional replicas of machinery and equipment that are moved about on a scale-model area. The relationship of distance is important, and the templates are manipulated until a desirable layout is obtained. Such factors as noise generation, vibration, and lighting are also important but are not a part of the experimentation and must be considered separately.

Analogue Models. Analogue comes from the Greek word *analogia*, which means proportion. This explains the concept of an *analogue model*; the focus is on similarity in relations. Analogues are usually meaningless from the visual standpoint.

Analogue models can be physical in nature, such as where electric circuits are used to represent mechanical systems, hydraulic systems, or even economic systems. Analogue computers use electronic components to model power distribution systems, chemical processes, and the dynamic loading of structures. The analogue is represented by physical elements. When a digital computer is used as a model for a system, the analogue is more abstract. It is represented by symbols in the computer program and not by the physical structure of the computer components.

The analogue may be a partial subsystem, or it may be an almost complete representation of the system under study. For example, the tail assembly design being tested in a wind tunnel may be complete in detail but incomplete in the properties being studied. The wind tunnel test may examine only the aerodynamic properties and not the structural, weight, or cost characteristics of the assembly. From this it is evident that only those features of an analogue model that serve to describe reality should be considered. These models, like other types, suffer from certain inadequacies.

Schematic Models. A schematic model is developed by reducing a state or event to a chart or diagram. The schematic model may or may not look like the real-world situation it represents. It is usually possible to achieve a much better understanding of the real-world system described by the model through use of an explicit coding process employed in the construction of the model. The execution of a football play may be diagrammed on a game board with a simple code. It is the idealized aspect of this schematic model that permits this insight into the football play.

An organization chart is a common schematic model. It is a representation of the state of formal relationships existing between various members of the organization. A human–machine chart is another example of a schematic model. It is a model of an event, that is, the time-varying interaction of one or more people and one or more machines over a complete work cycle. A flow process chart is a schematic model that describes the order or occurrence of several events that constitute an objective, such as the assembly of an automobile from a multitude of component parts.

In each case, the value of the schematic model lies in its ability to describe the essential aspects of the existing situation. It does not include all extraneous actions and relationships but rather concentrates on a single facet. Thus, the schematic model is not in itself a solution but only facilitates a solution. After the model has been carefully analyzed, a proposed solution can be defined, tested, and implemented.

Mathematical Models.   A mathematical model employs the language of mathematics and, like other models, may be a description and then an explanation of the system it represents. Although its symbols may be more difficult to comprehend than verbal symbols, they do provide a much higher degree of abstraction and precision in their application. Because of the logic it incorporates, a mathematical model may be manipulated in accordance with established mathematical procedures.

Almost all mathematical models are used either to predict or to control. Such laws as Boyle's law, Ohm's law, and Newton's laws of motion are formulated mathematically and may be used to predict certain outcomes when dealing with physical phenomena. Outcomes of alternative courses of action may also be predicted if a measure of evaluation is adopted. For example, a linear programming model may predict the profit associated with various production quantities within a multiproduct process. Mathematical models may be used to control an inventory. In quality control, a mathematical model may be employed to monitor the proportion of defects that will be accepted from a supplier. Such models maintain control over a state of reality.

Mathematical models directed to the study of systems differ from those traditionally used in the physical sciences in two important ways. First, because the system being studied usually involves social and economic factors, these models must often incorporate probabilistic elements to explain their random behavior. Second, mathematical models formulated to explain existing or planned operations incorporate two classes of variables: those under the control of a decision maker and those not directly under control. The objective is to select values for controllable variables so that some measure of effectiveness is optimized. Thus, these models are of great benefit in systems engineering and systems analysis.

2.2 Models and Indirect Experimentation

Models and the process of simulation provide a convenient means of obtaining factual information about a system being designed or a system in being. In component design, it is customary and feasible to build several prototypes, test them, and then modify the design based on the test results. This is often not possible in systems engineering because of the cost involved and the length of time required over the system life cycle. A major part of the design process requires decisions based on a model of the system rather than decisions derived from the system itself.

Direct and Indirect Experimentation.   In direct experimentation, the object, state, or event, and/or the environment are subject to manipulation, and the results are observed. For example, a couple might rearrange the furniture in their living room by this method. Essentially, they move the furniture and observe the results. This process may then be repeated with a second move and perhaps a third, until all logical alternatives have been exhausted. Eventually, one such move is subjectively judged best; the furniture is returned to this position, and the experiment is completed. Direct experimentation, such as this, may be applied to the rearrangement of equipment in a factory. Such a procedure is time-consuming, disruptive, and costly. Hence, simulation or indirect experimentation is employed with templates representing the equipment to be moved.

Direct experimentation in aircraft design would involve constructing a full-scale prototype that would be flight tested under real conditions. Although this is an essential step in the evolution of a new design, it would be costly as the first step. The usual procedure is evaluating several proposed configurations by building a model of each and then testing in a wind tunnel. This is the process of indirect experimentation or *simulation*. It is extensively used in situations where direct experimentation is not economically feasible.

In systems analysis, indirect experimentation is effected through the formulation and manipulation of decision models. This makes it possible to determine how changes in those aspects of the system under control of the decision maker affect the modeled system. Indirect experimentation enables the systems analyst to evaluate the probable outcome of a given decision without changing the operational system itself. In effect, indirect experimentation in the study of operations provides a means for making quantitative information available to the decision maker without disturbing the operations under his or her control.

Simulation Through Indirect Experimentation. In most design and operational situations, the objective sought is the optimization of an effectiveness or performance measure. Rarely, if ever, can this be done by direct experimentation with a system under development or a system in being. Also, there is no available theory by which the best model for a given system simulation can be selected. The choice of an appropriate model is determined as much by the experience of the systems analyst as the system itself.

The primary use of simulation in systems engineering is to explore the effects of alternative system characteristics on system performance without actually producing and testing each candidate system. Most models used will fit the classification given earlier, and many will be mathematical. The type used will depend on the questions to be answered. In some instances, simple schematic diagrams will suffice. In others, mathematical or probabilistic representations will be needed. In many cases, simulation with the aid of an analogue or digital computer will be required.

In most systems engineering and analysis undertakings, several models are usually formulated. These models form a hierarchy ranging from considerable aggregation to extreme detail. At the start of a systems project, knowledge of the system is sketchy and general. As the design progresses, this knowledge becomes more detailed and, consequently, the models used for simulation should be detailed.

2.3 Formulating and Validating Decision Models

In formulating a mathematical decision model, one attempts to consider all components of the system that are relevant to the system's effectiveness and cost. Because of the impossibility of including all factors in constructing the evaluation function, it is common practice to consider only those on which the outcome is believed to depend significantly. This necessary viewpoint sometimes leads to the erroneous conclusion that segments of the situation are actually isolated from each other. Although it may be feasible to consider only those relationships that are significantly pertinent, one should remember that all system elements are interdependent.

Manipulation of the model can lead to model modification to reduce the misfit between the model and the real world. This *validation* process has a recurrent pattern analogous to the scientific method. It consists of three steps: (1) postulate a model, (2) test the model prediction or explanation against measurements or observations, and (3) modify the model to reduce the misfit.

The validation process should continue until the model is supported reasonably by evidence from measurement and observation. As a model evolves to a state of validity, it provides postulates of reality that can be depended on. A model may have achieved a state of validity even though it gives biased results. Here the model is valid because it is consistent; that is, it gives results that do not vary. A decision model cannot be classified as accurate or inaccurate in any absolute sense. It may be considered to be accurate if it is an idealized substitute of the actual system under study. If manipulation of the model would yield the identical result that manipulation of reality would have yielded, the model would be true. If, however, one knew what the manipulation of reality would have yielded, the process of simulation would be unnecessary. Hence, it is difficult to test a decision model except by an intuitive check for reasonableness.

Engineers and managers are becoming increasingly aware that experience, intuition, and judgment are insufficient for the effective pursuit of design and operational objectives. Models are useful in design and operations because they take the decision maker part way to the point of decision.

3 DECISION EVALUATION THEORY

Decision evaluation is an important part of systems engineering and analysis. Evaluation is needed as a basis for choosing among the alternatives that arise during the activities of system design, as well as for optimizing systems already in operation. In either case, equivalence provides the common evaluation measure on which choice can be based.

This section presents two general categories of decision evaluation models. The first is based on money flows over time and designated *money flow modeling*. The second often embraces money flows but always incorporates optimization. It is designated *economic optimization modeling*.

3.1 Evaluation by Money Flow Modeling

Economic equivalence is expressed as the present equivalent (PE), annual equivalent (AE), or future equivalent (FE) amount, as well as the internal rate of return and the payback period.

A general equivalence function subsuming each equivalence expression may be stated as

$$\text{PE, AE, or FE} = f(F_t, i, n) \tag{1}$$

where $t = 0, 1, 2, \ldots, n$ and where

F_t = positive or negative money flow at the end of year t
i = annual rate of interest
n = number of years

The life cycle is the underlying money flow generator over its acquisition and utilization phases as shown for a hypothetical 17-year life cycle in Figure 1. Here acquisition spans several years, with expenditures composed of F_0 through F_7 (F_0–F_2 for conceptual and prelimi-

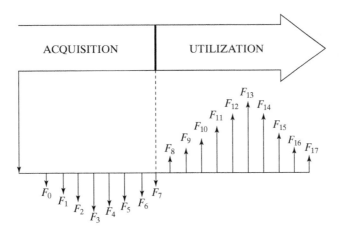

Figure 1 Money flows for acquisition and utilization.

nary design, F_3 and F_4 for detail design and development, and F_5–F_7 for production and/or construction). Net benefits or revenues occur at the end of each year for 10 years, F_8–F_{17}. Salvage value or cost, if any, is included in F_{17}.

The present equivalent, annual equivalent, and future equivalent amounts (or equivalent amounts at other points in time) are consistent bases for the evaluation of a single alternative or for the comparison of mutually exclusive alternatives. Any one of these may be used in accordance with the general economic equivalence function of Equation 1. In addition, Equation 1 provides a general structure for evaluation by the internal rate of return or payout period approaches.

3.2 Evaluation by Economic Optimization Modeling

Decision evaluation often requires a combination of both money flow modeling and economic optimization approaches. When investment cost, periodic costs, or project life is a function of one or more decision variables, it is important to optimize over these variables as a prerequisite to the determination of economic equivalence. This optimization is linked to decision evaluation through one or more money flows, which, in turn, are used in calculating a measure of economic equivalence. Optimization requires that an evaluation measure be derived from an economic optimization model.

An economic optimization function is a mathematical model formally linking an evaluation measure, E, with controllable decision variables, X, and system parameters, Y, which cannot be directly controlled by the decision maker. It provides a means for testing decision variables in the presence of system parameters. This test is an indirect experiment performed mathematically, which results in an optimized value for E. The functional relationship, in its unconstrained form, may be expressed as

$$E = f(X, Y) \tag{2}$$

As an example of the optimization function structure for an unconstrained decision situation, consider the determination of an optimal procurement quantity for inventory operations. Here, the evaluation measure is cost and the objective is to choose a procurement quantity in the face of demand, procurement cost, and holding cost so that the total cost is minimized. The procurement quantity is the variable directly under the control of the decision maker. Demand, procurement cost, and holding cost are not directly under his or her control. Use of the optimization function allows the decision maker to arrive at a value for the variable under his or her control that allows for a trade-off of conflicting cost elements.

The optimization function may be extended to both operational and design decisions involving alternatives. This extension involves the identification and isolation of design- or decision-dependent system parameters, Y_d, from design- or decision-independent system parameters, Y_i. Accordingly, Equation 2 can be restated in unconstrained form as

$$E = f(X, Y_d, Y_i) \tag{3}$$

As an example of the application of this version of the decision evaluation function, consider the establishment of a procurement and inventory system to meet the demand for an item that is available from one of several sources. Decision variables are the procurement level and the procurement quantity. For each source under consideration, there exists a set of source-dependent parameters. These are the item cost per unit, the procurement cost per procurement, the replenishment rate, and the procurement lead time. Uncontrollable (source-independent) system parameters include the demand rate, the holding cost per unit per period, and the shortage penalty cost. The objective is to determine the procurement level and procurement quantity and choose the procurement source so that total system cost will be minimized.

In design decision making, consider the deployment of a population of repairable equipment units to meet a demand (multiple-entity population system). Three decision variables may be identified: the number of units to deploy, the number of maintenance channels to provide, and the age at which units should be retired. Controllable (design-dependent) system parameters include the reliability, maintainability, energy efficiency, and design life. Uncontrollable (design-independent) system parameters include the cost of energy, the time value of money, and the penalty cost incurred when there are insufficient units operational to meet demand. Because the controllable parameters are design dependent, the objective is to develop design alternatives in the face of decision variables so that the design alternative that will minimize total system cost can be identified.

The design-dependent parameter approach facilitates the comprehensive life-cycle evaluation of alternative system designs. This approach involves separation of the system design space (represented by design-dependent parameters) from the optimization space (represented by design variables). Terms are defined as follows:

1. *Design-dependent parameters (Y_d)*. These are factors with values under the control of the designer(s) and that are impacted by the specialty disciplines during the development process. Every instance of the design-dependent parameter set represents a distinct candidate system or design alternative, because these parameters bound the design space. Examples of design-dependent parameters include reliability, producibility, maintainability, time to altitude, throughput, and so on.

2. *Design-independent parameters (Y_i)*. These are factors beyond control of the designer(s) but that impact the effectiveness of all candidate systems or design alternatives, and can significantly alter their relative "goodness" or desirability. Examples of design-independent parameters include labor rates, material cost, energy cost, inflation and interest factors, and so on.

3. *Design variables (X)*. These are factors that define the design optimization space. Each candidate system is optimized over the set of design variables before being compared with the other alternatives. In this way, equivalence is assured.

A summary of the evolution and development of the decision evaluation function is given in Table 1.

3.3 Models in Design and Operations

The place and role of Equations 2 and 3 differ in systems analysis and design evaluation. When considered in terms of the life cycle, each applies to a major segment as exhibited in Figure 2. Equation 2 is applicable to the optimization of operations already in being, whereas Equation 3 is useful for choosing from among mutually exclusive design alternatives based on design-dependent parameters.

Consider Table 1 and Figure 2 jointly as a way of elaborating upon the application of models in design and operations. Also herein is a mathematical basis for contrasting the use of models in systems engineering and in systems analysis. The first two entries in Table 1 and the operations phase in Figure 2 are decision model formulations applicable to operations in general, specifically within systems analysis as per Equation 2. Design-dependent parameters are not invoked. In the third entry of Table 1, a specific

TABLE 1 Forms of the Decision Evaluation Function

References	Functional Form	Application
Churchman, Ackoff, and Arnoff (1957)	$E = f(x_i, y_j)$ $E =$ system effectiveness $x_i =$ variables under direct control $y_j =$ variables not subject to direct control	Operations
Fabrycky, Ghare, and Torgersen (1984)	$E = f(X, Y); g(X) \lesseqgtr B$ $E =$ evaluation measure $X =$ controllable variables $Y =$ uncontrollable variables	Operations
Banks and Fabrycky (1987)	$E = f(X, Y_d, Y_i); g(X, Y_d) \lesseqgtr C$ $E =$ evaluation measure $X =$ procurement level and procurement quantity $Y_d =$ source-dependent parameters $Y_i =$ source-independent parameters	Procurement operations
Fabrycky and Blanchard (1991)	$E = f(X, Y_d, Y_i); g(X, Y_d) \lesseqgtr C$ $E =$ evaluation measure $X =$ design variables $Y_d =$ design-dependent parameters $Y_i =$ design-independent parameters	Design optimization

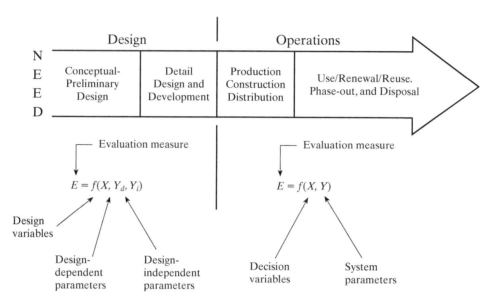

Figure 2 Economic optimization in design and operations.

operational situation is considered, that of procurement operations. In this application, system parameters are partitioned into source-dependent and source-independent subsets as per Equation 3. By so doing, one may utilize that equation to optimize across the decision variables for each instance of the parameter set, one instance for each source alternative. Albeit recognizing partitioned parameters, this is an application in systems analysis and not in systems engineering because the parameters are not design dependent as defined earlier.

In systems engineering, design-dependent parameters are invoked. Referring again to Table 1 (the fourth entry) and Figure 2 (the design phase), note that design-dependent parameters are explicitly partitioned. Values, or instances, of these parameters are TPMs and they make possible discerning the differences between mutually exclusive design alternatives. Note that Equation 3 collapses to Equation 2, indicating systems analysis in support of systems engineering. Systems analysis may serve to optimize operations in the face of a fixed design, or it may be utilized to help stumble through the design space always optimizing on the design variable set.

Out of the stumbling comes a focus on mutually exclusive design alternative after mutually exclusive design alternative. Because there are specific target values of the design-dependent parameters desired by the customer (TPMs[*]), these alternatives must be evaluated in the multiple criteria domain in such a way that the TPM deviations from desired values can easily be observed. Section 4 gives several classical methods for doing this and Section 5 gives a graphical method of displaying alternatives for consideration.

Models for design decisions and for operational decision making are abstractions of the system under study. Like all abstractions, models involve several assumptions: assumptions about the operating characteristics of components, about the behavior of people, and about the nature of the environment. These assumptions must be fully understood and evaluated when models are used for decision making in design and operations.

4 DECISIONS INVOLVING MULTIPLE CRITERIA

Rarely, if ever, is there only a single criterion present in a decision situation. In most cases, a decision must be made in the face of multiple criteria that jointly influence the relative desirability of the alternatives under consideration. The decision should be made only after considering all relevant criteria, recognizing that some are quantifiable and others are only qualitative in nature. Accordingly, this section presents some formal approaches for dealing with multiple criteria that incorporate both qualitative knowledge and judgment.

4.1 Multiple Criteria Considerations

Choice is not easy when multiple criteria are present, and this difficulty is perpetuated if a decision maker sidesteps basic concepts and simple guidelines. Therefore, before developing and presenting some formal and structured methods for choosing from among multiple alternatives, a number of useful concepts should be recognized and considered. Background concepts are in this section as a basis for the formal methods and approaches to follow.

Selecting the Criteria to Be Considered. It is important that alternatives be compared only after criteria of importance are identified, selected, and characterized. Furthermore, the process of identification and selection is often a useful step in converging toward a choice of the best alternative.

The most important concern in the selection of criteria or attributes is that they be *independent* of each other. Two criteria or attributes are assumed to be independent if the preference order and the trade-off for different levels of the criteria do not depend on the levels at which all other criteria occur.

Although it is not easy to achieve independence of criteria in practice, independence should be sought as a first step. Then any perceived lack of independence can be accommodated by combining interdependent criteria or by adjusting the final result.

The number of criteria or attributes finally selected should be kept to a reasonable minimum. Those judged to have an insignificant effect on the alternatives being considered can be safely eliminated from the analysis. But it is also recognized that too few criteria may mean that some important ones are being overlooked. Unfortunately, there is no way to determine the ideal number of criteria that should be considered in a given decision situation.

One good way to reduce the number of criteria under consideration is to seek those for which each alternative is equally good or bad regarding that criterion or attribute. These criteria contribute nothing to the overall evaluation and can be eliminated. Only differences are important in the analysis.

Rather than reducing a large number of criteria to a manageable but significant subset, an opposite approach can be used. It is to start with one, or a few, criteria and subject them to preliminary evaluation for the purpose of gaining insight into the decision situation. Then other criteria can be added incrementally. This process should continue only as long as the time and cost incurred are justified by the added information and insight gained regarding the choice to be made. This is an application of the law of diminishing returns in the face of an uncertain outcome.

Differences Are the Basis for Decision. Decisions between alternatives should be made only on the basis of their *true difference* across all criteria. Only true differences are important in decision making. All criteria or attributes that are identical may be canceled out. In this process, care must be exercised to be sure that canceled criteria are truly identical and of the same significance. Unless it is clear that criteria considered for cancelation are identical, it is best to carry them all the way through the analysis process.

In many studies, only small elements of a whole enterprise are considered. For example, studies are often made to evaluate the consequences of the purchase of a single tool or machine in a complex of many facilities. In such cases, it would be desirable to isolate the element from the whole by some means. To do this, for example, with respect to a machine being considered for purchase, it would be necessary to identify all the receipts and all the disbursements that would arise from the machine. If this could be done, the disbursements could be subtracted from the receipts. This difference would represent profit or gain, from which a rate of return could be calculated. But this should not be the end of the evaluation. Multiple criteria often exist, and differences in noneconomic elements should be examined closely.

As an example of a decision involving multiple criteria, consider the analysis and evaluation of alternatives to cross a river. The criteria may include the initial capital investment for a bridge, tunnel, barge, or other means of crossing, the crossing time, sustainable traffic volume, safety, environmental impact, and maintenance cost.

In these situations, one of several structured techniques that combine economic and noneconomic criteria can be effectively applied. This combination is then presented to the decision maker, who can be expected to add intuition and judgment, especially to perceived differences in the qualitative aspects of the situation.

4.2 Direct Ranking Methods

A direct method for choosing from among alternatives is simply to present them to the decision maker for ranking. But first, it is generally best to present the important criteria and ask that these be ranked. In this way, the decision maker will obtain better insight into the criteria upon which the selection of the best alternative depends. Accordingly, ranking methods may be applied to criteria or to alternatives, or first to criteria and then to alternatives. Ranking methods are difficult to apply to both simultaneously.

A procedure that may make the task of ranking easier and more effective is called the *method of paired comparisons*. The method of paired comparisons suggests the submission of criteria and/or alternatives to the decision maker two at a time. An additional benefit is that the paired comparisons provide a check on the internal consistency of value judgments provided by the decision maker.

As an example, consider five important criteria or attributes that pertain to the design of an office appliance. The criteria, designated 1–5, are to be rank ordered: (1) better, (2) cheaper, (3) faster, (4) repairable, and (5) disposable. These criteria are to be used in a subsequent analysis to help in identifying the best mutually exclusive design for the appliance.

When there are N criteria or alternatives to be ranked, $N(N - 1)/2$ pairs must be compared. For the five appliance design criteria, assume that the results of the paired comparisons are as follows (where $>$ means "preferred to" and $=$ means "equally preferred to"):

$1 < 2$	$2 > 3$	$3 > 4$	$4 > 5$
$1 > 3$	$2 > 4$	$3 = 5$	
$1 > 4$	$2 > 5$		
$1 = 5$			

A display of the pairwise comparisons is given in Table 2. A P is entered for each pair where the row criterion is preferred to the column criterion. There should not be any en-

TABLE 2 Exhibit of Preference Comparisons

Criteria	1	2	3	4	5	Times Preferred
1	–		P	P	=	2+
2	P	–	P	P	P	4
3			–	P	=	1+
4				–	P	1
5	=		=		–	+ +

tries on the diagonal in that a given criterion cannot be compared with itself. To verify that all pairs of criteria have been considered, check to see if there is an entry for each pair, with two entries being made for pairs that are equally ranked.

The number of times a given criterion in each row is preferred appears in the right-hand column of Table 2. A simple rule for ranking is to count the number of times one criterion is preferred over others. For this example, the rank order of preference is $2 > 1 > 3 > 4 = 5$. Thus, Cheaper is most important, with Better ranked second. Faster is ranked ahead of Repairable and Disposable, with these attributes being ranked in last place and considered to be approximately equally preferred.

The method presented assumes *transitivity* of preferences. That is, if $1 > 2$ and $2 > 3$, then 1 must be > 3. If two or more criteria are preferred an equal number of times (except for ties), there is evidence of a lack of consistency. There is no evidence of a lack of consistency in this example. If there was, the remedy would be to question and reconsider the preference choices pertaining to the affected criteria.

4.3 Systematic Elimination Methods

Systematic elimination methods are among the simplest approaches available for choosing from among alternatives in the face of multiple criteria. These methods are applicable when values and/or outcomes can be specified for all criteria and all alternatives. The values should be measurable (scalar) or at least rank orderable (ordinal).

Elimination methods have two limitations. First, they do not consider *weights* that might be applicable to the criteria or attributes. Second, they are *noncompensatory* in that they do not consider possible trade-offs among the criteria across alternatives.

An example will be used to illustrate three systematic elimination approaches. Consider Alternatives *A–D* in Table 3, with each having an estimated or predicted scalar or ordinal value (outcome). The right-hand columns specify ideal and minimum standard values for each criterion.

TABLE 3 Estimated Criterion Values for Alternatives

Criterion	Alternative				Ideal	Minimum Standard
	A	*B*	*C*	*D*		
1. Better	40	35	50	30	50	30
2. Cheaper	90	80	75	60	100	70
3. Faster	6	5	8	6	10	7
4. Gone	*G*	*P*	*VG*	*E*	*E*	*F*

In Table 3, Criteria 1–3 are estimated in scalar terms with higher numbers being better. Criterion 4 is rank ordered with E being excellent, VG being very good, G being good, F being fair, and P being poor. Three systematic elimination methods will be presented next.

Comparing Alternatives Against Each Other.   The most obvious elimination method is to check for dominance by making mutual comparisons across alternatives. Dominated alternatives can be safely eliminated from further consideration. If an alternative is better than or equal to another alternative across all criteria, then the other alternative is *dominated*.

For the decision situation in Table 3, note that Alternative *A* dominates Alternative *B*. This is due to 40 > 35 for Better, 90 > 80 for Cheaper, 6 > 5 for Faster, and G > P for Gone. No other pair can be found in which dominance occurs.

Dominance is only an elimination technique. It will not necessarily lead to selection of a preferred alternative unless other criteria are applied. This will be introduced in subsequent sections.

Comparing Alternatives Against a Standard.   Comparisons against a standard (also called *satisficing*) are commonplace and easy to explain. These comparison methods may be applied at one of two extremes, or at any point in between. Extremes provide the following rules:

Rule 1: An alternative may be retained (not eliminated) only if it meets the standard for at least one criterion.

Rule 2: An alternative may be retained (not eliminated) only if it meets the standard for all criteria.

Application of these rules and variants thereof may be accomplished by setting up a table to show the derived results as in Table 4. An *X* is used to indicate the criterion minimum values (standards) that are violated by each alternative.

Inspection of the results in Table 4 leads to the following conclusions about the alternatives under consideration:

1. All alternatives meet the standard for at least one criterion and, under Rule 1, all may be retained for further evaluation.
2. Alternative *C* meets the standard for all criteria and, under Rule 2, is the only one that may be retained.

Comparing Criteria Across Alternatives.   Comparisons across alternatives can be made in two ways after the relevant criteria are ranked in importance:

1. For the most important criterion, choose the alternative (if any) that is best. A tie between two or more alternatives is broken by using the second most important

TABLE 4   Alternatives Not Meeting Criteria Standards

Alternative	Criterion			
	1	2	3	4
A			*X*	
B			*X*	*X*
C				
D		*X*	*X*	

criterion as a basis for choice. Continue until a single alternative survives or until all criteria have been considered.

2. Examine one criterion at a time, making comparisons among the alternatives. Eliminate alternatives that do not meet the minimum standard value. Continue until all alternatives except one have been eliminated, or until all criteria have been examined.

For the example of this section, assume that the importance ranking for criteria is $2 > 1 > 3 > 4$, as determined by the direct ranking method. Using the minimum standards in Table 3, the comparison rules give the following results:

1. Criterion 2 (Cheaper) is most important and Alternative A is best when evaluated on this criterion alone. However, this alternative does not meet the minimum standard for one of the other criteria (Faster).

2. Examination of one criterion at a time and eliminating those that do not meet the minimum standard results in the following:

Criterion	Eliminate	Remaining
1	None	A, B, C, D
2	D	A, B, C
3	A, B, D	C
4	None	C

From the aforementioned method, under the first rule, Alternative A could be chosen if violation of the minimum for Criterion 3 (Faster) could be overlooked. Otherwise, Alternative C could be chosen as determined by application of the second rule.

4.4 Weighting Methods of Evaluation

Weighted importance ratings can extend the ranking and elimination methods just presented. This extension will explicitly recognize the higher importance or priority that some criteria or attributes should assume. Also, if an additive weighting method is used, adding the results together for different criteria permits strength on one criterion to compensate for weakness on another. Thus, strength may compensate for weakness in the final evaluation.

In this section, direct as well as subjective techniques for assigning weights or importance ratings to criteria are considered. Then, both tabular additive and graphical additive methods are presented. These techniques are effective aids to the selection of the preferred alternative. They are based on a combination of criterion weights and outcome ratings of the alternatives.

Weighting of Criteria or Attributes. Regardless of whether a direct assignment of weights is made or whether a subjective assignment of importance rating is used, the resulting weights for criteria apply to all alternatives. Also, it is essential that the resulting weights sum to 1.00 or 100% for each criterion.

Consider a version of the example, where the criteria of importance were Better, Cheaper, Faster, and so on. Assume that there are two mutually exclusive design alternatives being evaluated with respect to these criteria. Now limit the scope of the evaluation to only Better, Cheaper, and Faster by subsuming Supportable and Disposable under Better. This limitation makes the first criterion (Better) more complex, but it simplifies the analysis by giving only three criteria to consider for each design alternative.

TABLE 5 Additive Weights from Importance Rating

Criterion	Importance Rating	Additive Weight
Better	7	7/20 = 0.35
Cheaper	9	9/20 = 0.45
Faster	4	4/20 = 0.20
		1.00

Assume that Better is weighted 35%, Cheaper is weighted 50%, and Faster is weighted 15%, with the sum equal to 100%. If the weight of one criterion is changed, the others must be changed too. Thus, directly assigning weights may be thought of as the same as allocating 100 points to the criteria or attributes under consideration. These weighted criteria will apply to all design alternatives.

Alternatively, subjective importance ratings may be assigned to each criterion using a 10-point scale. Suppose that Better is ranked 7, Cheaper is ranked 9, and Faster is ranked 4. An analysis can then provide additive weights, as given in Table 5. These may differ some from the direct assignment of weights if done independently.

The Tabular Additive Method. Table 6 illustrates a tabular additive method based on input from the prior section. It identifies criteria, weights (W), the performance rating (R) of each alternative over each criterion, the product of weight and rating ($W \times R$), and the aggregate totals of $W \times R$. These aggregate totals may be used in comparing the alternatives.

The simple comparative scores in Table 6 are derived from a computation that combines criterion weight (directly assigned or from Table 5) and ratings of the anticipated outcome of each alternative on each criterion. Since the weights sum to 100% (or 1.00) and the best is 10 for all criteria, the weighted sum is an indication of how close each alternative comes to the ideal.

The Graphical Additive Method. A graphical version of the information given in Table 6 is shown in Figure 3. This is a stacked bar chart, with the total height of each bar corresponding to the total score for each alternative.

Each bar in Figure 3 is made up of components that are the product of criterion weights and performance ratings. The contribution of each component to the total for each alternative is exhibited. If additional alternatives are to be considered, the bar chart can quickly reveal the role of each component in making up the total score. In addition, this graphical method can be used to display the results of sensitivity analysis, particularly as it leads to possible decision reversal.

TABLE 6 Tabular Additive Method

Criterion	Weight (W)	Alternative A		Alternative B	
		Rating (R)	$W \times R$	Rating (R)	$W \times R$
Better	0.35	6	2.10	7	2.45
Cheaper	0.45	10	4.50	6	2.70
Faster	0.20	5	1.00	3	0.60
	1.00		7.60		5.75

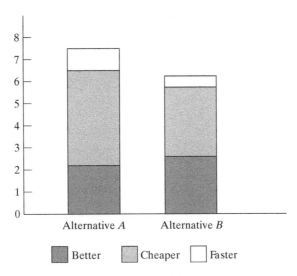

Figure 3 Bar chart for weighted evaluation.

5 THE DECISION EVALUATION DISPLAY

The decision evaluation display method of evaluation differs significantly from the methods presented in the prior sections. Some decision makers consider ranking, elimination, weighting, rating, and similar selection rules to be impediments to the effective application of insight, intuition, and judgment. An alternative is to put the emphasis on communicating only the differences upon which a decision depends, leaving the remaining path to a decision to the decision maker.

The decision evaluation display is based on the premise that *differences between alternatives* and the *degree of compliance* with multiple criteria, appropriately exhibited, are all that most decision makers need or desire. Experienced decision makers possess an inherent and acquired ability to process information needed to trade off competing criteria. Accordingly, the decision evaluation display is recommended as a means for simultaneously exhibiting the differences that multiple alternatives create in the face of multiple criteria.

The general form and basic structure of the decision evaluation display is illustrated in Figure 4. It is one way of simultaneously displaying the elements pertaining to one or more mutually exclusive alternatives in the face of economic and noneconomic criteria. These elements are as follows:

1. *Alternatives (A, B, C).* Two or more alternatives appear as vertical lines in the field of the decision evaluation display.
2. *Equivalent cost or profit.* The horizontal axis represents present equivalent, annual equivalent, or future equivalent cost or profit. Specific cost or profit values are indicated on the axis for each alternative displayed, with cost or profit increasing from left to right. In this way, equivalent economic differences between alternatives are made visible.
3. *Other criteria (X, Y, Z).* Vertical axes on the left represent one or more criteria, usually of a noneconomic nature. Each axis has its own scale depending upon the nature of the factor represented.

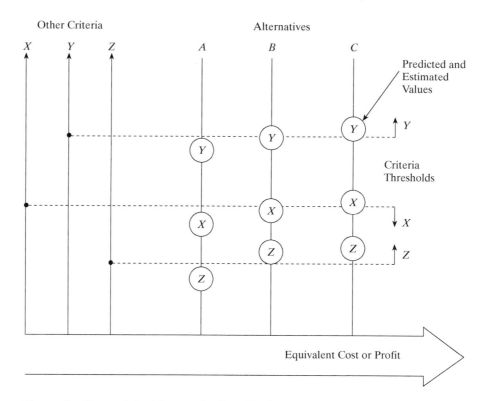

Alternatives and Models in Decision Making

Figure 4   General decision evaluation display.

4. *Other criteria thresholds.* Horizontal lines emanating from the vertical axes represent threshold or limiting values for noneconomic criteria (less than, equal to, or greater than).

5. *Predicted and/or estimated values.* Anticipated outcomes for each alternative (based on prediction or estimation) are entered in circles placed above, on, or below the thresholds. In this way, differences between desired and anticipated outcomes for each alternative are made visible.

Equivalent cost or profit from Equation 1, or from Equation 2 or 3, shown on the horizontal axis of Figure 4, is an objective measure. The aim in decision making is to select the mutually exclusive alternative with the lowest equivalent cost (or maximum equivalent profit) that adequately satisfies the other criteria.

6 DECISIONS UNDER RISK AND UNCERTAINTY

A particular decision can result in one of several outcomes, depending on which of several future events occurs. For example, a decision to go sailing can result in a high degree of satisfaction if the day turns out to be sunny, or in a low degree of satisfaction if it rains. These levels of satisfaction would be reversed if the decision were made to stay home. Thus, for the two states of nature, sun and rain, there are different payoffs depending on the alternative chosen.

In this section, a decision evaluation matrix is introduced to accommodate decisions under assumed certainty, risk, and uncertainty. Then each of these decision situations is treated separately, with emphasis on dealing with risk and uncertainty.

6.1 The Decision Evaluation Matrix

A decision evaluation matrix is a formal way of exhibiting the interaction of a finite set of alternatives and a finite set of possible futures (or states of nature). In this usage, alternatives have the meaning presented in Section 1; that is, they are courses of action from among which a decision maker expects to choose. The states of nature are normally not natural events, such as rain, sleet, or snow, but are a wide variety of future outcomes over which the decision maker has no direct control.

The general decision evaluation matrix is a model depicting the positive and negative results that may occur for each alternative under each possible future. In abstract form, this model is structured as shown in Figure 5. Its symbols are defined as follows:

A_i = an alternative available for selection by the decision maker, where $i = 1, 2, \ldots, m$

F_j = a future not under control of the decision maker, where $j = 1, 2, \ldots, n$

P_j = the probability that the jth future will occur, where $j = 1, 2, \ldots, n$

E_{ij} = evaluation measure (positive or negative) associated with the ith alternative and the jth future determined from Equation 1 or 3

Several assumptions underlie the application of this decision evaluation matrix model to decision making under assumed certainty, risk, and uncertainty. Foremost among these is the presumption that all viable alternatives have been considered and all possible futures have been identified. Possible futures not identified can significantly affect the actual outcome relative to the planned outcome.

Evaluation measures in the matrix model are associated with outcomes that may be either objective or subjective. The most common case is one in which the outcome values are objective and, therefore, subject to quantitative expression in cardinal form. For example, the payoffs may be profits expressed in dollars, yield expressed in pounds, costs (negative payoffs) expressed in dollars, or other desirable or undesirable measures. Subjective outcomes,

P_j		P_1	P_2	$\cdots$	P_n
F_j		F_1	F_2	$\cdots$	F_n
A_i					
A_1		E_{11}	E_{12}	$\cdots$	E_{1n}
A_2		E_{21}	E_{22}	$\cdots$	E_{2n}
$\vdots$		$\vdots$	$\vdots$		$\vdots$
A_m		E_{m1}	E_{m2}	$\cdots$	E_{mn}

Figure 5   The decision evaluation matrix.

conversely, are those that are valued on an ordinal or ranking scale. Examples are expressions of preference, such as a good corporate image being preferred to a poor image, higher quality outputs being preferred to those of lower quality, and so forth.

Following are the other assumptions of importance in the evaluation matrix representation of decisions:

1. The occurrence of one future precludes the occurrence of any other future (futures are mutually exclusive).
2. The occurrence of a specific future is not influenced by the alternative selected.
3. The occurrence of a specific future is not known with certainty, even though certainty is often assumed for analysis purposes.

6.2 Decisions under Assumed Certainty

In dealing with physical aspects of the environment, physical scientists and engineers have a body of systematic knowledge and physical laws on which to base their reasoning. Such laws as Boyle's law, Ohm's law, and Newton's laws of motion were developed primarily by collecting and comparing many comparable instances and by the use of an inductive process. These laws may then be applied with a high degree of certainty to specific instances. They are supplemented by many models for physical phenomena that enable conclusions to be reached about the physical environment that match the facts with narrow limits. Much is known with certainty about the physical environment.

Much less, particularly of a quantitative nature, is known about the environment within which operational decisions are made. Nonetheless, the primary aim of operations research and management science is to bring the scientific approach to bear to a maximum feasible extent. This is done with the aid of conceptual simplifications and models of reality, the most common being the assumption of a single known future. It is not claimed that knowledge about the future is in hand. Rather, the suppression of risk and uncertainty is one of the ways in which the scientific approach simplifies reality to gain insight. Such insight can assist greatly in decision making, provided that its shortcomings are recognized and accommodated.

The evaluation matrix for decision making under assumed certainty is not a matrix at all. It is a vector with as many evaluations as there are alternatives, with the outcomes constituting a single column. This decision vector is a special case of the matrix of Figure 5. It appears as in Figure 6 with the payoffs represented by E_i, where $i = 1, 2, \ldots, m$. The single future, which is

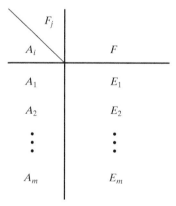

Figure 6   Decision evaluation vector.

assumed to occur with certainty, actually carries a probability of unity in the matrix. All other futures are suppressed by carrying probabilities of zero ($P = 0.0$).

When the outcomes, E_i, are stated in monetary terms (cost or profit), the decision rule or principle of choice is simple. If the alternatives are equal in all other respects, one would choose the alternative that minimizes cost or maximizes profit. In the case of cost, one would choose

$$\min_i \{E_i\} \quad \text{for } i = 1, 2, \ldots, m$$

For profit, one would choose

$$\max_i \{E_i\} \quad \text{for } i = 1, 2, \ldots, m$$

It is often not possible to accept the premise that only the cost or the profit differences are important, with intangibles and irreducibles having little or no effect. Unquantifiable nonmonetary factors may be significant enough to outweigh calculated costs or profit differences among alternatives. In other cases, the outcome is not easily expressed in monetary terms, or even in quantitative terms of some other evaluation measure, such as time, percentage of market, and so forth. Valid qualitative comparisons may be made when the quantitative outcomes cannot stand alone and when the outcomes are nonquantitative.

6.3 Decision Making under Risk

There is usually little assurance that anticipated or predicted futures will coincide with actual futures. The physical and economic elements on which a course of action depends may vary from their estimated values because of chance causes. Not only are the estimates of future costs problematical but also the anticipated future worth of most ventures is known only with a degree of assurance. This lack of certainty about the future makes decision making one of the most challenging tasks faced by individuals, industry, and government.

Decision making under risk occurs when the decision maker does not suppress acknowledged ignorance about the future but makes it explicit through the assignment of probabilities. Such probabilities may be based on experimental evidence, expert opinion, subjective judgment, or a combination of these.

Consider the following example. A computer systems firm has the opportunity to bid on two related contracts being advertised by a municipality. The first pertains to the selection and installation of hardware for a central computing facility together with required software. The second involves the development of a distributed computing network involving the selection and installation of hardware and software. The firm may be awarded either contract C_1 or contract C_2, or both contract C_1 and C_2. Thus, there are three possible futures.

Careful consideration of the possible approaches leads to the identification of five alternatives. The first is for the firm to subcontract the hardware selection and installation, but to develop the software itself. The second is for the firm to subcontract the software development, but to select and install the hardware itself. The third is for the firm to handle both the hardware and software tasks itself. The fourth is for the firm to bid jointly with a partner firm on both the hardware and software projects. The fifth alternative is for the firm to serve only as project manager, subcontracting all hardware and software tasks.

With the possible futures and various alternatives identified, the next step is to determine payoff values. Also to be determined are the probabilities for each of the three futures, where the sum of these probabilities must be unity. Suppose that these determinations lead to the profits and probabilities given in Table 7.

Alternatives and Models in Decision Making

TABLE 7 Decision Evaluation Matrix (Profit in Thousands of Dollars)

		(0.3)	(0.2)	(0.5)
Probability:		C_1	C_2	$C_1 + C_2$
Future:				
Alternative	A_1	100	100	400
	A_2	−200	150	600
	A_3	0	200	500
	A_4	100	300	200
	A_5	−400	100	200

Table 7 is structured in accordance with the format of the decision evaluation matrix model exhibited in Figure 5. It is observed that the firm anticipates a profit of $100,000 if Alternative A_1 is chosen and contract C_1 is secured. If contract C_2 is secured, the profit would also be $100,000. However, if both contract C_1 and C_2 are secured, the profit anticipated is $400,000. Similar information is exhibited for the other alternatives, with each row of the matrix representing the outcome expected for each future (column) for a particular alternative.

Before proceeding to the application of criteria for the choice from among alternatives, the decision evaluation matrix should be examined for dominance. Any alternatives that are clearly not preferred, regardless of the future that occurs, may be dropped from consideration. If the outcomes for alternative x are better than the outcomes for alternative y for all possible futures, alternative x is said to *dominate* alternative y, and y can be eliminated as a possible choice.

The computer systems firm, facing the evaluation matrix of Table 7, may eliminate A_5 from consideration because it is dominated by all other alternatives. This means that the possible choice of serving only as project manager is inferior to each and every one of the other alternatives, regardless of the way in which the projects are awarded. Therefore, the matrix can be reduced to that given in Table 8. The decision criteria in the sections that follow may be used to assist in the selection from among Alternatives A_1–A_4.

Aspiration Level Criterion. Some form of aspiration level exists in most personal and professional decision making. An aspiration level is some desired level of achievement such as profit, or some undesirable result level to be avoided, such as loss. In decision making under risk, the aspiration level criterion involves selecting some level of achievement that is to be met, followed by a selection of that alternative which maximizes the probability of achieving the stated aspiration level.

The computer systems firm is now faced with selecting from among Alternatives A_1–A_4, as presented in the reduced matrix of Table 8. Under the aspiration level criterion, management must set a minimum aspiration level for profit and possibly a maximum aspiration level

TABLE 8 Reduced Decision Evaluation Matrix (Profit in Thousands of Dollars)

		(0.3)	(0.2)	(0.5)
Probability:		C_1	C_2	$C_1 + C_2$
Future:				
Alternative	A_1	100	100	400
	A_2	−200	150	600
	A_3	0	200	500
	A_4	100	300	200

for loss. Suppose that the profit level is set to be at least $400,000 and the loss level is set to be no more than $100,000. Under these aspiration level choices, Alternatives A_1–A_3 qualify as to profit potential, but Alternative A_2 fails the loss test and must be eliminated. The choice could now be made between A_1 and A_3 by some other criterion, even though both satisfy the aspiration level criterion.

Most Probable Future Criterion.   A basic human tendency is to focus on the most probable outcome from among several that could occur. This approach to decision making suggests that all except the most probable future be disregarded. Although somewhat equivalent to decision making under certainty, this criterion works well when the most probable future has a significantly high probability so as to partially dominate.

Under the most probable future criterion, the computer systems firm would focus its selection process from among the four alternatives on the profits associated with the future designated $C_1 + C_2$ (both contracts awarded). This is because the probability of this future occurring is 0.5, the most probable possibility. Alternative A_2 is preferred by this approach.

The most probable future criterion could be applied to select between A_1 and A_3, as identified under the aspiration level criterion. If this is done, the firm would choose Alternative A_3.

Expected Value Criterion.   Many decision makers strive to make choices that will maximize expected profit or minimize expected loss. This is ordinarily justified in repetitive situations where the choice is to be made over and over again with increasing confidence that the calculated expected outcome will be achieved. This criterion is viewed with caution only when the payoff consequences of possible outcomes are disproportionately large, making a result that deviates from the expected outcome a distinct possibility.

The calculation of the expected value requires weighing all payoffs by their probabilities of occurrence. These weighed payoffs are then summed across all futures for each alternative. For the computer systems firm, Alternatives A_1–A_4 yield the following expected profits (in thousands):

$$A_1: \ \ \$\,100(0.3) \ + \ \$100(0.2) \ + \ \$400(0.5) \ = \ \$250$$

$$A_2: -\$200(0.3) \ + \ \$150(0.2) \ + \ \$600(0.5) \ = \ \$270$$

$$A_3: \ \ \ \$\,0(0.3) \ + \ \$200(0.2) \ + \ \$500(0.5) \ = \ \$290$$

$$A_4: \ \ \$\,100(0.3) \ + \ \$300(0.2) \ + \ \$200(0.5) \ = \ \$190$$

From this analysis it is clear that Alternative A_3 would be selected. Further, if this criterion were to be used to resolve the choice of either A_1 or A_3 under the aspiration level approach, the choice would be Alternative A_3.

Comparison of Decisions.   It is evident that there is no one best selection when these criteria are used for decision making under risk. The decision made depends on the decision criterion adopted by the decision maker. For the example of this section, the alternatives selected under each criterion were

Aspiration level criterion: A_1 or A_3
Most probable future criterion: A_2
Expected value criterion: A_3

If the application of the latter two criteria to the resolution of A_1 or A_3 chosen under the aspiration level criterion is accepted as valid, then A_3 is preferred twice and A_2 once. From this it might be appropriate to suggest that A_3 is the best alternative arising from the use of these three criteria.

6.4 Decision Making under Uncertainty

It may be inappropriate or impossible to assign probabilities to the several futures identified for a given decision situation. Often no meaningful data are available from which probabilities may be developed. In other instances, the decision maker may be unwilling to assign a subjective probability, as is often the case when the future could prove to be unpleasant. When probabilities are not available for assignment to future events, the situation is classified as decision making under uncertainty.

As compared with decision making under certainty and under risk, decisions under uncertainty are made in a more abstract environment. In this section, several decision criteria will be applied to the example of Table 7 to illustrate the formal approaches that are available.

Laplace Criterion. Suppose that the computer systems firm is unwilling to assess the futures in terms of probabilities. Specifically, the firm is unwilling to differentiate between the likelihood of acquiring contract C_1, contract C_2, and contracts C_1 and C_2. In the absence of these probabilities, one might reason that each possible state of nature is as likely to occur as any other. The rationale of this assumption is that there is no stated basis for one state of nature to be more likely than any other. This is called the *Laplace principle* or the *principle of insufficient reason* based on the philosophy that nature is assumed to be indifferent.

Under the Laplace principle, the probability of the occurrence of each future state of nature is assumed to be $1/n$, where n is the number of possible future states. To select the best alternative, one would compute the arithmetic average for each. For the decision matrix of Table 8, this is accomplished as shown in Table 9. Alternative A_3 results in a maximum profit of \$233,000 and would be selected.

Maximin and Maximax Criteria. Two simple decision rules are available for dealing with decisions under uncertainty. The first is the *maximin* rule, based on an extremely pessimistic view of the outcome of nature. The use of this rule would be justified if it is judged that nature will do its worst. The second is the *maximax* rule, based on an extremely optimistic view of the future. Use of this rule is justified if it is judged that nature will do its best.

Because of the pessimism embraced by the maximin rule, its application will lead to the alternative that assures the best of the worst possible outcomes. If E_{ij} is used to represent the payoff for the ith alternative and the jth state of nature, the required computation is

$$\max_i \{\min_j E_{ij}\}$$

TABLE 9 Computation of Average Profit (Thousands of Dollars)

Alternative	Average Payoff		
A_1	(\$100 + \$100 + \$400)	÷ 3 =	\$200
A_2	(−\$200 + \$150 + \$600)	÷ 3 =	\$183
A_3	(\$0 + \$200 + \$500)	÷ 3 =	\$233
A_4	(\$100 + \$300 + \$200)	÷ 3 =	\$200

TABLE 10 Profit by the Maximin Rule (Thousands of Dollars)

Alternative	$\underset{j}{\text{Min }} E_{ij}(\$)$
A_1	100
A_2	-200
A_3	0
A_4	100

Consider the decision situation described by the decision matrix of Table 8. The application of the maximin rule requires that the minimum value in each row be selected. Then the maximum value is identified from these and associated with the alternative that would produce it. This procedure is illustrated in Table 10. Selection of either Alternative A_1 or A_4 assures the firm of a profit of at least $100,000 regardless of the future.

The optimism of the maximax rule is in sharp contrast to the pessimism of the maximin rule. Its application will choose the alternative that assures the best of the best possible outcomes. As before, if E_{ij} represents the payoff for the ith alternative and the jth state of nature, the required computation is

$$\underset{i}{\max}\ \{\underset{j}{\max}\ E_{ij}\}$$

Consider the decision situation of Table 8 again. The application of the maximax rule requires that the maximum value in each row be selected. Then the maximum value is identified from these and associated with the alternative that would produce it. This procedure is illustrated in Table 11. Selection of Alternative A_2 is indicated. Thus, the decision maker may receive a profit of $600,000 if the future is benevolent.

A decision maker who chooses the maximin rule considers only the worst possible occurrence for each alternative and selects the alternative that promises the best of the worst possible outcomes. In the example where A_1 was chosen, the firm would be assured of a profit of at least $100,000, but it could not receive a profit any greater than $400,000. Or, if A_4 were chosen, the firm could not receive a profit any greater than $300,000. Conversely, the firm that chooses the maximax rule is optimistic and decides solely on the basis of the highest profit offered for each alternative. Accordingly, in the example in which A_2 was chosen, the firm faces the possibility of a loss of $200,000 while seeking a profit of $600,000.

TABLE 11 Profit by the Maximax Rule (Thousands of Dollars)

Alternative	$\underset{j}{\text{Max }} E_{ij}\ (\$)$
A_1	400
A_2	600
A_3	500
A_4	300

6.5 Hurwicz Criterion

Because the decision rules presented previously are extreme, they are shunned by many decision makers. Most people have a degree of optimism or pessimism somewhere between the extremes. A third approach to decision making under uncertainty involves an index of relative optimism and pessimism. It is called the *Hurwicz rule*.

A compromise between optimism and pessimism is embraced in the Hurwicz rule by allowing the decision maker to select an index of optimism, α, such that $0 \le \alpha \le 1$. When $\alpha = 0$, the decision maker is pessimistic about nature, while an $\alpha = 1$ indicates optimism about nature. Once α is selected, the Hurwicz rule requires the computation of

$$\max_i \{\alpha[\max_j E_{ij}] + (1 - \alpha)[\min_j E_{ij}]\}$$

where E_{ij} is the payoff for the ith alternative and the jth state of nature.

As an example of the Hurwicz rule, consider the payoff matrix of Table 8 with $\alpha = 0.2$. The required computations are shown in Table 12 and Alternative A_1 would be chosen by the firm.

Additional insight into the Hurwicz rule can be obtained by graphing each alternative for all values of α between zero and one. This makes it possible to identify the value of α for which each alternative would be favored. Such a graph is shown in Figure 7. It may be observed that Alternative A_1 yields a maximum expected profit for all values of $\alpha \le \frac{1}{2}$. Alternative A_3 exhibits a maximum for $\frac{1}{2} \le \alpha \le \frac{2}{3}$, and alternative A_2 gives a maximum for $\frac{2}{3} \le \alpha \le 1$. There is no value of α for which Alternative A_4 would be best except at $\alpha = 0$, where it is as good an alternative as A_1.

When $\alpha = 0$, the Hurwicz rule gives the same result as the maximin rule, and when $\alpha = 1$, it is the same as the maximax rule. This may be shown for the case where $\alpha = 0$ as

$$\max_i \{0[\max_j E_{ij}] + (1 - 0)[\min_j E_{ij}]\} = \max_i [\min_j E_{ij}]$$

For the case where $\alpha = 1$,

$$\max_i \{1[\max_j E_{ij}] + (1 - 1)[\min_j E_{ij}]\} = \max_i [\max_j E_{ij}]$$

Thus, the maximin rule and the maximax rule are special cases of the Hurwicz rule.

The philosophy behind the Hurwicz rule is that the focus on the most extreme outcomes or consequences bounds or brackets the decision. By use of this rule, the decision maker may weight the extremes in such a manner as to reflect their relative importance.

TABLE 12 Profit by the Hurwicz Rule with $\alpha = 0.2$ (Thousands of Dollars)

Alternative	$\alpha[\max E_{ij}] + (1 - \alpha)[\min E_{ij}]$
A_1	$0.2(\$400) + 0.8(\$100) = \$160$
A_2	$0.2(\$600) + 0.8(\$-200) = \$-40$
A_3	$0.2(\$500) + 0.8(0) = \$100$
A_4	$0.2(\$300) + 0.8(\$100) = \$140$

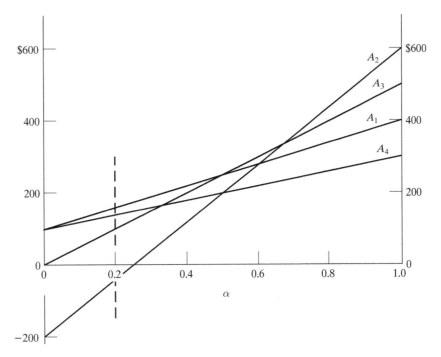

Figure 7 Values for the Hurwicz rule for four alternatives.

Comparison of Decisions. As was the case for decision criteria applied for decision making under risk, it is evident that there is no one best criterion for decision making under uncertainty. The decision made depends on the decision criterion applied by the decision maker. For the preceding examples, the alternatives selected were

Laplace criterion: A_3
Maximin criterion: A_1 or A_4
Maximax criterion: A_2
Hurwicz criterion ($\alpha = 0.2$): A_1

Examination of the selections recommended by the five decision rules indicates that each has its own merit. Several factors may influence a decision maker's choice of a rule in a given decision situation. The decision maker's attitude toward uncertainty (pessimistic or optimistic) and his or her personal utility function are important influences. Thus, the choice of a particular decision rule for a given decision situation must be based on subjective judgment.

7 SUMMARY AND EXTENSIONS

The delay in the development of models to explain and make decisions about systems may be attributed to an early preoccupation with the physical and life sciences. This is understandable because during much of history, the limiting factors in the satisfaction of human wants were predominantly physical and biological. But with the accumulation of scientific knowledge, human beings have been able to design complex systems to meet needs.

The centerpiece and foundation for this chapter is modeling and decision models, with full realization that models are not an end in themselves. Models for systems analysis and for the evaluation of system designs are focused on alternatives as "input" and informed decisions as "output"; that is, models are employed to bring a measure of rigor to decision making in systems engineering. Only a limited treatment of this important subject was possible herein. Accordingly, two references are offered as a means for extending the material covered: (1) *Systems Analysis and Modeling: A Macro–Micro Approach with Multidisciplinary Applications*, by D. W. Boyd, Academic Press, 2001 and, (2) "Modeling and Indirect Experimentation in System Design Evaluation," by W. J. Fabrycky, Inaugural Issue of *Systems Engineering*, Vol. 1, No. 1, 1994.

Decision evaluation theory was introduced as an overarching model-based construct for decision making involving multiple alternatives and multiple criteria. It was based upon engineering economics and operations research, two well-developed bodies of systematic knowledge. These bodies of knowledge are made useful to the systems engineering process through adaptation. In this connection, the most noteworthy paradigm is the design-dependent parameter approach coupled with the decision evaluation function. Then, a graphical decision evaluation display was presented conceptually as the suggested means for addressing multiple criteria. This display method does not lead to formal resolution and selection of a preferred alternative, but leaves weighting and trade-off with the decision maker. The technique assumes that the decision maker, or decision-making group, will apply the needed trade-off through subjective evaluation. More formal trade-off between life-cycle cost and effectiveness measures would require the application of preference functions and utility theory. This was not presented herein, but the interested reader is referred to the classic entitled *Decisions with Multiple Objectives: Preferences and Value Trade-Offs*, by R. L. Kenney and H. Raiffa, John Wiley and Sons, Inc., 1976.

This chapter ends with a section combining decision making under certainty, risk, and uncertainty, based on the decision evaluation matrix. A single certain future is often assumed, but rarely encountered. Accordingly, risk is often addressed with the utilization of probability measures and a number of selection methods. Decision making under uncertainty is also addressed through the application of selection rules, but without the use of probability statements about future outcomes. Selection of the appropriate formal method or model for a given decision situation depends upon the nature of the decision and the preference of the decision maker. There is no rule or set of guidelines that may be applied in all cases. However, it is generally accepted that use of structured methods and techniques can minimize the omission of options and other factors of importance.

In the real world, criteria are numerous and many facts may be missing. Even with the availability of methods for dealing jointly with multiple alternatives and multiple criteria, it is usually the case that alternatives cannot be completely delineated in accurate quantitative terms. Thus, the choice must be made on the basis of the judgment of one or more individuals. An important aim of decision analysis is to marshal the facts so that reason may be used to the fullest extent in arriving at a decision. In this way, judgment can be reserved for parts of situations where factual knowledge is absent. Whatever it is called, it is inescapable that judgment must always be the final step in arriving at a decision about the future. There is no other way if action is to be taken. There appears to be a marked difference in people's abilities to come to sound conclusions when some facts relative to a situation are missing. Those who possess sound judgment are richly rewarded. But as effective as intuition, hunch, or judgment may sometimes be, this type of thinking should be reserved for those areas where facts on which to base a decision are missing.

Where a diligent search uncovers insufficient information to reason the outcome of a course of action, the problem is to render as accurate a decision as the lack of facts permits.

In such situations, there is a decided tendency, on the part of many, to make little logical use of the data that are available in coming to a conclusion. The thinking is that because rough estimating has to be done on some elements of the situation, the estimate might as well embrace the entire situation. But an alternative may usually be subdivided into parts, and the available data are often adequate for a complete or nearly complete evaluation of several of the parts. The segregation of the known and unknown parts is in itself additional knowledge. Also, the unknown parts, when subdivided, frequently are recognized as being similar to parts previously encountered and thus become known.

After all the data that can be brought to bear on a decision situation have been considered, some areas of uncertainty may be expected to remain. If a decision is to be made, these areas of uncertainty must be bridged by the recognition and evaluation of intangibles. Some call the type of evaluation *intuition*, while others call it a *hunch* or *judgment*. In any event, the decision maker should be taken part way to the point of decision with the aid of models and simulation through indirect experimentation, as introduced in this chapter.

QUESTIONS AND PROBLEMS

1. A complete and all-inclusive alternative rarely emerges in its final state. Explain.

2. Should decision making be classified as an art or a science?

3. Contrast limiting and strategic factors.

4. Discuss the various meanings of the word model.

5. Describe briefly physical models, schematic models, and mathematical models.

6. How do mathematical models directed to decision situations differ from those traditionally used in the physical sciences?

7. Contrast direct and indirect experimentation.

8. Write the general form of the evaluation function for money flow modeling and define its symbols.

9. Identify a decision situation and indicate the variables under the control of the decision maker and those not directly under his or her control.

10. Contrast the similarities and differences in the economic optimization functions given by Equations 2 and 3 (see Figure 2).

11. Why is it not possible to formulate a model that accurately represents reality?

12. Under what conditions may a properly formulated model become useless as an aid in decision making?

13. Explain the nature of the cost components that should be considered in deciding how frequently to review a dynamic environment.

14. What caution must be exercised in the use of models?

15. Discuss several specific reasons why models are of value in decision making.

16. Identify a multiple-criteria decision situation with which you have experience. Select the three to five most important criteria.

17. Discuss the degree to which you think the criteria you selected in Question 16 are truly independent. Weight each criterion, check for consistency, and then normalize the weight so that the total sums to 100. Use the method of paired comparisons to rank the criteria in order of decreasing importance.

18. Extend Question 16 to include two or three alternatives. Evaluate how well each alternative ranks against each criterion on a scale of 1–10. Then compute the product of the ratings and

the criterion weights and sum them for each criterion to determine the weighted evaluation for each alternative.

19. The values for three alternatives considered against four criteria are given (with higher values being better). What can you conclude using the following systematic elimination methods?

    (a) Comparing the alternatives against each other (dominance).
    (b) Comparing the alternatives against a standard (Rules 1 and 2).
    (c) Comparing criteria across alternatives (criteria ranked 2 > 3 > 1 > 4).

| | **Alternative** | | | | |
Criterion	A	B	C	Ideal	Minimum Standard
1.	6	5	8	10	7
2.	90	80	75	100	70
3.	40	35	50	50	30
4.	G	P	VG	E	F

20. A specialty software development firm is planning to offer one of four new software products and wishes to maximize profit, minimize risk, and increase market share. A weight of 65% is assigned to annual profit potential, 20% to profitability risk, and 15% to market share. Use the tabular additive method for this situation and identify the product that would be best for the firm to introduce.

New Product	Profit Potential ($)	Profit Risk ($)	Market Share
SW I	100K	40K	High
SW II	140K	35K	Medium
SW III	150K	50K	Low
SW IV	130K	45K	Medium

21. Convert the tabular additive results from Problem 20 into a stacked bar chart display.

22. Rework the example in Section 5 if the importance ratings for Better, Cheaper, and Faster in Table 5 change from 7, 9, 4 to 9, 7, 4, respectively.

23. Sketch a decision evaluation display that would apply to making a choice from among three automobile makes in the face of the three top criteria of your selection.

24. Superimpose another source alternative (remanufacture) on the decision evaluation display in Section 5 to illustrate its expandability. Now discuss the criteria values that this source alternative would require to make it the preferred alternative.

25. Formulate an evaluation matrix for a hypothetical decision situation of your choice.

26. Formulate an evaluation vector for a hypothetical decision situation under assumed certainty.

27. Develop an example to illustrate the application of paired outcomes in decision making among a number of nonquantifiable alternatives.

28. What approaches may be used to assign probabilities to future outcomes?

29. What is the role of dominance in decision making among alternatives?

30. Give an example of an aspiration level in decision making.

31. When would one follow the most probable future criterion in decision making?

32. What drawback exists in using the most probable future criterion?

33. How does the Laplace criterion for decision making under uncertainty actually convert the situation to decision making under risk?

34. Discuss the maximin and the maximax rules as special cases of the Hurwicz rule.

35. The cost of developing an internal training program for office automation is unknown but described by the following probability distribution:

Cost ($)	Probability of Occurrence
80,000	0.20
95,000	0.30
105,000	0.25
115,000	0.20
130,000	0.05

What is the expected cost of the course? What is the most probable cost? What is the maximum cost that will occur with a 95% assurance?

36. Net profit has been calculated for five investment opportunities under three possible futures. Which alternative should be selected under the most probable future criterion? Which alternative should be selected under the expected value criterion?

	(0.3) $F_1(\$)$	(0.2) $F_2(\$)$	(0.5) $F_3(\$)$
A_1	100,000	100,000	380,000
A_2	−200,000	160,000	590,000
A_3	0	180,000	500,000
A_4	110,000	280,000	200,000
A_5	400,000	90,000	180,000

37. Daily positive and negative payoffs are given for five alternatives and five futures in the following matrix. Which alternative should be chosen to maximize the probability of receiving a payoff of at least 9? What choice would be made by using the most probable future criterion?

	(0.15) F_1	(0.20) F_2	(0.30) F_3	(0.20) F_4	(0.15) F_5
A_1	12	8	−4	0	9
A_2	10	0	5	10	16
A_3	6	5	10	15	−4
A_4	4	14	20	6	12
A_5	−8	22	12	4	9

38. The following matrix gives the payoffs in utiles (a measure of utility) for three alternatives and three possible states of nature:

	State of Nature		
	S_1	S_2	S_3
A_1	50	80	80
A_2	60	70	20
A_3	90	30	60

Which alternative would be chosen under the Laplace principle? The maximin rule? The maximax rule? The Hurwicz rule with $\alpha = 0.75$?

39. The following payoff matrix indicates the costs associated with three decision options and four states of nature:

	State of Nature			
	S_1	S_2	S_3	S_4
O_1	20	25	30	35
O_2	40	30	40	20
O_3	10	60	30	25

Select the decision option that should be selected under the maximin rule; the maximax rule; the Laplace rule; the minimax regret rule; and the Hurwicz rule with $\alpha = 0.2$. How do the rules applied to the cost matrix differ from those that are applied to a payoff matrix of profits?

40. A cargo aircraft is being designed to operate in different parts of the world where external navigation aids vary in quality. The manufacturer has identified five areas where the aircraft is likely to fly and the likelihood that the aircraft will be in each area. The aircraft will be in Areas 1–5 with the frequency given in the following table. The table also gives the five navigation systems under consideration, N_1, N_2, N_3, N_4, and N_5. The manufacturer has estimated the typical navigation errors to be expected in each area and these are shown in the table. Navigation errors are in nautical miles (nm). Small navigation errors are good; large navigation errors are not good.

Navigation Systems	**Area 1 (0.15)**	**Area 2 (0.20)**	**Area 3 (0.30)**	**Area 4 (0.20)**	**Area 5 (0.15)**
N_1	0.09	0.15	0.20	0.30	0.10
N_2	0.09	0.30	0.25	0.09	0.06
N_3	0.15	0.14	0.09	0.06	0.25
N_4	0.21	0.07	0.05	0.12	0.06
N_5	0.28	0.03	0.08	0.19	0.10

a) Which system(s) will maximize the probability of achieving a navigation error of 0.10 nm or less?

b) What is the probability of achieving this navigation error?

c) Which choice would be made by using the most probable future criterion?

41. The following matrix gives the expected profit in thousands of dollars for five marketing strategies and five potential levels of sales:

		Levels of Sales				
		L_1	L_2	L_3	L_4	L_5
	M_1	10	20	30	40	50
	M_2	20	25	25	30	35
Strategy	M_3	50	40	5	15	20
	M_4	40	35	30	25	25
	M_5	10	20	25	30	20

Which marketing strategy would be chosen under the maximin rule? The maximax rule? The Hurwicz rule with $\alpha = 0.4$?

42. Graph the Hurwicz rule for all values of α using the payoff matrix of Problem 40.

43. The following decision evaluation matrix gives the expected savings in maintenance costs (in thousands of dollars) for three policies of preventive maintenance and three levels of operation of equipment. Given the probabilities of each level of operation, $P_1 = 0.3$, $P_2 = 0.25$, and $P_3 = 0.45$, determine the best policy based on the most probable future criterion.

Policy	Level of Operation		
	L_1	L_2	L_3
M_1	10	20	30
M_2	22	26	26
M_3	40	30	15

Also, determine the best policy under uncertainty, using the Laplace rule, the Maximax rule, and the Hurwicz rule with $\alpha = 0.2$.

44. What should be done with those facets of a decision situation that cannot be explained by the model being used?

Models for Economic Evaluation

From Chapter 80 of *Systems Engineering and Analysis,* Fifth Edition, Benjamin S. Blanchard, Wolter J. Fabrycky. Copyright © 2011 by Pearson Education, Inc. Published by Pearson Prentice Hall. All rights reserved.

Models for Economic Evaluation

Economic considerations are important in systems engineering, the suggested means for bringing technical or engineered systems into being. For systems already in existence, economic considerations often provide a basis for the analysis and improvement of system operation. There are numerous examples of systems and their products that exhibit excellent physical design, but have little economic merit in relation to their worth.

Down through history, the limiting factor has been predominantly physical, but with the development of science and technology engineered systems and services that may not have economic utility have become physically possible. Fortunately, engineers can readily extend their inherent ability for analysis and evaluation to embrace economic factors. The primary aim of this chapter is to help achieve that goal.

Upon completion of this chapter, the reader will have gained essential insight into interest and interest formulas and the determination of economic equivalence. Economic equivalence will then be demonstrated as a means to evaluate single and multiple alternatives by five classical methods. Insight into multiple alternatives and multiple criteria will be obtained from applications of the decision evaluation display. Finally, break-even analysis extends the economic analysis methods to recognize fixed and variable costs.

The consideration and study of several fundamentals in this chapter will impart a capability and an understanding of the following:

- Interest formula derivations and their use in determining equivalence for various money flow patterns over time;
- Determining economic equivalence by classical measures, as well as by rate-of-return and payout measures, as bases for selecting from among alternatives;
- Finding the annual equivalent cost of an asset by incorporating both depreciation cost and the cost of invested capital;
- Evaluating single and multiple alternatives by classical methods, as well as by rate-of-return and payout methods;
- Evaluating multiple alternatives based on economic and noneconomic criteria using the decision evaluation display;

- Applying the concept of break-even analysis to evaluate alternatives where fixed and variable costs are present; and
- Considering multiple uncertain futures, as well as risk, in the evaluation of economic alternatives.

The intent of this chapter is to provide prerequisite material upon which subsequent chapters can rely whenever economic evaluations are indicated. A companion purpose is to prepare the reader for a study of design for affordability and for life-cycle cost-effectiveness analysis.

1 INTEREST AND INTEREST FORMULAS

The time value of money in the form of an interest rate is an important element in most decision situations involving money flow over time. Because money can earn at a certain interest rate, it is recognized that a dollar in hand at present is worth more than a dollar to be received at some future date. A lender may consider interest received as a gain or profit, whereas a borrower usually considers interest to be a charge or cost.

The interest rate is the ratio of the borrowed money to the fee charged for its use over a period, usually a year. This ratio is expressed as a percentage. For example, if $100 is paid for the use of $1,000 for 1 year, the interest rate is 10%. In compound interest, the interest earned at the end of the interest period is either paid at that time or earns interest upon itself. This compounding assumption is usually used in the economic evaluation of alternatives.

A schematic model for money flow over time is shown in Figure 1. It is the basis for the derivation of interest factors and may be applied to any phase of the system life cycle for the purpose of life-cycle cost or income analysis. Let

i = nominal annual rate of interest

n = number of interest periods, usually annual

P = principal amount at a time assumed to be the present

A = single amount in a series of n equal amounts at the end of each interest period

F = amount, n interest periods hence, equal to the compound amount of P, or the sum of the compound amounts of A, at the interest rate i

1.1 Single-Payment Compound-Amount Formula

When interest is permitted to compound, the interest earned during each interest period is added to the principal amount at the beginning of the next interest period. Using the terms defined, the relationship among F, P, n, and i can be developed as shown in Table 1. The resulting factor, $(1 + i)^n$, is known as the *single-payment compound-amount factor* and is designated as

$$\left(^{F/P,\, i,\, n}\right)$$

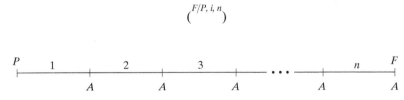

Figure 1   The money flow model.

TABLE 1 Single-Payment Compound-Amount Formula

Year	Amount at Beginning of Year	Interest Earned During Year	Compound Amount at End of Year
1	P	Pi	$P + Pi = P(1 + i)$
2	$P(1 + i)$	$P(1 + i)i$	$P(1 + i) + P(1 + i)i = P(1 + i)^2$
3	$P(1 + i)^2$	$P(1 + i)^2 i$	$P(1 + i)^2 + P(1 + i)^2 i = P(1 + i)^3$
n	$P(1 + i)^{n-1}$	$P(1 + i)^{n-1}i$	$P(1 + i)^{n-1} + P(1 + i)^{n-1}i = P(1 + i)^n = F$

This factor may be used to express the equivalence between a present amount, P, and a future amount, F, at an interest rate i for n years. The formula is

$$F = P(1 + i)^n$$

or

$$F = P(\overset{F/P,\, i,\, n}{\qquad})$$

(1)

1.2 Single-Payment Present-Amount Formula

The single-payment compound-amount formula may be solved for P and expressed as

$$P = F\left[\frac{1}{(1 + i)^n}\right]$$

The resulting factor, $1/(1 + i)^n$, is known as the *single-payment present-amount factor* and is designated

$$(\overset{P/F,\, i,\, n}{\qquad})$$

This factor may be used to express the equivalence between a future amount, F, and a present amount, P, at an interest rate i for n years. The formula is

$$P = F(\overset{P/F,\, i,\, n}{\qquad})$$

(2)

1.3 Equal-Payment Series Compound-Amount Formula

In some situations, a series of receipts or disbursements occurring uniformly at the end of each year may be encountered. The sum of the compound amounts of this series may be determined by reference to Figure 1.

The A dollars deposited at the end of the nth year will earn no interest and will contribute only A dollars to F. The A dollars deposited at the end of period $n - 1$ will earn interest in the amount of Ai, and $A(1 + i)$ will be contributed to the sum. The amount at the end of period $n - 2$ will contribute $A(1 + i)^2$. The sum of this series will be

$$F = A(1) + A(1 + i) + A(1 + i)^2 + \cdots + A(1 + i)^{n-2} + A(1 + i)^{n-1}$$

Multiplying this series by $(1 + i)$ gives

$$F(1 + i) = A[(1 + i) + (1 + i)^2 + (1 + i)^3 + \cdots + (1 + i)^{n-1} + (1 + i)^n]$$

Subtracting the first expression from the second gives

$$F(1 + i) - F = A[(1 + i)^n - 1]$$

$$Fi = A[(1 + i)^n - 1]$$

$$F = A\left[\frac{(1 + i)^n - 1}{i}\right]$$

The resulting factor, $[(1 + i)^n - 1]/i$, is known as the *equal-payment series compound-amount factor* and is designated

$$\left(^{F/A,\ i,\ n}\right)$$

This factor may be used to express the equivalence between an equal-payment series, A, and a future amount F, at an interest rate i for n years. The formula is

$$F = A\left(^{F/A,\ i,\ n}\right) \tag{3}$$

1.4 Equal-Payment Series Sinking-Fund Formula

The equal-payment series compound-amount formula may be solved for A and expressed as

$$A = F\left[\frac{i}{(1 + i)^n - 1}\right]$$

The resulting factor, $i/[(1 + i)^n - 1]$, is known as the *equal-payment series sinking-fund factor* and is designated

$$\left(^{A/F,\ i,\ n}\right)$$

This factor may be used to express the equivalence between a future amount, F, and an equal-payment series, A, at an interest rate i for n years. The formula is

$$A = F\left(^{A/F,\ i,\ n}\right) \tag{4}$$

1.5 Equal-Payment Series Capital-Recovery Formula

The substitution of $P(1 + i)^n$ for F in the equal-payment series sinking-fund formula results in

$$A = P(1 + i)^n\left[\frac{i}{(1 + i)^n - 1}\right]$$

$$= P\left[\frac{i(1 + i)^n}{(1 + i)^n - 1}\right]$$

The resulting factor, $i(1 + i)n/[(1 + i)^n - 1]$, is known as the *equal-payment series capital-recovery factor* and is designated

$$\left(^{A/P,\ i,\ n}\right)$$

This factor may be used to express the equivalence between future equal-payment series, A, and a present amount, P, at an interest rate i for n years. The formula is

$$A = P(^{A/P,\ i,\ n})\tag{5}$$

1.6 Equal-Payment Series Present-Amount Formula

The equal-payment series capital-recovery formula can be solved for P and expressed as

$$P = A\left[\frac{(1 + i)^n - 1}{i(1 + i)^n}\right]$$

The resulting factor, $[(1 + i)^n - 1]/[i(1 + i)^n]$, is known as the *equal-payment series present-amount factor* and is designated

$$(^{P/A,\ i,\ n})$$

This factor may be used to express the equivalence between future equal-payment series, A, and a present amount, P, at an interest rate i for n years. The formula is

$$P = A(^{P/A,\ i,\ n})\tag{6}$$

1.6 Summary of Interest Formulas

The interest factors derived in the previous paragraphs express relationships among P, A, F, i, and n. In any given application, an equivalence between F and P, P and F, F and A, A and F, P and A, or A and P is expressed for an interest rate i and a number of years n. Table 2 may be used to select the interest formula needed in a given situation. The factor designations summarized in the last column make it possible to set up a problem symbolically before determining the values of the factors involved.

TABLE 2 Summary of Interest Formulas

Formula Name	Function	Formula	Designation
Single-payment compound-amount	Given P Find F	$F = P(1 + i)^n$	$F = P(^{F/P,\ i,\ n})$
Single-payment present-amount	Given F Find P	$P = F\left[\dfrac{1}{(1 + i)^n}\right]$	$P = F(^{P/F,\ i,\ n})$
Equal-payment series compound-amount	Given A Find F	$F = A\left[\dfrac{(1 + i)^n - 1}{i}\right]$	$F = A(^{F/A,\ i,\ n})$
Equal-payment series sinking-fund	Given F Find A	$A = F\left[\dfrac{i}{(1 + i)^n - 1}\right]$	$A = F(^{A/F,\ i,\ n})$
Equal-payment series present-amount	Given A Find P	$P = A\left[\dfrac{(1 + i)^n - 1}{i(1 + i)^n}\right]$	$P = A(^{P/A,\ i,\ n})$
Equal-payment series capital-recovery	Given P Find A	$A = P\left[\dfrac{i(1 + i)^n}{(1 + i)^n - 1}\right]$	$A = P(^{A/P,\ i,\ n})$

Interest tables based on the designations summarized in Table 2 make it unnecessary to remember the name of each formula. Such tables are given in Appendix E for a range of values for i and n. Each column in the table is headed by a designation that identifies the function of the tabular value as established by the money flow diagram. Each table is based on $1 for a specific value of i, a range of n, and all six factors.

1.7 Geometric-Gradient-Series Formula

In some situations, annual money flows increase or decrease over time by a constant percentage. If g is used to designate the percentage change in the magnitude of the money flows from one period to the next, the magnitude of the tth flow is related to flow F_1 as

$$F_t = F_1(1 + g)^{t-1} \quad t = 1, 2, \ldots, n$$

When g is positive, the series will increase as illustrated in Figure 2. When g is negative, the series will decrease.

To derive an expression for the present amount, P, the relationship between F_1 and F_t can be used, together with the single-payment present-amount factor, as

$$P = F_1\left[\frac{(1 + g)^0}{(1 + i)^1}\right] + F_1\left[\frac{(1 + g)^1}{(1 + i)^2}\right] + F_1\left[\frac{(1 + g)^2}{(1 + i)^3}\right] + \cdots + F_1\left[\frac{(1 + g)^{n-1}}{(1 + i)^n}\right]$$

Multiply each term by $(1 + g)/(1 + g)$ and simplify

$$P = \frac{F_1}{1 + g}\left[\frac{(1 + g)^1}{(1 + i)^1} + \frac{(1 + g)^2}{(1 + i)^2} + \frac{(1 + g)^3}{(1 + i)^3} + \cdots + \frac{(1 + g)^n}{(1 + i)^n}\right]$$

Let

$$\frac{1}{1 + g'} = \frac{1 + g}{1 + i}$$

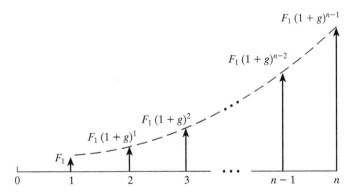

Figure 2 A geometric-gradient series with $g > 0$.

where g' is the *growth-free rate*, a rate that incorporates both g and i and can be used in place of i in the formulas and tables. Substituting for each item gives

$$P = \frac{F_1}{1 + g} \left[\frac{1}{(1 + g')^1} + \frac{1}{(1 + g')^2} + \frac{1}{(1 + g')^3} + \cdots + \frac{1}{(1 + g')^n} \right]$$

The terms within the brackets constitute another way of calculating the equal-payment series present-amount factor for n years. Therefore, substituting the original equal-payment series present-amount factor,

$$P = \frac{F_1}{1 + g} \left[\frac{(1 + g')^n - 1}{g'(1 + g')^n} \right]$$

or

$$P = F_1 \left[\frac{\overset{P/A, g', n}{()}}{1 + g} \right] \tag{7}$$

The factor within brackets is the *geometric-gradient series factor*. Its use requires finding g' by solving for g' in

$$\frac{1}{1 + g'} = \frac{1 + g}{1 + i}$$

$$g' = \frac{1 + i}{1 + g} - 1 \tag{8}$$

2 DETERMINING ECONOMIC EQUIVALENCE

If two or more situations are to be compared, their characteristics must be placed on an equivalent basis. Two things are said to be equivalent when they have the same effect. For instance, the torques produced by applying forces of 100 and 200 pound, 2 and 1 feet, respectively, from the fulcrum of the lever are equivalent because each produces a torque of 200 feet-pound.

Two monetary amounts are equivalent when they have the same value in exchange. Three factors are involved in the equivalence of sums of money: (1) the amount of the sums, (2) the time of occurrence of the sums, and (3) the interest rate. In this connection, the general equivalence function applies. It is directly applicable to alternatives described by money flows over time according to the convention established in Figure 1. Simple equivalence calculations, geometric-gradient-series calculations, and equivalence function diagrams are presented in this section.

2.1 Interest Formula Equivalence Calculations

Equivalence calculations based on simple applications of the interest formulas derived in Section 1 are illustrated in this section. More comprehensive applications are presented later. In each case, the determination of economic equivalence is in accordance with the money flow convention in Figure 1 and based on $1.

At an interest rate of 10% with $n = 8$ years, a P of $1 is equivalent to an F of $2.144. This may be stated as

$$F = \$1(\overset{F/P,\,10,\,8}{2.1440}) = \$2.1440$$

A practical application of this statement of equivalence is that $1 spent today must result in a revenue receipt or avoid a cost of $2.144, 8 years hence, if the interest rate is 10%.

A reciprocal situation is where $i = 12\%$, $n = 10$ years, and $F = \$1$. The P that is equivalent to $1 is

$$P = \$1(\overset{P/F,\,12,\,10}{0.3220}) = \$0.3220$$

A practical application of this statement of equivalence is that no more than $0.322 can be spent today to secure a revenue receipt or to avoid a cost of $1, 10 years hence, if the interest rate is 12%.

If $i = 8\%$ and $n = 20$ years, the F that is equivalent to an A of $1 is

$$F = \$1(\overset{F/A,\,8,\,20}{45.7620}) = \$45.7620$$

A practical application of this equivalence statement is that $1 spent each year for 20 years must result in a revenue receipt or it must avoid a cost of $45.762, 20 years hence, if the interest rate is 8%.

A reciprocal situation is where $i = 12\%$, $n = 6$ years, and $F = \$1$. The A that is equivalent to $1 is

$$A = \$1(\overset{A/F,\,12,\,6}{0.1232}) = \$0.1232$$

A practical application of this equivalence statement is that $0.1232 must be received each year for 6 years to be equivalent to the receipt of $1, 6 years hence.

If $i = 9\%$, $n = 10$ years, and A is $1, the P that is equivalent to $1 is

$$P = \$1(\overset{P/A,\,9,\,10}{6.4177}) = \$6.4177$$

A practical application of this statement of equivalence is that an investment of $6.4177 today must yield an annual benefit of $1 each year for 10 years if the interest rate is 9%.

A reciprocal situation is where $i = 14\%$, $n = 7$ years, and $P = \$1$. The A that is equivalent to $1 is

$$A = \$1(\overset{A/P,\,14,\,7}{0.1987}) = \$0.1987$$

A practical application of this statement of equivalence is that $1 can be spent today to capture an annual saving of $0.1987 per year over 7 years if the interest rate is 14%.

2.2 Geometric-Gradient-Series Equivalence Calculations

The geometric-gradient series was shown for an increasing series in Figure 2 and derived in Equations 7 and 8, but it may also be used for decreasing gradient situations. In this case, g will be negative, giving a positive value for g'.

As an example, suppose that a revenue flow of $1 during the first year is expected to decline by 10% per year over the next 8 years. The present equivalent of the anticipated revenue at an interest rate of 17% may be found as follows:

$$g' = \frac{1 + 0.17}{1 - 0.10} - 1 = 0.30 \quad \text{or} \quad 30\%$$

$$P = \$1 \frac{\overset{P/A,\,30,\,8}{(2.9247)}}{1 - 0.10} = \$3.25$$

In this example, the value for the P/A factor was available from the 30% interest table.

2.3 Equivalence Function Diagrams

A useful technique in the determination of equivalence is to plot present value as a function of the interest rate. For example, what value of i will make a P of $1,500 equivalent to an F of $5,000 if $n = 9$ years? Symbolically,

$$\$5,000 = \$1,500(\overset{F/P,\,i,\,9}{\quad})$$

The solution is illustrated graphically in Figure 3, showing that i is slightly less than 13%.

Present value may also be plotted as a function of n as a useful technique for determining equivalence. For example, what value of n will make a P of $4,000 equivalent to an F of $8,000 if $i = 8\%$? Symbolically,

$$\$8,000 = \$4,000(\overset{F/P,\,8,\,n}{\quad})$$

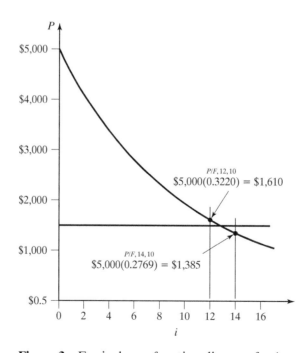

Figure 3 Equivalence-function diagram for i.

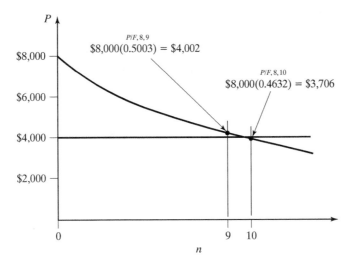

Figure 4 Equivalence-function diagram for n.

The solution is illustrated graphically in Figure 4, showing that n is between 9 and 10 years.

Equivalence functions can be plotted for A or F as a function of i or n if desired. In these cases, equivalence would be found between the two quantities A or F as a function of i or n.

3 EVALUATING A SINGLE ALTERNATIVE

In most cases, a decision is reached by selecting one option from among two or more available alternatives. Sometimes, however, the decision is limited to acceptance or rejection of a single alternative. In such a case, the decision will be based on the relative merit of the alternative and other opportunities believed to exist, even though none of the latter has been crystallized into definite proposals.

When only one specified possibility exists, it should be evaluated within a framework that will permit its desirability to be compared with other opportunities that may exist but are unspecified. The most common bases for comparison are the present-equivalent evaluation, the annual-equivalent evaluation, the future-equivalent evaluation, the rate-of-return evaluation, and the service-life evaluation.

The following example illustrates the several bases of evaluation of a single alternative under certainty about future outcomes. Assume that a waste heat recirculation system is contemplated for a small building. The anticipated costs and savings of this project are given in Table 3. Consider an interest rate of 12% as the cost of money and take January 1, 2005, to be the present.

For convenience, costs and savings that may occur during the year are assumed to occur at the end of that year or the start of the next year, considered to be the same point. There is some small error in the practice of considering money flows to be year-end amounts. This error is insignificant, however, in comparison with the usual errors in estimates, except under high interest rates. The costs and savings can be represented by the money flow diagram shown in Figure 5.

TABLE 3 Disbursements and Savings for a Single Alternative

Item	Date	Disbursements ($)	Savings ($)
Initial cost	1-1-2005	28,000	—
Saving, first year	1-1-2006	—	9,500
Saving, second year	1-1-2007	—	9,500
Overhaul cost	1-1-2007	2,500	—
Saving, third year	1-1-2008	—	9,500
Saving, fourth year	1-1-2009	—	9,500
Salvage value	1-1-2009	—	8,000

3.1 Present Equivalent Evaluation

The present equivalent evaluation is based on finding a present equivalent amount, PE, that represents the difference between present equivalent savings and present equivalent costs for an alternative at a given interest rate. Thus, the present equivalent amount at an interest rate i over n years is obtained by applying Equation 2.

$$PE(i) = F_0(\overset{P/F, i, 0}{}) + F_1(\overset{P/F, i, 1}{}) + F_2(\overset{P/F, i, 2}{}) + \cdots + F_n(\overset{P/F, i, n}{}) = \sum_{t=0}^{n} F_t(\overset{P/F, i, t}{}) \quad (9)$$

But because $(\overset{P/F, i, t}{}) = (1 + i)^{-t}$,

$$PE(i) = \sum_{t=0}^{n} F_t(1 + i)^{-t} \quad (10)$$

For the waste heat recirculation system example, the present equivalent amount at 12% using Equation 9 is

$$PE(12) = -\$28,000(\overset{P/F, 12, 0}{1.0000}) + \$9,500(\overset{P/F, 12, 1}{0.8929}) + \$7,000(\overset{P/F, 12, 2}{0.7972})$$
$$+ \$9,500(\overset{P/F, 12, 3}{0.7118}) + \$17,500(\overset{P/F, 12, 4}{0.6355})$$
$$= \$3,946$$

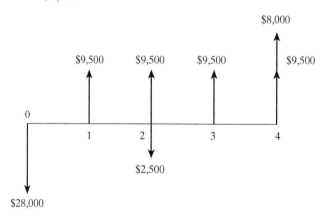

Figure 5 Money flow diagram from Table 3.

Or, by using Equation 10

$$PE(12) = -\$28{,}000(1.12)^0 + \$9{,}500(1.12)^{-1} + \$7{,}000(1.12)^{-2}$$
$$+ \$9{,}500(1.12)^{-3} + \$17{,}500(1.12)^{-4}$$
$$= \$3{,}946$$

Because the present equivalent amount is greater than zero, this is a desirable undertaking at 12%.

3.2 Annual Equivalent Evaluation

The annual equivalent evaluation is similar to the present equivalent evaluation except that the difference between savings and costs is now expressed as an annual equivalent amount, AE, at a given interest rate as follows:

$$AE(i) = PE(i)(\overset{A/P,\,i,\,n}{}) \tag{11}$$

But because

$$PE(i) = \sum_{t=0}^{n} F_t(1 + i)^{-t}$$

and

$$(\overset{A/P,\,i,\,n}{}) = \left[\frac{i(1 + i)^n}{(1 + i)^n - 1}\right]$$

$$AE(i) = \left[\sum_{t=0}^{n} F_t(1 + i)^{-t}\right]\left[\frac{i(1 + i)^n}{(1 + i)^n - 1}\right] \tag{12}$$

For the waste-heat recirculation example, the annual equivalent evaluation using Equation 11 gives

$$AE(12) = \$3{,}946(\overset{A/P,\,12,\,4}{0.3292}) = \$1{,}299$$

Or, by using Equation 12,

$$AE(12) = [-\$28{,}000(1.12)^0 + \$9{,}500(1.12)^{-1}$$
$$+ \$7{,}000(1.12)^{-2} + \$9{,}500(1.12)^{-3}$$
$$+ \$17{,}500(1.12)^{-4}]\left[\frac{0.12(1.12)^4}{(1.12)^4 - 1}\right]$$
$$= \$1{,}299$$

These results mean that if \$28,000 is invested on January 1, 2010, a 12% return will be received plus an equivalent of \$1,299 on January 1, 2012, 2013, 2014, and 2015.

3.3 Future Equivalent Evaluation

The future equivalent basis for comparison is based on finding an equivalent amount, FE, that represents the difference between the future equivalent savings and future equivalent costs for an alternative at a given interest rate. Thus, the future equivalent amount at an interest rate i over n years is obtained by applying Equation 1

$$FE(i) = F_0(\overset{F/P,\,i,\,n}{}) + F_1(\overset{F/P,\,i,\,n-1}{}) + \cdots + F_{n-1}(\overset{F/P,\,i,\,1}{}) + F_n(\overset{F/P,\,i,\,0}{})$$

$$= \sum_{t=0}^{n} F_t(\overset{F/P,\,i,\,n-t}{}) \tag{13}$$

But because

$$(\overset{F/P,\,i,\,n-t}{}) = (1 + i)^{n-t}$$

$$FE(i) = \sum_{t=0}^{n} F_t(1 + i)^{n-t} \tag{14}$$

or,

$$FE(i) = PE(i)(\overset{F/P,\,i,\,n}{}) \tag{15}$$

For the waste-heat recirculation example, the future equivalent evaluation using Equation 13 gives

$$FE(12) = -\$28,000(\overset{F/P,\,12,\,4}{1.5740}) + \$9,500(\overset{F/P,\,12,\,3}{1.4050}) + \$7,000(\overset{F/P,\,12,\,2}{1.2542})$$

$$+ \$9,500(\overset{F/P,\,12,\,1}{1.1208}) + \$17,500(\overset{F/P,\,12,\,0}{1.0000})$$

$$= \$6,211$$

Or using Equation 15,

$$FE(12) = \$3,946(\overset{F/P,\,12,\,4}{1.5740}) = \$6,211$$

Because the future equivalent amount is greater than zero, this is a desirable venture at 12%.

3.4 Rate-of-Return Evaluation

The rate-of-return evaluation is probably the best method for comparing a specific proposal with other opportunities believed to exist but not delineated. The rate of return is a universal measure of economic success because the returns from different classes of opportunities are usually well established and generally known. This permits comparison of an alternative against accepted norms. This characteristic makes the rate-of-return comparison well adapted to the situation where the choice is to accept or reject a single alternative.

Rate of return is a widely accepted index of profitability. It is defined as the interest rate that causes the equivalent receipts of a money flow to be equal to the equivalent disbursements of that money flow. The interest rate that reduces the $PE(i)$, $AE(i)$, or $FE(i)$ of a series of receipts and disbursements to zero is another way of defining the rate of return. Mathematically, the rate of return for an investment proposal is the interest rate $i*$ that satisfies the equation

$$0 = PE(i*) = \sum_{t=0}^{n} F_t(1 + i*)^{-t} \tag{16}$$

where the proposal has a life of n years.

As an example of the rate-of-return evaluation, the energy saving system can be evaluated on the basis of the rate of return that would be secured from the invested funds. In effect, a rate of interest will be specified that makes the receipts and disbursements equivalent. This can be done either by equating present equivalent, annual equivalent, or future equivalent amounts.

Equating the present equivalent amount of receipts and disbursements at $i = 15\%$ gives

$$[\$9,500(\overset{P/A,\,15,\,4}{2.8850}) + \$8,000(\overset{P/F,\,15,\,4}{0.5718})] = [\$28,000 + \$2,500(\overset{P/F,\,15,\,2}{0.7562})]$$

$$[\$27,408 + \$4,574] = [\$28,000 + \$1,890]$$

$$\$31,982 = \$29,800$$

At $i = 20\%$

$$[\$9,500(\overset{P/A,\,20,\,4}{2.5887}) + \$8,000(\overset{P/F,\,20,\,4}{0.4823})] = [\$28,000 + \$2,500(\overset{P/F,\,20,\,2}{0.6945})]$$

$$[\$24,592 + \$3,858] = [\$28,000 + \$1,736]$$

$$\$28,450 = \$29,736$$

Interpolating,

$$i = 15\% + (5)\left[\frac{[31,982 - 29,800] - 0}{[31,982 - 29,800] - [28,450 - 29,736]}\right]$$

$$i = 18.15\%$$

From these results, a present equivalent graph can be developed as shown in Figure 6. From the graph it is evident that the rate of return on the venture is just over 18%. This means that the investment of \$28,000 in the system should yield a 18.15% rate of return over the 4-year period.

3.5 Payout Evaluation

Often a proposed system can be evaluated in terms of how long it will take the system to pay for itself from benefits, revenues, or savings. Systems that tend to pay for themselves quickly are desirable because there is less uncertainty with estimates of short duration.

The payout period is the amount of time required for the difference in the present value of receipts (savings) to equal the present value of the disbursements (costs); the annual equivalent receipts and disbursements may also be equated. For the present equivalent approach

$$0 \le \sum_{t=0}^{n*} F_t(1 + i)^{-t} \tag{17}$$

The smallest value of $n*$ that satisfied the preceding expression is the payout duration.

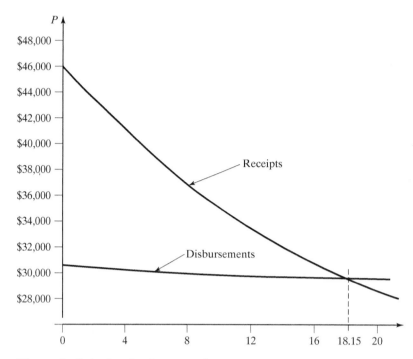

Figure 6 Solution for the rate of return.

For a duration of 3 years the equation for present equivalent of savings and disbursements is

$$[\$9,500(\overset{P/A,\,12,\,3}{2.4018}) + \$8,000(\overset{P/F,\,12,\,3}{0.7118})] = [\$28,000 + \$2,500(\overset{P/F,\,12,\,2}{0.7972})]$$

$$[\$22,817 + \$5,694] = [\$28,000 + \$1,993]$$

$$\$28,511 \neq \$29,993$$

For a duration of 4 years, the equation for present equivalent of savings and disbursements is

$$[\$9,500(\overset{P/A,\,12,\,4}{3.0374}) + \$8,000(\overset{P/F,\,12,\,4}{0.6355})] = [\$28,000 + \$2,500(\overset{P/F,\,12,\,2}{0.7972})]$$

$$[\$28,855 + \$5,084] = [\$28,000 + \$1,993]$$

$$\$33,939 \neq \$29,993$$

Interpolating gives

$$n = 3 + \frac{[28,511 - 29,993] - 0}{[28,511 - 29,993] - [33,939 - 29,993]}$$

$$n = 3.3 \text{ years}$$

From these results, a present equivalent graph can be developed as shown in Figure 7. From this graph it is evident that the payout period on the energy-saving system is just over 3 years.

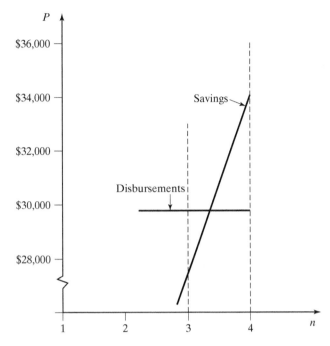

Figure 7   Solution for the payout period.

3.6 Annual Equivalent Cost of an Asset

A useful application of the annual equivalent evaluation approach pertains to the cost of an asset. The cost of any asset is made up of two components, the cost of depreciation and the cost of interest on the undepreciated balance. It can be shown that the annual equivalent cost of an asset is independent of the depreciation function used to represent the value of the asset over time.

As an example of the cost of depreciation plus the cost of interest on the undepreciated balance, consider the following example based on straight line depreciation. An asset has a first cost of $5,000, a salvage value of $1,000, and a service life of 5 years, and the interest rate is 10%. The annual costs are shown in Figure 8.

The present equivalent cost of depreciation plus interest on the undepreciated balance from Equation 9 is

$$PE(10) = \$1,300(\overset{P/F,\,10,\,1}{0.9091}) + \$1,220(\overset{P/F,\,10,\,2}{0.8265}) + \$1,140(\overset{P/F,\,10,\,3}{0.7513})$$
$$+ \$1,060(\overset{P/F,\,10,\,4}{0.6830}) + \$980(\overset{P/F,\,10,\,5}{0.6209})$$
$$= \$4,379$$

and the annual equivalent cost of the asset from Equation 11 is

$$AE(10) = \$4,379(\overset{A/P,\,10,\,5}{0.2638}) = \$1,155$$

Regardless of the depreciation function that describes the reduction in value of a physical asset over time, the annual equivalent cost of an asset may be expressed as the annual equivalent first cost minus the annual equivalent salvage value. This annual equivalent cost is

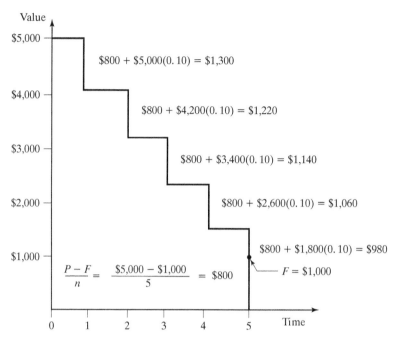

Figure 8 Value–time function.

the amount an asset must earn each year if the invested capital is to be recovered along with a return on the investment. The annual equivalent cost is derived as follows:

$$A = P(\overset{A/P,\,i,\,n}{\qquad}) - F(\overset{A/F,\,i,\,n}{\qquad})$$
$$(\overset{A/F,\,i,\,n}{\qquad}) = (\overset{A/P,\,i,\,n}{\qquad}) - i$$

But because

$$A = P(\overset{A/P,\,i,\,n}{\qquad}) - F[(\overset{A/P,\,i,\,n}{\qquad}) - i]$$
$$A = (P - F)(\overset{A/P,\,i,\,n}{\qquad}) + F(i) \tag{18}$$

This means that the annual equivalent cost of any asset may be found from knowledge of its first cost, P, its anticipated service life, n, its estimated salvage value, F, and the interest rate, i. For example, an asset with a first cost of $5,000, a service life of 5 years, and a salvage value of $1,000 will lead to an annual equivalent cost of

$$(\$5,000 - \$1,000)(\overset{A/P,\,10,\,5}{0.2638}) + \$1,000(0.10) = \$1,155$$

if the interest rate is 10%.

The annual equivalent cost formula is useful in a "should cost" study. For example, suppose that a contractor is to use equipment, tooling, and so on, for a certain defense contract that will cost $3,250,000 and will depreciate to $750,000 at the end of the 4-year production contract. If money costs the contractor 10%, the annual equivalent cost of the contractor's equipment should be

$$(\$3{,}250{,}000 - \$750{,}000)(\overset{A/P,\,10,\,4}{0.3155}) + \$750{,}000(0.10) = \$863{,}750$$

This annual equivalent amount is only one component of the total contract "should cost." Other components are the annual cost of labor, materials, and overhead.

4 EVALUATING MULTIPLE ALTERNATIVES

When several mutually exclusive alternatives provide service of equal value, it is desirable to compare them directly with each other. Where service of unequal value is provided by multiple alternatives, each alternative must be evaluated as a single alternative and is accepted or rejected on the basis of one or more of the comparisons suggested in Section 3. In many cases, however, the available alternatives do provide outputs that are identical or equal in value. Under this condition, the objective is to select the alternative that provides the desired service at the least cost.

Assume that a defense contractor is considering the acquisition of new test centers. Semiautomatic test equipment will cost $110,000 for each test center and can be expected to last 6 years, with a salvage value of $10,000. Operating costs will be $28,000 per year. Fully automatic equipment will cost $160,000, should last 6 years, and will have a salvage value of $14,000. The operating costs will be $12,000 per year. Here the services provided by the equipment are assumed to be identical. With a desired interest rate of 14%, the alternative that meets the criterion of least cost should be selected.

This is an example of acquisition by purchase. The general equivalence function will be used as a basis for comparing the alternatives.

4.1 Present Equivalent Evaluation

Under this method, the two alternatives may be compared on the basis of equivalent cost at a time taken to be the present. The present equivalent cost of the semiautomatic test equipment at 14% is

$$PE(14) = \$110{,}000 + \$28{,}000(\overset{P/A,\,14,\,6}{3.8887}) - \$10{,}000(\overset{P/F,\,14,\,6}{0.4556}) = \$214{,}328$$

The present equivalent cost of the fully automatic equipment is

$$PE(14) = \$160{,}000 + \$12{,}000(\overset{P/A,\,14,\,6}{3.8887}) - \$14{,}000(\overset{P/F,\,14,\,6}{0.4556}) = \$200{,}286$$

This comparison shows the present equivalent cost of the fully automatic test equipment to be less than the present equivalent cost of the semiautomatic test equipment by $14,042 ($214,328 − $200,286).

4.2 Annual Equivalent Evaluation

The annual equivalent costs are taken as an equal-cost series over the life of the assets. The annual equivalent cost of the semiautomatic test equipment is

$$AE(14) = \$110{,}000(\overset{A/P,\,14,\,6}{0.2572}) + \$28{,}000 - \$10{,}000(\overset{A/F,\,14,\,6}{0.1175}) = \$55{,}118$$

and the annual equivalent cost of the fully automatic test equipment is

$$AE(14) = \$160{,}000(\overset{A/P,\,14,\,6}{0.2572}) + \$12{,}000 - \$14{,}000(\overset{A/F,\,14,\,6}{0.1175}) = \$51{,}507$$

The annual equivalent difference of \$55,118 minus \$51,507 or \$3,611 is the annual equivalent cost superiority of the fully automatic test equipment. As a verification, the annual equivalent amount of the present equivalent difference is

$$\$14{,}042(\overset{A/P,\,14,\,6}{0.2572}) = \$3{,}611$$

4.3 Rate-of-Return Evaluation

The previous cost comparisons indicated that the fully automatic test equipment was more desirable at an interest rate of 14%. At some higher interest rate, the two alternatives will be identical in equivalent cost, and beyond that interest rate, the semiautomatic test equipment will be less expensive because of its lower initial cost.

The interest rate at which the costs of the two alternatives are identical can be determined by setting the present equivalent amounts for the alternatives equal to each other and solving for the interest rate, i. Thus,

$$\$110{,}000 + \$28{,}000(\overset{P/A,\,i,\,6}{\quad}) - \$10{,}000(\overset{P/F,\,i,\,6}{\quad})$$
$$= \$160{,}000 + \$12{,}000(\overset{P/A,\,i,\,6}{\quad}) - \$14{,}000(\overset{P/F,\,i,\,6}{\quad})$$
$$\$16{,}000(\overset{P/A,\,i,\,6}{\quad}) + \$4{,}000(\overset{P/F,\,i,\,6}{\quad}) = \$50{,}000$$

For $i = 20\%$

$$\$16{,}000(\overset{P/A,\,20,\,6}{3.3255}) + \$4{,}000(\overset{P/F,\,20,\,6}{0.3349}) = \$54{,}548$$

For $i = 25\%$

$$\$16{,}000(\overset{P/A,\,25,\,6}{2.9514}) + \$4{,}000(\overset{P/F,\,25,\,6}{0.2622}) = \$48{,}271$$

Then, by interpolation,

$$i = 20 + (5)\frac{\$[54{,}548 - \$50{,}000]}{\$[54{,}548 - \$48{,}271]}$$
$$i = 23.6\%$$

When funds will earn less than 23.6% elsewhere, the fully automatic test equipment will be most desirable. If funds can earn more than 23.6%, the semiautomatic equipment would be preferred.

4.4 Payout Evaluation

The service life of 6 years for each of the two test equipment alternatives is only the result of estimates and may be in error. If the services are needed for shorter or longer periods and if the assets are capable of providing the service for a longer period of time, the advantage may pass from one alternative to the other.

Just as there is an interest rate at which the two alternatives may be equally desirable, there may be a service life at which the equivalent costs may be identical. This service life may be obtained by setting the alternatives equal to each other and solving for the life, n. Thus, for an interest rate of 14%,

$$\$110{,}000 + \$28{,}000(\overset{P/A,\,14,\,n}{}) - \$10{,}000(\overset{P/F,\,14,\,n}{})$$

$$= \$160{,}000 + \$12{,}000(\overset{P/A,\,14,\,n}{}) - \$14{,}000(\overset{P/F,\,14,\,n}{})$$

$$\$16{,}000(\overset{P/A,\,14,\,n}{}) + \$4{,}000(\overset{P/F,\,14,\,n}{}) = \$50{,}000$$

For $n = 4$ years,

$$\$16{,}000(\overset{P/A,\,14,\,4}{2.9137}) + \$4{,}000(\overset{P/F,\,14,\,4}{0.5921}) = \$48{,}988$$

For $n = 5$ years,

$$\$16{,}000(\overset{P/A,\,14,\,5}{3.4331}) + \$4{,}000(\overset{P/F,\,14,\,5}{0.5194}) = \$57{,}007$$

Then, by interpolation,

$$n = 4 + (1)\frac{[\$50{,}000 - \$48{,}988]}{[\$57{,}007 - \$48{,}988]}$$

$$n = 4.12 \text{ years}$$

Thus, if the desired service were to be used for less than 4.12 years, the semiautomatic equipment would be the most desirable. The advantage passes to the automatic equipment if test center use is to exceed 4.12 years.

5 EVALUATION INVOLVING MULTIPLE CRITERIA

Multiple criteria must be considered when both economic and noneconomic factors are involved in making a decision.

As an example of the evaluation of multiple alternatives under certainty involving multiple criteria, suppose that there are two competing design concepts pertaining to the acquisition of an automated severe weather warning system. The costs (in millions of dollars) are estimated over the life cycle as shown in Figure 9. In addition to these projected costs, there are other criteria to be considered in the evaluation, as given below.

For dependability, the MTBF is required to be greater than 2,500 hours. In this example, the MTBF under Design A is predicted to be 3,000 hours, whereas for Design B it is predicted to be only 1,500 hours. The higher maintenance and operating cost for Design B ($3,000,000) reflects the more frequent maintenance situation when compared with Design A, which is only $1,000,000. All other effectiveness and performance factors are assumed to be identical for Designs A and B.

Another criterion is that of budget constraints during acquisition and utilization, specified as follows:

1. The capital requirement in any year of the acquisition phase should not exceed $7,300,000.
2. The maintenance and operating cost in any year during utilization should not exceed $2,000,000.

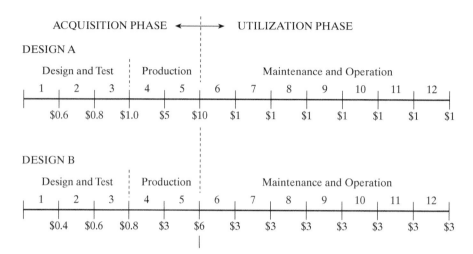

Figure 9 Two design alternatives for system acquisition.

Although other choices are available, the design concepts may be compared by finding their equivalent costs at the end of the acquisition phase (the point when the system goes into operation). For Design A, this equivalent life-cycle cost at 10% is

$$\$10 + \$5(\overset{F/P,\,10,\,1}{1.100}) + \$1(\overset{F/P,\,10,\,2}{1.210}) + \$0.8(\overset{F/P,\,10,\,3}{1.331}) + \$0.6(\overset{F/P,\,10,\,4}{1.464})$$
$$+ \$1(\overset{P/A,\,10,\,7}{4.8684}) = \$23.52 \text{ million}$$

And for Design B, the equivalent life-cycle cost is

$$\$6 + \$3(\overset{F/P,\,10,\,1}{1.100}) + \$0.8(\overset{F/P,\,10,\,2}{1.210}) + \$0.6(\overset{F/P,\,10,\,3}{1.331}) + \$0.4(\overset{F/P,\,10,\,4}{1.464})$$
$$+ \$3(\overset{P/A,\,10,\,7}{4.8684}) = \$26.26 \text{ million}$$

If life-cycle cost were the only consideration, Design A would be preferred to Design B because \$23.52 is less than \$26.26. However, multiple criteria exist and will be considered with the aid of a decision evaluation display as shown in Figure 10.

Although Design A exhibits the lowest equivalent life-cycle cost, it violates the acquisition budget constraint. Design B violates both the MTBF requirement and the annual budget constraint for maintenance and operations. Therefore, the decision maker must either accept Design A with the acquisition budget violation, or authorize continuation of the design effort to search for an acceptable alternative. The choice has to do with preference based on subjective evaluation.

6 MULTIPLE ALTERNATIVES WITH MULTIPLE FUTURES

An engineering firm is engaged in the design of a unique machine tool. Because the machine tool is technologically advanced, the firm is considering setting up a new production facility separate from its other operations.

Because the firm's own finances are not adequate to set up this new facility, there is a need to borrow capital from financial sources. The firm has identified three sources, namely, A, B, and C, which would lend money for this venture.

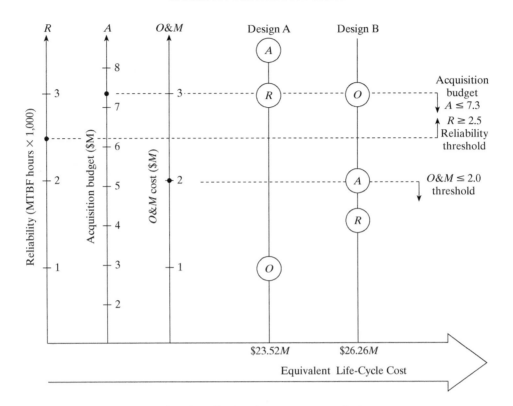

Figure 10 Decision evaluation display for two system designs.

Each of these financial sources will lend money based on the anticipated sales of the product and will use a varying interest rate based on the anticipated demand for the product. If the product demand is anticipated to be low over the next few years, the financial backers will charge a high interest rate. This may be because the backers think that the firm may not borrow money in the future if the product fails because of low demand.

Conversely, if the financial sources feel that the demand will be high, they will be willing to lend the money at a lower interest rate. This is because of the possibility of future business with the firm owing to its present product being popular.

The engineering firm has fairly accurate estimates of the receipts and disbursements associated with each demand level and the financial sources under consideration. The firm has knowledge about the initial outlay and anticipated receipts over 3 years for each demand level as is summarized in Table 4. The interest rates quoted by each financial source are summarized in Table 5.

The engineering firm wishes to select a financing source in the face of three demand futures, not knowing which demand future will occur. Accordingly, a decision evaluation matrix is developed as a first step. Using the present equivalent approach, nine evaluation values are derived and displayed in Table 6, one for each financing source for each demand level. For example, if financing comes from Source A, and the demand level is low, the present equivalent payoff is

$$PE(15) = -\$500 + \$400(\overset{P/A, 15, 3}{2.2832}) = \$413$$

TABLE 4 Anticipated Disbursements and Receipts for
Machine Tool Production (Thousands of Dollars)

Demand Level	Initial Outlay ($)	Annual Receipts ($)
Low	500	400
Medium	1,300	700
High	2,000	900

TABLE 5 Anticipated Interest Rates for Production Financing

Financing Source	Demand Level	Interest Rate
A	Low	15
	Medium	13
	High	7
B	Low	14
	Medium	12
	High	8
C	Low	15
	Medium	11
	High	6

If the firm applied probabilities to each of the demand futures, the financing decision becomes one under risk. If the probability of a low demand is 0.3, the probability of a medium demand is 0.2, and the probability of a high demand is 0.5, the expected present equivalent amount for each financing source is calculated as follows:

$$\text{Source A: } \$413(0.3) + \$353(0.2) + \$362(0.5) = \$376$$
$$\text{Source B: } \$429(0.3) + \$382(0.2) + \$320(0.5) = \$365$$
$$\text{Source C: } \$343(0.3) + \$411(0.2) + \$406(0.5) = \$388$$

Accordingly, financing Source C is selected by this decision rule.

In the event the firm has no basis for assigning probabilities to demand futures, the financing decision must be made under uncertainty. Under the Laplace criterion, each future is assumed to be equally likely and the present equivalent payoffs are calculated as follows:

$$\text{Source A: } \$(413 + 353 + 362) \div 3 = \$376$$
$$\text{Source B: } \$(429 + 382 + 320) \div 3 = \$377$$
$$\text{Source C: } \$(343 + 411 + 406) \div 3 = \$386$$

TABLE 6 Present Equivalent Payoff for Three Financing Sources

| Source | Demand Level | | |
	Low ($)	Medium ($)	High ($)
A	413	353	362
B	429	382	320
C	343	411	406

Financing Source C would be selected by this decision rule.

By the maximin rule, the firm calculates the minimum present equivalent amount that could occur for each financing source and then selects the source that provides a maximum. The minimums for each source are

Source A: $353

Source B: $320

Source C: $343

Financing Source A would be selected by this decision rule as the one that will maximize the minimum present equivalent amount.

By the maximax rule, the firm calculates the maximum present equivalent amount that could occur for each financing source and then selects the source that provides a maximum. The maximums for each source are

Source A: $413

Source B: $429

Source C: $411

Financing Source B would be selected by this decision rule as the one that will maximize the present equivalent amount.

A summary of the sources selected by each decision rule is given in Table 7.

7 BREAK-EVEN ECONOMIC EVALUATIONS

Break-even analysis may be graphical or mathematical in nature. This economic evaluation technique is useful in relating fixed and variable costs to the number of hours of operation, the number of units produced, or other measures of operational activity. In each case, the break-even point is of primary interest in that it identifies the range of the decision variable within which the most desirable economic outcome may occur.

When the cost of two or more alternatives is a function of the same variable, it is usually useful to find the value of the variable for which the alternatives incur equal cost. Several examples will be presented in this section.

7.1 Make-or-Buy Evaluation

Often a manufacturing firm has the choice of making or buying a certain component for use in the product being produced. When this is the case, the firm faces a make-or-buy decision.

TABLE 7 Summary of Financing Source Selections

Decision Rule	Financing Source		
	A	B	C
Expected value			X
Laplace			X
Maximin	X		
Maximax		X	

Suppose, for example, that a firm finds that it can buy the electric power supply for the system it produces for $8 per unit from a vendor. Alternatively, suppose that it can manufacture an equivalent unit for a variable cost of $4 per unit. It is estimated that the additional fixed cost in the plant would be $12,000 per year if the unit is manufactured. The number of units per year for which the cost of the two alternatives breaks even would help in making the decision.

First, the total annual cost is formulated as a function of the number of units for the make alternative. It is

$$TC_M = \$12{,}000 + \$4N$$

And the total annual cost for the buy alternative is

$$TC_B = \$8N$$

Break-even occurs when $TC_M = TC_B$ or

$$\$12{,}000 + \$4N = \$8N$$

$$\$4N = \$12{,}000$$

$$N = 3{,}000 \text{ units}$$

These cost functions and the break-even point are shown in Figure 11. For requirements in excess of 3,000 units per year, the make alternative would be more economical. If the rate of use is likely to be less than 3,000 units per year, the buy alternative should be chosen.

If the production requirement changes during the course of the production program, the break-even choice in Figure 11 can be used to guide the decision of whether to make or to buy the power unit. Small deviations below and above 3,000 units per year make little

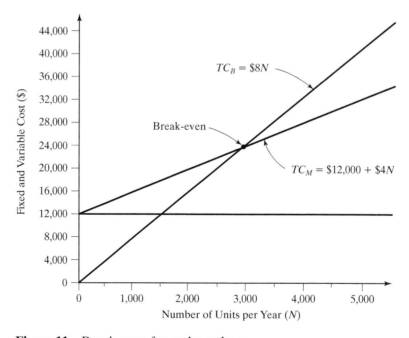

Figure 11 Break-even for make or buy.

difference. However, the difference can be significant when the production requirement is well above or below 3,000 units per year.

7.2 Lease-or-Buy Evaluation

As another example of break-even analysis, consider the decision to lease or buy a piece of equipment. Assume that a small electronic computer is needed for data processing in an engineering office. Suppose that the computer can be leased for $50 per day, which includes the cost of maintenance. Alternatively, the computer can be purchased for $25,000.

The computer is estimated to have a useful life of 15 years with a salvage value of $4,000 at the end of that time. It is estimated that annual maintenance cost will be $2,800. If the interest rate is 9% and it costs $50 per day to operate the computer, how many days of use per year are required for the two alternatives to break even?

First, the annual cost if the computer is leased is

$$TC_L = (\$50 + \$50)N$$
$$= \$100N$$

And the annual equivalent total cost if the computer is bought is

$$TC_B = (\$25,000 - \$4,000)(\overset{A/P,9,15}{0.1241}) + \$4,000(0.09) + \$2,800 + \$50N$$
$$= \$2,606 + \$360 + \$2,800 + \$50N$$
$$= \$5,766 + \$50N$$

The first three terms represent the fixed cost and the last term is the variable cost. Break-even occurs when $TC_L = TC_B$ or

$$\$100N = \$5,766 + \$50N$$
$$\$50N = \$5,766$$
$$N = 115 \text{ days}$$

A graphical representation of this decision situation is shown in Figure 12. For all levels of use exceeding 115 days per year, it would be more economical to purchase the computer. If the level of use is anticipated to be below 115 days per year, the computer should be leased.

7.3 Equipment Selection Evaluation

Suppose that a fully automatic controller for a load center can be fabricated for $140,000 and that it will have an estimated salvage value of $20,000 at the end of 4 years. Maintenance cost will be $12,000 per year, and the cost of operation will be $85 per hour.

As an alternative, a semiautomatic controller can be fabricated for $55,000. This device will have no salvage value at the end of a 4-year service life. The cost of operation and maintenance is estimated to be $140 per hour.

With an interest rate of 10%, the annual equivalent total cost for the fully automatic installation as a function of the number of hours of use per year is

$$TC_A = (\$140,000 - \$20,000)(\overset{A/P,10,4}{0.3155}) + \$20,000(0.10) + \$12,000 + \$85N$$
$$= \$51,800 + \$85N$$

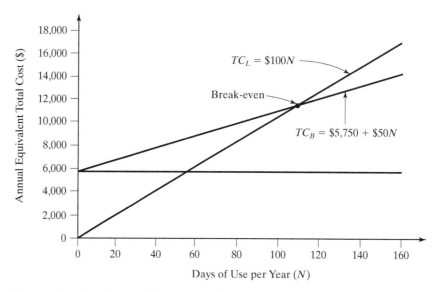

Figure 12 Break-even for lease or buy.

And the annual equivalent total cost for the semiautomatic installation as a function of the number of hours of use per year is

$$TC_S = \$55{,}000(\overset{A/P,10,4}{0.3155}) + \$140N$$
$$= \$17{,}400 + \$140N$$

Break-even occurs when $TC_A = TC_S$, or

$$\$51{,}800 + \$85N = \$17{,}400 + \$140N$$

$$\$55N = \$34{,}400$$

$$N = 625 \text{ hours}$$

Figure 13 shows the two cost functions and the break-even point. For rates of use exceeding 625 hours per year, the automatic controller would be more economical. However, if it is anticipated that the rate of use will be less than 625 hours per year, the semiautomatic controller should be fabricated.

7.4 Profitability Evaluation

There are two aspects of production operations. One consists of assembling the production system of facilities, material, and people, and the other consists of the sale of the goods produced. The economic success of an enterprise depends upon its ability to carry on these activities to the end that there may be a net difference between receipts and the cost of production.

Linear break-even analysis is useful in evaluating the effect on profit of proposals for new operations not yet implemented and for which no data exist. Consider a proposed activity consisting of the manufacturing and marketing of a certain plastic item for which the sale price per unit is estimated to be $30,000. The machine required in the operation will cost $140,000 and will have an estimated life of 9 years. It is estimated that the cost of production, including

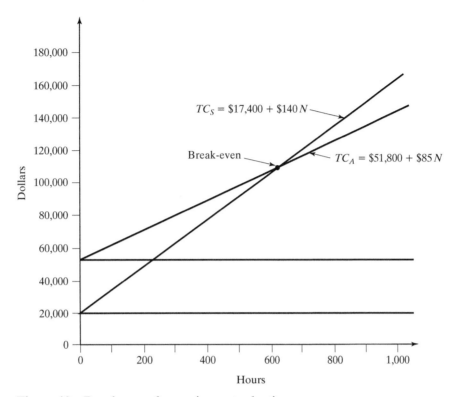

Figure 13 Break-even for equipment selection.

power, labor, space, and selling expense, will be $21 per unit sold. Material will cost $9.50 per unit. An interest rate of 12% is considered necessary to justify the required investment. The estimated costs associated with this activity are as given in Table 8.

The difficulty of making a clear-cut separation between fixed and variable costs becomes apparent when attention is focused on the item for repair and maintenance. In practice, it is very difficult to distinguish between repairs that are a result of deterioration that takes place with the passage of time and those that result from wear and tear. However, in theory, the separation can be made as shown in this example and is in accord with fact, with the exception, perhaps, of the assumption that repairs from wear and tear

TABLE 8 Fixed and Variable Manufacturing Costs

	Fixed Costs ($)	Variable Costs ($)
Capital recovery with return $\overset{A/P, 12, 8}{\$140,000(0.1877)}$	26,278	
Insurance and taxes	2,000	
Repairs and maintenance	1,722	0.50/unit
Materials		9.50/unit
Labor, electricity, space		11.00/unit
Total	30,000	21.00/unit

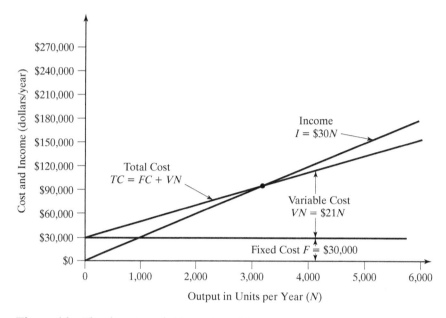

Figure 14 Fixed cost, variable cost, and income per year.

will be in direct proportion to the number of units manufactured. To be in accord with actualities, depreciation also undoubtedly should have been separated so that a part would appear as variable cost.

In this example, $F = \$30,000$, $TC = \$30,000 + \$21N$, and $I = \$30N$. If F, TC, and I are plotted as N varies from 0 to 6,000 units, the results will be as shown in Figure 14.

The cost of producing the plastic item will vary with the number of units made per year. The production cost per unit is given by $F/N + V$. If production cost per unit, variable cost per unit, and income per unit are plotted as N varies from 0 to 6,000, the results will be as shown in Figure 15. It will be noted that the fixed cost per unit may be infinite. Thus, in determining unit costs, fixed cost has little meaning unless the number of units to which it applies is known.

Income for most enterprises is directly proportional to the number of units sold. However, the income per unit may easily be exceeded by the sum of the fixed and the variable cost per unit for low production volumes. This is shown by comparing the total cost per unit curve and the income per unit curve in Figure 15.

8 BREAK-EVEN EVALUATION UNDER RISK

In Section 7, the make-or-buy analysis was aided by Figure 11 showing the break-even point on which a make-or-buy decision is based. The analysis was based on certainty about the demand for the electric power supply. Ordinarily, the decision maker does not know with certainty what the demand level will be.

Consider demand ranging from 1,500 units to 4,500 units with probabilities as given in Table 9. Entries for the cost to make and the cost to buy are calculated from the equations in Section 7 as was exhibited in Figure 11.

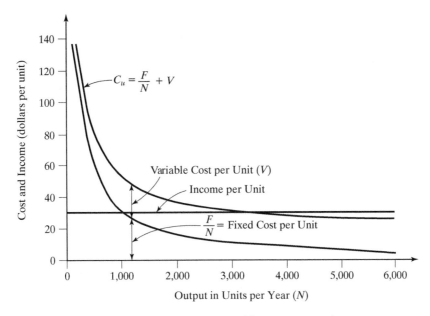

Figure 15 Fixed cost, variable cost, and income per unit.

The expected demand may be calculated from the probabilities associated with each demand level as given in Table 9.

$$1{,}500(0.05) + 2{,}000(0.10) + 2{,}500(0.15) + 3{,}000(0.20) + 3{,}500(0.25)$$
$$+ 4{,}000(0.15) + 4{,}500(0.10) = 3{,}175$$

From the equations for this decision situation, the expected total cost to make is

$$TC_M = \$12{,}000 + \$4(3{,}175)$$
$$= \$24{,}700$$

And the expected total cost to buy is

$$TC_B = \$8(3{,}175)$$
$$= \$25{,}400$$

TABLE 9 Demand Range for Make or Buy

	Demand in Units						
	1,500	2,000	2,500	3,000	3,500	4,000	4,500
Probability	0.05	0.10	0.15	0.20	0.25	0.15	0.10
Cost to make ($)	18,000	20,000	22,000	24,000	26,000	28,000	30,000
Cost to buy ($)	12,000	16,000	20,000	24,000	28,000	32,000	36,000

Accordingly, the power supply should be made. Alternatively, this decision might be based on the most probable future criterion. From Table 9 it is evident that a demand level of 3,500 is most probable. Therefore, the cost to make of $26,000 is compared with cost to buy of $28,000, and the decision would be to make the item. This decision is the same as that for the expected demand approach.

9 SUMMARY AND EXTENSIONS

In an increasingly competitive global marketplace, economic feasibility has become a consideration of primary importance. The engineering approach to the alteration of physical factors to create utility in the economic domain has advanced and broadened to the extent that market success often depends upon the ability to integrate economic and physical factors. In general, the essential prerequisite of successful engineering application is economic feasibility.

Most design and operational alternatives may be described in terms of their receipts and disbursements over time. When this is the case, reducing these monetary amounts to a common economic base is essential to valid decision making. In this chapter, the money flow model is used for all derivations, examples, and applications. It is applied using a family of interest formulas and a unique interest factor designation scheme. Money flow modeling is also the basis for considering the time-value of money over the system life cycle. Accordingly, life-cycle engineering economics is the central theme of this chapter.

This chapter developed only the basic concepts and methods of analysis in preparation for evaluating the economic feasibility of systems, products, and services in terms of worth and cost. The underlying body of knowledge goes beyond this chapter and comprises the well-known field of engineering economics. Further study to extend the compact presentation of this chapter may be enabled by reference to G. J. Thuesen and W. J. Fabrycky, *Engineering Economy*, 9th ed., Prentice-Hall, 2001. Also, 12 learning modules entitled Economy 2.0 are available from *http://www.a2i2.com* under Products. It is suggested that these computer-based modules be downloaded and exercised during the study of this chapter.

Both the American Society for Engineering Education and the Institute of Industrial Engineers have engineering economy divisions (ASEE EED and IIE EED) as part of their society organizations. These divisions work closely together to advance the art and practice of engineering economics. Specific information about the objectives and activities of these divisions may be obtained from two websites. For ASEE EED, visit *http://www.asee.org* and for IIE EED visit *http://www.iienet2.org*.

A quarterly journal entitled *The Engineering Economist* is published jointly by ASEE EED and IIE EED. This journal is the best source of technical literature on the subject of capital investment economics. Individuals interested in extending their understanding may view copies in most technical libraries, or may request subscription information from the Institute of Industrial Engineers, 3577 Parkway Lane, Suite 200, Norcross, GA 30092.

QUESTIONS AND PROBLEMS

1. How much money must be invested to accumulate $10,000 in 8 years at 6% compounded annually?

2. What amount will be accumulated by each of the following investments?

   a. $8,000 at 7.2% compounded annually over 10 years.

   b. $52,500 at 8% compounded annually over 5 years.

3. What is the present equivalent amount of a year-end series of receipts of $6,000 over 5 years at 8% compounded annually?

4. What is the present equivalent of a year-end series of receipts starting with a first-year base of $1,000 and increasing by 8% per year to year 20 with an interest rate of 8%?

5. What is the present equivalent of a year-end series of receipts starting with a first-year base of $1 million and decreasing by 25% per year to year 4 with an interest rate of 6%?

6. What interest rate compounded annually is involved if $4,000 results in $10,000 in 6 years?

7. How many years will it take for $4,000 to grow to $7,000 at an interest rate of 10% compounded annually?

8. What interest rate is necessary for a sum of money to double itself in 8 years? What is the approximate product of i and n (i as an integer) that establishes the doubling period? How accurate is this product of i and n for estimating the doubling period?

9. An asset was purchased for $52,000 with the anticipation that it would serve for 12 years and be worth $6000 as scrap. After 5 years of operation, the asset was sold for $18,000. The interest rate is 14%.

   a. What was the anticipated annual equivalent cost of the asset?

   b. What was the actual annual equivalent cost of the asset?

10. An epoxy mixer purchased for $33,000 has an estimated salvage value of $5,000 and an expected life of 3 years. An average of 200 pounds per month will be processed by the mixer.

    a. Calculate the annual equivalent cost of the mixer with an interest rate of 8%.

    b. Calculate the annual equivalent cost per pound mixed with an interest rate of 12%.

11. The following table shows the receipts and disbursements for a given venture. Determine the desirability of the venture for a 14% interest rate, based on the present equivalent comparison and the annual equivalent comparison.

End of the Year	Receipts ($)	Disbursements ($)
0	0	20,000
1	6,000	0
2	5,000	4,000
3	5,000	0
4	12,000	1,000

12. A microcomputer-based controller can be installed for $30,000 and will have a $3,000 salvage value after 10 years and is expected to decrease energy consumption cost by $4,000 per year.

    a. What rate of return is expected if the controller is used for 10 years?

    b. For what life will the controller give a return of 15%?

13. Transco plans on purchasing a bus for $75,000 that will have a capacity of 40 passengers. As an alternative, a larger bus can be purchased for $95,000 that will have a capacity of 50 passengers. The salvage value of either bus is estimated to be $8,000 after a 10-year life. If an annual net profit of $400 can be realized per passenger, which alternative should be recommended using a management-suggested interest rate of 15%? Using the actual cost of money at 7.5%?

14. An office building and its equipment are insured to $7,100,000. The present annual insurance premium is $0.85 per $100 of coverage. A sprinkler system with an estimated life of

20 years and no salvage value can be installed for $180,000. Annual maintenance and operating cost is estimated to be $3,600. The premium will be reduced to $0.40 per $100 coverage if the sprinkler system is installed.

a. Find the rate of return if the sprinkler system is installed.

b. With interest at 12%, find the payout period for the sprinkler system.

15. The design of a system is to be pursued from one of two available alternatives. Each alternative has a life-cycle cost associated with an expected future. The costs for the corresponding futures are given in the following table (in millions of dollars). If the probabilities of occurrence of the futures are 30%, 50%, and 20%, respectively, which alternative is most desirable from an expected cost viewpoint. Use an interest rate of 10%?

Design 1	Years											
Future	1	2	3	4	5	6	7	8	9	10	11	12
Optimistic	0.4	0.6	5.0	7.0	0.8	0.8	0.8	0.8	0.8	0.8	0.8	0.8
Expected	0.6	0.8	1.0	5.0	10.0	1.0	1.0	1.0	1.0	1.0	1.0	1.0
Pessimistic	0.8	0.9	1.0	7.0	10.0	1.2	1.2	1.2	1.2	1.2	1.2	1.2

Design 2	Years											
Future	1	2	3	4	5	6	7	8	9	10	11	12
Optimistic	0.4	0.4	0.4	1.0	3.0	2.5	2.5	2.5	2.5	2.5	2.5	2.5
Expected	0.6	0.8	1.0	3.0	6.0	3.0	3.0	3.0	3.0	3.0	3.0	3.0
Pessimistic	0.6	0.8	1.0	5.0	6.0	3.1	3.1	3.1	3.1	3.1	3.1	3.1

16. Prepare a decision evaluation matrix for the design alternatives in Problem 15 and then choose the alternative that is best under the following decision rules: Laplace, maximax, maximin, and Hurwicz with $\alpha = 0.6$. Assume that the choice is under uncertainty.

17. A campus laboratory can be climate conditioned by piping chilled water from a central refrigeration plant. Two competing proposals are being considered for the piping system, as outlined in the table. On the basis of a 10-year life, find the number of hours of operation per year for which the cost of the two systems will be equal if the interest rate is 9%.

	6-inch System	8-inch System
Motor size in horsepower	6	3
Installed cost of pump and pipe	$32,000	$44,000
Installed cost of motor	4,500	3,000
Salvage value of the system	5,000	6,000
Energy cost per hour of operation	$3.20	$2.00

18. Replacement fence posts for a cattle ranch are currently purchased for $4.20 each. It is estimated that equivalent posts can be cut from timber on the ranch for a variable cost of $1.50 each, constituting the value of the timber plus labor cost. Annual fixed cost for required equipment is estimated to be $1,200. If 1,000 posts will be required each year, what will be the annual saving if posts are cut?

19. An equipment operator can buy a maintenance component from a supplier for $960 per unit delivered. Alternatively, operator can rebuild the component for a variable cost of

$460 per unit. It is estimated that the additional fixed cost would be $80,000 per year if the component is rebuilt. Find the number of units per year for which the cost of the two alternatives will break even.

20. A marketing company can lease a fleet of automobiles for its sales personnel for $35 per day plus $0.18 per mile for each vehicle. As an alternative, the company can pay each salesperson $0.45 per mile to use his or her own automobile. If these are the only costs to the company, how many miles per day must a salesperson drive for the two alternatives to break even?

21. An electronics manufacturer is considering the purchase of one of two types of laser trimming devices. The sales forecast indicated that at least 8,000 units will be sold per year. Device A will increase the annual fixed cost of the plant by $20,000 and will reduce variable cost by $5.60 per unit. Device B will increase the annual fixed cost by $5,000 and will reduce variable cost by $3.60 per unit. If variable costs are now $20 per unit produced, which device should be purchased?

22. Machine A costs $20,000, has zero salvage value at any time, and has an associated labor cost of $1.15 for each piece produced on it. Machine B costs $36,000, has zero salvage value at any time, and has an associated labor cost of $0.90. Neither machine can be used except to produce the product described. If the interest rate is 10% and the annual rate of production is 20,000 units, how many years will it take for the cost of the two machines to break even?

23. An electronics manufacturer is considering two methods for producing a circuit board. The board can be hand-wired at an estimated cost of $9.80 per unit and an annual fixed equipment cost of $10,000. A printed equivalent can be produced using equipment costing $180,000 with a service life of 8 years and salvage value of $12,000. It is estimated that the labor cost will be $3.20 per unit and that the processing equipment will cost $4,000 per year to maintain. If the interest rate is 8%, how many circuit boards must be produced each year for the two methods to break even?

24. It is estimated that the annual sales of an energy saving device will be 20,000 the first year and increase by 10,000 per year until 50,000 units are sold during the fourth year. Proposal A is to purchase manufacturing equipment costing $120,000 with an estimated salvage value of $15,000 at the end of 4 years. Proposal B is to purchase equipment costing $280,000 with an estimated salvage value of $32,000 at the end of 4 years. The variable manufacturing cost per unit under proposal A is estimated to be $8.00, but is estimated to be only $2.60 under proposal B. If the interest rate is 9%, which proposal should be accepted for a 4-year production horizon?

25. The fixed operating cost of a machine center (capital recovery, interest, maintenance, space charges, supervision, insurance, and taxes) is F dollars per year. The variable cost of operating the center (power, supplies, and other items, but excluding direct labor) is V dollars per hour of operation. If N is the number of hours the center is operated per year, TC the annual total cost of operating the center, TC_h the hourly cost of operating the center, t the time in hours to process 1 unit of product, and M the center cost of processing 1 unit, write expressions for (a) TC, (b) TC_h, and (c) M.

26. In Problem 25, $F = $60,000 per year, $t = 0.2$ hour, $V = $50 per hour, and N varies from 1,000 to 10,000 in increments of 1,000.

   a. Plot values of M as a function of N.
   b. Write an expression for the total cost of direct labor and machine cost per unit, TC_u, using the symbols in Problem 25 and letting W equal the hourly cost of direct labor.

27. A certain firm has the capacity to produce 800,000 units per year. At present, it is operating at 75% of capacity. The income per unit is $0.10 regardless of the output. Annual fixed

costs are $28,000, and the variable cost is $0.06 per unit. Find the annual profit or loss at this capacity and the capacity for which the firm will break even.

28. An arc welding machine that is used for a certain joining process costs $90,000. The machine has a life of 5 years and a salvage value of $10,000. Maintenance, taxes, insurance, and other fixed costs amount to $5,000 per year. The cost of power and supplies is $28.00 per hour of operation and the total operator cost (direct and indirect) is $65.00 per hour. If the cycle time per unit of product is 60 minutes and the interest rate is 8%, calculate the cost per unit if (a) 200, (b) 600, and (c) 1,800 units of output are needed per year.

29. A certain processing center has the capacity to assemble 650,000 units per year. At present, it is operating at 65% of capacity. The annual income is $416,000. Annual fixed cost is $192,000 and the variable cost is $0.38 per unit assembled.

   a. What is the annual profit or loss attributable to the center?

   b. At what volume of output does the center break even?

   c. What will be the profit or loss at 70%, 80%, and 90% of capacity on the basis of constant income per unit and constant variable cost per unit?

30. Chemco operates two plants, A and B, which produce the same product. The capacity of plant A is 60,000 gallons while that of B is 80,000 gallons. The annual fixed cost of plant A is $2,600,000 per year and the variable cost is $32 per gallon. The corresponding values for plant B are $2,800,000 and $39 per gallon. At present, plants A and B are being operated at 35% and 40% of capacity, respectively.

   a. What would be the total cost of production of plants A and B?

   b. What are the total cost and the average unit cost of the total output of both plants?

   c. What would be the total cost to the company and cost per gallon if all production were transferred to plant A?

   d. What would be the total cost to the company and cost per gallon if all production were transferred to plant B?

Optimization in Design and Operations

The design of complex systems that appropriately incorporate optimization in the design process is an important challenge facing the systems engineer. A parallel but lesser challenge arises from the task of optimizing the operation of systems already in being. In the former case, optimum values of design variables are sought for each instance of the design-dependent parameter set. In the latter, optimum values for policy variables are desired. The modeling approaches are essentially the same. This distinction is important only as a means for contrasting system design and systems analysis.

In this chapter, the general approach to the formulation and manipulation of mathematical models is presented. The specific mathematical methods used vary in degree of complexity and depend on the system under study. Examples are used to illustrate design situations for static and dynamic systems. Both unconstrained and constrained examples are presented. Classical optimization techniques are illustrated first, followed by some nonclassical approaches, including linear programming methods.

These fundamental areas are developed in detail in this chapter by the integration of economic modeling and optimization, with hypothetical examples from design and operations. Consideration and study of this material will impart a general understanding of optimization as a means to an end, not as an end in itself. Specifically, the reader should become proficient in several areas:

- Utilizing calculus-based classical optimization concepts and methods for both univariate and multivariate applications in design and operations;
- Understanding that the utilization of optimization as presented in this chapter is traceable back to and inspired by the field of operations research;
- Applying decision theory and economic evaluation methods jointly to expand and realize the broader benefit of equivalence;
- Structuring unconstrained situations requiring optimization as a prelude to addressing life-cycle-based money flow representations;

- Considering investment cost optimization as a step on the way to determining life-cycle complete evaluations of design and operations;
- Applying optimization in the face of constraints as an extension of the unconstrained situations and examples;
- Evaluating situations involving optimization and multiple criteria, employing the decision evaluation display; and
- Understanding and applying constrained optimization by both the graphical and the simplex methods of linear programming.

The final section of this chapter provides a summary of the broad domain of optimization with selected example applications in design and operations. It also provides some recommended sources of additional material useful in extending insight and improving understanding.

1 CLASSICAL OPTIMIZATION THEORY

Optimization is the process of seeking the best. In systems engineering and analysis, this process is applied to each alternative in accordance with decision evaluation theory. In so doing, the general system optimization function of Equation A.1 as well as the decision evaluation function given by Equation A.2 are exercised and illustrated by hypothetical examples. Before presenting specific derivations and examples of optimization in design and operations, this section provides a review of calculus-based methods for optimizing functions with one or more decision variables.

1.1 The Slope of a Function

The slope of a function $y = f(x)$ is defined as the rate of change of the dependent variable, y, divided by the rate of change of the independent variable, x. If a positive change in x results in a positive change in y, the slope is positive. Conversely, a positive change in x resulting in a negative change in y indicates a negative slope.

If $y = f(x)$ defines a straight line, the difference between any two points x_1 and x_2 represents the change in x, and the difference between any two points y_1 and y_2 represents the change in y. Thus, the rate of change of y with respect to x is $(y_1 - y_2)/(x_1 - x_2)$, the slope of the straight line. This slope is constant for all points on the straight line; $\Delta y / \Delta x = $ constant.

For a nonlinear function, the rate of change of y with respect to changes in x is not constant but changes with changes in x. The slope must be evaluated at each point on the curve. This can be done by assuming an arbitrary point on the function p, for which the x and y values are x_0 and $f(x_0)$, as shown in Figure 1. The rate of change of y with respect to x at point p is equal to the slope of a line tangent to the function at that point. It is observed that the rate of change of y with respect to x differs from that at p at other points on the curve.

1.2 Differential Calculus Fundamentals

Differential calculus is a mathematical tool for finding successively better and better approximations for the slope of the tangent line shown in Figure 1. Consider another point on $f(x)$ designated q, situated at an x distance from point p equal to Δx and situated at a y distance from point p equal to Δy. Then the slope of a line segment through points p and q would have a slope $\frac{\Delta y}{\Delta x}$. But this is the average slope of $f(x)$

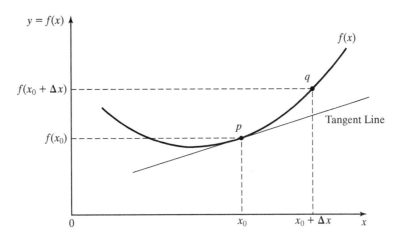

Figure 1   Slope of a function.

between the designated points. In classical optimization, it is the instantaneous rate of change at a given point that is sought.

The instantaneous rate of change in $f(x)$ at $x = x_0$ can be found by letting $\Delta x:\ 0$. Referring to Figure 1, this can be stated as

$$\underset{\Delta x:\ 0}{\text{limit}} \frac{f(x_0 + \Delta x) - f(x_0)}{\Delta x} \tag{1}$$

It is noted that as Δx becomes smaller and smaller, the line segment passing through points p and q approaches the tangent line to point p. Thus, the slope of the function at x_0 is given by Equation 1 when Δx is infinitesimally small.

Equation 1 is an expression for slope of general applicability. It is called a *derivative*, and its application is a process known as *differentiation*. For the function $y = f(x)$, the symbolism dy/dx or $f'(x)$ is most often used to denote the derivative.

As an example of the process of differentiation utilizing Equation 1, consider the function $y = f(x) = 8x - x^2$. Substituting into Equation 1 gives

$$\underset{\Delta x:\ 0}{\text{limit}} = \frac{[8(x + \Delta x) - (x + \Delta x)^2] - [8x - x^2]}{\Delta x}$$

$$= \frac{8x + 8\Delta x - x^2 - 2x\Delta x - \Delta x^2 - 8x + x^2}{\Delta x}$$

$$= \frac{8\Delta x - 2x\Delta x - \Delta x^2}{\Delta x}$$

$$= 8 - 2x - \Delta x$$

As $\Delta x:\ 0$ the derivative, $dy/dx = 8 - 2x$, the slope at any point on the function. For example, at $x = 4$, $dy/dx = 0$, indicating that the slope of the function is zero at $x = 4$.

An alternative example of the process of differentiation, not tied to Equation 1, might be useful to consider. Begin with a function of some form such as $y = f(x) = 3x^2 + x + 2$.

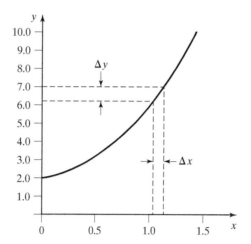

Figure 2 Function $3x^2 + x + 2$.

Let x increase by an amount Δx. Then y will increase by an amount Δy as shown in Figure 2, giving

$$y + \Delta y = 3(x + \Delta x)^2 + (x + \Delta x) + 2$$
$$= 3[x^2 + 2x(\Delta x) + (\Delta x)^2] + (x + \Delta x) + 2$$
$$\Delta y = 3[x^2 + 2x(\Delta x) + (\Delta x)^2] + (x + \Delta x) + 2 - y$$

But $y = 3x^2 + x + 2$, giving

$$\Delta y = 3x^2 + 6x(\Delta x) + 3(\Delta x)^2 + x + \Delta x + 2 - 3x^2 - x - 2$$
$$= 6x(\Delta x) + 3(\Delta x)^2 + \Delta x$$

The average rate of change of Δy with respect to Δx is

$$\frac{\Delta y}{\Delta x} = \frac{6x(\Delta x)}{\Delta x} + \frac{3(\Delta x)^2}{\Delta x} + \frac{\Delta x}{\Delta x}$$
$$= 6x + 3(\Delta x) + 1$$

But because

$$\lim_{\Delta x:\ 0} = \frac{dy}{dx} = f'(x)$$

the instantaneous rate of change is $6x + 1$.

Instead of proceeding as earlier for each case encountered, certain rules of differentiation have been derived for a range of common functional forms. Some of these are as follows:

1. Derivative of a constant: If $f(x) = k$, a constant, then $dy/dx = 0$.
2. Derivative of a variable: If $f(x) = x$, a variable, then $dy/dx = 1$.

3. Derivative of a straight line: If $f(x) = ax + b$, a straight line, then $dy/dx = a$.

4. Derivative of a variable raised to a power: If $f(x) = x^n$, a variable raised to a power, then $dy/dx = nx^{n-1}$.

5. Derivative of a constant times a function: If $f(x) = k[g(x)]$, a constant times a function, then

$$\frac{dy}{dx} = k\left[\frac{dg(x)}{dx}\right]$$

6. Derivative of the sum or difference of two functions: If $f(x) = g(x) \pm h(x)$, a sum or difference of two functions, then

$$\frac{dy}{dx} = \frac{dg(x)}{dx} \pm \frac{dh(x)}{dx}$$

7. Derivative of the product of two functions: If $f(x) = g(x)h(x)$, a product of two functions, then

$$\frac{dy}{dx} = g(x)\frac{dh(x)}{dx} + h(x)\frac{dg(x)}{dx}$$

8. Derivative of an exponential function: If $f(x) = e^x$, an exponent of e, then

$$\frac{dy}{dx} = e^x$$

9. Derivative of the natural logarithm: If $f(x) = \ln x$, natural log of x, then

$$\frac{dy}{dx} = \frac{1}{x}$$

The methods presented previously may be used to find *higher order* derivatives. When a function is differentiated, another function results that may also be differentiated if desired. For example, the derivative of $8x - x^2$ was $8 - 2x$. This was the slope of the function. If the slope is differentiated, the higher order derivative is found to be -2. This is the rate of change in dy/dx and is designated d^2y/dx^2 of $f''(x)$. The process of differentiation can be extended to even higher orders as long as there remains some function to differentiate.

1.3 Partial Differentiation

Partial differentiation is the process of finding the rate of change whenever the dependent variable is a function of more than one independent variable. In the process of partial differentiation, all variables except the dependent variable, y, and the chosen independent variable are treated as constants and differentiated as such. The symbolism $\partial y/\partial x_1$ is used to represent the partial derivative of y with respect to x_1.

As an example, consider the following function of two independent variables, x_1 and x_2:

$$4x_1^2 + 6x_1x_2 - 3x_2$$

The partial derivative of y with respect to x_1 is

$$\frac{\partial y}{\partial x_1} = 8x_1 + 6x_2 - 0$$

255

The partial derivative of y with respect to x_2 is

$$\frac{\partial y}{\partial x_2} = 0 + 6x_1 - 3$$

Partial differentiation permits a view of the slope, along one dimension at a time, of an n-dimensional function. By combining this information from such a function, the slope at any point on the function can be found. Of particular interest is the point on the function where the slope along all dimensions is zero. For the previous function, this point is found by setting the partial derivatives equal to zero and solving for x_1 and x_2 as follows:

$$8x_1 + 6x_2 = 0$$

$$6x_1 - 3 = 0$$

from which

$$x_1 = 1/2 \text{ and } x_2 = -2/3$$

The meaning of this point is explained in the following section.

1.4 Unconstrained Optimization

Unconstrained optimization simply means that no constraints are placed on the function under consideration. Under this assumption, a *necessary* condition for x^* to be an optimum point on $f(x)$ is that the first derivative be equal to zero. This is stated as

$$\left. \frac{df(x)}{dx} \right|_{x=x^*} = 0 \tag{2}$$

The first derivative being zero at the stationary point x^* is not *sufficient*, because it is possible for the derivative to be zero at a point of inflection on $f(x)$. It is sufficient for x^* to be an optimum point if the second derivative at x^* is positive or negative; x^* being a minimum if the second derivative is positive and a maximum if the second derivative is negative.

If the second derivative is also zero, a higher order derivative is sought until the first nonzero one is found at the nth derivative as

$$\left. \frac{d^n f(x)}{dx^n} \right|_{x=x^*} = 0 \tag{3}$$

If n is odd, then x^* is a point of inflection. If n is even and if

$$\left. \frac{d^n f(x)}{dx^n} \right|_{x=x^*} < 0 \tag{4}$$

then x^* is a local maximum. But if

$$\left. \frac{d^n f(x)}{dx^n} \right|_{x=x^*} > 0 \tag{5}$$

then x^* is a local minimum.

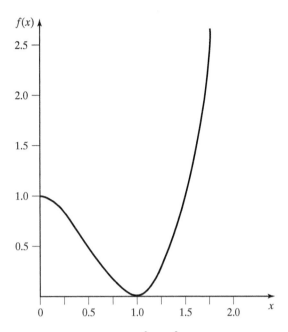

Figure 3   Function $2x^3 - 3x^2 + 1$.

As an example, suppose that the minimum value of x is sought for the function $f(x) = 2x^3 - 3x^2 + 1$ exhibited in Figure 3. The necessary condition for a minimum to exist is

$$\frac{df(x)}{dx} = 6x^2 - 6x = 0$$

Both $x = 0$ and $x = 1$ satisfy the previous condition. Accordingly, the second derivative is taken as

$$\frac{d^2f(x)}{dx^2} = 12x - 6$$

and evaluated at $x = 0$ and $x = 1$. At $x = 0$, the second derivative is -6, indicating that this is a maximum point. At $x = 1$, the second derivative is 6, indicating that this is the minimum sought. These points are shown in Figure 3.

For functions of more than one independent variable, the vector $x^* = [x_1, x_2, \ldots, x_n]$ will have a stationary point on a function $f(x)$ if all elements of the vector of first partial derivatives evaluated at x^* are zero. Symbolically,

$$F(x) = \left[\frac{\partial F(x)}{\partial x_1}, \frac{\partial F(x)}{\partial x_2}, \ldots, \frac{\partial F(x)}{\partial x_n} \right] = 0 \qquad (6)$$

Consider a function in two variables, x_1 and x_2, given as

$$f(x_1, x_2) = x_1^2 - 8x_2 + 2x_2^2 - 6x_1 + 30$$

The necessary condition requires that

$$\frac{\partial f(x_1, x_2)}{\partial x_1} = 2x_1 - 6 = 0$$

and

$$\frac{\partial f(x_1, x_2)}{\partial x_2} = 4x_2 - 8 = 0$$

The optimal value x_1^* is found to be 3 and the optimal value x_2^* is found to be 2. These may be substituted into $f(x_1, x_2)$ to give the optimal value:

$$f(x_1, x_2) = (3)^2 - 8(2) + 2(2)^2 - 6(3) + 30 = 13$$

To determine whether this value for $f(x_1, x_2)$ is a maximum or a minimum, the second partials are required. These are

$$\frac{\partial(2x_1 - 6)}{\partial x_1} = 2 \quad \text{and} \quad \frac{\partial(4x_2 - 8)}{\partial x_2} = 4$$

Each is greater than zero, indicating that a minimum may have been found. To verify this, the following relationship must be satisfied:

$$\left(\frac{\partial^2 y}{\partial x_1^2}\right)\left(\frac{\partial^2 y}{\partial x_2^2}\right) - \left(\frac{\partial^2 y}{\partial x_1 \partial x_2}\right)^2 > 0 \tag{7}$$

where

$$\frac{\partial^2 y}{\partial x_1 \partial y_2} = \frac{\partial(\partial y/\partial x_1)}{\partial x_2}$$

Because

$$\frac{\partial y}{\partial x_1} = 2x_1 - 6$$

$$\frac{\partial^2 y}{\partial x_1 \partial x_2} = \frac{\partial(2x_1 - 6)}{\partial x_2} = 0$$

Substituting back into Equation 7 gives $(2)(4) - 0 = 8$, verifying that the value found for $f(x_1, x_2)$ is truly a minimum.

2 UNCONSTRAINED CLASSICAL OPTIMIZATION

There are several decision situations in design and operations to which optimization approaches must be applied before alternatives can truly be compared equivalently. Many economic decisions are characterized by two or more cost factors that are affected differently by common design or policy variables. Certain costs may vary directly with an increase in the value of a variable, whereas others may vary inversely. When the total cost of an alternative is a function of increasing and decreasing cost components, a value may exist for the common variable, or variables, which will result in a minimum cost for the alternative. This section presents selected examples from classical situations that illustrate the application of unconstrained optimization theory in design and operations.

2.1 Optimizing Investment or First Cost

Investment (or first cost) has a significant effect on the equivalent life-cycle cost of a system. Accordingly, it is important that investment cost be minimized before a life-cycle cost analysis is performed. In this section the classical situation of bridge design will be used to illustrate the appropriate approach, not as an end in itself, but to illustrate a general principle.[1]

Before engaging in detail design, it is important to optimally allocate the anticipated capital investment to the bridge superstructure and to its piers. This can be accomplished by recognizing that there exists an inverse relationship between the cost of superstructure and the number of piers. As the number of piers increases, the cost of superstructure decreases. Conversely, the cost of superstructure increases as the number of piers decreases. Pier cost for the bridge is directly related to the number specified. This is a classical design situation involving increasing and decreasing cost components, the sum of which will be a minimum for a certain number of piers. Figure 4 illustrates two bridge superstructure designs, with the span between piers indicated by S.

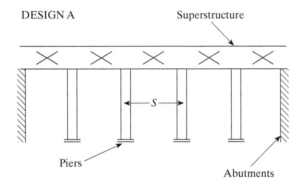

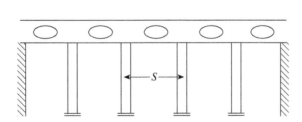

Figure 4 Two bridge superstructure design alternatives.

[1]This preliminary design example illustrates the selection of a preferred superstructure configuration requiring determination of the optimal pier spacing as an essential first step.

A general mathematical model may be derived for the evaluation of investment in superstructure and piers. The decision evaluation function of Equation A.1 applies as

$$E = f(X, Y)$$

where

E = evaluation measure (total first cost)
X = design variable of the span between piers
Y = system parameters of the bridge length, the superstructure weight, the erected cost of superstructure per pound, and the cost of piers per pier.

Equation A.2 is the design evaluation function applicable to the evaluation of alternative design configurations, where design-dependent parameters are key to selecting the best alternative. Let

L = bridge length (feet)
W = superstructure weight (pounds per foot)
S = span between piers (feet)
C_S = erected cost of superstructure (dollars per pound)
C_P = installed cost of piers (dollars per pier)

Assume that the weight of the superstructure is linear over a certain span range, $W = AS + B$, with the parameters A and B having been established by a sound empirical procedure. Accordingly, the superstructure cost (SC) will be

$$\text{SC} = (AS + B)(L)(C_S) \tag{8}$$

The total cost of piers, PC, will be

$$\text{PC} = \left(\frac{L}{S} + 1\right)(C_P) \tag{9}$$

where two abutments are included as though they were piers.

The total first cost of the bridge is expressed as

$$
\begin{aligned}
\text{TFC} &= \text{SC} + \text{PC} \\
&= (AS + B)(L)(C_S)\left(\frac{L}{S} + 1\right)(C_P) \\
&= ASLC_S + BLC_S + \frac{LC_P}{S} + C_P
\end{aligned}
\tag{10}
$$

To find the optimum span between piers, differentiate Equation 10 with respect to S and equate the result to zero as follows:

$$\frac{d(\text{TFC})}{dS} = ALC_S - \frac{LC_P}{S^2} = 0$$

$$S^* = \sqrt{\frac{C_P}{AC_S}} \tag{11}$$

The minimum total first cost for the bridge is found by substituting Equation 11 for S in Equation 10 to obtain

$$\text{TFC}^* = 2\sqrt{AC_P L^2 C_S} + BLC_S + C_P \tag{12}$$

Equation 11 can be used to find the optimal pier spacing for a given bridge design. However, to evaluate alternative bridge designs, Equation 12 would be used first, followed by a single application of Equation 11. This procedure is illustrated by examples below.

Single Design Alternative. Assume that a pedestrian bridge to serve a 1,000-foot scenic marshland crossing is to be fabricated from steel with a certain girder design configuration (see Figure 4). For this design alternative, the weight of the superstructure in pounds per foot is estimated to be linear (over a limited span range) and expressed as

$$W = 16S + 600$$

Also, assume that the superstructure is expected to cost \$0.65 per pound erected. Piers are anticipated to cost \$80,000 each in place, and this amount will also be used as the estimated cost for each abutment.

From Equation 11, the optimum span between piers is found to be

$$S^* = \sqrt{\frac{\$80,000}{16 \times \$0.65}} = 87.7 \text{ feet}$$

This result is a theoretical spacing and must be adjusted to obtain an integer number of piers. The required adjustment gives 12 piers (11 spans) for a total cost from Equation 10 of \$2,295,454. The span will be 90.9 feet, slightly greater than the theoretical minimum. A check can be made by considering 13 piers (12 spans) with each span being 83.3 feet. In this case, the total cost is higher (\$2,296,667), so the 12-pier design would be adopted.

Total cost as a function of the pier spacing (and the number of piers) is summarized in Table 1. Note that a minimum occurs when 12 piers are specified; actually, 10 piers and 2 abutments. The optimal pier spacing is 90.9 feet.

TABLE 1   TFC as a Function of the Span Between Piers

Span (feet)	Number of Piers	Pier Cost ($)	Superstructure Cost ($)	Total First Cost ($)
142.8	8	640,000	1,875,714	2,515,714
111.1	10	800,000	1,545,555	2,345,555
100.0	11	880,000	1,430,000	2,310,000
90.9	12	960,000	1,335,454	2,295,454
83.3	13	1,040,000	1,256,666	2,296,667
76.9	14	1,120,000	1,190,000	2,310,000
71.4	15	1,200,000	1,132,857	2,332,857
66.6	16	1,280,000	1,083,333	2,363,333
58.8	18	1,440,000	1,001,765	2,441,765

Multiple Design Alternatives. To introduce optimal design for multiple alternatives, assume that there is another superstructure configuration under consideration for the bridge design described in the preceding section (see Figure 4). The weight in pounds per foot for the alternative configuration is estimated from parametric methods to be

$$W = 22S + 0$$

Assume that all other factors are the same as for the previous design.

A choice between the design alternatives is made by finding the optimum total cost for each alternative utilizing Equation 12. This gives \$2,294,281 for Design A and \$2,219,159 for Design B. Thus, Design B would be chosen.

For Design B, the optimum span between piers is found from Equation 11 to be

$$S^* = \sqrt{\frac{\$80,000}{22 \times \$0.65}} = 74.8 \text{ feet}$$

and the lowest cost integer number of piers is found to be 14 (13 spans) with a span of 76.9 feet.

Table 2 summarizes this design decision by giving the total cost as a function of the number of piers for each alternative. The theoretical total cost function for each alternative is exhibited in Figure 5, along with an indication of the number of piers.

2.2 Optimizing Life-Cycle Cost

Investment cost was optimized in the derivations and example applications just presented. When the annual cost of maintenance and/or operations differs between design alternatives, it is important to formulate a life-cycle cost function including both first cost and these costs as a basis for choosing the best alternative.

Bridge Evaluation Including Maintenance. Equation 12 gave the total first cost of the bridge. It can be augmented by Equation 8.7 to give the present equivalent life-cycle cost (PELCC) as

$$\text{PELCC} = [2\sqrt{AC_PL^2C_S} + BLC_S + C_P] + M\left[\frac{\overset{P/A, g', n}{(\quad)}}{1 + g}\right] \qquad (13)$$

TABLE 2 TFC for Two Bridge Design Alternatives

Number of Piers	Pier Cost ($)	Superstructure Cost		Total First Cost	
		Design A ($)	Design B ($)	Design A ($)	Design B ($)
8	640,000	1,875,714	2,042,857	2,515,714	2,682,857
10	800,000	1,545,555	1,588,889	2,345,555	2,388,889
11	880,000	1,430,000	1,430,000	2,310,000	2,310,000
12	960,000	1,335,454	1,300,000	2,295,454	2,260,000
13	1,040,000	1,256,666	1,191,666	2,296,667	2,231,666
14	1,120,000	1,190,000	1,100,000	2,310,000	2,220,000
15	1,200,000	1,132,857	1,021,438	2,332,857	2,221,438
16	1,280,000	1,083,333	953,333	2,363,333	2,233,333
18	1,440,000	1,001,765	841,176	2,441,765	2,281,176

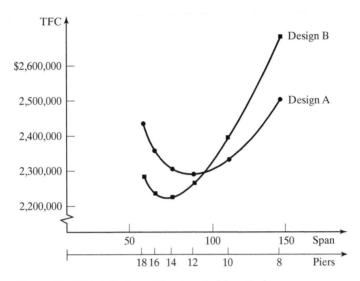

Figure 5   Total first cost for two bridge design alternatives.

where terms not previously defined are

n = anticipated service life of the bridge (years)
M = maintenance cost of the bridge in the first year
g = percentage increase in maintenance cost in each subsequent year

As an example of the application of Equation 13, assume that the two bridge design alternatives in Figure 4 differ in the maintenance cost to be experienced. Design A is estimated to cost $100,000 the first year with a 5% increase in each subsequent year. Design B will cost $90,000 to maintain in the first year with a 7% increase each subsequent year. If the interest rate is 10%, the PELCC for Design A as a function of its life, n, is

$$\text{PELCC}_A = [2\sqrt{16(\$80{,}000)(1{,}000)^2(0.65)} + 600(1{,}000)(\$0.65) + \$80{,}000]$$
$$+ \$100{,}000\left[\frac{\overset{P/A,\,g',\,n}{(\overline{\quad\quad})}}{1 + 0.05}\right]$$

where

$$g' = \frac{1 + i}{1 + g} - 1 = \frac{1 + 0.10}{1 + 0.05} - 1 = 4.76\%$$

The PELCC for Design B as a function of n is

$$\text{PELCC}_B = [2\sqrt{22(\$80{,}000)(1{,}000)^2(\$0.65)} + \$80{,}000] + \$90{,}000\left[\frac{\overset{P/A,\,g',\,n}{(\overline{\quad\quad})}}{1 + 0.07}\right]$$
$$g' = \frac{1 + i}{1 + g} - 1 = \frac{1 + 0.10}{1 + 0.07} - 1 = 2.80\%$$

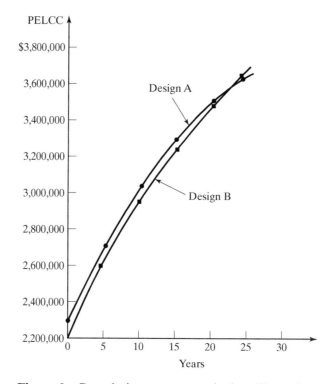

Figure 6 Cumulative present equivalent life-cycle cost for two bridge design alternatives.

Assuming a service life of 40 years ($n = 40$), the present equivalent cost for Design A is

$$\text{PELCC}_A = \$2,294,281 + \$100,000 \left[\frac{\overset{P/A,\ 4.76,\ 40}{(17.7382)}}{1.05} \right] = \$3,983,633$$

And for Design B, it is

$$\text{PELCC}_B = \$2,219,159 + \$90,000 \left[\frac{\overset{P/A,\ 2.80,\ 40}{(23.8807)}}{1.07} \right] = \$4,227,816$$

From this, it appears that Design A is best.

A break-even illustration of the present equivalent life-cycle cost of Designs A and B is exhibited in Figure 6. From this illustration, it is evident that Design A becomes better than Design B for a utilization period exceeding 22 years. Thus, the choice of Design A is confirmed.[2]

---

[2]Note that Design B was preferred when only investment cost was considered (see Figure 5 and Table 2). The inclusion of maintenance cost leads to a clear preference for Design A, illustrating the importance of life-cycle thinking.

2.3 Optimum Equipment Life

The optimum service life of an asset may be projected from knowledge of its first cost and by estimating maintenance and operation costs. Generally, the objective is to minimize the annual equivalent cost over the service life. This optimum life is also known as the minimum-cost service life or the optimum replacement interval.

In this section, a general model is developed for finding the optimum life of an asset. This model is then applied to a determination of the economic life of an existing asset and the selection of an asset from among competing design alternatives. In each case, the life-cycle cost provides the basis for decision.

Evaluation by Mathematical Approach. Operating and maintenance costs generally increase with an increase in the age of equipment. This rising trend is offset by a decrease in the annual equivalent cost of capital recovery with return. Thus, for some age, there will be a minimum life-cycle cost or a minimum-cost life.

If the time value of money is neglected, the average annual life-cycle cost of an asset with constantly increasing maintenance cost is

$$\text{ALCC} = \frac{P}{n} + O + (n - 1)\frac{M}{2} \tag{14}$$

where

$\text{ALCC} = $ average annual life-cycle cost
$P = $ first cost of the asset
$O = $ annual constant portion of operating cost (equal to first-year operation cost, of which maintenance is a part)
$M = $ amount by which maintenance costs increase each year
$n = $ life of the asset (years)

The minimum-cost life can be found mathematically using differential calculus as

$$\frac{d\text{ALCC}}{dn} = -\frac{P}{n^2} + \frac{M}{2} = 0$$

$$n^* = \sqrt{\frac{2P}{M}} \tag{15}$$

Therefore, minimum average life-cycle cost is given by

$$\text{ALCC}^* = P\sqrt{\frac{M}{2P}} + O + \left(\sqrt{\frac{2P}{M}} - 1\right)\frac{M}{2}$$

$$= \sqrt{\frac{PM}{2}} + O + \sqrt{\frac{PM}{2}} - \frac{M}{2}$$

$$= 2\sqrt{\frac{PM}{2}} + O - \frac{M}{2}$$

$$= \sqrt{2PM} + O - M/2 \tag{16}$$

As an example of the application of Equation 16, consider an asset with a first cost of $18,000, a salvage value of zero at any age, a first year operating and maintenance cost of $4,000, and maintenance costs increasing by $1,000 in each subsequent year. The minimum-cost life from Equation 15 is

$$n^* = \sqrt{\frac{2(\$18,000)}{\$1,000}} = 6 \text{ years}$$

The minimum total cost at this life is found from Equation 16 as

$$\text{ALCC}^* = \sqrt{2(\$18,000)(\$1,000)} + \$4,000 - \$500 = \$9,500$$

This situation can also be illustrated by tabulating the cost components and the average life-cycle cost as shown in Table 3. The minimum life-cycle cost occurs at a life of 6 years. Figure 7 illustrates the nature of the increasing and decreasing cost components and the optimum life of 6 years.

Consider an alternative equipment configuration that is more reliable than the previous configuration, leading to operating and maintenance costs in the first year of $3,200 and with an increase each subsequent year equal to $800. This configuration has a first cost of $25,600. The minimum-cost life from Equation 15 is

$$n^* = \sqrt{\frac{2(\$25,600)}{\$800}} = 8 \text{ years}$$

The minimum life-cycle cost at this age is found from Equation 16 as

$$\text{ALCC}^* = \sqrt{2(\$25,600)(\$800)} + \$3,200 - \$400 = \$9,200$$

This situation can be tabulated as before to illustrate the cost components and the average life-cycle cost. This is shown in Table 4. The minimum total cost occurs at a life of 8 years and is $9,200. Figure 8 illustrates the nature of the increasing and decreasing cost components and the optimum life of 8 years. This is the preferred configuration from an average life-cycle cost standpoint and from the standpoint of service life.

TABLE 3 Average Annual Life-Cycle Cost of an Asset (First Configuration)

End of Year	Maintenance Cost at End of Year ($)	Cumulative Maintenance Cost ($)	Average Maintenance Cost ($)	Average Capital Cost ($)	Average Life-Cycle Cost ($)
A	B	$C = \Sigma B$	$D = C/A$	$E = \$18,000/A$	$F = D + E$
1	4,000	4,000	4,000	18,000	22,000
2	5,000	9,000	4,500	9,000	13,500
3	6,000	15,000	5,000	6,000	11,000
4	7,000	22,000	5,500	4,500	10,000
5	8,000	30,000	6,000	3,600	9,600
6	9,000	39,000	6,500	3,000	9,500
7	10,000	49,000	7,000	2,571	9,571
8	11,000	60,000	7,500	2,250	9,750
9	12,000	72,000	8,000	2,000	10,000

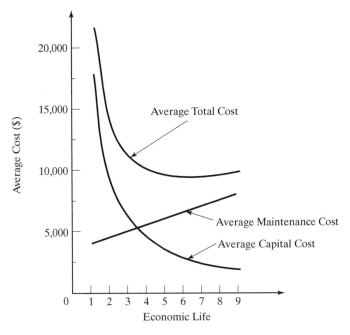

Figure 7 Component and LCC for an asset (first configuration).

Evaluation by Tabular Approach. Often it is not possible to find the minimum-cost life mathematically. When nonlinearities enter and the time value of money is taken into consideration, a tabular approach must be used. Consider the following example. The economic future of an asset with a first cost of $15,000, with linearly decreasing salvage values and with operating costs beginning at $1,000, increasing by $300 in the second year

TABLE 4 Average Annual Life-Cycle Cost of an Asset (Alternative Configuration)

End of Year	Maintenance Cost at End of Year ($)	Cumulative Maintenance Cost ($)	Average Maintenance Cost ($)	Average Capital Cost ($)	Average Life-Cycle Cost ($)
A	B	C = ΣB	D = C/A	E = $25,600/A	F = D − E
1	3,200	3,200	3,200	25,600	28,800
2	4,000	7,200	3,600	12,800	16,400
3	4,800	12,000	4,000	8,533	12,533
4	5,600	17,600	4,400	6,400	10,800
5	6,400	24,000	4,800	5,120	9,920
6	7,200	31,200	5,200	4,267	9,467
7	8,000	39,200	5,600	3,657	9,257
8	8,800	48,000	6,000	3,200	9,200
9	9,600	57,600	6,400	2,844	9,244
10	10,400	68,000	6,800	2,560	9,360
11	11,200	79,200	7,200	2,327	9,527
12	12,000	91,200	7,600	2,133	9,733

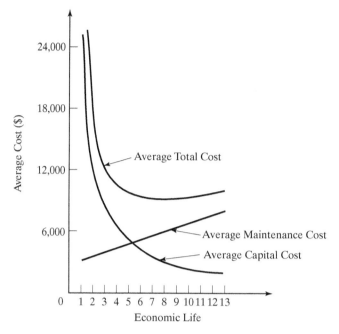

Figure 8 Component and LCC for an asset
(alternative configuration).

and thereafter (the increase each year is $55 more than the increase in the preceding years)
as shown in Figure 9.

To find the optimum life of the asset it is necessary to break out the money flows,
from those shown in Figure 9, that would apply for keeping the asset 1, 2, 3, 4, or 5 years. For
example, if the asset is to be retired after 3 years, the annual equivalent capital cost at
12% is

$$(\$15{,}000 - \$6{,}000)\overset{A/P, 12, 3}{(0.4164)} + \$6{,}000(0.12) = \$4.468$$

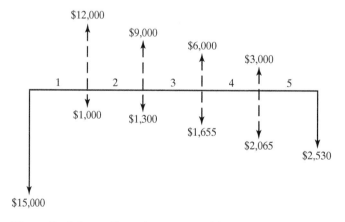

Figure 9 Money flows for an asset (first configuration).

268

TABLE 5 Tabular Calculations for Optimum Life

End of Year	Salvage Value When Asset Retired ($)	Operating Costs ($)	Annual Equivalent Cost of Asset ($)	Annual Equivalent Operating Cost ($)	Annual Equivalent LCC ($)
A	B	C	D	E	F
1	12,000	1,000	4,800	1,000	5,800
2	9,000	1,300	4,630	1,142	5,772
3	6,000	1,655	4,468	1,294	5,762
4	3,000	2,065	4,310	1,455	5,765
5	0	2,530	4,161	1,624	5,785

And the annual equivalent operation cost is

$$[(\$1,000)\overset{P/F,\,12,\,1}{(0.8929)} + \$1,300\overset{P/F,\,12,\,2}{(0.7972)} + \$1,655\overset{P/F,\,12,\,3}{(0.7116)}]\overset{A/P,\,12,\,3}{(0.4164)} = \$1,294$$

The annual equivalent life-cycle cost is $4,468 + $1,294 = $5,762 as shown in Column F of Table 5. This is the optimum and may be used for comparison with other alternatives.

Suppose that the need for the asset discussed and tabulated in Table 5 can be met by an alternative physical configuration. The initial investment needed for the alternative asset is $12,500, again with linearly decreasing salvage values and with operating costs beginning at $1,600, increasing by $200 in the second year and thereafter where the increase each year is $100 more than the increase in the preceding year, and with an interest rate of 12%. To find this asset's optimum life, it is necessary to identify, as before, the relevant money flows. These money flows are depicted in Figure 10 and are the basis for the annual equivalent life-cycle cost calculations shown in Table 6.

In this example, both the first case and the alternative configuration have an optimum life of 3 years. However, the alternative is preferred in that the annual equivalent life-cycle cost is lower by $5,762 − $5,538 = $224.

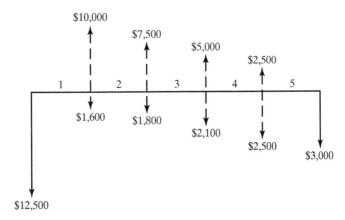

Figure 10   Money flows for an asset (alternative configuration).

TABLE 6 Tabular Calculations for Optimum Life (Alternative)

End of Year	Salvage Value When Asset Retired ($)	Operating Costs ($)	Annual Equivalent Cost of Asset ($)	Annual Equivalent Operating Cost ($)	Annual Equivalent LCC ($)
A	B	C	D	E	F
1	10,000	1,600	4,000	1,600	5,600
2	7,500	1,800	3,859	1,694	5,553
3	5,000	2,100	3,723	1,815	5,538
4	2,500	2,500	3,592	1,958	5,550
5	0	3,000	3,468	2,122	5,590

2.4 Optimizing Procurement and Inventory Operations

The bridge design and equipment selection examples illustrated the application of classical optimization theory to systems having structure without activity. In this section, classical optimization theory is applied to inventory systems with single-item, single-source, and multi-source characteristics. In this system with activity, the objective is to determine the optimum procurement and inventory policy so that the system will operate at minimum cost over time.

When to procure and how much to procure are policy variables of importance in the single-source system. The decision evaluation function of Equation A.1 applies as

$$E = f(X, Y)$$

where

E = evaluation measure (total system cost)
X = policy variables of when to procure and how much to procure
Y = system parameters of demand, procurement lead time, replenishment rate, item cost, procurement cost, holding cost, and shortage cost.

Equation A.2 is the decision evaluation function applicable to the multisource system, where source-dependent parameters are the key to determining the best procurement source. Both versions of the function are used in their unconstrained forms in this section.

Inventory Decision Evaluation Model. A mathematical model can be formulated for the general deterministic case (demand and lead time constant) from the inventory flow geometry shown in Figure 11. The following symbolism will be adopted:

TC = total system cost per period
L = procurement level
Q = procurement quantity
D = demand rate in units per period
T = lead time in periods
N = number of periods per cycle
R = replenishment rate in units per period
C_i = item cost per unit

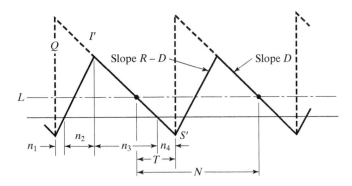

Figure 11 General deterministic inventory system geometry.

C_p = procurement cost per procurement
C_h = holding cost per unit per period
C_s = shortage cost per unit short per period

It is assumed that the replenishment rate is greater than the demand rate and that unsatisfied demand is not lost. From Figure 11, the number of periods per cycle can be expressed as

$$N = \frac{Q}{D} \tag{17}$$

Also, the following relationships are evident:

$$(n_1 + n_2)(R - D) = (n_3 + n_4)D \tag{18}$$

$$n_1 + n_2 = \frac{Q}{R} \tag{19}$$

and

$$n_3 + n_4 = \frac{I' + DT - L}{D} \tag{20}$$

From Equations 18–20,

$$I' = Q\left(1 - \frac{D}{R}\right) + L - DT \tag{21}$$

The total number of unit periods of stock on hand during the inventory cycle, I, is

$$I = \frac{I'}{2}(n_2 + n_3)$$

$$= \frac{I'^2}{2(R - D)} + \frac{I'^2}{2D} \tag{22}$$

Substituting Equation 21 for I' gives

$$I = \frac{[Q(1 - D/R) + L - DT]^2}{2} \left(\frac{1}{R - D} + \frac{1}{D} \right)$$ (23)

The total number of unit periods of shortage during the cycle, S, is

$$S = \frac{S'}{2} (n_1 + n_4)$$

$$= \frac{S'^2}{2(R - D)} + \frac{S'^2}{2D}$$

But because $S' = DT - L$,

$$S = \frac{(DT - L)^2}{2} \left(\frac{1}{R - D} + \frac{1}{D} \right)$$ (24)

The total system cost per period will be a summation of the item cost for the period, the procurement cost for the period, the holding cost for the period, and the shortage cost for the period, or

$$TC = IC + PC + HC + SC$$ (25)

Item cost for the period will be the product of the item cost per unit and the demand rate in units per period, or

$$IC = C_i D$$ (26)

The procurement cost for the period will be the procurement cost per procurement divided by the number of periods per inventory cycle, or

$$PC = \frac{C_p}{N}$$

$$= \frac{C_p D}{Q}$$ (27)

Holding cost for the period will be the product of the holding cost per unit per period and the average number of units on hand during the period, or

$$HC = \frac{C_h I}{N}$$

$$= \frac{C_h D}{Q} \left\{ \frac{[Q(1 - D/R) + L - DT]^2}{2} \left(\frac{1}{R - D} + \frac{1}{D} \right) \right\}$$

Note the existence of the following relationship:

$$\frac{D}{Q} \left(\frac{1}{R - D} + \frac{1}{D} \right) \frac{1}{Q(1 - D/R)}$$ (28)

Using Equation 34, the holding-cost component becomes

$$HC = \frac{C_h}{2Q(1 - D/R)}\left[Q\left(1 - \frac{D}{R}\right) + L - DT\right]^2 \tag{29}$$

Shortage cost for the period will be the product of the shortage cost per unit short per period and the average number of units short during the period, or

$$SC = \frac{C_s S}{N}$$

$$= \frac{C_s D}{Q}\left[\frac{(DT - L)^2}{2}\left(\frac{1}{R - D} + \frac{1}{D}\right)\right] \tag{30}$$

Equation 30 may be written as

$$SC = \frac{C_s(DT - L)^2}{2Q(1 - D/R)} \tag{31}$$

The total system cost per period will be a summation of the four cost components developed above and may be expressed as

$$TC = C_i D + \frac{C_p D}{Q} + \frac{C_h}{2Q(1 - D/R)}\left[Q\left(1 - \frac{D}{R}\right) + L - DT\right]^2$$

$$+ \frac{C_s(DT - L)^2}{2Q(1 - D/R)} \tag{32}$$

The minimum-cost procurement level and procurement quantity may be found by setting the partial derivatives equal to zero and solving the resulting equations. Modifying Equation 32 gives

$$TC = C_i D + \frac{C_p D}{Q} + \frac{C_h Q(1 - D/R)}{2} - C_h(DT - L)$$

$$+ \frac{C_h(DT - L)^2}{2Q(1 - D/R)} + \frac{C_s(DT - L)^2}{2Q(1 - D/R)} \tag{33}$$

By taking the partial derivative of Equation 33 with respect to Q and then with respect to $DT - L$, and setting both equal to zero, one obtains

$$\frac{\partial TC}{\partial Q} = -\frac{C_p D}{Q^2} + \frac{C_h(1 - D/R)}{2}$$

$$- \frac{C_h(DT - L)^2}{2Q^2(1 - D/R)} - \frac{C_s(DT - L)^2}{2Q^2(1 - D/R)} = 0 \tag{34}$$

$$\frac{\partial TC}{\partial(DT - L)} = -C_h + \frac{C_h(DT - L)}{Q(1 - D/R)} + \frac{C_s(DT - L)}{Q(1 - D/R)} = 0 \tag{35}$$

Equation 35 may be expressed as

$$\frac{DT - L}{Q} = \frac{C_h(1 - D/R)}{C_h + C_s} \tag{36}$$

Substituting Equation 36 into Equation 34 gives

$$-\frac{C_pD}{Q^2} + \frac{C_h(1 - D/R)}{1} - \frac{C_h^2(1 - D/R)}{2(C_h + C_s)^2} - \frac{C_sC_h^2(1 - D/R)}{2(C_h + C_s)^2} = 0$$

$$\frac{C_pD}{Q^2} = \frac{C_hC_s(1 - D/R)}{2(C_h + C_s)}$$

Solving for Q gives

$$Q^* = \sqrt{\frac{1}{1 - D/R}}\sqrt{\frac{2C_pD}{C_h} + \frac{2C_pD}{C_s}} \tag{37}$$

Substituting Equation 37 into Equation 36 gives

$$L^* = DT - \frac{C_h(1 - D/R)}{C_h + C_s}\sqrt{\frac{1}{1 - D/R}}\sqrt{\frac{2C_pD}{C_h} + \frac{2C_pD}{C_s}}$$

$$L^* = DT - \sqrt{1 - \frac{D}{R}}\sqrt{\frac{2C_pD}{C_s(1 + C_s/C_h)}} \tag{38}$$

Equations 37 and 38 may now be substituted back into Equation 33 to give an expression for the minimum total system cost.[3] After several steps, the result is

$$TC^* = C_iD + \sqrt{1 - \frac{D}{R}}\sqrt{\frac{2C_pC_hC_sD}{C_h + C_s}} \tag{39}$$

Single-Source Alternative.   As an example of an inventory system with only one procurement source alternative, consider a situation in which an item having the following system parameters is to be manufactured:

$D = 10$ units per period
$R = 20$ units per period
$T = 8$ periods
$C_i = \$4.82$ per unit
$C_p = \$100.00$ per procurement
$C_h = \$0.20$ per unit per period
$C_s = \$0.10$ per unit per period

---

[3]The policy variables, Q^* and L^*, supposedly occur at the minimum of the total cost function. Actually, there are three possibilities that may occur when the first partials of a function are set equal to zero and solved: a maximum, a minimum, or a point of inflection may be found. Although the nature of the total cost function derived here is such that a minimum is always found, a formal test could be applied using Equation 7 to verify that the minimum has been found.

TABLE 7 Total System Cost as a Function of L and Q

L \ Q	242	243	244	245	246	247	248
-6	56.399	56.394	56.390	56.386	56.383	56.380	56.377
-5	56.387	56.383	56.380	56.377	56.374	56.372	56.371
-4	56.378	56.375	56.372	56.370	56.368	56.367	56.366
-3	56.371	56.369	56.367	56.366	56.365	56.364	56.364
-2	56.366	56.365	56.364	56.363	56.363	56.364	56.364
-1	56.364	56.364	56.364	56.364	56.365	56.366	56.367
0	56.364	56.365	56.366	56.367	56.368	56.370	56.373
1	56.367	56.369	56.370	56.372	56.374	56.377	56.380
2	56.373	56.375	56.377	56.380	56.383	56.386	56.390
3	56.381	56.383	56.386	56.390	56.394	56.398	56.403
4	56.391	56.394	56.398	56.403	56.407	56.442	56.418

Total system cost may be tabulated as a function of Q and L by use of Equation 32. The resulting values in the region of the minimum-cost point are given in Table 7. The minimum-cost Q and L may be found by inspection. As in the previous examples, the surface generated by the TC values is seen to be rather flat. Also, the selection of integer values of Q and L does not affect the total cost appreciably.

The minimum-cost procurement quantity and minimum-cost procurement level may be found directly by substituting into Equations 37 and 38 as follows:

$$Q^* = \sqrt{\frac{1}{1 - 10/20}} \sqrt{\frac{2(\$100.00)(10)}{\$0.20} + \frac{2(\$100.00)(10)}{\$0.10}} = 244.95$$

$$L^* = 10(8) - \sqrt{1 - \frac{10}{20}} \sqrt{\frac{2(\$100.00)(10)}{\$0.10(1 + \$0.10/\$0.20)}} = -1.65$$

The resulting total system cost at the minimum-cost procurement quantity and procurement level may be found from Equation 39 as

$$TC^* = \$4.82(10) + \sqrt{1 - \frac{10}{20}} \sqrt{\frac{2(\$100.00)(\$0.20)(\$0.10)(10)}{\$0.20 + \$0.10}} = \$56.363$$

Accordingly, the item should be manufactured in lots of 245 units. Procurement action should be initiated when the stock on hand is negative; that is, when the backlog is two units. The total system cost will be $56.36 per period under these optimum operating conditions.

Single-Source Alternative for Purchase. As an example of an inventory system with only one procurement source, consider purchasing the needed item, a special case of the manufacture alternative presented in the previous section. In this case, $R = \infty$ (instantaneous replenishment) and $C = \infty$ (no shortages permitted). With these modifications, Figure 11 now appears as shown in Figure 12.

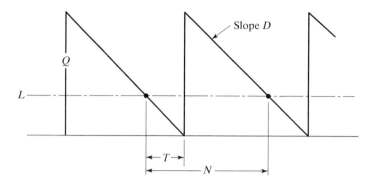

Figure 12 Deterministic inventory geometry with $R = \infty$ and $C = \infty$

The total system cost can be found from Equation 32 with the substitution of ∞ for both R and C_s. This gives

$$TC = C_iD + \frac{C_pD}{Q} + \frac{C_hQ}{2} \tag{40}$$

The optimal purchase quantity is found by taking the derivative of TC with respect to Q and setting the result equal to zero as

$$\frac{dTC}{dQ} = -\frac{C_pD}{Q^2} + \frac{C_h}{2} = 0 \tag{41}$$

Solving for the optimal value of Q gives

$$Q^* = \sqrt{\frac{2C_pD}{C_h}} \tag{42}$$

This could also have been found from Equation 37 with R and C_s set equal to ∞.

The optimum value of the procurement level, L^*, is simply D times T plus safety stock, where the cost of holding safety stock is justified by the contribution it makes to reducing shortages.

As an example of the application of Equation 42, assume that the annual demand for a spare part is 1,000 units. The cost per unit is \$6.00 delivered in the field. Procurement cost per order placed is \$10.00 and the cost of holding one unit in inventory for 1 year is estimated to be \$1.32.

The economic procurement quantity may be found by substituting the appropriate values into Equation 42 as

$$Q^* = \sqrt{\frac{2(\$10)(1,000)}{\$1.32}} = 123 \text{ units}$$

Total cost may be expressed as a function of Q by substituting the costs and various values of Q into Equation 40. The result is shown in Table 8. The tabulated total cost value

TABLE 8 Inventory System Cost as a Function of Q

Q	C_iD (\$)	$\dfrac{C_pD}{Q}$ (\$)	$\dfrac{C_hQ}{2}$ (\$)	TC (\$)
0	6,000	∞	0	∞
50	6,000	200	33	6,233
100	6,000	100	66	6,166
123	6,000	81	81	6,162
150	6,000	67	99	6,165
200	6,000	50	132	6,182
400	6,000	25	264	6,289
600	6,000	17	396	6,413

for $Q = 123$ is the minimum-cost quantity for the condition specified. Total cost as a function of Q is illustrated in Figure 13.[4]

Multiple Source Alternatives. Consider the following situation. A single-item inventory is maintained to meet demand. When the number of units on hand falls to a predetermined level, action is initiated to procure a replenishment quantity from one of

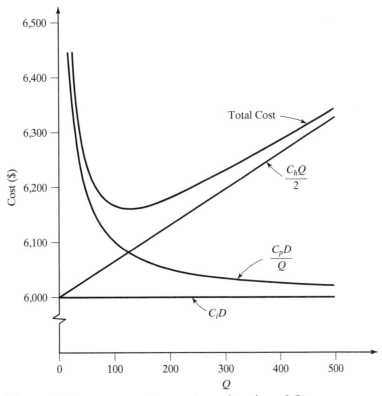

Figure 13 Inventory system cost as a function of Q.

[4]The total cost found does not include the cost of safety stock, which must be separately justified by trading off the cost of holding additional inventory against the cost of being short.

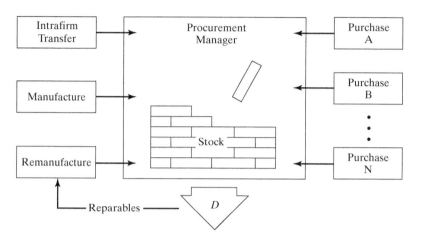

Figure 14 Multiple replenishment sources.

several possible sources, as is illustrated in Figure 14. The object is to determine the procurement level, the procurement quantity, and the procurement source in the light of system and cost parameters so that the sum of all costs associated with the system is minimized.

The procurement lead time, rate of replenishment, item cost, and procurement cost are source dependent. Thus, the decision evaluation function of Equation A.2 applies with the named source-dependent parameters. All other system and cost parameters are source independent. Explicit recognition of source-dependent parameters permits the use of Equation 39 to find the minimum-cost procurement source. The procedure is to compare TC^* for all sources and to choose that source that results in a minimum TC^*. Equations 37 and 38 may then be applied to find the minimum-cost procurement quantity and minimum-cost procurement level.

An example of the procedure is the situation in which a system manager is experiencing a demand of four units per period for an item that may be either manufactured or purchased from one of two vendors. Holding cost per unit per period is $0.24 and shortage cost per unit short per period is $0.17. Values for source-dependent parameters are given in Table 9.

The procurement source resulting in a minimum total system cost can be found from Equation 39. For the manufacturing alternative, it is

$$TC^* = \$19.85(4) + \sqrt{1 - \frac{4}{12}} \sqrt{\frac{2(\$17.32)(\$0.24)(\$0.17)(4)}{\$0.24 + \$0.17}}$$

$$= \$82.43$$

TABLE 9 Source-Dependent Parameters for Multisource System

Parameter	Manufacture	Vendor A	Vendor B
R	12.00	∞	∞
T	6.00	3.00	4.00
C_i ($)	19.85	17.94	18.33
C_p ($)	17.32	18.70	17.50

For the alternative designated Vendor A, it is

$$TC^* = \$17.94(4) + \sqrt{1 - \frac{4}{\infty}} \sqrt{\frac{2(\$18.70)(\$0.24)(\$0.17)(4)}{\$0.24 + \$0.17}}$$

$$= \$75.62$$

For the alternative designated Vendor B, it is

$$TC^* = \$18.33(4) + \sqrt{1 - \frac{4}{\infty}} \sqrt{\frac{2(\$17.50)(\$0.24)(\$0.17)(4)}{\$0.24 + \$0.17}}$$

$$= \$77.05$$

On the basis of this analysis, the alternative designated Vendor A would be chosen as the minimum-cost procurement source.

The minimum-cost procurement quantity for this source may be found from Equation 37 to be

$$Q^* = \sqrt{1 - \frac{4}{\infty}} \sqrt{\frac{2(\$18.70)(4)}{\$0.24} + \frac{2(\$18.70)(4)}{\$0.17}}$$

$$= 38.78$$

The minimum-cost procurement level for this source, from Equation 38, is

$$L^* = 4(3) - \sqrt{1 - \frac{4}{\infty}} \sqrt{\frac{2(\$18.70)(4)}{\$0.17(1 + \$0.17/\$0.24)}}$$

$$= -10.69$$

Thus, the system manager would initiate procurement action when the stock level falls to -10.69 units, for a procurement quantity of 38.78 units, from the procurement source designated Vendor A.

3 CONSTRAINED CLASSICAL OPTIMIZATION

In design and operations alike, physical and economic limitations often exist that act to limit system optimization globally. These limitations arise for a variety of reasons and generally cannot be removed by the decision maker. Accordingly, there may be no choice except to find the best or optimum solution subject to the constraints. In this section, the example applications already presented will be revisited to illustrate two physical constraint situations: span-constrained bridge design and space-constrained inventory system operations.

3.1 Span-Constrained Bridge Design

Table 2 gave total costs as a function of the number of piers for two bridge design alternatives. Suppose that there exists a constraint on the pier spacing, expressed as a minimum spacing of 110 feet to permit the safe passage of barge traffic. With this requirement, it is evident from Table 2 and Figure 15 that Design A now replaces Design B as the minimum-cost choice.

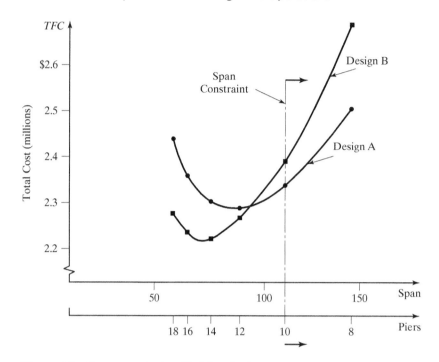

Figure 15   Span-constrained bridge design.

This particular constraint leads to a design decision reversal and to a penalty in total system cost. Although a reversal will not always occur, a penalty in total cost usually will. In this example, the penalty is found from Table 2 to be $2,388,889 − $2,345,555 or $43,334.

3.2 Space-Constrained Inventory Operations

Warehouse space is often a scarce resource. It may be expressed in cubic units designated W. Each unit in stock consumes a certain amount of space designated w. The units on hand require a total amount of space that must not exceed the amount available. In general, this may be formally stated as $I'w \leq W$, where I' is the maximum accumulation of stock.

The Space-Constrained Inventory Model.   Consider the inventory operations of Figure 11. The total cost for that model was given by Equation 32 and the maximum inventory, I', was given by Equation 21. If $I'w \leq W$, then $I' \leq W/w$, or

$$I' = Q\left(1 - \frac{D}{R}\right) + L - DT \leq \frac{W}{w}$$

If the warehouse restriction is active, then $I' = W/w$, or

$$I' = Q\left(1 - \frac{D}{R}\right) + L - DT - \frac{W}{w} = 0 \tag{43}$$

For an active warehouse restriction, the minimum total cost will occur along the intersection of the surface defined by Equation 32 and the plane defined by Equation 43. The new *TC* equation is determined by substituting Equation 43 into Equation 32 giving

$$TC = C_iD + \frac{C_pD}{Q} + \frac{C_h(W/w)^2}{2Q(1 - D/R)} + \frac{C_s[Q(1 - D/R) - W/w]^2}{2Q(1 - D/R)} \qquad (44)$$

After some manipulation,

$$TC = a_1 + \frac{a_2}{Q} = a_3Q \qquad (45)$$

where

$$a_1 = C_iD - C_s\left(\frac{W}{w}\right)$$

$$a_2 = C_pD + \frac{(C_h + C_s)(W/w)^2}{2(1 - D/R)}$$

and

$$a_3 = \frac{C_s(1 - D/R)}{2}$$

The optimal value of Q can be obtained by differentiating Equation 45 with respect to Q, setting the result equal to zero, and solving for Q^* as

$$\frac{\partial TC}{\partial Q} = -\frac{a_2}{Q^2} + a_3 = 0$$

and

$$Q^* = \sqrt{\frac{a_2}{a_3}}$$

which can be written in terms of the original parameters as

$$Q^* = \sqrt{\frac{2C_pD}{C_s(1 - D/R)} + \frac{(C_h + C_s)(W/w)^2}{C_s(1 - D/R)^2}} \qquad (46)$$

The optimal L^* can be determined from Equation 43 with $Q = Q^*$ as

$$L^* = DT + \frac{W}{w} - Q^*\left(1 - \frac{D}{R}\right) \qquad (47)$$

Finally, the minimum total cost is given by substituting the optimal values of Q into Equation 44, or using Equation 45 with

$$TC^* = a_1 + \frac{a_2}{Q^*} + a_3 Q^* \tag{48}$$

Space-Constrained Single-Source Alternative. As an example of the application of Equations 43–48, suppose that a procurement manager will purchase an item having the following parameters:

$D = 6$ units per period
$R = \infty$ units per period
$T = 7$ periods
$C_i = \$34.75$ per unit
$C_p = \$23.16$ per procurement
$C_h = \$0.30$ per unit per period
$C_s = \$0.30$ per unit per period
$W = 100$ cubic units of space
$w = 24$ cubic units per item

With no restrictions, the optimal $Q^* = 43.04$ using Equation 37, $L^* = 20.48$ using Equation 38, and $TC^* = \$214.96$ using Equation 39. If $R = \infty$, Equation 43 can be written as

$$I'^* = Q^* + L^* - DT \tag{49}$$

When the optimal values are used, Equation 49 yields

$$I'^* = 43.04 + 20.48 - 6(7) = 21.52$$

Because $W/w = 100/24 < 21.52 = I'^*$, the warehouse restriction is active.

Next, determine the optimal policy values for the constrained warehouse where $I'^* = W/w = 100/24 = 4.17$. From Equation 46 with $1 - D/R :$ unity,

$$Q^* = \sqrt{\frac{2(\$23.16)(6)}{\$0.30} + \frac{(\$0.30 + \$0.30)(4.17)^2}{\$0.30}} = 30.57$$

Next, using Equation 47 with $1 - D/R :$ unity, determine L^* as

$$L^* = 6(7) + 4.17 - 30.57 = 15.60$$

Finally, determine the optimal total cost, using Equation 44 with $1 - D/R :$ unity, as

$$TC^* = \$34.75(6) + \frac{\$23.16(6)}{30.57} + \frac{\$0.30(4.17)^2}{2(30.57)} + \frac{\$0.30[6(7) - 15.17]^2}{2(30.57)} = \$216.55$$

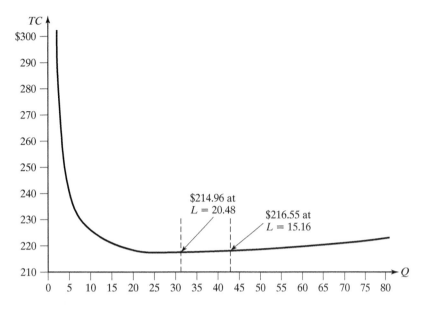

Figure 16   Total cost with restriction.

The difference in TC at the optimum is approximately \$1.60 per period. The total cost as a function of the procurement quantity both without and with the warehouse restriction is shown in Figure 16.

Space-Constrained Multiple-Source Alternatives.   The explicit recognition of source-dependent parameters permits the use of models for the constrained system to be used for the single-item, multisource system. The procedure is to evaluate each source and to find that source that will result in a minimum total cost subject to the restriction.

As an example, suppose that the system is constrained by a total warehouse space of 1,000 cubic units. Also suppose that the item requires 120 cubic units of space. Determine if the restriction is active for Vendor A using Equation 49 as

$$I'^* = 38.7806 - 10.6917 - 4(3) = 16.0889$$

Because $W/w = 1,000/120 < 16.0889 = I'^*$, the warehouse restriction is active.

Next, determine the optimal policy values for the constrained situation with $I'^* = 1,000/120 = 8.3333$. From Equation 46 with $1 - D/R$: unity, Q^* is found as

$$Q^* = \sqrt{\frac{2(\$18.70)4}{\$0.17} + \frac{(\$0.24 + \$0.17)(8.3333)^2}{\$0.17}} = 32.3648$$

Using Equation 47 with $1 - D/R$: unity, L^* is found to be

$$L^* = 4(3) + 8.3333 - 32.3648 = -12.0315$$

TABLE 10 Optimal Policies for Restricted Multisource System

Source	L^*	Q^*	TC^* ($)
Manufacture	5.6710	40.0000	82.5158
Vendor A	−12.0315	32.3648	75.8454
Vendor B	−7.1470	31.4820	77.2550

As the last step, determine the optimal total cost using Equation 44 with $1 - D/R:$ unity, as

$$TC^* = \$17.94(4) + \frac{\$18.70(4)}{32.3648} + \frac{\$0.24(8.3333)^2}{2(32.3648)}$$

$$+ \frac{\$0.17[(32.3648) - 8.3333]^2}{2(32.3648)} = \$75.85$$

A penalty of $75.85 less $75.62 = $0.23 occurs due to the warehouse constraint.

It is possible for the restricted source choice to differ from the source choice when no restriction exists. Accordingly, the other source possibilities in the example should be evaluated in the face of the warehouse restriction. The results of this evaluation are given in Table 10. Inspection reveals that Vendor A remains the best choice for the case where the warehouse capacity is 1,000 cubic units.

4 OPTIMIZATION INVOLVING MULTIPLE CRITERIA

Optimization is not an end in itself. It is only a means for bringing mutually exclusive alternatives into comparable (or equivalent) states. Like money flow modeling, it is an essential step in comparing alternatives equivalently. There is no need to go further if only a single criterion is involved.

When multiple criteria are present in a decision situation, neither economic optimization nor money flow analysis is sufficient. Although necessary, these steps must be augmented with information about the degree to which each alternative meets (or exceeds) specific criteria. One means for consolidating and displaying this information is through the decision evaluation display approach. In this section, two examples of the approach are presented based on situations presented in Sections 2 and 3.

4.1 Superstructure Design Preference

Consider the superstructure design alternatives illustrated in Figure 4. If the public (customer) is indifferent to the superstructure configuration for the bridge, then a choice can be made on the basis of economic factors alone. Recall that Design B is preferred for the unconstrained case and Design A for the constrained. This was based only on first cost considerations.

Now suppose that the Design A configuration has some undesirable aesthetic and safety characteristics in comparison with Design B. Specifically, suppose that the aesthetics committee rates Design A at only 5 on a 10-point scale versus a 9$^+$ for Design B. Further, the safety officer is concerned that Design A will be more risky to youth who venture to

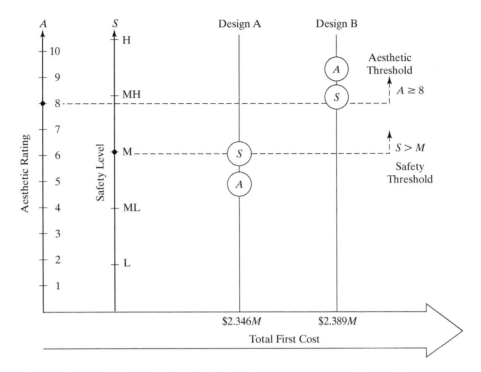

Figure 17 Decision evaluation display for superstructure alternatives.

climb within the superstructure. On an L/ML/M/MH/H scale, this design is rated M and Design B is rated MH.

The ratings for aesthetics and safety can now be exhibited along with total first cost as in Figure 17. Here the decision evaluation display shows the economic advantages identified and rated. The total first cost difference of $43,334 is no longer the only basis for the decision. Further, if aesthetics must rate at least 8 and safety greater than M, then Design A fails on the aesthetics measure. It is now left to negotiation in the presence of subjective evaluation as the basis for choice.

4.2 Procurement Source Preference

Manufacturing, or procurement from one of two vendors, is identified as a possible procurement source by the source-dependent parameters in Table 10. The total cost for each source was found from Equation 39 and it differs. However, suppose there are other considerations.

Manufacturing is judged to provide a source of supply without interruption and is rated 9$^+$ on a 10-point scale; Vendors A and B are not likely to be as dependable and ratings of 6 and 7 are applied to A and B, respectively. As to quality, Vendor A is judged to be able to provide a quality level just above H, Vendor B a level of MH$^+$, and manufacturing a level of just below MH. These values and the total costs are displayed in Figure 18.

Suppose that a dependability rating of at least 9 is desired, with a quality rating of at least MH. The decision is now reduced to a trade-off of two noneconomic factors against

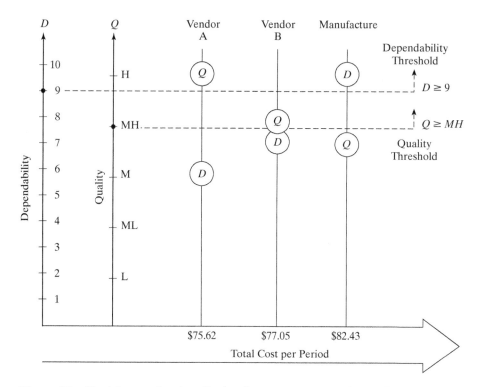

Figure 18   Decision evaluation display for procurement alternatives.

total cost. Choice can only be made by a knowledgeable person or group by the application of subjective evaluation and negotiation, aided by the decision evaluation display.

5 CONSTRAINED OPTIMIZATION BY LINEAR PROGRAMMING[5]

The general linear programming model (LP) may be stated mathematically as that of optimizing the evaluation function,

$$E = \sum_{j=1}^{n} e_j x_j \tag{50}$$

subject to the constraints

$$\sum_{j=1}^{n} a_{ij} x_j = c_i \quad i = 1, 2, \ldots, m$$

$$x_j \geq 0 \quad j = 1, 2, \ldots, n$$

---

[5]The mathematical formulation of the general linear programming model, together with the simplex optimization algorithm, was developed by George Dantzig in 1947. Simplex and its subsequent versions of LP are in contrast with the calculus methods of classical optimization presented in Section 1, based mainly on its being restricted in application to linear systems.

Optimization requires either maximization or minimization, depending on the measure of evaluation involved. The decision maker has control of the vector of variables, x_j. Not directly under his or her control is the vector of evaluation coefficients, e_j, the matrix of constraints, a_{ij}, and the vector of constants, c_i. This structure is in accordance with the constrained version of the decision evaluation function in Equation A.1.

5.1 Graphical Optimization Methods

The mathematical statement of the general linear programming model may be explained in graphical terms. There exist n variables that define an n-dimensional space. Each constraint corresponds to a hyperplane in this space. These constraints surround the region of feasible solution by hypersurfaces so that the region is the interior of a convex polyhedron. Because the evaluation function is linear in the n variables, the requirement that this function have some constant value gives a hyperplane that may or may not cut through the polyhedron. If it does, one or more feasible solutions exist. By changing the value of this constant, a family of hyperplanes parallel to each other is generated. The distance from the origin to a member of this family is proportional to the value of the evaluation function.

Two limiting hyperplanes may be identified: one corresponds to the largest value of the evaluation function for which the hyperplane just touches the polyhedron; the other corresponds to the smallest value. In most cases, the limiting hyperplanes just touch a single vertex of the limiting polyhedron. This outermost limiting point is the solution that optimizes the evaluation function.

Although this is a graphical description, a graphical solution is not convenient when more than three variables are involved. The following examples illustrate the general linear programming model and its graphical optimization method for operations having two and three activities.

Graphical Maximization for Two Activities. As an example of the case, where two activities compete for scarce resources, consider the following production system. Two products are to be manufactured. A single unit of product A requires 2.4 minutes of punch press time and 5.0 minutes of assembly time. The profit for product A is $0.60 per unit. A single unit of product B requires 3.0 minutes of punch press time and 2.5 minutes of welding time. The profit for product B is $0.70 per unit. The capacity of the punch press department available for these products is 1,200 minutes per week. The welding department has idle capacity of 600 minutes per week and the assembly department can supply 1,500 minutes of capacity per week. The manufacturing and marketing data for this production system are summarized in Table 11.

In this example, two products compete for the available production time. The objective is to determine the quantity of product A and the quantity of product B to produce so that total profit will be maximized. This will require maximizing

$$TP = \$0.60A + \$0.70B$$

TABLE 11 Manufacturing and Marketing Data for Two Products

Department	Product A	Product B	Capacity
Punch press	2.4	3.0	1,200
Welding	0.0	2.5	600
Assembly	5.0	0.0	1,500
Profit	$0.60	$0.70	

subject to

$$2.4A + 3.0B \leq 1,200$$

$$0.0A + 2.5B \leq 600$$

$$5.0A + 0.0B \leq 1,500$$

$$A \geq 0 \quad \text{and} \quad B \geq 0$$

The graphical equivalent of the algebraic statement of this two-product system is shown in Figure 19. The set of linear restrictions defines a region of feasible solutions. This region lies below $2.4A + 3.0B = 1,200$ and is restricted further by the requirements that $B \leq 240$, $A \leq 300$, and that both A and B be nonnegative. Thus, the scarce resources determine which combinations of the activities are feasible and which are not feasible.

The production quantity combinations of A and B that fall within the region of feasible solutions constitute feasible production programs. The combination or combinations of A and B that maximize profit are sought. The relationship between A and B is $1.167A = B$. This relationship is based on the relative profit of each product. The total profit realized will depend on the production quantity combination chosen. Thus, there is a family of isoprofit lines, one of which will have at least one point in the region of the feasible production quantity combinations and be a maximum distance from the origin. The member that satisfies this condition intersects the region of feasible solutions at the extreme point $A = 300$ and $B = 160$. This is shown as a dashed line in Figure 19 and represents a total profit of $\$0.60(300) + \$0.70(160) = \$292$. No other production quantity combination would result in a higher profit.

Alternative production programs with the same profit might exist in some cases. This would occur when the isoprofit line lies parallel to one of the limiting restrictions. For

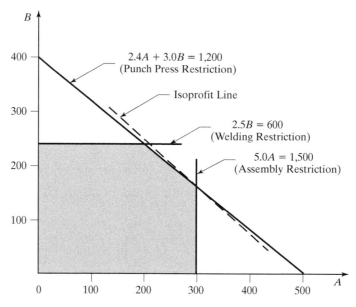

Figure 19 Graphical optimization for a two-product production operation.

example, if the relative profits of product A and product B were $A = 1.25B$, the isoprofit line in Figure 19 would coincide with the restriction $2.4A + 3.0B = 1,200$. In this case, the isoprofit line would touch the region of feasible solutions along a line instead of at a point. All production quantity combinations along the line would maximize profit.

Graphical Maximization for Three Activities.   When three activities compete for scarce resources, a three-dimensional space is involved. Each constraint is a plane in this space, and all constraints taken together identify a volume of feasible solutions. The evaluation function is also a plane, its distance from the origin being proportional to its value. The optimum value for the evaluation function occurs when this plane is located so that it is at the extreme point of the volume of feasible solutions.

As an example, suppose that the production operations for the previous system are to be expanded to include a third product, designated C. A single unit of product C will require 2.0 minutes of punch press time, 1.5 minutes of welding time, and 2.5 minutes of assembly time. The profit associated with product C is $0.50 per unit. Manufacturing and marketing data for this revised production situation are summarized in Table 12.

In this example, three products compete for the available production time. The objective is to determine the quantity of product A, the quantity of product B, and the quantity of product C to produce so that total profit will be maximized. This will require maximizing

$$TP = \$0.60A + \$0.70B + \$0.50C$$

subject to

$$2.4A + 3.0B + 2.0C \leq 1,200$$

$$0.0A + 2.5B + 1.5C \leq 600$$

$$5.0A + 0.0B + 2.5C \leq 1,500$$

$$A \geq 0, \quad B \geq 0, \quad \text{and} \quad C \geq 0$$

The graphical equivalent of the algebraic statement of this three-product production situation is shown in Figure 20. The set of restricting planes defines a volume of feasible solutions. This region lies below $2.4A + 3.0B + 2.0C = 1,200$ and is restricted further by the requirement that $2.5B + 1.5C \leq 600$, $5.0A + 1.5C \leq 1,500$, and that A, B, and C be nonnegative. Thus, the scarce resources determine which combinations of the activities are feasible and which are not feasible.

The production quantity combinations of A, B, and C that fall within the volume of feasible solutions constitute feasible production programs. The combination or combinations of A, B, and C that maximize total profit is sought. The expression $0.60A + 0.70B + 0.50C$ gives the relationship among A, B, and C based on the relative profit of each product. The total

TABLE 12   Manufacturing and Marketing Data for Three Products

Department	Product A	Product B	Product C	Capacity
Punch press	2.4	3.0	2.0	1,200
Welding	0.0	2.5	1.5	600
Assembly	5.0	0.0	2.5	1,500
Profit ($)	0.60	0.70	0.50	

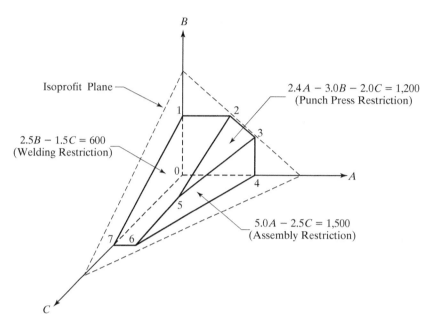

Figure 20 Graphical optimization for a three-product production operation.

profit realized will depend upon the production quantity combination chosen. Thus, there exists a family of isoprofit planes, one for each value of total profit. One of these planes will have at least one point in the volume of feasible solutions and will be a maximum distance from the origin. The plane that maximizes profit will intersect the volume at an extreme point. This calls for the computation of total profit at each extreme point as given in Table 13. The coordinates of each extreme point were found from the restricting planes, and the associated profit was calculated from the total profit equation.

Inspection of the total profit values in Table 13 indicates that profit is maximized at point 5, which has the coordinates 180, 96, and 240. This means that if 180 units of product A,

TABLE 13 Total Profit Computations at Extreme Points of Figure 14

	Coordinate			
Point	A	B	C	Profit ($)
0	0	0	0	0
1	0	240	0	168.00
2	200	240	0	288.00
3	300	160	0	292.00
4	300	0	0	180.00
5	180	96	240	295.20
6	100	0	400	260.00
7	0	0	400	200.00

96 units of product B, and 240 units of product C are produced, profit will be maximized. No other production quantity combination will result in a higher profit. Also, no alternative optimum solutions exist, because the total profit plane intersects the volume of feasible solutions at only a single point.

Addition of a third product to the two-product production system increased total profit from $292.00 per week to $295.20 per week. This increase in profit results from reallocations of the available production time and from greater utilization of the idle capacity of the welding department. The number of units of product A was reduced from 200 to 180 and the number of units of B was reduced from 240 to 96. This made it possible to add 240 units of product C to the production program, with the resulting increase in profit.

5.2 The Simplex Optimization Algorithm

Many problems to which the general linear programming model may be applied are n dimensional in that n activities compete for scarce resources. In these cases, graphical optimization methods cannot be applied and a numerical optimization technique known as the simplex algorithm can be used.

The *simplex optimization algorithm* is an iterative technique that begins with a feasible solution, tests for optimality, and proceeds toward an improved solution. It can be shown that the algorithm will finally lead to an optimal solution if such a solution exists. In this section, the simplex method will be applied to the three-product production problem presented above. Reference to the graphical solution will explain certain facets of the computational procedure.

The Simplex Matrix. The three-product production problem required the maximization of a total profit equation subject to certain constraints. These constraints must be converted to equalities of the form specified by the general linear programming model. This requires the addition of three "slack" variables to remove the inequalities. Thus, the constraints become

$$2.4A + 3.0B + 2.0C + S_1 = 1,200$$

$$0.0A + 2.5B + 1.5C + S_2 = 600$$

$$5.0A + 0.0B + 2.5C + S_3 = 1,500$$

The amount of departmental time not used in the production program is represented by a slack variable. Thus, each slack variable takes on whatever value is necessary for the equality to exist. If nothing is produced, the slack variables assume values equal to the total production time available in each department. This gives an initial feasible solution expressed as $S_1 = 1,200$, $S_2 = 600$, and $S_3 = 1,500$. Each slack variable is one of the x_j variables in the model. However, since no profit can be derived from idle capacity, total profit may be expressed as

$$TP = \$0.60A + \$0.70B + \$0.50C + \$0S_1 + \$0S_2 + \$0S_3$$

The initial matrix required by the simplex algorithm may now be set up as shown in Table 14. The first column is designated e_i and gives the coefficients applicable to the initial feasible solution. These are all zero, since the initial solution involves the allocation of all production time to the slack variables. The second column is designated *Sol* and gives the

TABLE 14   Initial Matrix for a Three-Product Production Problem

e_j			0	0	0	0.60	0.70	0.50	
e_i	Sol	b	S_1	S_2	S_3	A	B	C	θ
0	S_1	1,200	1	0	0	2.4	3.0	2.0	400
0	S_2	600	0	1	0	0	2.5	1.5	240
0	S_3	1,500	0	0	1	5.0	0	2.5	∞
E_j		0	0	0	0	0	0	0	
$E_j - e_j$			0	0	0	−0.60	−0.70	−0.50	

$$k$$

variables in the initial solution. These are the slack variables that were introduced. The third column is designated b and gives the number of minutes of production time associated with the solution variables of the previous column. These reflect the total production capacity in the initial solution. Each of the next three columns is headed by slack variables with elements of zero or unity, depending upon which equation is served by which slack variable. The e_j heading for these columns carries an entry of zero corresponding to a zero profit. The last three columns are headed by the activity variables, with elements entered from the restricting equations. The e_j heading for these columns is the profit associated with each activity variable. The last column, designated θ, is used during the computational process.

Testing for Optimality.   After an initial solution has been obtained, it must be tested to see if a program with a higher profit can be found. The optimality test is accomplished with the aid of the last two rows in Table 14. The required steps are as follows:

1. Enter values in the row designated E_j from the expression $E_j = \sum e_i a_{ij}$, where a_{ij} are the matrix elements in the ith row and the jth column.
2. Calculate $E_j - e_j$, for all positions in the row designated $E_j - e_j$.
3. If $E_j - e_j$ is negative for at least one j, a better program is possible.

Application of the optimality test to the initial feasible solution is shown in the last two rows of Table 14. The first element in the E_j row is calculated as $0(1,200) + 0(600) + 0(1,500) = 0$. The second is $0(1) + 0(0) = 0$. All values in this row will be zero because all e_i values are zero in the initial feasible solution. The first element in the $E_j - e_j$ row is $0 - 0 = 0$, the second is $0 - 0 = 0$, the third is $0 - 0 = 0$, the fourth is $0 - 0.60 = -0.60$, and so forth. Because $E_j - e_j$ is negative for at least one j, this initial solution is not optimal.

Iteration Toward an Optimal Program.   If the optimality test indicates that an optimal program has not been found, the following iterative procedure may be employed:

1. Find the minimum value of $E_j - e_j$ and designate this column k. The variable at the head of this column will be the incoming variable.
2. Calculate entries for the column designated θ from $\theta_i = b_i/a_{ik}$.
3. Find the minimum positive value of θ_i and designate this row r. The variable to the left of this row will be the outgoing variable.

TABLE 15 First Iteration for a Three-Product Production Problem

e_j			0	0	0	0.60	0.70	0.50	
e_j	Sol	b	S_1	S_2	S_3	A	B	C	θ
0	S_1	480	1	−1.20	0	2.40	0	0.20	200
0.70	B	240	0	0.40	0	0	1	0.60	∞
0	S_3	1,500	0	0	1	5.00	0	2.50	300
E_j		168	0	0.28	0	0	0.70	0.42	
$E_j − e_j$			0	0.28	0	−0.60	0	−0.08	
						k			

4. Set up a new matrix with the incoming variable substituted for the outgoing variable. Calculate new elements, a'_{ij}, as $a'_{rj} = a_{rj}/a_{rk}$ for $i = r$ and $a'_{ij} = a_{ij} − a_{ik}a'_{rj}$ for $i \neq r$.
5. Perform the optimality test.

Apply steps 1–3 of the foregoing procedure to the initial matrix. In Table 14, step 1 designates B as the incoming variable. Values for θ_i are calculated from step 2. Step 3 designates S_2 as the outgoing variable. The affected column and row are marked with a k and an r, respectively, in Table 14.

Steps 4 and 5 require a new matrix, as shown in Table 15. The incoming variable B, together with its associated profit, replaces the outgoing variable S_2 with its profit. All other elements in this row are calculated from the first formula of step 4. Elements in the remaining two rows are calculated from the second formula of step 4. The optimality test indicates that an optimal solution has not yet been reached. Note that after this iteration, the profit at point 1 of Figure 20 appears. Comparison of the results in Tables 14 and 15 with the total profit computations in Table 13 indicates that the isoprofit plane, which began at the origin, has now moved away from this initial position to point 1. The gain from this iteration was $168 − $0 = $168.

Because the first iteration did not yield an optimal solution, it is necessary to repeat steps 1–5. Steps 1–3 are applied to Table 15, designating A as the incoming variable and S_1 as the outgoing variable. The incoming variable, together with its associated profit, replaces the outgoing variable as shown in Table 16. All other elements in this new matrix are calculated from the formulas in step 4. Applications of the optimality test indicate that the

TABLE 16 Second Iteration for a Three-Product Production Problem

e_j			0	0	0	0.60	0.70	0.50	
e_j	Sol	b	S_1	S_2	S_3	A	B	C	θ
0.60	A	200	0.416	−0.50	0	1	0	0.084	2,381
0.70	B	240	0	0.40	0	0	1	0.60	400 r
0	S_3	500	−2.08	2.50	1	0	0	2.08	240
E_j		288	0.25	−0.02	0	0.60	0.70	0.47	
$E_j − e_j$			0.25	−0.02	0	0	0	−0.03	
								k	

TABLE 17 Third Iteration for a Three-Product Production Problem

e_j			0	0	0	0.60	0.70	0.50	
e_i	Sol	b	S_1	S_2	S_3	A	B	C	θ
0.60	A	180	0.50	−0.60	−0.04	1	0	0	
0.70	B	96	0.60	−0.32	−0.29	0	1	0	
0.50	C	240	−1	1.20	0.48	0	0	1	
E_j		295.20	0.22	0.016	0.016	0.60	0.70	0.50	
$E_j - e_j$			0.22	0.016	0.016	0	0	0	

solution indicated is still not optimal. Table 13 and Figure 18 show that the isoprofit plane is now at point 2. The gain from this iteration was $288 − $168 = $120.

Table 16 did not yield an optimal solution, requiring the reapplication of steps 1–5. Steps 1–3 designate C as the incoming variable and S_3 as the outgoing variable. This incoming variable, together with its associated profit, replaces the outgoing variable as shown in Table 17. All other elements in the matrix are calculated from the formulas of step 4. Application of the optimality test indicates that the solution exhibited by Table 17 is optimal. Table 17 and Figure 20 indicate that the isoprofit plane is now at point 5. The gain from this iteration was $295.20 − $288 = $7.20.

Minimizing by the Simplex Algorithm. The computational algorithm just presented may be used without modification for problems requiring minimization if the signs of the cost coefficients are changed from positive to negative. The principle that maximizing the negative of a function is the same as minimizing the function then applies. If these coefficients are entered in the simplex matrix with their negative signs, the value of the solution will decrease as the computations proceed.

5.3 Optimizing Operations and Design by LP

The derivation and examples of linear programming in Section 5 are based on the decision evaluation function of Equation A.1. They apply to the optimization of operations. To apply linear programming to design optimization, leading to the selection from among design alternatives, the decision evaluation function of Equation A.2 should be utilized. This is illustrated by the placement of Equations A.1 and A.2 on the life-cycle domains of operations and design in Figure A.1. Focus is on evaluation of design alternatives by Equation A.2, where design-dependent parameters are identified and partitioned from design-independent parameters.

For an overview of the foregoing utilizing LP, refer to the evaluation function of Equation 50 and consider the right-hand side of the constraint set in connection with the example of a production system. The right-hand side provides capacity metrics; specifically, it refers to the capacity of production equipment types wherein reliability and maintainability are design-dependent parameters. These parameters combine to determine capacity availability. Accordingly, the design of a production system can be optimized by LP for each mutually exclusive equipment design alternative based on the prediction of design targets for both MTBF and MTTR. Highly reliable and easily maintainable equipment will provide greater capacity at the processing center than the obverse. The cost thereof can be traded off against increased revenue traceable to greater production throughput.

Other design-dependent parameter applications utilizing Equation A.2 and LP exist. For example, the time consumed by each product on each of the equipment centers depends upon the design of that product. For each product design alternative, the LP algorithm could be used to return an optimal (maximum) profit. The product design alternative producing the greatest maximum profit could then be identified.

6 SUMMARY AND EXTENSIONS

Optimization in design and operations is the title and focus of this chapter. This theme was purposely chosen to promulgate the concept of a dual role for the familiar subject of optimization.

This chapter opened with the basics of classical optimization theory, emanating from the differential calculus. Unconstrained classical optimization involving single and multiple alternatives was then presented and illustrated by utilizing several hypothetical applications in design and operations. This was followed by example-based presentations of constrained classical optimizations. An extension of two examples when multiple criteria are present was then offered, utilizing the decision evaluation display. The final topic was that of constrained optimization by graphical and simplex methods of linear programming.

All topics abstracted above were intended to enable the implementation of equivalence in its full meaning; that is, for integrating money flow modeling and economic optimization modeling. Specific cases were offered to show how and why this integration is important in systems engineering and analysis. And, others are offered to highlight the correct method for comparing alternatives on a "fair" basis under both single and multiple criteria situations.

There are several general principles that came to light in the development of the example applications in this chapter. They deserve special visibility and are listed below for review:

1. From a strictly theoretical perspective, application of the differential calculus to determine the optimum value of an objective function requires a second derivative test, as governed by Equation 7. However, all examples in this chapter have well-behaved evaluation functions and the indicated maximum or minimum is obvious. For example, see Footnote 3 pertaining to Equation 39.

2. Money flow modeling, applied in accordance with the evaluation function of Equation A.3, should be preceded by optimization of investment (or first cost) utilizing Equation A.1 (see Equation 12). This provides assurance that the full intent of the concept of equivalence is incorporated before alternatives are compared with each other.

3. Comparing alternatives on the basis of first cost only may lead to an incorrect selection when life-cycle cost is considered. This is demonstrated for a specific example by comparing Figures 5 and 6 where decision reversal was shown to occur.

4. Life-cycle cost optimization in asset procurement can reveal the best allocation of scarce resources to acquisition (or first cost) and to maintenance and operations. An optimization example, demonstrating the desirability of greater acquisition investment to avoid higher operational cost, may be studied by comparing Table 4 with Table 3. However, this was without interest cost, so a companion example was presented in Tables 5 and 6 with interest cost incorporated. The higher investment in acquisition was not justified in this instance.

5. Reference was made in Figure 14 to the extension of inventory models to determine the best procurement source. This was done by the identification of source-dependent parameters and the repeated application of Equation 39 to the three source possibilities in the example of Table 9. This evaluation parallels the optimization of mutually exclusive design alternatives by the identification of design-dependent parameters in other examples.

6. Section 3 revisits two prior unconstrained examples by adding constraints and observing the effect. In the first example, the constraint leads to design decision reversal and a higher total system cost. This was illustrated graphically in Figure 15. In the second, the space constraint leads to a higher total cost per period shown in Figure 16. Penalties always occur when constrained and unconstrained situations are compared.

7. In the two examples of evaluation involving optimization and multiple criteria (Section 4), it is clear that the choice cannot be made on the basis of cost alone. Application of the decision evaluation display and involvement of the customer are required to select the preferred alternative for each example.

8. Optimization of a linear evaluation function subject to linear constraints is the last topic in this chapter. Its application to operations is legendary. The final subsection, 5.3, is an attempt to place linear programming into the design domain through design characteristics of the output product as well as by the role of reliability and maintainability (design-dependent parameters) in the availability of the producing recourse.

Extension of the operations material in this chapter is available from any good operations research textbook. However, those interested in the mapping of operations research models on the Churchman, Ackoff, and Arnoff paradigm of 1957 are encouraged to refer to W. J. Fabrycky, P. M. Ghare, and P. E. Torgersen, *Applied Operations Research and Management Science*, Prentice Hall, 1984. The suggested mapping will take the reader to Equation A.1, but not beyond, for many of the areas covered in this chapter and elsewhere.

QUESTIONS AND PROBLEMS

1. Specify the dimensions of the sides of a rectangle of perimeter p so that the area it encloses will be maximum.

2. The cost per unit produced at a certain facility is represented by the function

$$UC = 2x^2 - 10x + 50$$

where x is in thousands of units produced. For what value of x would unit cost be minimized (other than zero)? What is the minimum cost at this volume? Show that the value found is truly a minimum.

3. Advertising expenditures have been found to relate to profit approximately in accordance with the function

$$P = x^3 - 100x^2 + 3125x$$

where x is the expenditure in thousands of dollars. What advertising expenditure would produce the maximum profit? What profit is expected at this expenditure? Show that the derived result is truly a maximum.

4. The cost of producing and selling a certain item is $220x + $15,000$ for the first 1,000 units, $120x + $115,000$ for a production range between 1,000 and 2,500 units, and $205x − $97,500$ for more than 2,500 units, where x is the number of units produced. If the selling price is $200 per unit, and all units produced are sold, find the level of production that will maximize profit.

5. Ethyl acetate is made from acetic acid and ethyl alcohol. Let x = pounds of acetic acid input, y = pounds of ethyl alcohol input, and z = pounds of ethyl acetate output. The relationship of output to input is

$$\frac{z^2}{(1.47x − z)(1.91y − z)} = 3.9$$

   (a) Determine the output of ethyl acetate per pound of acetic acid, where the ratio of acetic acid of ethyl alcohol is 2.0, 1.0, and 0.67, and graph the result.

   (b) Graph the cost of material per pound of ethyl acetate for each of the ratios given and determine the ratio for which the material cost per pound of ethyl acetate is a minimum if acetic acid costs $0.80 per pound and ethyl alcohol costs $0.92 per pound.

6. It has been found that the heat loss through the ceiling of a building is 0.13 Btu per hour per square foot of area for each 1°F of temperature difference. If the 2,200 feet^2 ceiling is insulated, the heat loss in Btu per hour per degree temperature difference per square foot of area is taken as

$$\frac{1}{(1/0.13) + (t/0.27)}$$

   where t is the thickness in inches. The in-place cost of insulation 2, 4, and 6 inches thick is $0.36, $0.60, and $0.88 per square foot, respectively. The building is heated to 72°F, 3,000 hours per year, by a gas furnace with an efficiency of 50%. The mean outside temperature is 45°F and the natural gas used in the furnace costs $8.80 per 1,000 cubic feet and has a heating value of 1,020 Btu per cubic feet. What thickness of insulation, if any, should be used if the interest rate is 10% and the resale value of the building 6 years hence is enhanced $2,000 if insulation is added, regardless of the thickness?

7. An overpass is being considered for a certain crossing. The superstructure design under consideration will be made of reinforced concrete and will have a weight per foot depending on the span between piers in accordance with $W = 32(S) + 1,850$. Piers will be made of steel and will cost $250,000 each. The superstructure will be erected at a cost of $3.20 per pound. If the number of piers required is to be one less than the number of spans, find the number of piers that will result in a minimum total cost for piers and superstructure if $L = 600$ feet.

8. Two girder designs are under consideration for a bridge for a 1,200-foot crossing. The first is expected to result in a superstructure weight per foot of $22(S) + 800$, where S is the span between piers. The second should result in superstructure weight per foot of $20(S) + 1,000$. Piers and two required abutments are estimated to cost $220,000 each. The superstructure will be erected at a cost of $5.40 per pound. Choose the girder design that will result in a minimum cost and specify the optimum number of piers.

9. What is the cost advantage of choosing the best girder design for the bridge described in Problem 8? If the number of piers is determined from the best girder design alternative, but the other design alternative is adopted, what cost penalty is incurred?

10. A used automobile can be purchased by a student to provide transportation to and from school for $5,500 as is (i.e., the auto will have no warranty). First-year maintenance cost is

expected to be $350 and the maintenance costs will increase by $100 per year thereafter. Operation costs for the automobile will be $1,200 for every year the auto is used and its salvage value decreases by 15% per year.

(a) What is the economic life without considering the time value of money?

(b) With interest at 10%, what is the economic life?

11. As an alternative to the used automobile in Problem 10, the student can purchase a new "utility" model for $13,600 with a 3-year warranty. First-year maintenance cost is expected to be $50 and the maintenance cost will increase by $50 per year thereafter. Operation costs for this new automobile are expected to be $960 for each year of use and its salvage value decreases by 20% per year. What is the economic advantage of the new automobile without interest; with interest at 10%?

12. Special equipment can be designed and built for $80,000. This equipment will have a salvage value of $70,000 in year 1, which will decrease by $10,000 through year 8. Operation and maintenance costs will start at $18,000 in year 1, and increase by 5% per year through year 8. The organization proposing this design can borrow money with a 9% annual interest rate. Using the tabular approach, find the economic life of this equipment.

13. Company X is considering using an impact wrench with a torque-sensitive clutch to fasten bolts on one of their assembly lines. The impact wrench, which can be purchased for $950, has a maximum life of 4 years. The impact wrench will have a salvage value of $600 in year 1, and will decrease in value by $200 through year 4. The impact wrench will have operation and maintenance costs of $100 in year 1, $250 in year 2, $325 in year 3, and $400 in year 4. Company X has an MARR of 25%. Using the tabular approach, find the impact wrench's economic value. Show the impact wrench's economic value graphically.

14. Plot an inventory flow diagram similar to Figure 11 if $R = \infty$. Derive optimum values for Q, L, and TC under this assumption and verify the result by substituting into Equations 38–40.

15. Plot an inventory flow diagram similar to Figure 11 if $R = \infty$ and $C_s = \infty$. Derive optimum values of Q, L, and TC under these assumptions and verify the result by substituting into Equations 38–40.

16. An engine remanufacturer requires 82 pistons per day in its assembly operations. No shortages are to be allowed. The machine shop can produce 500 pistons per day. The cost associated with initiating manufacturing action is $400 and the holding cost is $0.45 per piston per day. The manufacturing cost is $105 per piston.

(a) Find the minimum-cost production quantity.

(b) Find the minimum-cost procurement level if production lead time is 8 days.

(c) Calculate the total system cost per day.

17. A subcontractor has been found who can supply pistons to the remanufacturer described in Problem 16. Procurement cost will be $90 per purchase order. The cost per unit is $108.

(a) Calculate the minimum total system cost per day for purchasing from the subcontractor.

(b) What is the economic advantage of adopting the minimum-cost source?

18. The demand for a certain item is 12 units per period. No shortages are to be allowed. Holding cost is $0.02 per unit per period. Demand can be met either by purchasing or manufacturing, with each source described by the data given in the following table.

	Purchase	**Manufacture**
Procurement lead time	18 periods	13 periods
Item cost	$11.00	$9.60
Procurement cost	$20.00	$90.00
Replenishment rate	∞	25 units/period

(a) Find the minimum-cost procurement source and calculate its economic advantage over its alternative source.

(b) Find the minimum-cost procurement quantity.

(c) Find the minimum-cost procurement level.

19. An electronic component manufacturing firm has a demand of 250 units per period. It costs $400 to initiate manufacturing action to produce at the rate of 600 units per period. The unit production cost is $90. The holding cost is $0.15 per unit and the shortage cost is $3.25 per unit short per period for unsatisfied demand. Determine (a) the minimum-cost manufacturing quantity, (b) the minimum-cost procurement level if production lead time is 12 periods, and (c) the total system cost per period.

20. Derive expressions for the minimum-cost procurement quantity and minimum-cost procurement level when C_h is assumed to be infinite. Name a real-world situation where such a model would apply.

21. If the span between piers must be at least 70 feet in Problem 7, what cost penalty is incurred for this constraint?

22. Suppose that no more than five piers can be utilized in the bridge design of Problem 7. What is the cost penalty incurred for this constraint if these piers cost $300,000 each?

23. An item with a demand of 300 units per period is to be purchased and no shortages are allowed. The item costs $1.30, holding cost is $0.02 per unit per period, and it costs $28.00 to process a purchase order. Each item consumes 2 cubic feet of warehouse space. The warehouse contains 1,500 cubic feet of space. Find the minimum-cost procurement quantity and procurement level under this restriction if the lead time is 2 periods. What is the cost penalty due to the warehouse restriction?

24. The demand for a certain item is 20 units per period. Unsatisfied demand causes a shortage cost of $0.60 per unit per period. The cost of initiating manufacturing action is $48.00 and the holding cost is $0.04 per unit per period. Production cost is $7.90 per unit and the item may be produced at the rate of 60 units per period. Each item consumes 3 cubic units of warehouse space. The warehouse space reserved for this item is limited to 300 cubic units. Find the minimum-cost procurement quantity and procurement level under this restriction if the lead time is 3 periods. What is the cost penalty per period due to the warehouse restriction?

25. Solve graphically for the values of x and y that maximize the function

$$Z = 2.2x + 3.8y$$

subject to the constraints

$$2.4x + 3.2y \leq 140$$
$$0.0x + 2.6y \leq 80$$
$$4.1x + 0.0y \leq 120$$
$$x \geq 0 \quad \text{and} \quad y \geq 0$$

26. A small shop has capability in turning, milling, drilling, and welding. The machine capacity is 16 hours per day in turning, 8 hours per day in milling, 16 hours per day in drilling, and 8 hours per day in welding. Two products, designated A and B, are under consideration. Each will yield a net profit of $0.35 per unit and will require the amount of machine time shown in the following table. Solve graphically for number of units of each product that should be scheduled to maximize profit.

	Product A	Product B
Turning	0.046	0.124
Milling	0.112	0.048
Drilling	0.040	0.000
Welding	0.000	0.120

27. Solve graphically for the values of x, y, and z that maximize the function

$$P = 7.8x + 9.4y + 2.6z$$

subject to the constraints

$$4.2x + 11.7y + 3.5z \leq 1,800$$

$$0.8x + 4.3y + 1.9z \leq 2,700$$

$$12.7x + 3.8y + 2.5z \leq 950$$

$$x \geq 0, \quad y \geq 0, \quad \text{and} \quad z \geq 0$$

28. Evaluate a production facility designed for higher availability than that given in Table 12. Specifically, assume that higher MTBF and lower MTTR design values lead to the capacity in minutes per week to be 1,320 in punch press, 680 in welding, and 1,600 in assembly, as shown in the following table. These higher design-dependent parameter values will be more costly, reducing profit per unit to $0.585 for product A, $0.68 for product B, and $0.49 for product C. Determine the optimal production mix and the profit per week that would be possible if this facility design with higher availability were adopted. Should this alternative be chosen over the baseline design?

Department	Product A	Product B	Product C	Capacity
Punch press	2.4	3.0	2.0	1,320
Welding	0.0	2.5	1.5	680
Assembly	5.0	0.0	2.5	1,600
Profit	$0.585	$0.68	$0.49	

Appendix

$$E = f(X, Y) \tag{A.1}$$

$$E = f(X, Y_d, Y_i) \tag{A.2}$$

$$PE, AE, \text{ or } FE = f(F_t, i, n) \tag{A.3}$$

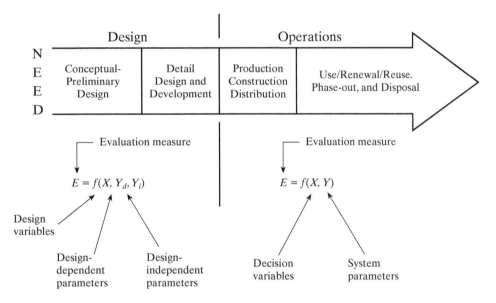

Figure A.1 Economic optimization in design and operations.

Queuing Theory and Analysis

The *queuing* or *waiting-line system* under consideration in this chapter may be described as follows. A facility of group of facilities is maintained to meet the demand for service created by a population of individuals or entities. These units form a queue or waiting line and receive service in accordance with a predetermined waiting-line discipline. In most cases, the serviced units rejoin the population and again become candidates for service. In other cases, the individuals form a waiting line at the next stage in the system.

Systems with these characteristics are common in operations where people, materials, equipment, or vehicles form waiting lines. The public forms waiting lines at cafeterias, doctors' offices, and theaters. In production, the flow of items in process forms a waiting line at each processing center. In maintenance, equipment to be repaired waits for service at maintenance facilities. In transportation, waiting lines form at toll gates, traffic signals, and loading docks. In each case, the objective is to optimize across a set of decision variables in the face of system parameters, whether these parameter values are fixed or subject to determination through design.

The fundamental areas abstracted above are developed with recognition that there are both infinite and finite population domains to consider. First, the general nature of the queuing system is presented. It is then represented by a Monte Carlo analysis. Single-channel models are derived followed by the derivation of models for multiple channels, both under several assumptions about the arrival- and service-time distributions. The chapter closes with a treatment of finite population queuing models. Consideration of these topics should give the reader proficiency in several areas:

- Understanding the components and elements that comprise queuing systems made up of arrival, waiting, and service mechanisms;
- Performing a Monte Carlo analysis of a queuing process and analyzing the results;
- Understanding the derivation of a number of common metrics for single- and multiple-channel queuing systems with exponential arrival and service times;
- Replacing the exponential service-time assumption with various nonexponential service-time distributions and understanding the effects;
- Contrasting the similarities and differences in infinite population queuing theory and finite population theory;

From Chapter 10 of *Systems Engineering and Analysis,* Fifth Edition, Benjamin S. Blanchard, Wolter J. Fabrycky. Copyright © 2011 by Pearson Education, Inc. Published by Pearson Prentice Hall. All rights reserved.

- Applying the Finite Queuing Tables to service channel optimization, service-time control, and alteration of the service factor; and
- Utilizing decision evaluation theory for the evaluation of infinite population and finite population queuing system analysis and design.

The final section of this chapter provides a summary of the domain of queuing theory and analysis and emphasizes key points for further study. It also provides some recommended material useful in extending the understanding of queuing systems.

1 THE QUEUING SYSTEM

A multiple-channel queuing system is illustrated schematically in Figure 1. It exists because the population shown requires service. In satisfying the demand on the system, the decision maker must establish the level of service capacity to provide. This will involve increasing or decreasing the service capacity by altering the service rate at existing channels or by adding or deleting channels. The following sections describe the components of the waiting-line system and indicate their importance in the decision evaluation process.

1.1 The Arrival Mechanism

The demand for service is the primary stimulus on the waiting-line system and the justification for its existence. As previously indicated, the waiting-line system may exist to meet the service demand created by people, materials, equipment, or vehicles. The characteristics of the arrival pattern depend on the nature of the population giving rise to the demand for service.

Waiting-line systems come into being because there exists a population of individuals or units requiring service from time to time. Usually, the arrival population is best thought of as a group of items, some of which depart and join the waiting line. For example, if the population is composed of all airborne aircraft, then flight schedules and random occurrences will determine the number of aircraft that will join the landing pattern of a given airport during a given time interval. If the population is composed of telephone subscribers, then the time of day and the day of the week, as well as many other

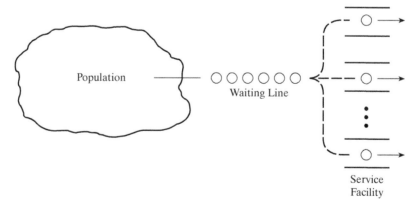

Figure 1 A multiple-channel queuing system.

factors, will determine the number of calls placed on an exchange. If the population consists of production machines, the deterioration, wear-out, and use rates will determine the departure mechanism causing the machines to join a waiting line of machines requiring repair service.

The arrival population, although always finite in size, may be considered infinite under certain conditions. If the departure rate is small relative to the size of the population, the number of units that potentially may require service will not be seriously depleted. Under this condition, the population may be considered infinite. Models used to explain the behavior of such systems are much easier to formulate than are models for a finite population. Examples of populations that may usually be treated as infinite are automobiles that may require passage over a bridge, customers who may potentially patronize a theater, telephone subscribers who may place a call, and production orders that may require processing at a specific machine center.

In some cases, the proportion of the population requiring service may be fairly large when compared with the population itself. In these cases, the population is seriously depleted by the departure of individuals to the extent that the departure rate will not remain stable. Because models used to explain waiting-line systems depend on the stability of the arrival rate, finite cases must be given special treatment.

Examples of waiting-line operations that might be classified as finite are equipment items that may require repair, semiautomatic production facilities that require operator attention, and company cafeterias that serve a captive population. Queuing systems for the case where the infinite population assumption does not hold is treated in Section 6.

1.2 The Waiting Line

In any queuing system, a departure mechanism exists that governs the rate at which individuals leave the population and join the queue. This departure mechanism is responsible for the formation of the waiting line and the need to provide service. Formation of the queue is a discrete process. Individuals or items joining the waiting line do so as integer values. The number of units in the waiting line at any point is an integer value. Rarely, if ever, is the queuing process continuous.

Individuals or items becoming a part of the waiting line take a position in the queue in accordance with a certain waiting-line discipline. The most common discipline is that of first come, first served. Other priority rules that may exist are the random selection process; the relative urgency rule; first come, last served; and disciplines involving a combination of these. In addition, individuals may remain in the queue for a period and then rejoin the population. This behavior is called *reneging*.

When a unit joins the waiting line, or is being serviced, a waiting cost is incurred. Waiting cost per unit per period will depend on the units in question. If expensive equipment waits for operator attention, or requires maintenance, the loss of profit may be sizable. Vehicles waiting in queue at a toll gate incur a waiting cost due to interruption of trip progress. Customers waiting at a checkout counter become irritated and the proprietor suffers a loss of goodwill.

Increasing the service capacity will cause a decrease in both the length of the waiting line and the time required for each service completion. As a result, the waiting time will be decreased. Because waiting cost to the system is a product of the number of units waiting and the time duration involved, this action will decrease this cost component. But since increasing the service facility capacity increases the service cost, it is appropriate to seek a reduction in waiting cost for the system only up to the point where the saving justifies the added facility cost.

1.3 The Service Mechanism

The rate at which units requiring service are serviced is assumed to be a variable directly under the control of the decision maker. This variable can be assigned a specific value to create a minimum-cost waiting-line system.

Service is the process of providing the activities required by the units in the waiting line. It may consist of collecting a toll, filling an order, providing a necessary repair, or completing a manufacturing operation. In each case, the act of providing the service causes a unit decrease in the waiting line. The service mechanism, like the arrival mechanism, is discrete, because items are processed on a unit basis.

The service facility may consist of a single channel, or it may consist of several channels in parallel, as in Figure 1. If it consists of only a single channel, all arrivals must eventually pass through it. If several channels are provided, items may move from the waiting line into the first channel that becomes empty. The rate at which individuals are processed depends on the service capacity provided at the individual channels and the number of channels in the system.

The service may be provided by human beings only, by human beings aided by tools and equipment, or by equipment alone. For example, collecting a fee is essentially a clerk's task, which requires no tools or equipment. Repairing a vehicle, on the other hand, requires a mechanic aided by tools and equipment. Processing a phone call dialed by the subscriber seldom requires human intervention and is usually paced by the automatic equipment. These examples indicate that service facilities can vary widely with respect to the person–machine mix used to provide the required service.

Each channel of the service facility represents a capital investment plus operating and maintenance costs. In addition, wages for personnel may be involved together with associated overhead rates. The capability of the channel to process units requiring service is a function of the resources expended at the channel. For example, the channel may consist of a single repairperson with modest tools, or it may be a crew of technicians with complex tools and equipment. The cost of providing such a facility will depend on the characteristics of the personnel and equipment employed.

Because increasing the service capacity will result in a reduction in the waiting line, it is appropriate to adjust service capacity so that the sum of waiting cost and service cost is a minimum. The general structure of evaluation models directed to this objective will be presented next.

1.4 Queuing System Evaluation

For a queuing system already in being, an evaluation function is applicable. Specifically,

$$E = f(X, Y)$$

where

E = measure of evaluation
X = policy variable concerning the level of service capacity to provide
Y = system parameters of the arrival rate, the service rate, the waiting cost, and the service facility cost

If the queuing system is being designed, an evaluation function is applicable. Specifically,

$$E = f(X, Y_d, Y_i)$$

where

E = measure of evaluation

X = design variable concerning the level of service capacity to specify

Y_d = design-dependent parameters (usually reliability and/or maintainability of the units requiring service)

Y_i = design-independent parameters of waiting cost and service facility cost

The following sections are devoted to developing decision evaluation models with the preceding characteristics. Let

TC = total system cost per period

A = number of periods between arrivals

S = number of periods to complete one unit of service

C_w = cost of waiting per unit per period

C_f = service facility cost for servicing one unit

Additional notation will be adopted and defined as required for deriving specific decision models.

2 MONTE CARLO ANALYSIS OF QUEUING

Decision models for probabilistic waiting-line systems are usually based on certain assumptions regarding the mathematical forms of the arrival- and service-time distributions. Monte Carlo analysis, however, does not require that these distributions obey certain theoretical forms. Waiting-line data are produced as the system is simulated over time. Conclusions can be drawn from the output statistics, whatever be the form of the underlying distributions. In addition, the detailed numerical description that results from Monte Carlo analysis assists greatly in understanding the probabilistic queuing process. This section illustrates the application of Monte Carlo analysis to an infinite population, single-channel waiting-line system (see Appendix: Probability Theory and Analysis, Section 3).

2.1 Arrival- and Service-Time Distributions

The probabilistic waiting-line system usually involves both an arrival-time distribution and a service-time distribution. Monte Carlo analysis requires that the form and parameters of these distributions be specified. The cumulative distributions may then be developed and used as a means for generating arrival- and service-time data.

For example, assume that the time between arrivals, A_x, has an empirical distribution with a mean of 6.325 periods ($A_m = 6.325$). Service time, S_x, will be assumed to have a normal distribution with a mean of 5.000 periods ($S_m = 5.000$) and a standard deviation of 1 period. These distributions are exhibited in Figure 2. The probabilities associated with each value of A_x and S_x are indicated. By summing these individual probabilities from left to right, the

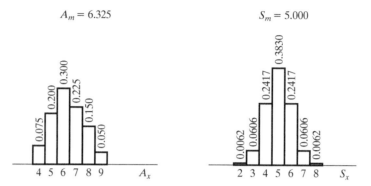

Figure 2 Arrival- and service-time distributions.

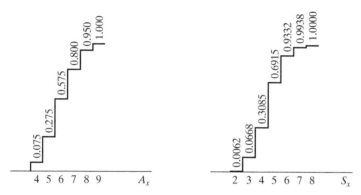

Figure 3 Cumulative arrival- and service-time distributions.

cumulative distributions of Figure 3 results. These distributions may be used with a table of random rectangular variates to generate arrival- and service-time random variables or with a computer-based algorithm.

2.2 The Monte Carlo Analysis

The queuing process under study is assumed to begin when the first arrival occurs. A unit will move immediately into the service facility if it is empty. If the service facility is not empty, the unit will wait in the queue. Units in the waiting line will enter the service facility on a first-come, first-served basis. The objective of the Monte Carlo analysis is to simulate this process over time. The number of unit periods of waiting in the queue and in service may be observed for each of several service rates. The service rate resulting in a minimum-cost system may then be adopted.

The waiting-line process resulting from the arrival- and service-time distributions of Figure 2 is shown in Figure 4. The illustration reads from left to right, with the second line being a continuation of the first, and so forth. The interval between vertical lines represents two periods, the heavy dots represent arrivals, the slanting path a unit in service, and the arrows a service completion. When a unit cannot move directly into the service channel it waits in the queue, which is represented by a horizontal path.

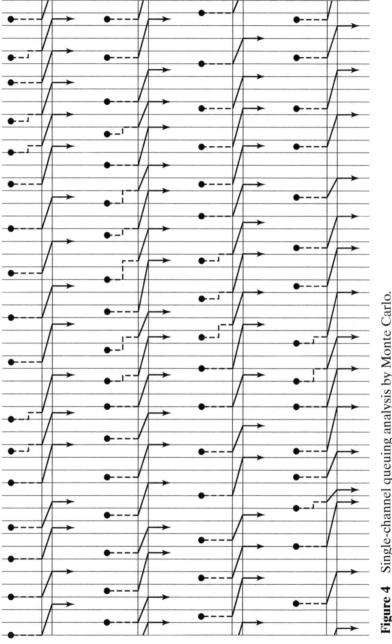

Figure 4 Single-channel queuing analysis by Monte Carlo.

The probabilistic waiting-line process illustrated in Figure 4 involves 400 periods and was developed in the following manner. First, the sequence of arrivals was established by the use of random rectangular variates and the cumulative arrival distribution of Figure 3. Next, each arrival was moved into the service channel if it was available. Channel availability is a function of the arrival pattern and the service durations selected with the aid of random rectangular variates and the cumulative service-time distribution of Figure 3.

Specifically, the Monte Carlo analysis proceeded as follows. Random rectangular variates were chosen as 5668, 3513, 2060, 7804, 0815, 2639, 9845, 6549, 6353, 7941, and so on. These correspond to arrival intervals of 6, 6, 5, 7, 5, 5, 9, 7, 7, 7, and so forth. Next, random rectangular variates were chosen as 323, 249, 404, 275, 879, 404, 740, 779, 441, 384, and so on. These correspond to service durations of 5, 4, 5, 4, 6, 5, 6, 6, 5, 5, and so forth. These service times determine the time an arrival enters the service channel and the time it is discharged. By proceeding in this manner, the results of Figure 4 are obtained.

2.3 Economic Evaluation of Results

The 400 periods simulated produced a waiting pattern involving 337 unit periods of waiting in service and 23 unit periods of waiting in the queue. The total number of unit periods of waiting for the 400-period sample was 360.

Suppose that waiting cost per unit per period is \$9.60 and that it costs \$16.10 per period to provide the service capability indicated by the service-time distribution of Figure 4. The total system cost for the study period is, therefore, \$9.60(360) + \$16.10(400) = \$9,896. This total system cost may be compared with the total system cost for alternative service policies by performing a Monte Carlo simulation for the alternative policies. Although this process is time consuming, it is applicable to many situations that cannot be treated by mathematical means.

3 SINGLE-CHANNEL QUEUING MODELS

Assume that the population of units that may demand service is infinite, with the number of arrivals per period a random variable with a Poisson distribution. It is also assumed that the time required to service each unit is a random variable with an exponential distribution. Events are recognized to have occurred at the time of arrival of a unit or at the time of completion of a service.

Under the assumption of Poisson arrivals and exponential service, it can be shown that the probability of the occurrence of an event (arrival or service completion) during a specific interval does not depend on the time of the occurrence of the immediately preceding event of the same kind. The expected number of arrivals per period may be expressed as $1/A_m$, defined as λ, and the expected number of service completions per period may be expressed as $1/S_m$, defined as μ.

3.1 Probability of n Units in the System

Under the foregoing assumptions, the probability that an arrival occurs between time t and time $t + \Delta t$ is $\lambda \Delta t$. Similarly, the probability that a service completion occurs between time t and time $t + \Delta t$, given that a unit is being serviced at time t, is $\mu \Delta t$. Let

n = number of units in the system at time t, including the unit being served, if any

$P_n(t)$ = probability of n units in the system at time t

Because the time interval Δt is small, it can be assumed that the probability of more than one arrival or service completion during the interval is negligible. Consider the event that there are n units in the system at time $t + \Delta t$ with $n \geq 1$ expressed as

Event {n units in the system at time $t + \Delta t$}
 = Event {n units in the system at time t, no arrivals during interval Δt, and no service completions during interval Δt}, or
 Event {$n + 1$ units in the system at time t, no arrivals during interval Δt, and one service completion during interval Δt}, or
 Event {$n - 1$ units in the system at time t, one arrival during interval Δt, and no service completion during interval Δt}

The probability of the event n units in the system at time $t + \Delta t$ can be written as the sum of the probabilities of these three mutually exclusive events as

$$
\begin{aligned}
P_n(t + \Delta t) &= \{P_n(t)[1 - \lambda\Delta t][1 - \mu\Delta t]\} \\
&\quad + \{P_{n+1}(t)[1 - \lambda\Delta t]\mu\Delta t\} + \{P_{n-1}(t)\lambda\Delta t[1 - \mu\Delta t]\} \\
&= P_n(t) - (\lambda + \mu)P_n(t)\Delta t + \lambda\mu P_n(t)(\Delta t)^2 + \mu P_{n+1}(t)\Delta t \\
&\quad - \lambda\mu P_{n+1}(t)(\Delta t)^2 + \lambda P_{n-1}(t)\Delta t - \lambda\mu P_{n-1}(t)(\Delta t)^2
\end{aligned} \tag{1}
$$

Terms involving $(\Delta t)^2$ can be neglected. Subtracting $P_n(t)$ from both sides and dividing by Δt yields

$$
\frac{P_n(t + \Delta t) - P_n(t)}{\Delta t} = -(\lambda + \mu)P_n(t) + \mu P_{n+1}(t) + \lambda P_{n-1}(t)
$$

In the limit,

$$
\lim_{\Delta t:\, 0} \frac{P_n(t + \Delta t) - P_n(t)}{\Delta t} = \frac{d}{dt}P_n(t) = -(\lambda + \mu)P_n(t) + \mu P_{n+1}(t) + \lambda P_{n-1}(t) \tag{2}
$$

For the special case $n = 0$,

Event {0 units in the system at time $t + \Delta t$}
 = Event {0 units in the system at time t and no arrivals during the interval Δt}, or
 Event {1 unit in the system at time t, no arrivals during the interval Δt, and one service completion during the interval Δt}

The probability of no units in the system at time $t + \Delta t$ can be written as the sum of the probabilities of these two mutually exclusive events as

$$
\begin{aligned}
P_0(t + \Delta t) &= \{P_0(t)[1 - \lambda\Delta t]\} + \{P_1(t)[1 - \lambda\Delta t]\mu\Delta t\} \\
&= P_0(t) - \lambda P_0(t)\Delta t + \mu P_1(t)\Delta t - \lambda\mu P_1(t)(\Delta t)^2
\end{aligned} \tag{3}
$$

Again neglecting terms involving $(\Delta t)^2$, subtracting $P_0(t)$ from both sides, and dividing by Δt, we obtain

$$
\frac{P_0(t + \Delta t) - P_0(t)}{\Delta t} = -\lambda P_0(t) + \mu P_1(t)
$$

In the limit,

$$\frac{d}{dt} P_n(t) = \lim_{\Delta t:\ 0} \frac{P_0(t + \Delta t) - P_0(t)}{\Delta t} = -\lambda P_0(t) + \mu P_1(t) \tag{4}$$

Equations 2 and 4 are called the *governing equations* of a Poisson arrival and exponential service single-channel queue. The differential equations constitute an infinite system for which the general solution is rather difficult to obtain. Figure 5 is an example of the nature of the solution for a particular case.

As long as the probabilities $P_n(t)$ are changing with time, the queue is considered to be in a transient state. From Figure 5 it should be noted that this change in $P_n(t)$ becomes smaller and smaller as the time increases. Eventually, there will be little change in $P_n(t)$, and the queue will have reached a steady state. In the steady state, the rate of change $dP_n(t)/dt$ can be considered to be zero and the probabilities considered to be independent of time. The steady-state governing equations can be written as

$$(\lambda + \mu)P_n = \mu P_{n+1} + \lambda P_{n-1} \tag{5}$$

and

$$\lambda P_0 = \mu P_1 \tag{6}$$

Equations 5 and 6 constitute an infinite system of algebraic equations that can be solved by substituting $P_{n+1} = P_n \cdot \rho$ into Equation 5 giving

$$(\lambda + \mu)\rho P_{n-1} = (\rho^2 \mu + \lambda)P_{n-1}$$

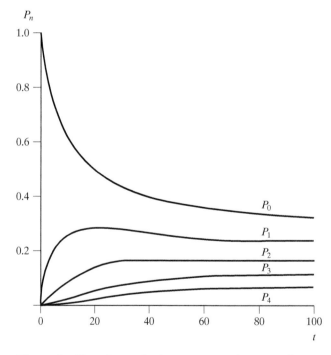

Figure 5 Transient solution to governing equations.

or

$$(\lambda + \mu)\rho = \rho^2\mu + \lambda$$

where

$$\rho = \frac{\lambda}{\mu}$$

Substituting in Equation 6 gives the same result. Using this substitution, the general solution can be written as

$$P_1 = \rho P_0$$

and

$$P_n = \rho P_{n-1}$$
$$= \rho^n P_0 \qquad (7)$$

However, because $\sum_{n=0}^{\infty} P_n = 1$,

$$1 = \sum_{n=0}^{\infty} P_0 \rho^n = P_0 \sum_{n=0}^{\infty} \rho^n$$
$$= P_0\left(\frac{1}{1-\rho}\right) \qquad (8)$$

Hence,

$$P_0 = 1 - \frac{\lambda}{\mu}$$

and

$$P_n = \left(1 - \frac{\lambda}{\mu}\right)\left(\frac{\lambda}{\mu}\right)^n \qquad (9)$$

The requirement for the convergence of the sum $\sum_{n=0}^{\infty}(\lambda/\mu)^n$ is that λ/μ be less than 1. This means that the arrival rate λ must be less than the service rate μ for the queue to reach steady state.

As an example of the significance of Equation 9 in waiting-line operations, suppose that a queue is experiencing Poisson arrivals with a mean rate of 1/10 unit per period and that the service duration is distributed exponentially with a mean of 4 periods. The service rate is, therefore, 1/4, or 0.25 unit per period. Probabilities associated with each value of n may be calculated as follows:

$$P_0 = (0.6)(0.4)^0 = 0.600$$

$$P_1 = (0.6)(0.4)^1 = 0.240$$

$$P_2 = (0.6)(0.4)^2 = 0.096$$

$$P_3 = (0.6)(0.4)^3 = 0.039$$

$$P_4 = (0.6)(0.4)^4 = 0.015$$

$$P_5 = (0.6)(0.4)^5 = 0.006$$
$$P_6 = (0.6)(0.4)^6 = 0.003$$
$$P_7 = (0.6)(0.4)^7 = 0.001$$

Figure 6 exhibits the probability distribution of n units in the system. Certain important characteristics of the waiting-line system can be extracted from this distribution. For example, the probability of 1 or more units in the system is 0.4, the probability of no units in the system is 0.6, the probability of more than 4 units in the system is 0.01, and so forth. Such information as this is useful when there is a restriction on the number of units in the system. By altering the arrival population or the service rate or both, the probability of the number of units in the system exceeding a specified value may be controlled.

3.2 Mean Number of Units in the System

The mean number of units in the system may be expressed as

$$n_m = \sum_{n=0}^{\infty} nP_n = \sum_{n=0}^{\infty} n(1 - \rho)\rho^n$$
$$= (1 - \rho) \sum_{n=0}^{\infty} n\rho^n$$

Let $g = \sum_{n=0}^{\infty} n\rho^n$, then

$$\rho g = \sum_{n=0}^{\infty} n\rho^{n+1} = \sum_{n=1}^{\infty} (n - 1)\rho^n$$

Subtracting ρg from g,

$$(1 - \rho)g = \sum_{n=0}^{\infty} n\rho^n - \sum_{n=1}^{\infty} (n - 1)\rho^n$$
$$= \sum_{n=1}^{\infty} n\rho^n - \sum_{n=1}^{\infty} n\rho^n + \sum_{n=1}^{\infty} \rho^n$$
$$= \sum_{n=1}^{\infty} \rho^n = \rho \sum_{n=0}^{\infty} \rho^n = \frac{\rho}{1 - \rho}$$

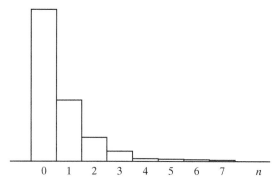

Figure 6   Probability distribution of n units in the system.

Hence,

$$n_m = (1 - \rho)g = \frac{\rho}{1 - \rho}$$

Substituting $\rho = \lambda/\mu$ yields

$$n_m = \frac{\lambda}{\mu - \lambda} \tag{10}$$

For the example given previously, the mean number of units in the system is

$$n_m = \frac{0.10}{0.25 - 0.10} = 0.667$$

3.3 Average Length of the Queue

The average length of the queue m_m can be expressed as the average number of units in the system less the average number of units being serviced:

$$m_m = \frac{\lambda}{\mu - \lambda} - \frac{\lambda}{\mu}$$

$$= \frac{\lambda^2}{\mu(\mu - \lambda)} \tag{11}$$

For the previous example, the average length of queue is

$$m_m = \frac{(0.10)^2}{0.25(0.25 - 0.10)} = 0.267$$

That is, the average length of a nonempty waiting line. The probability that the queue is nonempty is given by

$$P(m > 0) = 1 - P_0 - P_1$$
$$= 1 - (1 - \rho) - (1 - \rho)\rho = \rho^2 \tag{12}$$

And the average length of the nonempty queue is

$$(m|m > 0)_m = \frac{m_m}{P(m > 0)} \tag{13}$$

$$= \frac{\lambda^2/\mu(\mu - \lambda)}{\rho^2}$$

$$= \frac{\mu}{\mu - \lambda}$$

For the previous numerical example, the probability that the queue is nonempty is

$$P(m > 0) = \rho^2 = \frac{(0.10)^2}{(0.25)^2} = 0.16$$

and the average length of the nonempty queue is

$$\frac{0.25}{0.25 - 0.10} = 1.667$$

3.4 Distribution of Waiting Time

In a probabilistic queuing system, waiting time spent by a unit before it goes into service is a random variable that depends on the status of the system at the time of arrival and also on the time required to service the units already waiting for service. In the case of a single-channel system, an arriving unit can go immediately into service only if there are no other units in the system. In all other cases the arriving unit will have to wait.

Two different events can be identified under the Poisson arrival and exponential service assumption for the waiting time, W:

1. Event $\{w = 0\}$ is identical to Event $\{0$ units in the system$\}$.
2. Event {waiting time is in the interval w and $w + \Delta w$} is a composite event of there being n units in the system at the time of arrival, $n - 1$ services being completed during time w, and the last service being completed within the interval w and $w + \Delta w$.

The probability of the first event occurring is given by

$$p(w = 0) = P_0$$
$$= 1 - \frac{\lambda}{\mu} \tag{14}$$

The second event is illustrated in Figure 7. There will be one such event for every n. Furthermore,

$$P(w \leq \text{ waiting time } \leq w + \Delta w) = f(w)\Delta w$$
$$= \sum_{n=1}^{\infty} \{P_n \cdot P[(n - 1) \text{ services in times } w]\}$$
$$\cdot P(\text{ one service completion in time } \Delta w)$$
$$= \sum_{n=1}^{\infty} \left(1 - \frac{\lambda}{\mu}\right)\left(\frac{\lambda}{\mu}\right)^n \left[\frac{(\mu w)^{n-1} e^{-\mu w}}{(n - 1)!}\right]\mu\Delta w$$
$$= \sum_{n=1}^{\infty} \left(\frac{\lambda}{\mu}\right)^n \left[\frac{(\mu w)^{n-1}}{(n - 1)!}\right] e^{-\mu w}\left(1 - \frac{\lambda}{\mu}\right)\mu\Delta w$$

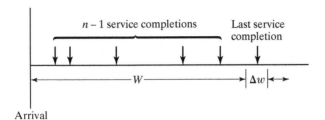

Figure 7 Waiting time when there are n units in the system.

Let $k = n - 1$. Then

$$f(w)\,\Delta w = \sum_{k=0}^{\infty}\left(\frac{\lambda}{\mu}\right)^k \frac{\mu^k w^k}{k!}\left(\frac{\lambda}{\mu}\right)e^{-\mu w}\left(1 - \frac{\lambda}{\mu}\right)\mu\,\Delta w \qquad (15)$$

However,

$$\sum_{k=0}^{\infty}\left(\frac{\lambda}{\mu}\right)^k \frac{\mu^k w^k}{k!} = \sum_{k=0}^{\infty}\left(\frac{\lambda w}{k!}\right)^k = e^{\lambda w} \qquad (16)$$

Substituting Equation 16 into Equation 15 gives

$$f(w)\,\Delta w = e^{\lambda w}\left(\frac{\lambda}{\mu}e^{-\mu w}\right)\mu\,\Delta w\left(1 - \frac{\lambda}{\mu}\right)$$

$$= \lambda\left(1 - \frac{\lambda}{\mu}\right)e^{-(\mu-\lambda)w}\,\Delta w$$

or

$$f(w) = \lambda\left(1 - \frac{\lambda}{\mu}\right)e^{-(\mu-\lambda)w} \qquad (17)$$

Equations 14 and 17 describe the waiting-time distribution for a single-channel queue with Poisson arrivals and exponential service. This distribution is partly discrete and partly continuous.

The mean time an arrival spends waiting for service can be obtained from the waiting-time distribution as

$$w_m = 0 \cdot P(w = 0) + \int_{w>0} w\lambda\left(1 - \frac{\lambda}{\mu}\right)e^{-(\mu-\lambda)w}\,dw$$

After several steps,

$$w_m = \frac{\lambda}{\mu(\mu - \lambda)} \qquad (18)$$

For the numerical example presented previously,

$$w_m = \frac{0.10}{0.25(0.25 - 0.10)}$$
$$= 2.667 \text{ periods}$$

The average time that a unit spends in the waiting-line system is composed of the average waiting time and the average time required for service, or

$$t_m = \frac{\lambda}{\mu(\mu - \lambda)} + \frac{1}{\mu}$$
$$= \frac{\lambda + \mu - \lambda}{\mu(\mu - \lambda)} = \frac{1}{\mu - \lambda} \tag{19}$$

For the preceding example, the average time spent in the system will be

$$t_m = \frac{1}{0.25 - 0.10} = 6.667 \text{ periods}$$

3.5 Minimum-Cost Service Rate

The expected total system cost per period is the sum of the expected waiting cost per period and the expected facility cost per period, that is,

$$TC_m = WC_m + FC_m$$

The expected waiting cost per period is obtained as the product of cost of waiting per period, the expected number of units arriving per period, and the average time each unit spends in the system, or

$$WC_m = C_w(\lambda) \frac{1}{\mu - \lambda}$$
$$= \frac{C_w \lambda}{\mu - \lambda} \tag{20}$$

Alternatively, the expected waiting cost per period can be obtained as the product of cost of waiting per period and the mean number of units in the system during the period, or

$$WC_m = C_w(n_m)$$
$$= \frac{C_w \lambda}{\mu - \lambda} \tag{21}$$

The expected service cost per period is the product of the cost of servicing one unit and the service rate in units per period, or

$$FC_m = C_f(\mu) \tag{22}$$

The expected total system cost per period is the sum of these cost components and may be expressed as

$$TC_m = \frac{C_w \lambda}{\mu - \lambda} + C_f(\mu) \tag{23}$$

A minimum-cost service rate may be found by differentiating with respect to μ, setting the result equal to zero, and solving for μ as follows:

$$\frac{dTC_m}{d\mu} = -C_w\lambda(\mu - \lambda)^{-2} + C_f = 0$$

$$(\mu - \lambda)^2 C_f = \lambda C_w$$

$$\mu = \lambda + \sqrt{\frac{\lambda C_w}{C_f}} \qquad (24)$$

As an application of the preceding optimization model, consider the following Poisson arrival and exponential service-time situation. The mean time between arrivals is eight periods, the cost of waiting is \$0.10 per unit per period, and the facility cost for serving one unit is \$0.165. The expected waiting cost per period, the expected facility cost per period, and the expected total system cost per period are exhibited as a function of μ in Table 1. The expected waiting cost per period is infinite when $\mu = \lambda$ and decreases as μ increases. The expected facility cost per period increases with increasing values of μ. The minimum expected total system cost occurs when μ is 0.4 unit per period.

The minimum-cost service rate may be found directly by substituting into Equation 24 as follows:

$$\mu = 0.125 + \sqrt{\frac{(0.125)(\$0.10)}{\$0.165}}$$

$$= 0.125 + 0.275 = 0.400 \text{ unit per period}$$

4 MULTIPLE-CHANNEL QUEUING MODELS

In the previous sections, arriving units were assumed to have been serviced through a single service facility. In many practical situations, however, there are several alternative service facilities. One example is the toll plaza on the turnpike, where several toll booths may serve the arriving traffic.

In the multiple-channel case the service facility will have c service channels, each capable of serving one unit at a time. An arriving unit will go to the first available service channel that is not busy. If all channels are busy, additional arrivals will form a single queue.

TABLE 1 Cost Components for Exponential Service Example

μ	WC_m (\$)	FC_m (\$)	TC_m (\$)
0.125	∞	0.0206	∞
0.150	0.5000	0.0248	0.5248
0.200	0.1667	0.0330	0.1997
0.250	0.1000	0.0413	0.1413
0.300	0.0714	0.0495	0.1209
0.400	0.0455	0.0660	0.1115
0.500	0.0333	0.0825	0.1158
0.600	0.0263	0.0990	0.1253
0.800	0.0185	0.1320	0.1505
1.000	0.0143	0.1650	0.1793

As soon as any busy channel completes service and becomes available, it accepts the first unit in the queue for service. The steady-state probabilities in such a system are defined as

$$P_{m,n}(t) = \text{probability that there are } n \text{ units waiting in queue}$$
$$\text{and } m \text{ channels are busy at time } t$$

It must be noted that m can only be an integer between 0 and c, and that n is zero unless $m = c$.

The determination of the steady-state probabilities follows the logic used in the single-channel case, although the solution becomes quite involved. The results are given below for the Poisson arrival and exponential service situation. Define

$$\rho = \frac{\lambda}{c\mu}$$

Then,

$$P_{c,n} = P_{0,0}\left(\frac{\lambda}{\mu}\right)^c \frac{1}{c!} \rho^n$$

$$P_{m,0} = P_{0,0}\left(\frac{\lambda}{\mu}\right)^m \frac{1}{m!}$$

$$P_{0,0} = \frac{1}{(\lambda/\mu)^c (1/c!)[1/(1-\rho)] + \displaystyle\sum_{r=0}^{r=c-1} (\lambda/\mu)^r (1/r!)} \tag{25}$$

As an example, consider a three-channel system with Poisson arrivals at a mean rate of 0.50 units per period and exponential service at each channel with a mean service rate of 0.25 units per period. Under these conditions, ρ is $[0.50/(3 \times 0.25)] = 2/3$ and

$$P_{0,0} = \frac{1}{(0.50/0.25)^3 (1/3!)[1/(1-2/3)] + \displaystyle\sum_{0}^{2}(0.50/0.25)^r (1/r!)}$$

$$= \frac{1}{4+1+2+2} = \frac{1}{9}$$

4.1 Average Length of the Queue

The average queue length is obtained from the expression

$$m_m = P_{0,0} \frac{(\lambda/\mu)^{c+1}}{(c-1)!(c-\lambda/\mu)^2} \tag{26}$$

For the example considered,

$$m_m = \frac{1}{9}\left[\frac{(0.50/0.25)^4}{2!(3-0.50/0.25)^2}\right]$$

$$= \frac{1}{9}\left[\frac{(2)^4}{2}\right] = \frac{8}{9} = 0.89 \text{ units}$$

320

4.2 Mean Number of Units in the System

The mean number of units in the system is

$$n_m = m_m + \frac{\lambda}{\mu} \qquad (27)$$

For the example considered,

$$n_m = 0.89 + \frac{0.50}{0.25} = 2.89 \text{ units}$$

4.3 Mean Waiting Time

The mean waiting time can be obtained from the expression for m_m as

$$w_m = \frac{m_m}{\lambda} \qquad (28)$$

In this example, $w_m = 0.89/0.50$ or 1.78 periods.

4.4 Average Delay or Holding Time

The average delay is obtained as the sum of waiting and service times as

$$d_m = w_m + \frac{1}{\mu} \qquad (29)$$

In the example, average delay is $1.78 + 4$ or 5.78 periods.

4.5 Probability That an Arriving Unit Must Wait

The probability of a delay is the same as the probability that all channels are occupied. This is

$$Pr(w > 0) = \sum_{0}^{\infty} P_{c,n}$$
$$= P_{0,0}\left(\frac{\lambda}{\mu}\right)^c \frac{1}{c!(1 - \rho)} \qquad (30)$$

In the example, the probability that an arriving unit has to wait is

$$P(w > 0) = \frac{1}{9}\left(\frac{0.50}{0.25}\right)^3 \frac{1}{3!(1 - 2/3)}$$
$$= \frac{1}{9}(8)\frac{1}{2} = \frac{4}{9} = 0.444$$

5 QUEUING WITH NONEXPONENTIAL SERVICE

The assumption that the number of arrivals per period obeys a Poisson distribution has a sound practical basis. Although it cannot be said that the Poisson distribution always adequately describes the distribution of the number of arrivals per period, much evidence exists to indicate that this is often the case. Intuitive considerations add support to this assumption because arrival rates are usually independent of time, queue length, or any other property of the waiting-line system. Evidence in support of the exponential distribution of service durations is not as strong. Often this distribution is assumed for mathematical convenience, as in previous sections.

When the service-time distribution is nonexponential, the development of decision models is somewhat more difficult. This section will present models with nonexponential service without proof.

5.1 Poisson Arrivals with Constant Service Times

When service is provided automatically by mechanical means, or when the service operation is mechanically paced, the service duration might be constant. Under these conditions, the service-time distribution has a variance of zero. The mean number of units in the system is given by

$$n_m = \frac{(\lambda/\mu)^2}{2[1 - (\lambda/\mu)]} + \frac{\lambda}{\mu} \tag{31}$$

and the mean waiting time is

$$w_m = \frac{\lambda/\mu}{2\mu[1 - (\lambda/\mu)]} + \frac{1}{\mu} \tag{32}$$

The expected total system cost per period is the sum of the expected waiting cost per period and the expected facility cost per period

$$TC_m = WC_m + FC_m$$

The expected waiting cost per period is the product of the cost of waiting per unit per period and the mean number of units in the system during the period, or

$$WC_m = C_w(n_m)$$
$$= C_w\left\{\frac{(\lambda/\mu)^2}{2[1 - (\lambda/\mu)]} + \frac{\lambda}{\mu}\right\}$$

The expected facility cost per period is the product of the cost of servicing one unit and the service rate in units per period, or

$$FC_m = C_f(\mu)$$

The expected total system cost per period is the sum of these cost components and may be expressed as

$$TC_m = C_w\left\{\frac{(\lambda/\mu)^2}{2[1 - (\lambda/\mu)]} + \frac{\lambda}{\mu}\right\} + C_f(\mu) \tag{33}$$

TABLE 2 Cost Components for Constant Service Example

μ	WC_m (\$)	FC_m (\$)	TC_m (\$)
0.1250	∞	0.0206	∞
0.1500	0.2913	0.0248	0.3161
0.2000	0.1145	0.0330	0.1475
0.2500	0.0750	0.0413	0.1163
0.3000	0.0566	0.0495	0.1061
0 4000	0.0383	0.0660	0.1043
0.5000	0.0292	0.0825	0.1117
0.6000	0.0236	0.0990	0.1226
0.8000	0.0170	0.1320	0.1490
1.0000	0.0134	0.1650	0.1785

As an application of the foregoing model, consider the example of the previous section. Instead of the parameter μ being the expected value from an exponential distribution, assume it to be a constant. The expected waiting cost per period, the expected service cost per period, and the expected total system cost per period are exhibited as a function of μ in Table 2. Although the expected waiting-cost function differs from the previous example, the minimum-cost service interval is still 0.4 units per period.

The examples may be more easily compared by graphing the expected total cost functions as shown in Figure 8. The upper curve is the expected total system cost function when μ is an expected value from an exponential distribution. The lower curve is the expected total system cost when μ is a constant. No significant difference in the minimum-cost policy is evident for the example considered.

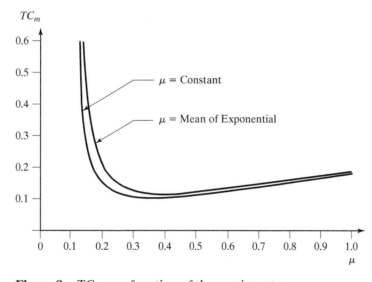

Figure 8 TC_m as a function of the service rate.

5.2 Poisson Arrivals with Any Service-Time Distribution

For further generality, it is desirable to have expressions for pertinent system characteristics regardless of the form of the service-time distribution. If σ^2 is the variance of the service-time distribution, the mean number of units in the system is given by

$$n_m = \frac{(\lambda/\mu)^2 + \lambda^2\sigma^2}{2[1 - (\lambda/\mu)]} + \frac{\lambda}{\mu} \tag{34}$$

and the mean waiting time is

$$w_m = \frac{(\lambda/\mu^2) + \lambda\sigma^2}{2[1 - (\lambda/\mu)]} + \frac{1}{\mu} \tag{35}$$

Equation 34 reduces to Equation 31 and Equation 35 reduces to Equation 32 when $\sigma^2 = 0$. In addition, because the variance of an exponential distribution is $(1/\mu)^2$, Equation 34 reduces to Equation 10 and Equation 35 reduces to Equation 23 when this substitution is made.

The expected total system cost per period is the sum of the expected waiting cost per period and the expected facility cost per period; that is,

$$TC_m = WC_m + FC_m$$

The expected waiting cost per period is the product of the cost of waiting per unit per period and the mean number of units in the system during the period. The expected facility cost per period may be taken as the product of the cost of serving one unit and the service rate in units per period. Therefore, the expected total system cost per period is

$$TC_m = C_w\left\{\frac{(\lambda/\mu)^2 + \lambda^2\sigma^2}{2[1 - (\lambda/\mu)]} + \frac{\lambda}{\mu}\right\} + C_f(\mu) \tag{36}$$

As an example of the application of this model, consider the following situation: The number of arrivals per hour has a Poisson distribution with a mean of 0.2 unit. The cost of waiting per unit per hour is \$2.10 and the cost of servicing one unit is \$4.05. The decision maker may choose one of two service policies. The first will result in a service rate of 0.4 units per hour with a service-time variance of 3 hours. The second will result in a service rate of 0.5 units per hour with a service-time variance of 4 hours.

The first policy will result in an expected total system cost of

$$TC_m = \$2.10\left\{\frac{(0.2/0.4)^2 + (0.2)^2(3)}{2[1 - (0.2/0.4)]} + \frac{0.2}{0.4}\right\} + \$4.05(0.4)$$

$$= \$1.83 + \$1.62 = \$3.45$$

The second policy will result in an expected total system cost of

$$TC_m = \$2.10\left\{\frac{(0.2/0.5)^2 + (0.2)^2(4)}{2[1 - (0.2/0.5)]} + \frac{0.2}{0.5}\right\} + \$4.05(0.5)$$

$$= \$1.40 + \$2.03 = \$3.43$$

From these results, it is evident that it makes little difference which policy is adopted.

6 FINITE POPULATION QUEUING MODELS

Finite waiting-line models must be applied to those waiting-line systems where the population is small relative to the arrival rate. In these systems, units leaving the population significantly affect the characteristics of the population and the arrival probabilities. It is assumed that both the time between calls for service for a unit of the population and the service times are distributed exponentially.

6.1 Finite Queuing Theory

Units leave the population when they fail, and the number of these failed units can be determined by an approach similar to that used in Section 3. Let

N = number of units in the population
M = number of service channels in the repair facility
λ = failure rate of an item, 1/MTBF
μ = repair rate of a repair channel, 1/MTTR
n = number of failed items
P_n = steady-state probability of n failed items
P_0 = probability that no items failed
$M\mu$ = maximum possible repair rate
λ_n = failure rate when n items already failed
μ_n = repair rate when n items already failed

The failure rate of an item is expressed as $\lambda = 1/$ MTBF and the failure rate of the entire population when n items already failed can be expressed as $\lambda_n = (N - n)\lambda$, where $N - n$ is the number of operational units, each of which fails at a rate of λ. Similarly, the repair rate of a repair channel is expressed as $\mu = 1/$ MTTR, and the repair rate of the entire repair facility when n items have already failed can be expressed as

$$\mu_n = \begin{cases} n\mu & \text{if } n \in 1, 2, \ldots, M-1 \\ M\mu & \text{if } n \in M, M + !, \ldots, N \end{cases} \tag{37}$$

An analysis using birth–death processes is employed to determine the probability distribution, P_n, for the number of failed items. In the birth–death process, the state of the system is the number of failed items (state $= 0, 1, 2, \ldots, N$). The rates of change between the states are the breakdown rate, λ_n, and the repair rate, μ_n. This gives

If steady-state operation of the system is assumed, it yields

$$N\lambda P_0 = \mu P_1$$

$$N\lambda P_0 + 2\mu P_2 = [\mu + (N-1)\lambda]P_1$$

$$(N-1)\lambda P_1 + 3\mu P_3 = [2\mu + (N-2)\lambda]P_2$$

$$\vdots$$

$$(N-M+2)\lambda P_{M-2} + M\mu P_M = [(M-1)\mu + (N-M+1)\lambda]P_{M-1}$$

$$\vdots$$

$$2\lambda P_{N-2} + M\mu P_N = (M\mu + \lambda)P_{N-1}$$

$$\lambda P_{N-1} = M\mu P_N$$

In addition,

$$\sum_{n=0}^{N} P_n = 1$$

Solving these balance equations gives

$$P_0 = \left(\sum_{n=0}^{N} C_n\right)^{-1} \tag{38}$$

where

$$C_n = \begin{cases} \dfrac{N!}{(N-n)!\,n!}\left(\dfrac{\lambda}{\mu}\right)^n & \text{if } n = 0, 1, 2, \ldots, M \\[3mm] \dfrac{N!}{(N-n)!\,M!\,M^{n-M}}\left(\dfrac{\lambda}{\mu}\right)^n & \text{if } n = M+1, M+2, \ldots, N \end{cases} \tag{39}$$

Equations 38 and 39 can now be used to find the steady-state probability of n failed units as $P_n = P_0 C_n$ for $n = 0, 1, 2, \ldots, N$.

For example, assume that a finite population of 10 units exists, with each unit having a mean time between failures of 32 hours. Also, assume that the mean time to service an item in a single-service channel is 8 hours.

From these data, the arrival and the service rates are found to be

$$\lambda = 1/32 \text{ and } \mu = 1/8$$

from which

$$\frac{\lambda}{\mu} = 0.25$$

First, compute C_n for $n = 0$ and 1 using the first expression in Equation 39 as

$$C_0 = \frac{10! \ (0.25)^0}{10! \ 0!} = 1$$

$$C_1 = \frac{10! \ (0.25)^1}{9! \ 1!} = 2.5$$

Continue computing C_n for $n = 2, 3, \ldots, 10$ using the second expression in Equation 39 as

$$C_2 = \frac{10! \ (0.25)^2}{8! \ 1! \ 1^1} = 5.625$$

$$C_3 = \frac{10! \ (0.25)^3}{7! \ 1! \ 1^2} = 11.250$$

$$C_4 = \frac{10! \ (0.25)^4}{6! \ 1! \ 1^3} = 19.6875$$

$$C_5 = \frac{10! \ (0.25)^5}{5! \ 1! \ 1^4} = 29.53125$$

$$C_6 = \frac{10! \ (0.25)^6}{4! \ 1! \ 1^5} = 36.9140625$$

$$C_7 = \frac{10! \ (0.25)^7}{3! \ 1! \ 1^6} = 36.9140625$$

$$C_8 = \frac{10! \ (0.25)^8}{2! \ 1! \ 1^7} = 27.685547$$

$$C_9 = \frac{10! \ (0.25)^9}{1! \ 1! \ 1^8} = 13.842773$$

$$C_{10} = \frac{10! \ (0.25)^{10}}{0! \ 1! \ 1^9} = 3.460693$$

$$\sum_{n=0}^{10} C_n = 188.410888$$

From Equation 38,

$$P_0 = \frac{1}{\displaystyle\sum_{n=0}^{10} C_n} = \frac{1}{188.410888} = 0.0053076$$

P_n for $n = 0, 1, 2, \ldots, N$ can now be computed from $P_n = P_0 C_n = 0.0053076(C_n)$ as follows:

$$P_0 = 0.0053076 \times 1 \qquad = 0.0053076$$
$$P_1 = 0.0053076 \times 2.5 \qquad = 0.0132690$$
$$P_2 = 0.0053076 \times 5.625 \qquad = 0.0298553$$
$$P_3 = 0.0053076 \times 11.250 \qquad = 0.0597105$$
$$P_4 = 0.0053076 \times 19.6875 \qquad = 0.1044934$$
$$P_5 = 0.0053076 \times 29.53125 \qquad = 0.1567400$$

$$P_6 = 0.0053076 \times 36.9140625 = 0.1959251$$

$$P_7 = 0.0053076 \times 36.9140625 = 0.1959251$$

$$P_8 = 0.0053076 \times 27.685547 \ \ = 0.1469438$$

$$P_9 = 0.0053076 \times 13.842773 \ \ = 0.0734719$$

$$P_{10} = 0.0053076 \times 3.460693 \ \ \ = 0.0183680$$

The steady-state probabilities of n failed units calculated previously can now be used to find the mean number of failed units from

$$\sum_{n=0}^{10} n \times P_n$$

as

$$0 \times 0.0053076 + 1 \times 0.0132690 + \cdots + 10 \times 0.0183680 = 6.023 \text{ units}$$

6.2 The Finite Queuing Tables[1]

The finite queuing derivations in the previous section and the appropriate numerical applications provide complete probabilities for n units having failed, P_n. From these probabilities, the expected number of failed units was obtained.

In this section, an approach based on the Finite Queuing Tables is presented. For convenience, the notation used in these tables will be adopted. Let

T = mean service time

U = mean time between calls for service

H = mean number of units being serviced

L = mean number of units waiting for service

J = mean number of units running or productive

Appendix: Finite Queuing Tables gives a portion of the Finite Queuing Tables (for populations of 10, 20, and 30 units). Each set of values is indexed by N, the number of units in the population. Within each set, data are classified by X, the service factor, and M, the number of service channels. Two values are listed for each value of N, X, and M. The first is D, the probability of a delay, expressing the probability that an arrival will have to wait. The second is F, an efficiency factor needed in the calculation of H, L, and J.

The service factor is a function of the mean service time and the mean time between calls for service,

$$X = \frac{T}{T + U} \tag{40}$$

The mean number of units being serviced is a function of the efficiency factor, the number of units in the population, and the service factor,

$$H = FNX \tag{41}$$

[1]L. G. Peck and R. N. Hazelwood, *Finite Queuing Tables* (New York: John Wiley & Sons, Inc., 1958).

328

The mean number of units waiting for service is a function of the number of units in the population and the efficiency factor,

$$L = N(1 - F) \tag{42}$$

Finally, the mean number of units running or productive is a function of the number of units in the population, the efficiency factor, and the service factor,

$$J = NF(1 - X) \tag{43}$$

A knowledge of N, T, and U for the waiting-line system under study, the expressions given previously, and a set of Finite Queuing Tables makes it easy to find mean values for important queuing parameters. For example, the mean number of failed units calculated in Section 6.1 to be 6.023 can be found from Equations 40–42 as

$$X = \frac{8}{8 + 32} = 0.20$$

$$H = 0.497(10)(0.20) = 0.994$$

$$L = 10(1 - 0.497) = 5.03$$

This gives $H + L = 6.024$ as the mean number failed. Other examples are given in the sections that follow.

6.3 Number of Service Channels Under Control

Assume that a population of 20 units exists, with each unit having a mean time between required services of 32 minutes. Each service channel provided will have a mean service time of 8 minutes. Both the time between arrivals and the service interval are distributed exponentially. The number of channels to be provided is under management control. The cost of providing one channel with a mean service-time capacity of 8 minutes is $10 per hour. The cost of waiting is $5 per unit per hour. The service factor for this system is

$$X = \frac{T}{T + U} = \frac{8}{8 + 32} = 0.20$$

Table 3 provides a systematic means for finding the minimum-cost number of service channels. The values in columns A and B are entered from Appendix: Finite Queuing Tables, Table 2, with $N = 20$ and $X = 0.20$. The mean number of units being serviced is found from

TABLE 3 Cost as a Function of the Number of Service Channels

M (A)	F (B)	H (C)	L (D)	$H + L$ (E)	Waiting Cost (F) ($)	Service Cost (G) ($)	Total Cost (H) ($)
8	0.999	4.00	0.02	4.02	20.10	80	100.00
7	0.997	3.99	0.06	4.05	20.25	70	90.25
6	0.988	3.95	0.24	4.19	20.95	60	80.95
5	0.963	3.85	0.74	4.59	22.95	50	72.95
4	0.895	3.58	2.10	5.68	28.40	40	68.40
3	0.736	2.94	5.28	8.22	41.10	30	71.10
2	0.500	2.00	10.00	12.00	60.00	20	80.00

Equation 41 and entered in column C. The mean number of units waiting for service is found from Equation 42 and is entered in column D. The mean number of units waiting in queue and in service is given in column E. The data of columns A and E may be multiplied by their respective costs to give the total system cost.

Multiplying $5 per unit per hour by the mean number of units waiting gives the waiting cost per hour in column F. Next, multiplying $10 per channel per hour by the number of channels gives the service cost per hour in column G. Finally, adding the expected waiting cost and the service cost gives the expected total system cost in column H. The minimum cost number of channels is found to be four.

In this example, the cost of waiting was taken to be $5 per unit per hour. If this is due to lost profit, resulting from unproductive units, the same solution may be obtained by maximizing profit. As before, the values in columns A and B of Table 4 are entered from Appendix: Finite Queuing Tables, Table 2, with $N = 20$ and $X = 0.20$. The mean number of units running or productive is found from Equation 43 and is entered in column C.

The profit per hour in column D is found by multiplying the mean number of productive units by $5 profit per productive unit per hour. The cost of service per hour in column E is obtained by multiplying the number of channels by $10 per channel per hour. Finally, the net profit in column F is found by subtracting the service cost per hour from the profit per hour. As before, the number of channels that should be used is four. This example illustrates that either the minimum-cost or the maximum-profit approach may be used with the same results.

6.4 Mean Service Time Under Control

Assume that a population of 10 units is to be served by a single-service channel. The mean time between calls for service is 30 minutes. If the mean service rate is 60 units per hour, the service cost will be $100 per hour. The service cost per hour is inversely proportional to the time in minutes to service one unit, expressed as $100/T$. Both the time between calls for service and the service duration are distributed exponentially. Lost profit because of units waiting in the system is $15 per hour.

Column A in Table 5 gives the capacity of the channel expressed as the mean service time in minutes per unit processed. The service factor for each service time is found from Equation 38 and entered in column B. The efficiency factors in column C are found by interpolation in Appendix: Finite Queuing Tables, Table 1, for $N = 10$ and the respective service factors of column B. The mean number of units running is found from Equation 41 and entered in column D. The data given in column A and column D may now be used to find the service capacity that results in a maximum net profit.

TABLE 4 Profit as a Function of the Number of Service Channels

Net Profit (F)	Service Cost (E)	Gross Profit (D)	J (C) ($)	F (B) ($)	M (A) ($)
8	0.999	15.98	79.90	80	−0.10
7	0.997	15.95	79.75	70	9.75
6	0.988	15.81	79.05	60	19.05
5	0.963	15.41	77.05	50	27.05
4	0.895	14.32	71.60	40	31.60
3	0.736	11.78	58.90	30	28.90
2	0.500	8.00	40.00	20	20.00

TABLE 5 Profit as a Function of the Mean Service Time

T(A)	X(B)	F(C)	J(D)	Gross Profit (E) ($)	Service Cost (F) ($)	Net Profit (G) ($)
1	0.032	0.988	9.56	143.20	100.00	43.20
2	0.062	0.945	8.86	132.90	50.00	82.90
3	0.091	0.864	7.85	117.75	33.33	84.42
4	0.118	0.763	6.73	101.00	25.00	76.00
5	0.143	0.674	5.77	86.51	20.00	66.51

The expected profit per hour in column E is found by multiplying the mean number of units running by $15 per hour. The cost of service per hour is found by dividing $100 by the value for T in column A. These costs are entered in column F. By subtracting the cost of service per hour from the expected gross profit per hour, the expected net profit per hour in column G is found. The mean service time resulting in an expected maximum profit is three periods.

6.5 Service Factor Under Control

Suppose that a population of production equipment is under study with the objective of deriving minimum-cost maintenance policy. It is assumed that both the time between calls for maintenance for a unit of the population and the service times are distributed exponentially. Two parameters of the system are subject to management control: (1) by increasing the repair capability (reducing T) the average machine downtime will be reduced; and (2) alternative policies of preventive maintenance will alter the mean time between breakdowns, U. Therefore, the problem of machine maintenance reduces to one of determining the service factor, X, that will result in a minimum-cost operation. This section will present methods for establishing and controlling the service factor in maintenance operations.

As machines break down, they become unproductive with a resulting economic loss. This loss may be reduced by reducing the service factor. But a decrease in the service factor requires either a reduction in the repair time or a more expensive policy of preventive maintenance, or both. Therefore, the objective is to find an economic balance between the cost of unproductive machines and the cost of establishing a specific service factor.

The analysis of this situation is facilitated by developing curves giving the percentage of machines not running as a function of the service factor. Figure 9 gives curves for selected populations when one service channel is provided. Each cure is developed from Equation 41 and the Finite Queuing Tables. As was expected, the percentage of machines not running increases as the service factor increases.

As an example of the determination of the minimum-cost service factor, suppose that eight machines are maintained by a mechanic. Each machine produces a profit of $22.00 per hour while it is running. The mechanic costs the company $28.80 per hour. Three policies of preventive maintenance are under consideration. The first will cost $80 per hour. After considering the increase in the mean time between breakdowns and the effect on service time, it is estimated that the resulting service factor will be 0.04. The second policy of preventive maintenance will cost only $42 per hour but will result in a service factor of 0.10. The third alternative involves no preventive maintenance at all; hence, it will cost nothing, but a service factor in excess of 0.2 will result. The time between calls for service and the service times are distributed exponentially. By reference to Figure 9, the results of Table 6 are developed. From

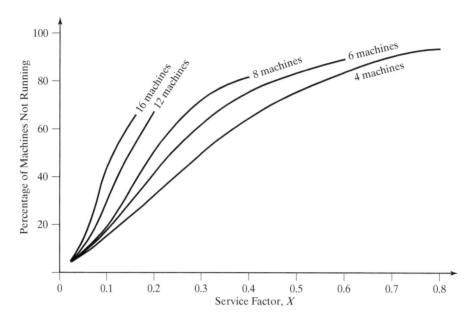

Figure 9 Percentage of machines not running as a function of the service factor.

TABLE 6 Three Policies of Preventive Maintenance

Maintenance Policy	Service Factor, X	Machines Not Running	Cost of Lost Profit ($)	Cost of Maintenance ($)	Cost of Mechanic ($)	Total Cost ($)
1	0.04	0.5	11.00	80	28.80	119.80
2	0.10	1.6	35.20	42	28.80	106.00
3	0.20	4.1	90.20	0	28.80	119.00

the last column, it is evident that the second preventive maintenance alternative should be adopted.

7 SUMMARY AND EXTENSIONS

The phenomenon of queuing is pervasive. It must be included in any serious study of systems analysis and engineering. But the modeling and analysis methods are more extensive than could be presented in this chapter. Accordingly, individuals are encouraged to consult one or both of the following books, depending upon the assumption being made about the underlying population:

1. For the infinite population assumption, see R. W. Hall, *Queuing Methods for Service and Manufacturing*, Prentice Hall, Upper Saddle River, NJ, 2001.
2. For the finite population case, see L. G. Peck and R. N. Hazelwood, *Finite Queuing Tables*, John Wiley and Sons, Inc., Hoboken, NJ, 1958.

Sections 1–5 of this chapter cover models and analyses in which the infinite assumption about the underlying population is made, even though there are no known infinite populations. Only Section 6 addresses the finite population case. There is relatively little in the queuing literature on the finite population topic.

The queuing system and its evaluation is the focus of Section 1. After defining and describing the arrival mechanism, the waiting line, and the service mechanism, attention is turned to the theoretical basis for queuing system evaluation. Again, the generic evaluation paradigm is applicable. Then, before proceeding further, Monte Carlo analysis of queuing is introduced and exercised. This adds a measure of realism to an otherwise abstract topic with the elements already introduced.

Single-channel queuing models are the easiest to derive, and this was done in Section 3. The assumption was that of Poisson arrivals and exponential service, producing metrics for the mean number of units in the system, the average length of the queue, and the distribution of waiting time. For multiple-channel queuing systems, models are presented without proof for the average length of the queue, the mean number of units in the system, the mean waiting time, the average holding time, and the probability that an arriving unit must wait. Relaxation of the exponential service-time distribution is then presented, but again without proof. The cases covered are those with constant service times and with any service-time distribution.

The finite population case has been dealt with at the end of this chapter. It was derived based on the birth–death process. Here, the Finite Queuing Tables, referenced above, are all-important. They are abstracted in Appendix: Finite Queuing Tables and then used in the examples that follow the derivation. The first is an example where the number of service channels is to be selected to minimize system cost. Then, control of the service factor to choose a minimum-cost service policy is presented by an example. Both examples depend upon the design-dependent parameters of reliability and maintainability.

QUESTIONS AND PROBLEMS

1. Under what conditions is it necessary or desirable to use Monte Carlo analysis in the study of a queuing system?
2. Without changing the standard deviation of S_x in Figure 2, change its mean from 5 to 6 and then redo 100 periods of the Monte Carlo analysis. Discuss the observed difference.
3. Why is it essential that μ be greater than λ in a probabilistic waiting-line process?
4. Suppose that the time between arrivals is distributed exponentially with a mean of 6 minutes and that service time is constant and equal to 5 minutes. Service is provided on a first-come, first-served basis. No units are in the system at 8:00 A.M. Use Monte Carlo analysis to estimate the total number of unit minutes of waiting between 8:00 A.M. and 12:00 noon.
5. Use Monte Carlo analysis to verify Equation 9 if the number of arrivals per period has a Poisson distribution with a mean of 0.10 and if the service duration is distributed exponentially with a mean of four periods. Plot the histogram of the number of units in the system and compare it with Figure 6.
6. Suppose that arrivals are distributed according to the Poisson distribution with a mean of 0.125 unit per period and that the service duration is distributed exponentially with a mean of five periods. Develop the probability distribution of n units in the system. What is the probability of there being more than four units in the system?

7. The arrival rate for a certain waiting-line system obeys a Poisson distribution with a mean of 0.5 unit per period. It is required that the probability of one or more units in the system not exceed 0.20. What is the minimum service rate that must be provided if the service duration is to be distributed exponentially?

8. What is the expected number of units in the system and the expected waiting time for the conditions of Problem 6?

9. The expected number of units in a waiting-line system experiencing Poisson arrivals with a rate of 0.4 unit per period must not exceed 8. What is the minimum service rate that must be provided if the service duration will be distributed exponentially? What will be the expected waiting time?

10. If the cost of waiting in Problem 7 is $5.00 per unit per period, and the service facility costs $2.50 per unit served, what is the total system cost? If the service rate is doubled at a total cost of $3.50 per unit served, what is the total system cost?

11. Plot the mean number of units in the system and the mean waiting time as a function of λ/μ if the number of arrivals per period obeys the Poisson distribution and if the service duration is distributed exponentially. What is the significance of this illustration?

12. The number of arrivals per period is distributed according to the Poisson with an expected value of 0.75 unit per period. The cost of waiting per unit per period is $3.20. The facility cost for serving one unit per period is $5.15. What expected service rate should be established if the service duration is distributed exponentially? What is the expected total system cost?

13. The expected waiting time in a waiting-line system with Poisson arrivals at a rate of 1.5 units per hours must not exceed 6 hours. What is the minimum constant service rate that must be provided? What is the expected number in the system?

14. Trucks arrive at a loading dock in a Poisson manner at the rate of 3.5 per day. The cost of waiting per truck per day is $320. A two-person crew that can load at a constant rate of four trucks per day costs $440 per day. Compute the total system cost for this operation.

15. Customers arrive at a bank at the rate of 0.2 per period and the service rate is 0.4 customers per period. If there is one teller, what is the average length of the line? If the arrival rate increases to 0.6 per period and the number of tellers is increased to two, how does the average length of the line change? Assume Poisson arrivals and an exponential service-time distribution.

16. Vehicles arrive at a toll plaza at the rate of 60 per hour. There are four booths, each capable of servicing 30 vehicles per hour. Assuming a Poisson distribution for arrivals and an exponential distribution for the service time, determine (a) the probability of no vehicles in the toll plaza, (b) the average length of the waiting line, and (c) the average time spent by a vehicle in the plaza.

17. In Problem 16, if it costs $55 per hour to operate a toll booth and the waiting cost per vehicle is $0.75 per minute spent at the plaza, how many booths should be operated to minimize total cost?

18. In a three-channel queuing system, the arrival rate is 1 unit per time period and the service rate is 0.6 unit per time period. Determine the probability that an arriving unit does not have to wait.

19. Use Equation 39 to extend the example in Section 6 to the case where there are two service channels. Compare the mean number waiting with the single-channel assumption.

20. Each truck in a fleet of 30 delivery trucks will return to a warehouse for reloading at an average interval of 150 minutes. An average of 30 minutes is required by the driver and one warehouse person to load the next shipment. If warehouse personnel are busy loading previous arrivals, the driver must wait in line. Both the time between arrivals and the loading

time are distributed exponentially. The cost of waiting in the system is \$48 per hour per truck and the total cost per warehouse person is \$40 per hour. Find the minimum cost number of warehouse people to employ.

21. A population of 10 cargo aircraft each produces a profit of \$4,000 per 24-hour period when not waiting to be unloaded. The time between arrivals is distributed exponentially with a mean of 144 hours. The unloading time at the ramp is distributed exponentially with a mean of 18 hours. It costs \$150 per hour to lease a ramp with this unloading capacity. How many ramps should be leased?

22. Each unit in a 10-unit population returns for service at an average interval of 20 minutes. If the cost of waiting is \$32 per hour per unit and the cost of service is \$50/T per unit per hour, find the optimum service time if there are two service channels. Assume Poisson arrivals and exponential service times.

23. A population of 30 chemical processing units is to be sampled by a two-person crew. The mean time between calls for this operation is 68 minutes. If the mean service rate is one unit per minute, the cost of the crew and instruments will be \$16 per minute. This cost will decrease to \$12, \$9, and \$7 for service intervals of 2, 3, and 4 minutes, respectively. The time between calls for service and the service duration are distributed exponentially. If lost profit is \$28 per hour for each unit that is idle, find the minimum cost service interval.

24. Plot the percentage of machines not running as a function of the service factor for a population of 20 machines with exponential arrivals and services (a) if a single channel is employed and (b) if two channels are employed.

25. A population of 10 cells is to be served by an unknown number of service channels. The mean time between service calls is 15 minutes. The cost per service channel is \$60 per hour at a mean service rate of 60 cells per hour and the service cost per channel per hour is inversely proportional to the service rate. Both the time between calls for service and the service times are distributed exponentially. If the net profit from each cell is \$10 per hour, determine the optimum mix of service channels and service times.

26. A group of 12 machines are repaired by a single repairperson when they break down. Each machine yields a profit of \$22 per hour while running. The mechanic costs the firm \$92 per hour, including overhead. At the present time no preventive maintenance is used. It is proposed that one technician be employed to perform certain routine maintenance and adjustment tasks. This will cost \$16.60 per hour but will reduce the service factor from 0.13 to 0.08. What is the economic advantage of implementing preventive maintenance?

Appendix

Appendix:

Probability Theory
and Analysis

Some models will give satisfactory results if variation is not incorporated. But, these models usually apply to physical phenomena where certainty is generally observed. Models formulated to analyze and evaluate human-made systems must incorporate probabilistic elements to be useful in the system engineering process. Accordingly, this appendix presents probability concepts and theory, probability distribution models, and an introduction to Monte Carlo analysis.

1 PROBABILITY CONCEPTS AND THEORY

If one tosses a coin, the outcome will not be known with certainty until either a head or a tail is observed. Prior to the toss, one can assign a probability to the outcome from knowledge of the physical characteristics of the coin. One may know that the diameter of an acorn ranges between 0.80 and 3.20 cm, but the diameter of a specific acorn to be selected from an oak tree will not be known until the acorn is measured. Experiments such as tossing a coin and selecting an acorn provide outcomes called *random events*. Most events in the decision environment are random and probability theory provides a means for quantifying these events.

1.1 The Universe and the Sample

The terms *universe* and *population* are used interchangeably. A *universe* consists of all possible objects, stages, and events within an arbitrarily defined boundary. A universe may be finite or it may be infinite. If it is finite, the universe may be very large or very small. If the universe is large, it may sometimes be assumed to be infinite for computational purposes. A universe need not always be large; it may be defined as a dozen events or as only one object. The relative usefulness of the universe as an entity will be paramount in its definition.

A *sample* is a part or portion of a universe. It may range in size from 1 to 0 less than the size of the universe. A sample is drawn from the population, and observations are made. This is done either because the universe is infinite in size or scope or because the population is large and/or inaccessible as a whole. The sample is used because it is smaller, more accessible, and more economical, and because it suggests certain characteristics of the population.

It is usually assumed that the sample is typical of the population in regard to the characteristics under consideration. The sample is then assessed, and inferences are made in regard to the population as a whole. To the extent that the sample is representative of the population, these inferences may be correct. The problem of selecting a representative sample from a population is an area in statistics to which an entire chapter might be devoted.

Subsequent discussion assumes that the sample is a *random sample*, that is, one in which each object or state or event that constitutes the population has an equally likely chance or probability of being selected and represented in the sample. It is rather simple to state this definition; it may be much more difficult to implement it in practice.

1.2 The Probability of an Event

A measure of the relative certainty of an event, before the occurrence of the event, is its probability. The usual representation of a probability is a number $P(A)$ assigned to the outcome A. This number has the following property: $0 \le P(A) \le 1$, with $P(A) = 0$ if the event is certain not to occur and $P(A) = 1$ if the event is certain to occur.

Because probability is only a measure of the certainty (or uncertainty) associated with an event, its definition is rather tenuous. The concept of relative frequency is sometimes employed to establish the number $P(A)$. Sometimes probabilities are established a priori. Other times they are simply a subjective estimate. Consider the example of tossing a fair coin. In a lengthy series of tosses, the coin may have come up heads as often as tails. Then the limiting value of the relative frequency of a head will be 0.5 and will be stated as $P(H) = 0.5$.

Two definitions pertaining to events are needed in the development of probability theorems:

1. Events A and B are said to be *mutually exclusive* if both cannot occur at the same time.
2. Event A is said to be *independent* of event B if the probability of the occurrence of A is the same regardless of whether or not B has occurred.

The probability of the occurrence of either one or another of a series of mutually exclusive events is the sum of probabilities of their separate occurrences. If a fair coin is tossed and success is defined as the occurrence of either a head or a tail, then the probability of a head or a tail is

$$P(H + T) = P(H) + P(T)$$

$$= 0.5 + 0.5 = 1.0 \tag{1}$$

The key to use of the addition theorem is the proper definition of mutually exclusive events. Such events must be distinct from one another. If one event occurs, it must be impossible for the second to occur at the same time. For example, assume that the probability of having a flat tire during a given period on each of four tires on an automobile is 0.3. Then the probability of having a flat tire on any of the four tires during this time period is not given by the addition of these four probabilities. If $P(T_1) = P(T_2) = P(T_3) = P(T_4) = 0.3$ are the respective probabilities of failure for each of the four tires, then

$$P(T_1 + T_2 + T_3 + T_4) \ne P(T_1) + P(T_2) + P(T_3) + P(T_4)$$

$$\ne 0.3 + 0.3 + 0.3 + 0.3 = 1.2$$

This cannot be true because the failure of tires is not mutually exclusive. During the time period established, two or more tires may fail, whereas in the example of coin tossing, it is not possible to obtain a head and a tail on the same toss.

1.3 The Multiplication Theorem

The probability of occurrence of independent events is the product of the probabilities of their separate events. Implicit in this theorem is the successful occurrence of two events simultaneously or in succession. Thus, the probability of the occurrence of two heads in two tosses of a coin is

$$P(H \cdot H) = P(H)P(H)$$
$$D = (0.5)(0.5) = 0.25 \tag{2}$$

The tire-failure problem can now be resolved by considering the probabilities of each tire not failing. The probability of each tire not failing is given by $P(\overline{T}_i) = 0.7$. The probability of no tire failing is then given by

$$P[(\overline{T}_1)(\overline{T}_2)(\overline{T}_3)(\overline{T}_4)] = P(\overline{T}_1)\, P(\overline{T}_2)\, P(\overline{T}_3)\, P(\overline{T}_4)$$
$$= (0.7)(0.7)(0.7)(0.7) = 0.2401$$

Thus, the probability of a tire failing, or of one or more tires failing, is

$$P(T_1 + T_2 + T_3 + T_4) = 1 - 0.2401 = 0.7599$$

This approach is valid, since the probability of one tire not failing is independent of the success or failure of the other three tires.

1.4 The Conditional Theorem

The probability of the occurrence of two dependent events is the probability of the first event times the probability of the second event, given that the first has occurred. This may be expressed as

$$P(W_1 \cdot W_2) = P(W_1)P(W_2|W_1) \tag{3}$$

This theorem is similar to the multiplication theorem, except that consideration is given to the lack of independence between events.

As an example, consider the probability of selecting two successive white balls from an urn containing three white and two black balls. This problem reduces to a calculation of the product of the probability of selecting a white ball times the probability of selecting a second white ball, given that the first attempt has been successful, or

$$P(W_1 \cdot W_2) = \left(\frac{3}{5}\right)\left(\frac{2}{4}\right) = \frac{3}{10}$$

The conditional theorem makes allowances for a change in probabilities between two successive events. This theorem will be helpful in constructing finite discrete probability distributions.

1.5 The Central Limit Theorem

Although many real-world variables are normally distributed, this assumption cannot be universally applied. However, the distribution of the means of samples or the sums of random variables approximates the normal distribution provided certain assumptions hold. The Central Limit Theorem states: If x has a distribution for which the moment-generating function exists, then the variable $\bar{x}$ has a distribution that approaches normality as the size of the sample tends toward infinity. The sample size required for any desired degree of convergence is a function of the shape of the parent distribution. Fairly good results have been demonstrated with a sample of $n = 4$ for both the rectangular and triangular distributions.

2 PROBABILITY DISTRIBUTION MODELS

The pattern of the distribution of probabilities over all possible outcomes is called a probability distribution. *Probability distribution models* provide a means for assigning the likelihood of occurrences of all possible values. Variables described in terms of a probability distribution are conveniently called *random variables*. The specific value of a random variable is determined by the distribution.

A probability distribution is completely defined when the probability associated with every possible outcome is defined. In most instances, the outcomes themselves are represented by numbers or different values of a variable, such as the diameter of an acorn. When the pattern of the probability distribution is expressed as a function of this variable, the resulting function is called a *probability distribution function*.

An example empirical probability distribution function may be developed as follows. A maintenance mechanic attends four machines and his services are needed only when a machine fails. He would like to estimate how many machines will fail each shift. From previous experience, and using the relative frequency concept of probability, the mechanic knows that 40% of the time only one machine will fail at least once during the shift. Further, 30% of the time two machines will fail, three machines will fail 20% of the time, and all four will fail 10% of the time.

The probability distribution of the number of failed machines may be expressed as $P(1) = 0.4$, $P(2) = 0.3$, $P(3) = 0.2$, and $P(4) = 0.1$. This probability distribution is exhibited in Figure 1.

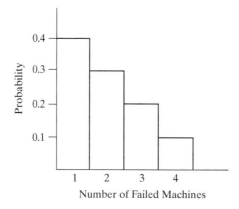

Figure 1 A probability distribution of the number of failed machines.

The probability distribution function for this case may be defined as

$$P(x) = \frac{5-x}{10} \qquad \text{if } x = 1, 2, 3, 4$$

$$P(x) = 0 \qquad \text{otherwise}$$

Although the function $P(x) = (5-x)/10$ uniquely represents the probability distribution pattern for the number of failed machines, the function itself belongs to a wider class of functions of the type $P(x) = (a-x)/b$. All functions of this type indicate similar patterns, yet each pair of numbers (a, b) uniquely defines a specific probability distribution. These numbers (a, b) are called *parameters*.

In the sense that they serve to define the probability distribution function, it is possible to look upon parameters as properties of the distribution function. The choice of representation of parameters is not unique, and the most desirable representation would reflect a measure of the properties of the universe under study. Two most commonly sought measures are the *mean*, an indication of central tendency, and the *variance*, a measure of dispersion.

The probability distribution just presented is discrete in that it assigns probabilities to an event that can only take on integer values. Continuous probability distributions are used to define the probability of the occurrence of an event that may take on values over a continuum. Under certain conditions, it may be desirable to use a continuous probability distribution to approximate a discrete probability distribution. By so doing, tedious summations may be replaced by integrals. In other instances, it may be desirable to make a continuous distribution discrete as when calculations are to be performed on a digital computer. Several discrete and continuous probability distribution models are presented subsequently.

2.1 The Binomial Distribution

The binomial distribution is a basic discrete sampling distribution. It is applicable where the probability is sought of exactly x occurrences in n trials of an event that has a constant probability of occurrence p. The requirement of a constant probability of occurrence is satisfied when the population being sampled is infinite in size, or where replacement of the sampled unit takes place.

The probability of exactly x occurrences in n trials of an event that has a constant probability of occurrence p is given as

$$P(x) = \frac{n!}{x!(n-x)!}\, p^x q^{n-x} \qquad 0 \leq x \leq n \tag{4}$$

where $q = 1 - p$. The mean and variance of this distribution are given by np and npq, respectively.

As an example of the application of the binomial distribution, assume that a fair coin is to be tossed five times. The probability of obtaining exactly two heads is

$$P(2) = \frac{5!}{2!(5-2)!}\, (0.5)^2 (1-0.5)^3$$

$$= 10(0.03125) = 0.3125$$

A probability distribution may be constructed by solving for the probability of exactly zero, one, two, three, four, and five heads in five tosses. If $p = 0.5$, as in this example, the resulting

distribution is symmetrical. If the distribution is skewed to the right; if the distribution is skewed to the left.

2.2   The Uniform Distribution

The uniform or rectangular probability distribution may be either discrete or continuous. The continuous form of this simple distribution is

$$f(x) = \frac{1}{a} \qquad 0 \le x \le a \tag{5}$$

The discrete form divides the interval 0 to a into $n + 1$ cells over the range 0 to n, with $1/(n + 1)$ as the unit probabilities. The mean and variance of the rectangular probability distribution are given as $a/2$ and $a^2/12$ for the continuous case, and as $n/2$ and $n^2/12 + n/6$ for the discrete case.

    The general form of the rectangular probability distribution is shown in Figure 2. The probability that a value of x will fall between the limits 0 and a is equal to unity. One may determine the probability associated with a specific value of x, or a range of x, by integration for the continuous case. The probability associated with a specific value of x for the discrete distributions of the previous section was found from the functions given. Determination of the probability associated with a range of x required a summation of individual probabilities. This is a fundamental difference in dealing with discrete and continuous probability distributions.

    Values are drawn at random from the rectangular distribution with x allowed to take on values ranging from 0 through 9. These random rectangular variates may be used to randomize a sample or to develop values drawn at random from other probability distributions as is illustrated in the last section of this appendix.

2.3   The Poisson Distribution

The Poisson is a discrete distribution useful in its own right and as an approximation to the binomial. It is applicable when the opportunity for the occurrence of an event is large, but when the actual occurrence is unlikely. The probability of exactly x occurrences of an event of probability p in a sample n is

$$P(x) = \frac{(\mu)^x e^{-\mu}}{x!} \qquad 0 \le x \le \infty \tag{6}$$

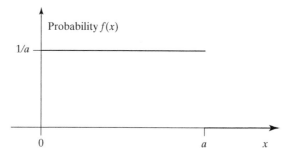

Figure 2   The general form of the rectangular distribution.

The mean and variance of this distribution are equal and given by μ, where $\mu = np$.

As an example of the application of the Poisson distribution, assume that a sample of 100 items is selected from a population of items which are 1% defective. The probability of obtaining exactly three defectives in the sample is found from Equation 6 as

$$P(3) = \frac{(1)^3(2.72)^{-1}}{3!} = 0.061$$

The Poisson distribution may be used as an approximation to the binomial distribution. Such an approximation is good when n is relatively large, p is relatively small, and in general, $pn < 5$. These conditions were satisfied in the previous example.

2.4 The Exponential Distribution

The exponential probability distribution is given by

$$f(x) = \frac{1}{a}e^{-x/a} \qquad 0 \le x \le \infty \tag{7}$$

The mean and variance of this distribution are given by a and a^2, respectively. Its form is illustrated in Figure 3.

As an example of the application of the exponential probability distribution, consider the selection of a light bulb from a population of light bulbs whose life is known to be exponentially distributed with a mean $\mu = 1,000$ hours. The probability of the life of this sample bulb not exceeding 1,000 hours would be expressed as $P(x \le 1,000)$. This would be the proportional area under the exponential function over the range $x = 0$ to $x = 1,000$, or

$$P(x \le 1,000) = \int_0^{1,000} f(x)dx$$

$$D = \int_0^{1,000} \frac{1}{1,000} e^{-x/1,000} dx$$

$$= -e^{-x/1,000} \Big|_0^{1,000}$$

$$= 1 - e^{-1} = 0.632$$

Note that 0.632 is that proportion of the area of an exponential distribution to the left of the mean. This illustrates that the probability of the occurrence of an event exceeding the mean value is only $1 - 0.632 = 0.368$.

2.5 The Normal Distribution

The normal or Gaussian probability distribution is one of the most important of all distributions. It is defined by

$$f(x) = \frac{1}{\sigma\sqrt{2\pi}} e^{[-(x-\mu)^2/2\sigma^2]} \qquad -\infty \le x \le +\infty \tag{8}$$

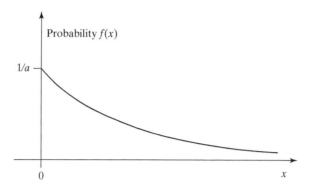

Figure 3 The general form of the exponential distribution.

The mean and variance are μ and σ^2, respectively. Variation is inherent in nature, and much of this variation appears to follow the normal distribution, the form of which is given in Figure 4.

The normal distribution is symmetrical about the mean and possesses some interesting and useful properties regarding its shape. Where distances from the mean are expressed in terms of standard deviations, σ, the relative areas defined between two such distances will be constant from one distribution to another. In effect, all normal distributions, when defined in terms of a common value of μ and σ, will be identical in form, and corresponding probabilities may be tabulated. Normally, cumulative probabilities are given from $-\infty$ to any value expressed as standard deviation units. This table gives the probability from $-\infty$ to Z, where Z is a standard normal variate defined as

$$Z = \frac{x - \mu}{\sigma} \tag{9}$$

This is shown as the shaded area in Figure 4.

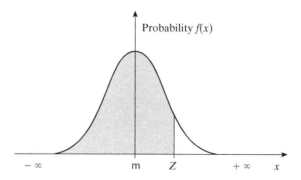

Figure 4 The normal probability distribution.

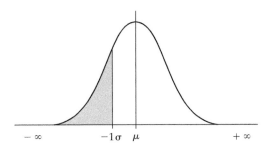

Figure 5 The area from $-\infty$ to -1σ under the normal distribution.

The area from $-\infty$ to -1σ is indicated as the shaded area in Figure 5. The probability of x falling in this range is 0.1587. Likewise, the area from $-\infty$ to $+2\sigma$ is 0.9773. If the probability of a value falling in the interval -1σ to $+2\sigma$ is required, the following computations are made.

$$P(\text{area } -\infty \text{ to } +2\sigma) = 0.9773$$

$$-P(\text{area } -\infty \text{ to } -1\sigma) = 0.1587$$

$$P(\text{area } -1\sigma \text{ to } +2\sigma) = 0.8186$$

This situation is shown in Figure 6.

2.6 The Lognormal Distribution

The lognormal probability distribution is related to the normal distribution. If a random variable $Y = \ln X$ is normally distributed with mean μ and variance σ^2, then the random variable X follows the lognormal distribution. Accordingly, the probability distribution function of the random variable X is defined as

$$f(x) = \frac{1}{x\sigma\sqrt{2\pi}} e^{[-(\ln x - \mu)^2/2\sigma^2]}, \qquad x > 0 \qquad (10)$$

where $\mu \in (-\infty, \infty)$ is called the scale parameter and $\sigma > 0$ is called the shape parameter. The mean and variance of the lognormal distribution are $e^{\mu + \sigma^2/2}$ and $e^{2\mu + \sigma^2}(e^{\sigma^2} - 1)$ respectively.

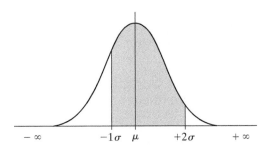

Figure 6 The area from -1σ to $+2\sigma$ under the normal distribution.

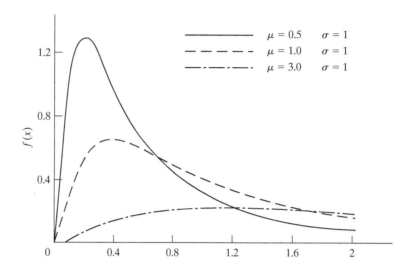

Figure 7   Lognormal distribution with constant shape parameter.

Figure 7 shows a lognormal distribution with a constant shape parameter and various scale parameters. Figure 8 shows a lognormal distribution with a constant scale parameter and various shape parameters.

If $X \sim \ln(\mu, \sigma^2)$ then $\ln X \sim N(\mu, \sigma^2)$. This implies that if n data points $x_1, x_2, \ldots, x_n$ are lognormal, then the logarithms of these data points, $\ln x_1, \ln x_2, \ldots, \ln x_n$ will be normally distributed and may be used for parameter estimation, goodness-of-fit testing, and hypothesis testing.

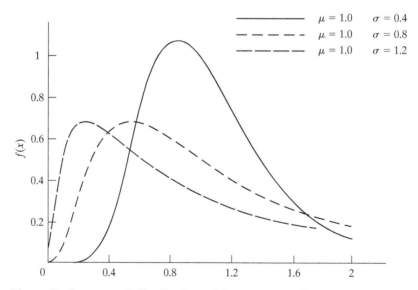

Figure 8   Lognormal distribution with constant scale parameter.

2.7 The Weibull Distribution

The probability distribution function of a random variable X, which follows the Weibull distribution, is given by

$$f(x) = \frac{\alpha}{\beta^\alpha} x^{\alpha-1} e^{-x/\beta)^\alpha}, \qquad x \geq 0 \qquad (11)$$

where $\alpha > 0, \beta > 0$ are the shape and scale parameters respectively, and defined on $(0, \infty)$. A Weibull variate X has mean $\dfrac{\beta}{\alpha} \Gamma\left(\dfrac{1}{\alpha}\right)$ and variance $\dfrac{\beta^2}{\alpha}\left\{ 2\Gamma\left(\dfrac{2}{\alpha}\right) - \dfrac{1}{\alpha}\left[\Gamma\left(\dfrac{1}{\alpha}\right)\right]^2 \right\}$, where $\Gamma()$ is the gamma function

$$\Gamma(z) = \int_0^\infty t^{z-1} e^{-t} dt$$

Weibull distributions with different shape and scale parameters are illustrated in Figures 9 and 10, respectively. The Weibull distribution has some interesting characteristics:

1. For $\alpha = 1$, the Weibull distribution is the same as the exponential distribution with parameter β.
2. For $\alpha = 3.4$, the Weibull distribution approximates the normal distribution.
3. If $X \sim$ Weibull (α, β) then $X^a \sim$ exponential $(\beta\alpha)$.

The Weibull distribution is frequently used as a time-to-failure model and in reliability analysis, especially in instances where failure data cannot be fitted by the exponential distribution.

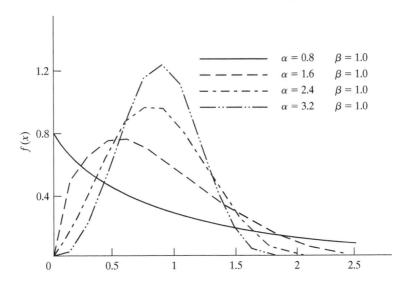

Figure 9 Weibull distribution with various shape parameters.

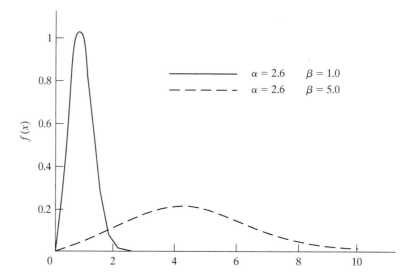

Figure 10 Weibull distribution with various scale parameters.

3 MONTE CARLO ANALYSIS

The decision environment is made up of many random variables. Thus, models used to explain operational systems must often incorporate probabilistic elements. In some cases, formal mathematical solutions are difficult or impossible to obtain from these models. Under such conditions it may be necessary to use a method known as *Monte Carlo analysis*. When applied to an operational system, Monte Carlo analysis provides a powerful means of simulation.

3.1 A Simple Monte Carlo Example

As an introduction to the idea of Monte Carlo analysis, consider its application to the determination of the area of a circle with a diameter of 1 inch. Proceed as follows:

1. Enclose the circle of a 1-inch square as shown in Figure 11.
2. Divide two adjoining sides of the square into tenths, or hundredths, or thousandths, and so on, depending on the accuracy desired.
3. Secure a sequence of pairs of random rectangular variates.
4. Use each pair of rectangular variates to determine a point within the square and possibly within the circle. This process is illustrated in Table 1 for 100 trials.
5. Compute a ratio of the number of times a point falls within the circle to the total number of trials. The value of this ratio is an approximate area for the circle expressed as a fraction of the 1 inch^2 represented by the square. It is 79/100, or 0.79, in this example.

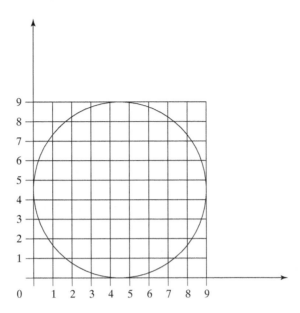

Figure 11 Area of circle by Monte Carlo analysis.

The example just presented has a well-known mathematical solution as follows:

$$A = \pi r^2 = 3.1416(0.50)^2 = 0.7854.$$

TABLE 1 Determining the area of a circle

Trial	Random number	In	Out
1	73	x	
2	26	x	
3	19		x
4	84	x	
5	81	x	
6	47	x	
7	18		x
8	44	x	
$\vdots$	$\vdots$	$\vdots$	$\vdots$
100	35	x	
Total		79	21

3.2 Steps in the Monte Carlo Procedure

Monte Carlo analysis may be implemented in accordance with a step-by-step procedure that is usually independent of the application. The steps are:

1. **Formalize the system logic**—The system chosen for study usually operates in accordance with a certain logical pattern. Therefore, before beginning the actual process of Monte Carlo analysis, it is necessary to formalize the operational procedure by the

construction of a model. This may require the development of a step-by-step flow diagram outlining the logic. If the actual simulation process is to be performed on a digital computer, it is mandatory to prepare an accurate logic diagram. From this, the computer can be programmed to pattern the process under study.

2. **Determine the probability distributions**—Each random variable in the operation refers to an event in the system being studied. Therefore, an important step in Monte Carlo analysis is determining the behavior of these random variables. This involves the development of empirical frequency distributions to describe the relevant variables by the collection of historical data. Once this is done, the frequency distribution for each variable may be studied statistically to ascertain whether it conforms to a known theoretical distribution.

3. **Develop the cumulative probability distributions**—Convert each probability distribution to its cumulative equivalent, with the cumulative probability exhibited on the ordinate ready to receive random number inputs. It is the cumulative probability distributions that serve to convert random rectangular variates to values drawn at random from the underlying probability distribution.

4. **Perform the Monte Carlo process**—The Monte Carlo process proceeds by exercising the system logic and saving the results of each trial. Analysis of a sample of trials gives insight into the estimated behavior of the system under study. The validity of the estimate depends upon the fidelity of the logic and the number of trials in the sample.

Appendix

Appendix

Finite Queuing Tables

TABLES 1–3 FINITE QUEUING FACTORS

The probability of a delay, D, and the efficiency factor, F, are given for populations of 10, 20, and 30 units. Each set of values is keyed to the service factor, X, and the number of channels, M. (These tabular values are adapted with permission from L. G. Peck and R. N. Hazelwood, *Finite Queuing Tables* [New York: John Wiley & Sons, Inc., 1958].)

TABLE 1 Finite Queuing Factors—Population 10

X	M	D	F	X	M	D	F	X	M	D	F
0.008	1	0.072	0.999		2	0.177	0.990		3	0.182	0.986
0.013	1	0.117	0.998		1	0.660	0.899		2	0.528	0.921
0.016	1	0.144	0.997	0.085	3	0.037	0.999		1	0.954	0.610
0.019	1	0.170	0.996		2	0.196	0.988	0.165	4	0.049	0.997
0.021	1	0.188	0.995		1	0.692	0.883		3	0.195	0.984
0.023	1	0.206	0.994	0.090	3	0.043	0.998		2	0.550	0.914
0.025	1	0.224	0.993		2	0.216	0.986		1	0.961	0.594
0.026	1	0.232	0.992		1	0.722	0.867	0.170	4	0.054	0.997
0.028	1	0.250	0.991	0.095	3	0.049	0.998		3	0.209	0.982
0.030	1	0.268	0.990		2	0.237	0.984		2	0.571	0.906
0.032	2	0.033	0.999		1	0.750	0.850		1	0.966	0.579
	1	0.285	0.988	0.100	3	0.056	0.998	0.180	5	0.013	0.999
0.034	2	0.037	0.999		2	0.258	0.981		4	0.066	0.996
	1	0.302	0.986		1	0.776	0.832		3	0.238	0.978
0.036	2	0.041	0.999	0.105	3	0.064	0.997		2	0.614	0.890
	1	0.320	0.984		2	0.279	0.978		1	0.975	0.549
0.038	2	0.046	0.999		1	0.800	0.814	0.190	5	0.016	0.999
	1	0.337	0.982	0.110	3	0.072	0.997		4	0.078	0.995
0.040	2	0.050	0.999		2	0.301	0.974		3	0.269	0.973
	1	0.354	0.980		1	0.822	0.795		2	0.654	0.873
0.042	2	0.055	0.999	0.115	3	0.081	0.996		1	0.982	0.522
	1	0.371	0.978		2	0.324	0.971	0.200	5	0.020	0.999
0.044	2	0.060	0.998		1	0.843	0.776		4	0.092	0.994
	1	0.388	0.975	0.120	4	0.016	0.999		3	0.300	0.968
0.046	2	0.065	0.998		3	0.090	0.995		2	0.692	0.854
	1	0.404	0.973		2	0.346	0.967		1	0.987	0.497
0.048	2	0.071	0.998		1	0.861	0.756	0.210	5	0.025	0.999
	1	0.421	0.970	0.125	4	0.019	0.999		4	0.108	0.992
0.050	2	0.076	0.998		3	0.100	0.994		3	0.333	0.961
	1	0.437	0.967		2	0.369	0.962		2	0.728	0.835
0.052	2	0.082	0.997		1	0.878	0.737		1	0.990	0.474
	1	0.454	0.963	0.130	4	0.022	0.999	0.220	5	0.030	0.998
0.054	2	0.088	0.997		3	0.110	0.994		4	0.124	0.990
	1	0.470	0.960		2	0.392	0.958		3	0.366	0.954
0.056	2	0.094	0.997		1	0.893	0.718		2	0.761	0.815
	1	0.486	0.956	0.135	4	0.025	0.999		1	0.993	0.453
0.058	2	0.100	0.996		3	0.121	0.993	0.230	5	0.037	0.998
	1	0.501	0.953		2	0.415	0.952		4	0.142	0.988
0.060	2	0.106	0.996		1	0.907	0.699		3	0.400	0.947
	1	0.517	0.949	0.140	4	0.028	0.999		2	0.791	0.794
0.062	2	0.113	0.996		3	0.132	0.991		1	0.995	0.434
	1	0.532	0.945		2	0.437	0.947	0.240	5	0.044	0.997
0.064	2	0.119	0.995		1	0.919	0.680		4	0.162	0.986
	1	0.547	0.940	0.145	4	0.032	0.999		3	0.434	0.938
0.066	2	0.126	0.995		3	0.144	0.990		2	0.819	0.774
	1	0.562	0.936		2	0.460	0.941		1	0.996	0.416
0.068	3	0.020	0.999		1	0.929	0.662	0.250	6	0.010	0.999
	2	0.133	0.994	0.150	4	0.036	0.998		5	0.052	0.997
	1	0.577	0.931		3	0.156	0.989		4	0.183	0.983
0.070	3	0.022	0.999		2	0.483	0.935		3	0.469	0.929
	2	0.140	0.994		1	0.939	0.644		2	0.844	0.753
	1	0.591	0.926	0.155	4	0.040	0.998		1	0.997	0.400
0.075	3	0.026	0.999		3	0.169	0.987	0.260	6	0.013	0.999
	2	0.158	0.992		2	0.505	0.928		5	0.060	0.996
	1	0.627	0.913		1	0.947	0.627		4	0.205	0.980
0.080	3	0.031	0.999	0.160	4	0.044	0.998		3	0.503	0.919
	2	0.866	0.732		4	0.533	0.906		7	0.171	0.982

TABLE 1 Finite Queuing Factors—Population 10 *(Continued)*

X	M	D	F	X	M	D	F	X	M	D	F
	1	0.998	0.384		3	0.840	0.758		6	0.413	0.939
0.270	6	0.015	0.999		2	0.986	0.525		5	0.707	0.848
	5	0.070	0.995	0.400	7	0.026	0.998		4	0.917	0.706
	4	0.228	0.976		6	0.105	0.991		3	0.991	0.535
	3	0.537	0.908		5	0.292	0.963	0.580	8	0.057	0.995
	2	0.886	0.712		4	0.591	0.887		7	0.204	0.977
	1	0.999	0.370		3	0.875	0.728		6	0.465	0.927
0.280	6	0.018	0.999		2	0.991	0.499		5	0.753	0.829
	5	0.081	0.994	0.420	7	0.034	0.998		4	0.937	0.684
	4	0.252	0.972		6	0.130	0.987		3	0.994	0.517
	3	0.571	0.896		5	0.341	0.954	0.600	9	0.010	0.999
	2	0.903	0.692		4	0.646	0.866		8	0.072	0.994
	1	0.999	0.357		3	0.905	0.700		7	0.242	0.972
0.290	6	0.022	0.999		2	0.994	0.476		6	0.518	0.915
	5	0.093	0.993	0.440	7	0.045	0.997		5	0.795	0.809
	4	0.278	0.968		6	0.160	0.984		4	0.953	0.663
	3	0.603	0.884		5	0.392	0.943		3	0.996	0.500
	2	0.918	0.672		4	0.698	0.845	0.650	9	0.021	0.999
	1	0.999	0.345		3	0.928	0.672		8	0.123	0.988
0.300	6	0.026	0.998		2	0.996	0.454		7	0.353	0.954
	5	0.106	0.991	0.460	8	0.011	0.999		6	0.651	0.878
	4	0.304	0.963		7	0.058	0.995		5	0.882	0.759
	3	0.635	0.872		6	0.193	0.979		4	0.980	0.614
	2	0.932	0.653		5	0.445	0.930		3	0.999	0.461
	1	0.999	0.333		4	0.747	0.822	0.700	9	0.040	0.997
0.310	6	0.031	0.998		3	0.947	0.646		8	0.200	0.979
	5	0.120	0.990		2	0.998	0.435		7	0.484	0.929
	4	0.331	0.957	0.480	8	0.015	0.999		6	0.772	0.836
	3	0.666	0.858		7	0.074	0.994		5	0.940	0.711
	2	0.943	0.635		6	0.230	0.973		4	0.992	0.571
0.320	6	0.036	0.998		5	0.499	0.916	0.750	9	0.075	0.994
	5	0.135	0.988		4	0.791	0.799		8	0.307	0.965
	4	0.359	0.952		3	0.961	0.621		7	0.626	0.897
	3	0.695	0.845		2	0.998	0.417		6	0.870	0.792
	2	0.952	0.617	0.500	8	0.020	0.999		5	0.975	0.666
0.330	6	0.042	0.997		7	0.093	0.992		4	0.998	0.533
	5	0.151	0.986		6	0.271	0.966	0.800	9	0.134	0.988
	4	0.387	0.945		5	0.553	0.901		8	0.446	0.944
	3	0.723	0.831		4	0.830	0.775		7	0.763	0.859
	2	0.961	0.600		3	0.972	0.598		6	0.939	0.747
0.340	7	0.010	0.999		2	0.999	0.400		5	0.991	0.625
	6	0.049	0.997	0.520	8	0.026	0.998		4	0.999	0.500
	5	0.168	0.983		7	0.115	0.989	0.850	9	0.232	0.979
	4	0.416	0.938		6	0.316	0.958		8	0.611	0.916
	3	0.750	0.816		5	0.606	0.884		7	0.879	0.818
	2	0.968	0.584		4	0.864	0.752		6	0.978	0.705
0.360	7	0.014	0.999		3	0.980	0.575		5	0.998	0.588
	6	0.064	0.995		2	0.999	0.385	0.900	9	0.387	0.963
	5	0.205	0.978	0.540	8	0.034	0.997		8	0.785	0.881
	4	0.474	0.923		7	0.141	0.986		7	0.957	0.777
	3	0.798	0.787		6	0.363	0.949		6	0.995	0.667
	2	0.978	0.553		5	0.658	0.867	0.950	9	0.630	0.938
0.380	7	0.019	0.999		4	0.893	0.729		8	0.934	0.841
	6	0.083	0.993		3	0.986	0.555		7	0.994	0.737
	5	0.247	0.971	0.560	8	0.044	0.996				

TABLE 2 Finite Queuing Factors—Population 20 *(Continued)*

X	M	D	F	X	M	D	F	X	M	D	F
0.005	1	0.095	0.999		1	0.837	0.866		3	0.326	0.980
0.009	1	0.171	0.998	0.052	3	0.080	0.998		2	0.733	0.896
0.011	1	0.208	0.997		2	0.312	0.986		1	0.998	0.526
0.013	1	0.246	0.996		1	0.858	0.851	0.100	5	0.038	0.999
0.014	1	0.265	0.995	0.054	3	0.088	0.998		4	0.131	0.995
0.015	1	0.283	0.994		2	0.332	0.984		3	0.363	0.975
0.016	1	0.302	0.993		1	0.876	0.835		2	0.773	0.878
0.017	1	0.321	0.992	0.056	3	0.097	0.997		1	0.999	0.500
0.018	2	0.048	0.999		2	0.352	0.982	0.110	5	0.055	0.998
	1	0.339	0.991		1	0.893	0.819		4	0.172	0.992
0.019	2	0.053	0.999	0.058	3	0.105	0.997		3	0.438	0.964
	1	0.358	0.990		2	0.372	0.980		2	0.842	0.837
0.020	2	0.058	0.999		1	0.908	0.802	0.120	6	0.022	0.999
	1	0.376	0.989	0.060	4	0.026	0.999		5	0.076	0.997
0.021	2	0.064	0.999		3	0.115	0.997		4	0.219	0.988
	1	0.394	0.987		2	0.392	0.978		3	0.514	0.950
0.022	2	0.070	0.999		1	0.922	0.785		2	0.895	0.793
	1	0.412	0.986	0.062	4	0.029	0.999	0.130	6	0.031	0.999
0.023	2	0.075	0.999		3	0.124	0.996		5	0.101	0.996
	1	0.431	0.984		2	0.413	0.975		4	0.271	0.983
0.024	2	0.082	0.999		1	0.934	0.768		3	0.589	0.933
	1	0.449	0.982	0.064	4	0.032	0.999		2	0.934	0.748
0.025	2	0.088	0.999		3	0.134	0.996	0.140	6	0.043	0.998
	1	0.466	0.980		2	0.433	0.972		5	0.131	0.994
0.026	2	0.094	0.998		1	0.944	0.751		4	0.328	0.976
	1	0.484	0.978	0.066	4	0.036	0.999		3	0.661	0.912
0.028	2	0.108	0.998		3	0.144	0.995		2	0.960	0.703
	1	0.519	0.973		2	0.454	0.969	0.150	7	0.017	0.999
0.030	2	0.122	0.998		1	0.953	0.733		6	0.059	0.998
	1	0.553	0.968	0.068	4	0.039	0.999		5	0.166	0.991
0.032	2	0.137	0.997		3	0.155	0.995		4	0.388	0.968
	1	0.587	0.962		2	0.474	0.966		3	0.728	0.887
0.034	2	0.152	0.996		1	0.961	0.716		2	0.976	0.661
	1	0.620	0.955	0.070	4	0.043	0.999	0.160	7	0.024	0.999
0.036	2	0.168	0.996		3	0.165	0.994		6	0.077	0.997
	1	0.651	0.947		2	0.495	0.962		5	0.205	0.988
0.038	3	0.036	0.999		1	0.967	0.699		4	0.450	0.957
	2	0.185	0.995	0.075	4	0.054	0.999		3	0.787	0.860
	1	0.682	0.938		3	0.194	0.992		2	0.987	0.622
0.040	3	0.041	0.999		2	0.545	0.953	0.180	7	0.044	0.998
	2	0.202	0.994		1	0.980	0.659		6	0.125	0.994
	1	0.712	0.929	0.080	4	0.066	0.998		5	0.295	0.978
0.042	3	0.047	0.999		3	0.225	0.990		4	0.575	0.930
	2	0.219	0.993		2	0.595	0.941		3	0.879	0.799
	1	0.740	0.918		1	0.988	0.621		2	0.996	0.555
0.044	3	0.053	0.999	0.085	4	0.080	0.997	0.200	8	0.025	0.999
	2	0.237	0.992		3	0.257	0.987		7	0.074	0.997
	1	0.767	0.906		2	0.643	0.928		6	0.187	0.988
0.046	3	0.059	0.999		1	0.993	0.586		5	0.397	0.963
	2	0.255	0.991	0.090	5	0.025	0.999		4	0.693	0.895
	1	0.792	0.894		4	0.095	0.997		3	0.938	0.736
0.048	3	0.066	0.999		3	0.291	0.984		2	0.999	0.500
	2	0.274	0.989		2	0.689	0.913	0.220	8	0.043	0.998
	1	0.815	0.881		1	0.996	0.554		7	0.115	0.994
0.050	3	0.073	0.998	0.095	5	0.031	0.999		6	0.263	0.980
	2	0.293	0.988		4	0.112	0.996		5	0.505	0.943

TABLE 2 Finite Queuing Factors—Population 20 (*Continued*)

X	M	D	F	X	M	D	F	X	M	D	F
	4	0.793	0.852		4	0.998	0.555	0.500	14	0.033	0.998
	3	0.971	0.677	0.380	12	0.024	0.999		13	0.088	0.995
0.240	9	0.024	0.999		11	0.067	0.996		12	0.194	0.985
	8	0.068	0.997		10	0.154	0.989		11	0.358	0.965
	7	0.168	0.989		9	0.305	0.973		10	0.563	0.929
	6	0.351	0.969		8	0.513	0.938		9	0.764	0.870
	5	0.613	0.917		7	0.739	0.874		8	0.908	0.791
	4	0.870	0.804		6	0.909	0.777		7	0.977	0.698
	3	0.988	0.623		5	0.984	0.656		6	0.997	0.600
0.260	9	0.039	0.998		4	0.999	0.526	0.540	15	0.023	0.999
	8	0.104	0.994	0.400	13	0.012	0.999		14	0.069	0.996
	7	0.233	0.983		12	0.037	0.998		13	0.161	0.988
	6	0.446	0.953		11	0.095	0.994		12	0.311	0.972
	5	0.712	0.884		10	0.205	0.984		11	0.509	0.941
	4	0.924	0.755		9	0.379	0.962		10	0.713	0.891
	3	0.995	0.576		8	0.598	0.918		9	0.873	0.821
0.280	10	0.021	0.999		7	0.807	0.845		8	0.961	0.738
	9	0.061	0.997		6	0.942	0.744		7	0.993	0.648
	8	0.149	0.990		5	0.992	0.624		6	0.999	0.556
	7	0.309	0.973	0.420	13	0.019	0.999	0.600	16	0.023	0.999
	6	0.544	0.932		12	0.055	0.997		15	0.072	0.996
	5	0.797	0.848		11	0.131	0.991		14	0.171	0.988
	4	0.958	0.708		10	0.265	0.977		13	0.331	0.970
	3	0.998	0.536		9	0.458	0.949		12	0.532	0.938
0.300	10	0.034	0.998		8	0.678	0.896		11	0.732	0.889
	9	0.091	0.995		7	0.863	0.815		10	0.882	0.824
	8	0.205	0.985		6	0.965	0.711		9	0.962	0.748
	7	0.394	0.961		5	0.996	0.595		8	0.992	0.666
	6	0.639	0.907	0.440	13	0.029	0.999		7	0.999	0.583
	5	0.865	0.808		12	0.078	0.995	0.700	17	0.047	0.998
	4	0.978	0.664		11	0.175	0.987		16	0.137	0.991
	3	0.999	0.500		10	0.333	0.969		15	0.295	0.976
0.320	11	0.018	0.999		9	0.540	0.933		14	0.503	0.948
	10	0.053	0.997		8	0.751	0.872		13	0.710	0.905
	9	0.130	0.992		7	0.907	0.785		12	0.866	0.849
	8	0.272	0.977		6	0.980	0.680		11	0.953	0.783
	7	0.483	0.944		5	0.998	0.568		10	0.988	0.714
	6	0.727	0.878	0.460	14	0.014	0.999		9	0.998	0.643
	5	0.915	0.768		13	0.043	0.998	0.800	19	0.014	0.999
	4	0.989	0.624		12	0.109	0.993		18	0.084	0.996
0.340	11	0.029	0.999		11	0.228	0.982		17	0.242	0.984
	10	0.079	0.996		10	0.407	0.958		16	0.470	0.959
	9	0.179	0.987		9	0.620	0.914		15	0.700	0.920
	8	0.347	0.967		8	0.815	0.846		14	0.867	0.869
	7	0.573	0.924		7	0.939	0.755		13	0.955	0.811
	6	0.802	0.846		6	0.989	0.651		12	0.989	0.750
	5	0.949	0.729		5	0.999	0.543		11	0.998	0.687
	4	0.995	0.588	0.480	14	0.022	0.999	0.900	19	0.135	0.994
0.360	12	0.015	0.999		13	0.063	0.996		18	0.425	0.972
	11	0.045	0.998		12	0.147	0.990		17	0.717	0.935
	10	0.112	0.993		11	0.289	0.974		16	0.898	0.886
	9	0.237	0.981		10	0.484	0.944		15	0.973	0.833
	8	0.429	0.954		9	0.695	0.893		14	0.995	0.778
	7	0.660	0.901		8	0.867	0.819		13	0.999	0.722
	6	0.863	0.812		7	0.962	0.726	0.950	19	0.377	0.981
	5	0.971	0.691		6	0.994	0.625		18	0.760	0.943

TABLE 3 Finite Queuing Factors—Population 30 (*Continued*)

X	M	D	F	X	M	D	F	X	M	D	F
0.004	1	0.116	0.999		1	0.963	0.772		3	0.426	0.976
0.007	1	0.203	0.998	0.044	4	0.040	0.999		2	0.847	0.873
0.009	1	0.260	0.997		3	0.154	0.996	0.075	5	0.069	0.998
0.010	1	0.289	0.996		2	0.474	0.977		4	0.201	0.993
0.011	1	0.317	0.995		1	0.974	0.744		3	0.486	0.969
0.012	1	0.346	0.994	0.046	4	0.046	0.999		2	0.893	0.840
0.013	1	0.374	0.993		3	0.171	0.996	0.080	6	0.027	0.999
0.014	2	0.067	0.999		2	0.506	0.972		5	0.088	0.998
	1	0.403	0.991		1	0.982	0.716		4	0.240	0.990
0.015	2	0.076	0.999	0.048	4	0.053	0.999		3	0.547	0.959
	1	0.431	0.989		3	0.189	0.995		2	0.929	0.805
0.016	2	0.085	0.999		2	0.539	0.968	0.085	6	0.036	0.999
	1	0.458	0.987		1	0.988	0.689		5	0.108	0.997
0.017	2	0.095	0.999	0.050	4	0.060	0.999		4	0.282	0.987
	1	0.486	0.985		3	0.208	0.994		3	0.607	0.948
0.018	2	0.105	0.999		2	0.571	0.963		2	0.955	0.768
	1	0.513	0.983		1	0.992	0.663	0.090	6	0.046	0.999
0.019	2	0.116	0.999	0.052	4	0.068	0.999		5	0.132	0.996
	1	0.541	0.980		3	0.227	0.993		4	0.326	0.984
0.020	2	0.127	0.998		2	0.603	0.957		3	0.665	0.934
	1	0.567	0.976		1	0.995	0.639		2	0.972	0.732
0.021	2	0.139	0.998	0.054	4	0.077	0.998	0.095	6	0.057	0.999
	1	0.594	0.973		3	0.247	0.992		5	0.158	0.994
0.022	2	0.151	0.998		2	0.634	0.951		4	0.372	0.979
	1	0.620	0.969		1	0.997	0.616		3	0.720	0.918
0.023	2	0.163	0.997	0.056	4	0.086	0.998		2	0.984	0.697
	1	0.645	0.965		3	0.267	0.991	0.100	6	0.071	0.998
0.024	2	0.175	0.997		2	0.665	0.944		5	0.187	0.993
	1	0.670	0.960		1	0.998	0.595		4	0.421	0.973
0.025	2	0.188	0.996	0.058	4	0.096	0.998		3	0.771	0.899
	1	0.694	0.954		3	0.288	0.989		2	0.991	0.664
0.026	2	0.201	0.996		2	0.695	0.936	0.110	7	0.038	0.999
	1	0.718	0.948		1	0.999	0.574		6	0.105	0.997
0.028	3	0.051	0.999	0.060	5	0.030	0.999		5	0.253	0.988
	2	0.229	0.995		4	0.106	0.997		4	0.520	0.959
	1	0.763	0.935		3	0.310	0.987		3	0.856	0.857
0.030	3	0.060	0.999		2	0.723	0.927		2	0.997	0.605
	2	0.257	0.994		1	0.999	0.555	0.120	7	0.057	0.998
	1	0.805	0.918	0.062	5	0.034	0.999		6	0.147	0.994
0.032	3	0.071	0.999		4	0.117	0.997		5	0.327	0.981
	2	0.286	0.992		3	0.332	0.986		4	0.619	0.939
	1	0.843	0.899		2	0.751	0.918		3	0.918	0.808
0.034	3	0.083	0.999	0.064	5	0.038	0.999		2	0.999	0.555
	2	0.316	0.990		4	0.128	0.997	0.130	8	0.030	0.999
	1	0.876	0.877		3	0.355	0.984		7	0.083	0.997
0.036	3	0.095	0.998		2	0.777	0.908		6	0.197	0.991
	2	0.347	0.988	0.066	5	0.043	0.999		5	0.409	0.972
	1	0.905	0.853		4	0.140	0.996		4	0.712	0.914
0.038	3	0.109	0.998		3	0.378	0.982		3	0.957	0.758
	2	0.378	0.986		2	0.802	0.897	0.140	8	0.045	0.999
	1	0.929	0.827	0.068	5	0.048	0.999		7	0.115	0.996
0.040	3	0.123	0.997		4	0.153	0.995		6	0.256	0.987
	2	0.410	0.983		3	0.402	0.979		5	0.494	0.960
	1	0.948	0.800		2	0.825	0.885		4	0.793	0.884
0.042	3	0.138	0.997	0.070	5	0.054	0.999		3	0.979	0.710
	2	0.442	0.980		4	0.166	0.995	0.150	9	0.024	0.999

TABLE 3 Finite Queuing Factors—Population 30 (*Continued*)

X	M	D	F	X	M	D	F	X	M	D	F
	8	0.065	0.998		7	0.585	0.938		7	0.901	0.818
	7	0.155	0.993		6	0.816	0.868		6	0.981	0.712
	6	0.322	0.980		5	0.961	0.751		5	0.999	0.595
	5	0.580	0.944		4	0.998	0.606	0.290	14	0.023	0.999
	4	0.860	0.849	0.230	12	0.023	0.999		13	0.055	0.998
	3	0.991	0.665		11	0.056	0.998		12	0.117	0.994
0.160	9	0.036	0.999		10	0.123	0.994		11	0.223	0.986
	8	0.090	0.997		9	0.242	0.985		10	0.382	0.969
	7	0.201	0.990		8	0.423	0.965		9	0.582	0.937
	6	0.394	0.972		7	0.652	0.923		8	0.785	0.880
	5	0.663	0.924		6	0.864	0.842		7	0.929	0.795
	4	0.910	0.811		5	0.976	0.721		6	0.988	0.688
	3	0.996	0.624		4	0.999	0.580		5	0.999	0.575
0.170	10	0.019	0.999	0.240	12	0.031	0.999	0.300	14	0.031	0.999
	9	0.051	0.998		11	0.074	0.997		13	0.071	0.997
	8	0.121	0.995		10	0.155	0.992		12	0.145	0.992
	7	0.254	0.986		9	0.291	0.981		11	0.266	0.982
	6	0.469	0.961		8	0.487	0.955		10	0.437	0.962
	5	0.739	0.901		7	0.715	0.905		9	0.641	0.924
	4	0.946	0.773		6	0.902	0.816		8	0.830	0.861
	3	0.998	0.588		5	0.986	0.693		7	0.950	0.771
0.180	10	0.028	0.999		4	0.999	0.556		6	0.993	0.666
	9	0.070	0.997	0.250	13	0.017	0.999	0.320	15	0.023	0.999
	8	0.158	0.993		12	0.042	0.998		14	0.054	0.998
	7	0.313	0.980		11	0.095	0.996		13	0.113	0.994
	6	0.546	0.948		10	0.192	0.989		12	0.213	0.987
	5	0.806	0.874		9	0.345	0.975		11	0.362	0.971
	4	0.969	0.735		8	0.552	0.944		10	0.552	0.943
	3	0.999	0.555		7	0.773	0.885		9	0.748	0.893
0.190	10	0.039	0.999		6	0.932	0.789		8	0.901	0.820
	9	0.094	0.996		5	0.992	0.666		7	0.977	0.727
	8	0.200	0.990	0.260	13	0.023	0.999		6	0.997	0.625
	7	0.378	0.973		12	0.056	0.998	0.340	16	0.016	0.999
	6	0.621	0.932		11	0.121	0.994		15	0.040	0.998
	5	0.862	0.845		10	0.233	0.986		14	0.086	0.996
	4	0.983	0.699		9	0.402	0.967		13	0.169	0.990
0.200	11	0.021	0.999		8	0.616	0.930		12	0.296	0.979
	10	0.054	0.998		7	0.823	0.864		11	0.468	0.957
	9	0.123	0.995		6	0.954	0.763		10	0.663	0.918
	8	0.249	0.985		5	0.995	0.641		9	0.836	0.858
	7	0.446	0.963	0.270	13	0.032	0.999		8	0.947	0.778
	6	0.693	0.913		12	0.073	0.997		7	0.990	0.685
	5	0.905	0.814		11	0.151	0.992		6	0.999	0.588
	4	0.991	0.665		10	0.279	0.981	0.360	16	0.029	0.999
0.210	11	0.030	0.999		9	0.462	0.959		15	0.065	0.997
	10	0.073	0.997		8	0.676	0.915		14	0.132	0.993
	9	0.157	0.992		7	0.866	0.841		13	0.240	0.984
	8	0.303	0.980		6	0.970	0.737		12	0.392	0.967
	7	0.515	0.952		5	0.997	0.617		11	0.578	0.937
	6	0.758	0.892	0.280	14	0.017	0.999		10	0.762	0.889
	5	0.938	0.782		13	0.042	0.998		9	0.902	0.821
	4	0.995	0.634		12	0.093	0.996		8	0.974	0.738
0.220	11	0.041	0.999		11	0.185	0.989		7	0.996	0.648
	10	0.095	0.996		10	0.329	0.976	0.380	17	0.020	0.999
	9	0.197	0.989		9	0.522	0.949		16	0.048	0.998
	8	0.361	0.974		8	0.733	0.898		15	0.101	0.995

TABLE 3 Finite Queuing Factors—Population 30 (*Continued*)

X	M	D	F	X	M	D	F	X	M	D	F
	14	0.191	0.988		16	0.310	0.977		22	0.038	0.998
	13	0.324	0.975		15	0.470	0.957		21	0.085	0.996
	12	0.496	0.952		14	0.643	0.926		20	0.167	0.990
	11	0.682	0.914		13	0.799	0.881		19	0.288	0.980
	10	0.843	0.857		12	0.910	0.826		18	0.443	0.963
	9	0.945	0.784		11	0.970	0.762		17	0.612	0.936
	8	0.988	0.701		10	0.993	0.694		16	0.766	0.899
	7	0.999	0.614		9	0.999	0.625		15	0.883	0.854
0.400	17	0.035	0.999	0.500	20	0.032	0.999		14	0.953	0.802
	16	0.076	0.996		19	0.072	0.997		13	0.985	0.746
	15	0.150	0.992		18	0.143	0.992		12	0.997	0.690
	14	0.264	0.982		17	0.252	0.983		11	0.999	0.632
	13	0.420	0.964		16	0.398	0.967	0.600	23	0.024	0.999
	12	0.601	0.933		15	0.568	0.941		22	0.059	0.997
	11	0.775	0.886		14	0.733	0.904		21	0.125	0.993
	10	0.903	0.823		13	0.865	0.854		20	0.230	0.986
	9	0.972	0.748		12	0.947	0.796		19	0.372	0.972
	8	0.995	0.666		11	0.985	0.732		18	0.538	0.949
0.420	18	0.024	0.999		10	0.997	0.667		17	0.702	0.918
	17	0.056	0.997	0.520	21	0.021	0.999		16	0.837	0.877
	16	0.116	0.994		20	0.051	0.998		15	0.927	0.829
	15	0.212	0.986		19	0.108	0.994		14	0.974	0.776
	14	0.350	0.972		18	0.200	0.988		13	0.993	0.722
	13	0.521	0.948		17	0.331	0.975		12	0.999	0.667
	12	0.700	0.910		16	0.493	0.954	0.700	25	0.039	0.998
	11	0.850	0.856		15	0.663	0.923		24	0.096	0.995
	10	0.945	0.789		14	0.811	0.880		23	0.196	0.989
	9	0.986	0.713		13	0.915	0.827		22	0.339	0.977
	8	0.998	0.635		12	0.971	0.767		21	0.511	0.958
0.440	19	0.017	0.999		11	0.993	0.705		20	0.681	0.930
	18	0.041	0.998		10	0.999	0.641		19	0.821	0.894
	17	0.087	0.996	0.540	21	0.035	0.999		18	0.916	0.853
	16	0.167	0.990		20	0.079	0.996		17	0.967	0.808
	15	0.288	0.979		19	0.155	0.991		16	0.990	0.762
	14	0.446	0.960		18	0.270	0.981		15	0.997	0.714
	13	0.623	0.929		17	0.421	0.965	0.800	27	0.053	0.998
	12	0.787	0.883		16	0.590	0.938		26	0.143	0.993
	11	0.906	0.824		15	0.750	0.901		25	0.292	0.984
	10	0.970	0.755		14	0.874	0.854		24	0.481	0.966
	9	0.994	0.681		13	0.949	0.799		23	0.670	0.941
	8	0.999	0.606		12	0.985	0.740		22	0.822	0.909
0.460	19	0.028	0.999		11	0.997	0.679		21	0.919	0.872
	18	0.064	0.997		10	0.999	0.617		20	0.970	0.832
	17	0.129	0.993	0.560	22	0.023	0.999		19	0.991	0.791
	16	0.232	0.985		21	0.056	0.997		18	0.998	0.750
	15	0.375	0.970		20	0.117	0.994	0.900	29	0.047	0.999
	14	0.545	0.944		19	0.215	0.986		28	0.200	0.992
	13	0.717	0.906		18	0.352	0.973		27	0.441	0.977
	12	0.857	0.855		17	0.516	0.952		26	0.683	0.953
	11	0.945	0.793		16	0.683	0.920		25	0.856	0.923
	10	0.985	0.724		15	0.824	0.878		24	0.947	0.888
	9	0.997	0.652		14	0.920	0.828		23	0.985	0.852
0.480	20	0.019	0.999		13	0.972	0.772		22	0.996	0.815
	19	0.046	0.998		12	0.993	0.714		21	0.999	0.778
	18	0.098	0.995		11	0.999	0.655	0.950	29	0.226	0.993
	17	0.184	0.989	0.580	23	0.014	0.999		28	0.574	0.973

Control Concepts
and Methods

From Chapter 11 of *Systems Engineering and Analysis,* Fifth Edition, Benjamin S. Blanchard, Wolter J. Fabrycky. Copyright © 2011 by Pearson Education, Inc. Published by Pearson Prentice Hall. All rights reserved.

Control Concepts
and Methods

Most systems are deployed and then operate in an environment that changes over time. Except for static systems, a changing environment can lead to instability unless control action is applied. The value of dynamic systems may be enhanced through control action applied during operations. Control of a portion or all of a system can help maintain system performance within specified tolerances or can increase the worth of system output.

The concern for control began with physical systems and servo theory. Engineers who turned their attention to large-scale systems in the 1950s became interested in servomechanisms and other control devices. A large-scale, human-made system is one in which there are many states, control variables, and constraints. Classical control theory cannot be used directly to optimize outputs for most systems of this type. Nevertheless, classical concepts can provide a basis for structuring system descriptions and internal relationships so that feedback and adaptive phenomena may be incorporated during system design.

The fundamental concepts abstracted above are developed in this chapter, recognizing that there are numerous control situations, models, and methods that could be considered. Those chosen for inclusion are statistical process controls for variables and attributes, optimum policy control, two project control methods, and four aspects from the general domain of total quality control. Study of these topics should give the reader an understanding of the following concepts and ideas:

- The concept and elements of a control system, types of control systems, and information as the medium of control;
- Statistical process control and its application through control charts for variables and control charting methods for attributes;
- Monitoring an optimum solution obtained from a decision model through methods of optimum policy control;
- The control of project activities and schedules, utilizing critical path methods (CPM) and program evaluation and review techniques (PERT);

- The role of experimental design (ED), parameter design (PD), and quality engineering (QE) in the broad domain of total quality control (TQC); and
- Parameter design per Taguchi and the design-dependent parameter approach of this textbook; their similarities and differences.

The last section of this chapter provides a summary of the diverse control topics listed above. It also emphasizes certain key points for further consideration. Study of the material in this chapter assumes that the reader is proficient in working with probability theory and statistical analysis.

1 SOME BASIC CONTROL CONCEPTS

A control problem may arise from the need to regulate speed, temperature, quality, or quantity or to determine the trajectory of an aircraft or space vehicle. In each case, the allocation of scarce resources over time is involved. Control variables must be related in some way to state variables describing the system characteristic or the condition being controlled.

The classical example of control is the determination of a missile trajectory. Control variables are the amount, timing, and direction of the thrust applied to the missile, where the thrust available is constrained by fuel availability. State variables describe the trajectory incorporating mass, position, and velocity. The thrusts and the state variables are related by differential equations in the light of the effectiveness function to be optimized. The objective is usually to maximize the payload that can be delivered to a given destination.

Control problems also arise in large-scale human-equipment systems of activity. In air traffic control, state variables are the number of aircraft en route, the number of aircraft waiting to land at available airports, the number of aircraft waiting on the ground, the available number of communication channels, and so on. Control variables in air traffic control are the instructions given by the traffic controllers.

1.1 The Elements of a Control System

Every control system has four basic elements. These elements always occur in the same sequence and have the same relationship with each other. They are as follows:

1. A controlled characteristic or condition.
2. A sensory device or method for measuring the characteristic or condition.
3. A control device that will compare measured performance with planned performance.
4. An activating device that will alter the system to bring about a change in the output characteristic or condition being controlled.

Figure 1 illustrates the relationships among the four elements of a control system. The output of the system (Figure 1, Block 1) is the characteristic or condition to be measured. This output may be speed, temperature, quality, or any other characteristic or condition of the system under consideration.

The sensory device (Figure 1, Block 2) measures output performance. System design must incorporate such devices as tachometers, thermocouples, thermometers, transducers, inspectors, and other physical or human sensors. Information gathered from a sensory device is essential to the operation of a control system.

The third element in a control system is the control device (Figure 1, Block 3). This element exists to determine the need for control action based on the information provided by

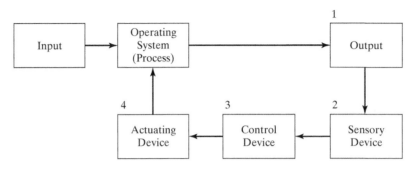

Figure 1   Control system elements and relationships.

the sensory device. It may be a minicomputer or microcomputer, but need be no more than a visual or hand-calculated comparison depending upon the situation. It is important that this element in the control system be able to detect significant differences between planned output and actual output.

Any out-of-control signal received from the control device must be acted upon by the actuating device (Figure 1, Block 4). This device is the fourth element in a control system. Its role is that of implementation. Actuating devices may be mechanical, electromechanical, hydraulic, pneumatic, or they may be human, as in the case where a machine operator changes the setting on a machine to alter the dimension of a part being produced.

1.2 Types of Control Systems

There are two basic types of control systems, open-loop and closed-loop. Each has its own characteristic. The basic difference is whether or not the control device is an integral part of the system it controls.

In *open-loop control*, the optimal control action is completely specified at the initial time in the system cycle. In *closed-loop control*, the optimal control action is determined both by the initial conditions and by the current state-variable values. Whereas in open-loop control all decisions are made in advance, in closed-loop control decisions are revised in the light of new information received about the system state.

Common examples of open-loop and closed-loop control are the laundry dryer and the home thermostat. In the laundry dryer the length of the drying cycle is set once by human action. A home thermostat, on the other hand, provides control signals to the furnace whenever temperature falls below a predetermined level. When a system is regulated by a person, as in the case of a laundry dryer, attention must be paid to see that the desired output is obtained. In this open-loop example, the drying time must be determined by experience and be properly set into the timer.

If control is exercised in terms of system operation, as with a thermostat, it is closed-loop control. The significant difference in this type of control system is that the control device is an element of the controlled system it serves. In closed-loop control the four elements of a control system, shown in Figure 1, all act together to achieve the purpose. Open-loop control requires only initial measurements and is preferred, unless there are random elements either in measurement or in system operation.

1.3 Feedback and Feed-Forward Control

An essential part of a closed-loop system is feedback. *Feedback* is a process in which the output of a system is measured continuously or at predetermined intervals as in sample-data control presented next. Generally, costs are incurred in measuring a characteristic or condition. These costs may be in terms of added workload, energy, weight, and so on.

The *feed-forward* control structure provides a way to augment feedback by measuring or predicting the disturbance before it influences the process under control. This feed-forward modification of feedback control must conform to specific design criteria. The measured disturbance must: (1) be measurable, (2) detect the occurrence of an important disturbance, (3) not be causally affected by the final element, and (4) have disturbance dynamics not any faster than the compensation dynamics.

Usually, feedback control is retained in a system because feed-forward only compensates for the measured disturbances and because the prediction model employed is not perfect. Like all models, it is only an approximation of reality. The fidelity of the feed-forward control action can be no better than the fidelity of the underlying prediction.

1.4 Information as the Medium of Control

To control every characteristic or condition of a system's output is usually not possible. Accordingly, it is important to make a wise choice of the characteristic to be measured. It is important that the successful control of the characteristic chosen lead to successful control of the system. This requirement is usually met if the controlled characteristic is stated in the language to be used in the corrective loop. Before information is gathered, one must determine what is to be transmitted.

It is the flow of measurement information which, when converted into corrective information, makes it possible for a characteristic or condition of a system to be controlled. Information to be compared with the control standard must be expressed in the same terms as the standard to facilitate a decision regarding the control status of the system. This information can often be secured by sampling the output. Of course, a sample is only an estimate of a group or population. Statistical sampling based on probability theory can, however, provide valid control information subject to some error.

The sensor element in Figure 1 provides information that becomes the basis for control action. Output information is compared with the characteristic or condition to be controlled. In feedback control, significant deviations are noted and corrective information is provided to the activating device.

Information received from the control device causes the actuating device to respond. However, every control system has a finite delay in its action time that can cause problems of overcorrection. This difficulty can be partially overcome by reducing the time lag between output measurement and input correction, or by introducing a lead time in the control system. Feed-forward control is another way of reducing the delay in control action but it introduces problems of its own because of its future focus.

2 STATISTICAL PROCESS CONTROL

Variation is inherent in the output characteristic or condition of most processes. Patterns of statistical variation exist in the number of units processed through maintenance per time period, the dimension of a machined part, the procurement lead time for an item, and the number of defects per unit quantity of product produced. Statistical process control is a methodology for testing to see if an operating system is in control.

Measurement is the key to evaluation in Statistical Process Control. This measurement is applied to system output for the purpose of detecting a difference between actual values and ideal or target values. By applying statistical control to a process properly centered on product specifications, considerable faith can be placed in the quality of the output product. Statistical control permits producers to identify and eliminate assignable causes of variation, while reducing random or unassignable causes. Furthermore, output produced under statistical control generally need not be subjected to sampling inspection. Sampling inspection seeks to find and eliminate defective items, a practice that is proving to be too disruptive and costly.

A stable or probabilistic steady-state pattern of variation exists when the parameters of the statistical distribution describing an operation remain constant over time. Steady-state variation of this type is normally an exception rather than the rule. Many operations will produce a probabilistic nonsteady-state pattern of variation over time. When this is the case, the mean or the variance of the distribution describing the pattern change with time. Sometimes, the form of the distribution will also undergo change.

Control limits may often be placed about an initial stable pattern of variation to detect subsequent changes from that pattern; a statistical inference is made when a sample value is considered. If the sample falls within the control limits, the process under study is said to be *in control*. If the sample falls outside the limits, the process is deemed to have changed and is said to be *out of control*.

Figure 2 illustrates control limits placed at a distance $k\sigma$ from the mean of the initial stable pattern of variation. If a sample value falls outside the control limits, and the process has not changed, a Type I error of probability α has been made. Conversely, if a sample value falls within these limits and the process has changed, a Type II error of probability β has been made. In the first case, the null hypothesis has been rejected in error. In the second, the null hypothesis has not been rejected, with a resulting error.

If the control limits are set relatively far apart, it is unlikely that a Type I error will be made. The control model, however, is then not likely to detect small shifts in the parameter. In this case the probability, β, of a Type II error is large. If the other approach is taken, and the limits are placed relatively close to the initial stable pattern of variation, the value of α will increase. The advantage gained is greater sensitivity of the model for the shifts that may occur. The ultimate criterion will be the costs associated with the making of Type I and Type II errors. The limits should be established to minimize the sum of these two costs.

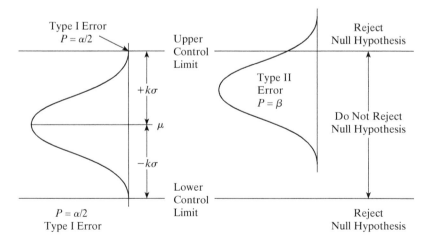

Figure 2 A stable and a changed pattern of variation.

Statistical process control is based on the assumption of a stable process. If stable, one can measure a product characteristic that reflects the behavior of the process. The measured characteristic will have a steady-state distribution (in a probabilistic sense); that is, each sample of the product from the process will have the identical statistical distribution.

A stable process that yields a product that satisfies specifications may later become unstable. Instability may arise from special causes. Continued monitoring of the process using statistical process control can detect the transition to an unstable state and the presence of special causes. A process that passes the statistical test is not always stable. However, the probability is very small that an unstable process will continue to produce output that passes the statistical tests (refer to Figure 2).

Although process control methods were first applied to manufacturing, their use is not restricted to production. Statistical process control is also used to evaluate and improve engineering processes. It is a means to obtain useful information to feed back into product and production system design. In addition to control charts, Pareto curves, cause and effect diagrams, CPNI networks, and PERT are used to evaluate and improve engineering and production processes.

As real-time information becomes more readily available in design, production, and support, practitioners are beginning to measure not only product parameters but also changes in process control parameters. Such applications are closely related to control engineering and use many of its basic principles.

3 STATISTICAL CONTROL CHARTS

A statistical control chart is a control device (Figure 1, Block 3) that uses measurement information from a sensory device (Figure 1, Block 2) and provides system change information to an actuating device (Figure 1, Block 4). Control charts use samples taken from system output and control limits to determine if an operation is in control. When a sample value falls outside the control limits, a stable system of statistical variation probably no longer exists. Action may then be taken to find the cause of the "out-of-control" condition so that control may again be established, or compensation be made for the change that has occurred.

Charts are used in statistical process control for conducting stability tests. When the value of a characteristic can be measured, $\overline{X}$ and R charts are used; when the fraction of a two-valued characteristic is being measured, a p chart is appropriate; and when the overall number of defects is being measured, the c chart is applicable.[1]

3.1 Control Charts for Variables

A *control chart* is a graphical representation of a mathematical model used to monitor a random variable process to detect changes in a parameter of that process. Charting statistical data is a test of the null hypothesis that the process from which the sample came has not changed. A control chart is employed to distinguish between the existence of a stable pattern of variation and the occurrence of an unstable pattern. If an unstable pattern of variation is detected, action may be initiated to discover the cause of the instability. Removal of the assignable cause should permit the process to return to the stable state.

---

[1]These charting methods are attributed to the work of Walter Shewhart in the 1920s. Shewhart's approach to variation focuses more on process stability than on conformance to specifications. Engineers should define specifications and statistical tolerances as part of the design process with the process capability in mind.

Control charts for variables are used for continuous operations. Two charts are available for operations of this type. The $\overline{X}$ *chart* is a plot over time of sample means taken from a process. It is primarily employed to detect changes in the mean of the process from which the samples came. The *R chart* is a plot over time of the ranges of these same samples. It is employed to detect changes in the dispersion of the process. These charts are often employed together in control operations.

Constructing the $\overline{X}$ Chart. The $\overline{X}$ chart receives its input as the mean of a sample taken from the process under study. Usually, the sample will contain four or five observations, a number sufficient to make the Central Limit Theorem applicable.[2] Accepting an approximately normal distribution of the sample means allows the establishment of control limits with a predetermined knowledge of the probability of making a Type I error. It is not necessary to know the form of the distribution of the process.

The first step in constructing an $\overline{X}$ chart is to estimate the process mean, μ, and the process variance, σ^2. This requires taking m samples each of size n and calculating the mean, $\overline{X}$, and the range, R, for each sample. Table 1 illustrates the format that may be used in the calculations. The mean of the sample means, $\overline{\overline{X}}$ is used as an estimate of μ and is calculated as

$$\overline{\overline{X}} = \frac{\sum_{i=1}^{m} \overline{X}}{m} \tag{1}$$

and the mean of the sample ranges, $\overline{R}$, is calculated as

$$\overline{R} = \frac{\sum_{i=1}^{m} R}{m} \tag{2}$$

The expected ratio between the average range, $\overline{R}$, and the standard deviation of the process has been computed for various sample sizes, n. This ratio is designated d_2 and is expressed as

$$d_2 = \frac{\overline{R}}{\sigma}$$

TABLE 1   Computational Format for Determining $\overline{X}$ and R

Sample Number	Sample Values	Mean $\overline{X}$	Range R
1	$x_{11}, x_{12}, \ldots, x_{1n}$	$\overline{X}_1$	R_1
2	$x_{21}, x_{22}, \ldots, x_{2n}$	$\overline{X}_2$	R_2
$\vdots$	$\vdots \quad \vdots \quad \vdots$	$\vdots$	$\vdots$
M	$x_{m1}, x_{m2}, \ldots, x_{mn}$	$\overline{X}_m$	R_m

---

[2]The *Central Limit Theorem* states that the distribution of sample means approaches normality as the sample size approaches infinity, provided that the sample values are independent.

Therefore, σ can be estimated from the sample statistic $\overline{R}$ as

$$\sigma = \frac{\overline{R}}{d_2} \tag{3}$$

Values of d_2 as a function of the sample size may be found in Table 2.

The mean of the $\overline{X}$ chart is set at $\overline{\overline{X}}$. The control limits are normally set at $\pm 3\sigma_{\overline{X}}$, which results in the probability of making a Type I error of 0.0027. Because

$$\sigma_{\overline{X}} = \frac{\sigma}{\sqrt{n}}$$

substitution into Equation 3 gives

$$\sigma_{\overline{X}} = \frac{\overline{R}}{d_2\sqrt{n}}$$

and

$$3\sigma_{\overline{X}} = \frac{3\overline{R}}{d_2\sqrt{n}} \tag{4}$$

The factor $3/d_2\sqrt{n}$ has been tabulated as A_2 in Table 2. Therefore, the upper and lower control limits for the $\overline{X}$ chart may be specified as

$$UCL_{\overline{X}} = \overline{\overline{X}} + A_2\overline{R} \tag{5}$$

$$LCL_{\overline{X}} = \overline{\overline{X}} - A_2\overline{R} \tag{6}$$

Constructing the R Chart. The R chart is constructed in a manner similar to the $\overline{X}$ chart. If the $\overline{X}$ chart has already been completed, $\overline{R}$ can be calculated from Equation 2. Tabular values of 3σ control limits for the range have been compiled for varying sample

TABLE 2 Factors for the Construction of $\overline{X}$ and R Charts

Sample Size, n	$\overline{X}$ Chart		R Chart	
	d_2	A_2	D_3	D_4
2	1.128	1.880	0	3.267
3	1.693	1.023	0	2.575
4	2.059	0.729	0	2.282
5	2.326	0.577	0	2.115
6	2.534	0.482	0	2.004
7	2.704	0.419	0.076	1.924
8	2.847	0.373	0.136	1.864
9	2.970	0.337	0.184	1.816
10	3.078	0.308	0.223	1.777

sizes and are included in Table 2. The upper and lower control limits for the R chart are then specified as

$$UCL_R = D_4\overline{R} \tag{7}$$

$$LCL_R = D_3\overline{R} \tag{8}$$

Because $D_3 = 0$ for sample size of $n \leq 6$ in Table 2, the $LCL_R = 0$. Actually, 3σ limits yield a negative lower control limit that is recorded as zero. This means that with samples of six or fewer, it will be impossible for a value on the R chart to fall outside the lower limit. Thus, the R chart will not be capable of detecting reductions in the dispersion of the process output.

Application of the $\overline{X}$ and R Charts.   Once the control limits have been specified for each chart, the data used in constructing the limits are plotted. Should all values fall within both sets of limits, the charts are ready for use. Should one or more values fall outside one set of limits, however, further inquiry is needed. A value outside the limits on the $\overline{X}$ chart indicates that the process may have undergone some change in regard to its central tendency. A value outside the limits on the R chart is evidence that the process variability may now be out of control. In either case, one should search for the source of the change in process behavior. If one or two values fall outside the limits and an assignable cause can be found, then these one or two values may be discarded and revised control limits calculated. If the revised limits contain all the remaining values, the control chart is ready for implementation. If they do not, the procedure may be repeated before using the control chart.

Assume that control charts are to be established to monitor the weight in ounces of propellant loaded into canisters. The canisters should hold 10 ounces, and to achieve this weight, the process must be set to deliver slightly more. Samples of five have been taken every 30 minutes. The sample data, together with the sample means and sample ranges, are given in Table 3.

An R chart is first constructed from these data. Using Equation 2, $\overline{R}$ is calculated as 2.05. The control limits are then determined from Equations 7 and 8 as

$$UCL_R = 2.115(2.05) = 4.34$$
$$LCL_R = 0(2.05) = 0$$

These limits are used to construct the R chart of Figure 3. Since all values fall within the control limits, the R chart is accepted as a means of assessing subsequent process variation. Had a point fallen outside the calculated limits, that point would have had to be discarded and limits recalculated.

Attention should next be directed to the $\overline{X}$ chart. The mean of the sample means, $\overline{\overline{X}}$, is found from Equation 1 to be 19. The mean of the sample ranges, $\overline{R}$, has already been calculated as 2.05. Preliminary control limits for the $\overline{X}$ chart can now be calculated from Equations 5 and 6 as

$$UCL_{\overline{X}} = 12.19 + 0.577(2.05) = 13.37$$
$$LCL_{\overline{X}} = 12.19 - 0.577(2.05) = 11.01$$

These limits are used to construct the $\overline{X}$ chart shown in Figure 4.

TABLE 3 Weight of Propellant in Canisters (ounce)

Sample Number	Sample Values					Mean $\overline{X}$	Range R
1	11.3	10.5	12.4	12.2	12.0	11.7	1.9
2	9.6	11.7	13.0	11.4	12.8	11.7	3.4
3	11.4	12.4	11.7	11.4	12.4	11.9	1.0
4	12.0	11.9	13.2	11.9	12.2	12.2	1.3
5	12.4	11.9	11.7	11.6	10.5	11.6	1.9
6	13.8	12.5	13.9	11.9	11.4	12.7	2.5
7	13.3	11.6	13.2	10.7	11.4	12.0	2.6
8	11.1	11.3	13.2	12.8	12.0	12.1	2.1
9	12.5	11.9	13.8	11.6	13.0	12.6	2.2
10	12.1	11.7	12.0	11.7	12.9	12.1	1.2
11	11.7	12.6	12.3	11.2	10.8	11.7	1.8
12	13.8	12.3	12.4	14.1	11.3	12.8	2.8
13	10.6	11.8	13.1	12.8	11.7	12.0	2.5
14	12.0	11.2	12.1	11.7	12.1	11.8	0.9
15	11.5	13.1	13.9	11.9	10.7	12.2	3.2
16	13.4	12.6	12.4	11.9	11.8	12.4	1.6
17	12.1	13.1	14.1	11.4	12.3	12.6	2.7
18	11.5	13.2	12.4	12.6	12.2	12.4	1.7
19	13.8	14.2	13.5	13.2	12.8	13.5	1.4
20	11.5	11.4	13.1	11.6	10.8	11.7	2.3

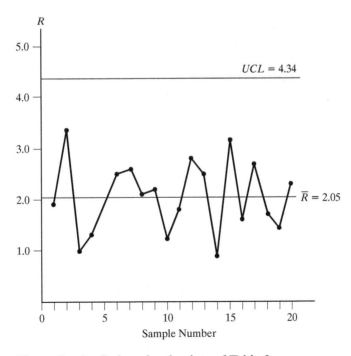

Figure 3 An R chart for the data of Table 3.

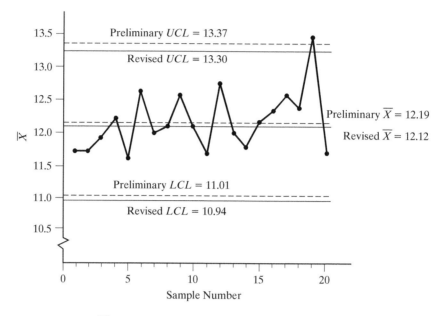

Figure 4 An $\overline{X}$ chart for the data of Table 3.

The 20 sample means may now be plotted. It is noted that the mean of sample 19 exceeds the upper control limit. This would indicate that at this point in time, the universe from which this sample was selected was not exhibiting a stable pattern of variation. Some change occurred between the time of selecting sample 18 and sample 19. It is further noted that after sample 19, the process returned to its original state. One may base action on the assumption that these statements are true, particularly if one thinks that some recognized assignable cause effected the change. Actually, the mean of sample 19 might have exceeded the control limit by chance, and a Type I error might have been made. Alternatively, the pattern of variation might have shifted some time before sample 18 or not returned after this time. Then a Type II error would have been made at these other points in time.

The data of sample 19 should now be discarded and the control limits recalculated for the remaining pattern of variation. The mean of the remaining sample means is 12.12. The range is not recalculated because it is assumed that the process variation did not change when sample 19 was selected. The control limits are now revised to

$$UCL_{\overline{X}} = 12.12 + 0.577(2.05) = 13.30$$

$$LCL_{\overline{X}} = 12.12 - 0.577(2.05) = 10.94$$

These are indicated on the control chart of Figure 4 and again checked to determine that no sample means exceed these limits. It is noted that no further changes are necessary and the control chart of Figure 4, with the revised limits, and the chart of Figure 3 may now be employed to monitor the process. This control model may be implemented to test future process variation to see that it does not change from that used to construct the chart.

3.2 Control Charts for Attributes

A system characteristic or condition is sometimes expressible in a two-value manner. A defense system may be operationally ready or it may be down for maintenance. The surface finish of a workpiece may be acceptable or not. A maintenance crew may be working or idle. When a two-value classification is made, the proportion of observations falling in one class during a predetermined time period may be monitored over time with a *p* chart.

Often observations yield a multivalue but still discrete classification. An employee may suffer none, one, two, or more lost-time accidents during a given time period. The number of demands on a supply system will be a discrete number during a specified period. A maintenance mechanic may perform an assigned task perfectly, or may have made one or more errors. When a discrete classification is made, a system characteristic or condition can be monitored over time with a *c* chart.

The *p* Chart. When a characteristic or condition is sampled and then placed into one of two defined classes, the proportion of units falling into one class may be controlled over time or from one sample to another with a *p* chart. The applicable probability distribution is the binomial. The mean of this distribution and its standard deviation may be expressed as:

$$\mu = np$$

$$\sigma = \sqrt{np(1 - p)}$$

These parameters may be expressed as proportions by dividing by the sample size, n. If $\bar{p}$ is then defined as an estimate of the proportion parameter μ/n, and s_p as an estimate of σ/n, then these statistics can be expressed as

$$\bar{p} = \frac{\text{total number in the class}}{\text{total number of observations}} \tag{9}$$

$$s_p = \sqrt{\frac{\bar{p}(1 - \bar{p})}{n}} \tag{10}$$

The application of the *p* chart will be illustrated with an example of an activity-sampling study. This activity measurement technique is used to obtain information about the activities of workers or machines, usually in lesser time and at a lower cost than by conventional means. Random and instantaneous observations are taken by classifying the activity at a point in time into one and only one category. In the simplest form, the categories *idle* and *busy* are used. The control chart is useful in work-sampling studies in that the observed proportions can be verified as in control and following a stable pattern of variation or out of control with an unstable pattern. In the latter case, a search can be undertaken for an assignable cause.

Consider the case of an activity-sampling study involving computer terminals in an office. The objective was to determine the proportion of time the terminals were in use as opposed to the time they were idle. One hundred observations were taken each day over all working days in a month. The number of times the terminals were in use and the proportion for each day is given in Table 4. From this table and Equation 9, $\bar{p}$ is obtained as

$$\bar{p} = \frac{546}{(21)(100)} = 0.260$$

TABLE 4 Number of Times a Day the Terminals Are in Use

Working Day	Times in Use	Proportion	Working Day	Times in Use	Proportion
1	22	0.22	12	46	0.46
2	33	0.33	13	31	0.31
3	24	0.24	14	24	0.24
4	20	0.20	15	22	0.22
5	18	0.18	16	22	0.22
6	24	0.24	17	29	0.29
7	24	0.24	18	31	0.31
8	29	0.29	19	21	0.21
9	18	0.18	20	26	0.26
10	27	0.27	21	24	0.24
11	31	0.31	Total	546	

The standard deviation of the data is calculated with Equation 10 as

$$s_p = \sqrt{\frac{(0.26)(0.74)}{100}} = 0.044$$

With $\bar{p} = 0.26$ as the best estimate of the population proportion, a control chart may now be constructed. Control limits in activity sampling are usually established at $\bar{p} \pm 2s_p$ and these limits will not vary from one day to another in this example since a constant sample size has been maintained for each day. With these control limits, the probability of making a Type I error may be defined as $\alpha = 0.0456$ if the normal distribution is used as an approximation. Such an approximation is realistic for this example since the binomial will approximate the normal distribution if $n > 50$ and $0.20 < p < 0.80$. If these requirements were not met or a more accurate estimate were needed, the binomial distribution would have had to be employed. The control limits for the p chart are defined and calculated as

$$UCL_p = \bar{p} + 2s_p \tag{11}$$
$$= 0.260 + 2(0.044) = 0.348$$
$$LCL_p = \bar{p} - 2s_p \tag{12}$$
$$= 0.260 - 2(0.044) = 0.172$$

The p chart is constructed as shown in Figure 5. A plot of the data indicates that day 12 was not typical of the pattern of use established by the rest of the month. Subsequent investigation reveals that personnel from another office were also using the terminals that one day because their equipment had preceded them in a move to another building. As this is an atypical situation, the sample is discarded and a revised mean and standard deviation are calculated as

$$\bar{p} = \frac{500}{(20)(100)} = 0.250$$

$$s_p = \sqrt{\frac{(0.25)(0.75)}{100}} = 0.043$$

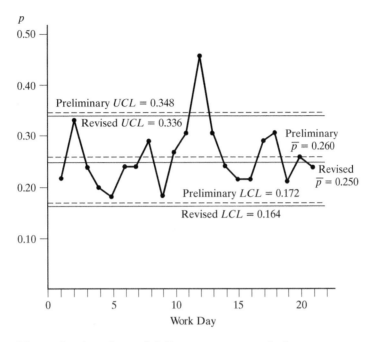

Figure 5 A *p* chart of daily computer terminal usage.

Revised control limits are now calculated and placed on the same control chart as

$$UCL_p = \bar{p} + 2s_p$$
$$= 0.250 + 2(0.043) = 0.336$$
$$LCL_p = \bar{p} - 2s_p$$
$$= 0.250 - 2(0.043) = 0.164$$

It is noted that no further days fall outside these limits and that the chart and data, without day 12, can be taken as a stable pattern of variation of terminal usage. Deviations from this in-control condition can be noted and appropriate control action be taken.

The *c* Chart. Some systems exhibit characteristics expressible numerically. For example, the number of arrivals per hour seeking service at a toll booth is of interest when deciding on the level of service capability to provide. If the number of arrivals per hour deviates from the stable pattern of variation, it may be necessary to compensate by either opening or closing certain toll booths.

The Poisson distribution is usually used to describe the number of arrivals per time period. Here, the opportunity for the occurrence of an event, n, is large but the probability of each occurrence, p, is small. The mean and the variance of the Poisson distribution are equal as $\mu = \sigma^2 = np$. These parameters can be estimated from the statistics with $\bar{c}$ and s_c^2 defined as these estimates. In many applications, values for n and p cannot be determined, but their product np can be established. Then the mean and variance can be estimated as

$$\bar{c} = s_c^2 = \frac{\Sigma(np)}{n} \tag{13}$$

TABLE 5 Number of Arrivals per Hour Seeking Service at a Toll Booth

Hour Number	np Number of Arrivals	Hour Number	np Number of Arrivals
1	6	11	6
2	4	12	4
3	3	13	2
4	5	14	2
5	4	15	4
6	6	16	8
7	5	17	2
8	4	18	3
9	2	19	5
10	5	20	4
		Total = 84	

Consider the application of a c chart to the arrival process previously described. Data for the past 20 hours have been collected and are presented in Table 5. The mean of the arrival population may be estimated from $\bar{c}$ with Equation 13 as

$$\bar{c} = \frac{84}{20} = 4.20$$

and the standard deviation may be estimated as

$$s_c = \sqrt{\bar{c}}$$
$$= \sqrt{4.20} = 2.05$$

With these estimates, the control chart may now be constructed. Control limits may be established as $\bar{c} \pm 3s_c$. The probability of making a Type I error can be determined from the cumulative values of Appendix: Probability and Statistical Tables, Table 1. For the example under consideration,

$$UCL_c = \bar{c} + 3s_c \tag{14}$$
$$= 4.2 + 3(2.05) = 10.35$$

$$LCL_c = \bar{c} - 3s_c \tag{15}$$
$$= 4.2 - 3(2.05) < 0$$

There is no lower control limit. The probability of making a Type I error is the probability of 11 or more arrivals in a given hour from a population with $\bar{c} = 4.2$. This is $1 - P$ (10 or less) or $1 - 0.994 = 0.006$. Alternatively, the control limit could have been defined as a probability limit. If it were thought desirable to define $\alpha \leq 0.01$, the control limit would have been specified as 9.5. Under this control policy, the probability of detecting 10 or more arrivals would satisfy $\alpha \leq 0.01$.

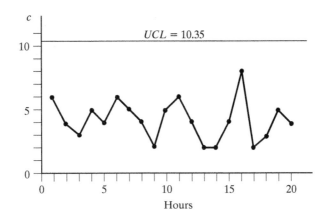

Figure 6 A *c* chart of the number of arrivals per hour.

In the *c* chart of Figure 6, all values fall within the control limits and the chart can be utilized as originally formulated. Action will be initiated to alter the service capability to meet the new demand pattern. At the same time, a new chart based upon the recent data will be constructed and implemented. In this way, the decision maker is informed of conditions in the system that requires modification of operating policy.

4 OPTIMUM POLICY CONTROL

A statistical control chart may be used to monitor the optimum solution obtained from a decision model. Most decision models require monitoring because they do not automatically adapt themselves to the instability of the decision environment. These models are usually formulated at a point in time for a specific set of input parameters. A solution derived from a decision model will be optimal only as long as the input parameters retain the values initially established.[3]

Optimum policy control will be illustrated for the case where an optimum service factor has been determined for a maintenance situation.[4] Once the service factor resulting in a minimum-cost operation has been established, it might be desirable to implement a control model for detecting the effects of a shift in the arrival rate or the service rate or both. This may be accomplished by constructing a control chart for the number of machines not running. The statistical control charts presented cannot be applied directly to this variable, since its distribution is badly skewed. The expected form of the distribution is known in advance; however, a factor usually missing in other control applications.

As an example of the application of control charts in waiting-line operations, consider a one worker/*N* machine situation.[5] Suppose that 20 automatic machines are to be run by one worker, and that the minimum-cost service factor is found to be 0.03. It is assumed that the machines require service at randomly distributed times, that the service times are

---

[3]In terms of $E = f(X, Y)$, this means that E will remain optimal for specific values of X only as long as Y values retain their value initially established.

[4]This example follows a computational procedure.

[5]This example was adapted from R. W. Llewellyn, "Control Charts for Queuing Applications," *Journal of Industrial Engineering*, Vol. 11, No. 4 (July–August 1960).

distributed exponentially, and that machines are serviced on a first-come, first-served basis. The probability of n machines in the queue and in service (not running) is given by

$$P_n = \frac{N!}{(N - n)!} \rho^n P_0 \tag{16}$$

where

P_n = probability that n machines are not running at any point in time

P_0 = probability that all machines are running at any point in time

ρ = ratio of λ/μ

N = number of machines assigned

n = number of machines not running at any point in time

Values for P_n are given in column C of Table 6 for $0 \leq n \leq N$. The first entry is found by dividing the first entry in column B by 2.30005387. The second results from dividing the second entry in column B by the same value, and so on. The results for P_n are plotted in Figure 7. The distribution of n is seen to be extremely skewed, being strictly convex. Unlike some control chart applications, the distribution of the variable to be controlled is known. It may be used to determine the control limits.

Because the probability that all machines are running is approximately 0.435, the lower control limit is obviously 0. The upper control limit is all that needs to be determined,

TABLE 6 Calculation of P_n and ΣP_n

T n (A)	P_n P_0 (B)	P_n (C)	ΣP_n (D)
0	1.000000000	0.434772422	0.434772422
1	0.600000000	0.260863453	0.695635875
2	0.342000000	0.148692168	0.844328043
3	0.184680000	0.080293771	0.924621814
4	0.094186800	0.040949823	0.965571637
5	0.045209664	0.019655915	0.985227552
6	0.020344349	0.008845162	0.994072714
7	0.008544627	0.003714968	0.997787682
8	0.003332405	0.001448838	0.999236520
9	0.001199666	0.000521582	0.999758102
10	0.000395890	0.000172122	0.999930224
11	0.000118767	0.000051637	0.999981861
12	0.000032067	0.000013942	0.999995803
13	0.000007696	0.000003346	0.999999149
14	0.000001616	0.000000703	0.999999852
15	0.000000291	0.000000127	0.999999979
16	0.000000044	0.000000019	0.999999998
17	0.000000005	0.000000002	1.000000000
18	0.000000000	0.000000000	1.000000000
19	0.000000000	0.000000000	1.000000000
20	0.000000000	0.000000000	1.000000000
	2.300053887		

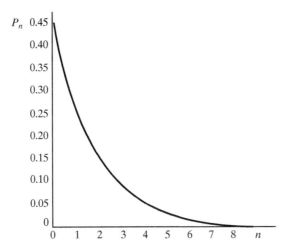

Figure 7 Probability distribution of the number of machines not running.

because the variable n can go out of control only at the top. Thus, the entire critical range will be at the upper end of the distribution. It will be desirable to set this limit so that it will not be violated too frequently. Because n is an integer, the magnitude of the critical range can be observed as a function of n from column D in Table 6. If six idle machines are chosen as a point in control and seven is a point out of control, the probability of designating the system out of control when it is really in control is

$$1 - \sum_{n=0}^{6} P_n = 0.0059$$

Therefore, a control limit set at 6.5 should be satisfactory. The control chart applied to a period of operation might appear as in Figure 8. It can be concluded that the service factor has not changed during this period of observation.

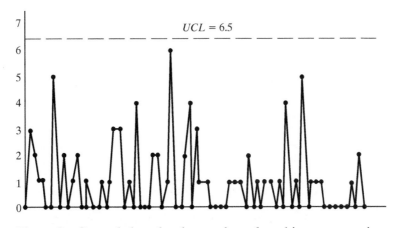

Figure 8 Control chart for the number of machines not running.

For the application of this control model to be valid, two conditions must be met. First, the observation must be made at random times. Second, the observations must be spaced far enough apart, so that the results are independent. For example, if the service time averages 5 minutes and if six machines are idle in an observation, then an observation taken 10 minutes later would probably indicate at least four idle machines. The readings should be far enough apart for a waiting line to be dissipated. To ensure both randomness and spacing, all numbers from 45 to 90 could be taken from a random number table with their order preserved: numbers from 00 to 44 and from 91 to 99 would be dropped. These numbers can then be used to space the observations. This would yield an average of eight random observations per day with a minimum spacing of 45 minutes and a maximum spacing of 90 minutes.

If the control chart indicates that n is no longer in control, corrective action must be taken. This will require investigation to determine whether the service factor has changed because of a change in the mean time between calls or a change in the mean service time, or both. Specific items that might be studied are the policy of preventive maintenance, the age of the machines, the capability of the operator, or material characteristics. Once the assignable cause for the out-of-control condition is located, it may be corrected so that the optimal operating condition may be restored.

5 PROJECT CONTROL WITH CPM AND PERT

A project is composed of a series of activities directed to the accomplishment of a desired objective. Most projects are nonrepetitive and confront the decision maker with a situation in which prior experience and information for control is nonexistent. Therefore, techniques for project control should meet two conditions: (1) During the planning phase of a project, they should make possible the determination of a logical, preferably optimal, project plan; and (2) during the execution of a project, they should make possible the evaluation of a project progress against the plan.

The planning and control of a large-scale project may be accomplished by utilizing an activity network as the planning and control medium. Project planning and control techniques may be classified as deterministic and designated CPM or probabilistic and designated PERT.

CPM and PERT are similar in their logical structure. Projects are represented by sets of required activities, with the principal difference being the treatment of estimated time for the completion of each activity. In PERT, activity time is considered to be a random variable with an assumed probability distribution.

5.1 Critical Path Methods

Many engineering projects involving design, development, and construction operations are nonrepetitive in nature. Although a particular project is to be executed only once, many interdependent activities are required. Some must be executed simultaneously. Each will require personnel and equipment, the cost of which will depend upon the resource commitment. The total project duration and its overall cost depend upon the activities on the *critical path* and the resources allocated to them. Critical path methods for dealing with this situation are presented in this section.

Activity and Event Networks. A network is the basic structural entity behind all critical path methods. It is used to portray the interrelationships among the activities associated with a project. Figure 9 illustrates a network that represents the activities and events in connection with the preparation of a foundation. Four events and three activities are shown.

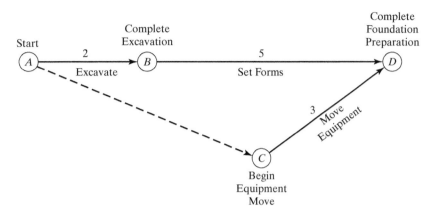

Figure 9 Activity–event network for preparing a foundation.

An event in CPM is represented by a circle. It indicates the completion of an activity. In the case of an initial event, it indicates project start or the initiation of the associated activities. Events A and C in Figure 9 indicate the start of excavation and the start of equipment move activities. Event D represents the completion of foundation preparation.

Activities in CPM are represented by arrows that interconnect events. These arrows symbolize the effort required to complete an event in units of time. Their direction specifies the order in which the events must occur. For example, activities BD (set forms) and CD (equipment move-in) are required for event D (complete preparation) to be achieved. The precedence relationships between events D and B and between event D and event C are established by the arrows. Events B and C must be realized before the foundation is prepared (event D).

In some network representations it is necessary or useful to create dummy activities to clarify precedence relationships. A dummy activity is illustrated in Figure 9 by a broken arrow that connects activity A with activity C, signifying that event C is preceded by event A. Even though dummy activities do not require time and do not consume resources, they may alter the completion time of a project by establishing the order in which events are performed.

Associated with each activity is an elapsed time estimate. This estimate is usually shown above the arrow in the network. The length of the arrows need not be proportional to the duration of the activity. In Figure 9 the time estimates are in weeks.

Two important absolute times for each event must be identified. The first is the earliest time, T_E, which indicates the calendar time at which an event can occur, provided all previous activities were completed at their earliest possible times. The second is the latest time, T_L, for each event, defined as the latest calendar time at which an event can be completed without delaying the initiation of the following activities and completion of the project.

The difference between the latest and the earliest times for each event is called the *slack time* for that event. Positive slack for an event indicates the maximum amount of time that it can be delayed without delaying subsequent events and the overall project. In the example, event B has no slack since the latest time for completion of excavation and the earliest time for setting the forms is the same point in time. Any delay in realizing event B would cause a delay in the project. From this, it is evident that the activities of excavation and setting the forms are critical. The activity of equipment move-in is noncritical to the realization of event D because the start of equipment move-in can occur at any time on or before 3 weeks before the completion of the project. Thus, 4 weeks is the slack time for event C.

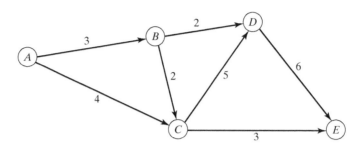

Figure 10 A network for the construction of
a certain structure.

Finding the Critical Path. In CPM all critical activities must be identified and given special attention. This is necessary because any delay in performing these activities may lead to a delay in subsequent activities and the project as a whole. Often it is possible to identify the critical path by tracing the activities along a sequence of events that have zero slack time. The sum of the activity times along this critical path gives the shortest possible elapsed time for project completion. This is 7 weeks in the foundation sample.

As an example of a systematic procedure for finding the critical path, consider a project requiring the construction of a certain structure. Five major events must be realized through seven activities as illustrated in Figure 10. The number appearing above each arrow is the estimated duration of the activity in months, with the arrow indicating the precedence relationship.

The procedure for finding the critical path starts by determining the earliest time and the latest time for each event. By subtracting T_E and T_L for each event, the slack time is found. To facilitate this process each event can be labeled (T_E, T_L, S), where S represents the slack time for the event. First, $T_E S$ are determined for each event and are labeled $(T_E,)$. Next, $T_L S$ are found and the labels become (T_E, T_L). Finally, S is found by subtracting T_E from T_L and completing the label (T_E, T_L, S) for each event.

The earliest time for each event is determined by adding the earliest time for the event immediately preceding to the activity duration connecting the two events. When there is more than one preceding event, the largest value of the sum is chosen as the earliest time for the event under consideration. The process for determining these earliest times proceeds forward from the initial event, which is assumed to have an earliest time of zero. This process is illustrated in Figure 11 and Table 7 for the example under consideration.

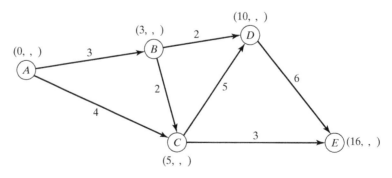

Figure 11 The earliest time for each event.

TABLE 7 Computation of Earliest Times

Event	Immediate Predecessor	T_E for Immediate Predecessor Plus Activity Time	Largest Sum = The Earliest Time
A	None	$0 + 0 = 0$	0
B	A	$0 + 3 = 3$	3
C	A	$0 + 4 = 4$	
	B	$3 + 2 = 5$ ⟵	5
D	B	$3 + 2 = 5$	
	C	$5 + 5 = 10$ ⟵	10
E	C	$5 + 3 = 8$	
	D	$10 + 6 = 16$ ⟵	16

The latest time for each event can be determined in a manner similar to that used in finding the earliest time except that the process starts with the final event and proceeds backward. The latest time for this final event is the same as its earliest time; if it were not, the project would be delayed. For each event, the latest time is determined by subtracting the duration time of an activity connecting the event and an event immediately succeeding from the latest time of the event immediately succeeding. If there is more than one event immediately succeeding, the smallest value of the difference is chosen to be the latest time for the event under consideration. A slack time for each event is then found by computing the difference between T_L and T_E, as illustrated in Figure 12 and Table 8.

The critical path can now be identified with the T_L, T_E, and S values entered in Figure 12. This is done by tracing all zero slack events and viewing them as a path. In this example, the critical path is $A \rightarrow B \rightarrow C \rightarrow D \rightarrow E$ with a minimum project duration of 16 months.

Economic Aspects of CPM. The CPM example above was presented under the assumption that the activities are performed in normal time with a normal allocation of resources. This is called a *normal schedule*. When one or more activities are performed with additional resources to shorten their time durations, a *crash schedule* is said to exist.

Two costs are associated with a project. Direct costs exist for each activity. These costs increase with an increase in the resources allocated for the purpose of decreasing the activity time. Indirect costs exist that are associated with the overall project duration. They increase in direct proportion to an increase in the total time required for project completion.

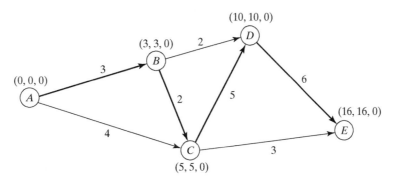

Figure 12 Labeling of events and identification of critical path.

TABLE 8 Computation of Latest Times

Event	Immediate Successor	T_L for Immediate Successor Minus Activity Time	Smallest Difference = The Latest Time
E	None	16 − 0 = 16	16
D	E	16 − 6 = 10	10
C	D	10 − 5 = 5 ⟵	5
	E	16 − 3 = 13	
B	C	5 − 2 = 3 ⟵	3
	D	10 − 2 = 8	
A	B	3 − 3 = 0 ⟵	0
	C	5 − 4 = 1	

This relationship should be self-evident, for overhead costs continue independent of activity levels and they depend on the passage of time.

In most projects, it is of interest to investigate the economic aspects of increasing direct costs as activities are expedited in the light of decreasing overhead costs. An optimum allocation of resources to each activity is sought so that the sum of direct and indirect costs will be a minimum.

The first step in finding the optimum project schedule is to find the critical path under normal resource conditions. This is the starting point. Next, the activity on the critical path that has the least impact on direct cost is shortened. When shortening this activity, other critical paths may appear. If they do, it is necessary to reduce the duration of activities on these paths an equal number of time units for each path. No further activity time reduction should be attempted either when a limit has been reached or when the indirect costs saved are not greater than the extra expenditure of direct resources. This process is repeated for all activities on the critical path or paths.

Consider the example project shown in Figure 13. Indirect costs of $1,000 can be saved for each day removed from the total project duration. Event times shown are those that would occur under normal conditions. The label for each event represents T_E, T_L, and S as defined earlier. The critical path is $A \rightarrow B \rightarrow D \rightarrow E$.

Data on activity durations under normal and crash conditions are given in Table 9. These data are daily costs computed from the assumption that each additional worker adds a direct cost of $50 per day. Additional equipment costs are those that are estimated to arise for each day reduced from the normal activity duration.

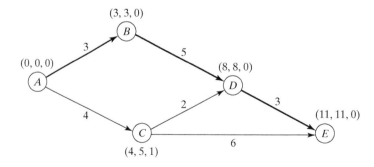

Figure 13 A CPM network with normal activities.

TABLE 9 Computation of Crash Costs

Activity	Number of Days Reduction	Additional Labor Cost			Additional Equipment Cost ($)	Total Crash Cost ($)
		Additional Workers	Working Days	Additional Costs ($)		
AB	1	2	2	200	600	800
AC	1	3	3	450	300	750
BD	1	1	4	200	500	700
	2	3	3	450	450	900
	3	3	2	300	1,700	2,000
CD	0	0	2	0	0	0
CE	1	0	5	0	950	950
	2	4	4	800	1,110	1,900
DE	0	0	3	0	0	0

The project depicted in Figure 13 will take 11 days to complete under the normal activity times shown. No crash cost is incurred and, therefore, there is no saving in indirect cost.

A 10-day schedule can be achieved by reducing BD by 1 day. From Table 9, a crash cost of $700 is found. A saving of $1,000 in indirect cost will result. No further reduction in activity BD should be attempted because the critical path will change. This 10-day schedule is shown in Figure 14.

Here two critical paths now exist: the original path and a new path $A \rightarrow C \rightarrow E$. Reducing activity BD by one more day yields a 9-day schedule. This leads to a crash cost of $900, which is less than the crash cost for shortening BD and AB by 1 day each. The duration of one activity in the critical path $A \rightarrow C \rightarrow E$ should also be shortened by 1 day at the same time. A 1-day reduction of AC costs $750, which is less than shortening CE by 1 day. A saving of $2,000 in indirect cost will result. This 9-day schedule is shown in Figure 15.

In the 9-day schedule, path $A \rightarrow B \rightarrow D \rightarrow E$ and path $A \rightarrow C \rightarrow E$ remain critical as in the 10-day schedule. Further equal time reduction should be sought in these critical paths.

An 8-day schedule is obtained by reducing AB by 1 day at a crash cost of $800 and BD by 2 days at a crash cost of $900 for a total of $1,700. This is better than a 3-day reduction in activity BD. It is necessary to reduce CE and AC by 1 day each at a crash cost of $950 plus $750 or $1,700. This is better than a 2-day reduction in CE. A saving of $3,000 is possible by this schedule. Three critical paths result: $A \rightarrow B \rightarrow D \rightarrow E$, $A \rightarrow C \rightarrow D \rightarrow E$, and $A \rightarrow C \rightarrow E$, each with a duration of 8 days. No further reductions are possible. The resulting schedule is shown in Figure 16.

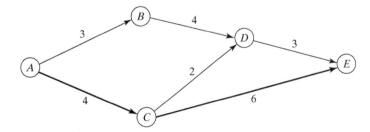

Figure 14 A 10-day schedule.

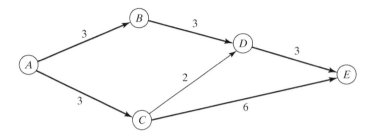

Figure 15 A 9-day schedule.

Finally, it is necessary to find the optimum schedule. This is accomplished by summarizing the crash costs and the savings in Table 10. A maximum net saving occurs for a schedule of 9 days.

5.2 Program Evaluation and Review Technique

PERT is ideally suited to early project planning where precise time estimates are not readily available. The advantage of PERT is that it offers a way of dealing with random variation, making it possible to allow for chance in the scheduling of activities. PERT may be used as a basis for computing the probability that the project will be completed on or before its scheduled date.

PERT Network Calculations. A PERT network is illustrated in Figure 17. One starts with an end objective (e.g., event *H* in Figure 17) and works backward until event *A* is identified. Each event is labeled, coded, and checked in terms of program time frame. Activities are then identified and checked to ensure that they are properly sequenced. Some activities can be performed on a concurrent basis, and others must be accomplished in series. For each completed network, there is one beginning event and one ending event, and all activities must lead to the ending event.

The next step in developing a PERT network is to estimate activity times and to relate these times to the probability of their occurrences. An example of calculations for a typical PERT network is presented in Table 11. The following steps are required:

1. Column 1 lists each event, starting from the last event and working downward to the start.
2. Column 2 lists all previous events that have been indicated prior to the event listed in column 1.
3. Columns 3–5 give the optimistic (t_a), the most likely (t_b), and the pessimistic (t_c) times, in weeks or months, for each activity. Optimistic time means that there is little

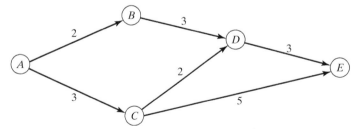

Figure 16 An 8-day schedule.

TABLE 10 Net Savings for Crash Schedules

Crash Schedule	Crash Cost ($)	Indirect Cost Saving ($)	Net Saving ($)
10-day	700	1,000	300
9-day	1,650	2,000	350
8-day	3,400	3,000	−400

chance that the activity can be completed before this time, while pessimistic time means that there is little likelihood that the activity will take longer. Time value (t) is defined as a random variable with a beta distribution and with range from a to b and mode m, as shown in Figure 18.

4. Column 6 gives the expected or mean time (t_e) from

$$t_e = \frac{t_a + 4t_b + t_c}{6}$$

5. Column 7 gives the variance, σ^2, from

$$\sigma^2 = \left(\frac{t_c - t_a}{6}\right)^2$$

6. Column 8 gives the earliest expected time (T_E) as the sum of all times (t_e) for each activity and the cumulative total of the expected times through the preceding event, remaining on the same path throughout the network. When several activities lead to an event, the highest time value (t_e) is used. For example, path $A \rightarrow D \rightarrow E \rightarrow H$ in Figure 17 totals 46.8.

7. Column 9 gives the latest allowable time (T_L) calculated by starting with the latest time for the last event, which equals T_E (where T_E equals 46.8). Then one works backward, subtracting the value in column 6 for each activity, staying in the same path.

8. Column 10 gives the slack time (S) as the difference between the latest allowable time (T_L) and the earliest expected time (T_E) from

$$S = T_L - T_E$$

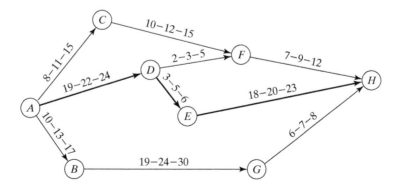

Figure 17   A PERT network with events and activity times.

TABLE 11 Example of PERT Calculations

Event Number (1)	Previous Events (2)	t_a (3)	t_b (4)	t_c (5)	t_e (6)	σ^2 (7)	T_E (8)	T_L (9)	S (10)	PT (11)	Probability (12)
H	G	6	7	8	7.0	0.108				46.0	0.266
	F	7	9	12	9.2	0.693					
	E	18	20	23	20.2	0.693	46.8	46.8	0.0		
G	B	19	24	30	24.2	3.349	37.7	39.8	2.1		
F	D	2	3	5	3.2	0.250					
	C	10	12	15	12.2	0.693	23.4	37.6	14.2		
E	D	3	5	6	4.8	0.250	26.6	26.6	0.0		
D	A	19	22	24	21.8	0.693	21.8	21.8	0.0		
C	A	8	11	15	11.2	1.369	11.2	25.4	14.2		
B	A	10	13	17	13.5	1.369	13.5	15.6	2.1		

9. Columns 11 and 12 give the required project time (PT) for the network. Assuming that PT is 46.0, it is necessary to determine the probability of meeting this requirement. This probability is determined from

$$Z = \frac{PT - T_L}{\sqrt{\Sigma \text{ variances}}}$$

Z is the area under the normal distribution curve and the variance is the sum of the individual variances applicable to the critical path activities in Figure 17 (path *A: D: E: H*). For this example,

$$Z = \frac{46.0 - 46.8}{\sqrt{0.693 + 0.250 + 0.693}} = -0.625$$

From Appendix: Probability and Statistical Tables, Table 3, the calculated value of -0.625 represents an area of approximately 0.266; that is, the probability of meeting the scheduled time of 46 weeks is 0.266.

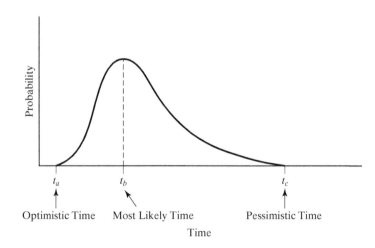

Figure 18 β-distribution of activity times.

When evaluating the resultant probability value (column 12 of Table 11), management must decide on the range of factors allowable in terms of risk. If the probability factor is too low, additional resources may be applied to the project to reduce the activity schedule times and improve the probability of success. Conversely, if the probability factor is too high, this may indicate that excess resources are being applied, some of which should be diverted to other projects.

PERT-Cost.   An extension of PERT to include economic considerations brings cost factors into project control decisions. A cost network can be superimposed on a PERT network by estimating the total cost and the cost slope for each activity time.

When implementing PERT-Cost, there is always the time-cost option that enables the decision maker to evaluate alternatives relative to the allocation of resources for activity accomplishment. In many instances, time can be saved by applying more resources or, conversely, cost may be reduced by extending the time to complete an activity. The time-cost option can be attained by applying the following steps:

1. Determine alternative time and cost estimates for all activities and select the lowest cost alternative in each instance.
2. Calculate the critical path time for the network. If the calculated value is equal to or less than the total time permitted, select the lowest-cost option for each activity and check to ensure that the total of the incremental activity times does not exceed the allowable overall program completion time. If the calculated value exceeds the program time permitted, review the activities along the critical path, select the alternative with the lowest cost slope, and then reduce the time value to be compatible with the program requirement.
3. After the critical path has been established in terms of the lowest-cost option, review all the network paths with slack time and shift activities to extend the times and reduce costs wherever possible. Activities with the steepest cost-time slopes should be addressed first.

The time-cost option can be illustrated by referring to two paths of the network presented in Figure 17. Data associated with each path are given in Table 12. In this example, path $A \rightarrow D \rightarrow E \rightarrow H$ is critical, and the alternative path $(A \rightarrow C \rightarrow F \rightarrow H)$ has a slack of 14.2. If one is concerned with the critical path (since it is the governing factor in the overall network relative to program success or failure), the shifting of resources from the alternative path to critical-path activities should be considered. For instance, assume that $200 is reallocated from activity HF to activity HE. As a result, activities HF and HE will now require 12.2 and 18.7 units of time, respectively. The total time along the critical path is reduced, and the project

TABLE 12   Time-Cost Options

Path $A \rightarrow D \rightarrow E \rightarrow H$				Path $A \rightarrow C \rightarrow F \rightarrow H$			
Event Number	Previous Event	t_e	Estimated Activity Cost ($)	Event Number	Previous Event	t_e	Estimated Activity Cost ($)
H	E	20.2	3,500	H	F	9.2	1,050
E	D	4.8	940	F	C	12.2	2,350
D	A	21.8	4,800	C	A	11.2	1,870
Total		46.8	$9,240	Total		32.6	$5,370

is completed in 45.3 units of time. Conversely, the time along the alternative path is increased to 35.6 units, and the slack is reduced to 9.7 units. The entire network must now be reevaluated to determine the effects of the changes on the critical path. This process may be continued by trading off resources against time until the most acceptable result is obtained.

6 TOTAL QUALITY CONTROL

Total quality control (TQC) is a management concept being stimulated by the need to compete in global markets where higher quality, lower cost, and more rapid development are essential to market leadership. Management for total quality is rapidly emerging as a means for improving the quality-cost characteristics of products, processes, and services.

Approaches now being promulgated emphasize that conformance to design specifications is necessary, but not sufficient. Quality losses begin to accumulate whenever a product parameter deviates from its nominal or optimal value. Accordingly, it is being recognized that quality must be designed into products and processes. Furthermore, optimal values for product and process parameters must be established and controlled throughout the product life cycle.

6.1 Experimental Design

Experimental design (ED) was first introduced in the 1920s by Sir Ronald Fisher. Its subsequent development led to gains in the efficiency of experimentation by changing factors, not one at a time, but together in a factorial design. Fisher introduced the concept of randomization (so that trends resulting from unknown disturbing factors would not bias results), the idea that a valid estimate of experimental error could be obtained from the design, and the method of blocking to eliminate systematic differences introduced by using different lots.

Experimental designs were introduced into industry in the 1930s. At that time, the Industrial and Agricultural Section of the Royal Statistical Society was inaugurated in London and papers on applications to glass manufacturing, light bulb production, textile spinning, and the like were presented and discussed. This led to new statistical methods such as variance component analysis to reduce variation in textile production.

During World War II, the need for designs that could screen large numbers of factors led to the introduction of fractional factorial designs and other orthogonal arrays. These designs have been widely applied in industry, and many successful industrial examples are described in papers and books dating from the 1950s. Also in the early 1950s, George Box developed response surface methods for the improvement and optimization of industrial processes by experimentation.

Modern engineering research and development involves considerable experimentation and analysis. Statistically designed experiments enable the engineer to achieve development objectives in less time with fewer resources. It is not unusual for development cycle time and cost to be reduced by one-half when experimental design methods are integrated into a concurrent engineering development process. When product quality and process design are addressed early in the life cycle, many decisions that affect product life, reliability, and producibility can be evaluated before the product is produced.

6.2 Parameter Design

Parameter design (PD) is based on the idea of controllables and uncontrollables in design and operations. Parameter optimization involves the selection of controllable values in the product and the processes of manufacture and maintenance so that some measure of merit will be improved (or optimized if possible).

Parameter optimization is traditionally used when the goal is to improve some performance measure. It is now being applied to variability reduction for increased quality and lower cost. Although parameter design to reduce variability may be accomplished by a variety of methods, there is one technique that is gaining recognition. This technique is attributable to Genichi Taguchi.

Taguchi uses the term *quality engineering* to describe an approach to achieving both improved quality and low cost. He identifies two methods of dealing with variability: parameter optimization and tolerance design. The first attempts to minimize the effects of variation and the second seeks to remove its causes.

There have been many examples of the successful application of parameter optimization. The most widely reported is the case of tile manufacturing in Japan. Similar improvements have been demonstrated in automobile manufacturing, electronic component production, computer operations, IC chip bonding processes, ultrasonic welding, and the design of disc brakes and engines.

The concept of a quality loss function is used in parameter optimization to capture the effect of variability. This function represents the loss to society resulting from variability of function as well as from harmful side effects during the production and use of a product. Loss is calculated from the view of society with an emphasis on the viewpoint of the customer. For example, the loss to an automobile owner when a part fails includes the cost of repair, the cost of lost wages, and the inconvenience of not having use of the car.

In many applications the quality loss can be approximated by a quadratic function of the important design parameter. The function is assumed to be analytic, and so it can be expanded in a Taylor series about some known optimal value T. Then the first two terms of the expansion can be assumed to be zero and the principal term becomes

$$L(y) = (y - T)^2 \frac{L''(T)}{2!}$$

or

$$L(y) = k(y - T)^2$$

$L(y) = $ the loss as a function of y

$T = $ the optimum or target value for parameter y

$L''(T) = $ the second derivative of L evaluated at T

$k = $ the loss coefficient for the particular application

Taguchi used the loss function to measure the effect of variation whether or not the specification is satisfied. He also used the orthogonal experimental design for efficiently evaluating the effect of individual parameter settings in the face of noise.

6.3 Quality Engineering

Quality engineering (QE) as promulgated by Taguchi involves the following activities:

1. *Plan an experiment.* Identify the main function, side effects, and failure modes. Identify noise factors and testing conditions for evaluating quality loss, the quality characteristic to be observed, the quality function to be optimized, and the controllable parameters and their most likely settings.

2. *Perform the experiment.* Conduct a limited number of experiments that simultaneously vary all of the parameters according to the pattern of an appropriate orthogonal matrix. Collect data on the results of the experiments.

3. *Analyze and verify the results.* Analyze the data, determine the optimum parameter settings, and predict the performance under these settings. Conduct a verification experiment to confirm the results obtained at the optimum settings.

As with statistical process control, the Taguchi method relies on a hypothesis. Part of the evaluation includes a series of tests to verify that the hypothesis should not be rejected based on experimental evidence. To compare the effects of changing the values of different parameters, Taguchi introduces the signal-to-noise ratio measure embracing a simple addition of the contribution of individual parameter settings. This additive property is the hypothesis to be tested.

The relationship of Taguchi parameter design and the design-dependent parameter approach is illustrated in Figure 19. A summary of the relationship follows:

1. Parameter Design Approach (Taguchi).
   (a) *Loss function.* Quadratic in the deviation of actual parameter value from target value and related to the specific design by k.
   (b) *Experimental design.* A factorial experiment relating control factors at two levels to noise factors at two or more levels.

2. Design-Dependent Parameter Approach.
   (a) *Design evaluation function.* A life-cycle cost function optimized on design variables for each instance of design-dependent parameters.
   (b) *Indirect experimentation.* The use of modeling, Monte Carlo sampling, and sensitivity analysis to evaluate the LCC penalty for various combinations of design-dependent parameter values in the face of variation in design-independent parameters.

Design-dependent parameter approaches and parameter design (Taguchi style) have much in common. Taguchi would call the concurrent design process and the design evaluation function approach *system design*. System design involves innovation and knowledge from the science and engineering fields. It includes the selection of materials, parts, and product parameter values in the product design stage and the selection of production equipment and values for process factors in the process design stage.

Tentative nominal values for design-dependent parameters are then tested over specified ranges. This step is Taguchi's parameter design. Its purpose it to determine the best combination of parameter levels or values. Parameter design determines product parameter values that are least sensitive to joint changes in design-independent parameter noise and environmental noise factors.

Indirect experimentation through modeling and simulated sampling is a means of determining the best combination of levels for design-dependent parameters. The loss function best used is the design evaluation function. Taguchi parameter design is an approach yet to be adopted to conceptual and preliminary design. Because the product and the production process does not yet exist, ED approaches are not feasible except on surrogate systems.

Tolerance design, as defined by Taguchi, is expensive and is, therefore, to be minimized. It is in this area that SPC can be most helpful. The objective, however, is to define design-dependent parameters in such a way that production, operations, and support capabilities are not adversely affected by design-dependent parameter drift. The

PARAMETER DESIGN APPROACH (TAGUCHI)

Loss Function

$$L = k(y - T)^2$$

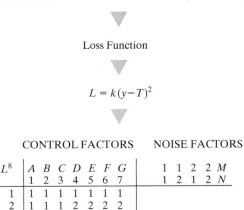

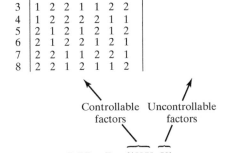

$$DEF = E = \overline{f(X, Y_d, Y_i)}$$

Design Evaluation Function

DESIGN-DEPENDENT PARAMETER APPROACH

Figure 19 Parameter design and design-dependent parameter relationships.

objective of the Taguchi and the design-dependent parameter approaches is to achieve robustness against noise factors, so that life-cycle cost may be minimized.

7 SUMMARY AND EXTENSIONS

The basic control concepts presented in Section 1 are not explained in great depth. Also, the breadth of topical coverage is limited. Although the material is adequate to get the subsequent sections underway, those readers interested to go deeper into the classical dimensions of the theory of control and its application may refer to *Introduction to System Science*, by G. M. Sandquist, Prentice Hall, 1985. This book offers a system-science embedded view of control theory. Others exist, but most are oriented strictly to physical systems.

Two large professional societies are engaged in the area of controls and control systems. These are as follows: (1) The Control Systems Society of IEEE (Institute of Electrical and Electronics Engineers) with website *www.ieeecss.org* and (2) IFAC (International Federation of Automatic Control) with website *www.ifac-control.org*. In addition, the American Society for Quality with website *www.asq.org* gives special attention to the dimension of control and related quality matters.

Three fundamental topics from probability theory underpin Section 2 on statistical process control:

1. The Central Limit Theorem—A theorem stating that the distribution of sample means approaches normality as the sample size increases, provided that the sample values are independent. This provides the basis for control charts for variables.

2. The Binomial Distribution—A theoretical probability distribution providing the basis for control charts for attributes, where the sampled characteristic falls into one of two classes.

3. The Poisson Distribution—A theoretical probability distribution providing the basis for control charts for attributes, where the sampled characteristic is a numerical count. This distribution also underpins Section 3 on Optimum Policy Control.

A single comprehensive book, *Introduction to Statistical Quality Control*, 5th ed., by D. C. Montgomery, John Wiley & Sons, Inc., 2005, is recommended to those wishing to obtain greater understanding of statistical quality control. In addition, the website for the American Society for Quality at *http://www.asq.org* is an excellent source of information.

Project control is very important in systems engineering. Accordingly, this chapter developed the basic theory behind both CPM and PERT, including PERT-Cost. The relationship and importance of these control methods to systems engineering management will not be discussed in this chapter.

The final section on total quality control revisits statistical process control from a historical and somewhat more basic perspective. It then touches on experimental design from a historical perspective and also from a modern view with a focus on life-cycle outcome predictability. Finally, parameter design and quality engineering are presented to close this chapter. These are topics closely related to the design-dependent parameter approach promulgated in this text.

QUESTIONS AND PROBLEMS

1. Speed is one characteristic or condition of an automobile that must be controlled. Discuss the role of each element of a control system for speed control.

2. Sketch a diagram such as Figure 1 that would apply to a thermostatically controlled heating system. Discuss each block.

3. Give an example of open-loop control and an example of closed-loop control.

4. What is the relationship among an unstable pattern of variation, control limits, and a Type I error? A Type II error?

5. Samples of $n = 10$ were taken from a process for a period of time. The process average was estimated to be $\overline{X} = 0.0250$ inch and the process range was estimated as $\overline{R} = 0.0020$ inch. Specify the control limits for an $\overline{X}$ chart and for an R chart.

6. Control charts by variables are to be established on the tensile strength in pounds of a yarn. Samples of five have been taken each hour for the past 20 hours. These were recorded as shown in the table.

							Hour												
1	2	3	4	5	6	7	8	9	10	11	12	13	14	15	16	17	18	19	20
50	44	44	48	47	47	44	52	44	13	47	49	47	43	44	45	45	50	46	48
51	46	44	52	46	44	46	46	46	44	44	48	51	46	43	47	45	49	47	44
49	50	44	49	46	43	46	45	46	49	44	41	50	46	40	51	47	45	48	49
42	47	47	49	48	40	48	42	46	47	42	46	48	48	40	48	47	49	46	50
43	48	48	46	50	45	46	55	43	45	50	46	42	46	46	46	46	48	45	46

(a) Construct an $\overline{X}$ chart based on these data.

(b) Construct an R chart based on these data.

7. A lower specification limit of 42 pounds is required for the condition of Problem 6. Sketch the relationship between the specification limit and the control limits. What proportion, if any, of the yarn will be defective?

8. The total number of accidents during the long weekends and the number of fatalities for a 10-year period are given in the table. Assuming that the process is in control with regard to the proportion of fatalities, construct a p chart and record the data of the last 8 weekends.

Weekend	Number of Accidents	Number of Fatalities	Weekend	Number of Accidents	Number of Fatalities
1	2,378	426	16	3,943	523
2	3,375	511	17	3,950	557
3	3,108	498	18	4,358	536
4	3,756	525	19	4,217	533
5	3,947	564	20	3,959	547
6	2,953	475	21	4,108	554
7	3,075	490	22	4,379	579
8	3,173	504	23	4,455	598
9	3,479	528	24	4,753	585
10	3,545	555	25	4,276	543
11	3,865	537	26	3,868	507
12	3,747	529	27	3,947	523
13	4,011	569	28	3,665	575
14	3,108	470	29	4,078	569
15	3,207	510	30	4,025	578

9. During a 4-week inspection period, the number of defects listed in the table were found in a sample of 400 electronic components. Construct a c chart for these data. Does it appear as though there existed an assignable cause of variation during the inspection period?

Date	Number of Defects	Date	Number of Defects
1	7	11	6
2	8	12	8
3	9	13	16
4	8	14	2
5	3	15	4
6	9	16	2
7	5	17	6
8	6	18	5
9	15	19	3
10	9	20	7

10. A survey during a safety month showed the number of defective cars on a highway as given in the table. A sample of 100 cars was taken during each day. Construct a *c* chart for the data. Is there any assignable cause of variation during this period?

Date	Defective Cars	Date	Defective Cars	Date	Defective Cars
1	12	11	9	21	9
2	15	12	7	22	8
3	13	13	13	23	17
4	9	14	12	24	16
5	14	15	6	25	13
6	17	16	18	26	18
7	8	17	15	27	12
8	21	18	5	28	9
9	12	19	14	29	11
10	14	20	12	30	19

11. One operator is assigned to run 14 automatic machines. The minimum-cost service factor is 0.2. The machines require attention at random, and the service duration is distributed exponentially. Machines receive service on a first-come, first-served basis. Specify the upper control limit if the probability of designating the system out of control when it is really in control must not exceed 0.05.

12. Find the critical path in the network shown.

13. Eleven activities and eight events constitute a certain research and development project. The activities and their expected completion times are as follows :

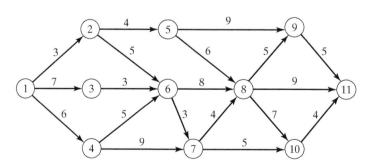

a. Represent the project in the form of an activity–event network.
b. Calculate T_E and T_L for each event and label each with T_E, T_L, and S.
c. Identify the critical path and calculate the shortest possible time for project completion.

Activity	Completion Time in Weeks
$A \rightarrow B$	5
$A \rightarrow C$	6
$A \rightarrow D$	3
$B \rightarrow E$	10
$B \rightarrow F$	7
$C \rightarrow E$	8
$D \rightarrow E$	2
$E \rightarrow F$	1
$E \rightarrow G$	2
$F \rightarrow H$	5
$G \rightarrow H$	6

14. In Problem 13, assume that event C must occur before event D occurs. Modify the activity–event network for this assumption and show how the critical path would be changed.

15. Nine activities and six events are required to execute and complete a certain construction program. The activities and their completion times in weeks under normal and under expedited conditions are as follows:

Activity	Normal		Expedited	
	Duration	Cost ($)	Duration	Cost ($)
AB	10	3,000	8	5,000
AC	5	2,500	4	3,600
AD	2	1,100	1	1,200
BC	6	3,000	3	12,000
BE	4	8,500	4	8,500
CE	7	9,800	5	10,200
CF	3	2,700	1	3,500
DF	2	9,200	2	9,200
EF	4	300	1	4,600

A linear crash cost-time relationship exists between the normal and expedited conditions. Overhead of $11,000 per week can be saved if the program is completed earlier than the normal schedule.

(a) Represent the program in the form of an activity–event network.
(b) Find the earliest time within which the program can be completed under normal conditions.
(c) Find the minimum-cost schedule for the program.

16. Eight maintenance activities, their normal durations in weeks, and the crew sizes for the normal condition are as follows:

Activity	Duration in Weeks	Crew Size
AC	6	8
BC	6	7
BE	2	6
CD	3	2
CE	4	1
DF	7	4
EF	4	6
FG	5	10

(a) Represent the maintenance project in the form of an activity–event network.

(b) Identify the critical path and find the minimum time for project completion.

(c) Extra crew members can be used to expedite activities *BC, CE, DF,* and *EF* at a cost given below:

Extra Crew	Weeks Saved	Crash Cost per Week ($)
1	1	100
2	2	120
3	2	200
4	3	250

If there is a penalty cost of $1,250 per week beyond the minimum maintenance time of 17 weeks, recommend the minimum-cost schedule.

17. Determine the critical path of the PERT network shown. What is the second most critical path?

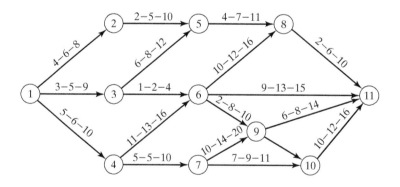

18. Calculate the probability of meeting a scheduled time of 50 units for the PERT network shown below:

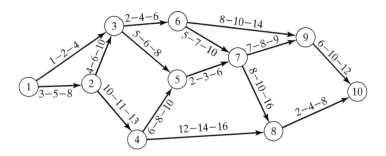

Appendix

Appendix:

Probability and Statistical Tables

TABLE 1 RANDOM RECTANGULAR VARIATES

Random variates from the rectangular distribution, $f(x) = \frac{1}{10}$, are presented. (These tables are reproduced with permission from the RAND Corporation, *A Million Random Digits with 100,000 Normal Deviates* [New York: The Free Press, 1955], pp. 130–131.)

TABLE 2 CUMULATIVE POISSON PROBABILITIES

Cumulative probabilities $\times$ 1,000 for the Poisson distribution are given for μ up to 24. The tabular values were computed from $\Sigma(\mu^x e^{-\mu}/x!)$. (These tables are reproduced from W. J. Fabrycky, P. M. Ghare, and P. E. Torgersen, *Applied Operations Research and Management Science* [Upper Saddle River, N.J.: Prentice Hall, Inc., 1984].)

TABLE 3 CUMULATIVE NORMAL PROBABILITIES

Cumulative probabilities are given from $-\infty$ to $Z = (x - \mu)/\sigma$ for the standard normal distribution. (Tabular values are adapted with permission from E. L. Grant and R. S. Leavenworth, *Statistical Quality Control*, 4th ed. [New York: McGraw-Hill Book Co., 1972].)

TABLE 1   Random Rectangular Variates

14541	36678	54343	94932	25238	84928	30668	34992	69955	06633
88626	98899	01337	48085	83315	33563	78656	99440	55584	54178
31466	87268	62975	19310	28192	06654	06720	64938	67111	55091
52738	52893	51373	43430	95885	93795	20129	54847	68674	21040
17444	35560	35348	75467	26026	89118	51810	06389	02391	96061
62596	56854	76099	38469	26285	86175	65468	32354	02675	24070
38338	83917	50232	29164	07461	25385	84838	07405	38303	55635
29163	61006	98106	47538	99122	36242	90365	15581	89597	03327
59049	95306	31227	75288	10122	92687	99971	97105	37597	91673
67447	52922	58657	67601	96148	97263	39110	95111	04682	64873
57082	55108	26992	19196	08044	57300	75095	84330	92314	11370
00179	04358	95645	91751	56618	73782	38575	17401	38686	98435
65420	87257	44374	54312	94692	81776	24422	99198	51432	63943
52450	75445	40002	69727	29775	32572	79980	67902	97260	21050
82767	26273	02192	88536	08191	91750	46993	02245	38659	28026
17066	64286	35972	32550	82167	53177	32396	34014	20993	03031
86168	32643	23668	92038	03096	51029	09693	45454	89854	70103
33632	69631	70537	06464	83543	48297	67693	63137	62675	56572
77915	56481	43065	24231	43011	40505	90386	13870	84603	73101
90000	92887	92668	93521	44072	01785	27003	01851	40232	25842
55809	70237	10368	58664	39521	11137	20461	53081	07150	11832
50948	64026	03350	03153	75913	72651	28651	94299	67706	92507
27138	59012	27872	90522	69791	85482	80337	12252	83388	48909
03534	58643	75913	63557	25527	47131	72295	55801	44847	48019
48895	34733	58057	00195	79496	93453	07813	66038	55245	43168
57585	23710	77321	70662	82884	80132	42281	17032	96737	93284
95913	24669	42050	92757	68677	75567	99777	49246	93049	79863
12981	37145	95773	92475	43700	85253	33214	87656	13295	09721
62349	64163	57369	65773	86217	00135	33762	72398	16343	02263
68193	37564	56257	50030	53951	84887	34590	22038	40629	29562
56203	82226	83294	60361	29924	09353	87021	08149	11167	81744
31945	23224	08211	02562	20299	85836	94714	50278	99818	62489
68726	52274	59535	80873	35423	05166	06911	25916	90728	20431
79557	25747	55585	93461	44360	18359	20493	54287	43693	88568
05764	29803	01819	51972	91641	03524	18381	65427	11394	37447
30187	66931	01972	48438	90716	21847	35114	91839	26913	68893
30858	43646	96984	80412	91973	81339	05548	49812	40775	14263
85117	38268	18921	29519	33359	80642	95362	22133	40322	37826
59422	12752	56798	31954	19859	32451	04433	62116	14899	38825
73479	91833	91122	45524	73871	77931	67822	95602	23325	37718
83648	66882	15327	89748	76685	76282	98624	71547	49089	33105
19454	91265	09051	94410	06418	34484	37929	61070	62346	79970
49327	97807	61390	08005	71795	49290	52285	82119	59348	55986
54482	51025	12382	35719	66721	84890	38106	44136	95164	92935
30487	19459	25693	09427	10967	36164	33893	07087	16141	12734
42998	68627	66295	59360	44041	76909	56321	12978	31304	97444
03668	61096	26292	79688	05625	52198	74844	69815	76591	35398
45074	91457	28311	56499	60403	13658	81838	54729	12365	24082
58444	99255	14960	02275	37925	03852	81235	91628	72136	53070
82912	91185	89612	02362	93360	20158	24796	38284	55328	96041

TABLE 1 Random Rectangular Variates (*Continued*)

44553	29642	20317	69470	57789	27631	68040	73201	51302	66497
01914	36106	71351	69176	53353	57353	42430	68050	47862	61922
00768	37958	69915	17709	31629	49587	07136	42959	56207	03625
29742	67676	62608	54215	97167	07008	77130	15806	53081	14297
07721	20143	56131	56112	23451	48773	38121	74419	11696	42614
99158	07133	04325	43936	83619	77182	55459	28808	38034	01054
97168	13859	78155	55361	04871	78433	58538	78437	14058	79510
07508	63835	83056	74942	70117	91928	10383	93793	31015	60839
68400	66460	67212	28690	66913	90798	71714	07698	31581	31086
88512	62908	65455	64015	00821	23970	58118	93174	02201	16771
94549	31145	62897	91582	94064	14687	47570	83714	45928	32685
02307	86181	44897	60884	68072	77693	83413	61680	55872	12111
28922	89390	66771	39185	04266	55216	91537	36500	48154	04517
73898	85742	97914	74170	10383	16366	37404	73282	20524	85004
66220	81596	18533	84825	43509	16009	00830	13177	54961	31140
64452	91627	21897	31830	62051	00760	43702	22305	79009	15065
26748	19441	87908	06086	62879	99865	50739	98540	54002	98337
61328	52330	17850	53204	29955	48425	84694	11280	70661	27303
89134	85791	73207	93578	62563	37205	97667	61453	01067	31982
91365	23327	81658	56441	01480	09677	86053	11505	30898	82143
54576	02572	60501	98257	40475	81401	31624	27951	60172	21382
39870	60476	02934	39857	06430	59325	84345	62302	98616	13452
82288	29758	35692	21268	35101	77554	35201	22795	84532	29927
57404	93848	87288	30246	34990	50575	49485	60474	17377	46550
22043	17104	49653	79082	45099	24889	04829	49097	58065	23492
61981	00340	43594	22386	41782	94104	08867	68590	61716	36120
96056	16227	74598	28155	23304	66923	07918	15303	44988	79076
64013	74715	31525	62676	75435	93055	37086	52737	89455	83016
59515	37354	55422	79471	23150	79170	74043	49340	61320	50390
38534	33169	40448	21683	82153	23411	53057	26069	86906	49708
41422	50502	40570	59748	59499	70322	62416	71408	06429	70123
38633	80107	10241	30880	13914	09228	68929	06438	17749	81149
48214	75994	31689	25257	28641	14854	72571	78189	35508	26381
54799	37862	06714	55885	07481	16966	04797	57846	69080	49631
25848	27142	63477	33416	60961	19781	65457	23981	90348	24499
27576	47298	47163	69614	29372	24859	62090	81667	50635	08295
52970	93916	81350	81057	16962	56039	27739	59574	79617	45698
69516	87573	13313	69388	32020	66294	99126	50474	04258	03084
94504	41733	55936	77595	55959	90727	61367	83645	80997	62103
67935	14568	27992	09784	81917	79303	08616	83509	64932	34764
63345	09500	40232	51061	09455	36491	04810	06040	78959	41435
87119	21605	86917	97715	91250	79587	80967	39872	52512	78444
02612	97319	10487	68923	58607	38261	67119	36351	48521	69965
69860	16526	41420	01514	46902	03399	12286	52467	80387	10561
27669	67730	53932	38578	25746	00025	98917	18790	51091	24920
59705	91472	01302	33123	35274	88433	55491	27609	02824	05245
36508	74042	44014	36243	12724	06092	23742	90436	33419	12301
13612	24554	73326	61445	77198	43360	62006	31038	54756	88137
82893	11961	19656	71181	63201	44946	14169	72755	47883	24119
97914	61228	42903	71187	54964	14945	20809	33937	13257	66387

TABLE 2 Cumulative Poisson Probabilities × 1,000

μ \ x	0	1	2	3	4	5	6	7	8	9	10	11	12	13	14
0.1	905	995	1,000												
0.2	819	982	999	1,000											
0.3	741	963	996	1,000											
0.4	670	938	992	999	1,000										
0.5	607	910	986	998	1,000										
0.6	549	878	977	997	1,000										
0.7	497	844	966	994	999	1,000									
0.8	449	809	953	991	999	1,000									
0.9	407	772	937	987	998	1,000									
1.0	368	736	920	981	996	999	1,000								
1.1	333	699	900	974	995	999	1,000								
1.2	301	663	879	966	992	998	1,000								
1.3	273	627	857	957	989	998	1,000								
1.4	247	592	833	946	986	997	999	1,000							
1.5	223	558	809	934	981	996	999	1,000							
1.6	202	525	783	921	976	994	999	1,000							
1.7	183	493	757	907	970	992	998	1,000							
1.8	165	463	731	891	964	990	997	999	1,000						
1.9	150	434	704	875	956	987	997	999	1,000						
2.0	135	406	677	857	947	983	995	999	1,000						
2.2	111	355	623	819	928	975	993	998	1,000						
2.4	091	308	570	779	904	964	988	997	999	1,000					
2.6	074	267	518	736	877	951	983	995	999	1,000					
2.8	061	231	469	692	848	935	976	992	998	999	1,000				
3.0	050	199	423	647	815	916	966	988	996	999	1,000				
3.2	041	171	380	603	781	895	955	983	994	998	1,000				
3.4	033	147	340	558	744	871	942	977	992	997	999	1,000			
3.6	027	126	303	515	706	844	927	969	988	996	999	1,000			
3.8	022	107	269	473	668	816	909	960	984	994	998	999	1,000		
4.0	018	092	238	433	629	785	889	949	979	992	997	999	1,000		
4.2	015	078	210	395	590	753	867	936	972	989	996	999	1,000		
4.4	012	066	185	359	551	720	844	921	964	985	994	998	999	1,000	
4.6	010	056	163	326	513	686	818	905	955	980	992	997	999	1,000	
4.8	008	048	143	294	476	651	791	887	944	975	990	996	998	999	1,000
5	007	040	125	265	440	616	762	867	932	968	986	995	998	999	1,000
6	002	017	062	151	285	446	606	744	847	916	957	980	991	996	1,000

TABLE 2 Cumulative Poisson Probabilities × 1,000 (Continued)

μ \ x	0	1	2	3	4	5	6	7	8	9	10	11	12	13	14
8	000	003	014	042	100	191	313	453	593	717	816	888	936	966	983
9	000	001	006	021	055	116	207	324	456	587	706	803	876	926	959
10		000	003	010	029	067	130	220	333	458	583	697	792	864	917
11		000	001	005	015	038	079	143	232	341	460	579	689	781	854
12		000	001	002	008	020	046	090	155	242	347	462	576	682	772
13			000	001	004	011	026	054	100	166	252	353	463	573	675
14				000	002	006	014	032	062	109	176	260	358	464	570
15				000	001	003	008	018	037	070	118	185	268	363	466

μ \ x	15	16	17	18	19	20	21	22	23	24	25	26	27	28	29
7	998	999	1,000												
8	992	996	998	999											
9	978	989	995	998	999	1,000									
10	951	973	986	993	997	998	999	1,000							
11	907	944	968	982	991	995	998	999	1,000						
12	844	899	937	963	979	988	994	997	999	999	1,000				
13	764	835	890	930	957	975	986	992	996	998	999	1,000			
14	669	756	827	883	923	952	971	983	991	995	997	999	999	1,000	
15	568	664	749	819	875	917	947	967	981	989	994	997	998	999	1,000

μ \ x	0	1	2	3	4	5	6	7	8	9	10	11	12	13	14
16					000	001	004	010	022	043	077	127	193	275	368
17					000	001	002	005	013	026	049	085	135	201	281
18						000	001	003	015	015	030	055	092	143	208
19						000	001	002	004	009	018	035	061	098	150
20							000	001	002	005	011	021	039	066	105
21								000	001	003	006	013	025	043	072
22								000	001	002	004	008	015	028	048
23									000	001	002	004	009	017	031
24									000	000	001	003	005	011	020

TABLE 2 Cumulative Poisson Probabilities × 1,000 (Continued)

x＼μ	15	16	17	18	19	20	21	22	23	24	25	26	27	28	29
16	467	566	659	742	812	868	911	942	963	978	987	993	996	998	999
17	371	468	564	655	736	805	861	905	937	959	975	985	991	995	997
18	287	375	469	562	651	731	799	855	899	932	955	972	983	990	994
19	215	292	378	469	561	647	725	793	849	893	927	951	969	980	988
20	157	221	297	381	470	559	644	721	787	843	888	922	948	966	978
21	111	163	227	302	384	471	558	640	716	782	838	883	917	944	963
22	077	117	169	232	306	387	472	556	637	712	777	832	877	913	940
23	052	82	123	175	238	310	389	472	555	635	708	772	827	873	908
24	034	056	087	128	180	243	314	392	473	554	632	704	768	823	868

x＼μ	30	31	32	33	34	35	36	37	38	39	40	41	42	43	44
16	999	1,000													
17	999	999	1,000												
18	997	998	999	1,000											
19	993	996	998	999	999										
20	987	992	995	997	999	1,000									
21	976	985	991	994	997	999	1,000								
22	959	973	983	989	994	998	999	999	1,000						
23	936	956	971	981	988	993	996	997	999	999	1,000				
24	904	932	953	969	979	987	992	995	997	998	999	999	1,000		

TABLE 3 Cumulative Normal Probabilities

Z	0.09	0.08	0.07	0.06	0.05	0.04	0.03	0.02	0.01	0.00
−3.5	0.00017	0.00017	0.00018	0.00019	0.00019	0.00020	0.00021	0.00022	0.00022	0.00023
−3.4	0.00024	0.00025	0.00026	0.00027	0.00028	0.00029	0.00030	0.00031	0.00033	0.00034
−3.3	0.00035	0.00036	0.00038	0.00039	0.00040	0.00042	0.00043	0.00045	0.00047	0.00048
−3.2	0.00050	0.00052	0.00054	0.00056	0.00058	0.00060	0.00062	0.00064	0.00066	0.00069
−3.1	0.00071	0.00074	0.00076	0.00079	0.00082	0.00085	0.00087	0.00090	0.00094	0.00097
−3.0	0.00100	0.00104	0.00107	0.00111	0.00114	0.00118	0.00122	0.00126	0.00131	0.00135
−2.9	0.0014	0.0014	0.0015	0.0015	0.0016	0.0016	0.0017	0.0017	0.0018	0.0019
−2.8	0.0019	0.0020	0.0021	0.0021	0.0022	0.0023	0.0023	0.0024	0.0025	0.0026
−2.7	0.0026	0.0027	0.0028	0.0029	0.0030	0.0031	0.0032	0.0033	0.0034	0.0035
−2.6	0.0036	0.0037	0.0038	0.0039	0.0040	0.0041	0.0043	0.0044	0.0045	0.0047
−2.5	0.0048	0.0049	0.0051	0.0052	0.0054	0.0055	0.0057	0.0059	0.0060	0.0062
−2.4	0.0064	0.0066	0.0068	0.0069	0.0071	0.0073	0.0075	0.0078	0.0080	0.0082
−2.3	0.0084	0.0087	0.0089	0.0091	0.0094	0.0096	0.0099	0.0102	0.0104	0.0107
−2.2	0.0110	0.0113	0.0116	0.0119	0.0122	0.0125	0.0129	0.0132	0.0136	0.0139
−2.1	0.0143	0.0146	0.0150	0.0154	0.0158	0.0162	0.0166	0.0170	0.0174	0.0179
−2.0	0.0183	0.0188	0.0192	0.0197	0.0202	0.0207	0.0212	0.0217	0.0222	0.0228
−1.9	0.0233	0.0239	0.0244	0.0250	0.0256	0.0262	0.0268	0.0274	0.0281	0.0287
−1.8	0.0294	0.0301	0.0307	0.0314	0.0322	0.0329	0.0336	0.0344	0.0351	0.0359
−1.7	0.0367	0.0375	0.0384	0.0392	0.0401	0.0409	0.0418	0.0427	0.0436	0.0446
−1.6	0.0455	0.0465	0.0475	0.0485	0.0495	0.0505	0.0516	0.0526	0.0537	0.0548
−1.5	0.0559	0.0571	0.0582	0.0594	0.0606	0.0618	0.0630	0.0643	0.0655	0.0668
−1.4	0.0681	0.0694	0.0708	0.0721	0.0735	0.0749	0.0764	0.0778	0.0793	0.0808
−1.3	0.0823	0.0838	0.0853	0.0869	0.0885	0.0901	0.0918	0.0934	0.0951	0.0968
−1.2	0.0985	0.1003	0.1020	0.1038	0.1057	0.1075	0.1093	0.1112	0.1131	0.1151
−1.1	0.1170	0.1190	0.1210	0.1230	0.1251	0.1271	0.1292	0.1314	0.1335	0.1357
−1.0	0.1379	0.1401	0.1423	0.1446	0.1469	0.1492	0.1515	0.1539	0.1562	0.1587
−0.9	0.1611	0.1635	0.1660	0.1685	0.1711	0.1736	0.1762	0.1788	0.1814	0.1841
−0.8	0.1867	0.1894	0.1922	0.1949	0.1977	0.2005	0.2033	0.2061	0.2090	0.2119
−0.7	0.2148	0.2177	0.2207	0.2236	0.2266	0.2297	0.2327	0.2358	0.2389	0.2420
−0.6	0.2451	0.2483	0.2514	0.2546	0.2578	0.2611	0.2643	0.2676	0.2709	0.2743
−0.5	0.2776	0.2810	0.2843	0.2877	0.2912	0.2946	0.2981	0.3015	0.3050	0.3085
−0.4	0.3121	0.3156	0.3192	0.3228	0.3264	0.3300	0.3336	0.3372	0.3409	0.3446
−0.3	0.3483	0.3520	0.3557	0.3594	0.3632	0.3669	0.3707	0.3745	0.3783	0.3821
−0.2	0.3859	0.3897	0.3936	0.3974	0.4013	0.4052	0.4090	0.4129	0.4168	0.4207
−0.1	0.4247	0.4286	0.4325	0.4364	0.4404	0.4443	0.4483	0.4522	0.4562	0.4602
−0.0	0.4641	0.4681	0.4721	0.4761	0.4801	0.4840	0.4880	0.4920	0.4960	0.5000

TABLE 3 Cumulative Normal Probabilities   (*Continued*)

Z	0.00	0.01	0.02	0.03	0.04	0.05	0.06	0.07	0.08	0.09
+0.0	0.5000	0.5040	0.5080	0.5120	0.5160	0.5199	0.5239	0.5279	0.5319	0.5359
+0.1	0.5398	0.5438	0.5478	0.5517	0.5557	0.5596	0.5636	0.5675	0.5714	0.5753
+0.2	0.5793	0.5832	0.5871	0.5910	0.5948	0.5987	0.6026	0.6064	0.6103	0.6141
+0.3	0.6179	0.6217	0.6255	0.6293	0.6331	0.6368	0.6406	0.6443	0.6480	0.6517
+0.4	0.6554	0.6591	0.6628	0.6664	0.6700	0.6736	0.6772	0.6808	0.6844	0.6879
+0.5	0.6915	0.6950	0.6985	0.7019	0.7054	0.7088	0.7123	0.7157	0.7190	0.7224
+0.6	0.7257	0.7291	0.7324	0.7357	0.7389	0.7422	0.7454	0.7486	0.7517	0.7549
+0.7	0.7580	0.7611	0.7642	0.7673	0.7704	0.7734	0.7764	0.7794	0.7823	0.7852
+0.8	0.7881	0.7910	0.7939	0.7967	0.7995	0.8023	0.8051	0.8079	0.8106	0.8133
+0.9	0.8159	0.8186	0.8212	0.8238	0.8264	0.8289	0.8315	0.8340	0.8365	0.8389
+1.0	0.8413	0.8438	0.8461	0.8485	0.8508	0.8531	0.8554	0.8577	0.8599	0.8621
+1.1	0.8643	0.8665	0.8686	0.8708	0.8729	0.8749	0.8770	0.8790	0.8810	0.8830
+1.2	0.8849	0.8869	0.8888	0.8907	0.8925	0.8944	0.8962	0.8980	0.8997	0.9015
+1.3	0.9032	0.9049	0.9066	0.9082	0.9099	0.9115	0.9131	0.9147	0.9162	0.9177
+1.4	0.9192	0.9207	0.9222	0.9236	0.9251	0.9265	0.9279	0.9292	0.9306	0.9319
+1.5	0.9332	0.9345	0.9357	0.9370	0.9382	0.9394	0.9406	0.9418	0.9429	0.9441
+1.6	0.9452	0.9463	0.9474	0.9484	0.9495	0.9505	0.9515	0.9525	0.9535	0.9545
+1.7	0.9554	0.9564	0.9573	0.9582	0.9591	0.9599	0.9608	0.9616	0.9625	0.9633
+1.8	0.9641	0.9649	0.9656	0.9664	0.9671	0.9678	0.9686	0.9693	0.9699	0.9706
+1.9	0.9713	0.9719	0.9726	0.9732	0.9738	0.9744	0.9750	0.9756	0.9761	0.9767
+2.0	0.9773	0.9778	0.9783	0.9788	0.9793	0.9798	0.9803	0.9808	0.9812	0.9817
+2.1	0.9821	0.9826	0.9830	0.9834	0.9838	0.9842	0.9846	0.9850	0.9854	0.9857
+2.2	0.9861	0.9864	0.9868	0.9871	0.9875	0.9878	0.9881	0.9884	0.9887	0.9890
+2.3	0.9893	0.9896	0.9898	0.9901	0.9904	0.9906	0.9909	0.9911	0.9913	0.9916
+2.4	0.9918	0.9920	0.9922	0.9925	0.9927	0.9929	0.9931	0.9932	0.9934	0.9936
+2.5	0.9938	0.9940	0.9941	0.9943	0.9945	0.9946	0.9948	0.9949	0.9951	0.9952
+2.6	0.9953	0.9955	0.9956	0.9957	0.9959	0.9960	0.9961	0.9962	0.9963	0.9964
+2.7	0.9965	0.9966	0.9967	0.9968	0.9969	0.9970	0.9971	0.9972	0.9973	0.9974
+2.8	0.9974	0.9975	0.9976	0.9977	0.9977	0.9978	0.9979	0.9979	0.9980	0.9981
+2.9	0.9981	0.9982	0.9983	0.9983	0.9984	0.9984	0.9985	0.9985	0.9986	0.9986
+3.0	0.99865	0.99869	0.99874	0.99878	0.99882	0.99886	0.99889	0.99893	0.99896	0.99900
+3.1	0.99903	0.99906	0.99910	0.99913	0.99915	0.99918	0.99921	0.99924	0.99926	0.99929
+3.2	0.99931	0.99934	0.99936	0.99938	0.99940	0.99942	0.99944	0.99946	0.99948	0.99950
+3.3	0.99952	0.99953	0.99955	0.99957	0.99958	0.99960	0.99961	0.99962	0.99964	0.99965
+3.4	0.99966	0.99967	0.99969	0.99970	0.99971	0.99972	0.99973	0.99974	0.99975	0.99976
+3.5	0.99977	0.99978	0.99978	0.99979	0.99980	0.99981	0.99981	0.99982	0.99983	0.99983

Design for Reliability

From Chapter 12 of *Systems Engineering and Analysis,* Fifth Edition, Benjamin S. Blanchard, Wolter J. Fabrycky. Copyright © 2011 by Pearson Education, Inc. Published by Pearson Prentice Hall. All rights reserved.

Design for Reliability

One of the most significant objectives in fulfilling the requirements for system operational feasibility is achieved through *design for reliability*. Reliability is the ability of a system to perform its intended mission when operating for a designated period of time, or through a planned mission scenario (or series of scenarios), in a realistic operational environment. The system, when operating in a true user's environment, is expected to be able to satisfy all of the operational objectives desired and specified by the customer.

History is replete with instances where systems operating in the field have not measured up to their expectations. Failures have occurred prematurely (for one reason or another), the requirements for maintenance have been high, and the costs throughout the system life cycle have been excessive. In addition, many of the systems currently in operational use have been unable to accomplish the mission for which they were designed.

When addressing cause-and-effect relationships, the inability of systems to realize their desired operational objectives has been primarily due to lack of proper consideration of reliability and its characteristics in the design from the beginning. Reliability must be properly specified during conceptual design in meaningful quantitative terms as one of the *design-to* requirements, when the overall requirements (i.e., *technical performance measures* – TPMs) for the system are being specified. These top-level requirements must then be allocated to the applicable subsystems, and the appropriate characteristics (attributes) must be incorporated into subsystem and component designs to ensure compliance with initially specified mission-related requirements.

The objective of this chapter is to provide a comprehensive understanding of reliability and its importance in the system design process. The intent is to accomplish this through presentation of the following topics:

- The definition and explanation of reliability;
- Measures of reliability—the reliability function, the failure rate, and component relationships;
- Reliability in the system life cycle—system requirements, reliability models, reliability allocation, component selection and application, redundancy in design, design participation, and design review and evaluation;

- Reliability analysis methods—failure mode, effects, and criticality analysis (FMECA); fault-tree analysis (FTA); stress–strength analysis; reliability prediction; reliability growth analysis; and
- Reliability test and evaluation.

Reliability requirements must be considered within and throughout the process and as a major parameter in accomplishing the design functions. Accordingly, the primary aim of this chapter is to provide the needed technical material in order to obtain understanding of, and capability with, the implementation of reliability in design. A summary, selected references and websites, and extensions for further study are given in the last section.

1 DEFINITION AND EXPLANATION OF RELIABILITY

Reliability may be defined simply as the probability that a system or product will accomplish its designated mission in a satisfactory manner or, specifically, the probability that the entity will perform in a satisfactory manner for a given period when used under specified operating conditions. Inherent within this definition are the elements of *probability, satisfactory performance, time or mission-related cycle,* and *specified operating conditions.* Reliability must be directly related to one or more specific mission scenarios, or to the system operational requirements.

Probability, the primary element in the reliability definition, is usually stated in quantitative terms as representing a fraction or a percent specifying the number of times that one can expect an event to occur in a total number of trials. For example, a statement that the probability of survival of an entity operating 80 hours is 0.75 (or 75%) indicates that one can expect it to function properly for at least 80 hours 75 times out of 100. When there are several supposedly identical items operating under similar conditions, one may specify that 75% of the items will operate in a satisfactory manner when required for the specified duration of the designated mission. Failures are expected to occur at different points of time; thus, failures are described in probabilistic terms, or probabilistically. The fundamentals of reliability are dependent on the subject of probability theory, as introduced in Appendix: Probability Theory and Analysis.

The second element in the reliability definition is *satisfactory performance,* indicating that specific criteria must be established that describe what is considered to be satisfactory. This relates to the definition of system operational requirements, the functions to be accomplished, and the TPMs. *What must the system do in order to satisfy the needs of the customer?*

The third element, *time,* is perhaps the most important because it represents a measure against which the degree of system operational performance can be determined. One must know the time period within which to assess the probability of completing a mission, or a designated function, as scheduled. Of particular interest is the ability to predict the probability of an entity surviving (without failure) for a designated interval of time. Because time is critical in reliability, it is often expressed in terms of *mean time between failure* (MTBF) or *mean time to failure* (MTTF).[1]

---

[1] Reliability may also be specified in terms of the number of cycles of operation or the number of successful functions completed. In the world of software, failures may be related to calendar time, processor time, the number of transactions per period, the number of faults per line item or module of code, and so on. The specific TPMs, or measures of reliability, must be related to the type of system and its makeup.

The *specified operating conditions* under which a system or product is expected to function constitute the fourth significant element of the reliability definition. These conditions include environmental factors, such as the geographical location where the system is expected to operate and the anticipated period of time, the operational profile, and the potential impacts resulting from temperature cycling, humidity, vibration and shock, and so on. Such factors must be the conditions not only during the period when the system is actually operating but also during the period when the system (or a portion thereof) is in a storage mode or is being transported from one location to the next. Experience has indicated that the transportation, handling, and storage modes are sometimes more critical from a reliability standpoint than are the conditions experienced during actual system operational use.[2]

The four elements given here are critical in determining the reliability of a system or product. As reliability is an inherent characteristic of design, it is essential that these elements be adequately considered at project inception. Reliability requirements must be addressed in the definition of operational requirements for the system, the appropriate reliability TPMs must be identified and prioritized, and the resultant DDPs must be *designed in* from the beginning.

2 MEASURES OF RELIABILITY

The evaluation of any system or product in terms of reliability is based on precisely defined reliability concepts and measures. Accordingly, this section is concerned with the development of selected reliability measures and terms. A basic understanding of these is required before discussing reliability program functions as related to the system/product design. Terms such as the reliability function, failure rate, probability density function, reliability model, and so on, are addressed and defined.

2.1 The Reliability Function

The *reliability function*, also known as the *survival function*, is determined from the probability that a system (or product) will be successful for at least some specified time t. The reliability function, $R(t)$, is defined as

$$R(t) = 1 - F(t) \tag{1}$$

where $F(t)$ is the probability that the system will fail by time t. $F(t)$ is basically the *failure distribution function* or *unreliability function*. If the random variable t has a density function of $f(t)$, the expression for reliability is

$$R(t) = 1 - F(t) = \int_t^\infty f(t)dt \tag{2}$$

If the time to failure is described by an exponential density function, then, from Appendix: Probability Theory and Analysis, Section 2,

$$f(t) = \frac{1}{\theta} e^{-t/\theta} \tag{3}$$

---

[2]Functional analysis should aid in the identification and definition of operating functions, transportation functions, and so on; that is, specified operating conditions.

where θ is the mean life and t the time period of interest. The reliability at time t is

$$R(t) = \int_t^\infty \frac{1}{\theta} e^{-t/\theta} dt = e^{-t/\theta} \tag{4}$$

Mean life (θ) is the arithmetic average of the lifetimes of all items considered, which for the exponential function is MTBF. Thus,

$$R(t) = e^{-t/M} = e^{-\lambda t} \tag{5}$$

where λ is the instantaneous failure rate and M the MTBF.

If an item has a constant failure rate, the reliability of that item at its mean life is approximately 0.37. Thus, there is a 37% probability that a system will survive its mean life without failure.[3] Mean life and failure rate are related as

$$\lambda = \frac{1}{\theta} \tag{6}$$

Figure 1 shows the exponential reliability function, where time is given in units of t/M. The illustration focuses on the reliability function for the exponential distribution, which may be used in those applications where the failures are random.

The failure characteristics of different entities are not necessarily the same. There are several well-known probability distribution functions that have been found in practice to describe the failure characteristics of different equipments. These include the binomial, exponential, normal, Poisson, and Weibull distributions. Thus, one should not assume that any one distribution is applicable in all situations (refer to Appendix: Probability Theory and Analysis, Section 2).

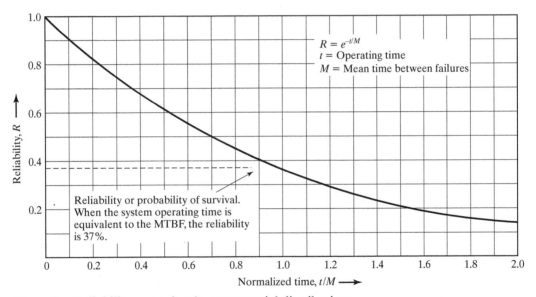

Figure 1 Reliability curve for the exponential distribution.

[3]Refer to Appendix: Probability Theory and Analysis (Section 2) for the derivation of the mean value of an exponential distribution. Also see Section 2.2 and Figure 1.

2.2 The Failure Rate

The rate at which failures occur in a specified time interval is called the *failure rate* for that interval. The failure rate per hour is expressed as

$$\lambda = \frac{\text{number of failures}}{\text{total operating hours}} \tag{7}$$

Failure rate may be expressed in terms of failures per hour, percentage of failures per 1,000 hours, or failures per million hours. As an example, suppose that 10 components were tested for 600 hours under specified operating conditions. The components (which are not repairable) failed as follows: component 1 failed after 75 hours, component 2 failed after 125 hours, component 3 failed after 130 hours, component 4 failed after 325 hours, and component 5 failed after 525 hours. Thus, there were five failures and the total operating time was 4,180 hours. Using Equation 7, the calculated failure rate per hour is

$$\lambda = \frac{5}{4,180} = 0.001196$$

As another example, suppose that the operating cycle for a given system is 169 hours, as illustrated in Figure 2. During that time, six failures occur at the points indicated. A failure is defined as an instance when the system is not operating within a specified set of parameters. The failure rate, or corrective maintenance frequency, per hour is

$$\lambda = \frac{\text{number of failures}}{\text{total mission time}} = \frac{6}{142} = 0.04225$$

Assuming an exponential distribution, the system mean life or the mean time between failure (MTBF) is

$$\text{MTBF} = \frac{1}{\lambda} = \frac{1}{0.04225} = 23.6686 \text{ hours}$$

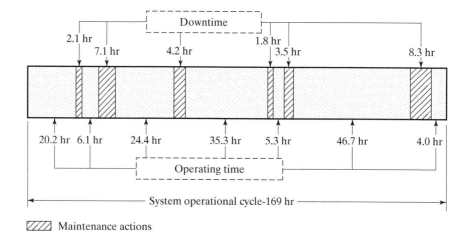

Figure 2   A system operational cycle.

Figure 3 presents a reliability nomograph (for the exponential failure distribution) that facilitates calculations of MTBF, λ, R(t), and operating time. If the MTBF is 200 hours (failure rate = 0.005) and the operating time is 2 hours, the nomograph gives a reliability value of 0.99.

When assuming the negative exponential distribution, the failure rate is considered to be relatively constant during normal system operation if the system design is mature. That is,

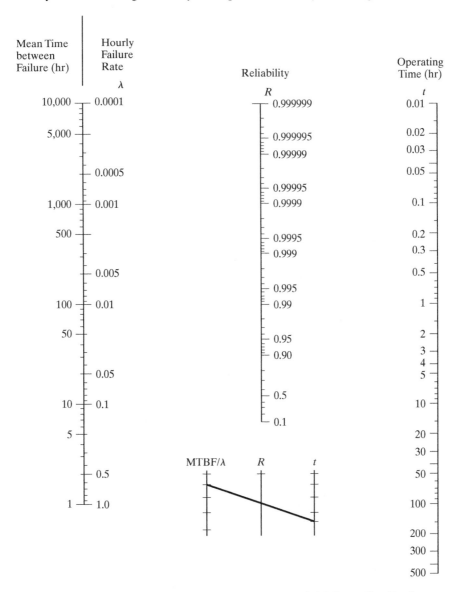

Figure 3 Reliability nomograph for the exponential failure distribution. *Source*: NAVAIR 00-65-502/NAVORD OD 41146, *Reliability Engineering Handbook*, Naval Air Systems Command and Naval Ordnance Systems Command.

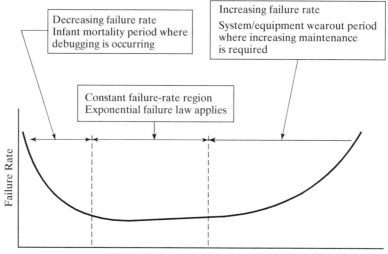

Bathtub Curve Based on Time-Dependent Failure Rate

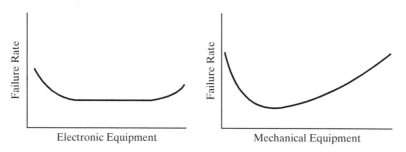

Figure 4 Typical failure-rate curve relationships.

when equipment is produced and the system is initially distributed for operational use, there are usually a higher number of failures resulting from component variations and mismatches, manufacturing processes, and so on. The initial failure rate is higher than anticipated, but gradually decreases and levels off, as illustrated in Figure 4, during the *debugging* period. Similarly, when the system reaches a certain age, there is a *wearout* period when the failure rate increases. The relatively level portion of the curve in Figure 4 is the constant failure-rate region, where the exponential failure law applies.

Figure 4 illustrates certain relative relationships. Actually, the curve may vary considerably depending on the type of system and its operational profile. Further, if the system is continually being modified for one reason or another, the failure rate may not be constant. In any event, the illustration does provide a good basis for discussion when considering relative failure-rate trends.

In the domain of software, failures may be related to calendar time, processor time, the number of transactions per period, the number of faults per module of code, and so on. Expectations are usually based on an operational profile and criticality to the mission. Thus, an accurate description of the mission scenario(s) is required. As the system evolves from the detail design and development phase to the operational use phase, the ongoing maintenance of software often becomes a major issue. Although the failure rate of equipment generally as-

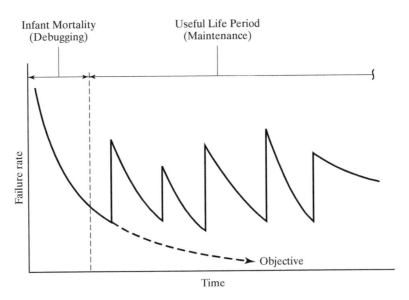

Figure 5 Failure-rate curve with maintenance (software application).

sumes the profiles in Figure 4, the maintenance of software often has a negative effect on the overall system reliability. The performance of software maintenance on a continuing basis, along with the incorporation of system changes in general, usually impacts the overall failure rate, as shown in Figure 5. When a change or modification is incorporated, "bugs" are usually introduced, and it takes a while for these to be "worked out."

To this point, the definition of failures and their associated characteristics have been related to both equipment and software primary failures, based on practices and experiences in the past. The emphasis has been directed more to the components of a system than to the system as an entity. However, when defining failures (and failure rates) from a total *systems* perspective, whenever the system is not functioning properly, a failure has occurred. Failures may be classified as inherent "primary" or "catastrophic" component failures, "dependent" or "secondary" failures, manufacturing defects, operator-induced failures, maintenance-induced failures, failures resulting from component wearout, material damage due to handling, and so on. The total overall failure rate (in instances per hour or instances per mission) is of concern herein. Table 1 illustrates how the various "causes" of failure contribute to the total failure rate for two different systems. Referring to the table, it should be noted that depending on the system type, complexity, human–equipment makeup, and so on, the individual contributions may vary from one system to another. For example, the number of potential failures resulting from operator-induced actions is greater for System *B* than for System *A*.

When relating these factors to experiences in the user's operational environment, one often encounters the situation that is illustrated in Figure 6.[4] In this example, an aerospace system has completed a designated number of missions. Through this series of mission scenarios, 100 system failures were reported (i.e., the operator indicated that a failure had occurred and

---

[4]The issues pertaining to operator-induced and maintenance-induced failures are not detailed here.

TABLE 1 Composite Failure Rates for Two Operating Systems

Contributor	Failure Rate (Instances/hour)	
	System A	System B
1. Inherent (primary) defects	0.000392	0.000298
2. Dependent (secondary) defects	0.000072	0.000061
3. Manufacturing defects	0.000002	0.000004
4. Operator-induced failures	0.000003	0.000134
5. Maintenance-induced failures	0.000012	0.000015
6. Equipment/material handling failures	0.000005	0.000004
7. Equipment/material wearout failures	0.000001	0.000002
Total	0.000487	0.000518

notified the maintenance organization accordingly). For each reported failure, an unscheduled maintenance action (MA) was initiated. Some of the maintenance actions, or 75 in this case, resulted in the removal and replacement of system components at the organizational level which, in turn, resulted in a requirement for spares/repair parts. Some of the items that were returned to the intermediate shop for higher-level maintenance were verified as being faulty (48 in this case), some were sent on to the factory for a higher level of maintenance (12), and some were checked out and found to be fully operational (15). In this example, there were 58 unscheduled maintenance actions that resulted in confirmed and "traceable" failures. On the other hand, there were 100 reported failures, either real or perceived by the operator. While the emphasis in the past has been directed to the 58 instances, a total *systems* perspective requires managers to address the total; that is, 100 instances. In other words, in the design of systems, one needs to incorporate the appropriate characteristics that will prevent instances such as those identified in Table 1.

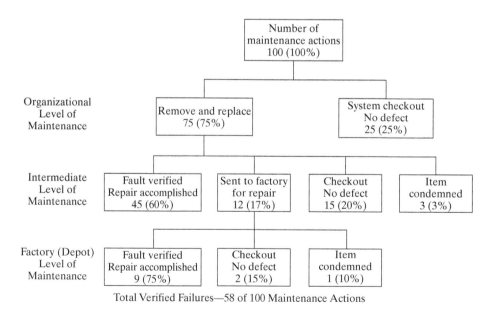

Figure 6 System *XYZ* unscheduled maintenance actions (system failures).

2.3 Component Relationships

After defining the basic reliability function and some of the measures associated with system failures, it is appropriate to consider their application in design. Components are selected and combined in a serial manner, in parallel, or in combinations thereof. Block diagrams are developed to show component relationships and for accomplishing reliability analysis.[5]

Series Networks. The series relationship is probably the most commonly used and is the simplest to analyze. It is illustrated in Figure 7. In a *series network*, all components must operate in a satisfactory manner if the system is to function properly. Assuming that a system includes subsystems A–C, the reliability of the system is the product of the reliabilities for the individual subsystems expressed as

$$R = (R_A)(R_B)(R_C). \qquad (8)$$

As an example, suppose that an electronic system includes a transmitter, a receiver, and a power supply. The transmitter reliability is 0.8521, the receiver reliability is 0.9712, and the power supply reliability is 0.9357. The overall reliability for the electronic system is

$$R = (0.8521)(0.9712)(0.9357) = 0.7743$$

If a series system is expected to operate for a specified time period, its required overall reliability can be derived. Substituting Equation 5 into Equation 8 gives

$$R_s = (e^{-\lambda_1 t})(e^{-\lambda_2 t})\dots(e^{-\lambda_n t})$$

for a series with n components. This may be expressed as

$$R_s = e^{-(\lambda_1 + \lambda_2 + \dots + \lambda_n)t} \qquad (9)$$

Suppose that a series system consists of four subsystems and is expected to operate for 1,000 hours. The four subsystems have the following MTBFs: subsystem A, MTBF = 6,000 hours; subsystem B, MTBF = 4,500 hours; subsystem C, MTBF = 10,500 hours; subsystem D, MTBF = 3,200 hours. The objective is to determine the overall reliability of the series network where

$$\lambda_A = \frac{1}{6,000} = 0.000167 \text{ failure/hour}$$

$$\lambda_B = \frac{1}{4,500} = 0.000222 \text{ failure/hour}$$

$$\lambda_C = \frac{1}{10,500} = 0.000095 \text{ failure/hour}$$

$$\lambda_D = \frac{1}{3,200} = 0.000313 \text{ failure/hour}$$

Figure 7 A series network.

[5]The development of reliability block diagrams should evolve directly from the functional analysis, as should the development of functional block diagrams.

The reliability of the series network is found from Equation 9 as

$$R = e^{-(0.000797)(1,000)} = 0.4507$$

This means that the probability of the system surviving (i.e., reliability) for 1,000 hours is 45.1%. If the requirement were reduced to 500 hours, the reliability would increase to about 67%.

Parallel Networks. A pure *parallel network* is one where several of the same components are in parallel and where all the components must fail to cause total system failure. A parallel network with two components is illustrated in Figure 8. Assuming that components A and B are identical, the system will function if either A or B, or both, are working. The reliability is expressed as

$$R = R_A + R_B - (R_A)(R_B) \tag{10}$$

Consider next a network with three components in parallel, as shown in Figure 9. The network reliability is expressed as

$$R = 1 - (1 - R_A)(1 - R_B)(1 - R_C) \tag{11}$$

If components A–C are identical, the reliability expression can be simplified to

$$R = 1 - (1 - R)^3$$

for a system with three parallel components. For a system with n identical components, the reliability is

$$R = 1 - (1 - R)^n \tag{12}$$

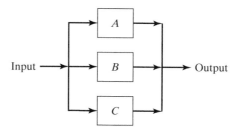

Figure 9 Parallel network with three components.

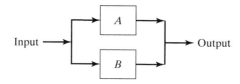

Figure 8 A parallel network.

Parallel redundant networks are used primarily to improve system reliability, as Equations 11 and 12 indicate mathematically. Consider the following example: a system includes two identical subsystems in parallel and the reliability of each subsystem is 0.95. The reliability of the system is found from Equation 10 as

$$R = 0.95 + 0.95 - (0.95)(0.95) = 0.9975$$

Suppose that the reliability of the preceding system needs improvement beyond 0.9975. By adding a third identical subsystem in parallel, the system reliability is found from Equation 11 to be

$$R = 1 - (1 - 0.95)^3 = 0.999875$$

Note that there is a reliability improvement of 0.002375 over the previous configuration.

If the subsystems are not identical, Equation 10 can be used. For example, a parallel redundant network with two subsystems, $R_A = 0.75$ and $R_B = 0.82$, gives a system reliability of

$$R = 0.75 + 0.82 - (0.75) = 0.955$$

Redundancy can be applied in design at different hierarchical indenture levels of the system. At the subsystem level, it may be appropriate to incorporate parallel functional capabilities, where the system will continue to operate if one path fails to function properly. The flight control capability (incorporating electronic, digital, and mechanical alternatives) in an aircraft is an example where there are alternate paths in case of a failure in any one. At the detailed piece-part level, redundancy may be incorporated to improve the reliability of critical functions, particularly in areas where the accomplishment of maintenance is not feasible. For example, in the design of many electronic circuit boards, redundancy is often built in for the purpose of improving reliability, while at the same time the accomplishment of maintenance is not practical.

The application of redundancy in design is a key area for evaluation. Although redundancy per se does improve reliability, the incorporation of extra components in the design requires additional space and weight and the costs are higher. This leads to a few questions: *Is redundancy really required in terms of criticality to system operation and accomplishment of the mission? At what level should redundancy be incorporated? What type of redundancy should be considered ("active" or "standby")? Should maintainability provisions be considered? Are there any alternative methods for improving reliability (e.g., improved part selection, part derating)?* In essence, there are many interesting and related concerns that require further investigation.

Combined Series–Parallel Networks.　Various levels of reliability can be achieved through the application of a combination of series and parallel networks. Consider the three examples illustrated in Figure 10.

The reliability of the first network in Figure 10 is given by

$$R_a = R_A(R_B + R_C - R_BR_C) \tag{13}$$

For the second network the reliability is given by

$$R_b = [1 - (1 - R_A)(1 - R_B)][1 - (1 - R_C)(1 - R_D)] \tag{14}$$

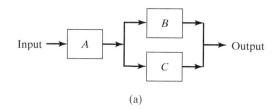

(a)

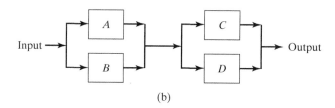

(b)

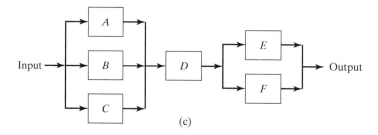

(c)

Figure 10 Some combined series–parallel networks.

And for the third network the reliability is given by

$$R_c = [1 - (1 - R_A)(1 - R_B)(1 - R_C)][R_D][R_E + R_F - (R_E)(R_F)] \qquad (15)$$

Combined series–parallel networks such as those in Figure 10 require that the analyst first evaluate the redundant elements to obtain unit reliability. Overall system reliability is then determined by finding the product of all unit reliabilities.

3 RELIABILITY IN THE SYSTEM LIFE CYCLE

Reliability, as an inherent characteristic of design, must be addressed in the overall systems engineering process beginning in the conceptual design phase. Referring to Figure 11, qualitative and quantitative reliability requirements are developed through the accomplishment of feasibility analysis, the development of operational requirements and the maintenance concept, and the identification and prioritization of TPMs. The applicable measures of reliability must be established, their importance with respect to other system metrics must be delineated, and the requirements for design must be identified through the development of the appropriate TPMs.

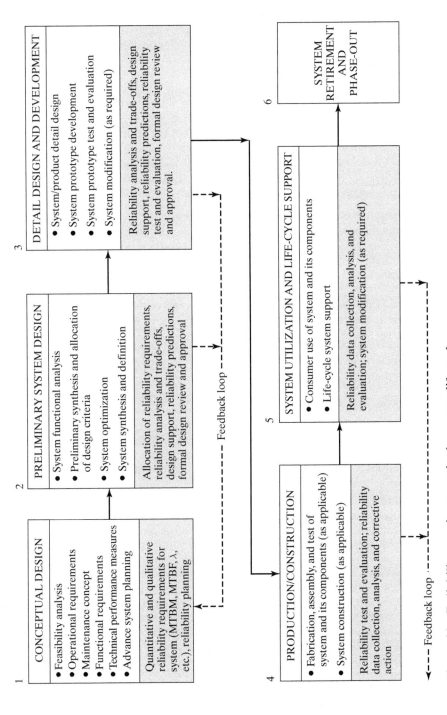

Figure 11 Reliability requirements in the system life cycle.

Given the specification of reliability requirements in Block 1 of the figure, the next step is to develop a reliability model for the purpose of allocating top-level requirements down to the subsystem level and below as part of the system allocation process. Having defined the basic requirements, one evolves through the iterative process of system synthesis, analysis, and evaluation. To facilitate this process, there are several different analysis methods/tools that can be effectively applied in evolving from conceptual design to preliminary and detail design (i.e., Blocks 2 and 3 of Figure 11). As the design progresses and models of the system are developed (both analytical and physical), the ongoing evaluation process occurs and reliability testing is included as part of the overall system test and evaluation activity. From this point on, the evaluation effort continues through the production/construction and system utilization phases (Blocks 4 and 5), with the appropriate feedback and system modification for corrective action or for product improvement as required. In essence, there are reliability requirements in each phase of the life cycle as illustrated in Figure 11.

3.1 System Requirements

Every system is developed in response to a customer need to fulfill some anticipated function. The effectiveness with which the system fulfills this function is the ultimate measure of its utility and its value to the customer (i.e., user). System effectiveness is a composite of many factors, with reliability being a major contributor in determining the ultimate usefulness of a system.

Reliability requirements, specified both in quantitative and qualitative terms, are defined within the context of the system operational requirements and the maintenance concept. This includes the following:

1. Definition of system performance and effectiveness factors, mission profile(s), and system functional requirements (use conditions, duty cycles, and how the system is to be operated).

2. Definition of the operational life cycle (the anticipated time that the system will be in the inventory and in operational use).

3. Definition of the environment in which the system is expected to operate and be maintained (temperature, humidity, shock and vibration, levels of noise and toxicity, etc.). This should include a range of values as applicable and should cover all operational, transportation and handling, maintenance and support, and storage modes.

4. Definition of the operational and supporting interfaces likely to impact the system as it performs its mission(s) throughout its planned life cycle. This should include the system operating within some system-of-systems (SOS) configuration and the internal and external effects from other systems operating within the same SOS structure.

The basic question at this point is, *What reliability should the system have (exhibit) in order to accomplish its intended mission successfully, throughout the specified life cycle, and in the environment defined?* If the operational requirements specify that the system must function 24 hours a day for 360 days a year without failure, the system reliability requirements may be rather stringent. Conversely, if the system is only required to operate 2 hours per day for 260 days per year, the specified requirements may be different. In any event, quantitative and qualitative reliability requirements must be defined for the system based on the foregoing considerations.

As presented in Section 2, quantitative requirements are usually expressed in terms of $R(t)$, MTBF, MTTF, λ, successful operational cycles per period, or various combinations thereof. For software, reliability may be specified as the number of errors per module of software or lines of code, number of processor errors, the time to first failure, or some equivalent measure impacting system failure. For a system configuration like a river crossing bridge, reliability requirements may be expressed in terms of degrees of degradation, or the number of maintenance actions required, over a specified time period given the expected traffic flow patterns. In any event, the specific measure of reliability is related to one or more operational parameters.

3.2 Reliability Models

The system is defined, in "functional" terms, through accomplishment of the functional analysis and the development of functional flow block diagrams. The results of the functional analysis lead to the development of a reliability block diagram and a model that can serve as the basis for accomplishing reliability allocation, reliability prediction, stress–strength analysis, and subsequent design analysis and evaluation tasks. An example of a reliability model is illustrated in Figure 12.

Series–parallel relationships are established and developed further as the design definition evolves. Components are selected to fulfill the functional requirements for each block, reliability characteristics are identified for each component and for each block, and the results are combined in the accomplishment of an early top-down allocation of reliability requirements, and subsequently in the accomplishment of a bottom-up reliability prediction for the purpose of design evaluation. The reliability model serves as a baseline for the identification of possible weak areas and where possible design improvements can be introduced. In addition, the reliability model serves as an input in the accomplishment of reliability prediction, FMECA, FTA, maintenance, and related analyses.

3.3 Reliability Allocation

Top-level reliability requirements are specified for the system, and these requirements are then allocated to subsystem level, unit level, and down to the level needed to provide a meaningful input to the design (i.e., establishment of the right input design criteria at the proper level).

The approach used in the allocation of reliability requirements is not "fixed," may be somewhat subjective at times, and will vary depending on the system type and complexity. Referring to Figure 13, it is assumed that at the system level there is an MTBF requirement of 450 hours. The question is, *What should be specified as a reliability requirement for Units A, B, and C?* In response, the following steps are appropriate:

1. Identify the elements of the system where the design is known and where reliability data are available or can be readily assessed. For example, through the selection of commercial off-the-shelf (COTS) items, it is expected that field data may be available from the manufacturer and that one can acquire a relatively close assessment of item reliability based on past experience. Such experience factors should be assigned to

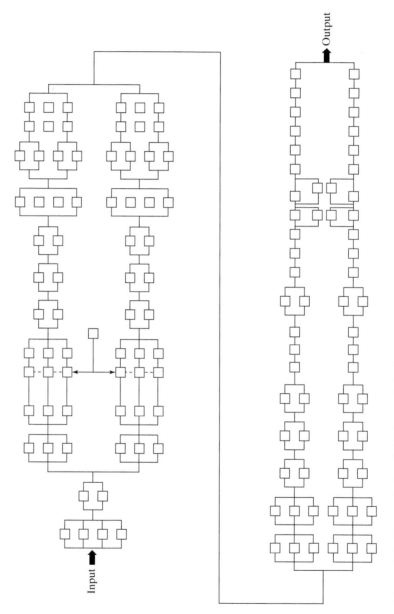

Figure 12 Expanded reliability model of system.

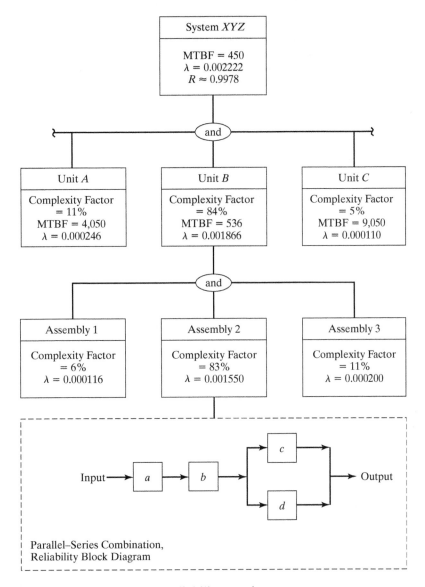

Figure 13 Allocation of reliability requirements.

the appropriate elements of the system, and their respective contribution to the 450-hour requirement should be determined.[6]

2. Identify the areas that are new and where design information is not available. Assign complexity or weighting factors to each applicable block. Complexity factors may be

---

[6]When evaluating field data from past applications of COTS items or equivalent, care must be taken to ensure that the use factors, duty cycles, stresses, and so forth are similar. A given COTS item can demonstrate a high degree of reliability in one application but may not in another application where the mission scenarios and stresses are different.

based on an estimate of the number and relationship of component parts, the duty cycle, whether it is anticipated that an item will be subjected to temperature extremes, and so on. Experience of the designer and his/her knowledge of similar applications in the past will help in this area. The portion of the system reliability requirement that is not already allocated to the areas of known design is allocated using the assigned weighting factors.

The end result should constitute a series of lower-level values that, when combined, will represent the system reliability requirement initially specified. The reliability factors established for the three units in Figure 13 serve as design criteria and are included in the appropriate development or product specifications used in the acquisition of the units.

In the event that System *XYZ* in Figure 13 is operating within a *SOS* configuration and that there is a "common" unit or assembly required for successful system operation, the allocation process must deal with the common element, in accordance with either the first option above (for an items with a known reliability) or the second option (for a newly developed item where there has been no previous operational experience).[7]

3.4 Component Selection and Application

The reliability of a system depends largely on the reliability of its component parts, and the selection of parts must be compatible with the requirements of the particular application of those parts. The process of procuring what is advertised as being a reliable part is not adequate in itself and does not guarantee a reliable system. The specific application of the component is of prime importance, particularly when considering factors such as part tolerances and drift characteristics, electrical and environmental stresses, and so on.

In electrical and electronic systems, part tolerances, drift characteristics, electrical stresses, and environmental stresses can have a major impact on system reliability and on the individual failure modes of various system components. For mechanical and structural items, the same concern prevails when considering part tolerances, component stress–strength factors, and material fatigue characteristics. Consequently, a fundamental approach to attaining a high level of reliability is to select and apply those components and materials of known reliabilities and capable of meeting system requirements. Major emphasis in the *design for reliability* should consider several factors:

1. The selection of *standardized* components and materials to the greatest extent possible. The utilization of common and standard components, with known physical characteristics, known reliabilities, and so on, should be preferred over the selection of new and nonstandard items that have not previously been qualified for operational use.
2. The test and evaluation of all components and materials prior to design acceptance. This includes the evaluation of component operating features, physical tolerances, sensitivity to certain stresses, physics-of-failure characteristics, and other specific characteristics of the component(s) related to its intended application.

The challenge in design is to select and utilize only those components that, when combined, are capable of meeting the overall reliability requirements for the system. This objective can be facilitated through the combination of theoretical analysis, a reliability model

---

[7]As the design progresses, it may become known that a given reliability requirement cannot be met (e.g., the MTBF of 536 hours for Unit *B*). In this case, it may be necessary to authorize a *reallocation* where the requirement for Unit *B* is "relaxed." However, if this is implemented, the reliability requirements for Units *A* and *C* will have to be "tightened" if the overall system requirement is to be met. Some trade-offs study is necessary at this point to arrive at a optimum (minimum cost) allocation of the reliability requirements.

evaluating application and relationships, and laboratory analysis evaluating the results from component testing.

3.5 Redundancy in Design

Under certain conditions in system design it may be necessary to consider the use of redundancy to enhance system reliability by providing two or more functional paths (or channels of operation) in areas that are critical for successful mission accomplishment. However, the application of redundancy per se will not necessarily solve all problems because it usually implies increased weight and space, increased power consumption, greater complexity, and higher cost. Conversely, the use of redundancy may be the only solution for reliability improvement in specific situations.

Redundancy can be applied at several levels, as illustrated in Figure 14. At the system level, Block G is redundant with the other blocks in the network and is at a different level than Block C, which is redundant with Block D. From the block diagram, the paths that will result in successful system operation are $A, B, C, E; A, B, C, F; A, B, D, E; A, B, D, F;$ and G.

The probability of success, or the reliability, of each path may be calculated using the multiplication rule expressed in Equation 8. The calculation of system reliability (considering all paths combined) requires knowledge of the type of redundancy used and the individual block reliabilities. For operating (or active) redundancy, where all blocks are fully energized during an operational cycle, the appropriate equations in Section 2 may be used to calculate system reliability. Equations 10–12 and the related examples illustrate the gain in reliability that can be obtained through redundancy.

When design problems become somewhat more complex than the illustrations presented in Section 2.3, the number of possible events becomes greater. For example, in Figure 9, the following possibilities exist:

1. Subsystem $A, B,$ and C are all operating.
2. Subsystems A and B are operating while C is failed.
3. Subsystems A and C are operating while B is failed.
4. Subsystems B and C are operating while A is failed.

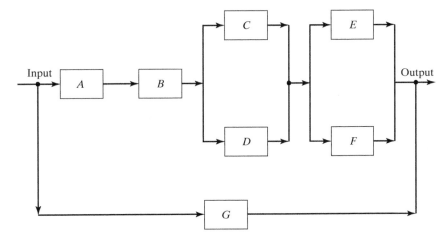

Figure 14   Reliability block diagram illustrating redundancy at system and subsystem levels.

5. Subsystem A is operating while B and C are failed.
6. Subsystem B is operating while A and C are failed.
7. Subsystem C is operating while A and B are failed.
8. Subsystems A, B, and C are all failed.

In the interest of simplicity, let R represent the reliability and $1 - R = Q$, the unreliability. Then,

$$R^3 \text{ represents } A, B, \text{ and } C \text{ operating}$$

$$3R^2Q \begin{cases} \text{represents } A \text{ and } B \text{ operating, } C \text{ failed} \\ \text{represents } A \text{ and } C \text{ operating, } B \text{ failed} \\ \text{represents } B \text{ and } C \text{ operating, } A \text{ failed} \end{cases}$$

$$3RQ^2 \begin{cases} \text{represents } A \text{ operating, } B \text{ and } C \text{ failed} \\ \text{represents } B \text{ operating, } A \text{ and } C \text{ failed} \\ \text{represents } C \text{ operating, } A \text{ and } B \text{ failed} \end{cases}$$

$$Q^3 \text{ represents } A, B, \text{ and } C \text{ failed}$$

Because the sum of $R^3, 3R^2Q, 3RQ^2$, and Q^3 represents all possible events,

$$R^3 + 3R^2Q + 3RQ^2 + Q^3 = 1 \tag{16}$$

Referring to Figure 9, it is assumed that the reliability of each block (i.e., blocks A, B, and C) is 0.95. Then, the reliability of the network is determined as

$$R = R^3 + 3R^2Q + 3RQ^2$$
$$= (0.95)^3 + 3(0.95)^2(0.05) + 3(0.95)(0.05)^2 = 0.999875$$

Note that this value agrees with the results of Section 2.3 and Equation 11.

Referring to Figure 14, suppose that it is desired to calculate the reliability for the network as illustrated. The approach is to first calculate the reliability for the redundant subsystems C–F; apply the product rule for A, B, and the resulting networks in this path (i.e., C–F); and then determine the combined reliability of the two overall redundant paths (to include subsystem G). It is assumed that the reliability of each of the individual subsystems is as follows:

$$\text{Subsystem } A = 0.97$$
$$\text{Subsystem } B = 0.98$$
$$\text{Subsystem } C = 0.92$$
$$\text{Subsystem } D = 0.92$$
$$\text{Subsystem } E = 0.93$$
$$\text{Subsystem } F = 0.90$$
$$\text{Subsystem } G = 0.99$$

The reliability of the redundant network including subsystems C and D is

$$R_{CD} = R_C + R_D - (R_C)(R_D) = 0.9936$$

The reliability of the redundant network including subsystems E and F is

$$R_{EF} = R_E + R_F - (R_E)(R_F) = 0.9930$$

The reliability of the path is

$$R_{ABCDEF} = (R_A)(R_B)(R_{CD})(R_{EF}) = 0.9379$$

The reliability of the combined network in Figure 14 is

$$R_{\text{system}} = R_{ABCDEF} + R_G - (R_{ABCDEF})(R_G) = 0.999379$$

Thus far, the discussion has addressed only one form of redundancy, operating redundancy, where all subsystems are fully energized throughout the system operating cycle. In actual practice, however, it is often preferable to use standby redundancy. Figure 15 provides a simple illustration of standby redundancy where subsystem A is operating full time, and subsystem B is standing by to take over operation if subsystem A fails. The standby unit (i.e., subsystem B) is not operative until a failure-sensing device senses a failure in subsystem A and switches operation to subsystem B, either automatically or through manual selection. Because of the fact that subsystem B is not operating unless a failure of subsystem A occurs, the system reliability for such a configuration is higher than that for a comparable system where both subsystems A and B are operating continuously.

When determining the reliability of standby systems, the Poisson distribution may be used because standby systems display the constant λt characteristic of this distribution (refer to Appendix: Probability Theory and Analysis, Section 2 for discussion of the Poisson process). In essence, the probability of no failure is represented by the first term, $e^{-\lambda t}$; the probability of one failure is $(\lambda t)(e^{-\lambda t})$; and so on.

Referring to Figure 15, where one operating subsystem and one standby subsystem are grouped together, one must consider the probability that no failure or one failure will occur (with one subsystem remaining in satisfactory condition). This combined probability is expressed as

$$P(\text{one standby}) = e^{-\lambda t} + (\lambda t)e^{-\lambda t} \tag{17}$$

where λt is the expected number of failures. If one operating subsystem and two standby subsystems are grouped together, then the combined probability is

$$P(\text{two standbys}) = e^{-\lambda t} + (\lambda t)e^{-\lambda t} + \frac{(\lambda t)^2 e^{-\lambda t}}{2!}$$

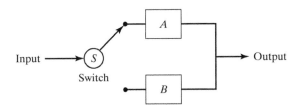

Figure 15 Standby redundant network.

An additional term in the Poisson distribution is added for each subsystem in standby.

As an example, suppose that one must determine the system reliability for a configuration consisting of one operating subsystem and one identical standby operating for a period of 200 hours. This configuration is illustrated in Figure 15, and it is assumed that the reliability of the switch is 100%. The failure rate (λ) for each subsystem is 0.002 failure per hour. Using Equation 17, the reliability is

$$R = e^{-\lambda t}(1 + \lambda t) \text{ and } \lambda t = (0.002)(200) = 0.4$$
$$= e^{-0.4}(1 + 0.4)$$
$$= (0.67032)(1.4) = 0.9384$$

To illustrate the difference between operating redundancy and standby redundancy, assume that both of the subsystems in Figure 15 are operating throughout the mission. The reliability of the configuration is determined as

$$R = 1 - (1 - 0.67032)^2 = 0.8913$$

As anticipated, the reliability of the standby system is higher (0.9384) than the reliability of the system using operating redundancy (0.8913).

3.6 Design Review and Evaluation

The design configuration and the characteristics of the various system components are evaluated in terms of compliance with the initially specified reliability requirements for the system. If the requirements appear to be fulfilled, the design is approved as is. If not, the appropriate changes are initiated for corrective action.

In conducting a reliability review, one may wish to develop a *design review checklist*, including questions of both a generic and a specific nature. An example of some key areas of interest may be noted through the questions below:

1. Have reliability quantitative and qualitative requirements for the system been adequately defined from the beginning?
2. Have these requirements been properly allocated to the various subsystems (and downward) as applicable? Is there a top-down/bottom-up "traceability" of these requirements?
3. Are the reliability requirements realistic? Are they compatible with other system requirements?
4. Has system design complexity been minimized; for example, number of components/parts?
5. Have system failure modes and effects been identified?
6. Are system, subsystem, unit, and component-part failure rates known?
7. Are the failure characteristics (i.e., physics of failure) known for each applicable component part?
8. Has the system or product wearout period been defined?
9. Have component parts with excessive failure rates been identified?
10. Have all critical-useful-life items been identified and eliminated where possible?

11. Have fail-safe characteristics been incorporated where applicable (i.e., protection against secondary/dependent failures resulting from primary failures)?
12. Has the utilization of adjustable components been minimized (if not eliminated)?
13. Have cooling provisions been incorporated in design "hot-spot" areas?
14. Have all hazardous conditions been eliminated?
15. Have all system reliability requirements been met?

The items covered are certainly not all-inclusive, but merely reflect a sample of possible questions that one might ask in a design review (it should be noted that the answer to these questions should be *yes*). Response to these questions is primarily dependent on the results of the ongoing reliability analyses and predictions accomplished as part of the day-to-day design participation process.

4 RELIABILITY ANALYSIS METHODS

Throughout design (and as part of the iterative process of system synthesis, analysis, and evaluation), there are a select number of reliability tools that can be effectively used in support of the objectives described throughout this text. Of specific interest in this chapter are the *failure mode, effects, and criticality analysis (FMECA); fault-tree analysis (FTA); stress–strength analysis; reliability prediction; and reliability growth modeling.* These tools can be applied in any phase of the system life cycle, but must be tailored to the depth of design definition and the application at hand.

4.1 Failure Mode, Effects, and Criticality Analysis (FMECA)

The failure mode, effects, and criticality analysis (FMECA) is a design technique that can be applied to identify and investigate potential system (product or process) weaknesses. It includes the necessary steps for examining all ways in which a system failure can occur, the potential effects of failure on system performance and safety, and the seriousness of these effects. The FMECA can be used initially during the conceptual and preliminary design stages, and can be subsequently applied as system definition evolves through detail design and development. Although the analysis is best used to impact "before-the-fact" enhancements to system design, it can also be used as an "after-the-fact" tool to evaluate and improve existing systems on a continuing basis.

FMECA can be applied to either a functional entity or a physical entity. For instance, Figure 16 conveys a partial illustration of the major functions of a package handling plant, or a series of activities for the processing of packages for distribution. The question is, *What is likely to happen if Function A.3.4 were to fail, and what are the possible consequences as a result of such a failure?* Conversely, one might ask the same question when referring to a physical entity. Thus, the FMECA needs to address both the *product* and the *process*.

The approach used in conducting a FMECA is shown in Figure 17. A brief description of the steps follows:

1. *Define system (product or process) requirements.* Describe the system in question, the expected outcomes, and the relevant technical performance measures (TPMs). Figure 18 shows an example where the FMECA can be applied to the manufacturing

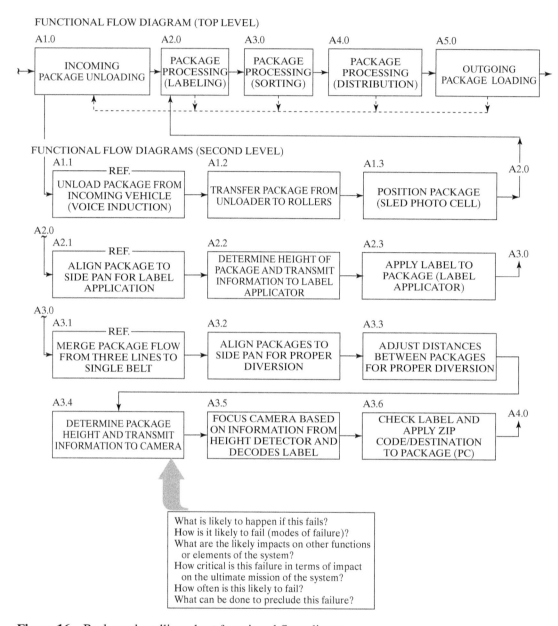

Figure 16 Package-handling plant functional flow diagram.

process for a gasket, as well as addressing the gasket within the automobile. One can first conduct a FMECA on the "stamping operation," determine the modes and effects of failure on the other operations within the process, and assess the impacts on the gasket. Then, a product-oriented FMECA can be accomplished assessing the effects of gasket failure on the engine block and on the automobile as an entity.

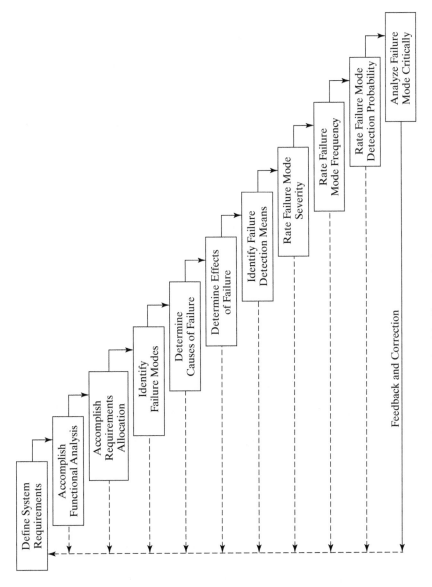

Figure 17 Failure mode, effects, and criticality analysis process.

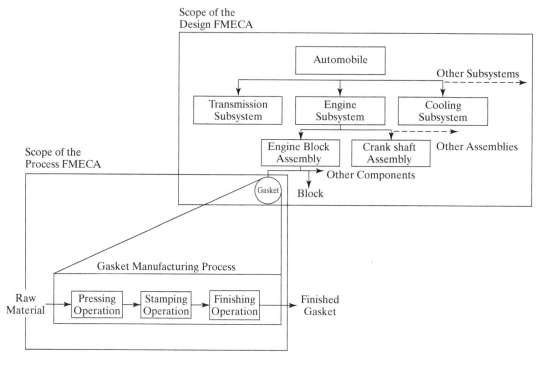

Figure 18   Sample design and process Failure Mode, Effects, and Criticality Analysis.

2. *Accomplish functional analysis.* This involves defining the system in functional terms. A system may be broken down into functional entities early in the life cycle and subsequently into a physical packaging scheme (refers to Figure 18).

3. *Accomplish requirements allocation.* This is a top-down breakout of system-level requirements.

4. *Identify failure modes.* A "failure mode" is the manner in which a system element fails to accomplish its function. For example, a switch may fail in an "open" position; a pipe may "rupture;" a given material may "shear" because of stress; a document may fail to be delivered on time; and so on.

5. *Determine causes of failure.* This involves analyzing the process or product to determine the actual cause(s) responsible for the occurrence of a failure. Typical causes might include abnormal equipment stresses during operation, aging and wearout, a software coding error, poor workmanship, defective materials, damage because of transportation and handling, or operator- and maintenance-induced faults. Although experience with similar systems, or the availability of good data from the field, is preferred, using an Ishikawa "cause-and-effect" diagram can prove to be highly effective in delineating potential failure causes. Figure 19 illustrates such a diagram.

6. *Determine the effects of failure.* Failures impact, often in multiple ways, the performance and effectiveness of not only the associated functional element but also the overall system. It is important to consider the effects of failure on other elements at

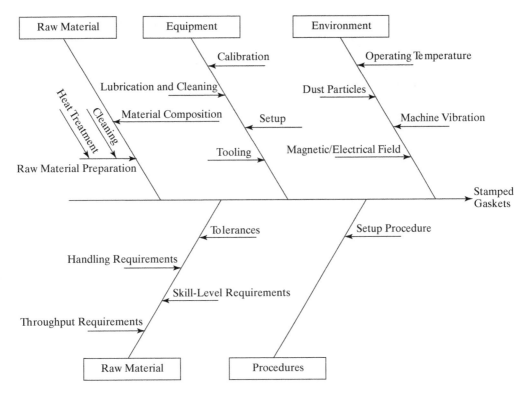

Figure 19 Ishikawa cause-and-effect "fishbone" diagram.

the same level in the system hierarchical structure, at the next higher level, and on the overall system. Referring to Figure 18, if the stamping operation fails, what are the effects on the finishing operation, on the gasket itself, on the automobile, and ultimately on the customer?

7. *Identify failure detection means.* For a process-oriented FMECA, it refers to the current process controls which may detect the occurrence of failures or defects. However, when the FMECA has a design focus, it refers to the existence of any design features, aids, gauges, readout devices, condition monitoring provisions, or evaluation procedures that will result in the detection of potential failures.

8. *Rate failure mode severity.* This refers to the *seriousness* of the effect or impact of a particular failure. If a failure occurs, will this cause the death of the operator and the system to be destroyed, or will it cause only a slight degradation in performance? For the purpose of illustration, the degree of severity may be expressed quantitatively (using a checklist approach) on a scale of 1–10 with *minor* effects being 1, *low* effects being 2–3, *moderate* effects being 4–6, *high* effects being 7–8, and *very high* effects being 9–10. The level of severity can be related to issues pertaining to safety or the degree of customer dissatisfaction.

9. *Rate failure mode frequency.* Given that a function or physical component within the system may fail in a variety of ways, this step addresses the frequency of occurrence of each individual failure mode. The sum of all modal failure frequencies for a system element must equal its failure rate. For the purposes of quantification, a scale of 1–10

may be used with *remote* (failure is unlikely) being 1, *low* (relatively few failures) being 2–3, *moderate* (occasional failures) being 4–6, *high* (repeated failures) being 7–8, and *very high* (failure is almost inevitable) being 9–10. These ratings can be based on the expected number of failures per unit of time, or equivalent.

10. *Rate failure mode detection probability.* This pertains to the probability that process controls, design features/aids, verification procedures, and so on, will detect potential failures in time to prevent a major system catastrophe. For a process application, this refers to the probability that a set of process controls currently in place will be in a position to detect and verify a failure before its effects are transferred to a subsequent process, or to the end consumer/customer. For the purposes of quantification, a scale of 1–10 may be used with *very high* being 1–2, *high* being 3–4, *moderate* being 5–6, *low* being 7–8, *very low* being 9, and *absolute certainty of nondetection* being 10.

11. *Analyze failure mode criticality.* The objective is to consolidate the preceding information in an effort to delineate the more critical aspects of system design. Criticality, in this context, is a function of *severity, frequency,* and *probability of detection,* and may be expressed in terms of a *risk priority number* (RPN). RPN can be determined from

$$RPN = (\text{severity rating})(\text{frequency rating})(\text{probability of detection rating})$$

(18)

The RPN reflects failure mode criticality. On inspection, one can see that a failure mode that has a high frequency of occurrence, has significant impact on system performance, and is difficult to detect is likely to have a very high RPN.

12. *Initiate recommendations for product/process improvement.* This pertains to the iterative process of identifying areas with high RPNs and evaluating the causes, and the subsequent initiation of recommendations for product/process improvement (refer to the feedback loop in Figure 17). A Pareto analysis, such as illustrated in Figure 20, can be accomplished to make visible the high-priority items that need to be addressed.

Figure 21 shows a partial example of a popular format that is used for recording the results of the FMECA. The information was derived from the functional flow analysis and expanded to include the results from the steps presented in Figure 17.

4.2 Fault-Tree Analysis (FTA)

The fault-tree analysis (FTA) is a deductive approach involving the graphical enumeration and analysis of different ways in which a particular failure can occur and the probability of its occurrence. It may be applied during the early stages of design, is oriented to specific failure modes, and is developed using a top-down fault-tree structure, such as illustrated in Figure 22. A separate fault tree is developed for every critical failure mode.

The first step is to identify a top-level event. One needs to be specific in defining this event, and it must be clearly observable and measurable. For example, it could be delineated as the "system catches fire" rather than the "system fails." Once the top-level event has been clearly defined, it is necessary to construct a *causal* hierarchy in the form of a fault tree. The causes for the top event are determined using a technique such as Ishikawa's *cause-and-effect* diagram (refer to Figure 19); each of these causes is next investigated for its causes; and so on. The next step is to determine the reliability of the top-level event by determining the probabilities of all relevant input events and the subsequent consolidation of these probabilities in accordance with the underlying logic of the tree. If the reliability of the top-level event turns out to be unacceptable, then corrective action is required.

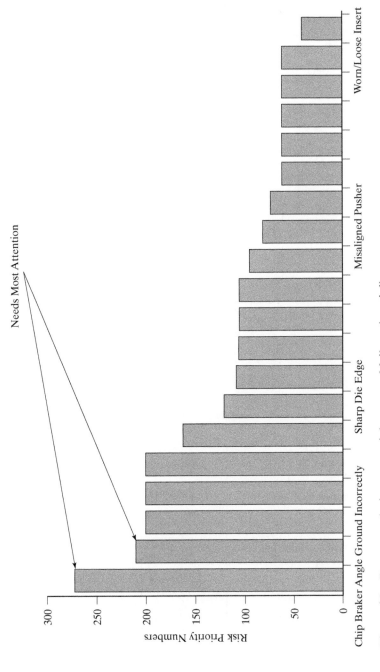

Figure 20 Pareto analysis—potential causes of failure (partial).

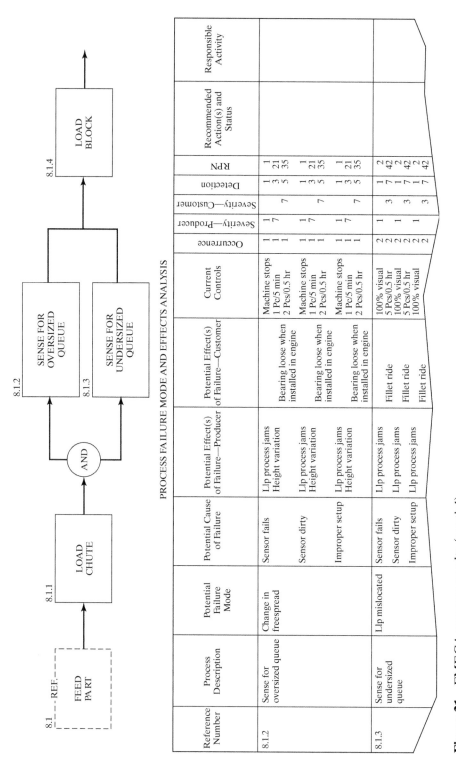

Figure 21 FMECA process results (partial).

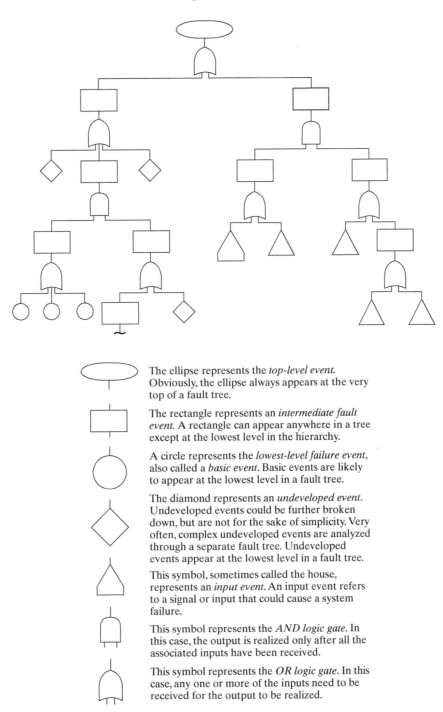

The ellipse represents the *top-level event*. Obviously, the ellipse always appears at the very top of a fault tree.

The rectangle represents an *intermediate fault event*. A rectangle can appear anywhere in a tree except at the lowest level in the hierarchy.

A circle represents the *lowest-level failure event*, also called a *basic event*. Basic events are likely to appear at the lowest level in a fault tree.

The diamond represents an *undeveloped event*. Undeveloped events could be further broken down, but are not for the sake of simplicity. Very often, complex undeveloped events are analyzed through a separate fault tree. Undeveloped events appear at the lowest level in a fault tree.

This symbol, sometimes called the house, represents an *input event*. An input event refers to a signal or input that could cause a system failure.

This symbol represents the *AND logic gate*. In this case, the output is realized only after all the associated inputs have been received.

This symbol represents the *OR logic gate*. In this case, any one or more of the inputs need to be received for the output to be realized.

Figure 22   Sample fault-tree analysis format and symbology.

The FTA can be effectively applied in the early phases of design when potential problems are suspected. It is narrow in focus and often easier to accomplish than the FMECA, requiring less input data to complete. For large and complex systems, which are highly software intensive and where there are many interfaces, the use of the FTA may be preferred in lieu of the FMECA. Conversely, accomplishing the FMECA provides more information for design evaluation, the establishing of "cause-and-effect" relationships across the spectrum of the system, and for pinpointing high-cost (high-risk) contributors.

4.3 Stress–Strength Analysis

Of major concern in the design for system reliability are the stress and strength characteristics of its components. Component parts are designed and manufactured to operate in a specified manner when utilized under nominal conditions. If additional stresses are imposed because of electrical loads, temperature, vibration, shock, humidity, and so on, then unexpected failures will occur and the reliability of the system will be less than anticipated. Also, if materials are used in a manner where nominal strength characteristics are exceeded, fatigue occurs and the materials may fail much earlier than expected. In any event, overstress conditions will result in reliability degradation and under stress conditions may be costly as a result of overdesign; that is, incorporating more than what is actually necessary to do the job.

A stress–strength analysis is often undertaken to evaluate the probability of identifying a situation(s) where the value of stress is much larger than (or the strength much less than) the nominal value. Such an analysis may be accomplished through the following steps:

1. For selected components, determine nominal stresses as a function of loads, temperature, vibration, shock, physical properties, and time.
2. Identify factors affecting maximum stress, such as stress concentration factors, static and dynamic load factors, stresses as a result of manufacturing and heat treating, environmental stress factors, and so on.
3. Identify critical stress components and calculate critical mean stresses (e.g., maximum tensile stress and shear stress).
4. Determine critical stress distributions for the specified useful life. Analyze the distribution parameters and identify component safety margins. Applicable distributions may include normal, Poisson, Gamma, Weibull, log-normal, or variations thereof.
5. For those components that are critical and where the design safety margins are inadequate, corrective action must be initiated. This may constitute component-part substitution or a complete redesign of the system element in question.

Reliability models are used to facilitate the stress-strength analysis process (refer to Figure 12). On the basis of such an analysis, component-part failure rates (λ) are adjusted as appropriate to reflect the effects of the stresses of the parts involved.

4.4 Reliability Prediction

As engineering data become available, reliability prediction is accomplished as a check on design in terms of the system requirement and the factors specified through allocation. The predicted values of R, MTBF, and/or MTTF are compared against the requirement, and areas of incompatibility are evaluated for possible design improvement.

Predictions are accomplished at different times in the system development process and will vary somewhat depending on the type of data available. Basic prediction techniques are summarized as follows:

1. *Prediction may be based on the analysis of similar equipment.* This technique should only be used when the lack of data prohibits the use of more sophisticated techniques. The prediction uses MTBF values for similar equipments of similar degrees of complexity performing similar functions and having similar reliability characteristics. The reliability of the new equipment is assumed to be equal to that of the equipment that is most comparable in terms of performance and complexity. Part quantity and type, stresses, and environmental factors are not considered. This technique is easy to perform but not very accurate.

2. *Prediction may be based on an estimate of active element groups (AEG).* The AEG is the smallest functional building block that controls or converts energy. An AEG includes one active element (relay, transistor, pump, machine, etc.) and a number of passive elements. By estimating the number of AEGs and using complexity factors, one can predict MTBF.

3. *Prediction may be accomplished from an equipment parts count.* There are a variety of methods used that differ somewhat due to data source, the number of part type categories, and assumed stress levels. Basically, a design parts list is used and parts are classified in certain designated categories. Failure rates are assigned and combined to provide a predicted MTBF at the system level. A representative approach is illustrated in Table 2.

4. *Prediction may be based on a stress analysis (discussed earlier).* When detailed equipment design is relatively firm, the reliability prediction becomes more sophisticated. Part types and quantities are determined, failure rates are applied, and stress ratios and environmental factors are considered. The interaction effects between components are addressed. This approach is peculiar and varies somewhat with each particular system/product design. Computer methods are often used to facilitate the prediction process.

TABLE 2   Reliability Prediction Data Summary

Component Part	λ/Part (%/1,000 hours)	Quantity of Parts	(λ/Part) (Quantity)
Part *A*	0.161	10	1.610
Part *B*	0.102	130	13.260
Part *C*	0.021	72	1.512
Part *D*	0.084	91	7.644
Part *E*	0.452	53	23.956
Part *F*	0.191	3	0.573
Part *G*	0.022	20	0.440

Failure rate (λ) = 48.995%/1,000 hours Σ = 48.995%

$$\text{MTBF} = \frac{1,000}{0.48995} = 2,041 \text{ hours}$$

Source: MIL-HDBK-217, Military handbook, *Reliability and Failure Rate Data for Electronic Equipment* (Washington, DC: Department of Defense).

The figures derived through reliability prediction constitute a direct input to maintainability prediction data, supportability analysis, and the determination of specific support requirements (test and support equipment, spare and repair parts, etc.). Reliability basically determines the frequency of corrective maintenance and the quantity of maintenance actions anticipated throughout the life cycle; thus, it is imperative that reliability prediction results be as accurate as possible.

4.5 Reliability Growth Analysis

Referring to Appendix: Figures, Figure 1, the results from the reliabilty prediction prepared for the conceptual design review indicate a system MTBF of approximately 430 hours, while the overall system requirement is 485 hours (as a minimum). The question is, *What changes need to be incorporated in system design in order to realize the necessary "growth" required to meet the specified requirement of 485 hours?*

In response, a formalized plan needs to be implemented such that reliability *growth* will occur in order to correct the deficiency as early in the design process as practicable, as compared to waiting until the system test and evaluation phase and then discovering that the design will not meet the specified requirements. From the results of the reliability prediction, the designer can identify those elements of the system where the failure rates are relatively high and, in turn, tend to cause the low MTBF. Application of the FMECA may aid in determining cause-and-effect relationships and in identifying specific aspects of the design that can be improved for reliability. Given the identified "weak" area(s), a recommended design change is then introduced following the procedure shown in Figure A.3). Subsequently, a follow-on reliability prediction can be accomplished to determine the the overall impact of the change on the system MTBF. This process can be repeated until such time that one is confident in meeting the minimum MTBF requirement of 485 hours.

Referring to the figure, a design change is implemented during the preliminary design phase (Mod. 1), with the resultant predicted MTBF of approximately 445 hours; a second design change (Mod. 2) is initiated and there is some growth (with a MTBF of over 450 hours); and subsequently two more design changes (Mod. 3 and Mod. 4) are implemented during the detail design and devlopment phase so that the predicted reliability MTBF requirement of 485 hours is realized.

While the specific reliability growth plan illustrated in Figure 23 reflects a rather costly approach involving four different design changes (i.e., ECPs and their implementation), it may be feasible to identify a number of different design areas concurrently that contribute to the low MTBF and then implement the necessary changes at a single point in time. In any event, it is important that one acquire some degree of assurance that the system reliability requirements will be met, and that any suspected areas of deficiency should be corrected as early as possible in the overall design process. This is particularly relevant at this time as many of the systems being delivered and currently in operational use are not meeting the reliability requirements as initially specified.

5 RELIABILITY TEST AND EVALUATION

Reliability testing is conducted as part of the system test and evaluation effort. Specifically, reliability testing is accomplished as part of and included within the scope of Types 3 and 4 testing, described in Appendix: Figures, Figure 2.

As is required for any category of testing, there is a planning phase, a test preparation phase, the actual test and evaluation activity itself, the data collection and analysis phase,

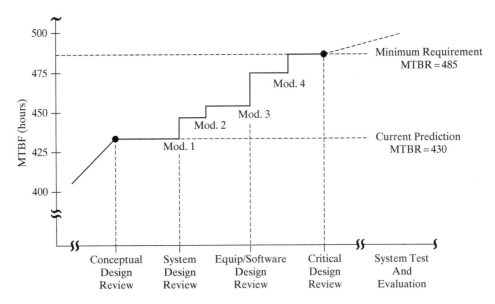

Figure 23 Reliability growth plan.

and test reporting. Only those procedures applicable to reliability test methods are covered in this chapter.

The ultimate objective of reliability testing is to determine whether the system (or product) under test meets the specified MTBF requirements. To accomplish this, the system is operated in a prescribed manner for a designated period and failures are recorded and evaluated as the test progresses. System acceptance is based on demonstrating a minimum acceptable life.

There are a number of different test methods and statistical procedures in practice today that are designed to measure system reliability. Many of these assume that the exponential distribution is applicable. The criteria for system acceptance (or rejection) are based on statistical assumptions involving test sample size, consumer–producer decision risk factors, test data confidence limits, and so on. These assumptions often vary from one application to the next, depending on the type of system, the mission that the system is expected to perform, and whether new design techniques are used in system development reflecting a potential high-risk area. In view of this, it is impossible to cover all facets of reliability testing within the confines of this chapter. However, in order to provide some understanding of the approaches used, it is intended to briefly cover reliability sequential qualification testing, reliability acceptance testing, and life testing.

5.1 Reliability Sequential Qualification Testing

Reliability qualification testing is conducted to provide an evaluation of system development progress, as well as the assurance that specified requirements have been met prior to proceeding to the next phase (i.e., the production or construction phase of life cycle). Initially, a reliability MTBF is established for the system, followed by allocation and the development of design criteria. System design is accomplished and reliability analyses and predictions are accomplished to evaluate (on an analytical basis) the design configuration relative to compliance with system requirements. If the predictions indicate compliance, the design progresses to the construction of a prototype, or preproduction model, of the system and qualification testing commences.

When a reliability sequential test is conducted (oriented primarily to electronics, electrical, electro-mechanical, and related systems/product), there are three possible decisions: (1) accept the system, (2) reject the system, or (3) continue to test. Figure 24 illustrates a typical sequential test plan. The system under test is operated in a manner reflecting actual customer utilization in a realistic environment. The objective is to simulate (to the extent possible) a mission profile, or at least to subject the system under test to conditions similar to those that will be present when the system is in operational use by the customer. The accomplishment of this objective generally involves an environmental testing facility and the operation of the system or equipment through different duty cycles and environmental conditions. Figure 25 illustrates a sample duty cycle. The system is operated through a series of these duty cycles for a designated period of time.

Referring to the sequential test plan in Figure 24, system operating hours are accumulated along the abscissa and failures are plotted against the ordinate. If enough operating time is acquired without too many failures, a decision is made to accept the system and testing is discontinued. Conversely, if there are a significant number of failures occurring early in test operations, then a reject decision may prevail and the system is unacceptable as is. A marginal condition results in a decision to continue testing until a designated point in time. In essence, the sequential test plan allows for an early decision. Highly reliable systems will be accepted with a minimum amount of required testing. If the system is unreliable, this will also be readily evident at an early point in time. In this respect, sequential testing is extremely beneficial. Conversely, if the system's inherent reliability is marginal, the amount of test time involved can be rather extensive and costly. An actual example of sequential testing is presented later.

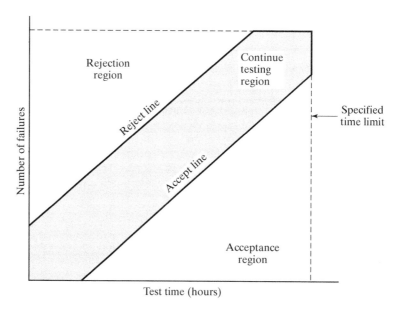

Figure 24   Sequential test plan.

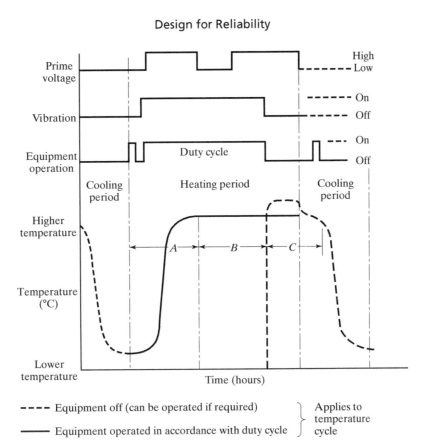

A = Time for facility to reach stabilization at higher temperature.

B = Time of system equipment operation at higher temperature.

C = Optional hot soak and hot startup checkout.

Figure 25 Sample environmental test cycle. *Source*: MIL-STD-781, *Reliability Design Qualification and Production Acceptance Tests: Exponential Distribution* (Washington, DC: Department of Defense).

Sequential test plans are highly influenced by the risks that both the producer and consumer are willing to accept in connection with decisions made as a result of testing. These risks are defined as follows:

1. *Producer's risk (α).* The probability of rejecting a system when the measured MTBF is equal to or better than the specified MTBF. In other words, this refers to the probability of rejecting a system when it really should be accepted, which constitutes a risk to the system manufacturer or producer (also known as a Type I error).
2. *User's or consumer's risk (β).* The probability of accepting a system when the measured MTBF is less than the specified MTBF. This refers to the probability of accepting a system that actually should be rejected, which constitutes a risk to the user (also known as a Type II error).

The probability of making an incorrect decision on the basis of test results must be addressed in a manner similar to hypothesis testing. An assumption is made and a test is accomplished to support (or disprove) that assumption. A null hypothesis (H_0) is established, that is, a statement or conjecture about a parameter such as "the true MTBF is equal to 100." The alternative hypothesis (H_1) is "the MTBF is not equal to 100." When testing an item (representing a sample of a total population), the question arises as to whether to accept H_0. The desired result is to accept when the null hypothesis is true and reject when false, or to minimize the chances of making an incorrect decision. The relationship of risks in sample testing is illustrated in Table 3. In selecting a sequential test plan, it is necessary to first identify two values of MTBF:

1. Specified MTBF representing the system requirement (assume that θ_0 represents this value).
2. Minimum MTBF that is considered to be acceptable based on the results of testing (assume that θ_1 represents this value).

Given θ_0 and θ_1, one must next decide on the values of producer's risk (α) and consumer's risk (β). Most test plans in use today accept risk values ranging between 5% and 25%. For example, a test where $\alpha = 0.10$ means that in 10 of 100 instances the test plan will reject items that should have been accepted. Risk values are generally negotiated in the test planning phase.

Referring to Figure 24, the design of the sequential test plan is based on the values of θ_0, θ_1, α and β. For instance, the *accept line* slope is based on the following expression:

$$t_1 = \frac{\ln(\theta_0/\theta_1)}{1/\theta_1 - 1/\theta_0}(r) - \frac{\ln[\beta/(1 - \alpha)]}{1/\theta_1 - 1/\theta_0} \tag{19}$$

where r is the number of failures (the accept line is plotted for assumed values of r). The ratio of θ_0 to θ_1 is known as the *discrimination ratio*, a standard test parameter. The *reject line* slope is based on Equation 20:

$$t_2 = \frac{\ln(\theta_0/\theta_1)}{1/\theta_1 - 1/\theta_0}(r) - \frac{\ln[(1 - \beta)/\alpha]}{1/\theta_1 - 1/\theta_0} \tag{20}$$

Determination of the specified time limit in Figure 24 is generally based on a multiple of θ_1.

TABLE 3 Risks in Sample Testing

True State of Affairs	Accept H_0 and Reject H_1 (i.e., MTBF = 100)	Reject H_0 and Accept H_1 (i.e., MTBF $\neq$ 100)
H_0 is true (i.e., MTBF = 100)	High probability $1 - \alpha$ (i.e., 0.90)	Low probability Error, α (i.e., 0.10)
H_0 is false and H_1 is true (i.e., MTBF $\neq$ 100)	Low probability Error, β (i.e., 0.10)	High probability $1 - \beta$ (i.e., 0.90)

Design for Reliability

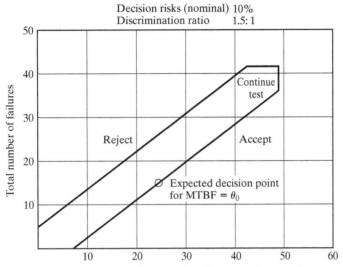

Decision risks (nominal) 10%
Discrimination ratio   1.5: 1

Total test time (in multiples of lower test MTBF, θ_1)

Number of failures	Total test time* Reject (equal or less)	Total test time* Accept (equal or more)	Number of failures	Total test time* Reject (equal or less)	Total test time* Accept (equal or more)
0	N/A	6.60	21	18.92	32.15
1	N/A	7.82	22	20.13	33.36
2	N/A	9.03	23	21.35	34.58
3	N/A	10.25	24	22.56	35.79
4	N/A	11.46	25	23.78	37.01
5	N/A	12.68	26	24.99	38.22
6	0.68	13.91	27	26.21	39.44
7	1.89	15.12	28	27.44	40.67
8	3.11	16.34	29	28.65	41.88
9	4.32	17.55	30	29.85	43.10
10	5.54	18.77	31	31.08	44.31
11	6.75	19.98	32	32.30	45.53
12	7.97	21.20	33	33.51	46.74
13	9.18	22.41	34	34.73	47.96
14	10.40	23.63	35	35.94	49.17
15	11.61	24.84	36	37.16	49.50
16	12.83	26.06	37	38.37	49.50
17	14.06	27.29	38	39.59	49.50
18	15.27	28.50	39	40.82	49.50
19	16.49	29.72	40	42.03	49.50
20	17.70	30.93	41	49.50	N/A

*Total test time is total unit hours of equipment on time and is expressed in multiples of the lower test MTBF.

Figure 26 Accept–reject criteria for a sample test plan. *Source*: MIL-STD-781, *Reliability Design Qualification and Production Acceptance Tests: Exponential Distribution* (Washington, DC: Department of Defense).

As indicated, there are a number of different sequential test plans that have been used for reliability testing. One example is illustrated in Figure 26. The established α and β decision risks are 10% (each), the discrimination ratio agreed on is 1.5:1, and the accept–reject criteria are as specified.

449

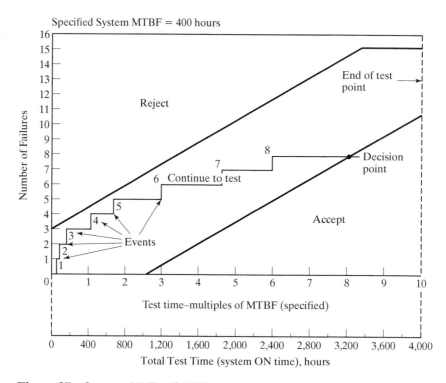

Figure 27 System *XYZ* reliability test plan.

Another approach is illustrated in Figure 27, representing actual experience in testing a specific system (to be designated as System *XYZ*). Referring to the figure, the specified MTBF for the system is 400 hours, and the maximum designated testing time for the sequential test plan used is 4,000 hours, or 10 times the specified MTBF. The test approach involves the selection of a designated quantity of equipments, operating the equipment under certain performance conditions over an extended period of time, and monitoring the equipment for failure. Failures are noted as events, corrected through the appropriate maintenance actions, and the applicable equipment is returned to full operational status for continued testing. An analysis of each event should determine the cause of failure, and trends may be established if more than one failure is traceable to the same cause. This may be referred to as a pattern failure, and in such cases, a change should be initiated to eliminate the occurrence of future failures of the same type.

Referring to Figure 27, the system is accepted after 3,200 hours of testing, and eight events are recorded. Thus, the specified MTBF requirement is fulfilled. In this instance, the test sample is six pieces of equipment. If a single model is designated for test with a minimum of 4,000 test hours required, the length of the test program (assuming continuous testing) will obviously extend beyond 6 months. This may not be feasible in terms of the overall program schedule and the delivery of qualified equipment. Conversely, the use of two or more models will result in a shorter test program. Sometimes the implementation of accelerated test conditions will result in an additional number of failures. These effects should be evaluated, and an optimum balance should be established between a smaller sample size and longer test times and a larger sample size and shorter test times.

5.2 Reliability Acceptance Testing

As indicated, reliability testing may be accomplished as part of qualification testing prior to commencing with full-scale production (which has been discussed) and during full-scale production on a sampling basis. To determine the effects of the production process on system reliability, it may be feasible to select a sample number of equipments/units or products from each production lot and test them in the same manner as described above. The sample may be based on a percentage of the total equipments spread over the entire production period, or a set number of equipment(s) selected during a given calendar time period (e.g., the two items of equipment per month throughout the production phase).

In any event, the selected equipment is tested and an assessed MTBF is derived from the test data. This value is compared against the specified MTBF and the measured value determined from earlier qualification testing. Growth MTBF trends (or negative trends) may be determined by plotting the resultant values as the testing activity progresses.

5.3 Reliability Life Testing

There are two basic forms of life testing in use: (1) life tests based on a fixed-test time, and (2) life tests based on the occurrence of a predetermined number of failures. The first approach to life testing (based on time) assumes that a computed fixed test time will be planned and a specified quantity of failures will be predetermined. System acceptance occurs if the actual number of failures during test is equal to (or less than) the predetermined quantity of failures at the end of the scheduled test time.

The second approach to life testing (based on failures) assumes that a test plan is developed specifying a predetermined quantity of failures and a computed test time dependent on an expected system failure rate. Testing continues until the specified quantity of failures occurs. System acceptance occurs if the test time is equal to (or is greater than) the computed time at the point of last failure. Reliability life tests are often accomplished at the detailed component level when a given item is being evaluated for possible incorporation into the design.

5.4 Operational Reliability Assessment

To this point, the measure of system reliability has been based on a combination of analytical studies, predictions, and the demonstration of prototype models of various system elements (e.g., equipment). The opportunity to observe the system being utilized in a realistic user environment has not been possible. Given the successful completion of all required reliability qualification and acceptance tests, the next step is to accomplish a true assessment of system reliability operating in the user's environment. Referring to Appendix: Figure, Figure 2, this can be accomplished by conducting Types 3 and 4 testing as part of the overall system test, evaluation, and validation.

As system tests are conducted, the data collection, analysis, and reporting capability must provide the right type and quantity of information, enabling the proper validation of reliability requirements. A failure analysis report should be prepared to cover each system failure, and the information provided must include the time of failure (and system operating characteristics when the failure occurred), symptom of failure, failure mode(s), effects of the failure on system operation, effects of the failure on other elements of the system, effects of failure on other systems within the same system-of-systems (SOS) configuration, and the actual cause(s) of failure. Enough data must be collected to positively determine *what*, *when*, *where*, and *why* it happened. This, of course, should lead to either the validation of a specified system MTBF (or equivalent) requirement, or a deficiency requiring corrective action and a possible design modification.

6 SUMMARY AND EXTENSIONS

This chapter provides an overview of the subject of reliability and stresses its importance in the systems engineering design and development process. This chapter commences with key definitions and continues through the initial establishment of reliability requirements for the system, development of the primary measures of reliability, description of selected reliability analysis models and techniques, application of reliability activities throughout the system life cycle, and validation of reliability in design through a formalized test and evaluation process. The intent is to emphasize the importance of reliability within the total spectrum of design-related requirements and activities needed to meet the objectives of systems engineering.

With regard to current practices from a historical perspective, emphasis has been placed primarily on determining the reliability requirements for the various individual elements of a system, rather than addressing the requirements for the system as entity whole. For example, there has been a great deal of effort and progress made through the years in defining the reliability requirements for equipment, more recently in deriving reliability metrics and requirements for software, and so on. A challenge for the future is to advance methodology for integrating these requirements at the system level. This includes the challenge of integrating reliability requirements pertaining to the human being, facilities, data/information processes, and the like. Progress beyond that reflected in Table 1 (composite failure rates, including inherent, dependent, and operator-induced) is needed, where the reliabilities of all of the elements of a system are properly specified, integrated, considered throughout the design process, and reflected in the ultimate system configuration. From a systems engineering perspective, the reliability of all of the elements of a system should be considered as an integrated entity.

Finally, there is always the challenge and need for acquiring good reliability data (e.g., component failure rates, material degradation factors, environmental impact factors, and the like) of a historical nature based on system operational experience. The complexity of systems, with the ongoing inclusion of new technologies, is increasing at a rapid pace, and acquiring the proper failure rates as a basis for reliability prediction constitutes a major problem. This is particularly true with respect a system operating within a system-of-systems (SOS) structure. The techniques discussed throughout this chapter appear to be valid, but the output results highly depend on the availability of good input data. While this is a continuing problem, the challenge relative to future improvement remains.

Up-to-date information pertaining to various facets of reliability can be acquired by visiting the following websites:

1. Annual Reliability and Maintainability Symposium (RAMS)—*http://www.rams.org.*
2. Reliability Information Analysis Center (RIAC)—*http://theRIAC.org.*
3. Society of Reliability Engineers (SRE)—*http://www.sre.org.*

QUESTIONS AND PROBLEMS

1. Define *reliability*. What are its major characteristics?
2. Why is reliability important in system design? When in the life-cycle process should it be considered? To what extent should reliability be emphasized in system design, and what are some of the factors that govern this?

3. What are the quantitative measures of reliability (discuss measures for hardware, software, personnel, facilities, and data)?

4. How would you define the overall *failure rate* for a system? What should be included (or excluded)? Why?

5. Refer to Figure 4. Reliability predictions/estimations are usually based on what portion of the "bathtub" curve? What is likely to happen if the system is delivered to the customer during the early "infant mortality" portion of the curve? What would you do to extend the flat portion of the curve?

6. Refer to Figure 5. Identify some of the possible "causes" for the jagged portion of the curve.

7. A system consists of four subassemblies connected in series. The individual subassembly reliabilities are as follows:

$$\text{Subassembly } A = 0.98$$

$$\text{Subassembly } B = 0.85$$

$$\text{Subassembly } C = 0.90$$

$$\text{Subassembly } D = 0.88$$

Determine the overall system reliability.

8. A system consists of three subsystems in parallel (assume operating redundancy). The individual subsystem reliabilities are as follows:

$$\text{Subsystem } A = 0.98$$

$$\text{Subsystem } B = 0.85$$

$$\text{Subsystem } C = 0.88$$

Determine the overall system reliability.

9. Refer to Figure 14. Determine the overall network reliability if the individual reliabilities of the subsystems are as follows:

$$\text{Subsystem } A = 0.95$$

$$\text{Subsystem } B = 0.97$$

$$\text{Subsystem } C = 0.92$$

$$\text{Subsystem } D = 0.94$$

$$\text{Subsystem } E = 0.90$$

$$\text{Subsystem } F = 0.88$$

$$\text{Subsystem } G = 0.98$$

10. The failure rate (λ) of a device is 22 failures per million hours. Two standby units are added to the system. Determine the MTBF of the system.

11. Calculate the reliability of a system consisting of one operating unit and one identical standby unit operating for a period of 200 hours. The failure rate (λ) of each unit is 0.003 failure per hour and the failure sensing switch reliability is 1.0.

12. A system consists of five subsystems with the following MTBFs:

$$\text{Subsystem } A: \text{ MTBF} = 10{,}540 \text{ hours}$$

$$\text{Subsystem } B: \text{ MTBF} = 16{,}220 \text{ hours}$$

$$\text{Subsystem } C: \text{ MTBF} = 9{,}500 \text{ hours}$$

$$\text{Subsystem } D: \text{ MTBF} = 12{,}100 \text{ hours}$$

$$\text{Subsystem } E: \text{ MTBF} = 3{,}600 \text{ hours}$$

The five subsystems are connected in series. Determine the probability of survival for an operating period of 1,000 hours.

13. Ten components were tested for 500 hours, each within prescribed operating conditions. Component 1 failed after 30 hours; component 2 failed after 85 hours; component 3 failed after 220 hours; and component 4 failed after 435 hours. Determine the overall composite failure rate (λ) for the system.

14. Suppose that you have the following design data on an equipment item and that, based on these data, you may wish to accomplish a reliability prediction. What is the predicted MTBF?

Component	Quantity of Parts Used	Failure Rate (% 1,000 hours)
A	16	0.135
B	75	0.121
C	32	0.225
D	44	0.323
E	60	0.120
F	15	0.118
G	28	0.092

15. Assume that there is a requirement for a new system with a specified performance capability and a reliability of 70%. In response to an "invitation to bid," three supplier configurations have been proposed and are reflected in Figure 28. The component reliability factors are noted in the following table:

Component	Reliability	Component	Reliability	Component	Reliability
A	0.84	G	0.87	M	0.83
B	0.86	H	0.88	N	0.85
C	0.89	I	0.89	O	0.84
D	0.86	J	0.86	P	0.89
E	0.87	K	0.85	Q	0.89
F	0.82	L	0.86		

The overall cost associated with each of the supplier proposals is $57,000 for Configuration A, $39,000 for Configuration B, and $42,000 for Configuration C.

(a) Determine the system reliability for each of the three configurations.

(b) In evaluating the three alternatives, employing cost-effectiveness (CE) criteria, which configuration would you select as being preferred?

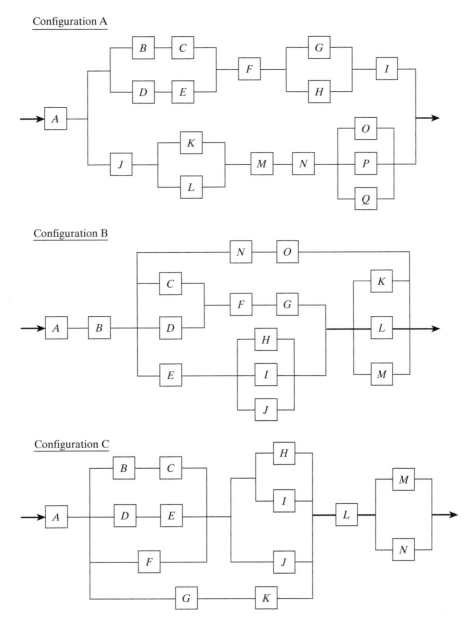

Figure 28 Three alternative configurations.

16. When designing a system for *reliability*, what characteristics (attributes) should be considered for incorporation into the ultimate configuration in order to attain the desired level of reliability? Identify at least five and describe how they would impact system reliability.

17. What is a reliability *model*? How would you develop a reliability model (i.e., what are the steps that you would take to do so)? Why is it needed? Describe some of the common applications of a model.

18. Describe the process that you would follow in identifying and developing the appropriate reliability requirements at the *system level*, and down to the *subsystem level, unit level,* and below as required.

19. What is meant by *redundancy*? Describe the various types of redundancy. How could the incorporation of redundancy in design benefit reliability? What are some of the possible negative impacts that could result by incorporating redundancy?

20. Describe each of the following, its purpose, application, and the information acquired:

(a) Failure mode, effects, and criticality analysis (FMECA)

(b) Fault-tree analysis (FTA)

(c) Stress–strength analysis

(d) Reliability prediction

(e) Reliability growth analysis

21. Select a system of your choice and describe the system in functional terms (construct a functional block diagram). Select an element of that system (i.e., product or process) and conduct a FMECA.

22. Assume that there is a 500-hour MTBF requirement for a new system being developed. At the last design review, the predicted MTBF was 400 hours. What steps would you take to improve the reliability and ensure that the requirement will be met? What techniques/tools would you employ to facilitate the objective?

23. At what stage in the system life cycle are the requirements for reliability test and evaluation determined? What steps would you take to ensure that these requirements will be met (i.e., validation steps)?

24. Describe some of the advantages of reliability *sequential* testing. Identify some of the disadvantages. What is meant by *life* testing? *Accelerated* testing?

25. Assume that you are planning to produce a multiple quantity of "like" products. What steps would you take to ensure that all of the products delivered will reflect the same reliability characteristics?

26. Define what is meant by *producer's* risk and *consumer's* risk.

27. A system is made up of five components. There is no redundancy in the system; that is, all system components are in series. The components have been experiencing an MTBF as shown in Column 2 of the following table. Design analysis has shown that all components can be redesigned to improve the individual MTBFs by a factor of 5. The estimated development cost to redesign each component is shown in Column 3. Total development cost is the sum of all redesign costs for those components (1–5) that the developer chooses to improve, but it is not known which of these components should be redesigned. A cost-effectiveness (CE) criterion has been defined to establish the priority for component redesign. Cost-effectiveness is defined as the improvement (reduction) in failure rate divided by the cost to implement the design change for each component. The component with the highest value of CE will have the highest priority for redesign. Other components will be listed in order of redesign priority, in accordance with its value of CE. The time basis for all reliabilities in this problem is 1,000 hours. Cumulative cost is the total cost to implement the design change for components 1–n.

(a) Calculate the MTBF of the entire system before redesign.

(b) Calculate the failure rate of each component before design.

(c) Calculate the CE of each redesign for each component.

(d) Prioritize these reliability improvements.

Component	MTBF per Component (hours)	Cost to Reduce Failures ($M)
1	9,100	1
2	12,500	2
3	5,000	3
4	14,300	4
5	25,000	5

(e) Calculate the reliability after each improvement is made (in order of cost-effectiveness), including the aggregate reliability if all components are redesigned.

(f) Plot total system reliability as a function of total development cost.

(g) Calculate the cumulative reliability over all the improvements.

(h) Determine how many components should be redesigned to achieve a reliability of at least 0.8.

(i) Determine how many components should be redesigned if the customer does not want to budget more than $10M to the effort.

Appendix: Figures

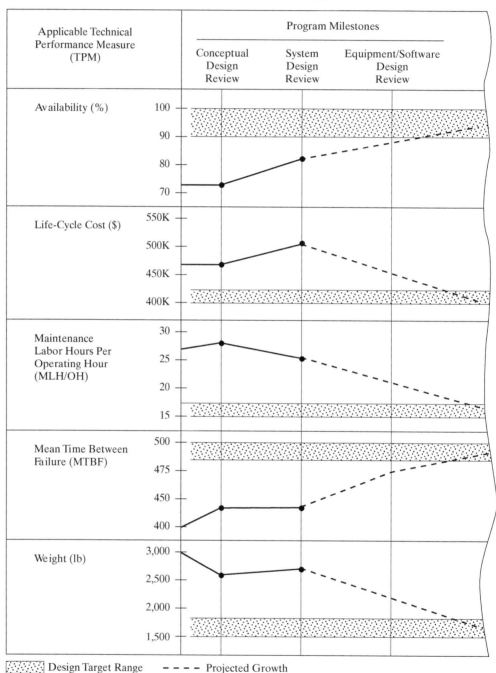

Design Target Range   – – – – Projected Growth

Figure 1 Parameter measurement and evaluation at design review ("tracking").

Design for Reliability

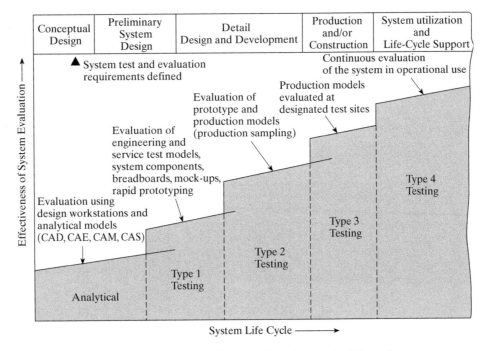

Figure 2 Stages of system test and evaluation during the life cycle.

Appendix:

Probability Theory and Analysis

Some models will give satisfactory results if variation is not incorporated. But, these models usually apply to physical phenomena where certainty is generally observed. Models formulated to analyze and evaluate human-made systems must incorporate probabilistic elements to be useful in the system engineering process. Accordingly, this appendix presents probability concepts and theory, probability distribution models, and an introduction to Monte Carlo analysis.

1 PROBABILITY CONCEPTS AND THEORY

If one tosses a coin, the outcome will not be known with certainty until either a head or a tail is observed. Prior to the toss, one can assign a probability to the outcome from knowledge of the physical characteristics of the coin. One may know that the diameter of an acorn ranges between 0.80 and 3.20 cm, but the diameter of a specific acorn to be selected from an oak tree will not be known until the acorn is measured. Experiments such as tossing a coin and selecting an acorn provide outcomes called *random events*. Most events in the decision environment are random and probability theory provides a means for quantifying these events.

1.1 The Universe and the Sample

The terms *universe* and *population* are used interchangeably. A *universe* consists of all possible objects, stages, and events within an arbitrarily defined boundary. A universe may be finite or it may be infinite. If it is finite, the universe may be very large or very small. If the universe is large, it may sometimes be assumed to be infinite for computational purposes. A universe need not always be large; it may be defined as a dozen events or as only one object. The relative usefulness of the universe as an entity will be paramount in its definition.

A *sample* is a part or portion of a universe. It may range in size from 1 to 0 less than the size of the universe. A sample is drawn from the population, and observations are made. This is done either because the universe is infinite in size or scope or because the population is large and/or inaccessible as a whole. The sample is used because it is smaller, more accessible, and more economical, and because it suggests certain characteristics of the population.

It is usually assumed that the sample is typical of the population in regard to the characteristics under consideration. The sample is then assessed, and inferences are made in regard to the population as a whole. To the extent that the sample is representative of the population, these inferences may be correct. The problem of selecting a representative sample from a population is an area in statistics to which an entire chapter might be devoted.

Subsequent discussion assumes that the sample is a *random sample*, that is, one in which each object or state or event that constitutes the population has an equally likely chance or probability of being selected and represented in the sample. It is rather simple to state this definition; it may be much more difficult to implement it in practice.

1.2 The Probability of an Event

A measure of the relative certainty of an event, before the occurrence of the event, is its probability. The usual representation of a probability is a number $P(A)$ assigned to the outcome A. This number has the following property: $0 \le P(A) \le 1$, with $P(A) = 0$ if the event is certain not to occur and $P(A) = 1$ if the event is certain to occur.

Because probability is only a measure of the certainty (or uncertainty) associated with an event, its definition is rather tenuous. The concept of relative frequency is sometimes employed to establish the number $P(A)$. Sometimes probabilities are established a priori. Other times they are simply a subjective estimate. Consider the example of tossing a fair coin. In a lengthy series of tosses, the coin may have come up heads as often as tails. Then the limiting value of the relative frequency of a head will be 0.5 and will be stated as $P(H) = 0.5$.

Two definitions pertaining to events are needed in the development of probability theorems:

1. Events A and B are said to be *mutually exclusive* if both cannot occur at the same time.
2. Event A is said to be *independent* of event B if the probability of the occurrence of A is the same regardless of whether or not B has occurred.

The probability of the occurrence of either one or another of a series of mutually exclusive events is the sum of probabilities of their separate occurrences. If a fair coin is tossed and success is defined as the occurrence of either a head or a tail, then the probability of a head or a tail is

$$P(H + T) = P(H) + P(T)$$
$$= 0.5 + 0.5 = 1.0 \tag{1}$$

The key to use of the addition theorem is the proper definition of mutually exclusive events. Such events must be distinct from one another. If one event occurs, it must be impossible for the second to occur at the same time. For example, assume that the probability of having a flat tire during a given period on each of four tires on an automobile is 0.3. Then the probability of having a flat tire on any of the four tires during this time period is not given by the addition of these four probabilities. If $P(T_1) = P(T_2) = P(T_3) = P(T_4) = 0.3$ are the respective probabilities of failure for each of the four tires, then

$$P(T_1 + T_2 + T_3 + T_4) \ne P(T_1) + P(T_2) + P(T_3) + P(T_4)$$
$$\ne 0.3 + 0.3 + 0.3 + 0.3 = 1.2$$

This cannot be true because the failure of tires is not mutually exclusive. During the time period established, two or more tires may fail, whereas in the example of coin tossing, it is not possible to obtain a head and a tail on the same toss.

1.3 The Multiplication Theorem

The probability of occurrence of independent events is the product of the probabilities of their separate events. Implicit in this theorem is the successful occurrence of two events simultaneously or in succession. Thus, the probability of the occurrence of two heads in two tosses of a coin is

$$P(H \cdot H) = P(H)P(H)$$
$$D = (0.5)(0.5) = 0.25 \tag{2}$$

The tire-failure problem can now be resolved by considering the probabilities of each tire not failing. The probability of each tire not failing is given by $P(\overline{T}_i) = 0.7$. The probability of no tire failing is then given by

$$P[(\overline{T}_1)(\overline{T}_2)(\overline{T}_3)(\overline{T}_4)] = P(\overline{T}_1) \, P(\overline{T}_2) \, P(\overline{T}_3) \, P(\overline{T}_4)$$
$$= (0.7)(0.7)(0.7)(0.7) = 0.2401$$

Thus, the probability of a tire failing, or of one or more tires failing, is

$$P(T_1 + T_2 + T_3 + T_4) = 1 - 0.2401 = 0.7599$$

This approach is valid, since the probability of one tire not failing is independent of the success or failure of the other three tires.

1.4 The Conditional Theorem

The probability of the occurrence of two dependent events is the probability of the first event times the probability of the second event, given that the first has occurred. This may be expressed as

$$P(W_1 \cdot W_2) = P(W_1)P(W_2|W_1) \tag{3}$$

This theorem is similar to the multiplication theorem, except that consideration is given to the lack of independence between events.

As an example, consider the probability of selecting two successive white balls from an urn containing three white and two black balls. This problem reduces to a calculation of the product of the probability of selecting a white ball times the probability of selecting a second white ball, given that the first attempt has been successful, or

$$P(W_1 \cdot W_2) = \left(\frac{3}{5}\right)\left(\frac{2}{4}\right) = \frac{3}{10}$$

The conditional theorem makes allowances for a change in probabilities between two successive events. This theorem will be helpful in constructing finite discrete probability distributions.

1.5 The Central Limit Theorem

Although many real-world variables are normally distributed, this assumption cannot be universally applied. However, the distribution of the means of samples or the sums of random variables approximates the normal distribution provided certain assumptions hold. The Central Limit Theorem states: If x has a distribution for which the moment-generating function exists, then the variable $\bar{x}$ has a distribution that approaches normality as the size of the sample tends toward infinity. The sample size required for any desired degree of convergence is a function of the shape of the parent distribution. Fairly good results have been demonstrated with a sample of $n = 4$ for both the rectangular and triangular distributions.

2 PROBABILITY DISTRIBUTION MODELS

The pattern of the distribution of probabilities over all possible outcomes is called a probability distribution. *Probability distribution models* provide a means for assigning the likelihood of occurrences of all possible values. Variables described in terms of a probability distribution are conveniently called *random variables*. The specific value of a random variable is determined by the distribution.

A probability distribution is completely defined when the probability associated with every possible outcome is defined. In most instances, the outcomes themselves are represented by numbers or different values of a variable, such as the diameter of an acorn. When the pattern of the probability distribution is expressed as a function of this variable, the resulting function is called a *probability distribution function*.

An example empirical probability distribution function may be developed as follows. A maintenance mechanic attends four machines and his services are needed only when a machine fails. He would like to estimate how many machines will fail each shift. From previous experience, and using the relative frequency concept of probability, the mechanic knows that 40% of the time only one machine will fail at least once during the shift. Further, 30% of the time two machines will fail, three machines will fail 20% of the time, and all four will fail 10% of the time.

The probability distribution of the number of failed machines may be expressed as $P(1) = 0.4$, $P(2) = 0.3$, $P(3) = 0.2$, and $P(4) = 0.1$. This probability distribution is exhibited in Figure 1.

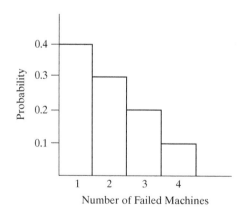

Figure 1 A probability distribution of the number of failed machines.

The probability distribution function for this case may be defined as

$$P(x) = \frac{5-x}{10} \qquad \text{if } x = 1, 2, 3, 4$$

$$P(x) = 0 \qquad\qquad \text{otherwise}$$

Although the function $P(x) = (5-x)/10$ uniquely represents the probability distribution pattern for the number of failed machines, the function itself belongs to a wider class of functions of the type $P(x) = (a-x)/b$. All functions of this type indicate similar patterns, yet each pair of numbers (a, b) uniquely defines a specific probability distribution. These numbers (a, b) are called *parameters*.

In the sense that they serve to define the probability distribution function, it is possible to look upon parameters as properties of the distribution function. The choice of representation of parameters is not unique, and the most desirable representation would reflect a measure of the properties of the universe under study. Two most commonly sought measures are the *mean*, an indication of central tendency, and the *variance*, a measure of dispersion.

The probability distribution just presented is discrete in that it assigns probabilities to an event that can only take on integer values. Continuous probability distributions are used to define the probability of the occurrence of an event that may take on values over a continuum. Under certain conditions, it may be desirable to use a continuous probability distribution to approximate a discrete probability distribution. By so doing, tedious summations may be replaced by integrals. In other instances, it may be desirable to make a continuous distribution discrete as when calculations are to be performed on a digital computer. Several discrete and continuous probability distribution models are presented subsequently.

2.1 The Binomial Distribution

The binomial distribution is a basic discrete sampling distribution. It is applicable where the probability is sought of exactly x occurrences in n trials of an event that has a constant probability of occurrence p. The requirement of a constant probability of occurrence is satisfied when the population being sampled is infinite in size, or where replacement of the sampled unit takes place.

The probability of exactly x occurrences in n trials of an event that has a constant probability of occurrence p is given as

$$P(x) = \frac{n!}{x!(n-x)!} p^x q^{n-x} \qquad 0 \le x \le n \qquad (4)$$

where $q = 1 - p$. The mean and variance of this distribution are given by np and npq, respectively.

As an example of the application of the binomial distribution, assume that a fair coin is to be tossed five times. The probability of obtaining exactly two heads is

$$P(2) = \frac{5!}{2!(5-2)!} (0.5)^2 (1-0.5)^3$$

$$= 10(0.03125) = 0.3125$$

A probability distribution may be constructed by solving for the probability of exactly zero, one, two, three, four, and five heads in five tosses. If $p = 0.5$, as in this example, the resulting

distribution is symmetrical. If the distribution is skewed to the right; if the distribution is skewed to the left.

2.2   The Uniform Distribution

The uniform or rectangular probability distribution may be either discrete or continuous. The continuous form of this simple distribution is

$$f(x) = \frac{1}{a} \qquad 0 \le x \le a \tag{5}$$

The discrete form divides the interval 0 to a into $n + 1$ cells over the range 0 to n, with $1/(n + 1)$ as the unit probabilities. The mean and variance of the rectangular probability distribution are given as $a/2$ and $a^2/12$ for the continuous case, and as $n/2$ and $n^2/12 + n/6$ for the discrete case.

   The general form of the rectangular probability distribution is shown in Figure 2. The probability that a value of x will fall between the limits 0 and a is equal to unity. One may determine the probability associated with a specific value of x, or a range of x, by integration for the continuous case. The probability associated with a specific value of x for the discrete distributions of the previous section was found from the functions given. Determination of the probability associated with a range of x required a summation of individual probabilities. This is a fundamental difference in dealing with discrete and continuous probability distributions.

   Values are drawn at random from the rectangular distribution with x allowed to take on values ranging from 0 through 9. These random rectangular variates may be used to randomize a sample or to develop values drawn at random from other probability distributions as is illustrated in the last section of this appendix.

2.3   The Poisson Distribution

The Poisson is a discrete distribution useful in its own right and as an approximation to the binomial. It is applicable when the opportunity for the occurrence of an event is large, but when the actual occurrence is unlikely. The probability of exactly x occurrences of an event of probability p in a sample n is

$$P(x) = \frac{(\mu)^x e^{-\mu}}{x!} \qquad 0 \le x \le \infty \tag{6}$$

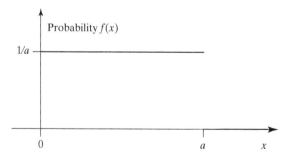

Figure 2   The general form of the rectangular distribution.

The mean and variance of this distribution are equal and given by μ, where $\mu = np$.

As an example of the application of the Poisson distribution, assume that a sample of 100 items is selected from a population of items which are 1% defective. The probability of obtaining exactly three defectives in the sample is found from Equation 6 as

$$P(3) = \frac{(1)^3(2.72)^{-1}}{3!} = 0.061$$

The Poisson distribution may be used as an approximation to the binomial distribution. Such an approximation is good when n is relatively large, p is relatively small, and in general, $pn < 5$. These conditions were satisfied in the previous example.

2.4 The Exponential Distribution

The exponential probability distribution is given by

$$f(x) = \frac{1}{a} e^{-x/a} \qquad 0 \leq x \leq \infty \tag{7}$$

The mean and variance of this distribution are given by a and a^2, respectively. Its form is illustrated in Figure 3.

As an example of the application of the exponential probability distribution, consider the selection of a light bulb from a population of light bulbs whose life is known to be exponentially distributed with a mean $\mu = 1,000$ hours. The probability of the life of this sample bulb not exceeding 1,000 hours would be expressed as $P(x \leq 1,000)$. This would be the proportional area under the exponential function over the range $x = 0$ to $x = 1,000$, or

$$P(x \leq 1,000) = \int_0^{1,000} f(x)dx$$

$$D = \int_0^{1,000} \frac{1}{1,000} e^{-x/1,000} dx$$

$$= -e^{-x/1,000} \Big|_0^{1,000}$$

$$= 1 - e^{-1} = 0.632$$

Note that 0.632 is that proportion of the area of an exponential distribution to the left of the mean. This illustrates that the probability of the occurrence of an event exceeding the mean value is only $1 - 0.632 = 0.368$.

2.5 The Normal Distribution

The normal or Gaussian probability distribution is one of the most important of all distributions. It is defined by

$$f(x) = \frac{1}{\sigma\sqrt{2\pi}} e^{[-(x-\mu)^2/2\sigma^2]} \qquad -\infty \leq x \leq +\infty \tag{8}$$

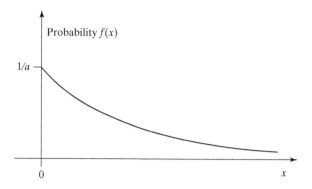

Figure 3 The general form of the exponential distribution.

The mean and variance are μ and σ^2, respectively. Variation is inherent in nature, and much of this variation appears to follow the normal distribution, the form of which is given in Figure 4.

The normal distribution is symmetrical about the mean and possesses some interesting and useful properties regarding its shape. Where distances from the mean are expressed in terms of standard deviations, σ, the relative areas defined between two such distances will be constant from one distribution to another. In effect, all normal distributions, when defined in terms of a common value of μ and σ, will be identical in form, and corresponding probabilities may be tabulated. Normally, cumulative probabilities are given from $-\infty$ to any value expressed as standard deviation units. This table gives the probability from $-\infty$ to Z, where Z is a standard normal variate defined as

$$Z = \frac{x - \mu}{\sigma} \tag{9}$$

This is shown as the shaded area in Figure 4.

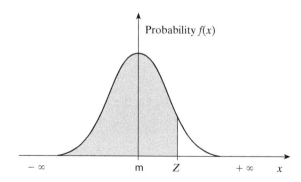

Figure 4 The normal probability distribution.

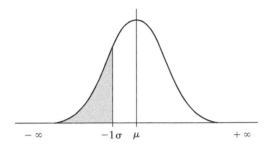

Figure 5   The area from $-\infty$ to -1σ
under the normal distribution.

The area from $-\infty$ to -1σ is indicated as the shaded area in Figure 5. The probability of x falling in this range is 0.1587. Likewise, the area from $-\infty$ to $+2\sigma$ is 0.9773. If the probability of a value falling in the interval -1σ to $+2\sigma$ is required, the following computations are made.

$$P(\text{area} -\infty \text{ to } +2\sigma) = 0.9773$$

$$-P(\text{area} -\infty \text{ to } -1\sigma) = 0.1587$$

$$P(\text{area} -1\sigma \text{ to } +2\sigma) = 0.8186$$

This situation is shown in Figure 6.

2.6   The Lognormal Distribution

The lognormal probability distribution is related to the normal distribution. If a random variable $Y = \ln X$ is normally distributed with mean μ and variance σ^2, then the random variable X follows the lognormal distribution. Accordingly, the probability distribution function of the random variable X is defined as

$$f(x) = \frac{1}{x\sigma\sqrt{2\pi}} e^{[-(\ln x - \mu)^2/2\sigma^2]}, \qquad x > 0 \tag{10}$$

where $\mu \in (-\infty, \infty)$ is called the scale parameter and $\sigma > 0$ is called the shape parameter. The mean and variance of the lognormal distribution are $e^{\mu + \sigma^2/2}$ and $e^{2\mu + \sigma^2}(e^{\sigma^2} - 1)$ respectively.

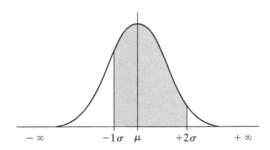

Figure 6   The area from -1σ to $+2\sigma$
under the normal distribution.

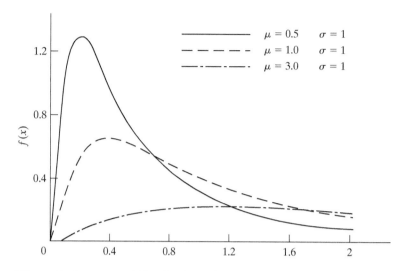

Figure 7 Lognormal distribution with constant shape parameter.

Figure 7 shows a lognormal distribution with a constant shape parameter and various scale parameters. Figure 8 shows a lognormal distribution with a constant scale parameter and various shape parameters.

If $X \sim \ln (\mu, \sigma^2)$ then $\ln X \sim N(\mu, \sigma^2)$. This implies that if n data points $x_1, x_2, \ldots, x_n$ are lognormal, then the logarithms of these data points, $\ln x_1, \ln x_2, \ldots, \ln x_n$ will be normally distributed and may be used for parameter estimation, goodness-of-fit testing, and hypothesis testing.

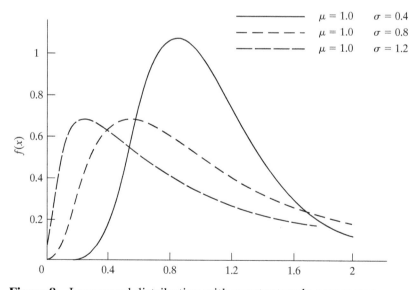

Figure 8 Lognormal distribution with constant scale parameter.

2.7 The Weibull Distribution

The probability distribution function of a random variable X, which follows the Weibull distribution, is given by

$$f(x) = \frac{\alpha}{\beta^\alpha} x^{\alpha-1} e^{-x/\beta)^\alpha}, \qquad x \geq 0 \tag{11}$$

where $\alpha > 0, \beta > 0$ are the shape and scale parameters respectively, and defined on $(0, \infty)$. A Weibull variate X has mean $\dfrac{\beta}{\alpha} \Gamma\left(\dfrac{1}{\alpha}\right)$ and variance $\dfrac{\beta^2}{\alpha} \left\{ 2\Gamma\left(\dfrac{2}{\alpha}\right) - \dfrac{1}{\alpha}\left[\Gamma\left(\dfrac{1}{\alpha}\right)\right]^2 \right\}$, where $\Gamma()$ is the gamma function

$$\Gamma(z) = \int_0^\infty t^{z-1} e^{-t} dt$$

Weibull distributions with different shape and scale parameters are illustrated in Figures 9 and 10, respectively. The Weibull distribution has some interesting characteristics:

1. For $\alpha = 1$, the Weibull distribution is the same as the exponential distribution with parameter β.
2. For $\alpha = 3.4$, the Weibull distribution approximates the normal distribution.
3. If $X \sim$ Weibull (α, β) then $X^\alpha \sim$ exponential $(\beta\alpha)$.

The Weibull distribution is frequently used as a time-to-failure model and in reliability analysis, especially in instances where failure data cannot be fitted by the exponential distribution.

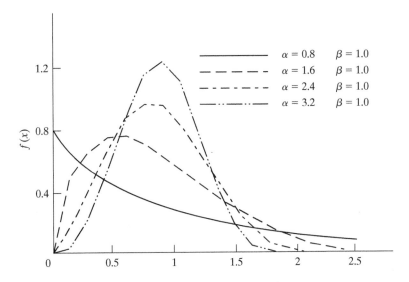

Figure 9 Weibull distribution with various shape parameters.

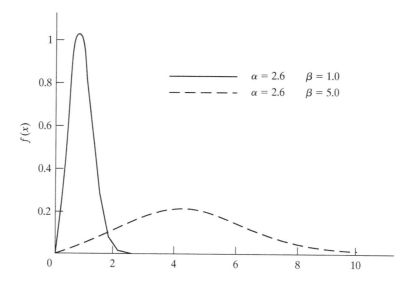

Figure 10 Weibull distribution with various scale parameters.

3 MONTE CARLO ANALYSIS

The decision environment is made up of many random variables. Thus, models used to explain operational systems must often incorporate probabilistic elements. In some cases, formal mathematical solutions are difficult or impossible to obtain from these models. Under such conditions it may be necessary to use a method known as *Monte Carlo analysis*. When applied to an operational system, Monte Carlo analysis provides a powerful means of simulation.

3.1 A Simple Monte Carlo Example

As an introduction to the idea of Monte Carlo analysis, consider its application to the determination of the area of a circle with a diameter of 1 inch. Proceed as follows:

1. Enclose the circle of a 1-inch square as shown in Figure 11.
2. Divide two adjoining sides of the square into tenths, or hundredths, or thousandths, and so on, depending on the accuracy desired.
3. Secure a sequence of pairs of random rectangular variates.
4. Use each pair of rectangular variates to determine a point within the square and possibly within the circle. This process is illustrated in Table 1 for 100 trials.
5. Compute a ratio of the number of times a point falls within the circle to the total number of trials. The value of this ratio is an approximate area for the circle expressed as a fraction of the 1 inch^2 represented by the square. It is 79/100, or 0.79, in this example.

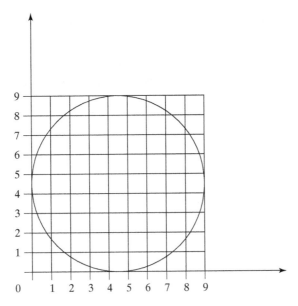

Figure C.11   Area of circle by Monte Carlo analysis.

The example just presented has a well-known mathematical solution as follows:

$$A = \pi r^2 = 3.1416(0.50)^2 = 0.7854.$$

TABLE C.1   Determining the area of a circle

Trial	Random number	In	Out
1	73	x	
2	26	x	
3	19		x
4	84	x	
5	81	x	
6	47	x	
7	18		x
8	44	x	
$\vdots$	$\vdots$	$\vdots$	$\vdots$
100	35	x	
Total		79	21

3.2   Steps in the Monte Carlo Procedure

Monte Carlo analysis may be implemented in accordance with a step-by-step procedure that is usually independent of the application. The steps are:

1. **Formalize the system logic**—The system chosen for study usually operates in accordance with a certain logical pattern. Therefore, before beginning the actual process of Monte Carlo analysis, it is necessary to formalize the operational procedure by the

construction of a model. This may require the development of a step-by-step flow diagram outlining the logic. If the actual simulation process is to be performed on a digital computer, it is mandatory to prepare an accurate logic diagram. From this, the computer can be programmed to pattern the process under study.

2. **Determine the probability distributions**—Each random variable in the operation refers to an event in the system being studied. Therefore, an important step in Monte Carlo analysis is determining the behavior of these random variables. This involves the development of empirical frequency distributions to describe the relevant variables by the collection of historical data. Once this is done, the frequency distribution for each variable may be studied statistically to ascertain whether it conforms to a known theoretical distribution.

3. **Develop the cumulative probability distributions**—Convert each probability distribution to its cumulative equivalent, with the cumulative probability exhibited on the ordinate ready to receive random number inputs. It is the cumulative probability distributions that serve to convert random rectangular variates to values drawn at random from the underlying probability distribution.

4. **Perform the Monte Carlo process**—The Monte Carlo process proceeds by exercising the system logic and saving the results of each trial. Analysis of a sample of trials gives insight into the estimated behavior of the system under study. The validity of the estimate depends upon the fidelity of the logic and the number of trials in the sample.

Design for Maintainability

From Chapter 13 of *Systems Engineering and Analysis,* Fifth Edition, Benjamin S. Blanchard, Wolter J. Fabrycky. Copyright © 2011 by Pearson Education, Inc. Published by Pearson Prentice Hall. All rights reserved.

Design for Maintainability

A primary objective in design for operational feasibility is to ensure that the system is available and operating in an effective and efficient manner as required in accomplishing its specified mission objective(s). The realization of an availability requirement depends on two factors: (1) the inherent reliability of the applicable system and, (2) maintainability—or the ability of that system to be maintained and/or repaired (in the event of failure) and returned to service rapidly and efficiently.

The purpose of this chapter is to cover the concepts, models, and methods useful in accomplishing *design for maintainability*. Maintainability is a design characteristic (a design dependent parameter) pertaining to ease, accuracy, safety, and economy in the performance of maintenance functions. Maintainability is the *ability* of a system to be maintained, whereas maintenance is a series of actions taken to restore or retain a system in an effective operational state. Maintainability must be inherent or "built into" the design, while maintenance is the result of design.

Many systems in use today are complex and do fulfill customer expectations when operating. However, past experience indicates that the reliability of some of these systems is marginal and that they are inoperative much of the time, requiring extensive and costly maintenance. Downtime and the waste of resources for maintenance stem from lack of the proper consideration of reliability and maintainability in design. Specific maintainability requirements must be specified in measurable terms and be incorporated during the system design and development process.

Maintainability is the counterpart of reliability. Both of these design-dependent parameters have to do with continuation of the operation and service expected from a system. Upon completion of this chapter, the reader will have obtained an understanding of maintainability and its importance in the system design process through consideration of the following:

- A definition and explanation of maintainability;
- Measures of maintainability—elapsed times, frequencies, labor hours, and cost;
- Availability and effectiveness factors;
- Maintainability in the system life cycle—system requirements, maintainability allocation, component selection and application, design participation, and design review;

- Maintainability analysis methods—reliability and maintainability trade-off evaluation, prediction, reliability-centered maintenance (RCM), level-of-repair analysis (LORA), maintenance task analysis (MTA), and total productive maintenance (TPM); and
- Maintainability demonstration.

Maintainability requirements must be considered throughout the system activities and interactions life cycle and as a major parameter in accomplishing the design of a system. Accordingly, the last section of this chapter provides a summary of how maintainability should be considered and gives and suggested extensions for study.

1 DEFINITION AND EXPLANATION OF MAINTAINABILITY

An objective in systems engineering is to design and develop a system or product that can be maintained effectively, safely, in the least amount of time, at the least cost, and with a minimum expenditure of support resources (e.g., people, materials, support equipment, and facilities) without adversely affecting the mission of that system. Maintainability is the *ability* of a system to be maintained, whereas maintenance constitutes a series of actions to be taken to restore or retain a system in an effective operational state. Maintainability is a design-dependent parameter. Maintenance is a result of design.

Maintainability, as a characteristic of design, can be expressed in terms of maintenance times, maintenance frequency factors, maintenance labor hours, and maintenance cost. These terms may be presented as different figures-of-merit; therefore, maintainability may be defined on the basis of a combination of factors, such as the following:

1. A characteristic of design and installation that is expressed as the probability that an item will be retained in or restored to a specified condition within a given period of time, when maintenance is performed in accordance with prescribed procedures and resources.
2. A characteristic of design and installation that is expressed as the probability that maintenance will not be required more than x times in a given period when the system is operated in accordance with prescribed procedures.
3. A characteristic of design and installation that is expressed as the probability that the maintenance cost for a system will not exceed y dollars per designated period when the system is operated and maintained in accordance with prescribed procedures.

Maintainability requirements must be specified not only for the prime mission-related elements of a system but for the various elements of the logistics and maintenance support infrastructure as well. Such requirements must be tailored to the applicable mission scenario(s). The appropriate metrics are developed through the identification and prioritization of the technical performance measures (TPMs).

2 MEASURES OF MAINTAINABILITY

Maintainability, defined in the broadest sense, can be measured in terms of a combination of different maintenance factors. From a system perspective, it is assumed that maintenance can be broken down into the following general categories:

1. *Corrective maintenance.* Unscheduled maintenance accomplished, as a result of failure, to *restore* a system or product to a specified level of performance. This includes the initial detection of failure(s), localization and fault isolation (diagnostics), disassembly (access), removal and replacement (or repair) of faulty component, reassembly, adjustment and/or alignment (as required), and final checkout and verification of proper system performance; that is, the corrective maintenance cycle.

2. *Preventive maintenance.* Scheduled maintenance accomplished to *retain* a system at a specified level of performance by providing systematic inspection, detection, servicing, or the prevention of impending failures through periodic item replacements.

When dealing with software, there are several additional categories that are often used. *Adaptive maintenance* refers to the continuing process of modifying software to be responsive to changing requirements in data or in the processing environment (but within the original functional structure). *Perfective maintenance* is a term often used to describe the modification of software for the purposes of enhancing performance, packaging, and other evolving needs of the consumer.

Of particular interest herein are the measures of maintainability pertaining to "downtime" (i.e., the elapsed time when the system is not operating because of maintenance), personnel labor hours (i.e., the total number of people, skill levels, and the labor hours expended in the accomplishment of maintenance), maintenance frequencies (i.e., the mean time between maintenance factors), maintenance cost, and related factors dealing with the various elements of support (i.e., spares/repair parts, test equipment, transportation and handling, maintenance facilities, and applicable supply chain factors). The measures most commonly used are described in this section.

2.1 Maintenance Elapsed-Time Factors

The elapsed-time category includes active corrective and preventive maintenance times, administrative and logistics delay times, and total maintenance downtime (MDT).

Mean Corrective Maintenance Time ($\overline{M}ct$). When a system fails, a series of steps is required to repair or restore the system to its full operational status. These steps include failure detection, fault isolation, disassembly to gain access to the faulty item, repair, and so on, as illustrated in Figure 1. Completion of these steps for a given failure constitutes a corrective maintenance cycle.

Throughout the system utilization phase, there will be a number of individual maintenance actions involving the series of steps illustrated in the figure. The mean corrective maintenance time ($\overline{M}ct$), or the mean time to repair (MTTR) that is equivalent, is a composite value representing the arithmetic average of these individual maintenance cycle times (Mct_i).

For the purpose of illustration, Table 1 includes data covering a sample of 50 corrective maintenance repair actions on a typical system. Each of the times indicated represents the

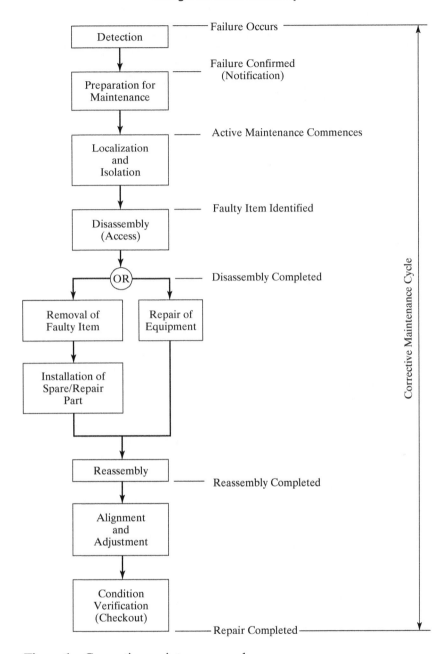

Figure 1   Corrective maintenance cycle.

completion of one corrective maintenance cycle. Based on the set of raw data presented, which constitutes a random sample, a frequency distribution table and frequency histogram may be prepared, as illustrated in Table 2 and Figure 2, respectively.

Referring to Table 2, the range of observations is between 30 and 97 minutes, or a total of 67 minutes. This range can be divided into class intervals, with a class interval width of

TABLE 1 Corrective Maintenance Times (Mct$_i$ in minutes)

40	58	43	45	63	83	75	66	93	92
71	52	55	64	37	62	72	97	76	75
75	64	48	39	69	71	46	59	68	64
67	41	54	30	53	48	83	33	50	63
86	74	51	72	87	37	57	59	65	63

TABLE 2 Frequency Distribution

Class Interval	Frequency	Cumulative Frequency
29.5–39.5	5	5
39.5–49.5	7	12
49.5–59.5	10	22
59.5–69.5	12	34
69.5–79.5	9	43
79.5–89.5	4	47
89.5–99.5	3	50

10 assumed for convenience. A logical starting point is to select class intervals of 20–29, 30–39, and so on. In such instances, it is necessary to establish the dividing point between two adjacent intervals, such as 29.5, 39.5, and so on.

Given the distribution of repair times, one can plot a histogram showing time values in minutes and the frequency of occurrence as in Figure 2. By determining the midpoint of each class interval, a frequency polygon can be developed as illustrated in Figure 3. This provides an indication of the form of the probability distribution applicable to repair times for this particular system.

As additional corrective maintenance actions occur and data points are plotted for the system in question, the curve illustrated in Figure 3 may take the shape of the normal distribution. The normal curve is defined by the arithmetic mean ($\overline{Mct}$) and the standard

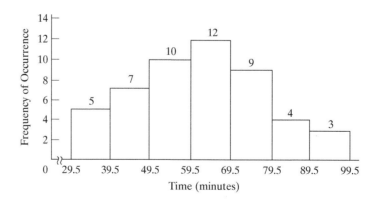

Figure 2 Histogram of maintenance times.

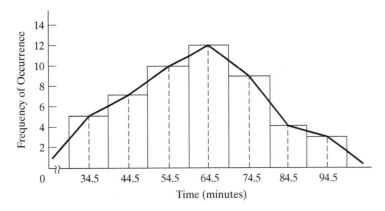

Figure 3   Frequency polygon.

deviation (σ). From the maintenance repair times presented in Table 1, the arithmetic mean is determined as

$$\overline{\text{Mct}} = \frac{\sum_{i=1}^{n} \text{Mct}_i}{n} = \frac{3{,}095}{50} = 61.9 \text{ minutes (assume 62)} \qquad (1)$$

where Mct_i is the total active corrective maintenance cycle time for each maintenance action and n is the sample size. Thus, the average value for the sample of 50 maintenance actions is 62 minutes.

The standard deviation (σ) measures the dispersion of maintenance time values. When a standard deviation is calculated, it is convenient to generate a table giving the deviation of each task time from the mean of 62. Table 3 illustrates this for only four individual task times, although all 50 tasks should be treated. The total value of 13,013 does cover all 50 tasks.

TABLE 3   Variance Data

Total	$\text{Mct}_i - \overline{\text{Mct}}$	$(\text{Mct}_i - \overline{\text{Mct}})^2$
40	-22	484
71	$+9$	81
75	$+13$	169
67	$+5$	25
etc.	etc.	etc.
Total		13,013

The standard deviation of the sample normal distribution curve can now be determined as

$$\sigma = \sqrt{\frac{\sum_{i=1}^{n}(Mct_i - \overline{Mct})^2}{n-1}} = \sqrt{\frac{13{,}013}{49}} = 16.3 \text{ minutes (assume 16)} \qquad (2)$$

One may wish to determine what percentage of the total repair actions falls within the range 46–78 minutes. This can be calculated by converting the range to multiples of standard deviation. In this case, the interval of 46–78 equals $\overline{Mct} \pm 1\sigma$, or 62 minutes ± the standard deviation of 16 minutes. When normality is assumed, it can be stated that 68% of the total population sampled falls within the range 46–78 minutes and that 99.7% of the sample population lies within the range $\overline{Mct} \pm 3\sigma$, or 14–110 minutes.

As a typical application, it may be desirable to determine the percentage of total population repair times that lies between 40 and 50 minutes. Graphically, this is represented in Figure 4. The problem is to find the percent represented by the shaded area. This can be calculated as follows:

1. Convert maintenance times of 40 and 50 minutes into standard values (Z) or the number of standard deviations above and below the mean of 62 minutes:

$$Z \text{ for 40 minutes} = \frac{X_1 - \overline{X}}{\sigma} = \frac{40 - 62}{16} = -1.37$$

$$Z \text{ for 50 minutes} = \frac{X_2 - \overline{X}}{\sigma} = \frac{50 - 62}{16} = -0.75$$

The maintenance times of 40 and 50 minutes represent −1.37 and −0.75 standard deviations below the mean (because the values are negative).

2. Point $X_1 (Z = -1.37)$ represents an area of 0.0853, and point $X_2 (Z = -0.75)$ represents an area of 0.2266, as given in Appendix D, Table D.3.

3. The shaded area A in Figure 4 represents the difference in area, or area $X_2 - X_1 = 0.2266 - 0.0853 = 0.1413$. Thus, 14.13% of the population of maintenance times are estimated to lie between 40 and 50 minutes.

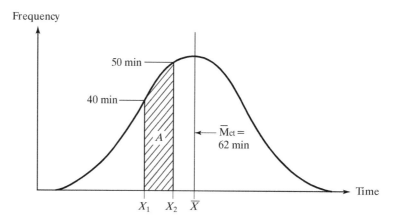

Figure 4 Normal distribution of repair times.

Next, confidence limits should be determined. Because the 50 maintenance tasks represent only a sample of all maintenance actions on the equipment being evaluated, it is possible that another sample of 50 maintenance actions on the same equipment could have a mean value either greater or less than 62 minutes. The original 50 tasks were selected at random, however, and statistically represent the entire population. Using the standard deviation, an upper and lower limit can be placed on the mean value ($\overline{\text{M}}$ct) of the population. For instance, if one is willing to accept a chance of being wrong 15% of the time (85% confidence limit), then

$$\text{Upper limit} = \overline{\text{M}}\text{ct} + Z\left(\frac{\sigma}{\sqrt{n}}\right) \tag{3}$$

where $\sigma/\sqrt{n}$ represents the standard error factor.

The Z value is obtained from Cumulative Normal Probabilities table, where 0.1492 is close to 15% and reflects a Z of 1.04. Thus,

$$\text{Upper limit} = 62 + (1.04)\left(\frac{16}{\sqrt{50}}\right) = 64.35 \text{ minutes}$$

This means that the upper limit is 64.4 minutes at a confidence level of 85%, or that there is an 85% chance that $\overline{\text{M}}$ct will be less than 64.4. Variations in risk and upper limit values are shown in Table 4. If a specified $\overline{\text{M}}$ct limit is established for the design of equipment (based on mission and operational requirements), and it is known (or assumed) that maintenance times are normally distributed, then one would have to compare the results of predictions or measurements (e.g., 64.35 minutes) accomplished during the development process with the specified value to determine the degree of compliance.

When considering probability distributions in general, the time dependency between probability of repair and the time allocated for repair can usually be expected to produce a probability density function in one of three common distribution forms (normal, exponential, and lognormal), as illustrated in Figure 5.[1]

1. The *normal* distribution applies to the relatively straightforward maintenance tasks and repair actions (simple removal and replacement tasks) that consistently require a fixed amount of time to complete with little variation.

TABLE 4 Risk-Upper-Limit Variations

Risk (%)	Confidence (%)	Z	Upper Limit (minutes)
5	95	1.65	65.72
10	90	1.28	64.89
15	85	1.04	64.35
20	80	0.84	63.89

[1]Although past experience has indicated that repair times usually follow the normal and lognormal distributions, sometimes the exponential distribution is assumed for the sake of modeling convenience, particularly in the area of reliability. The exponential distribution is assumed when dealing with failure rates. Such an assumption is sometimes misleading and presumes that zero repair times are the most frequent. Accordingly, the analyst must take care to ensure that the correct repair-time distribution is selected in analyzing maintainability data.

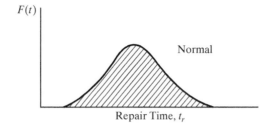

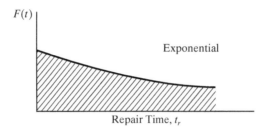

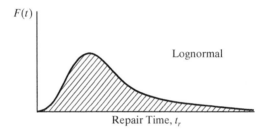

Figure 5   Repair-time distributions.

2. *The exponential* distribution applies to maintenance tasks involving part substitution methods of failure isolation in large systems that result in a constant repair rate.
3. *The lognormal* distribution applies to most maintenance tasks and repair actions comprised of several subsidiary tasks of unequal frequency and time duration.

As indicated, the maintenance task times for many systems and equipments do not always fit the normal curve. There may be a few representative maintenance actions where repair times are extensive, causing a skew to the right. This is particularly true for electronic equipment, where the distribution of repair times often follows a lognormal curve as shown in Figure 6. Derivation of the specific distribution curve for a set of maintenance task times is accomplished using the same procedure as given in the preceding paragraphs. A frequency table is generated, and a histogram is plotted.

A sample of 24 corrective maintenance repair actions for a typical electronic equipment item is presented in Table 5. Using the data in the table, the mean is determined as

$$\overline{\text{Mct}} = \frac{\sum_{i=1}^{n} \text{Mct}_i}{n} = \frac{1{,}637}{24} = 68.21 \text{ minutes}$$

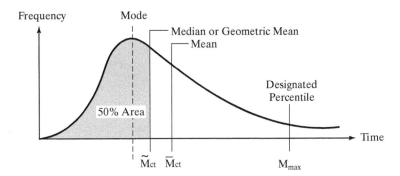

Figure 6 Lognormal distribution of repair times.

When determining the $\overline{\text{M}}$ct for a specific sample population of maintenance repair actions, the use of Equation 4, which has wider application, is appropriate and is

$$\overline{\text{M}}\text{ct} = \frac{\Sigma(\lambda_i)(\text{Mct}_i)}{\Sigma\lambda_i} \tag{4}$$

where λ_i is the failure rate of the individual (ith) element of the item being measured, usually expressed in failures per system operating hour. Equation 4 calculates $\overline{\text{M}}$ct as a weighted average using reliability factors.

It should be noted that $\overline{\text{M}}$ct considers only *active* maintenance time or the time that is spent working directly on the system. Logistics delay time (LDT) and administrative delay time (ADT) are not included. Although all elements of time are important, the $\overline{\text{M}}$ct factor is primarily oriented to a measure of the supportability characteristics in a particular product such as equipment, software, and so on.

Mean Preventive Maintenance Time ($\overline{\text{M}}$pt). Preventive maintenance refers to the actions required to retain a system at a specified level of performance. It may include such functions as periodic inspection, servicing, scheduled replacement of critical items, calibration, overhaul, and so on. $\overline{\text{M}}$pt is the mean (or average) elapsed time to perform preventive or scheduled maintenance on an item and is expressed as

$$\overline{\text{M}}\text{pt} = \frac{\Sigma(\text{fpt}_i)(\text{Mpt}_i)}{\Sigma\text{fpt}_i} \tag{5}$$

where fpt_i is the frequency of the individual (ith) preventive maintenance action in actions per system operating hour, and Mpt_i is the elapsed time required for the ith preventive maintenance action.

Preventive maintenance may be accomplished while the system is in full operation or could result in downtime. In this instance, the concern is for preventive maintenance actions

TABLE 5 Corrective Maintenance Repair Times (minutes)

55	28	125	47	58	53	36	88
51	110	40	75	64	115	48	52
60	72	87	105	55	82	66	65

TABLE 6   Calculation for $\widetilde{\text{M}}$ct

Mct_i	Log Mct_i	$(\text{Log Mct}_i)^2$	Mct_i	Log Mct_i	$(\text{Log Mct}_i)^2$
55	1.740	3.028	64	1.806	3.262
28	1.447	2.094	115	2.041	4.248
125	2.097	4.397	48	1.681	2.826
47	1.672	2.796	52	1.716	2.945
58	1.763	3.108	60	1.778	3.161
53	1.724	2.972	72	1.857	3.448
36	1.556	2.421	87	1.939	3.760
88	1.945	3.783	105	2.021	4.084
51	1.708	2.917	55	1.740	3.028
110	2.041	4.166	82	1.914	3.663
40	1.602	2.566	66	1.819	3.309
75	1.875	3.516	65	1.813	3.287
			Total	43.315	78.785

that result in system downtime. Again, $\overline{\text{M}}$pt includes only *active* system maintenance time, not logistic delay and administrative delay times.

Median Active Corrective Maintenance Time ($\widetilde{\text{M}}$ct).   The median maintenance time is the value that divides all of the downtime values so that 50% are equal to or less than the median and 50% are equal to or greater than the median. The median will usually give the best average location of the data sample. The median for a normal distribution is the same as the mean, whereas the median in a lognormal distribution is the same as the geometric mean, illustrated in Figure 6. $\widetilde{\text{M}}$ct (which is also equivalent to MTTR_g) is calculated as

$$\widetilde{\text{M}}\text{ct} = \text{antilog} \frac{\sum_{i=1}^{n} \log \text{Mct}_i}{n} = \text{antilog} \frac{\Sigma(\lambda_i)(\log \text{Mct}_i)}{\Sigma \lambda_i} \tag{6}$$

For illustrative purposes, the maintenance time values in Table 5 are presented in the format illustrated in Table 6. The median is computed as

$$\widetilde{\text{M}}\text{ct} = \text{antilog} \frac{\sum_{1}^{24} \log \text{Mct}_i}{24}$$

$$= \text{antilog} \frac{43.315}{24} = \text{antilog } 1.805 = 63.8 \text{ minutes}$$

Median Active Preventive Maintenance Time ($\widetilde{\text{M}}$pt).   The median active preventive maintenance time is determined using the same approach as for calculating $\widetilde{\text{M}}$ct. $\widetilde{\text{M}}$pt is expressed as

$$\widetilde{\text{M}}\text{pt} = \text{antilog} \frac{\Sigma(\text{fpt}_i)(\log \text{Mpt}_i)}{\Sigma \text{fpt}_i} \tag{7}$$

Mean Active Maintenance Time ($\overline{M}$). $\overline{M}$ is the mean or average elapsed time required to perform scheduled (preventive) and unscheduled (corrective) maintenance. It excludes logistics delay time and administrative delay time, and is expressed as

$$\overline{M} = \frac{(\lambda)(\overline{M}ct) + (fpt)(\overline{M}pt)}{\lambda + fpt} \tag{8}$$

where λ is the corrective maintenance rate or failure rate, and fpt is the preventive maintenance rate.

Maximum Active Corrective Maintenance Time (M_{max}). M_{max} can be defined as that value of maintenance downtime below which a specified percentage of all maintenance actions can be expected to be completed. M_{max} is related primarily to the lognormal distribution, and the 90th or 95th percentile point is generally taken as the specified value as shown in Figure 6. It is expressed as

$$M_{max} = antilog\ (\overline{\log Mct}) + Z\sigma_{\log Mct_i}) \tag{9}$$

where $\overline{\log Mct}$ is the mean of the logarithms of Mct_i, Z is the value corresponding to the specific percentage point at which M_{max} is defined (see Table 4, –1.65 for 95%), and

$$\sigma_{\log Mct_i} = \sqrt{\frac{\sum_{i=1}^{n} (\log Mct_i)^2 - \left(\sum_{i=1}^{n} \log Mct_i\right)^2 /n}{n-1}} \tag{10}$$

or the standard deviation of the sample logarithms of average repair times, Mct_i.

For example, M_{max} at the 95th percentile for the data sample in Table 5 is determined as

$$M_{max} = antilog\ [\log \overline{M}ct + (1.65)\sigma_{\log Mct_i}] \tag{11}$$

where, referring to Equation 10 and Table 6,

$$\sigma_{\log Mct_i} = \sqrt{\frac{78.785 - (43.315)^2/24}{23}} = 0.163$$

Substituting the standard deviation factor and the mean value into Equation 11 gives

$$M_{max} = antilog\ [\log \overline{M}ct + (1.65)(0.163)]$$
$$= antilog\ (1.805 + 0.269) = 119\ minutes$$

If maintenance times are distributed lognormally, M_{max} cannot be derived directly by using the observed maintenance values. However, by taking the logarithm of each repair value, the resulting distribution becomes normal, facilitating usage of the data in the manner identical to the normal case.

Logistics Delay Time (LDT). Logistics delay time refers to that maintenance downtime that is expended as a result of waiting for a spare part to become available, waiting for the availability of an item of test equipment in order to perform maintenance, waiting for transportation, waiting to use a facility required for maintenance, and so on. LDT does not include active maintenance time but does constitute a major element of total maintenance downtime (MDT).

Administrative Delay Time (ADT). Administrative delay time refers to that portion of downtime during which maintenance is delayed for reasons of an administrative nature (e.g., personnel assignment priority, labor strike, organizational constraint, etc.). ADT does not include active maintenance time but often constitutes a significant element of total maintenance downtime (MDT).

Maintenance Downtime (MDT). Maintenance downtime constitutes the total elapsed time required (when the system is not operational) to repair and restore a system to full operating status, or to retain a system in that condition. MDT includes $\overline{M}$, LDT, and ADT. The mean or average value is calculated from the elapsed times for each function and the associated frequencies (similar to the approach used in determining $\overline{M}$).

In summary, Figure 7 illustrates the relationships of the various downtime factors within the context of the overall time domain.

2.2 Maintenance Labor Hour Factors

The maintainability factors covered in the previous paragraphs relate to elapsed times. Although elapsed times are extremely important in the performance of maintenance, one must also consider the maintenance labor hours expended in the process. Elapsed times can

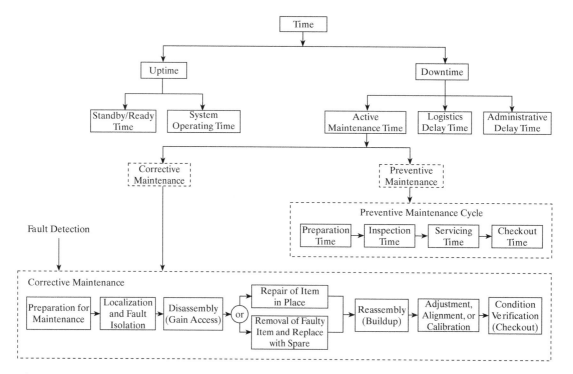

Figure 7 Composite view of uptime/downtime factors.

be reduced (in many instances) by applying additional human resources in the accomplishment of specific tasks. However, this may turn out to be an expensive trade-off, particularly when high skill levels are required to perform tasks that result in less overall clock time. In other words, maintainability is concerned with the *ease* and *economy* in the performance of maintenance. As such, an objective is to obtain the proper balance among elapsed time, labor time, and personnel skills at a minimum maintenance cost.

When considering measures of maintainability, it is not only important to address such factors as $\overline{M}ct$ and MDT, but it is also necessary to consider the labor-time element. Thus, some additional measures must be employed, such as the following:

1. Maintenance labor hours per system operating hour (MLH/OH)
2. Maintenance labor hours per cycle of system operation (MLH/cycle)
3. Maintenance labor hours per month (MLH/month)
4. Maintenance labor hours per maintenance action (MLH/MA)

Any of these factors can be specified in terms of mean values. For example, $\overline{\text{MLH}}_c$ is the mean corrective maintenance labor hours, expressed as

$$\overline{\text{MLH}}_c = \frac{(\Sigma \lambda_i)(\text{MLH}_i)}{\Sigma \lambda_i} \qquad (12)$$

where λ_i is the failure rate of the ith item (failures/hour), and MLH_i is the average maintenance labor hours necessary to complete repair of the ith item.

In addition, the values for mean preventive maintenance labor hours and mean total maintenance labor hours (to include preventive and corrective maintenance) can be calculated on a similar basis. These values can be predicted for each level of maintenance and are employed in determining specific support requirements and associated cost.[2]

2.3 Maintenance Frequency Factors

In terms of the measures of reliability, with MTBF and λ being key factors, it is obvious that reliability and maintainability are very closely related. The reliability factors, MTBF and λ, are the basis for determining the frequency of corrective maintenance. Maintainability deals with the characteristics in system design pertaining to minimizing the corrective maintenance requirements for the system when it assumes operational status. Thus, in this area, reliability and maintainability requirements for a given system must be compatible and mutually supportive.

In addition to the corrective maintenance aspect of system support, maintainability also deals with the characteristics of design that minimize (if not eliminate) preventive maintenance requirements. Sometimes, preventive maintenance requirements are added with the objective of improving system reliability (e.g., reducing failures by specifying selected component replacements at designated times). However, the introduction of

---

[2]Note that labor hours can be expended in the accomplishment of some preventive maintenance actions without causing any system downtime. Thus, it may be possible to accomplish certain scheduled maintenance activities on the system while it is in an operational mode.

preventive maintenance can turn out to be costly if not carefully controlled. Further, the accomplishment of too much preventive maintenance (particularly for complex systems or products) often has a degrading effect on system reliability as failures are frequently induced in the process. Hence, an objective of maintainability is to provide the proper balance between corrective maintenance and preventive maintenance at least overall cost.

Mean Time Between Maintenance (MTBM). MTBM is the mean or average time between all maintenance actions (corrective and preventive) and can be calculated as

$$\text{MTBM} = \frac{1}{1/\text{MTBM}_u + 1/\text{MTBM}_s} \tag{13}$$

where MTBM_u is the mean interval of unscheduled (corrective) maintenance and MTBM_s is the mean interval of scheduled (preventive) maintenance. The reciprocals of MTBM_u and MTBM_s constitute the maintenance rates in terms of maintenance actions per hour of system operation. MTBM_u should approximate MTBF, assuming that a combined failure rate is used that includes the consideration of primary inherent failures, dependent failures, manufacturing defects, operator- and maintenance-induced failures, and so on.

Mean Time Between Replacement (MTBR). MTBR, a factor of MTBM, refers to the mean time between item replacements and is a major parameter in determining spare part requirements. On many occasions, corrective and preventive maintenance actions are accomplished without generating the requirement for the replacement of a component part. In other instances, item replacements are required, which, in turn, necessitates the availability of a spare part and an inventory requirement. In addition, higher levels of maintenance support (i.e., intermediate and depot levels) may also be required.

In essence, MTBR is a significant factor, applicable in both corrective and preventive maintenance activities involving item replacement, and is a key parameter in determining logistic support requirements. A maintainability objective in system design is to maximize MTBR (or minimize the number of component replacements where possible).

2.4 Maintenance Cost Factors

For many systems/products, maintenance cost constitutes a major segment of total life-cycle cost. Further, experience has indicated that maintenance costs are significantly affected by design decisions made throughout the early stages of system development. Thus, it is essential that total life-cycle cost be considered as a major design parameter beginning with the definition of system operational requirements.

Of particular interest in this chapter is the aspect of *economy* in the performance of maintenance actions. Maintainability is directly concerned with the characteristics of system design that will ultimately result in the accomplishment of maintenance at minimum overall cost.

When considering maintenance cost, the following cost-related indices may be appropriate as criteria in system design:

1. Cost per maintenance action ($/MA)
2. Maintenance cost per system operating hour ($/OH)
3. Maintenance cost per month ($/month)

4. Maintenance cost per mission or mission segment ($/mission)
5. The ratio of maintenance cost to total life-cycle cost

2.5 Related Maintenance Factors

There are several additional factors that are closely related to and highly dependent on the maintainability measures described. These include various logistics factors, such as the following:

1. Supply responsiveness or the probability of having a spare part available when needed, spare part demand rates, supply lead times for given items, levels of inventory, and so on
2. Test and support equipment effectiveness (reliability and availability of test equipment), test equipment use, system test thoroughness, and so on
3. Maintenance facility availability and use
4. Transportation modes, times between maintenance facilities, and frequency
5. Maintenance organizational effectiveness and personnel efficiency
6. Data and information processing capacity, time, and frequency

There are numerous other logistics and supply chain factors that should also be specified, measured, and controlled if the ultimate mission of the system is to be fulfilled. For instance, specifying a 15-minute $\overline{M}ct$ requirement for a prime element of the system is highly questionable if there is a low probability of having a spare part available when required (resulting in a long logistics delay-time possibility); specifying specific maintenance labor hour requirements may not be appropriate if the maintenance organization is not properly staffed or available to perform the required function(s); specifying system test-time requirements may be inappropriate if the predicted reliability of the test equipment is less than the reliability of the item being tested; and so on.

There are many examples where the interactions between the prime system and its elements of support are critical, and both areas must be considered in the establishment of system requirements during conceptual design. In addition, the maintainability requirements for a given system must take into consideration the impact (both on other systems and from other systems on this system) when there are "common" functions and when the system is operating within a system-of-systems (SOS) configuration.

Maintainability, as a characteristic in design, is closely related to the area of system support since the results of maintainability directly affect maintenance requirements. Thus, when specifying maintainability factors, one should also address the qualitative and quantitative requirements for system support to determine the effects of one area on another.

Referring to Figure 8, there is a top-down and bottom-up relationship. To comply with the specified maintenance time requirements, the appropriate support infrastructure must be in place. This, in turn, may have a feedback effect on overall system effectiveness and cost. In addition, if the support infrastructure is not ideal, this may require some additional modifications in the design of the prime elements of the system (e.g., the incorporation of additional redundancy in design) or some added redundancies in the support structure (e.g., additional spares, personnel, and test equipment). The results could turn out to be costly.

Design for Maintainability

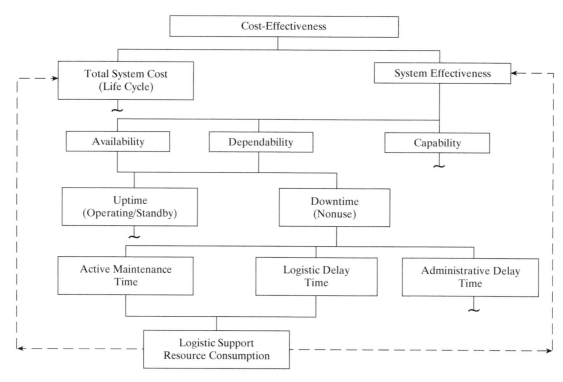

Figure 8 The relationship between maintenance downtime and logistic ansd factors.

3 AVAILABILITY AND EFFECTIVENESS MEASURES

From the reliability measures and the maintainability measures (described in Section 2), it is appropriate at this point to build upon these in addressing some higher-level measures for the overall system. More specifically, these may include measures of both system *availability* and *effectiveness*.[3]

Availability (A).   Availability may be expressed and defined in three ways.

1. *Inherent availability* (A_i). Inherent availability is "the probability that a system or equipment, when used under stated conditions in an *ideal* support environment (i.e., readily available tools, spares, maintenance personnel, etc.), will operate satisfactorily

[3]These system-level measures (as applicable) are initially derived during the conceptual design phase through the development of system operational requirements and the maintenance and support concept; they may also be specified as technical performance measures.

at any point in time as required." It excludes preventive or scheduled maintenance actions, logistics delay time, and administrative delay time, and is expressed as

$$A_i = \frac{\text{MTBF}}{\text{MTBF} + \overline{\text{M}}\text{ct}} \qquad (14)$$

where MTBF is the mean time between failure and $\overline{\text{M}}$ct is the mean corrective maintenance time. The term $\overline{\text{M}}$ct is equivalent to the mean time to repair (MTTR).

2. *Achieved availability* (A_a). Achieved availability is "the probability that a system or equipment, when used under stated conditions in an *ideal* support environment (i.e., readily available tools, spares, personnel, etc.), will operate satisfactorily at any point in time." This definition is similar to that for A_i except that preventive (i.e., scheduled) maintenance is included. It excludes logistics delay time and administrative delay time, and is expressed as

$$A_a = \frac{\text{MTBM}}{\text{MTBM} + \overline{\text{M}}} \qquad (15)$$

where MTBM is the mean time between maintenance and $\overline{\text{M}}$ is the mean active maintenance time. MTBM and $\overline{\text{M}}$ are functions of corrective (unscheduled) and preventive (scheduled) maintenance actions and times, respectively.

3. *Operational availability* (A_0). Operational availability is the "probability that a system or equipment, when used under stated conditions in an *actual* operational environment, will operate satisfactorily when called upon." It is expressed as

$$A_0 = \frac{\text{MTBM}}{\text{MTBM} + \text{MDT}} \qquad (16)$$

where MDT is the mean maintenance downtime. The reciprocal of MTBM is the frequency of maintenance, which, in turn, is a significant factor in determining logistic support requirements. MDT includes active maintenance time ($\overline{\text{M}}$), logistics delay time, and administrative delay time.

The term *availability* is used differently in different situations. If one is to impose an availability figure-of-merit as a *design requirement* for a given equipment supplier, and the supplier has no control over the operational environment in which that equipment is to function, then A_a or A_i might be appropriate figures-of-merit against which the supplier's equipment can be properly assessed. Conversely, if one is to assess a system in a realistic operational environment, then A_0 is a preferred figure-of-merit to employ for assessment purposes. Further, availability may be applied at any time in the overall mission profile representing a point estimate or may be more appropriately related to a specific segment of the mission where the requirements are different from other segments. Thus, one must define precisely what is meant by "availability" and how it is to be applied. In any event, reliability and maintainability are major factors in the determination of system availability.

System Effectiveness (SE). System effectiveness may be defined as "the probability that a system can successfully meet an overall operational demand within a given time when operated under specified conditions" or "the ability of a system to do the job for which it was intended." *System effectiveness*, like operational availability, is a term

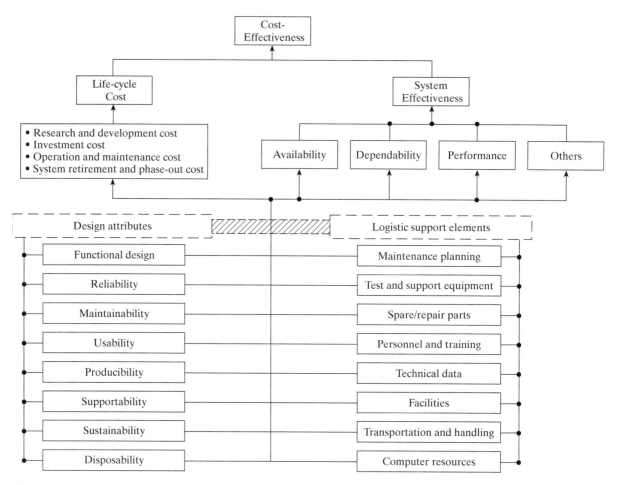

Design for Maintainability

Figure 9 The elements of effectiveness.

used in a broad context to reflect the technical characteristics of a system (i.e., performance, availability, supportability, dependability, etc.) and may be expressed differently depending on the specific mission application. Sometimes a single figure-of-merit is used to express system effectiveness and sometimes multiple figures-of-merit are employed. The objective is to reflect system design attributes and logistic support elements, as is illustrated in Figure 9.[4]

Cost-Effectiveness (CE). *Cost-effectiveness* relates to the measure of a system in terms of mission fulfillment (system effectiveness) and total life-cycle cost and can be expressed in various ways, depending on the specific mission or system parameters that

[4]The terms *system effectiveness* and *cost-effectiveness* are defined and applied differently depending on the system, its mission, and the metrics (TPMs) used to define the requirements. The discussion herein is *generic* and must be tailored to the specific system in question. Further, Figure 9 is presented to illustrate some overall relationships at the concept level.

one wishes to evaluate. In essence, cost-effectiveness includes the elements illustrated in Figure 9 and can be expressed as indicated in Equations 17–21. Reliability and maintainability are major factors in determining the cost effectiveness of a system.

$$\text{Figure-of-merit (FOM)} = \frac{\text{system benefits}}{\text{life-cycle cost}} \tag{17}$$

$$\text{FOM} = \frac{\text{system capacity}}{\text{life-cycle cost}} \tag{18}$$

$$\text{FOM} = \frac{\text{system effectiveness}}{\text{life-cycle cost}} \tag{19}$$

$$\text{FOM} = \frac{\text{availability}}{\text{life-cycle cost}} \tag{20}$$

$$\text{FOM} = \frac{\text{supportability}}{\text{life-cycle cost}} \tag{21}$$

4 MAINTAINABILITY IN THE SYSTEM LIFE CYCLE

Maintainability, like reliability, must be an inherent consideration within the overall systems engineering process beginning during the conceptual design phase. Referring to Figure 10, qualitative and quantitative maintainability requirements are developed through the accomplishment of feasibility analysis, the development of operational requirements and the maintenance concept, and the identification and prioritization of TPMs. The applicable measures of maintainability must be established, their importance with respect to other system metrics must be delineated, and the requirements for design must be identified through the development of the appropriate DDPs.

Given the specification of maintainability requirements, the next step is to evolve from the functional analysis (the identification of maintenance functions in particular) down to the subsystem level and below as part of the system allocation process. Having defined the basic requirements, one proceeds through the iterative process of system synthesis, analysis, and evaluation. To facilitate this process, there are different analysis methods/tools that can be effectively used in preliminary and detail design (refer to Blocks 2 and 3 of Figure 21). As design progresses and physical models are developed, the ongoing evaluation process occurs, and maintainability demonstration testing is accomplished as part of the overall system test and evaluation activity.

4.1 System Requirements

Every system is developed in response to a need or to fulfill some anticipated function. The effectiveness with which the system fulfills this function is the ultimate measure of its utility and its value to the customer. This effectiveness is a composite of performance, maintainability, and other factors. In essence, maintainability constitutes a major factor in determining the usefulness of the system.

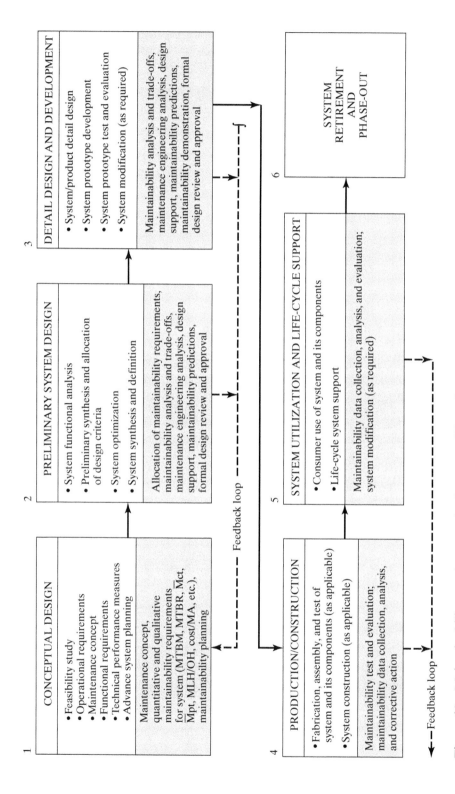

Figure 10 Maintainability requirements in the system life cycle.

Maintainability requirements for a system, specified in both quantitative and qualitative terms, are defined as part of the overall system operational requirements and the maintenance concept. Of particular interest are the following:

1. Definition of system performance factors, the mission profile, and system utilization requirements (use conditions, duty cycles, and just how the system is to be operated).
2. Definition of the operational life cycle (the anticipated time that the system will be in the inventory and in operational use).
3. Definition of the basic system maintenance and support concept (the anticipated levels of maintenance, maintenance responsibilities, major functions at each level, and the prime elements of logistic support at each level—type of spares, test equipment, personnel skills, facilities, etc.).
4. Definition of the environment in which the system is expected to operate and be maintained (temperature, humidity, vibration, arctic or tropics, mountainous or flat terrain, etc.). This may include a range of values and should also cover the environment experienced for all transportation, handling, and storage modes.

Given the foregoing information, one must determine the extent to which maintainability characteristics should be incorporated in the system design. If the operational requirements specify that the system must function 90% of the time and the estimated reliability is low, then the system maintainability requirements may be rather stringent in order to maintain the overall 90% availability. Conversely, if the estimated reliability is high (resulting in a very few anticipated failures), the specified maintainability requirements may be different. Further, if the maintenance concept dictates that only the organization and depot levels of maintenance are allowed, then the maintainability requirements may be different from what would likely be specified with three levels of maintenance planned (to include intermediate maintenance). In any event, quantitative and qualitative maintainability requirements must be defined for the system based on the foregoing considerations. Quantitative requirements are usually expressed in terms of MTBM, MTBR, $\overline{M}$, $\overline{M}$ct, $\overline{M}$pt, MDT, MLH/OH, \$/MA, or a combination thereof.

In specifying maintainability, one must first identify the major system requirements from the top-level functional flow diagram. These functions are expanded as necessary for the purposes of system definition, and a functional packaging scheme is identified. Maintenance functional diagrams are developed, significant maintenance functions are defined, and maintainability requirements are established for the appropriate maintenance functions and the various functional packages of the system. This is an iterative process of requirements definition.

4.2 Maintainability Allocation

After requirements for the system have been established, it is then necessary to translate these requirements into lower-level design criteria through maintainability allocation. This is accomplished as part of the system requirements allocation process.

For the purpose of illustration, it is assumed that System XYZ must be designed to meet an inherent availability requirement of 0.9989, a MTBF of 450 hours, and a MLH/OH (for corrective maintenance) of 0.2, and a need exists to allocate $\overline{M}ct$ and MLH/OH to the assembly level. From Equation 14, $\overline{M}ct$ is

$$\overline{M}ct = \frac{MTBF(1 - A_i)}{A_i} \qquad (22)$$

or

$$\overline{M}ct = \frac{450(1 - 0.9989)}{0.9989} = 0.5 \text{ hour}$$

Thus, the system's $\overline{M}ct$ requirement is 0.5 hour, and this requirement must be allocated to Units A–C and the assemblies within each unit. The allocation process is facilitated through the use of a format similar to that illustrated in Table 7.[5]

Referring to Table 7, each item type and the quantity (Q) of items per system configuration are indicated. Allocated reliability factors are specified in column 3 and the degree to which the failure rate of each unit contributes to the overall failure rate (represented by C_f) is entered in column 4. The average corrective maintenance time for each unit is estimated and entered in column 6. These times are ultimately based on the inherent characteristics of equipment design, which are not known at this point in the system life cycle. Thus, corrective maintenance times are initially derived using a complexity factor, which is indicated by failure rate. As a goal, the item that contributes the highest percentage to the anticipated total failures (Unit B in this instance) should require a low $\overline{M}ct$, and those with low contributions may require a higher $\overline{M}ct$. On certain occasions, however, the design costs associated with obtaining a low $\overline{M}ct$ for a complex item may lead to a modified approach, which is feasible as long as the end result ($\overline{M}ct$ at the system level) falls within the quantitative requirement.[6]

TABLE 7 Maintainability Allocation for System *XYZ*

1	2	3	4	5	6	7
Item	Quantity of Items per System (Q)	Failure Rate (λ) $\times$ 1,000 hours	Contribution of Total Failures $C_f = (Q)(\lambda)$	Percent Contribution $C_p = C_f \Sigma C_f \times 100$ (%)	Average Corrective Maintenance Time $\overline{M}ct$ (hour)	Contribution of Total Corrective Maintenance Time $C_t = (C_f)(\overline{M}ct)$
1. Unit A	1	0.246	0.246	11	0.9	0.221
2. Unit B	1	1.866	1.866	84	0.4	0.746
3. Unit C	1	0.110	0.110	5	1.0	0.110
Total			$\Sigma C_f = 2.222$	100		$\Sigma C_f = 1.077$

$$\overline{M}ct \text{ for System } XYZ = \frac{\Sigma C_t}{\Sigma C_f} = \frac{1.077}{2.222} = 0.485 \text{ hour (requirement: 0.5 hour)}$$

[5]MTBM and $\overline{M}pt$ may be allocated on a comparable basis with MTBF and $\overline{M}ct$, respectively.

[6]In any event, the maintainability parameters depend on the reliability parameters. Also, it will frequently occur that reliability allocations are incompatible with maintainability allocations (or vice versa). Hence, a close feedback relationship between these activities is mandatory.

The estimated value of C_t for each unit is entered in column 7, and the sum of the contributions for all units can be used to determine the overall system's $\overline{\text{Mct}}$ as

$$\overline{\text{Mct}} = \frac{\Sigma C_t}{\Sigma C_f} = \frac{1.077}{2.222} = 0.485 \text{ hour} \tag{23}$$

In Table 7, the calculated $\overline{\text{Mct}}$ for the system is within the requirement of 0.5 hour. The $\overline{\text{Mct}}$ values for the units provide corrective maintenance downtime criteria for design (i.e., DDPs), and the values are included in the applicable design specifications.

Once allocation is accomplished at the unit level, the resultant $\overline{\text{Mct}}$ values can be allocated to the next lower equipment indenture item. For instance, the 0.4 hour $\overline{\text{Mct}}$ value for Unit B can be allocated to Assemblies 1–3 and the procedure for allocation is the same as employed in Equation 23. An example of allocated values for the assemblies of Unit B is included in Table 8.

The $\overline{\text{Mct}}$ value covers the aspect of *elapsed* or *clock time* for restoration actions. Sometimes this factor, when combined with a reliability requirement, is sufficient to establish the necessary maintainability characteristics in design. On other occasions, specifying $\overline{\text{Mct}}$ by itself is not adequate because there may be several design approaches that will meet the $\overline{\text{Mct}}$ requirement but not necessarily in a cost-effective manner. Meeting a $\overline{\text{Mct}}$ requirement may result in an increase in the skill levels of personnel accomplishing maintenance actions, increasing the quantity of personnel for given maintenance functions, or incorporating automation for manual operations. In each instance there are costs involved; thus, one may wish to specify additional constraints, such as the skill level of personnel at each maintenance level and the maintenance labor hours per operating hour (MLH/OH) for significant items. In other words, a requirement may dictate that an item be designed such that it can be repaired within a specified elapsed time with a given quantity of personnel possessing skills of a certain level. This will influence design in terms of accessibility, packaging schemes, handling requirements, diagnostic provisions, and so on, and is perhaps more meaningful overall.

The MLH/OH factor is a function of task complexity and the frequency of maintenance. The system-level requirement is allocated on the basis of system operating hours, the anticipated quantity of maintenance actions, and an estimate of the number of labor hours per maintenance action. Experience data are used where possible.

Following the completion of quantitative allocations for each indenture level of equipment, all values are included in the functional breakdown. In the event that System

TABLE 8 Unit *B* Allocation

1	2	3	4	5	6	7
Assembly 1	1	0.116	0.116	6%	0.5	0.058
Assembly 2	1	1.550	1.550	83%	0.4	0.620
Assembly 3	1	0.200	0.200	11%	0.3	0.060
Total			1.866	100%		0.738

$$\overline{\text{Mct}} \text{ for Unit B} = \frac{\Sigma C_t}{\Sigma C_f} = \frac{0.738}{1.866} = 0.395 \text{ hour (requirement: 0.4 hour)}$$

XYZ in the figure is operating within a *SOS* configuration and there is a "common" unit or assembly required for successful operation of the system, the allocation process must deal with the common unit in a manner similar to that of reliability allocation.

4.3 Component Selection and Application

The extent to which maintainability characteristics are built into the design depends not only on the design of individual components themselves but also on the way in which they are mounted or located within the system structure. Specific objectives may include the following:

1. Select *standardized* components and materials where possible. The goal is to minimize the overall number and different types of components, reducing the quantity and variety of spares/repair part requirements (and associated inventories) while ensuring dependable sources of supply throughout the system life cycle.

2. For repairable items, select those that incorporate built-in *self-test features* and the level and depth of *diagnostics* that will facilitate completing the maintenance cycle in minimum time and with a high degree of confidence. Referring to Figure 1, the most time-consuming steps are often in the areas of localization and diagnostics, particularly for electronic equipment.

3. Select items that can be repaired using common and standard tools and test equipment, in readily available facilities, and without requiring highly skilled personnel with extensive training.

4. Ensure that the appropriate degree of *accessibility* is provided that will allow one to identify rapidly the item requiring removal (or gain access to the area requiring repair), enable quick removal and replacement of the faulty item with a spare, and accomplish the follow-on system checkout to ensure a proper operating state with a minimum of alignment and adjustment (without requiring calibration). Generally, items that are expected to require the most maintenance should be the most accessible (i.e., high failure-rate items or critical components).

5. Incorporate a modularized functional-packaging approach where, in the event of failure, a faulty item can be removed and replaced quickly, without requiring the removal and replacement of other items in the process. The interdependency of items on each other should be minimized. In addition, the faulty items should be easily removable, of a plug-in variety, and/or with quick-release fasteners. Complete *physical* and *functional interchangeability* is desired.

6. Avoid the selection of short-life components (i.e., critical-useful-life items) and the requirement for preventive maintenance. The goal is to eliminate all requirements for scheduled or preventive maintenance unless justified on the basis of reliability data/information. This can be facilitated through the accomplishment of an RCM analysis that is described in Section 5.

7. Incorporate the proper amount of labeling and identification of components in repairable items to aid the technician in completing his or her tasks in an effective and efficient manner. The above areas represent just a few of the concerns in designing for maintainability. The overall objective is to minimize maintenance times, high frequencies, and the logistic support resources required in the performance of maintenance tasks.

4.4 Design Review and Evaluation

Review of the extent to which maintainability is incorporated in system design is accomplished as an inherent part of the process. The characteristics of the system (and its elements) are evaluated in terms of the initially specified maintainability requirements for the system. If the requirements appear to have been met, the design is approved and the program enters the next phase. If not, the appropriate changes are initiated for corrective action.

In accomplishing a maintainability review, a checklist may be developed to facilitate the review process. An abbreviated representative list, not to be considered as being all inclusive, is presented below. It should be noted that the answer to those questions that are applicable should be *yes*.

1. Have maintainability quantitative and qualitative requirements for the system been adequately defined and specified?
2. Are the maintainability requirements compatible with other system requirements? Are they realistic?
3. Are the maintainability requirements compatible with the system maintenance concept?
4. Has the proper level of accessibility been provided in the design to allow for the easy accomplishment of repair or item replacement? Are access requirements compatible with the frequency of maintenance? Accessibility for items requiring frequent maintenance should be greater than that for items requiring infrequent maintenance.
5. Is standardization incorporated to the maximum extent practicable throughout the design? In the interest of developing an efficient supply support capability, the number of different types of spares should be held to a minimum.
6. Is functional packaging incorporated to the maximum extent possible? Interaction effects between modular packages should be minimized. It should be possible to limit maintenance to the removal of one module (the one containing the failed part) when a failure occurs and not require the removal of two, three, or four different modules.
7. Have the proper diagnostic test provisions been incorporated into the design? Is the extent or depth of testing compatible with the level-of-repair analysis?
8. Are modules and components having similar functions electrically, functionally, and physically interchangeable?
9. Are the handling provisions adequate for heavy items requiring transportation (e.g., hoist lugs, lifting provisions, handles, etc.)?
10. Have quick-release fasteners been used on doors and access panels? Have the total number of fasteners been minimized? Have fasteners been selected based on the requirement for standard tools in lieu of special tools?
11. Have adjustment, alignment, and calibration requirements been minimized (if not eliminated)?
12. Have servicing and lubrication requirements been held to a minimum (if not eliminated)?
13. Are assembly, subassembly, module, and component labeling requirements adequate? Are the labels permanently affixed and unlikely to come off during a maintenance action or as a result of environmental conditions?
14. Have all system maintainability requirements been met?

5 MAINTAINABILITY ANALYSIS METHODS

Within the context of maintainability analysis (accomplished as part of the iterative process of system synthesis, analysis, and evaluation), there are a number of tools/models that can be effectively used in support of the objectives described throughout this text. Of particular interest herein are the trade-offs between reliability and maintainability, maintainability prediction, reliability-centered maintenance (RCM), level-of-repair analysis (LORA), maintenance task analysis (MTA), and total productive maintenance (TPM). These are discussed in the paragraphs that follow.

5.1 Reliability And Maintainability Trade-Off Evaluation

Suppose that there is a requirement to replace existing equipment with a new one for improving operational effectiveness. The current need specifies that the equipment must operate 8 hours per day, 360 days per year, for 10 years. The existing equipment meets an availability of 0.961, a MTBF of 125 hours, and a $\overline{M}ct$ of 5 hours. The new system must meet an availability of 0.990, a MTBF greater than 300 hours, and a $\overline{M}ct$ not exceeding 5.0 hours. An anticipated quantity of 200 items of equipment is to be procured. Three different alternative design configurations are being considered to satisfy the requirement, and each configuration constitutes a modification of the existing equipment.

Figure 11 graphically shows the relationships among inherent availability (A_i), MTBF, and $\overline{M}ct$ and illustrates the allowable area for trade-off. The selected configuration must reflect the reliability and maintainability characteristics represented by the shaded area. The existing design is not compatible with the new requirement. Three alternative design configurations are being considered. Each configuration meets the availability requirement, with Configuration A having the highest estimated reliability MTBF and Configuration C reflecting the best maintainability characteristics with the lowest $\overline{M}ct$ value. The objective is to select the best of the three configurations on the basis of cost.

When considering cost, there are costs associated with research and development (R&D) activity, investment or manufacturing costs, and operation and maintenance (O&M) costs. For instance, improving reliability or maintainability characteristics in design will result in an increase in R&D and investment (manufacturing) cost. In addition, experience has indicated that such improvements will result in lower O&M cost, particularly in

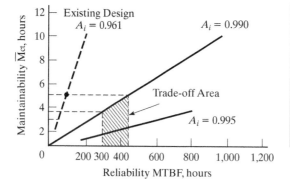

Conf.	A_i	MTBF	$\overline{M}ct$
Existing	0.961	125	5.0
Alt. A	0.991	450	4.0
Alt. B	0.990	375	3.5
Alt. C	0.991	320	2.8

Other configurations are eligible for consideration as long as the effectiveness parameters fall within the trade-off areas.

Figure 11 Reliability and maintainability trade-off.

TABLE 9 Cost Summary (Partial Costs)

Category	Configuration A	Configuration B	Configuration C	Remarks
R & D cost	$17,120	$15,227	$12,110	High-reliability parts,
Reliability design	2,109	4,898	7,115	packaging, accessibility
Maintainability design				
Investment cost	$3,422,400	$3,258,400	$3,022,200	$17,112/Equipment A;
manufacturing				$16,292/Equipment B;
(200 systems)				$15,111/Equipment C
O&M cost				
Maintenance personnel	$1,280,000	$1,536,000	$1,800,000	12,800 Maintenance
and support	342,240	325,840	302,220	Action/Equipment A;
Spare and repair parts				15,360 Maintenance
				Action/Equipment B;
				18,800 Maintenance
				Action/Equipment C;
				10% of manufacturing
				cost for spares
Total	$5,063,869	$5,140,365	$5,143,645	

the areas of maintenance personnel and support cost and the cost of spare/repair parts. Thus, initially the analyst may look at only these categories. If the final decision is close, it may be appropriate to investigate other categories. A summary of partial cost data is presented in Table 9.[7]

As shown in Table 9, the delta costs associated with the three alternative equipment configurations are included for R&D and investment. Maintenance personnel and support costs, included as part of the O&M cost, are based on estimated operating time for the 200 items of equipment throughout the required 10-year period of use (i.e., 200 items of equipment operating 8 hours per day, 360 days per year, for 10 years) and the reliability MTBF factor. Assuming that the average cost per maintenance action is $100, maintenance personnel and support costs are determined by multiplying this factor by the estimated quantity of maintenance actions, which is determined from total operating time divided by the MTBF value. Based on the delta values presented in Table 9, Configuration A satisfies the system availability, reliability, and maintainability requirements with the least cost.

5.2 Maintainability Prediction

Maintainability prediction involves an early assessment of the maintainability characteristics in system design and is accomplished periodically at different stages in the design process. Through the review of design data, predictions of the MTBM, $\overline{M}ct$, $\overline{M}pt$, MLH/OH, and so on, are made and are compared with the initially specified requirements identified in the maintainability allocation process. Areas of noncompliance are evaluated for possible design improvement.

Maintainability prediction, in the broadest sense, includes the early quantitative estimation of maintenance elapsed-time factors, maintenance labor hour factors, maintenance

---

[7]Only those *delta costs* considered significant for this evaluation are included.

frequency factors, and maintenance cost factors. The accomplishment of this requires that the engineer or analyst review design data, layouts, component-part lists, reliability factors, and supporting data with the intent of identifying anticipated maintenance tasks and the resources required for task completion. Maintainability prediction not only requires the estimation of task times and frequencies but also requires a qualitative assessment of the design characteristics for supportability.

Prediction of $\overline{\text{M}}$ct. Prediction of mean corrective maintenance time ($\overline{\text{M}}$ct) may be accomplished using a system element breakdown and determining maintenance tasks and the associated elapsed times in progressing from one element to another. The breakdown covers subsystems, units, assemblies, subassemblies, and parts. Maintainability subtasks such as localization, isolation, access, repair, and checkout, based on the characteristics incorporated in the design, are evaluated and identified with one of the functional levels and down to each component part. Times applicable to each part (assuming that every part will fail at some point) are combined to provide factors for the next higher level. A sample data format for an assembly is presented in Table 10.

Referring to Table 10, the frequency factors are represented by the failure rate (λ) for each part determined from reliability prediction data, and the elapsed times required for fault localization, isolation, disassembly to gain access, and so on, are noted. Maintenance task times are usually estimated from experience and data obtained on similar systems already in use. The summation of the various individual times constitutes a maintenance cycle time (i.e., Mct_i), as illustrated in Figure 1. Similar data prepared on each assembly in the system are combined, as illustrated in Table 11, and the factors are computed to arrive at the predicted $\overline{\text{M}}$ct.

Prediction of $\overline{\text{M}}$pt. Prediction of preventive maintenance time may be accomplished using a method similar to the corrective maintenance approach discussed earlier. Preventive maintenance tasks are estimated along with frequency and task times. An example is presented in Table 12.

TABLE 10 Maintainability Prediction Worksheet (Assembly 4)

Part Category	λ	N	$(N)(\lambda)$	Maintenance Times (hour)							$(N)(\lambda)(\text{Mct}_i)$
				Loc	Iso	Acc	Ali	Che	Int	Mct_i	
Part A	0.161	2	0.322	0.08	0.08	0.14	0.01	0.01	0.11	0.370	0.119
Part B	0.102	4	0.408	0.01	0.05	0.12	0.01	0.02	0.12	0.330	0.134
Part C	0.021	5	0.105	0.03	0.04	0.11	—	0.01	0.14	0.330	0.034
Part D	0.084	1	0.084	0.01	0.03	0.10	0.02	0.03	0.11	0.300	0.025
Part E	0.452	9	1.060	0.02	0.04	0.13	0.02	0.03	0.08	0.390	1.299
Part F	0.191	8	1.520	0.01	0.02	0.11	0.01	0.02	0.07	0.240	0.364
Part G	0.022	7	0.154	0.02	0.05	0.15	—	0.05	0.15	0.420	0.064
	Total		6.653							Total	2.039

N = Quantity of parts Iso = Isolation Che = Checkout
λ = Failure rate Acc = Access Int = Interchange
Loc = Localization Ali = Alignment Mct_i = Maintenance cycle time
For determination of $\overline{\text{MLH}}_c$, enter labor hours for maintenance times.

Design for Maintainability

TABLE 11 Maintainability Prediction Data Summary

Worksheet No.	Item designation	Worksheet Factor	
		$\Sigma(N)(\lambda)$	$\Sigma(N)(\lambda)(\text{Mct}_i)$
1	Assembly 1	7.776	3.021
2	Assembly 2	5.328	1.928
3	Assembly 3	8.411	2.891
4	Assembly 4	6.653	2.039
5	Assembly 5	5.112	2.576
⋮	⋮	⋮	⋮
13	Assembly 13	4.798	3.112
Total		86.476	33.118

$$\overline{\text{Mct}} = \frac{\Sigma(N)(\lambda)(\text{Mct}_i)}{\Sigma(N)(\lambda)} = \frac{33.118}{86.486} = 0.382 \text{ hour}$$

Prediction of Maintenance Resource Requirements.   Maintenance resources in this instance include the personnel and training requirements, test and support equipment, supply support (e g., spare and repair parts and associated inventories), transportation and handling requirements, facilities, computer software, and data needed in the accomplishment of maintenance actions. For instance, *What test equipment is required to accomplish diagnostics and fault isolation? What personnel types and skill levels are required to repair the faulty item? What type of handling equipment is necessary to transport the item to the intermediate maintenance shop?* In essence, the elapsed-time element alone does not provide an adequate prediction of the maintainability characteristics in design. One needs to predict the time element and the resources required.

5.3 Reliability-Centered Maintenance (RCM)

Reliability-Centered Maintenance (RCM) is a systematic approach to developing a focused, effective, and cost-efficient preventive maintenance program and control plan for a system or product. This technique is best initiated during the early design process and

TABLE 12 Preventive Maintenance Data Summary

Description of Preventive Maintenance Task	Task Frequency $(\text{fpt}_i)(N)$	Task Time (Mpt_i)	Product $(\text{fpt}_i)(N)(\text{Mpt}_i)$
1. Lubricate	0.115	5.511	0.060
2. Calibrate	0.542	4.234	0.220
⋮	⋮	⋮	⋮
31. Service	0.321	3.315	0.106
Total	13.260		31.115

$$\overline{\text{Mpt}} = \frac{\Sigma(\text{fpt}_i)(N)(\text{Mpt}_i)}{\Sigma(\text{fpt}_i)(N)} = \frac{31.115}{13.260} = 2.346 \text{ hours}$$

evolves as the system is developed, produced, and deployed. However, this technique can also be used to evaluate preventive maintenance programs for existing systems with the objective of continuous product/process improvement in mind.[8]

The RCM technique was developed in the 1960s, primarily through the efforts of the commercial airline industry. It can be accomplished using a structured decision tree that leads the analyst through a "tailored" logic approach to delineate the most applicable preventive maintenance tasks (their nature and frequency). Referring to Figure 12, a simplified top-down RCM decision–logic structure is presented, where system safety is a prime consideration along with performance and cost. By following the steps, one is led to either the specification of preventive maintenance or a recommendation for redesign.[9]

With the RCM analysis directed toward the establishment of a cost-effective preventive maintenance program, a necessary prerequisite is the accomplishment of the FMECA, one of the outputs from the FMECA could lead to the identification of a preventive maintenance requirement—one that can be justified on the basis of reliability information. Figure 13 shows how the RCM analysis model can be integrated with some of the other tools discussed herein.

5.4 Level-of-Repair Analysis (LORA)

In expanding the maintenance concept to establish criteria for system design, it is necessary to determine whether it is economically feasible to repair certain assemblies or to discard them when failures occur. If the decision is to accomplish repair, it is appropriate to determine the maintenance level at which the repair should be accomplished (i.e., intermediate maintenance or supplier/depot maintenance). This example illustrates a level-of-repair analysis based on life-cycle cost criteria, which may be accomplished during conceptual or preliminary system design and/or subsequently as feasible.

Suppose that a computer system is to be distributed in quantities of 65 throughout three major geographical areas. The system will be used to support both scientific and management functions within various industrial firms and government agencies. Although the actual system utilization will vary from one consumer organization to the next, an average use of 4 hours per day (for a 360-day year) is assumed.

The computer system is currently in the early development stage, should be in production in 18 months, and will be operational in 2 years. The full complement of 65 com-

[8]Quite often during system development, the designer, when selecting components, will not be knowledgeable of the specific reliability physics-of-failure characteristics of an item. So, to ensure the proper level of reliability for the system, the designer will select a component and indicate the need for preventive maintenance (e.g., a periodic item replacement) "just in case!" This has, for many systems in the past, resulted in the overspecification of maintenance requirements (i.e., the specification of too much preventive maintenance, which is unjustified from a reliability perspective due to its cost).

[9]A maintenance steering group (MSG) was formed in the 1960s that undertook the development of this technique. The result was a document called "747 Maintenance Steering Group Handbook: Maintenance Evaluation and Program Development (MSG-1)" published in 1968. This effort, focused on a particular aircraft, was next generalized and published in 1970 as "Airline/Manufacturer Maintenance Program Planning Document-MSG2." The MSG-2 approach was further developed and published in 1978 as "Reliability Centered Maintenance," Report Number A066-579, prepared by United Airlines, and in 1980 as "Airline/Manufacturer Maintenance Program Planning Document-MSG3." The MSG-3 report has been revised as "Airline/Manufacturer Maintenance Program Development Document (MSG-3), 1993." These reports are available from the Air Transport Association.

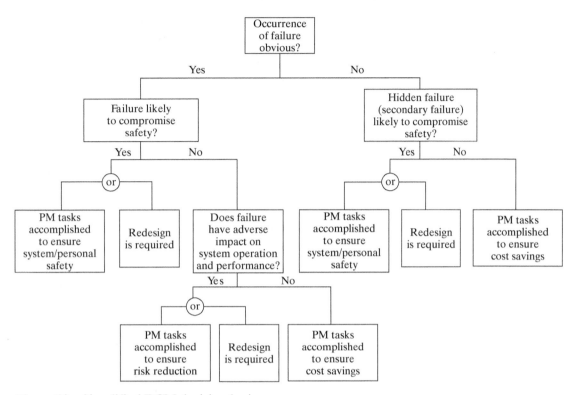

Figure 12 Simplified RCM decision logic.

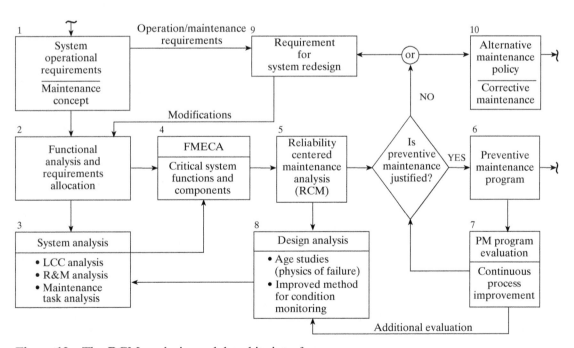

Figure 13 The RCM analysis model and its interfaces.

puter systems is expected to be in use in 4 years and will be available through the 8 years of the program before system phase-out commences. The system life cycle, for the purposes of the analysis, is 10 years.

Based on early design data, the computer system will be packaged in major units with a built-in test capability that will enable the isolation of faults to the unit level. Faulty units will be removed and replaced at the organizational level (i.e., customer's facility) and sent to the intermediate maintenance shop for repair. Unit repair will be accomplished through assembly replacement and assemblies will be either repaired or discarded. A total of 15 assemblies are being considered and the requirement is to justify the assembly repair or discard decision on the basis of life-cycle cost criteria. The operational requirements, maintenance concept, and program plan are illustrated in Figure 14.

The stated problem primarily pertains to the analysis of 15 major assemblies of the given computer system configuration to determine whether the assemblies should be repaired or discarded when failures occur. In other words, various assemblies will be individually evaluated in terms of (1) assembly repair at the intermediate level of maintenance, (2) assembly repair at the supplier or depot level of maintenance, and (3) disposing of the assembly. Life-cycle costs, as applicable to the assembly level, will be estimated and employed in the alternative selection process. Total overall computer system costs have been determined at a higher level, and are not included in this example.

Given the information in the problem statement, the next step is to develop a cost breakdown structure (CBS) and to establish evaluation criteria. The evaluation criteria include consideration of all costs in each applicable category of the CBS, but the emphasis is on operation and support (O&S) costs as a function of acquisition cost. Thus, the research and development cost and the production cost are presented as one element, whereas various segments of O&S costs are identified individually. Figure 15 presents evaluation criteria, cost data, and a brief description and justification supporting each category. The information shown in the figure covers only one of the 15 assemblies but is typical for each case.

The data presented in Figure 15 represent Assembly A-1, used in a manner illustrated in Figure 14. The next step is to employ the same criteria to determine the recommended repair-level decision for each of the other 14 assemblies (i.e., Assemblies 2–15). Although acquisition costs, reliability and maintainability factors, and certain logistics requirements are different for each assembly, many of the cost-estimating relationships are the same. The objective is to be *consistent* in analysis approach and in the use of input cost factors to the maximum extent possible and where appropriate. The summary results for all 15 assemblies are presented in Table 13.

Referring to Table 13, note that the decision for Assembly A-1 favors repair at the intermediate level; the decision for Assembly A-2 is repair at the supplier or depot level; the decision for Assembly A-3 is not to accomplish repair at all but to discard the assembly when a failure occurs; and so on. The table reflects recommended policies for each individual assembly. In addition, the overall policy decision, when addressing all 15 assemblies as an integral package, favors repair at the supplier's facility.

Before arriving at a final conclusion, the analyst should reevaluate each situation where the decision is close. Referring to Figure 15, it is clearly uneconomical to accept the discard decision; however, the two repair alternatives are relatively close. Based on the results of the various individual analyses, the analyst knows that repair-level decisions highly depend on the unit acquisition cost of each assembly and the total estimated number of replacements over the expected life cycle (i.e., maintenance actions based on assembly reliability). The trends are illustrated in Figure 16, where the decision tends to shift from discard

Design for Maintainability

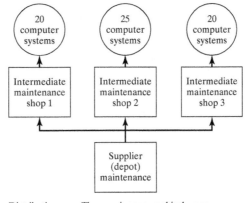

Distribution: Three major geographical areas
Utilization: Four hr/day throughout year (average)

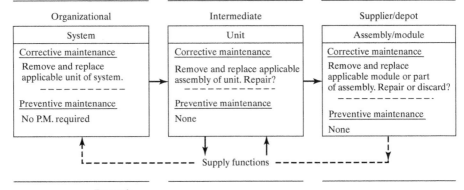

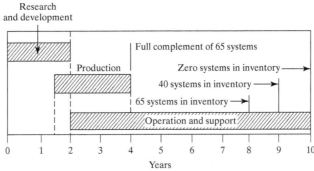

Figure 14 Basic system concepts.

to repair at the intermediate level as the unit acquisition cost and the number of replacements increase (or the reliability decreases).[10]

In instances where the individual analysis result lies close to the crossover lines in Figure 16, the analyst may wish to review the input data, the assumptions, and accomplish a

---

[10]The curves projected in Figure 16 are characteristic of this particular analysis and will vary with changes in operational requirements, system use, the maintenance concept, production requirements, and so on.

Evaluation Criteria	Repair at Intermediate Cost ($)	Repair at Supplier Cost ($)	Discard at Failure Cost ($)	Description and Justification
1. Estimated acquisition costs for assembly A-1 (to include R & D cost and production cost)	550/Assy. or 35,750	550/Assy. or 35,750	475/Assy. or 30,875	Acquisition cost includes all applicable costs allocated to assembly A-1 based on a requirement of 65 systems Assembly design and production are simplified in the discard area.
2. Unscheduled maintenance costs	6,480	8,100	Not applicable	Based in the 8-year useful system life, 65 systems, a utilization of 4 hrs/day, a failure rate (λ) of 0.00045 for assembly A-1, and a $\overline{M}ct$ of 2 hr, the expected number of maintenance actions is 270. When repair is accomplished, two technicians are required on a full-time basis. The labor rates are $12/hr for intermediate maintenance and $15/hr for supplier maintenance.
3. Supply support spare assemblies	3,300	4,950	128,250	For intermediate maintenance 6 spare assemblies are required to compensate for transportation time, the maintenance queue, TAT, etc. For supplier/depot maintenance 9 spare assemblies are required. 100% spares are required in the discard case.
4. Supply support spare modules or parts for assembly repair	6,750	6,750	Not applicable	Assume $25 for materials per repair action.
5. Supply support inventory management	2,010	2,340	25,650	Assume 20% of the inventory value (spare assemblies, modules, and parts).
6. Test and support equipment	5,001	1,667	Not applicable	Special test equipment is required in the repair case. The acquisition and support cost is $25,000 per installation. The allocation for assembly A-1 per installation is $1,667. No special test equipment is required in the discard case.
7. Transportation and handling	Not applicable	2,975	Not applicable	Transportation costs at the intermediate level are negligible. For supplier maintenance, assume 340 one-way trips at $175/100 lb. One assembly weighs 5 pounds.
8. Maintenance training	260	90	Not applicable	Delta training cost to cover maintenance of the assembly is based on the following: Intermediate—26 students, 2 hrs each, $200/student week. Supplier—9 students, 2 hrs each, $200/student week.
9. Maintenance facilities	594	810	Not applicable	From experience, a cost estimating relationship of $0.55 per direct maintenance labor-hour is assumed for the intermediate level, and $0.75 is assumed for the supplier level.
10. Technical data	1,250	1,250	Not applicable	Assume 5 pages for diagrams and text covering assembly repair at $250/page.
11. Disposal	270	270	2,700	Assume $10/assembly and $1/module or part as the cost of disposal.
Total estimated cost	61,665	64,952	187,475	

Figure 15   Repair versus discard evaluation (Assembly A-1).

sensitivity analysis involving the high-cost contributors. The purpose is to assess the risk involved and verify the decision. This is the situation for Assembly A-1, where the decision is close relative to repair at the intermediate level versus repair at the supplier's facility (refer to Figure 15).

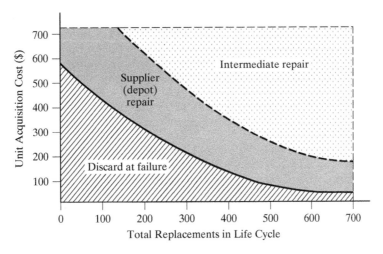

Figure 16 Economic screening criteria.

After reviewing the individual analyses of the 15 assemblies to ensure that the best possible decision is reached, the results in Table 13 are updated as required. Assuming that the decisions remain basically as indicated, the analyst may proceed in either of two ways. First, the decisions in the table may be accepted without change, supporting a *mixed* policy with some assemblies being repaired at each level of maintenance and other assemblies being discarded at failure. With this approach, the analyst should review the interaction effects that could occur (i.e., the effects on spares, use of test and support equipment, maintenance

TABLE 13 Summary of Repair-Level Costs

| Assembly Number | Maintenance Status | | | |
	Repair at Intermediate Cost ($)	Repair at Supplier Cost ($)	Discard at Failure Cost ($)	Decision
A-1	61,665	66,702	187,475	Repair—intermediate
A-2	58,149	51,341	122,611	Repair—supplier
A-3	85,115	81,544	73,932	Discard
A-4	85,778	78,972	65,071	Discard
A-5	66,679	61,724	95,108	Repair—supplier
A-6	65,101	72,988	89,216	Repair—intermediate
A-7	72,223	75,591	92,114	Repair—intermediate
A-8	89,348	78,204	76,222	Discard
A-9	78,762	71,444	89,875	Repair—supplier
A-10	63,915	67,805	97,212	Repair—intermediate
A-11	67,001	66,158	64,229	Discard
A-12	69,212	71,575	82,109	Repair—intermediate
A-13	77,101	65,555	83,219	Repair—supplier
A-14	59,299	62,515	62,005	Repair—intermediate
A-15	71,919	65,244	63,050	Discard
Policy cost	1,071,267	1,037,362	1,343,449	Repair—supplier

personnel use, etc.). In essence, each assembly is evaluated individually based on certain assumptions, the results are reviewed in the context of the whole, and possible feedback effects are assessed to ensure that there is no significant impact on the decision.

A second approach is to select the overall least-cost policy for all 15 assemblies treated as an entity (i.e., assembly repair at the supplier or depot level of maintenance). In this case, all assemblies are designated as being repaired at the supplier's facility, and each individual analysis is reviewed in terms of the criteria in Figure 15 to determine the possible interaction effects associated with the single policy. The result may indicate some changes to the values in Table 13.

Finally, the output of the repair-level analysis must be reviewed to ensure compatibility with the initially specified system maintenance concept. The analysis data may either directly support and be an expansion of the maintenance concept, or the maintenance concept will require change as a consequence of the analysis. If the latter occurs, other facets of system design may be significantly impacted. The consequences of such maintenance concept changes must be thoroughly evaluated before arriving at a final repair-level decision. The basic level-of-repair analysis procedure is illustrated in Figure 17.

5.5 Maintenance Task Analysis (MTA)

The MTA constitutes the process of evaluating a given system configuration with the following objectives in mind:

1. Identify the resources required for sustaining maintenance and support of the system throughout its planned life cycle. Such resource requirements may include the quantity and skill levels of personnel needed for maintenance, spares/repair parts and associated inventories, tools and test equipment, transportation and handling requirements, maintenance facilities and associated capital assets, information and technical data, and computer resources required in support of maintenance activities.

2. Provide an assessment of the configuration relative to the incorporation of maintainability characteristics in design, both in the design of the prime mission-related elements of the system and in the design of the maintenance and support infrastructure. The objective is to ensure that the system design configuration at the time is compatible with the TPMs and the DDPs, and that the required resource requirements for support are minimized. The system makeup should not only include the prime mission-related elements (i.e., operational hardware and software, operating personnel, operational facilities, and so on) but also the elements of the support network and infrastructure as well—there are two sides of the balance as conveyed in Figure 9.

Accordingly, the MTA is based on a given design configuration as it exists at the time of analysis and evaluation and serves as an excellent "assessment" tool in measuring the effectiveness of the support infrastructure as currently envisioned. With the MTA being accomplished iteratively throughout the design process and with a good "feedback" capability, it provides a good mechanism for the incorporation of design changes as needed for continuous product and process improvement.

The MTA may be accomplished at a gross level during the conceptual design phase when there is enough definition of specific elements of the system (evolving from the maintenance concept) or when an existing repairable item is being considered for incorpo-

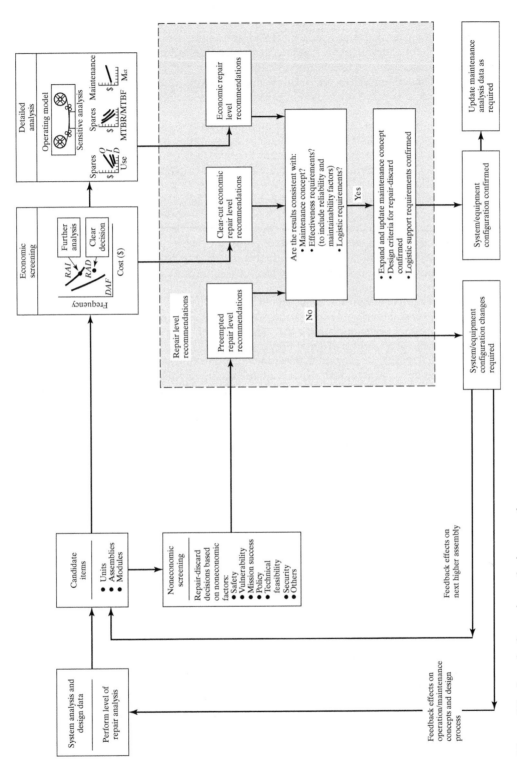

Figure 17 Level-of-repair analysis procedure.

ration into the overall design (i.e., a COTS item). As the design progresses during the preliminary and detail design phases and the system configuration becomes better defined, the analyst may accomplish a MTA through a review of available design data, drawings, component part and material lists, technical reports, and so on. The analyst will evaluate what he or she sees from the data, projected in the context of what is anticipated in the system utilization phase.

In conducting the MTA, the functional analysis provides the necessary foundation. For the purposes of illustration, a manufacturing plant has been selected and is described through the abbreviated functional block diagram shown in Figure 18. The appropriate metrics (i.e., TPMs) should be identified and allocated to each functional block. Each operational function can then be viewed in terms of *go/no-go* criteria, with the development of maintenance functions evolving from the operating functions. This process is illustrated in Figure 19.

The maintenance functions can then be broken down into subfunctions, duties, groups of activity, tasks, subtasks, and so on. With the "whats" described in terms of identified tasks, the analyst identifies and evaluates the various alternative approaches for task accomplishment. The results lead to the identification of the "hows" and the specific resources required for maintenance.

Referring to Figure 18, it is suspected that the "System Inspection and Test" activity (Function 13) constitutes a potential problem area and that it is desired to accomplish an MTA for the purposes of evaluation. Given that this function fails frequently, the analyst may commence with the preparation of an abbreviated logic flow diagram, as shown in Figure 20. Starting with a typical symptom of failure (one that occurs most often), a step-by-step *go/no-go* approach can be delineated, with the critical tasks being identified by the numbers just above each block. The selected "path" covers the corrective maintenance cycle, including diagnostics, repair, and condition verification.

The next step is to evaluate the critical tasks identified in Figure 20. Referring to Figure 21 (sheet 1 and sheet 2, which is an extension), each of the selected tasks is evaluated and the required maintenance resources are identified (i.e., task times and sequences in Blocks 12 and 13, task frequency in Block 14, personnel quantities and skill-level requirements in Blocks 15–18, spares and replacement parts in Blocks 8–11 (sheet 2), test and support equipment in Blocks 12–15, and facility requirements in Block 16).[11]

Given the information presented in Figure 21 (sheets 1 and 2), the analyst may wish to question a few specific areas of concern:

1. With the extensive resources required for the repair of Assembly A-7 (e.g., the variety of special test and support equipment, the necessity for a "clean-room" facility for maintenance, the extensive amount of time required for the removal and replacement of CB-1A5, etc.), it may be feasible to identify Assembly A-7 as being "nonrepairable." In other words, investigate the feasibility of whether the assemblies of Unit *B* should be classified as "repairable" or "discard at failure."

2. Referring to Tasks 1 and 2, a "built-in test" capability exists at the organizational level for fault isolation to the subsystem. However, fault isolation to the unit requires a special system tester (0-2310B) and it takes 25 minutes of testing plus a highly skilled (supervisory skill) individual to accomplish the function. In essence, one should investigate the feasibility of extending the built-in test down to the unit level and eliminate the need for the special system tester and the highly skilled individual.

---

[11]B. S. Blanchard, D. Verma, and E. L. Peterson, *Maintainability: A Key to Effective Serviceability and Maintenance Management* (New York: John Wiley & Sons, Inc., 1995).

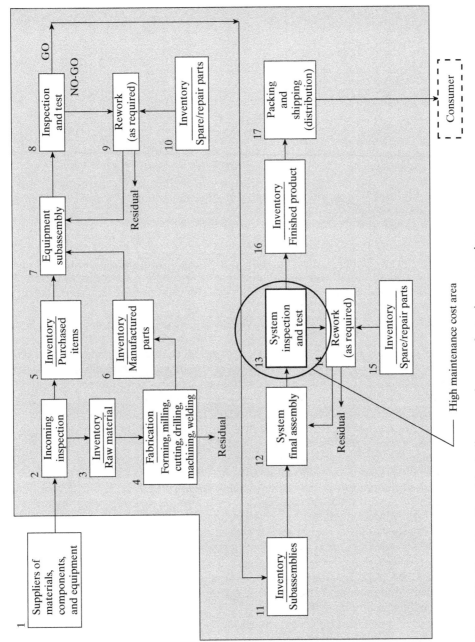

Figure 18 Function flow block diagram of a manufacturing operation.

515

Operational functions

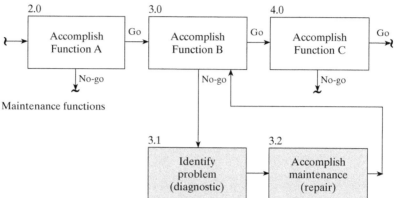

Figure 19 The identification of maintenance functions from operational functions.

3. The physical removal and replacement of Unit *B* from the system takes 15 minutes, which seems rather extensive. Although perhaps not a major item, it would be worthwhile investigating whether the removal/replacement time can be reduced (to less than 5 minutes, for example).

4. Referring to Tasks 10–15, a special clean-room facility is required for maintenance. Assuming that the various assemblies of Unit *B* are repaired (versus being classified as "discard at failure"), it would be worthwhile to investigate changing the design of these assemblies such that a clean-room environment is not required for maintenance. In other words, can the expensive maintenance facility requirement be eliminated?

5. There is an apparent requirement for a number of new "special" test equipment/tool items; that is, special system tester [0-2310B], special system tester [I-8891011-A], special system tester [I-8891011-8], C.B. test set [D-2252-A], special extractor tool [EX20003-4], and special extractor tool [EX45112-6]. Usually, these *special* items are limited as to general application for other systems and are expensive to acquire and maintain. Initially, one should investigate whether these items can be eliminated. If test equipment/tools are required, can *standard* items be used (in lieu of special items)? Also, if the various special testers are required, can they be integrated into a "single" requirement? In other words, can a single item be designed to replace the three special testers and the C.B. test set? Reducing the overall requirements for special test and support equipment is a major objective.

6. Referring to Task 9, there is a special handling container for the transportation of Assembly A-7. This may impose a problem in terms of the availability of the container at the time and place of need. It would be preferable if normal packaging and handling methods could be used.

7. Referring to Task 14, the removal and replacement of C8-1A5 takes 40 minutes and requires a highly skilled individual to accomplish the maintenance task. Assuming that Assembly A-7 is repairable, it would be appropriate to simplify the circuit board removal/replacement procedure by incorporating plug-in components or at least simplify the task to allow one with a basic skill level to accomplish.

Design for Maintainability

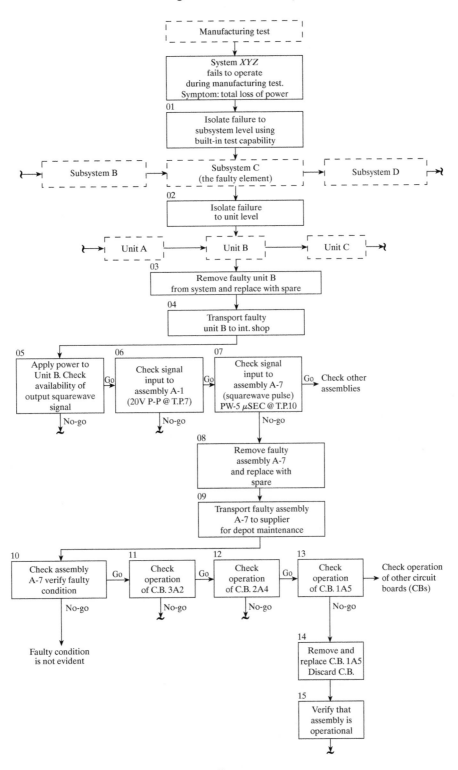

Figure 20   Abbreviated logic flow diagram

1. System XYZ
2. Item Name/Part No. Manufacturing Test/A4321
3. Next Higher Assy. Assembly and Test
4. Description of requirements During manufacturing and test of Product 12345 (Serial No. 654), System XYZ failed to operate. The symptom of failure was "loss of total power output." Requirements: troubleshoot and repair system.

5. Req. No. 01
6. Requirement Diag./Repair
7. Req. Freq. 0.00450
8. Maint. Level Org./Inter.
9. Ma. Cont. No. A12B100

10. Task Number	11. Task Description	12. Elapsed Time-Minutes	13. Total Elapsed Time	14. Task Freq.	15. B	16. I	17. S	18. Total
01	Isolate failure to subsystem level (Subsystem C is faulty)		5	0.00450	5	–	–	5
02	Isolate failure to unit level ①		25		–	25	–	25
	(Unit B is faulty) ②				–	–	25	25
03	Remove unit B from system and replace with a spare unit B (2nd cycle)		15		15	–	–	15
04	Transport faulty unit to int. shop		30		30	–	–	30
05	Apply power to faulty unit, check for output for squarewave signal (3rd cycle)		20		–	20	–	20
06	Check signal input to assembly A-1 (20v P-P @ T.P 7)		15		–	15	–	15
07	Check signal input to assembly A-7 (Squarewave, PW-5μsec @ T.P2) (4th cycle)		20		–	20	–	20
08	Remove faulty A-7 and replace		10		10	–	–	10
09	Transport faulty assembly A-7 to supplier for depot maintenance – 14 calendar days in transit –							
10	Check A-7 and verify faulty cond. (5th cycle)		25		–	–	25	25
11	Check operation of CB-3A2		15		–	–	15	15
12	Check operation of CB-2A4		10		–	–	10	10
13	Check operation of CB-1A5 (6th cycle)		20		–	–	20	20
14	Remove and replace faulty CB-1A5 (7th cycle)		40		–	40	–	40
	Discard faulty circuit board							
15	Verify that assembly is operational and return to inventory		15		–	–	15	15
	Total		**265**	**0.00450**	**60**	**120**	**110**	**290**

Elapsed Time-Minutes scale: 2 4 6 8 10 12 14 16 18 20 22 24 26 28 30 32 34 36 38

Figure 21 Maintenance task analysis (sheet 1).

Item Name/Part No. Manufacturing Test/A4321		1. Req. No. 01		3. Requirement Diagnostic Troubleshooting and Repair		4. Req. Freq. 0.00450	5. Maint. Level: Organization, Intermediate, Depot	6. Ma. Cont. No. A12B100
7. Task Number	8. Qty. per Assy.	9. Parts Nomenclature / 11. Part Number (Replacement Parts)	10. Rep. Freq.	12. Qty.	13. Item Nomenclature / 15. Item Part Number (Test and Support/Handling Equipment)	14. Use Time (min)	16. Description of Facility Requirements	17. Special Technical Data Instructions
01	–	– – – – –	–	1	Built-in test equip. A123456	5	– – – – –	Organizational maintenance
02	–	– – – – –	–	1	Special system tester O–2310B	2.5	– – – – –	
03	1	Unit B B180265X	0.01866	1	Standard tool kit STK–100–B	15	– – – – –	
04	–	– – – – –	–	1	Standard cart (M–10)	30	– – – – –	Intermediate maintenance
05	–	– – – – –	–	1	Special system tester I–8891011–A	20	– – – – –	
06	–	– – – – –	–	1	Special system tester I–8891011–A	15	– – – – –	
07	–	Assembly A–7 MO–2378A	–	1	Special system tester I–8891011–A	20	– – – – –	
08	1	– – – – –	0.00995	1	Special extractor tool EX20003–4	10	– – – – –	Refer to special removal instructions
09	–	– – – – –	–	1	Container, special handling T–300A	14 days	– – – – –	Normal trans. environment
10	–	– – – – –	–	1	Special system tester I–8891011–A	25	Clean room environment	Supplier (depot) maintenance
11	–	– – – – –	–	1	C.B. test set D–2252–A	15		
12	–	– – – – –	–	1	C.B. test set D–2252–A	10		
13	–	– – – – –	–	1	C.B. test set D–2252–A	20		
14	1	CB–1A5 GDA–221056C	0.00450	1	Special extractor tool/EX45112–6; standard tool kit STK–200	40		
15	–	– – – – –	–	1	Special system tester I–8891011–A	15		Return operating assy. to inventory

Figure 21 Maintenance task analysis (sheet 2).

The MTA evolves from the defined maintenance concept and the functional analysis and uses, as an input, data from the FMECA, RCM, reliability and maintainability predictions, and the level-of-repair analysis as necessary. Figure 22 includes a sample illustration showing the relationships of these various analyses. With the accomplishment of the MTA covering all of the repairable elements of the system, the total maintenance and support infrastructure can be further defined through the development of a detailed maintenance plan. From this plan, one can proceed with the necessary provisioning actions, procurement, and the subsequent acquisition of spare parts, test equipment, facilities, data, and so on. The MTA constitutes an essential input to support requirements.

5.6 Total Productive Maintenance (TPM)

Total productive maintenance, a concept introduced by the Japanese in the early 1970s, represents an "integrated life-cycle approach to the maintenance and support of a manufacturing plant." Experience indicated that many of the existing factories had been operating at less than full capacity, productivity was low in general, and the costs of factory operations were high. Further, a large portion of the total cost of doing business was due to "production losses" as a result of excessive factory downtime and maintenance requirements. These costs, in turn, were transferred to the cost of the product being manufactured. In other words, the cost of many of the products being delivered for consumer use was high because of the cost of manufacturing, which, in turn, had a negative impact in the commercial marketplace. Thus, an intensive effort was initiated in Japan to increase productivity and reduce the costs of manufacturing, with the issue of *maintenance* being a "target" of opportunity.[12]

Although the issue of maintenance was the initial "motivator," the implications relative to implementing the concepts and principles of TPM are more far-reaching. In essence, TPM is an approach for improving the overall effectiveness and efficiency of a manufacturing plant. More specifically, the objectives are to:

1. Maximize the overall effectiveness of manufacturing equipment and processes. This pertains to maximizing the availability of the production process through improvement of equipment reliability and maintainability, with the goal of minimizing downtime.

2. Establish a life-cycle approach in the accomplishment of preventive maintenance. This pertains to the application of a reliability-centered maintenance (RCM) approach to justify the requirements for preventive maintenance.

3. Involve all operating departments/groups within a manufacturing plant organization in the planning for and subsequent implementation of a maintenance program (i.e., representation from engineering, operations, testing, marketing, and maintenance). The objective is to gain full *commitment* throughout the manufacturing organization.

4. Involve employees from the plant manager to the workers on the floor. There must be commitment from the top down in the organizational hierarchy.

5. Initiate a program based on the promotion of maintenance through "motivation management" and the development of autonomous small-group activities (i.e., the

---

[12]S. Nakajima (ed.), *TPM Development Program: Implementing Total Productive Maintenance* (New York: English translation. Productivity Press, 1989). Subsequently, there have been numerous additional references on TPM published by Productivity Press. Also, refer to T. Wireman, *Total Productive Maintenance*, 2nd ed. (Boca Raton, FL: Industrial Press, 2003).

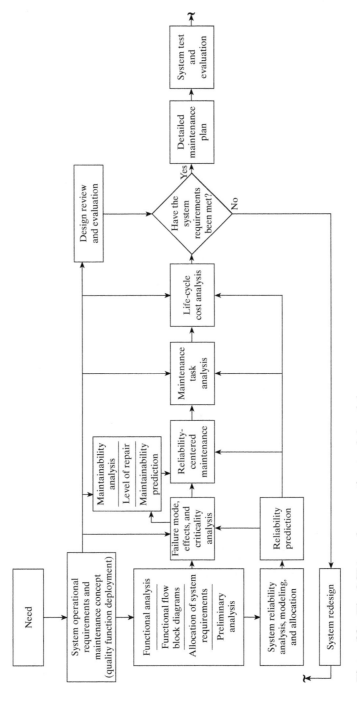

Figure 22 The MTA and supporting tools/models.

accomplishment of maintenance through a "team" approach). The operator must assume greater responsibility for the operation of his or her own equipment and must be trained to detect equipment abnormalities, understand the cause-and-effect relationships, and accomplish minor repair. The operator is then supported by other members of the team for the accomplishment of higher-level maintenance activities. The application of the FMECA and the Ishikawa diagram can help in identifying cause-and-effect relationships.

The measure of total productive maintenance (TPM) can be expressed in terms of *overall equipment effectiveness* (OEE), which is a function of *availability, performance rate*, and *quality rate* as

$$\text{OEE} = (\text{availability})(\text{performance rate})(\text{quality rate}) \tag{24}$$

where

$$\text{Availability } (A) = \frac{\text{loading time} - \text{downtime}}{\text{loading time}} \tag{25}$$

with *loading time* referring to the total time available for the manufacture of products (expressed in days, months, or minutes) and *downtime* referring to stoppages resulting from breakdown losses (failures), losses because of setups/adjustments, and so forth

and

$$\text{Performance rate } (P) = \frac{(\text{output})(\text{actual cycle time})}{(\text{loading time} - \text{downtime})} \times \frac{(\text{ideal cycle time})}{(\text{actual cycle time})} \tag{26}$$

with *output* referring to the number of products produced; *ideal cycle time* referring to the ideal or designed cycle time per product (e.g., minutes per product); and *actual cycle time* referring to the actual time experienced (minutes per product). This metric addresses the process and rate of performance

and

$$\text{Quality rate } (Q) = \frac{\text{input} - (\text{quality defects} + \text{startup defects} + \text{rework})}{\text{input}} \tag{27}$$

with *input* referring to the total number of products processed and *defects* referring to the total number of failures for one reason or another.

In essence, one is dealing with the manufacturing plant as a *system*, with the OEE factor being a measure of system effectiveness, as discussed in Section 3. As such, the manufacturing plant can be defined in functional terms, metrics can be allocated for each functional element, high-cost contributors can be identified, cause-and-effect relationships can be established, and recommendations for design improvement can be initiated as appropriate. Thus, although this subject is being addressed in this chapter (because of the fact that *maintenance* was the initial "driver"), many concepts and principles are applicable to the design, development, construction, operation, and support of a manufacturing plant.

6 MAINTAINABILITY DEMONSTRATION

Maintainability demonstration is conducted as part of the system test and evaluation effort. Specifically, maintainability demonstration is accomplished as part of Type 2 testing to verify that qualitative and quantitative maintainability requirements have been achieved. It also provides for the assessment of various logistic support factors related to, and having an impact on, maintainability parameters and item downtime (e.g., test and support equipment, spare/repair parts, technical data, personnel, and maintenance policies).

Maintainability demonstration is usually accomplished during the latter part of the detail design and development phase and should be conducted in an environment that simulates, as closely as practical, the operational and maintenance environment planned for the item. The maintainability demonstration may vary considerably depending on system requirements and the test objectives. Two representative approaches are described in this section to provide an idea of the steps involved.

6.1 Demonstration Method 1

Maintainability demonstration Method 1 follows a sequential test approach that is similar to a reliability test. Two different sequential test plans are employed to demonstrate $\overline{M}ct$ and M_{max} (for corrective maintenance). An accept decision for the equipment under test is reached when that decision can be made for both test plans. The test plans assume that the underlying distribution of corrective maintenance task times is lognormal. The sequential test plan approach allows for a quick decision when the maintainability of the equipment under test is either far above or far below the specified values of $\overline{M}ct$ and M_{max}.

Maintainability testing is accomplished by simulating faults in the system and observing the task times and logistic resources required to correct the situation. It involves the following steps:

1. A failure is induced in the equipment without knowledge of the test team. The induced failure should not be evident in any respect other than that normally resulting from the simulated mode of failure. In other words, the technician(s) scheduled to perform the maintenance demonstration shall not be given any hints (through visual or other evidence) as to where the failure is induced.

2. The maintenance technician will be called on to check out the equipment operationally. At some point in the checkout procedure, a symptom of malfunction is detected.

3. Once a malfunction has been detected, the maintenance technician(s) will proceed to accomplish the necessary corrective maintenance tasks (i.e., fault localization, isolation, disassembly, remove and adjustment and alignment, and system checkout—see Figure 1). In the performance of each step, the technician should follow approved maintenance procedures and use the proper test and support equipment. The maintenance tasks performed must be consistent with the maintenance concept and the specified levels of maintenance appropriate to the demonstration. Replacement parts required to perform repair actions being demonstrated shall be compatible with the spare and repair parts recommended for operational support.

4. While the maintenance technician is performing the corrective tasks (commencing with the identification of a malfunction and continuing until the equipment has been returned to full operational status), a test recorder collects data on task sequences, areas of task difficulty, and task times. In addition, the adequacies and inadequacies of

logistic support are noted. For example, Was the right type of support provided? Were there any test delays because of inadequacies? Was there an overabundance of certain items and/or a shortage of others? Did each specified element of support do the job for which it was intended in a satisfactory manner? Were the test procedures adequate? These and related questions should be in the mind of the data recorder during the observation of a test.

This maintainability test cycle is accomplished n times, where n is the selected sample size. For the sequential test, the number of demonstrations could possibly extend to 100, assuming that a continue-to-test decision prevails. Thus, in preparing for the test, a sample size of 100 demonstrations should be planned. The selected tasks should be representative and based on the expected percent contribution toward total maintenance requirements. Those items with high failure rates will fail more often and require more maintenance and logistic resources; hence, they should appear in the demonstration to a greater extent than items requiring less maintenance.

The task selection process is accomplished by proportionately distributing the 100 tasks among the major functional elements of a system. Assuming that three units compose a system, the 100 tasks may be allocated, as illustrated in Table 14. Referring to the table, the elements of the system and the associated failure rates (from reliability data) are listed. The percentage contribution of each item to the total anticipated corrective maintenance (column 5) is computed as

$$\text{Item percent contribution} = \frac{Q\lambda}{\Sigma Q\lambda} \times 100 \qquad (28)$$

This factor is used to allocate the tasks proportionately to each unit. In a similar manner, the 21 tasks within Unit A can be allocated to assemblies within that unit, and so on. When the allocation is completed, there may be one task assigned to a particular assembly and the assembly may contain several components, the failure of which reflects different failure modes (e.g., no output, erratic output, low output, etc.). Through a random process, one of the components in the assembly will be selected as the item where the failure is to be induced, and the method by which the failure is induced is specified.

With the tasks identified and listed in random order, the demonstration proceeds with the first task, then the second, third, and so on. The criteria for accept–reject decisions are illustrated in Figure 23. Task times (Mct_i) are measured and compared with the specified $\overline{\text{M}}\text{ct}$ and M_{max} values. When the demonstrated time exceeds the specified value, an

TABLE 14 Corrective Maintenance Task Allocation

(1) Item	(2) Quantity (Q)	(3) Failures/Item % 1,000 hours (λ)	(4) Total Failures (Q) (λ)	(5) % Contribution	(6) Allocated Maintenance Tasks for Demonstration
Unit A	1	0.48	0.48	21	21
Unit B	1	1.71	1.71	76	76
Unit C	1	0.06	0.06	3	3
Total			2.25	100	100

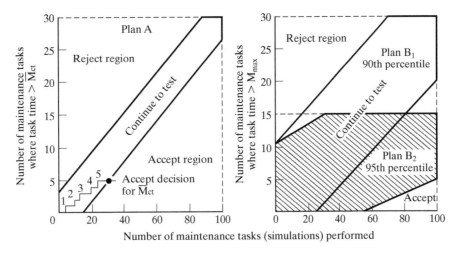

Figure 23 Graphical representation of maintainability demonstration plans. *Source:* MIL-STD-471, *Maintainability Verification, Demonstration, Evaluation* (Washington, DC: Department of Defense).

event is noted along the ordinate of the graph and problem areas are described. Testing then continues until the event line enters either the reject region or the accept region.

An example of the demonstration test score sheet is illustrated in Table 15. The accept–reject numbers support the decision lines in Figure 23 (refer to the $\overline{M}$ct curve). In this instance, 29 tasks were completed before an accept decision was reached. The sequential test requires that both the $\overline{M}$ct and the M_{max} criteria be met before the equipment is fully acceptable. M_{max} may be based on either the 90th or 95th percentile, depending on the specified system requirement and the test plan selected. If one test plan is completed with the event line crossing into the accept region, testing will continue until a decision is made in the other test plan.

Referring to Figure 23, the criteria for sequential testing specify that the minimum number of tasks possible for a quick decision for $\overline{M}$ct is 12 (test plan A). For M_{max} at the 90th percentile, the least number of tasks possible is 26 (test planB_1), whereas the figure is 57 for the 95th percentile (test plan B_2). Thus, if the maintainability of an item is exceptionally good, demonstrating the complete sample of 100 tasks may not be necessary, thus saving time and cost. Conversely, if the maintainability of an item is marginal and a continue-to-test decision prevails, the test program may require the demonstration of all 100 tasks. If truncation is reached, the equipment is acceptable for $\overline{M}$ct if 29 or fewer tasks do not exceed the specified $\overline{M}$ct value. Comparable factors for M_{max} are five or fewer for the 90th percentile and two or fewer for the 95th percentile.

6.2 Demonstration Method 2

This method is applicable to the demonstration of $\overline{M}$ct, $\overline{M}$pt, and $\overline{M}$. The underlying distribution of maintenance times is not restricted (no prior assumptions), and the sample size constitutes 50 corrective maintenance tasks for $\overline{M}$ct and 50 preventive maintenance tasks for $\overline{M}$pt. $\overline{M}$ is determined analytically from the test results for $\overline{M}$ct and $\overline{M}$pt. $\overline{M}_{max}$ can also be determined if the underlying distribution is assumed to be lognormal. This method offers the advantage of a fixed sample size, which facilitates the estimation of test costs.

TABLE 15 Demonstration Source Sheet

REQMT: $\overline{\text{Mct}} = 0.5$ hour = 30 minutes | Plan A

Maint. Task No.	Task Time Mct_i	Cum No. $\text{Mct}_i > \overline{\text{Mct}}$	Accept When Cum $\leq$ Than	Reject When Cum $>$ Than
1	12 minutes	0	—	—
2	6	0	—	—
3	18	0	—	—
4	32	1	—	—
5	19	1	—	5
6	27	1	—	6
7	108	2	—	6
8	6	2	—	6
9	14	2	—	7
10	47	3	—	7
11	28	3	—	7
12	19	3	0	7
13	4	3	0	8
14	24	3	0	8
15	78	4	1	8
⋮	⋮	⋮	⋮	⋮
24	20	4	3	11
25	127	5	4	11
26	21	5	4	12
27	13	5	4	12
28	28	5	4	12
29	8	5	5	12

Accept for $\overline{\text{Mct}}$

The test involves the selection and performance of maintenance tasks in a similar manner as described for demonstration Method 1. Tasks are selected based on their anticipated contribution to the total maintenance picture, and each task is performed and evaluated in terms of maintenance times and required logistic resources. Illustration of this method is best accomplished through an example. It is assumed that a system is designed to meet the following requirements and must be demonstrated accordingly:

$$\overline{M} = 75 \text{ minutes}$$
$$\overline{\text{Mct}} = 65 \text{ minutes}$$
$$\overline{\text{Mpt}} = 110 \text{ minutes}$$
$$M_{\text{max}} = 120 \text{ minutes}$$
$$\text{Producer's risk}(\alpha) = 20\%$$

This test is accomplished and the data collected are presented in Table 16. The determination of $\overline{\text{Mct}}$ (upper confidence limit) is based on the expression

$$\text{Upper limit} = \overline{\text{Mct}} + Z\left(\frac{\sigma}{\sqrt{n_c}}\right) \tag{29}$$

TABLE 16   Maintenance Test-Time Data

Demonstration Task Number	Observed Time Mct_i	$Mct_i - \overline{Mct}$ $(Mct_i - 62)$	$(Mct_i - \overline{Mct})^2$
1	58	−4	16
2	72	+10	100
3	32	−30	900
⋮	⋮	⋮	⋮
50	48	−14	196
Total	3,105		15,016

where

$$\overline{Mct} = \frac{\sum\limits_{i=1}^{n_c} Mct_i}{n_c} = \frac{3{,}105}{50} = 62.1 \text{ (assume 62)}$$

and $Z = 0.84$ from Table 4. Then,

$$\sigma = \sqrt{\frac{\sum\limits_{i=1}^{n_c} (Mct_i - \overline{Mct})^2}{n_c - 1}} = \sqrt{\frac{15{,}016}{49}} = 17.5$$

where n_c is the corrective maintenance sample size of 50.

$$\text{Upper limit} = 62 + \frac{(0.84)(17.5)}{\sqrt{50}} = 64.07 \text{ minutes}$$

The completed $\overline{Mct}$ statistic is compared to the corresponding accept–reject criteria, which is to accept if

$$\overline{Mct} + Z\left(\frac{\sigma}{\sqrt{n_c}}\right) \leq \overline{Mct} \text{ (specified)} \tag{30}$$

and reject if

$$\overline{Mct} + Z\left(\frac{\sigma}{\sqrt{n_c}}\right) > \overline{Mct} \text{ (specified)} \tag{31}$$

Applying demonstration test data, it can be seen that 64.07 minutes (the upper value of $\overline{Mct}$ derived by test) is less than the specified value of 65 minutes. Therefore, the system passes the $\overline{Mct}$ test and is accepted.

For preventive maintenance, the same approach is used. Fifty preventive maintenance tasks are demonstrated and task times (Mpt_i) are recorded. The sample mean preventive downtime is

$$\overline{Mpt} = \frac{\sum Mpt_i}{n_p} \tag{32}$$

The accept–reject criteria are the same as stated in Equations 30 and 31, except that preventive maintenance factors are used. That is, accept if

$$\overline{M}\text{pt} + Z\left(\frac{\sigma}{\sqrt{n_p}}\right) \le 110 \text{ minutes}$$

and reject if

$$\overline{M}\text{pt} + Z\left(\frac{\sigma}{\sqrt{n_p}}\right) > 110 \text{ minutes}$$

Given the test values for $\overline{M}\text{ct}$ and $\overline{M}\text{pt}$, the calculated mean maintenance time is

$$\overline{M} = \frac{(\lambda)(\overline{M}\text{ct}) + (\text{fpt})(\overline{M}\text{pt})}{\lambda + \text{fpt}} \tag{33}$$

where

λ = corrective maintenance rate or the expected number of corrective maintenance tasks occurring in a designated period

fpt = preventive maintenance rate or the expected number of preventive maintenance tasks occurring in the same period

Using test data, the resultant value of $\overline{M}$ should be equal to or less than 75 minutes. Finally, M_{max} is determined from

$$M_{max} = \text{antilog}\,(\overline{\log M\text{ct}} + Z\sigma_{\log M\text{ct}_i}) \tag{34}$$

The calculated value should be equal to or less than 120 for acceptance. An example of calculation for M_{max} is presented in Section 2.1. If all the demonstrated values are better than the specified values, following the criteria defined previously, the system is accepted. If not, some retest or redesign may be required, depending on the seriousness of the problem.

6.3 Maintainability Assessment

Given the successful completion of maintainability demonstration tests, as part of Type 2 testing, the next step is the accomplishment of a "true" assessment of system maintainability in the field. Referring to Figure A.1, this can be accomplished by conducting Types 3 and 4 testing as part of the overall system test, evaluation, and validation effort. As system tests are conducted, the data collection, analysis, and reporting process should include such coverage as all of the preventive and corrective maintenance tasks that are accomplished, task sequences, task frequencies, task times, personnel labor hours, and the supporting resources (i.e., test and support equipment, spares/repair parts, personnel, facilities, etc.) that are required in the accomplishment of system maintenance. This, of course, should lead either to the "validation" of a specified system MTBM, MDT, MLH/OH (or equivalent requirement), or to a noted deficiency requiring corrective action and a possible design modification.

7 SUMMARY AND EXTENSIONS

This chapter provides an overview of the subject of maintainability, emphasizing its relationship to reliability and its importance in the systems engineering design and development process. Beginning with some key definitions, the chapter includes the initial establishment of maintainability requirements, development of the primary measures of maintainability, the measures of system availability and effectiveness, description of selected maintainability analysis models and techniques, the application of maintainability activities throughout the system life cycle, and the validation of maintainability through a formalized demonstration and test.

A review of this chapter enables one to recognize that there is a very close relationship and interdependency between reliability and maintainability. Many of the requirements pertaining to maintainability in design depend on reliability data as an input, and many design decisions in maintainability can have a good feedback effect on reliability. The measures of reliability and maintainability, when combined, must be properly balanced to meet an overall system availability requirement. Accordingly, reliability and maintainability must be properly applied and integrated in order to attain true operational feasibility.

Looking ahead, a major challenge includes the application of maintainability principles and concepts across the system as an entity. As is the case with regard to reliability, history is replete with instances where the emphasis on maintainability has been directed to various elements of a system (e.g., equipment), and not viewing these elements on a total integrated basis. A system is made up of products and processes and includes various mixes of hardware, software, people, facilities, data, information, and so on. System failures may stem from different causes, so there must be more emphasis placed on maintainability requirements across the spectrum. Referring to Section 3, for example, we need to deal more with the measures of maintainability within the overall context of *operational availability* (A_o) and *system effectiveness*, as compared to specifying maintainability requirements based only on some of the lower-level factors.

The following websites are recommended as a source for additional information:

1. Annual Reliability and Maintainability Symposium (RAMS)—*http://www.rams.org*.
2. Reliability Information Analysis Center (RIAC)—*http://theRIAC.org*.
3. Society for Maintenance and Reliability Professionals (SMRP)—*http://www.smrp.org*.

On a more specific note, when dealing with the system as an entity, one needs to consider *all* of its elements to include the logistic support infrastructure. A significant part of this infrastructure includes various activities within the total supply chain (SC). In this day of international trade and exchange, the supply chain often constitutes a major element of support, and it is this element that frequently contributes to large segments of system downtime. Holdups as a result of holidays, labor strikes, issues of a cultural nature, and so on, often cause unexpected delays not earlier anticipated. Thus, when considering over system operational availability and system effectiveness, the issues of supply chain must be addressed.

QUESTIONS AND PROBLEMS

1. Define *maintainability*. How does it differ from *maintenance*? Provide some examples.

2. Why is maintainability important in system design? When in the life-cycle process should it be considered? Why?

3. What are the quantitative measures of maintainability (discuss measures for hardware, software, personnel, facilities, and for the logistic support infrastructure)?

4. What is the significant difference between MTBF and MTBM? Between MTBF and MTBR? Between MTBR and MTBM?

5. Corrective-maintenance task times were observed as given in the following table.

Task Time (minutes)	Frequency	Task Time (minutes)	Frequency
41	2	37	4
39	3	25	10
47	2	36	5
35	5	31	7
23	13	13	3
27	10	11	2
33	6	15	8
17	12	29	8
19	12	21	14

(a) What is the range of observations?

(b) Using a class interval width of four, determine the number of class intervals. Plot the data and construct a curve. What type of distribution is indicated by the curve?

(c) What is the $\overline{M}ct$?

(d) What is the geometric mean of the repair times?

(e) What is the standard deviation of the sample data?

(f) What is the M_{max} value? Assume 90% confidence level.

6. Corrective-maintenance task times were observed as given in Table 1.

(a) What is the range of observations?

(b) Assuming seven classes with a class interval width of four, plot the data and draw a curve. What type of distribution is indicated by the curve?

(c) What is the mean repair time?

Task Time (minutes)	Frequency	Task Time (minutes)	Frequency
35	2	25	12
17	6	19	10
12	2	21	12
15	4	23	13
37	1	29	8
27	10	13	3
33	3	9	1
31	6	—	—

(d) What is the standard deviation of the sample data?

(e) The system is required to meet a mean repair time of 25 minutes at a stated confidence level of 95%. Do the data reveal that the specification requirements will be met? Why?

7. With a specified inherent availability of 0.990 and a calculated MTBF of 400 hours, what is the $\overline{M}ct$?

8. Calculate as many of the following parameters as you can with the information given.

Determine		Given
A_i	MTBM	$\lambda = 0.004$
A_a	MTBF	Total operation time = 10,000 hours
A_o	$\overline{M}$	Mean downtime = 50 hours
$\overline{M}ct$	$MTTR_g$	Total number of maintenance actions = 50
M_{max}		Mean preventive maintenance time = 6 hours
		Mean logistics plus administrative

9. Given the following data, calculate the achieved availability.

(a) $\overline{M}ct = 0.5$ hour.

(b) $MTBM_u = 2.0$ hours.

(c) $\overline{M}pt = 2.0$ hours.

(d) $MTBM_s = 1,000$ hours.

10. What is maintainability allocation? What is its purpose? To what extent (level in the system hierarchy) should allocation be accomplished?

11. A $\overline{M}ct$ of 1 hour has been assigned to system ABC. Allocate this time to the assembly level of the system using the data given in the following table.

Assembly	Quantity	λ	$\overline{M}ct$
A	1	0.05	_____ ?
B	2	0.16	_____ ?
C	1	0.27	_____ ?
D	1	0.12	_____ ?

12. Assume that a system must operate 40 hours per week, 50 weeks per year, for 15 years, in response to a designated customer need. The system MTBF is 400 hours, the $\overline{M}ct$ is 2 hours, and two maintenance technicians are assigned for the duration of each maintenance action. Determine the MLH/OH.

13. What is the purpose of maintainability prediction? When are maintainability predictions accomplished in the life cycle?

14. Describe each of the following, its purpose, application, and the information acquired through its application:

(a) Reliability-centered maintenance analysis (RCM).

(b) Level-of-repair analysis (LORA).

(c) Maintenance task analysis (MTA).

15. In conducting a level-of-repair analysis on an assembly, the results indicate a clear-cut decision to accomplish *repair* (versus discard at failure). However, the data indicate a close decision on whether to accomplish repair at the intermediate level of maintenance or at the supplier (depot) level. In the performance of a sensitivity analysis, it was decided that system operating time would likely be significantly higher than initially projected. Thus, operating time was increased while assuming that other system characteristics will remain the same (unchanged). Given this change, how might this influence the repair decision (if at all)? Please explain.

16. Describe *total productive maintenance (TPM)*. What are the objectives of TPM? What are the measures associated with TPM?

17. Calculate the OEE given that loading time is 460 minutes/day; downtime is 100 minutes; ideal cycle time is 10 minutes per product; actual cycle time is 15 minutes per product; the number of products scheduled for production is 22; and the number of products actually produced is 20.

18. When designing a system for *maintainability*, what characteristics (attributes) should be considered for incorporation into the ultimate configuration in order to attain the desired level of maintainability? Identify at least five and describe how they would impact system maintainability.

19. Assume that you own the following items:

    (a) Television
    (b) Radio
    (c) Computer
    (d) Automobile
    (e) Humidifier
    (f) Printer
    (g) Lawnmower

    In the event of a failure, identify the maintenance approach that you would follow in restoring the operational capability for each item. What criteria would you use in making a decision?

20. How are preventive maintenance requirements determined? How are they justified? What is likely to occur if they are not properly justified?

21. How are the FMECA and the RCM analysis related (if at all)?

22. Select a system (or an element of one) of your choice, simulate or assume that a failure has occurred, and accomplish an MTA covering the corrective maintenance requirements (similar to the steps illustrated in Figures 20 and 21).

23. What is the purpose of maintainability demonstration? What elements are validated? When is it accomplished in the system life cycle?

24. In preparing for a maintainability demonstration,

    (a) What tasks are to be demonstrated and how are they selected?
    (b) What criteria are used in the selection of personnel for the accomplishment of the required maintenance tasks to be demonstrated?
    (c) What determines the requirements and how are the other resources (e.g., test and support equipment, spares/repair parts and inventories, facilities, and data) selected for the demonstration?

25. The $\overline{M}$ct requirement for an equipment item is 65 minutes and the established risk factor is 10%. A maintainability demonstration is accomplished and yields the results given in the following table for the 50 tasks demonstrated. (Task times are in minutes.)

39	57	70	51	74	63	66	42	85	75
42	43	54	65	47	40	53	32	50	73
64	82	36	63	68	70	52	48	86	36
74	67	71	96	45	58	82	32	56	58
92	91	75	74	67	73	49	62	64	62

Did the equipment item pass the maintainability demonstration?

26. The $\overline{M}$pt requirement for an equipment unit is 120 minutes. A maintainability demonstration is accomplished and yields the results given in the following table for the 50 tasks demonstrated. (Task times are in minutes.) Did the equipment item pass the maintainability demonstration? The risk factor is 10%.

150	120	133	92	89	115	122	69	172	161
144	133	121	101	114	112	181	78	112	91
82	131	122	159	135	108	95	67	118	103
78	93	144	152	136	86	113	102	65	115
113	101	94	129	148	118	102	106	117	115

27. Refer to maintainability demonstration method 2 and determine whether the system will meet the $\overline{M}$ct requirement if the assumed producer's risk is 5%.

28. Describe the relationships between reliability and maintainability; that is, how does reliability relate to maintainability and vice versa? Provide several examples of the trade-offs that may be required and accomplished in the design of a system.

Appendix: Figure

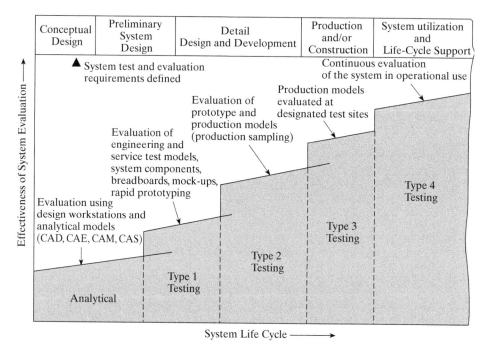

Figure A.1   Stages of system test and evaluation during the life cycle.

Design for Usability (Human Factors)*

From Chapter 14 of *Systems Engineering and Analysis,* Fifth Edition, Benjamin S. Blanchard, Wolter J. Fabrycky. Copyright © 2011 by Pearson Education, Inc. Published by Pearson Prentice Hall. All rights reserved.

Design for Usability (Human Factors)*

Much of the emphasis in the design and development of human-made systems is directed to their hardware and software components. Reliability, maintainability, and related design considerations are applied to equipment, software, and associated elements of the system. However, for the system to be effective and robust during service, the human element needs to be addressed from a usability and operability perspective. This includes an understanding of the interfaces between the human and other elements of the physical system, or *human systems integration* (HSI).

Hardware and software design alone will not guarantee good system usability (operability) no matter how well it is done. Accordingly, the objective of this chapter is to emphasize the importance of, and to provide guidance in, the *design for usability (human factors)*. Design considerations that pertain to the human and the human factors in the operation and maintenance of a system are the central focus. Good system design applies not only to the design for normal use but also to the design against misuse and abuse.

Safety and security issues require attention as well. Consideration must be given to the operator's (and maintainer's) anthropometric characteristics (i.e., human physical dimensions), sensory factors (i.e., capabilities relative to sight, sound, touch, and smell), physiological factors (i.e., impacts from the environment), psychological factors (i.e., personal attitude, expectations, and motivation), and the interrelationships among these. System design objectives pertain to maximizing the ease of use; minimizing the possibility of human-induced errors; improving the work environment and personal job enhancement capabilities; maximizing human safety; minimizing the risks of acute or chronic human illness, injury, or disability; reducing training requirements; and so on.

The intent of this chapter is to provide a general understanding of human factors. Upon completion of the material, the reader will have obtained an appreciation of the importance

---

*Equivalent terms may include *ergonomics, systems psychology, human engineering*, and *human systems integration* (HSI).

of including appropriate human factors requirements in the system design process. The chapter includes the following topics:

- A definition and explanation of human factors and human systems integration;
- The common measures in human factors;
- Human factors in the system life cycle—system requirements, requirements allocation, design participation, and design review;
- Human factors analysis methods—operator task analysis (OTA), operational sequence diagrams (OSDs), error analysis, and safety/hazard analysis;
- Personnel training requirements; and
- Personnel test and evaluation.

Consideration of the human being, as a major element of the system (along with hardware, software, facilities, data, etc.) early and throughout the system life cycle, is essential in designing for operational feasibility. Further, consideration of the human as an operator and/or user of the system or product makes a contribution to the usability of the human-made system.

1 DEFINITION AND EXPLANATION OF HUMAN FACTORS

Requirements for the *human*, as part of a system design, are initially derived through the definition of system operational requirements, the maintenance concept, and the accomplishment of a top-level functional analysis. A description of the mission(s) to be performed is an essential first step in the requirements definition process. Based on this, a functional analysis is completed where operational and maintenance functions are identified indicating the *whats*: what the system must do. Through the subsequent process of synthesis, analysis, and evaluation, a specific design approach may be selected for determining the *hows*: the manner by which the functions will be accomplished. This leads to the identification of *human* requirements for the system.[1]

Given the functions that have been identified and allocated to the human, the next step is to break these down into *job operations, duties, tasks, subtasks*, and so on, as illustrated in Figure 1. A brief description of each is presented as follows:

1. *Job operation.* Completion of a function normally includes a combination of duties and tasks. A job operation may involve one or more related groups of duties, and may require one or more individuals in its accomplishment (i.e., positions in a given specialty field). An example is operating a motor vehicle or accomplishing a maintenance requirement.

2. *Duty.* Defined as a set of related tasks within a given job operation. For instance, when considering the operating of a motor vehicle, a set of related tasks may include (a) driving the motor vehicle in traffic on a daily basis, (b) registering the motor vehicle yearly, (c) servicing the motor vehicle as required, and (d) accomplishing vehicle preventive maintenance on a periodic basis.

3. *Task.* Constitutes a composite of related activities (informational, decision, and control activities) performed by an individual in accomplishing a prescribed amount of work in a specified environment. A task may include a series of closely associated operations,

---

[1]Human activities and tasks may also evolve from the hardware and software life cycles as lower elements of the system are defined. It is through the analysis of a unit of hardware (or software) that a human requirement may be identified.

Design for Usability (Human Factors)

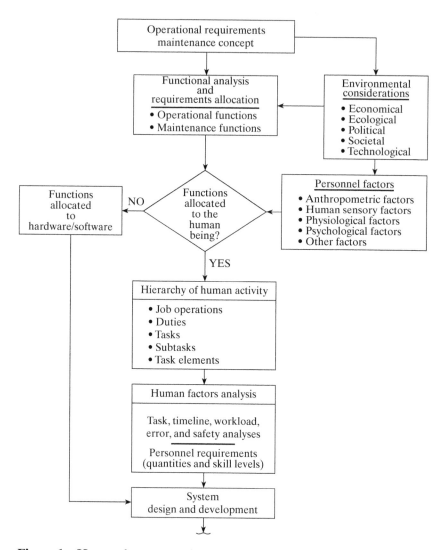

Figure 1 Human factors requirements.

maintenance inspections, and so on. Relative to driving a motor vehicle, examples of tasks are (a) applying the appropriate pressure on the accelerator in order to maintain the desired vehicle speed, (b) shifting gears when necessary in order to maintain engine rpm, or (c) turning the steering wheel as required enabling the motor vehicle to move in the desired direction.

4. *Subtask.* Depending on the complexity of the situation, a task may be broken down into subtasks to cover discrete actions of a limited nature. A subtask may constitute the shifting of gears from first to second (in the motor vehicle example), a machine adjustment, or a similar act.

5. *Task element.* Task elements may be categorized as per the smallest logically definable facet of activity (based on perceptions, decisions, and control actions) that requires individual behavioral responses in completing a task or a subtask. For example, the

identification of a specific signal level on a display, a decision pertaining to a single physical action, the actuation of a switch on a control panel, and the interpretation of a go/no-go signal might be classified as a task element. This is the lowest category of activity where job behavioral characteristics are identified and evaluated.

The hierarchy outlined represents a breakdown from the top system-level functions to the smallest element of activity involving human performance. In developing this breakdown, it is not always easy to separate functions from tasks, tasks from subtasks, and so on. However, one must proceed from the top-level function(s) down to the level necessary to establish the proper human–machine interface. As with an equipment hierarchy, this breakdown constitutes a logical division of human activity that serves as a basis in establishing human factors requirements in design.

With the functional analysis results and the basic hierarchy of human activity identified (at least on a preliminary basis), one needs to address some of the environmental and personnel factors that will ultimately influence work accomplishment. Figure 1 illustrates the relationship of these elements in the process of defining the overall human factors requirements for the system. The environmental considerations noted in the figure constitute those factors that impose a constraint on the system development activity as an entity. Personnel factors may be categorized in terms of anthropometric factors, human sensory factors, physiological factors, and psychological factors. These categories are briefly described subsequently.

1.1 Anthropometric Factors

When establishing basic design requirements for the system, particularly regarding human activities, one obviously must consider the physical dimensions of the human body. The weight, height, arm reach, hand size, and so on are critical when designing operator stations, consoles, control panels, accesses for maintenance purposes, and the like. Further, body dimensions will vary somewhat from a static position to a dynamic condition. *Static* measurements pertain to the human subject in a rigid standardized position, whereas *dynamic* measurements are made with the human in various working positions and undergoing continuous movement. As movement occurs, body measurements will change. Thus, when considering various system design alternatives, the designer must be aware of the different human profiles that are likely to occur in the performance of operator and maintenance activities associated with the system.

To aid the designer in obtaining information on body measurements, there are two basic sources of data to consider: (1) anthropometric surveys, in which measurements of a sample of the population have been made; and (2) experimental data derived from simulating the operating conditions peculiar to the system being developed. The designer may have access to either source of data, or a combination of both. In many instances, however, the acquisition of meaningful experimental data is quite costly. Hence, static measurements are often used for the purpose of design guidance.

Figures 2 and 3 illustrate examples of static body measurements in the standing and sitting positions, respectively.[2] Anthropometric measurements are usually provided in percentiles, ranges, and means (or medians). The data presented represent the 5th and 95th percentiles for men and women for a given sample population. Although the designer will not be able to cover all possible sizes or profiles, he or she can cover most situations as well as

---

[2]The factors presented in Figures 2 and 3 were taken from MIL-STD-1472, Military Standard, *Human Engineering Design Criteria for Military Systems, Equipment and Facilities* (Washington, DC: Department of Defense). Another good source for anthropometric data is M. S. Sanders and E. J. McCormick, *Human Factors in Engineering Design*, 7th ed. (New York: McGraw-Hill, 1993).

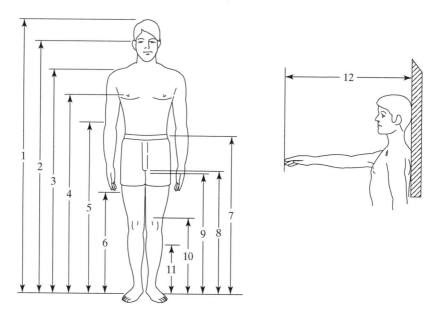

Factors	Percentile Values (cm)			
	5th Percentile		95th Percentile	
	Men	Women	Men	Women
Weight (kg)	57.4	46.4	91.6	76.5
Standing body dimensions				
1. Stature	163.8	152.4	185.6	172.2
2. Eye height	152.1	142.7	173.3	160.1
3. Shoulder height	133.6	123.0	154.2	143.4
4. Chest height		110.0		127.3
5. Elbow height	104.8		120.0	
6. Fingertip height	61.5		73.2	
7. Waist height	97.5	93.1	115.2	110.1
8. Crotch height	76.3	66.4	91.8	81.4
9. Gluteal furrow height		66.2		79.4
10. Kneecap height	47.5		58.6	
11. Calf height	31.1		40.6	
12. Functional reach	72.7	67.7	90.9	80.4

Figure 2 Anthropometric data—standing body dimensions.

decide where to cut off. The ultimate design configuration must accommodate most of the population if operator and maintenance activities are to be accomplished efficiently. Those individuals having extreme measurements (e.g., the small percentage of the population below the 5th percentile and above the 95th percentile) may be able to perform the necessary operator or maintenance functions; however, the results are likely to be inefficient due to the introduction of various human stresses causing early personal fatigue and possible system failure.

In designing a system, workspace requirements must be established for both operator personnel in the performance of operating functions and maintenance personnel in the accomplishment of maintenance tasks. Figures 4 and 5 provide examples of the design with anthropometric factors in mind.

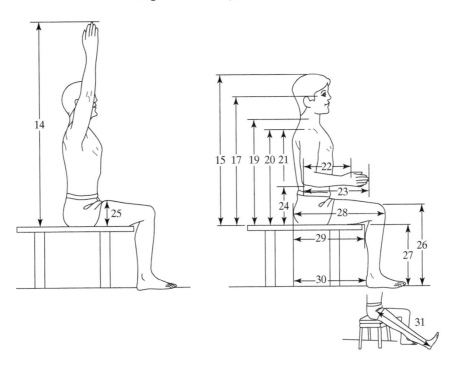

Factors	Percentile Values (cm)			
	5th Percentile		95th Percentile	
	Men	Women	Men	Women
Seated body dimensions				
14. Vertical arm reach, sitting	128.6		147.8	
15. Sitting height, erect	84.5	78.4	96.9	90.9
16. Sitting height, relaxed	82.5	76.9	94.8	89.7
17. Eye height, sitting erect	72.8	68.7	84.6	78.8
18. Eye height, sitting relaxed	70.8	67.2	82.5	77.6
19. Mid-shoulder height	57.1	53.7	67.7	62.5
20. Shoulder height, sitting	54.2		65.4	
21. Shoulder-elbow length	33.8	30.2	40.2	36.2
22. Elbow-grip length	32.6		37.9	
23. Elbow-fingertip length	44.3	38.9	51.9	45.7
24. Elbow rest height	17.5	18.7	28.0	26.9
25. Thigh clearance height		10.4		14.6
26. Knee height, sitting	49.7	43.7	58.7	51.6
27. Popliteal height	40.6	38.0	50.0	44.1
28. Buttock-knee length	54.9	52.0	64.3	61.9
29. Buttock-popliteal length	45.8	43.4	54.5	52.6
30. Buttock-heel length	46.7		56.4	
31. Buttock-heel length (diagonal)	103.9		120.4	

Figure 3 Anthropometric data—seated body dimensions.

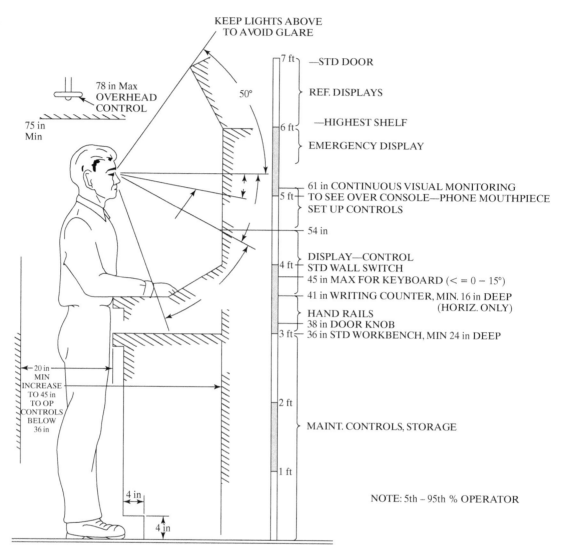

Figure 4 Suggested parameters for standard operator workplaces. *Source:* H. P. Van Cott and R. G. Kinkade, *Human Engineering Guide to Equipment Design*, revised ed. (Washington, DC: U.S. Government Printing Office, 1972).

The application of anthropometric data in design involves many considerations. The human body and workspace dimensions are significant. However, one should consult the literature before proceeding further because the limited material in this section is provided for illustrative purposes only. There exist criteria covering many facets of console and control panel design, workspace design, seat design, work surface design, and so on. Also, associated with body dimensions are the aspects of force and weight-lifting capacity. A human being can exert more force, with less fatigue, if the system (i.e., the segment of the system where the interface exists) is designed properly. The amount of force that can be exerted is determined by the position of the body and the members applying the force, the direction of application, and

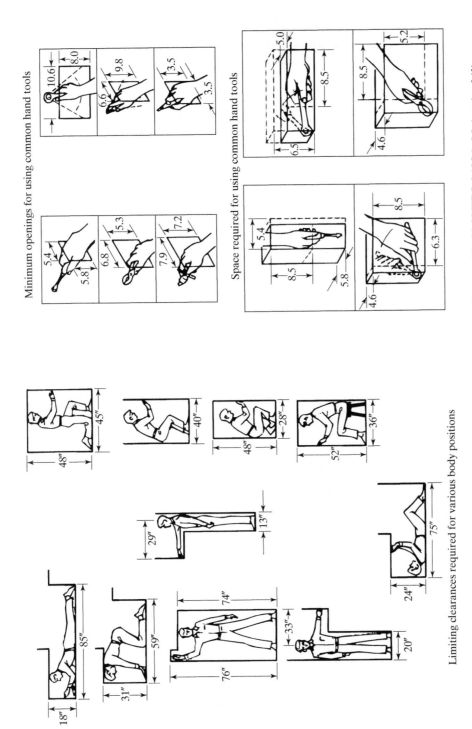

Minimum openings for using common hand tools

Space required for using common hand tools

Limiting clearances required for various body positions

Figure 5 Anthropometric data pertaining to work space requirements. *Source:* NAV-SHIPS 94324, *Maintainability Design Criteria Handbook for Designers of Shipboard Electronics Equipment* (Washington, DC: Naval Ship Systems Command, U.S. Navy, 1964).

543

the object to which the force is applied. Weight-lifting capacity, which is closely related to force, highly depends on the individual's weight, size, and position. The field of anthropometry deals primarily with the science and technique of human measurements; however, there are numerous other factors that are directly affected by design decisions.

1.2 Human Sensory Factors

In dealing with the human–machine interface in system design, one must be cognizant of certain human sensory capacities. Although most of the human senses may be affected through system design, factors pertaining to vision (sight) and hearing (noise) are of particular significance.

Vision. Vision or sight, as it pertains to system design, is usually limited to the desired vertical and horizontal fields, as illustrated in Figure 6. The designer should consider the specified degrees of eye and head rotation as the maximum allowable values in the design of operator consoles and control panels. Requirements outside of these recommended limits will result in operator fatigue, inefficiency, and system failure(s).

Within the broad field of view illustrated in the figure, the human eye may see different objects from different angles. Sight is stimulated by the electromagnetic radiation of certain wavelengths, or the visible portion of the electromagnetic spectrum. The eye sees different lines (i.e., parts of the spectrum) with varying degrees of brightness. Relative to color, one can usually perceive all colors while looking straight ahead. However, color perception begins to decrease as the viewing angle decreases. Therefore, when designing consoles (or panels) with color-coded meters or color warning light displays, one must consider the placement of such things relative to the operator's field of view. Figure 7 illustrates the limits of color vision.

Finally, the satisfactory performance of tasks highly depends on the level of illumination. Not only must the designer be concerned with field of view and color, but the proper level of illumination is also an obvious necessity. Illumination levels will vary somewhat depending on the task to be performed. Table 1 identifies suggested levels of illumination for designated items.

Hearing. In human activities, the designer needs to address both the requirements for oral communication and the aspects of noise. Of particular concern is the effect of noise on the performance of work. Noise is generally regarded as a distracter and a deterrent when considering efficiency of work accomplishment. As the noise level increases, a human being begins to experience discomfort, and both productivity and efficiency decrease. Further, oral communication becomes ineffectual or impossible. When the noise level approaches 120 dB, a human being will usually experience a physical sensation in some form. At levels above 130 dB, pain can occur.

Another major factor in determining the extent to which noise is a distracter is its character—whether steady or intermittent. In situations where the noise is steady, an individual can adapt to it, and work efficiency may not be significantly compromised. If the noise level is too high (even though steady), the individual will probably experience permanent injury through loss of hearing. Conversely, when the noise is intermittent, the individual is usually distracted regardless of the intensity. In this instance, a greater effort is required to maintain job efficiency, or else the possibility of early fatigue occurs.

The designer, when dealing with the human being as a component of the system, must identify the operator and maintenance tasks that are to be accomplished by people. These tasks must be evaluated in terms of the environment in which the tasks are to be performed, and the noise generated by the system (or generated externally) must be maintained at a level

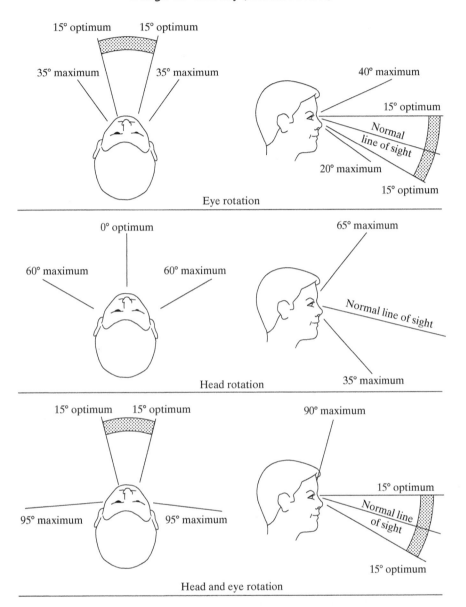

Figure 6 Vertical and horizontal visual field.

where human efficiency is maximized. The desired intensity level will likely fall within a range 50–80 dB; however, the designer should consult the available literature to establish the proper design criteria for the variety of applications that are likely to occur.

Other Senses Beyond that of vision and hearing, there are other senses that cannot be completely ignored in the design of systems. For example, in relatively closed environments where there is a very unpleasant odor, the results may lead to human nausea, which, in turn, results in decreased performance. Thus, the sense of *smell* becomes a major issue

Design for Usability (Human Factors)

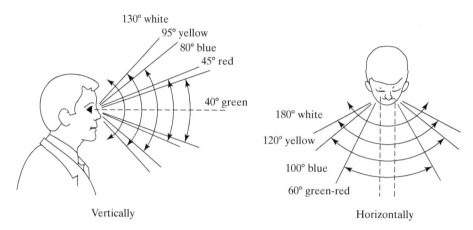

130° white
95° yellow
80° blue
45° red
40° green

Vertically

180° white
120° yellow
100° blue
60° green-red

Horizontally

Figure 7 Approximate limits of color differentiation.

TABLE 1 Specific Task Illumination Requirements

Illumination Levels

	Foot-Candles*	
Type of Task	Recommended	Minimum
1. Business machine operation	100	50
2. Console surface	50	30
3. Dials, gauges, and meters	50	30
4. Factory assembly—general	50	30
5. Factory assembly—precise	300	200
6. Inspection tasks—extra fine	300	200
7. Inspection tasks—rough	50	30
8. Office work, general	70	50
9. Ordinary seeing tasks	50	30
10. Panels	50	30
11. Passageways and stairways	20	10
12. Reading, large print	30	10
13. Reading, small type	70	50
14. Repair work—general	50	30
15. Repair work—instrument	200	100
16. Testing—extra fine	200	100
17. Testing—rough	50	30
18. Transcribing and tabulation	100	50

*As measured at the task object or 30 in above the floor.
This information (in part) was extracted from MIL-STD-1472, Military Standard, *Human Engineering Design Criteria for Military Systems, Equipment, and Facilities* (Washington, DC: Department of Defense).

and, when designing operator/maintainer stations, care must be taken to ensure the proper filtering and circulation of air. Further, the sense of *feeling (touch)* is important, particularly in the design of workstations and in the shape of controls and special keys on control panels. For example, a control knob that is pointed may indicate one thing while a circular control another. Of perhaps a more serious nature is the sense of *balance* (vestibular senses) that deals with the position of the body and its related motion(s). The effects of seasickness, airsickness, or equivalent can have a significant degrading impact on the performance of both operator and maintenance tasks.[3]

1.3 Physiological Factors

The study of physiology is obviously well beyond the scope of this text; however, some recognition must be given to the effects of environmental stresses on the human body while performing system tasks. *Stress* refers to any aspect of external activity or the environment acting on the individual (who is performing a system task) in such a manner as to cause a degrading effect. Stress may result in both physiological and psychological effects, and *strain* is often the consequence of stress measured in terms of one or more physical characteristics of the human body. Some of the causes of stress are as follows:

1. *Temperature extremes.* Experience has indicated that certain temperature extremes are detrimental to work efficiency. As the temperature increases above the comfort zone (i.e., 55°F–75°F), mental processes slow down, motor response is slower, and the likelihood of error increases. Conversely, as the temperature is lowered (e.g., 50°F and lower), physical fatigue and stiffening of the extremities begin. For this reason, the designer needs to be aware of the environmental profiles for the system, particularly with regard to anticipated system use in the arctic or in the tropics.

2. *Humidity.* Heat and humidity are usually significant factors in causing a reduction in the operational efficiency of personnel. A human being can tolerate much higher temperatures with dry air than if the air is humid. A high temperature (e.g., 90°F or higher), combined with a high humidity (90% or higher), could significantly result in a degradation of human performance.

3. *Vibration.* This refers to the alternating motion of a body or a surface with respect to some point of reference. Most transportation vehicles, machines, appliances, tools, and so on vibrate to some degree, and the degree of vibration (i.e., magnitude of the vibration frequency in Hz) will have different effects on various activities. For example, the equilibrium of a human body may be affected with a vibration of between 30 and 300 Hz, one's speech may be impacted with a vibration between 1 and 20 MHz, one's depth perception may be impacted with a vibration between 25 and 60 MHz, and so on. In any event, individual productivity will depend on the direction of the vibration, human work methods, climate conditions, and many other biological factors.

4. *Noise.* The consequences of noise and its impact on human efficiency are discussed earlier in the context of human sensory factors (i.e., hearing). However, this factor is listed again,

---

[3]Several good references covering the various factors pertaining to the human (i.e., sensory factors, physiological factors, and psychological factors) include (1) A. Chapanis, *Human Factors in Systems Engineering* (Hoboken, NJ: John Wiley & Sons, Inc., 1996), and (2) Sanders, M. S. and E. J. McCormick, *Human Factors in Engineering and Design*, 7th ed. (New York: McGraw-Hill, 1993).

since high steady or intermittent levels of noise may have a highly degrading effect on human performance.

5. *Other factors.* Several additional factors may (to varying degrees) cause stress on the human body. These include the effects of radiation, gas or toxic substances in the air, sand and dust, and so forth.

The external stress factors listed previously will normally result in individual human strain. Strain may, in turn, have an impact on any one or more of the human body systems (i.e., circulatory system, digestive system, nervous system, respiratory system, etc.). Measures of strain include such common parameters as pulse rate, blood pressure, body temperature, oxygen consumption, and the like.

1.4 Psychological Factors

Psychological factors pertain to the human mind and the aggregate of emotions, traits, and behavior patterns as they relate to job performance. All other conditions may be optimum from the standpoint of performing a task in an efficient manner; however, if the individual operator (or maintenance technician) lacks initiative, motivation, dependability, self-confidence, communication skills, and so on, the probability of performing in an efficient manner is low.

Psychological factors may be highly influenced by physiological factors. In general, one's attitude, initiative, motivation, and so on are dependent on the needs and expectations of the individual. Fulfillment of these needs and expectations is a function of the organizational environment within which the individual performs and the leadership characteristics of supervisory personnel in that organization. If the individual operator/maintainer does not perceive the opportunity for on-the-job growth, or if the managerial style of his or her immediate supervisor is not directly supportive of personal goals and objectives, job performance may deteriorate accordingly. The psychological aspects of human performance are many and varied, and the reader is advised to review the literature on behavioral science in order to gain some insight as to the cause-and-effect relationships pertaining to good versus poor performance.

With the preceding basic characteristics in mind, the human must now be viewed as a component of the system. A key issue here is that of *information processing*. Information may be generally classified as *static* or *dynamic* (i.e., that which is constant and fixed on a display or that which is continually changing). Further, there are different categories of information to include quantitative information, qualitative information, warning and signal information, identification information, graphic information, alphanumeric and symbolic information, time-phased information, and so on. The designer must be cognizant of the requirements for the system, and must understand the human's capacities and abilities in this area.

Figure 8 portrays a simple information-processing model where four basic human subsystems are identified. The *sensing* subsystem responds to specific types of energy identified through the human senses (i.e., vision, hearing, feeling, etc.). This provides the *stimulus* to initiate some form of action. The *information-processing* subsystem addresses the human's capacity to receive and process information. Of particular interest is the amount of information that an individual can transmit (often expressed in terms of "bits") and the rate at which he or she can transmit it. The *storage* subsystem refers to the human memory and its capacity, or the ability to retrieve data and facilitate the information processing. Finally, there is the *response* subsystem, which leads to the accomplishment of some function/task through a combination of physical motions (i.e., the output from the model). Inherent within this model is the *feedback*, which permits responses to be accurate in terms of the original input.[4]

[4]Figure 8 constitutes a modified version of Figure 2.1 in H. P. Van Cott and R. G. Kinkade, *Human Engineering Guide to Equipment Design*, revised ed. (Washington, DC: U.S. Government Printing Office, 1972).

Design for Usability (Human Factors)

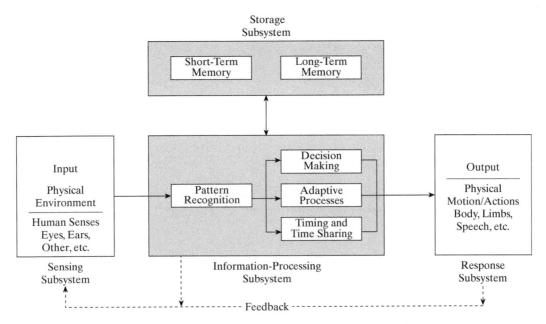

Figure 8   The human information-processing subsystem (simplified).

In summary, the designer needs to understand not only the characteristics dealing with anthropometry, sensory factors, and the impact of external factors on the human but also the capacity and abilities of the human relative to the information-processing requirement illustrated in Figure 8.

2 THE MEASURES IN HUMAN FACTORS

An overall objective is to design a system such that it can be operated and maintained in an effective and efficient manner, throughout its planned life cycle, in response to customer needs. The *effectiveness* side of the spectrum refers to the accomplishment of system operational and maintenance functions, in a specified manner, within the desired time frame, and without inducing errors in the process. The goal is to maximize performance, availability, dependability, and whatever other system-level goals that have been established. At the same time, to be *efficient* means that the functions must be accomplished at minimum total life-cycle cost. As with reliability and maintainability, the measures of interest herein impact both sides of this spectrum, and may include the following:

1. The quantity of personnel required for the *operation* of the system and for the accomplishment of its mission. A metric could be stated in terms of operator labor hours per hour of system operation (OLH/OH), or operator labor hours per mission segment or cycle of operation. Factors can be established for each personnel skill level (e.g., low, intermediate, and/or high skill level).

2. The quantity of personnel required for the *maintenance* of the system in a designated period of time, in terms of a specific mission scenario or in hours of system operation. A common measure is maintenance labor hours per operating hour (MLH/OH).

Factors can be established for each personnel skill level (e.g., low, intermediate, and/or high skill level).

3. The elapsed time that it takes to accomplish an *operational* mission, a segment of a mission, or an operational function (e.g., time/function).

4. The time that it takes to accomplish a *maintenance* function.

5. The number of human errors committed by the *operator* per mission, mission scenario, operator function/task, or period of time (e.g., errors/function).

6. The number of human errors committed by the *maintainer* per maintenance action, maintenance function, or period of time (e.g., errors/maintenance action).

7. Personnel training rate, or the number of operator and maintenance personnel trained per period of time (e.g., personnel trained/month).

8. The quantity of personnel training days per period of time (e.g., training days/month).

9. The cost of *operator* personnel per hour of system operation, per mission, or per mission segment (e.g., $/OH).

10. The cost of *maintenance* personnel per hour of system operation, per mission, or per maintenance action (e.g., $/MA).

11. The cost of *training* per individual, per organization, or per time period (e.g., $/person, $/month, or $/organization/month).

While there may be a number of different measures depending on the type of system and the nature of its mission, these represent some of the most common for many applications.

3 HUMAN FACTORS IN THE SYSTEM LIFE CYCLE

Human factors, like reliability and maintainability, must be an inherent consideration within and throughout the systems engineering process beginning with the conceptual design phase. Referring to Figure 9, qualitative and quantitative requirements pertaining to the human must be identified through the development of system operational requirements and the maintenance concept, as well as the identification and prioritization of TPMs. As the top-level functions are initially defined for the system, trade-off studies are conducted with the intent of identifying functions/tasks that will be accomplished (1) solely by the human; (2) automatically through the use of equipment and software where, and only where, there is no anticipated human involvement; or (3) using a combination of people, equipment, software, facilities, and the like. The requirements for people, as significant elements of the system, are identified and specified. These requirements may take the form of the measures, as applicable, described in Section 2.

Given the specification of requirements for human factors in Block 1 of Figure 9 (and also included in the System Specification—Type *A*), the next step is to translate these requirements from the system-level functional analysis down to the various subsystems and below through the process of allocation. The iterative process of synthesis, analysis, and evaluation is accomplished throughout the preliminary and detail design and development phases on a continuing basis. This process is facilitated through the effective utilization of selected methods/tools such as the operator task analysis (OTA), development of operational sequence diagrams (OSDs), safety/hazard analysis, maintenance task analysis (MTA), and so on. As the design progresses and physical models are developed, an ongoing

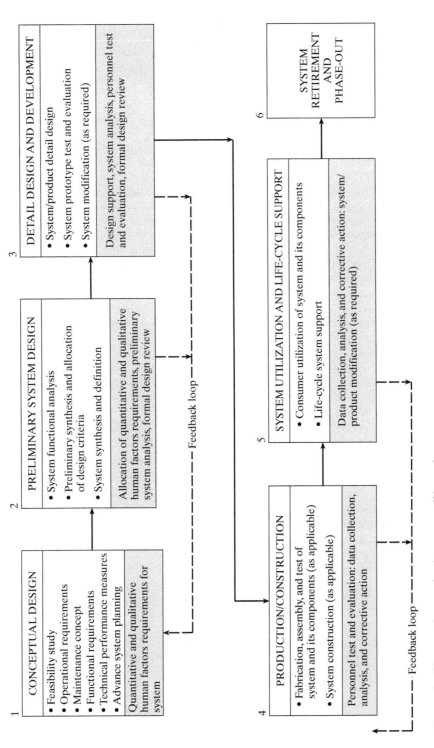

Figure 9 Human factors in the system life cycle.

assessment occurs through the personnel test and evaluation effort accomplished as part of the overall system test and evaluation activity. True and complete *human systems integration* is the overall objective.

3.1 System Requirements

The human, as an element of the system, is first visualized during the initial description of the need and the accomplishment of the feasibility analysis. The customer (or user) identifies his or her role as an inherent part of the system from the beginning, whether one is dealing with a transportation system, river crossing bridge system, communications system, health care system, or equivalent. Through development of the functional analysis and the completion of design trade-offs, the functions to be accomplished by the human are delineated further. Operational functions/tasks are first identified and then lead to the definition of maintenance functions/tasks.

At these early stages in design, one needs to establish some specific design-to requirements (as an input to design) for the human element of the system. These requirements, to be established as part of the definition of system requirements, should include the specification of whatever measures are applicable in order to operate and maintain the system easily, safely, economically, and without the introduction of errors. For example, it may be stated that *the system shall be designed such that it can be successfully operated by no more than* **x** *number of personnel (between the 95th percentile of men and the 5th percentile of women), with* **basic** *skills not to exceed* **y** *level, throughout the mission, in time* **z***, and without introducing any errors in the process.*

Some of the missing quantitative requirements can be derived from the measures presented in Section 2. When defining the appropriate human skill levels, the user's organization should serve as the source for defining the desired requirements. For example, an individual at a *basic* skill level may exhibit the following qualifications:[5]

1. Age: 18–21 years.
2. Education: high school graduate.
3. General reading/writing level: ninth grade.
4. Experience: no regular work experience prior to training.
5. General qualifications: after a limited amount (40 hours) of familiarization training, plus some on-the-job training (OJT)(the completion of three missions), the individual can accomplish normal functions, involving system actuation, manipulation of controls, communications, and the recording of data. The individual is able to follow clearly presented instructions where interpretation and decision making are not necessary. This individual will normally require close supervision.

The definition of higher skill levels (e.g., *intermediate* and *supervisory*) will, of course, include at the entry level a higher basic educational requirement, additional training, and a greater amount of experience. These entry levels, and associated requirements, are tailored to the specific organizations operating and maintaining the system. The prime human factors objective is to design the system for ease of use (operation and support), by individuals with basic skills, and without requiring an extensive amount of training.

---

[5]An example of the personnel skill levels defined by the U.S. Labor Department is included in A. Chapanis, *Human Factors in Systems Engineering* (Hoboken, NJ: John Wiley & Sons, Inc., 1996), pp. 155–156.

3.2 Requirements Allocation

Given the requirements for the overall system, the appropriate qualitative and quantitative design criteria can be established for the various elements of the system; that is, equipment, software, and facilities. The guidelines presented in Figures 2–7 provide examples as a start. Referring to Figure 4, in the design of rack-mounted consoles, specific criteria can be established covering the placement of components based on operational requirements and on the anthropometric and sensory factors associated with the human. In the design of control panels, there are specific guidelines pertaining to the type, sequencing, and location of switches, knobs, and other operator controls; the type of gauges, displays, and readout devices depending on the nature of the information desired; labeling and color coding; transfer of information through audio and/or visual means; and so on.

3.3 Design Review and Evaluation

Review of the design for the incorporation of human engineering characteristics is accomplished as an inherent part of the process. The characteristics of the system (and its elements) are evaluated in terms of the initially specified human factors requirements for the system. If the requirements appear to be fulfilled, the design is approved as is. If not, the appropriate changes are initiated for corrective action.

In support of conducting a human factors review, one may wish to develop a design review checklist, including questions such as those listed below. It should be noted that the answer to these questions that are applicable should be *yes*:

1. Have the qualitative and quantitative requirements for human factors in design been adequately defined and specified? Have they been allocated from the top down?
2. Has a top-level system analysis been accomplished to identify those functions that are to be completed by the human? Have these functions been broken down into job operations, duties, tasks, subtasks, and task elements to the extent practicable?
3. Have the appropriate interfaces been defined between the human and the other elements of the system (i.e., equipment, software, facilities, and elements of support)?
4. Does the system design adequately reflect the proper consideration of anthropometric, human sensory, physiological, and psychological factors?
5. Does the design reflect consideration of the abilities and capacities of the human in dealing with information-processing requirements (refer to Figure 8)?
6. Have those tasks to be accomplished by the human been justified through the OTA? Can they be traced from the functional analysis?
7. Have the personnel quantities and skill levels been defined for the system? Has the system been designed such that it can be successfully operated by an individual with *basic* skills? Have the number of operating personnel been optimized to the extent practicable?
8. Have the interfaces between the human and other elements of the system been properly defined and justified through the development of OSDs? Are the OTA and OSDs compatible? Do they justify and lead into the development of training requirements?
9. Has a system safety/hazard analysis been completed? Is it compatible with the FMECA?

10. Has an error analysis been conducted? Do the results feed back to the FMECA and the safety/hazard analysis? Have the results been provided as an input to the development of personnel training requirements?

11. Have the environments been adequately defined for each area where system functions/tasks are to be accomplished by the human? Are they optimal?

12. In the design of control panels and displays, are the controls standardized? Are the controls sequentially positioned? Are they placed according to frequency or criticality of use? Is control spacing adequate? Is control labeling adequate? Have the proper control and display relationships been incorporated? Are the proper types of panel switches used? Is control panel lighting adequate? Has the overall design been justified through the OSDs?

13. In the area of safety, have system/product hazards from heat, cold, thermal change, barometric change, humidity change, shock, vibration, light, mold, bacteria, corrosion, rodents, fungi, odors, chemicals, oils, greases, handling and transportation, and so on been eliminated?

14. Have fail-safe provisions been incorporated in the design?

15. Have protruding devices been eliminated or are they suitably protected?

The items covered are certainly not intended as being all-inclusive but merely represent a sample of possible interest areas. A good design review should address these and related issues.

4 HUMAN FACTORS ANALYSIS METHODS

Within the context of human factors analysis (accomplished as an iterative part of system synthesis, analysis, and evaluation), there are a few tools that can be effectively utilized in support of the objectives described throughout this text. Of particular interest herein are the operator task analysis (OTA), development of operational sequence diagrams (OSDs), error analysis, and system safety/hazard analysis. The relationships among these, and related, activities are illustrated in Figure 10.

4.1 Operator Task Analysis (OTA)

This facet of analysis involves a systematic study of the human behavior characteristics associated with the completion of system tasks. In accomplishing an operator task analysis (OTA), the following steps are applied and Figure 11 presents a sample data format that can facilitate implementation:

1. Identify system operator functions and establish a hierarchy of these functions in terms of job operations, duties, tasks, subtasks, and task elements as described in Section 1.

2. For each function involving the human element, determine the specific information necessary for operator personnel decisions. Such decisions may lead to the actuation of a control, the monitoring of a system condition, or equivalent. Information required for decision making may be presented in the form of a visual display or an audio signal of some type.

3. For each action, determine the adequacy of the information fed back to the human as a result of control activations, operational sequences, and so on.

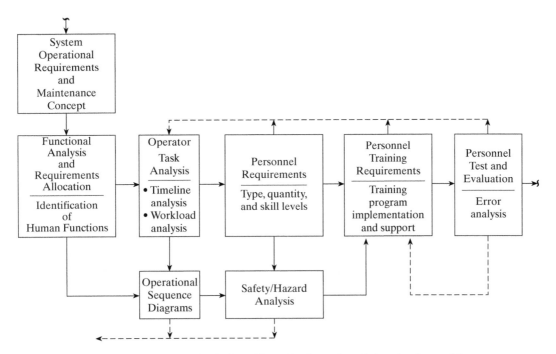

Figure 10 The application and relationships of selected tools/methods used for human factors in design.

4. Determine the time requirements, frequency of occurrence, accuracy requirements, and the criticality of each action (or series of actions) accomplished by the human.

5. Determine the impact of the environmental and personnel factors and constraints on the human activities identified in item 1.

6. Determine the human skill-level requirements for all operator personnel actions, group and assign tasks to a designated workstation, and determine the total quantity and skill levels of personnel required for the system.[6]

Utilizing the format in Figure 11, each subtask (or equivalent) is identified in column 3. This may involve operating, checking, adjusting, troubleshooting, or conducting some similar activity. The action stimulus, or event, that instigates performance of the subtask is noted in column 4. This may include initiating a command, an out-of-tolerance display indication, a symptom of failure, or equivalent. The required action, or response, is noted in column 5. This may include a control movement, a voice communication command, or some other human response. The feedback or indication of the adequacy of the response is included in column 6. Column 7 indicates the initial classification of the subtask along with other similar subtasks, and column 8 is used to enter potential sources of human error. Column 9 is divided into two sections, with *allowable* time being the period within which the subtask must be completed and *necessary* time being the actual time required. A time constraint may be initially specified

---

[6]The approach here is similar to what is accomplished for the assignment and grouping of maintenance tasks analyzed through the maintenance task analysis (MTA). In certain cases, it may be appropriate to accomplish the OTA and MTA jointly on an integrated basis, particularly when dealing with the determination of personnel quantities and skill levels.

Function (1)	Operate aircraft power plant and system controls
Task (2)	Control jet engine operation

Subtask (3)	Action stimulus (4)	Required action (5)	Feedback (6)	Task classification (7)	Potential errors (8)	Time (9) Allowable (9a)	Time (9) Necessary (9b)	Work station (10)	Skill level (11)
3.1 Adjust engine rpm	4.1 Engine rpm on tachometer	5.1 Depress throttle control downward	6.1 Increase in indicated tachometer rpm	7.1 Operator task, aircraft commander	8.1 a. Misread tachometer b. Fail to adjust throttle to proper rpm	9a.1 10 sec	9b.1 7 sec	10.1 aircraft commander seat	11.1 Low

Figure 11 Sample format for task allocation and analysis. *Source:* H. P. Van Cott and R. G. Kinkade, *Human Engineering Guide to Equipment Design*, revised ed. (Washington, DC: U.S. Government Printing Office, 1972).

through the allocation process discussed in Section 3.2. The geographical location or the prospective workstation where the subtask is to be accomplished is noted in column 10. Finally, the anticipated skill level required for subtask completion is included in column 11.

The purpose of the task analysis is to ensure that each *stimulus* is tied to a *response* (and that each response is directly related to a stimulus). Further, the individual human motions are analyzed on the basis of dexterity, mental and motor skill requirements, stress and strain characteristics of the human performing the task/subtask, and so on. The objective is to (1) identify those areas of system design where potential human–machine problems exist, (2) identify the necessary personnel quantities and skill-level requirements for operating the system, and (3) identify requirements for personnel training. The OTA format, as shown in Figure 11, can be extended to include the performance of a *timeline* analysis and a *workload* analysis. The timeline analysis addresses those tasks that can be accomplished sequentially and those in parallel, and the workload analysis leads to the identification of the quantity of personnel required.

4.2 Operational Sequence Diagram (OSD)

The OSD can be used to aid in evaluating the *flow of information* from the point when the operator first becomes involved with the system to the completion of the mission. Information flow in this instance pertains to operator decisions, operator control activities, and the transmission of data. Figure 12 presents an example of an OSD format.

Stemming from the functional analysis and the OTA, *operator* actions can be identified in a top-down sequential format as shown in the figure. In this instance, there are two system operators required, each assigned to a workstation. The objective is to show how the information flows between the two operators and their respective workstations, presented in time as one progresses from the top down. The symbology included in the figure reflects the "actions" that occur. This, in turn, leads to the requirements in design for specific types of displays, the incorporation of electrical or mechanical means for information processing, the incorporation of automation, and so on. Through an evaluation of these sequences of operation, and considering the human characteristics described in Section 1, one can assess the adequacy of design regarding the interfaces between the human and other elements of the system (i.e., control panel design in this instance). Results from the generation of OSDs can aid in the development of personnel training requirements.

4.3 Error Analysis

An error occurs when a human action exceeds some limit of acceptability, where the limits of acceptable performance have been defined. Errors may be broken down into errors of *omission*, when a human fails to perform a necessary task, and errors of *commission*, when a task is performed incorrectly (i.e., selection, sequence, or time errors). The possible causes of error are the following:

1. Inadequate workspace and work layout—poor workstation design relative to seating, available space, activity sequences, and accessibility to system elements.
2. Inadequate design of facilities, equipment, and control panels for human factors—inadequate displays and readout devices, poor layout of controls, and lack of proper labeling.
3. Poor environmental conditions—inadequate lighting, high or low temperature, and high noise level.

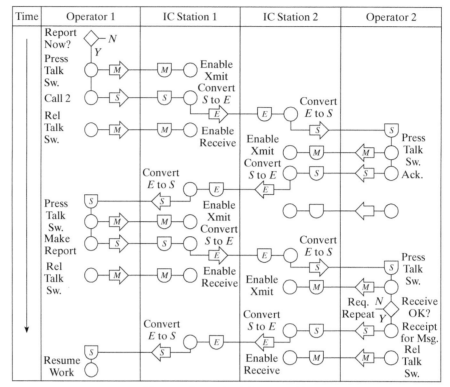

Figure 12 Operational sequence diagram (example). *Source:* MIL-H-46855, Military Specification, *Human Engineering Requirements for Military Systems, Equipment, and Facilities* (Washington, DC: Department of Defense).

4. Inadequate training, job aids, and procedures—lack of proper training, and poorly written operating and maintenance procedures.

5. Poor supervision—lack of communications, no feedback, and lack of good planning resulting in overtime.

The error analysis can be accomplished analytically in conjunction with the OTA, MTA, during the development of OSDs, and physically as part of the system test and evaluation effort. The objective is to select a human with the appropriate skills and training, simulate the operation of the system by undertaking a series of functions/tasks, and measure the number of errors that occur in the process. Using an Ishikawa "cause-and-effect" diagram approach can aid in determining the causes of the errors (i.e., *how was the error introduced in the system?*). In addition, it is important to note the effects of the error on other elements of the system, on the system as an entity, and on other systems within the same system-of-systems (SOS) configuration. Thus, it is important that the error analysis be closely integrated with both the FMECA and the System Safety/Hazard Analysis discussed subsequently.

4.4 Safety/Hazard Analysis

The safety/hazard analysis is closely aligned with the FMECA. Safety pertains to both personnel and the other elements of the system with *personnel* being emphasized herein. The safety/hazard analysis generally includes the following basic information:

1. *Description of hazard.* System operational requirements and the system maintenance concept are reviewed to identify possible hazardous conditions. Past experience on similar systems or products utilized in comparable environments serves as a good starting point. Hazardous conditions may include acceleration and motion, electrical shock, chemical reactions, explosion and fire, heat and temperature, radiation, pressure, moisture, vibration and noise, and toxicity.

2. *Cause of hazard.* Possible causes should be described for each identified hazard. In other words, *What events are likely to occur in creating the hazard?*

3. *Identification of hazard effects.* Describe the effects of each identified hazard on both personnel and equipment. Personnel effects may include injuries, such as cuts, bruises, broken bones, punctures, heat exhaustion, asphyxiation, trauma, and respiratory or circulatory damage.

4. *Hazard classification.* Hazards may be categorized according to their impact on personnel and equipment as follows:

   (a) *Negligible hazard (Category I).* Such conditions as environment, personnel error, characteristics in design, errors in procedures, or equipment failures that will not result in significant personnel injury or equipment damage.

   (b) *Marginal hazard (Category II).* Such conditions as environment, personnel error, characteristics in design, errors in procedures, or equipment failures that can be controlled without personnel injury or major system damage.

   (c) *Critical hazard (Category III).* Such conditions as environment, personnel error, characteristics in design, errors in procedures, or equipment failures that will cause personnel injury or major system damage, or that will require immediate corrective action for personnel or system survival.

   (d) *Catastrophic hazard (Category IV).* Such conditions as environment, personnel error, characteristics in design, errors in procedures, or equipment failures that will cause death or severe injury, or complete system loss.

5. *Anticipated probability of hazard occurrence.* Through statistical means, estimate the probability of occurrence of the anticipated hazard frequency in

terms of calendar time, system operation cycles, equipment operating hours, or equivalent.

6. *Corrective action or preventive measures.* Describe the action(s) that can be taken to eliminate or minimize (through control) the hazard. It is hoped that all hazardous conditions will be eliminated; however, in some instances it may only be possible to reduce the hazard level from Category IV to one of the lesser critical categories.

The safety/hazard analysis serves as an aid in initially establishing design criteria and as an evaluation tool for the subsequent assessment of design for safety. Although the format may vary somewhat, a safety analysis may be applied in support of requirements for both industrial safety and system/product safety.

5 PERSONNEL AND TRAINING REQUIREMENTS

Personnel requirements for the system may be categorized in terms of *operator* personnel and *maintenance* personnel. Operator personnel requirements are derived through the human factors analysis, particularly through the generation of the detailed operator task analysis and operational sequence diagrams. Maintenance personnel requirements evolve from the maintainability analysis, the supportability analysis, and the detail maintenance task analysis (MTA).

In defining these requirements for the system, personnel numbers and skill levels are determined for all human activities by function, job operation, duty, task, and so on. These requirements, initially identified in small increments, are combined on the basis of similarities and complexity level. Individual position requirements are identified, and worksheets are prepared specifying the duties and tasks to be performed by each individual. From this information, personnel quantities and skill-level requirements are identified for the system as an entity.

Given the personnel requirements (in terms of positions) for the system, the next step is to identify available resources and those individuals who can be employed to operate and maintain the system. These selected individuals are evaluated relative to their current skills as compared to the skill levels necessary for the system. The differences dictate the needs for training.

For the sake of illustration, let us assume that as a result of the human factors analysis, the system will require *x* individuals with a *basic* skill, *y* individuals with an *intermediate* skill, and *z* individuals with a *high* skill, and that these skill-level classifications are broadly defined as follows:[7]

1. *Basic skill level.* A basic skill level is assumed to require an individual between 18 and 21 years of age, a high school graduate with a ninth-grade general reading and writing level who has no regular work experience. After a limited amount of familiarization training on the system, this individual can accomplish simple operator functions involving system actuation, manipulation of controls, communications, and the recording of data. The individual is able to follow clearly presented instructions where interpretation and decision making are not required. Close supervision of this individual is normally required.

2. *Intermediate skill level.* An intermediate skill level normally requires an individual over 21 years of age, with approximately 2 years of college or equivalent course work in a

---

[7]As indicated, these classifications are defined in rather broad terms to illustrate a concept. In certain instances, it may be necessary to elaborate further to provide a good basis for determining specific training needs, particularly regarding complex functions.

technical institute, who has had some specialized training on similar systems in the field, and who has 2–5 years of work experience. Personnel in this classification can perform relatively complex tasks, where the interpretation of data and some decision making may be required, and can accomplish simple on-site preventive maintenance tasks. This individual requires little supervision.

3. *High skill level.* A high skill level normally requires an individual with 2–4 years of formal college or equivalent course work in a technical institute, who has taken several specialized training courses in various related fields and possesses 10 years or more of related on-the-job experience. An individual in this classification may be assigned to train and supervise basic and intermediate skill-level personnel, and is in a position to interpret procedures, accomplish complex tasks, and make major decisions affecting system operating policies. Further, this individual is qualified to accomplish (or supervise) all on-site preventive maintenance requirements for the system.

Given requirements for the system, it is assumed that all available resource personnel currently possess the necessary qualifications for entry into positions requiring a basic skill level; that is, with some general familiarization training on the system, they can usually perform the level of activity noted for an individual with a basic skill.

On the basis of these assumptions, it is necessary to develop a training program that will

1. Train entry-level personnel in the fundamentals of system operation to fill the positions where a *basic skill* level is required.
2. Train entry-level personnel in the performance of operator and maintenance functions to the extent necessary to satisfy the *intermediate skill-level* requirements for the system.
3. Train entry-level personnel in the performance of operator and maintenance functions to the extent necessary to satisfy the *high skill-level* requirements for the system.

Training may be accomplished through a combination of formal structured programs and (OJT). Training requirements include both the training of personnel initially assigned to the system and the training of replacement personnel throughout the life cycle as required due to attrition. In addition, training must cover both operator activities and maintenance activities.

In designing training programs, the specific requirements may vary considerably depending on the complexity of the human functions to be performed in the operation and maintenance of the system. The level of complexity is a function of system design as discussed in the earlier sections of this chapter. If there are many human functions of a highly complex nature, then the training requirements will likely be extensive and the associated cost will be high. Conversely, if the functions to be performed by the human being are relatively simple, then training requirements (and associated cost) will be minimal. In any event, a training program should be designed and tailored to meet the personnel needs for the system. The program content must be of the appropriate level to provide future system operators and maintenance personnel with the tools necessary to enable the performance of their respective functions in an effective and efficient manner.

With this objective in mind, a formal training plan is often developed for the system during the detail system/product design phase. This plan should include a statement of training objectives, a description of the training program in terms of modules and content, a proposed training schedule, a description of required training aids and data, and an estimate of anticipated training cost(s). The training plan must be implemented in time to support system operation and maintenance activities as the system is distributed for use.

6 PERSONNEL TEST AND EVALUATION

System *validation* commences with the application of analytical models (e.g., simulation) and extends through Types 1–4 testing. Within each phase of testing, the human element of the system is evaluated in one way or another. As engineering (physical) models, mock-ups, and training simulators are first developed, tests are conducted to evaluate certain facets of system operation and maintenance. The accomplishment of individual discrete tasks and the associated human requirements can be assessed. As part of Types 2 and 3 testing during the accomplishment of reliability test and maintainability demonstration, a certain degree of evaluation can be accomplished through the demonstration of maintenance tasks. During Types 3 and 4 test phases, evaluation of the system as an integrated entity takes place, and it is possible to conduct entire mission scenarios and to evaluate all of the elements of the system as a result.

As these various tests are conducted, the basic objective is to (1) monitor human performance in completing all operational and maintenance requirements; (2) assess operator and maintenance tasks in terms of elapsed times, step-by-step sequences, dexterity requirements, areas of difficulty, and human errors; (3) identify problem areas and initiate recommendations for corrective action; and (4) modify personnel selection and training (and data) requirements as necessary for compatibility with any system changes that may occur. In any event, the process is followed, but with emphasis on the human being and the interfaces between the human and other elements of the system.

7 SUMMARY AND EXTENSIONS

This chapter provides an overview of human factors and related considerations that must be addressed during the design process to ensure the proper implementation of requirements for *human system integration* (HSI). The content commences with some key terms and definitions and a description of some of the more significant characteristics of the typical human being (i.e., anthropometric, human sensory, physiological, and psychological factors). Subsequent topics cover measures of human factors, human factors in the system life cycle, human factors analysis methods, safety/hazard analysis, personnel training requirements, and personnel test and evaluation.

It should be added that the basic objective is not only to *design for system usability* but also to *design against misuse and abuse*. When addressing the issue of abuse, its classification as being the result of either an *intentional* or *unintentional* action is useful. Much emphasis has been on unintentionally introduced problems during the accomplishment of operator and maintenance functions that result in system failures and concerns for system safety. While the existence of these is certainly important, there is an added challenge in today's world—the intentional introduction of problems and the act of sabotage (or terrorism).

Accordingly, there needs to be an added dimension in and throughout the design process; that is, *design for security*. The following goals should be considered:

1. The development and incorporation of an external security alarm capability that will detect the presence of unauthorized personnel and prevent them from operating, maintaining, and/or gaining access to the system and its elements, and one that will ultimately lead to the prevention of an "outsider" from inducing a problem that will result in system damage or destruction.

2. The incorporation of a "condition-based monitoring" capability that will enable one to check the status of the system and its elements on a continuing basis. To accomplish this requires the appropriate sensors, readout devices, inspection methods, and so on to be included to verify that the system and its components are in the condition intended, and that the appropriate diagnostics be incorporated that will lead to the correction of any problem that may exist.

3. The incorporation of a built-in capability (mechanisms) that will detect and initiate an alarm in the event that a problem is detected. A design that will, in the event of a problem, prevent a subsequent chain reaction of failures leading to system damage or destruction is desirable.

The designer must address such issues as preventing unauthorized personnel from gaining access to the system, being able to determine the actual condition of the system at any given time, and being able to detect and subsequently prevent the intentional introduction of failures that will lead to system destruction and operator/maintainer death. The issue pertaining to the *design for security* can be an appropriate extension to the material in this chapter, and can be woven into the system safety/hazard analysis.

The purpose of this chapter is to illustrate and emphasize the importance of human factors within the total spectrum of design-related requirements and activities that are necessary to meet system engineering objectives. There are four helpful websites that serve as a current resource:

1. Human Factors and Ergonomics Society (HFES)—*http://www.hfes.org.*
2. National Association of Safety Professionals (NASP)—*http://www.naspweb.com.*
3. Occupational Safety and Health Administration (OSHA)—*http://www.osha.gov.*
4. System Safety Society (SSS)—*http://www.system-safety.org.*

QUESTIONS AND PROBLEMS

1. Describe what is meant by *human factors*. Why is it important to consider human factors in design? When in the system life cycle should it be considered and why?

2. Describe the process that leads to the identification of the *human* functional requirements in the performance of system operation and maintenance.

3. Describe each of the following in detail and provide an example of the relationship(s) to system design:

   (a) Anthropometric factors
   (b) Human sensory factors

    (c) Physiological factors

    (d) Psychological factors

4. Describe the possible impact(s) of physiological and psychological factors on system operation and the completion of a defined mission. Describe the possible impact(s) on each other. Provide some examples.

5. What effects are likely to occur in the performance of a function by a human if the operator/maintainer is undertrained for the job? Overtrained for the job? Why?

6. Refer to Figure 8. Describe some of the considerations pertaining to the human that need to be addressed in the design of a system.

7. Identify and describe some of the measures that you might apply as design-to requirements for the human elements of a system. Select a system of your choice, identify the major subsystems, and allocate these requirements to the subsystem level.

8. Describe the interrelationships between the human factor requirements (presented in this chapter), the reliability requirements, and maintainability requirements; that is, their impact on each other. Provide some examples.

9. Describe each of the following, along with its purpose, application, and the information acquired through its application:

    (a) Operator task analysis (OTA)

    (b) Operational sequence diagram (OSD)

    (c) Error analysis

    (d) Safety/hazard analysis

10. Select a system (or an element of one) of your choice and accomplish an OTA. Provide an illustration showing the top-down evolution of the OTA from an operational functional flow diagram, and then provide an expanded OTA.

11. From the OTA described in Problem 10, select a specific part of the system where there is a hardware–human interface (e.g., control panel requiring human–machine interaction) and develop an OSD.

12. How does the OTA relate to the MTA (if at all)?

13. How does the safety/hazard analysis relate to the FMECA (if at all)?

14. How are personnel quantities and skill-level requirements defined for a system? Identify the steps involved.

15. Select an organization with which you are familiar, assume that you are required to identify the entry-level requirements for the organization, and develop a set of descriptions for an individual at each skill level: *basic*, *intermediate*, and *supervisory*.

16. How are personnel training requirements determined? What input data are required?

17. What steps would you take in the design of a training program for an individual? For an organization?

18. How would you measure the effectiveness of a training program?

19. What is the purpose of personnel test and evaluation? When are the requirements for such tests initially defined? When in the system life cycle is personnel test and evaluation activities accomplished? Provide some examples of specific tests that are conducted where human factor requirements can be validated.

20. How does the personnel test and evaluation effort described in Section 6 relate to reliability testing and maintainability demonstration?

Design for Logistics and Supportability

Major systems have been planned, designed and developed, produced, and deployed with little consideration having been given to the need for logistics, maintenance, and sustaining support over their life cycles. When these essential activities are addressed, it has primarily been "after-the-fact;" that is, after the basic system design configuration has been established, often without the benefit of having had the desired design attributes built in and inherent within. Incorporating design changes at this late point in time is quite costly, largely because the development of an effective and efficient support infrastructure is no longer easily accomplished at this stage.

Given this situation, it is essential that (1) the logistics and maintenance support infrastructure be considered as a major element of the system; and (2) this be properly addressed during and throughout the system design process. If the system is to accomplish its mission in an effective and efficient manner, the needed logistics and maintenance support infrastructure must be reliable and put in place concurrently. It must be designed-in from the beginning, concurrently with other system elements.

This chapter provides the concepts, models, and methods useful in addressing the *design for logistics and supportability* early in the life cycle, leading to the specification and development of an effective and efficient system support infrastructure. This includes (1) the logistics and supply chain (SC) activities associated with initial procurement and acquisition, manufacture and/or production, and transportation and distribution of the system and its elements to customer (user) operational sites; and (2) the subsequent sustaining maintenance and support of the system throughout its planned life cycle. These areas of system support must be considered throughout the system engineering process.

The objective of this chapter is to provide an understanding of logistics and maintenance support from an engineering perspective, utilizing a total integrated approach, through a study of the following factors:

- The definition and explanation of logistics and supportability;
- Logistics and system support in the system-of-systems (SOS) environment;

- The elements of logistics and system support;
- The measures of logistics and supportability—supply chain, purchasing and material flow, transportation and packaging, warehousing and distribution, maintenance organization, spares/repair parts, test and support equipment, maintenance facilities, computer resources, and technical data factors;
- Logistics and maintenance support in the system life cycle—system requirements, requirements allocation, design participation, and design review;
- Supportability analysis (SA); and
- Supportability test and evaluation.

Developing the requirements and plans for logistics and system maintenance and support depends on the proper application of reliability, maintainability, human factors, and related principles and concepts. This application should be guided by the topics in this chapter. In addition, there are a number of useful websites recommended in the summary section of this chapter.

1 DEFINITION AND EXPLANATION OF LOGISTICS AND SUPPORTABILITY

The logistics and maintenance support infrastructure, as defined herein, includes (1) the logistics and supply chain activities associated with the initial purchasing and acquisition, manufacture and/or production, transportation and distribution, and installation of the system and its elements at the appropriate customer (user) operational site(s); and (2) the subsequent sustaining maintenance and support of the system throughout its entire life cycle. While these activities have normally been addressed and treated separately in the past, the objective herein is to view these on a total integrated life-cycle basis.

Referring to Figure 1, the various blocks reflect some of the major activities associated with the development, production/construction, operation, logistics, and maintenance support of a system. Initially, there is an identification of need, definition of system requirements, and the accomplishment of some early planning activity (Block 1). This leads into design and development, involving both the system developer and one or more suppliers (Blocks 2 and 3, respectively). Given an assumed design configuration, the production and/or construction process commences, involving a prime manufacturer and a number of different suppliers (Blocks 4 and 3, respectively). Subsequently, the system is transported and installed at the appropriate customer/user operational site(s), and different components of the system may either be distributed to some warehouse or directly to the operational site (Blocks 7 and 5, respectively). In essence, there is a *forward* (or *outward*) flow of activities; that is, the flow of activities from the initial identification of a need to the point when the system first becomes operational at the user's site, which is reflected by the shaded areas in Figure 1.

Within this forward flow of activities (i.e., Blocks 2–5 and 7 in Figure 1) are those functions that are primarily associated with the various aspects of the logistics *supply chain* (SC), emphasized throughout the commercial sector, which includes (1) the physical supply of items from the various applicable sources of supply to the manufacturer; (2) the materials handling, associated inventories, and flow of items throughout the production process; and (3) the transportation and physical distribution of finished goods from the manufacturer/ producer to the customer's operational site(s). Logistics in this context, which has been primarily oriented to the processing of relatively small items (e.g., consumables), can be

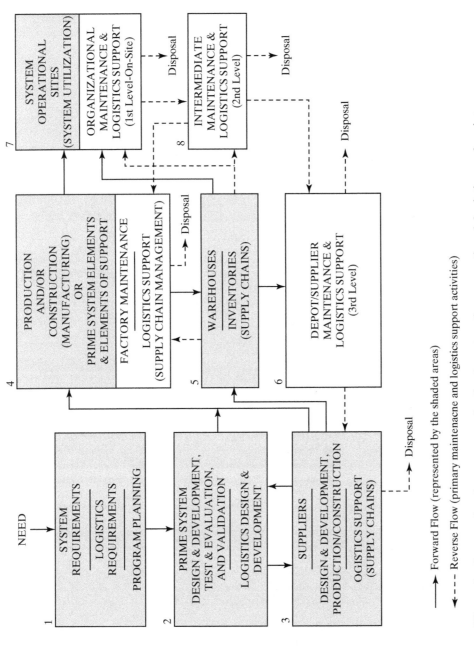

Figure 1 System development, production/construction, operational, logistics, and maintenance support activities.

defined as *that part of the supply chain process that plans, implements, and controls the efficient, effective forward and reverse flow and storage of goods, services, and related information between the point of origin and the point of consumption in order to meet customer requirements.*[1] Emphasis in the past has been primarily directed to the physical aspects of materials flow and handling as shown in Figure 2. Further, the spectrum of activities in the figure has been considered after-the-fact, whereas the emphasis in this text must also include those design-related activities pertaining to the development of the supply chain. In other words, those activities illustrated in Figure 2 must be initially addressed in the design and development process represented by Block 2 in Figure 1.

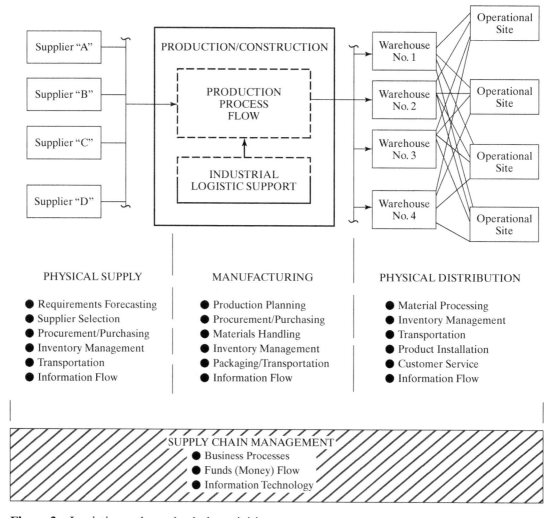

Figure 2 Logistics and supply chain activities.

[1]This definition was developed by the Council of Supply Chain Management Professionals (CSCMP), 333 East Butterfield Road, Suite 140, Lombard, IL 60148. Refer to website *http://cscmp.org*, "Glossary of Terms," October 2002.

During the past decade, the area of "business logistics" (i.e., logistics in the commercial sector) has been expanded significantly to include the introduction of the latest electronic commerce (EC) methods, information technology (IT), electronic data interchange (EDI), and the application of good business processes to the activities shown in Figure 2. Further, the trends toward more outsourcing, greater international competition, and increased globalization have resulted in the need for the establishment of coalitions and industry/government partnerships worldwide. This has resulted in development of the currently popular concepts of the SC and *supply chain management* (SCM). The *supply chain* refers to that group of organizations and activities that pertain to the overall flow of materials and services from the various supplier sources to the ultimate customer(s).

In the late 1990s, a group of researchers from the Massachusetts Institute of Technology (MIT) offered the following definition for SCM: *a process-oriented, integrated approach to procuring, producing, and delivering end products and services to customers. It includes sub-suppliers, suppliers, internal operations, trade customers, and end users. It covers the management of materials, information, and funds flow.*[2]

In 2001, a group of academicians conducted a survey in the field, identified and categorized six slightly different definitions of SCM, and arrived at the following definition by consensus: *the systemic, strategic coordination of the traditional business functions within a particular company and across businesses within the supply chain, for the purposes of improving the long-term performance of the individual companies and the supply chain as a whole.*[3]

SCM pertains to management of the supply chain, or a group of supply chains, with the objective of providing the required customer services, both effectively and efficiently. It requires a highly integrated approach, employing the appropriate resources (e.g., transportation, warehousing, inventory control, and information) and implementing the necessary business processes to ensure complete customer satisfaction. The Council of Supply Chain Management Professionals (CSCMP) adopted this definition at the time.

In addition to the forward (outward) flow of activities in Figure 1, there is also a *reverse (or backward)* flow, which covers the follow-on maintenance and support of the system after it has been initially installed and operational at the customer's (user's) site. Referring to the figure, this includes all activities associated with the accomplishment of organizational maintenance (Block 7), intermediate-level maintenance (Block 8), factory and/or depot-level maintenance (Blocks 4 and 6), supplier maintenance (Block 6), and replenishment of the necessary materials to support the maintenance actions at all levels (Blocks 3–5); for example, spares/repair parts and associated inventories, test and support equipment, personnel, facilities, information/data, and so on. This facet of activity, as part of the logistics and maintenance support infrastructure, is reflected by the operational and maintenance flow in Figure A.1.

Referring to Figure A.1, the activities shown include those within the spectrum of *integrated logistics support* (ILS), developed in the *defense* sector in the mid-1960s. The principles and concepts of ILS were developed further throughout the 1970s, 1980s, and 1990s, with ILS being defined as *a disciplined, unified, and iterative approach to the management and technical activities necessary to (1) integrate support considerations into system and*

[2]P. J. Metz, "Demystifying Supply Chain Management," *Supply Chain Management Review*, Winter 1998 (New York: Reed Elsevier, 1998).

[3]J. T. Mentzer (University of Tennessee), W. DeWitt (University of Maryland), J. S. Keebler (St. Cloud State University), Soonhong Min (Georgia Southern University), N. W. Nix (Texas Christian University), C. D. Smith (University of San Diego), and Z. G. Zacharia (Texas Christian University), "Defining Supply Chain Management," *Journal of Business Logistics*, Vol. 22, No. 2, 2001 (Lombard, IL: CSCMP).

equipment design; (2) develop support requirements that are related consistently to readiness objectives, to design, and to each other; (3) acquire the required support; and (4) provide the required support during the operational phase at minimum cost.[4]

The objectives in applying and implementing ILS are to (1) deal with logistics and maintenance support from a total systems perspective; (2) view the related requirements within the context of the entire system life cycle; and (3) address these system life-cycle requirements early in the system design and development process. As such, the implementation of ILS includes both the activities of "business logistics" illustrated in Figure 2 and those of system life-cycle maintenance and support shown in Figures A.1 and 1.

More recently, there has been an increased emphasis on logistics and maintenance support early in the system design and development process through the introduction of the concept of *acquisition logistics*, which is *a multifunctional technical management discipline associated with the design, development, test, production, fielding, sustainment, and improvement modifications of cost-effective systems that achieve the user's peacetime and wartime readiness requirements. The principal objectives of acquisition logistics are to ensure that support considerations are an integral part of the system's design requirements, that the system can be cost-effectively supported throughout its life cycle, and that the infrastructure elements necessary to the initial fielding and operational support of the system are identified and developed and acquired.*[5]

The emphasis on dealing with logistics and maintenance support in the design process is based on the fact that (through past experience) a significant portion of a system's life-cycle cost (LCC) can be attributed directly to the operation and support of the system in the field, and that much of this cost is based on design and management decisions made during the early stages of system development. In other words, early design decisions can have a large impact on the cost of those downstream activities associated with system operation and maintenance. Thus, it is essential that logistics and the *design for supportability* be addressed from the beginning.

To further stress this requirement, the Department of Defense introduced, more recently, the concept of *performance-based logistics* (PBL).[6] The objective is to emphasize the importance of and need for considering the logistics and maintenance support infrastructure in design (refer to Figures A.1 and 1) by establishing some specific measures (or metrics) and to include these as quantitative *design-to* performance requirements in the appropriate specifications. Such requirements may be covered within the spectrum of system-level requirements (e.g., operational availability (Ao), system effectiveness, life-cycle cost) and/or to a specific element of the logistics and maintenance support infrastructure (e.g., personnel effectiveness, quantitative goals for transportation times and cost, supply support effectiveness, facility utilization and throughput, information access and process time, and test equipment reliability). These logistics and support-related factors must be integrated with (and supportive of) the higher-level system performance factors as they pertain to the overall mission of the system (refer to the TPM identification and development process).

[4]DSMC, *Integrated Logistics Support Guide* (Fort Belvoir, VA: Defense Systems Management College, 1994). (DSMC is currently recognized as a major element within the Defense Acquisition University).
[5]MIL-HDBK-502, *Department of Defense Handbook on Acquisition Logistics* (Washington, DC: Department of Defense, May 1997).
[6]DOD 50002.2-R, *Mandatory Procedures for Major Defense Acquisition Programs* (MDAPS) and *Major Automated Information System* (MAIS) *Acquisition Programs* (Washington, DC: Department of Defense, April 5, 2002).

In covering the logistics and the maintenance support infrastructure, one needs to consider the entire spectrum of activity, including both the commercial and defense approaches (shown in Figures 1 and 2), on a total integrated life-cycle basis. The interactions among the various activities are numerous, and it is recommended that one have a good understanding of the flow of activities from system inception to utilization and back. In addition, one needs to consider this overall infrastructure (and associated activities) as a major element of a system, and this needs to be accomplished as an integral part of the design process from the beginning. As an inherent part of the system engineering design process, one must develop both the internal flow processes within each of the blocks in the two figures along with the external flow processes between the blocks to ensure that the end result represents an overall effective and efficient logistics (supply chain) and maintenance support capability. The objective in this chapter is to include all of these activities within the broad spectrum of supportability, with the *design for supportability* as a specific goal. In other words, addressing the issue of supportability is critical within and throughout the systems engineering process.

2 LOGISTICS IN THE SYSTEM-OF-SYSTEMS (SOS) ENVIRONMENT

When designing a new system within the context of a system-of-systems (SOS) network, one needs to ensure that (1) the specified logistics and maintenance support infrastructure for this new system is both effective and efficient and is completely responsive to the new system requirements; and (2) that the newly developed maintenance and support infrastructure is compatible and does not in any way degrade the equivalent capabilities of the other systems with the same SOS configuration. In this era of increased outsourcing globally, there are likely to be many different suppliers, representing many different and varied nationalities, providing support for more than just a single system in a given network. Referring to Figure 1, the basic supply chain is represented by Blocks 3–5 and 7. Projecting this further, one is likely to experience a configuration as shown in Figure 3, where one supplier (as essential element of the supply chain) may be directly supporting more that a single system; for example, supplier "b" serves as a critical element of the supply chain for all three of the systems in the figure.

Referring to the figure, the systems being supported (i.e., communications and the air and ground transportation systems) may have a completely different set of requirements for logistics and system support, and the suppliers providing the support may represent different nationalities with different language, cultural, work-oriented, and related unique capabilities. The working environment (labor requirements, local and national holidays, business and ethical practices, etc.) in one supplier's organization may be completely different from that in another. There have been instances when such variations have resulted in extensive logistics delays, not providing the right support at the place and time of need, and overall inefficiencies in logistics responsiveness in general. This, in turn, has resulted in degradation in the overall operational availability and effectiveness of the system being supported. Thus, it is essential that the numerous interfaces within a given SOS configuration, and system interoperability requirements, be addressed in the early design process when new systems are being introduced.

3 THE ELEMENTS OF LOGISTICS AND SYSTEM SUPPORT

Referring to the forward and reverse flows in Figure 1, it can be seen that the resource requirements for system logistics and maintenance support include personnel, transportation (ground, sea, and/or air), spares/repair parts and related inventories, test and support equip-

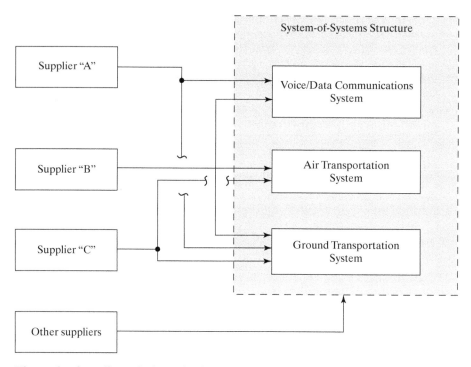

Figure 3   Supplier relationships in a system-of-systems (SOS) structure.

ment, facilities (maintenance, warehousing, utilities), information/data (documentation), computer software, and various combinations thereof. In the initial planning, design, and implementation of the logisitics and maintenance support infrastructure, it is essential to ensure that these requirements are fully integrated, as conveyed in Figure 4, since the interrelationships among these various elements are numerous. A brief description of the necessary planning activitiy and each of the elements in the figure is presented in the sections to follow.

Logistics and Maintenance Support Planning.   Inherent within the inner circle in Figure 4 are those ongoing iterative activities of planning, organization, and management necessary to ensure that the logistics and maintenance support requirements for any given system are properly coordinated and implemented from the beginning. Initial planning (during the conceptual design phase) leads to the establishment of the support requirements for the activities illustrated in Figure 1 (and expanded in Figures A.1 and 2). Included are the *design for supportability* requirements (as an input to the system design process), supplementing the overall system-level requirements included in the System Engineering Management Plan (SEMP). As the system design and development process evolves (during the preliminary system design and detail design and development phases), the planning activity continues and leads to the development of a comprehensive detailed *maintenance plan* covering the requirements for the sustaining maintenance and support of the system during and throughout its programmed life cycle.

Logistics, Maintenance, and Support Personnel.   Personnel required to perform unique logistics and system maintenance activities are included in this category. Such activities

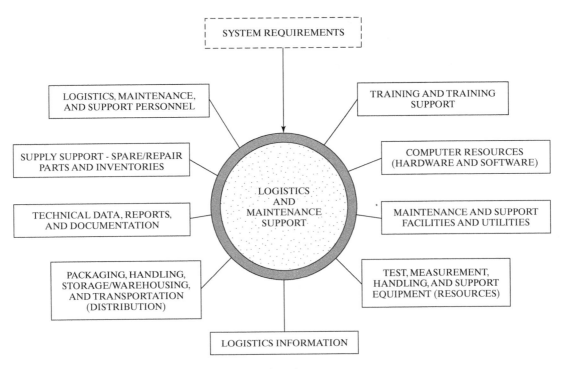

Figure 4 The basic elements of logistics and maintenance support.

cover the initial provisioning and procurement of items for system support, production-related logistics and supply chain functions shown in Figure 2, the initial installation and checkout of the system and its elements at the user's operational site(s), customer service and interim contractor logistics support (CLS) during the initial stages of system operation, the sustaining support of the system throughout its planned period of use, and those functions required for system retirement and the recycling/disposal of material as required. Personnel at all levels of maintenance are included. In the evaluation of a particular system (within a SOS configuration), it is important to include only those who can be directly attributed to the logistics and maintenance support of that system.

Training and Training Support. All personnel, materials and equipment, facilities, data, documentation, and associated resources necessary for the training of operator and maintenance personnel, to include both *initial* and *replenishment/replacement* training, are included. Training equipment (e.g., simulators, mock-ups, special displays and devices), training manuals, and computer resources (software) are developed and utilized as necessary to support the day-to-day on-site training, distance education (via the Internet), classroom education of a more formal nature, and/or various combinations thereof.

Supply Support: Spares/Repair Parts and Associated Inventories. This category includes all spares (repairable units, assemblies, and modules), repair parts (nonrepairable components), consumables (liquids, lubricants, fuels, and disposable items), special supplies, software modules, and supporting inventories necessary to maintain the prime mission-related elements of the system and the various elements of the logistics and maintenance support infrastructure throughout the operational use phase and as required during the retirement and material recycling/disposal phase.

Computer Resources: Hardware and Software. All computers, associated software, connecting components, networks, special facilities, and interfaces necessary to support the day-to-day flow of information for all logistics and system maintenance support functions throughout the system life cycle are included.

Technical Data, Reports, and Documentation. Technical data may include system installation and checkout procedures, operating and maintenance instructions, inspection and calibration procedures, overhaul instructions, facilities data, system modification instructions, engineering design data (specifications, drawings, materials and parts lists, CAD/CAM/CAS data, and special computerized information), logistics provisioning and procurement data, supplier data, special component tracking reports, system operational and maintenance data, and supporting databases. Included in this category is the ongoing and iterative process of data collection, analysis, and reporting covering the system throughout its life cycle (i.e., the maintenance data collection and assessment capability).

These data acquisition tasks can be facilitated through the recent introduction and application of automatic information technologies (AIT), active and passive radio frequency identification (RFID) tags, advances in bar coding methods, and global positioning system (GPS) technology. The recent thrust(s) in RFID applications, both in the commercial and defense sectors, should do much to enhance the logistics and support processes described throughout this chapter when such applications become more refined and standardized in the future.

Maintenance Support Facilities and Utilities. This category covers all special facilities that are unique and are required to support the logistics and maintenance activities at all levels (refer to Figures A.1 and 2). This includes physical plant, portable buildings and mobile vans, fixed maintenance shops, warehouses and storage buildings, personnel housing structures, calibration laboratories, and special repair shops. Capital equipment and utilities (heat, power, energy requirements, environmental controls, communications, safety and security provisions, etc.) are generally included as part of facilities.

Packaging, Handling, Storage/Warehousing, and Transportation (Distribution). This category includes all materials, equipment, special provisions, containers (reusable and disposable), and supplies necessary to support the packaging, safety and preservation, security, storage, handling, and/or transportation of the prime mission-related elements of the system and the various elements of the logistics and maintenance support infrastructure throughout the system life cycle. Referring to Figure 1, all of the transportation and handling functions represented by the lines (forward and reverse flows) between the blocks are included. The requirements pertaining to the primary modes of transportation (air, highway, railway, waterway, and pipeline) must be addressed.

Test, Measurement, Handling, and Support Equipment. All tools, condition monitoring equipment, diagnostic and checkout equipment, special test equipment, metrology and calibration equipment, maintenance fixtures and stands, and special handling equipment required to support operational and maintenance functions throughout the system life cycle are included. This covers the test and support equipment at each level of maintenance, shown in Figure A.1, and all of the equipment necessary for support of test equipment maintenance and calibration (i.e., essential secondary standards, transfer standards, and those elements required to ensure the *traceability* of requirements back to a primary standard).

Logistics Information. This item refers to the resources necessary to ensure that an effective and efficient logistics and maintenance information flow is provided throughout the system life cycle, and to the organizations responsible for all of the functions and activities reflected in Figure 1. This flow includes the necessary communication links among the customer, producer (prime contractor), subcontractors, suppliers, and supporting maintenance organizations. All of the activities within the applicable supply chain(s) and maintenance support infrastructures must be in the communications loop at all times. It is essential that the proper type and amount of information be provided to the appropriate organizational elements, in the proper format, and in a reliable and timely manner, with the necessary security provisions included. Inherent within this category is the utilization of the latest electronic commerce (EC) methods, electronic data interchange (EDI) capabilities, e-mail, and the Internet.

The objective is to provide the right *balance* of resources applied throughout the system support infrastructure illustrated in Figures A.1 and 2. A cost-effective approach must be reached that requires the proper mix of best commercial practices, supplemented by new developmental items where necessary and when justified. As new technologies are introduced and current processes are improved, the specific resource requirements at each level may shift somewhat. For example, the need for large spares/repair parts inventories may not be required as transportation times and costs decline. The availability of overnight express, combined with good communications processes, can help solve the high-cost large-inventory problem. Further, the increased application of both active and passive RFID technologies should result in reduced spares/repair parts inventories through better accountability and enhanced processing capabilities. The need for large data packages (e.g., technical manuals and drawings) at each maintenance location is reduced with the advent of new EDI data formats and processes. The development of computer-based information systems provides for faster and more accurate information and greater visibility about the type and location of various assets at a given point in time. The decision-making and communications processes may be enhanced accordingly. Numerous trade-off studies may be conducted involving various mixes and combinations of the logistics elements identified herein which will (hopefully) lead to the design and development of an effective and efficient support infrastructure.

4 THE MEASURES OF LOGISTICS AND SUPPORTABILITY

Inherent within the overall effectiveness measures of a system, and the degree to which the system is able to accomplish its mission, are those measures associated with the logistics and maintenance support infrastructure (as a major element of the system) and its availability when needed. Thus, it is important to address these logistics and maintenance support measures, some cause-and-effect relationships, and the effect of the system design process on the ultimate results.

The purpose herein is to define some of the system measures, concentrating on those directly associated with the various elements of logistics identified in Figure 4.

4.1 Supply Chain Factors

As described in Section 1, the supply chain includes those functions, associated primarily with the forward flow of activities in Figure 1, involved with the initial acquisition (procurement) of items from various sources of supply, the flow of materials throughout the production process, the transportation and distribution of products from the manufacturer to the customer (user), the sustaining onsite customer service as required, and all related business-oriented processes necessary to ensure that the entire flow is both effective and efficient. A simplified illustration of the flow is presented in Figure 5 (which is an extension of Figure 1). This figure shows different paths that may be utilized depending on the specific requirements; that is, materials shipped directly from supplier to manufacturer, supplier to customer, manufacturer to customer, and so on. Further, the reverse flow addresses returned materials and the requirements for maintenance.

With regard to the measures pertaining to the supply chain, once again there are two sides of the balance to consider. There are both technical and economic factors that can be applied. At the top level of the spectrum, for example, a technical measure of effectiveness can be stated in terms of the following, or the product thereof:[7]

1. *Capability*, or the ability to accomplish all of the functions required (e.g., purchasing, materials handling, transportation, warehousing, and maintenance). This is basically a measure of performance of the overall flow in Figure 5.

2. *Availability*, or the ability to respond to any or all of the requirements at any point in time when needed. This pertains to the reliability of the flow in Figure 5, and can also be stated in terms of *operational availability (Ao)*.

3. *Quality*, or the process responsiveness in terms of stated customer objectives (e.g., performance of the right function, in the right location, at the right time, with the right information, and at the right value). This term can be related to the *quality rate (Q)* utilized as one of the metrics in determining the *OEE* for a manufacturing capability. There are some underlying expectations that need to be realized from a customer satisfaction perspective, and any deviation here would constitute a subsystem defect.

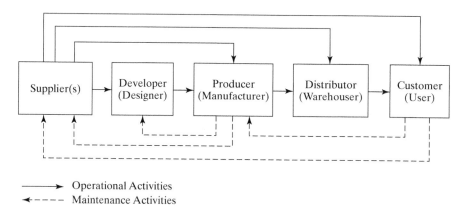

⟶ Operational Activities
◄ - - - - Maintenance Activities

Figure 5   Logistics activity flow (materials, information, and money).

---

[7]It should be noted that these factors may vary somewhat depending on the specific system configuration, its makeup, and its mission. The intent here is to illustrate an attempt to define the top-level metrics for a major subsystem; that is, the logistics and maintenance support infrastructure.

To complete the proper balance, one needs to deal with the issue of cost and, more specifically, the LCC associated with the infrastructure shown in Figure 5, and its impact on the LCC for the overall system as an entity. The total cost for the infrastructure may be broken down as shown in Equation 1:

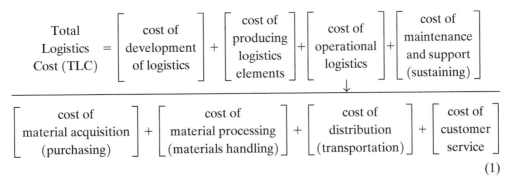

$$\text{(1)}$$

The term *quality* is used in a rather broad context here, applied at a high level, and can be related to the degree to which the system is able to attain (and maintain) a status reflecting complete fulfillment of all customer requirements. The critical issues include not only system performance and reliability but also the responsiveness of the logistics and maintenance support infrastructure in fulfilling these requirements. For example, it may be appropriate to establish some specific design-to goals such as the following:

1. The response time (i.e., time to react in response to an identified need) for the logistics and maintenance support infrastructure shall not exceed 4 hours.
2. The total processing time, from the identification of a need for a product (or service) to the delivery of the product to the point of consumption (or completion of the service) shall not exceed 24 hours.
3. The total cost of processing an item through the logistics and maintenance support infrastructure (i.e., packaging, transportation, handling, documentation) shall not exceed x dollars per action.
4. The process time for removing an obsolete item from the inventory shall not exceed 12 hours, and the cost per item processed shall not exceed x dollars.
5. The defect rate in terms of products delivered (services provided) shall not exceed 1% per designated interval of time.

Having established some specific design-to goals, along with those established for reliability and maintainability, and by providing some benchmarks for measuring response to customer needs, one can then measure and evaluate the system in terms of meeting these goals. Any deviations from the established benchmarks can be classified as *defects*.

4.2 Purchasing and Material Flow Factors

Within the overall spectrum of the supply chain is a wide variety of activities that are critical to the successful fulfillment of logistics objectives. Of specific interest are those factors that are associated with the initial *purchasing* and the subsequent *flow of materials* through the manufacturing and assembly of products (refer to Blocks 3 and 4 of Figure 1). In the area of *purchasing*, there may be a number of measures, such as

1. The time that it takes to initiate and process a purchase order.
2. The quantity of purchase orders processed within a designated time period.
3. The quality of the purchasing process.

Relative to the issue of *quality*, the following standards may be applied:[8]

1. Delivered complete—all items delivered in the quantities requested.
2. Delivered on time—using the customer's definition of on-time delivery.
3. Complete and accurate documentation (including packing slips, bills of lading, and invoices) to support the order.
4. Delivered in perfect condition and in the correct configuration to be used by the customer, faultlessly installed (as applicable).

With regard to the *flow of materials* through the product manufacturing and assembly process, such metrics may be specified in terms of the following:

1. The quantity of materials processed through the factory in a designated time.
2. The time it takes to process materials.
3. The quality of materials processed.
4. The cost of materials processed.

In the area of "quality," the emphasis is on the processing of materials on time, to the desired location, with the necessary supporting data/information as required, and in good shape (i.e., in perfect condition and without any damage).

4.3 Transportation and Packaging Factors

The element of *transportation* constitutes a major activity throughout the various facets of the logistics and maintenance support infrastructure. First, it is a key element in the supply chain and commercial logistics segment of the overall structure (refer to flow of activity between the blocks in Figure 1); second, it is a key element in the sustaining maintenance and support activity throughout the system utilization phase (refer to Figure A.1); and third, it is essential in the system retirement and material recycling/disposal phase.

Transportation requirements include the movement of humans and material resources, in support of both operational and maintenance activities, from one location to another. When evaluating the effectiveness of transportation, one must deal with the following factors:

1. Transportation route, both national and international (distances, number of nationalities, customs requirements, political and social factors, and so on).
2. Transportation capability or capacity (volume of goods transported, number of loads, ton-miles per year, frequency of transportation, modes of transportation, legal forms, and so on).
3. Transportation time (short-haul versus long-haul time, mean delivery time, time per transportation leg, and so on).
4. Transportation cost (cost per shipment, cost of transportation per mile/kilometer, cost of packaging and handling, and so on).

---

[8]LG803R1, *Supply Chain Management: A Recommended Performance Measurement Scorecard* (McLean, VA: Logistics Management Institute (LMI), June 1999).

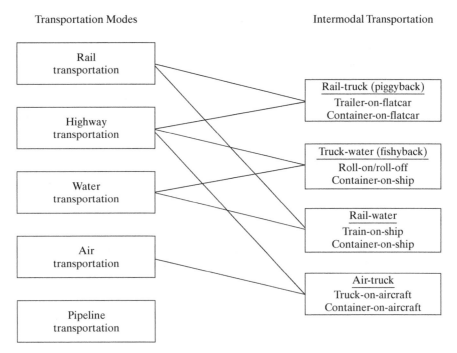

Figure 6 The various forms of transportation.

The basic modes of transportation include *railroad, highway, waterway, air,* and *pipeline*. In addition, there are various combinations of these modes that can be categorized as *intermodal*, as shown in Figure 6. Depending on the geographical location of system elements and the applicable suppliers (i.e., the need for transportation), the degree of urgency in terms of delivery-time requirements, and so on, alternative routing requirements can be identified. Through the process of conducting trade-offs considering time factors, available modes, and cost, a recommended transportation route is established. This may include transportation by a specific mode or through the use of an intermodal approach.

Given the possible modes of transportation and the proposed routes that have been identified to support the *outward* and *reverse* flows pertaining to operational and maintenance activities (refer to Figure 1), along with the various environments in which an item may be subjected when being transported from one location to another, the key issue becomes that of *packaging* design, or the design of an item for *transportability* or *mobility*. Products that are to be transported must be designed in such a way as to eliminate damage, possible degradation, and so on. The following questions should be addressed:[9]

1. Does the package (in which the item is being transported or contained) incorporate the desired strength-of-material characteristics?
2. Can the package stand rough handling and/or long-term storage without degradation?

---

[9]It is not uncommon to find that a product in transit will be subjected to a harsh or rigorous environment—one that is often more rigorous than will be experienced throughout the performance of the product's normal operational mission(s). Further, there is the ongoing concern about the possibility of terrorism and the intentional destruction of a given product while it is being transported from one location to another. The appropriate safety/security provisions must be incorporated.

3. Does the package provide adequate protection against various environmental conditions such as rain, temperature extremes, humidity, vibration and shock, sand/dust, salt spray?

4. Is the package compatible with existing transportation and handling methods; that is, does it possess good stacking qualities, can it be readily identified for tracking and tracing, and is it readily accessible for removal/replacement, if required?

5. Has the package been designed for safety and security; that is, to prevent pilferage, theft, and/or sabotage?

With regard to the metrics associated with the element of transportation, the key issues that should be considered include the following:

1. The *availability* of transportation, or the probability that the appropriate transportation capability will be available when required.

2. The *reliability* of transportation, or given that the appropriate transportation is available when required, the probability that it will complete its mission as planned.

3. The *time* that it takes to transport a product from one point to another.

4. The *maintainability* of a given transportation capability, or the probability that the applicable transportation capability can be repaired within a specified time and with the specified resources in the event of a failure.

5. The *cost* of transportation, or the cost per one-way trip.

6. The *LCC* of a given transportation capability for a designated period of time (in years).

4.4 Warehousing and Distribution Factors

Warehousing is an integral part of every logistics capability. The basic functions of warehousing are *movement, storage,* and *information transfer*. A major objective is to provide an ideal product flow and acceptable level of service between the producer and the customer by providing warehouses at designated locations with varying inventory levels based on local demand. The movement function can be broken down to include activities such as the following:

1. *Receiving:* the unloading of goods and products from an inbound carrier.

2. *Transfer/put away:* the physical movement of products into the warehouse for storage, movement to areas for specialized activities such as consolidation, and/or movement directly to an outbound dock for shipment.

3. *Order picking:* the selection of products from storage in response to customer orders/requirements.

4. *Cross-docking:* the movement of products directly from the inbound receiving dock to the outbound shipping dock, without requiring put-away, storage, and order-picking activities.

5. *Shipping:* the packing, loading on an outbound carrier, and shipment of products to the desired customer destination(s).

In recent years, the practice of cross-docking has become popular and is being adopted to an increasing degree in order to be more responsive to customer demands by getting the product to its destination faster and to reduce costs by not requiring put-away, storage, and order-picking activities. The implementation of such a policy will, of course, be dependent on the number of daily shipments required, the distance and geographical location(s) of the customer sites, the type and ease of movement of the product, and so on.

The storage function can be broken down to include *temporary storage* (storage of products that support short-term replenishment activities) and *long-term storage* (storage of products in excess of the requirement for normal replenishment in order to build up a buffer or safety stock). The information transfer function includes the necessary communications, databases, and so on, utilizing EC methods (or equivalent) to provide timely information pertaining to order requests, product stock-keeping locations, tracking of product in transit, current inventory levels, inbound and outbound shipments, facility space utilization, customer data, and so on.

The major categories of warehouses include both *private* and *public* warehouses. Private warehouses are those owned and operated primarily by production/manufacturing organizations to support their own logistics requirements. On the other hand, there are a number of different categories of public warehouses to serve a variety of needs; for example, *general merchandise warehouses, commodity warehouses, bulk-storage warehouses, bonded warehouses, temperature-controlled warehouses,* and *household goods warehouses.* Like common transportation carriers, public warehouses are obligated to provide the public with a wide variety of services. The most critical issues relative to the metrics associated with warehousing are the following:

1. The *time* that it takes to ship a product (from the initial notification of a requirement).
2. The *cost* of each product shipment (from storage to delivery at the customer site).
3. The *cost* of inventory holding and management (for the warehouse overall).
4. The *value* of the products shipped/value of the overall inventory.
5. The *percentage* of space utilization and the *cost* per area of utilization.
6. The *volume of products handled*, or the total number of products processed per year.

4.5 Maintenance Organization Factors

The measures associated with a maintenance organization are basically the same as those factors that are typical for any organization. Of particular interest relative to system support are the following:

1. The direct maintenance labor time for each personnel category, or skill level, expended in the performance of system maintenance activities. Labor time may be broken down to cover both unscheduled and scheduled maintenance individually. Labor time may be expressed in:
   (a) Maintenance labor hours per system operating hour (MLH/OH).
   (b) Maintenance labor hours per mission cycle (or segment of a mission).
   (c) Maintenance labor hours per month (MLH/Mo).
   (d) Maintenance labor hours per maintenance action (MLH/MA).

2. The indirect labor time required to support system maintenance activities (i.e., overhead factor).
3. The personnel attrition rate or turnover rate (in percent).
4. The personnel training rate or the worker-days of formal training per year of system operation and support.

5. The number of maintenance work orders processed per unit of time (e.g., week, month, or year), and the average time required for work-order processing.

6. The average administrative delay time, or the average time from when an item is initially received for maintenance to the point when active maintenance on that item actually begins.

When addressing the total spectrum of system maintenance (and the design for supportability), the organizational element is critical to the effective and successful life-cycle support of a system. The right personnel quantities and skills must be available when required, and the individuals assigned to the job must be properly trained and motivated. As in any organization, it is important to establish measures dealing with organizational effectiveness and productivity.

4.6 Spares, Repair Parts, and Related Inventory Factors

Spares and repair parts (and associated inventories) are necessary for the performance of all unscheduled (corrective) maintenance actions and for those scheduled maintenance actions where component replacements are required. Spare part requirements are initially based on the system maintenance concept, and specific types and quantities are identified for each level of maintenance, as illustrated in Figure A.1. Spares/repair part quantities are a function of demand rates and include consideration of the following:

1. Spares and repair parts covering items replaced as a result of corrective and preventive maintenance actions. Spares are major replacement items that are repairable, whereas repair parts are nonrepairable smaller components.

2. An additional stock level of spares to compensate for repairable items in the process of undergoing maintenance. If there is a backup (lengthy queue) of items in the intermediate maintenance shop or at the depot/manufacturer awaiting repair, these items obviously will not be available as recycled spares for subsequent maintenance actions; thus, the inventory is further depleted (beyond expectation), or a stock-out condition results. In addressing this problem, it becomes readily apparent that the test equipment capability, personnel, and facilities directly affect the maintenance turnaround times and the quantity of additional spare items needed.

3. An additional stock level of spares and repair parts to compensate for the procurement lead times required for item acquisition. For instance, prediction data may indicate that 10 maintenance actions requiring the replacement of a certain item will occur within a 6-month period, and it takes 9 months to acquire replacements from the supplier. One might ask: What additional repair parts will be necessary to cover the operational needs and yet compensate for the long supplier lead time? The added quantities will, of course, vary depending on whether the item is designated as repairable or will be discarded at failure.

4. An additional stock level of spares to compensate for the condemnation or scrappage of repairable items. Repairable items returned to the intermediate maintenance shop or depot are sometimes condemned (i.e., not repaired) because, through inspection, it is decided that the item was not economically feasible to repair. Condemnation will vary depending on equipment utilization, handling, environment, and organization capability. An increase in the condemnation rate will generally result in an increase in spare part requirements.

In reviewing the foregoing considerations, of particular significance is the determination of spares requirements as a result of item replacements in the performance of corrective maintenance. Major factors involved in this process are (1) the reliability of the item to be spared, (2) the quantity of items used, (3) the required probability that a spare will be available when needed, (4) the criticality of item application with regard to mission success, and (5) cost. Use of the reliability and probability factors are illustrated in the following examples:

Probability of Success with Spares Availability Considerations. Assume that a single component with a reliability of 0.8 (for time t) is used in a unique system application and that one backup spare component is purchased. Determine the probability of system success having a spare available in time t (given that failures occur randomly and are exponentially distributed).

This situation is analogous to the case of an operating component and a parallel component in standby (i.e., standby redundancy). The applicable expression is stated as

$$P = e^{-\lambda t} + (\lambda t)e^{-\lambda t} \tag{2}$$

With a component reliability of 0.8, the value of λt is 0.223. Substituting this value into Equation 2 gives a probability of success of

$$P = e^{-0.223} + (0.223)e^{-0.223}$$

$$= 0.8 + (0.223)(0.8) = 0.9784$$

If we next assume that the component is supported with two backup spares (where all three components are interchangeable), the probability of success during time t is determined from

$$P = e^{-\lambda t} + (\lambda t)e^{-\lambda t} + \frac{(\lambda t)^2 e^{-\lambda t}}{2!}$$

or

$$P = e^{-\lambda t}\left[1 + \lambda t + \frac{(\lambda t)^2}{2!}\right] \tag{3}$$

With a component reliability of 0.8 and a value of λt of 0.223, the probability of success is

$$P = 0.8\left[1 + 0.223 + \frac{(0.223)^2}{(2)(1)}\right]$$

$$= 0.8(1.2479) = 0.9983$$

Thus, adding another spare component results in one additional term in the Poisson expression. If two spare components are added, two additional terms are added, and so on.

The probability of success for a configuration consisting of two operating components, backed by two spares, with all components being interchangeable can be found from the expression

$$P = e^{-2\lambda t}\left[1 + 2\lambda t + \frac{(2\lambda t)^2}{2!}\right] \tag{4}$$

With a component reliability of 0.8 and $\lambda t = 0.223$,

$$P = e^{-0.446}\left[1 + 0.446 + \frac{(0.446)^2}{(2)(1)}\right]$$

$$= 0.6402[1 + 0.446 + 0.0995] = 0.9894$$

These examples illustrate the computations used in determining system success with spare parts for three simple component configuration relationships. Various combinations of operating components and spares can be assumed and the system success factors can be determined by using

$$1 = e^{-\lambda t} + (\lambda t)e^{-\lambda t} + \frac{(\lambda t)^2 e^{-\lambda t}}{2!} + \frac{(\lambda t)^3 e^{-\lambda t}}{3!} + \cdots + \frac{(\lambda t)^n e^{-\lambda t}}{n!} \tag{5}$$

Equation 5 can be simplified into a general Poisson expression:

$$f(x) = \frac{(\lambda t)^x e^{-\lambda t}}{x!} \tag{6}$$

The objective is to determine the probability that x failures will occur if an item is placed in operation for t hours, and each failure is corrected (through item replacement) as it occurs. With n items in the system, the number of failures in t hours will be $n\lambda t$, and the general Poisson expression becomes

$$f(x) = \frac{(n\lambda t)^x e^{-n\lambda t}}{x!} \tag{7}$$

To facilitate calculations, a cumulative Poisson probability graph is presented in Figure 7 derived from Equation 7. The ordinate value can be viewed as a confidence factor. Several simple examples are presented to illustrate the application of Figure 7.

Probability of Mission Completion. Suppose that one needs to determine the probability that a system will complete a 30-hour mission without a failure when the system has a known mean life of 100 hours. Let

$$\lambda = 1 \text{ failure per 100 hours, or 0.01 failure per hour}$$

$$t = 30 \text{ hours}$$

$$n = 1 \text{ system}$$

$$n\lambda t = (1)(0.01)(30) = 0.3$$

Enter Figure 7, where $n\lambda t$ is 0.3. Proceed to the intersection where r equals zero and read the ordinate scale, indicating a value of approximately 0.73. Thus, the probability that the system will complete a 30-hour mission is 0.73.

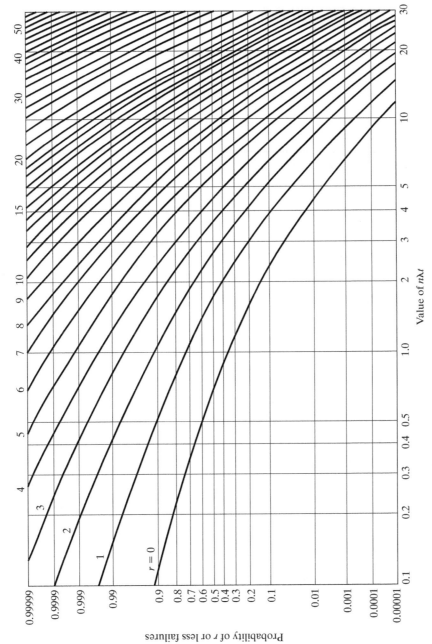

Figure 7 Spares determination using Poisson cumulative probabilities. *Source:* NAVAIR 00-65-502/NAVORD OD 41146, *Reliability Engineering Handbook* (Washington, DC: Naval Air Systems Command and Naval Ordnance Systems Command).

Assume that the identified system is installed in an aircraft and that 10 aircraft are scheduled for a 15-hour mission. Determine the probability that at least 7 systems will operate for the duration of the mission without failure. Let

$$n\lambda t = (10)(0.01)(15) = 1.5$$

and

$$r = 3 \text{ failures or fewer (allowed)}$$

Enter Figure 7, where $n\lambda t$ equals 1.5. Proceed to the intersection where r equals 3 and read the ordinate scale indicating a value of approximately 0.92. Thus, there is a 92% confidence that at least 7 systems out of 10 will operate successfully. If an 80% operational reliability is specified (i.e., 8 systems must operate without failure), the confidence factor decreases to about 82%.

Although the graph in Figure 7 provides a simplified solution, the use of Equation 5 is preferable for accurate results. In addition, there are many textbooks that contain tables covering the Poisson expansion.

Spare-Part Quantity Determination. Spare-part quantity determination is a function of a probability of having a spare part available when required, the reliability of the item in question, the quantity of items used in the system, and so on. An expression, derived from the Poisson distribution, useful for spare-part quantity determination is

$$P = \sum_{n=0}^{n=s} \left[\frac{R(-\ln R)^n}{n!} \right] \tag{8}$$

where

      P = probability of having a spare of a particular item available when required
      S = number of spare parts carried in stock
      R = composite reliability (probability of survival); $R = e^{-K\lambda t}$
      K = of parts used of a particular type
  $\ln R$ = natural logarithm of R

In determining spare-part quantities, one should consider the level of protection desired (safety factor). The protection level is the P value in Equation 8. This is the probability of having a spare available when required. The higher the protection level, the greater the quantity of spares required. This results in a higher cost for item procurement and inventory maintenance. The protection level, or safety factor, is a hedge against the risk of stock-out.

When determining spare-part quantities, system operational requirements should be considered (e.g., system effectiveness, availability) and the appropriate level at each location where corrective maintenance is performed should be established. Different levels of corrective maintenance may be appropriate for different items. For example, spares required to support prime equipment components that are critical to the success of a mission may be based on one factor; high-value or high-cost items may be handled differently from low-cost items; and so on. An optimum balance between stock level and cost is desired.

Figures 8a and 8b present a nomograph that simplifies the determination of spare-part quantities using Equation 8. The nomograph not only simplifies solutions for basic

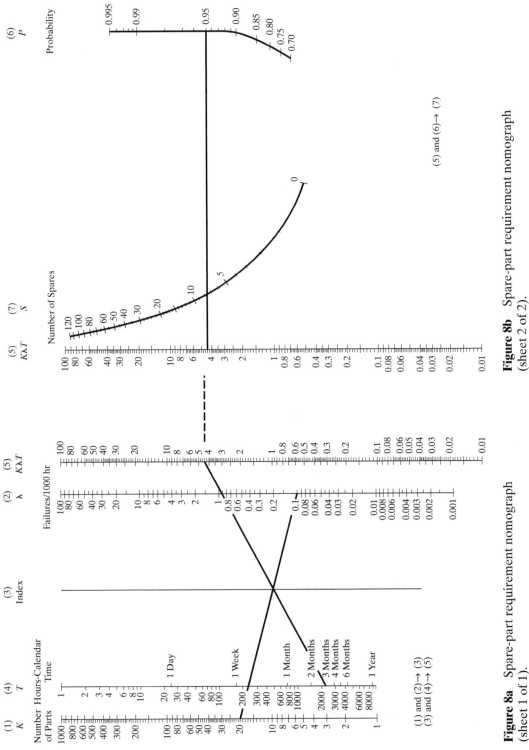

Figure 8b Spare-part requirement nomograph (sheet 2 of 2).

Figure 8a Spare-part requirement nomograph (sheet 1 of 1).

(1) and (2)→ (3)
(3) and (4) → (5)

(5) and (6)→ (7)

spare-part availability questions but also provides information that can aid in the evaluation of alternative design approaches in terms of spares and in the determination of provisioning cycles. The following examples illustrate the use of the nomograph.[10]

Suppose that a piece of equipment contains 20 parts of a specific type with a failure rate (λ) of 0.1 failure per 1,000 hours of operation. The equipment operates 24 hours a day, and spares are procured and stocked at 3-month intervals. How many spares should be carried in inventory to ensure a 95% chance of having a spare part when required? Let

$$K = 20 \text{ parts}$$

$$\lambda = 0.1 \text{ failure per 1,000 hours}$$

$$T = 3 \text{ months}$$

$$K\lambda T = (20)(0.0001)(24)(30)(3) = 4.32$$

$$P = 95\%$$

According to the nomograph in Figures 8a and 8b, approximately eight spares are required.

As a second example, suppose that a particular part is used in three different equipments (A–C). Spares are procured every 180 days. The number of parts used, the part failure rate, and the equipment operating hours per day are given in Table 1.

The number of spares that should be carried in inventory to ensure a 90% chance of having a spare available when required is calculated as follows:

1. Determine the product of K, λ, and T as

$$A = (25)(0.0001)(180)(12) = 5.40$$
$$B = (8)(0.00007)(180)(15) = 5.29$$
$$C = (35)(0.00015)(180)(20) = 18.90$$

2. Determine the sum of the $K\lambda T$ values as

$$\sum K\lambda T = 5.40 + 5.29 + 18.90 = 29.59$$

TABLE 1   Data for Spares Inventory

Item	K	Failures per 1,000 Hours (λ)	Operating Hours per Day (T)
Equipment A	25	0.10	12
Equipment B	28	0.07	15
Equipment C	35	0.15	20

[10]NAVSHIPS 94324, *Maintainability Design Criteria Handbook for Designers of Shipboard Electronic Equipment* (Washington, DC: Naval Ship Systems Command, Department of the Navy, 1964).

3. Using sheet 2 of the nomograph, Figure 8b, construct a line from the $K\lambda T$ value of 29.59 to the point where P is 0.90. The approximate number of spares required is 36.

Inventory System Considerations. The overall inventory requirements for spares and repair parts must be addressed in addition to evaluating specific demand situations. Although too much inventory on hand may be responsive in meeting the demand requirements and all related contingencies, the cost of maintaining the inventory may be high. There is a great deal invested (i.e., resources tied up) and the possibility of obsolescence may be great, particularly if system design changes are being introduced. Conversely, too little inventory increases the risk of stock depletion, the possibility of the system not being in an operational state, and as a result, high costs. An optimum balance (not too much or too little) must be sought between the inventory on hand, the procurement frequency, and the procurement quantity.[11]

When addressing the issues of *inventory*, one needs to commence with a review of the theory and concepts. Figure A.3 shows a general deterministic inventory model where the demand and the procurement lead time are constant. Referring to the figure, the traditional factors of *operating level, safety stock, procurement cycle, procurement lead time, order point,* and so on, are discussed. Further, the development of the *Economic Order Quantity* (EOQ) model and the relationships between inventory quantities and cost are discussed and, when presented in the context of Figure A.3, provide a theoretical representation of an inventory cycle for a given item.

Figure 9 presents a probabilistic inventory situation where the demand and procurement lead time are random variables. The necessity for safety stock is evident from this illustration. A stock-out condition may cause the system to be nonoperational, which may be quite costly. The overall objective is to have the needed amount and type of spares available for the lowest total system cost.

While the EOQ model may be generally applicable in instances where there are relatively large quantities of smaller common and standard repair parts, it may be preferable to utilize other models for the procurement of *high-value* items and/or components considered to be highly *critical* to mission success. High-value items are those components with an unusually high unit procurement price that should be purchased on a unit-to-unit basis. The dollar value of these components is usually significant and may even exceed the total value of the hundreds of other spares and repair parts in the inventory.

Relative to criticality, some items are considered to be more critical than others in terms of impact on mission success. For instance, the lack of a $100 item may cause the system to be inoperative, while the lack of a $10,000 item might not cause a major problem. The criticality of an item is generally based on its function in the system and not necessarily on its acquisition cost. The degree of criticality may be determined from the failure mode, effects, and criticality analysis (FMECA).

The spares acquisition process may also vary to some extent between items of a comparable nature if the usage rates (i.e., duty cycles) are significantly different. Fast-moving

---

[11]Although the value and objectives of the *just-in-time* approach to inventory are recognized, the practical applications of such approaches are not always possible when considering the distribution of system elements, criticality of need, available transportation, and so on.

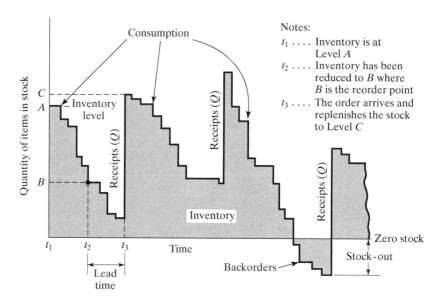

Figure 9 Representation of an actual inventory cycle.

items may be procured locally near the point of usage (such as an intermediate maintenance shop) whereas slower-moving items stocked at the depot may be acquired from a
remotely located supplier if the pipeline and procurement lead times are not as critical.

4.7 Test and Support Equipment Factors

The general category of test and support equipment may include a wide spectrum of items,
such as precision electronic test equipment, mechanical test equipment, ground handling
equipment, special jigs and fixtures, maintenance stands, and the like. These items, in varying
configurations and mixes, may be assigned to different levels of maintenance and geographically dispersed throughout the country (or world). However, regardless of the nature and
application, the objective is to provide the right item for the job intended, at the right time
and location, and in the quantity required.

Because of the likely diversification of the test and support equipment for any given
system, it is difficult to specify quantitative measures that can be universally applied.
Certain measures are appropriate for electronic test equipment, other measures are
applicable to ground-handling equipment, and so on. Further, the specific location and
application of a given item of test equipment may also result in different measures. For
instance, an item of electronic test equipment used in support of on-site organizational
maintenance may have different requirements than a similar item of test equipment used
for intermediate maintenance accomplished in a remote shop.

Although all the test and support equipment requirements at each level of maintenance are considered to be important relative to successful system operation, the testers
or test stations in the intermediate and depot maintenance facilities are of particular
concern, since these items are likely to support several system elements at different

customer locations. An intermediate maintenance facility may be assigned to provide the necessary corrective maintenance support for a large number of system elements dispersed throughout a wide geographical area. This means that a variety of items (all designated for intermediate-level maintenance) will arrive from different operational sites at different times.

When determining the specific test equipment requirements for a shop, one must define (1) the type of item that will be returned to the shop for maintenance; (2) the test functions to be accomplished, including the performance parameters to be measured as well as the accuracies and tolerances required for each item; and (3) the anticipated frequency of test functions per unit of time. The type and frequency of item returns (i.e., shop arrivals) is based on the maintenance concept and system reliability data. The distribution of arrival times for a given item is often a negative exponential with the number of items arriving within a given period following a Poisson distribution, as presented in Appendix C.2. As items arrive in the shop, they may be processed immediately or there may be a waiting line, or queue, depending on the availability of the test equipment and the personnel to perform the required maintenance functions.

When evaluating the test process itself, one should calculate the anticipated test equipment use requirements (i.e., the total amount of on-station time required per day, month, or year). This can be estimated by considering the repair-time distributions for the various items arriving in the shop. However, the ultimate elapsed times may be influenced significantly, depending on whether manual, semiautomatic, or automatic test methods are employed.

Given the test equipment utilization needs (from the standpoint of total test station time required for processing shop arrivals), it is necessary to determine the anticipated operational availability (Ao) of the test equipment configuration being considered for the application. Availability is a function of reliability and maintainability. Thus, one must consider the MTBM and MDT values for the test equipment itself. Obviously, the test equipment configuration should be more reliable than the system component being tested. Also, in instances where the complexity of the test equipment is high, the logistic resources required for the support of the test equipment may be extensive (e.g., the frequent requirement to calibrate an item of test equipment against a secondary or primary standard in a *clean-room* environment). There may be a requirement to determine the time that the test equipment will be available to perform its intended function.

The final determination of the requirements for test equipment in a maintenance facility is accomplished through an analysis of various alternative combinations of arrival rates, queue length, test station process times, or quantity of test stations. In essence, one is dealing with a single-channel or multichannel queuing situation. As the maintenance configuration becomes more complex, involving many variables (some of which are probabilistic in nature), Monte Carlo analysis may be appropriate. In any event, there may be a number of feasible servicing alternatives, and a preferred approach is sought.

4.8 Maintenance Facility Factors

Special facilities are required to support all activities pertaining to the accomplishment of active maintenance tasks, the storage of spares/repair parts and test and support equipment necessary for maintenance, the space and housing of personnel for related administrative activities, and so on. Referring to Figure A.1, this primarily includes the maintenance shops at the intermediate and supplier/manufacturer/depot levels of maintenance. Although the specific quantitative measures associated with facilities may vary significantly from one system to the next, the following factors are considered to be relevant in most cases:

1. Item process time or turnaround time (TAT); that is, the elapsed time necessary to process an item for maintenance, returning it to full operational status.
2. Facility utilization; that is, the ratio of the time utilized to the time available for use, percent utilization in terms of space occupancy, and so on.
3. Energy utilization in the performance of maintenance; that is, unit consumption of energy per maintenance action, cost of energy consumption per increment of time or per maintenance action, and so on.
4. Total facility cost for system operation and support; that is, total cost per month, cost per maintenance action, and so on.

4.9 Computer Resources and Maintenance Software Factors

For many systems, software has become a major element of support. This is particularly true where automation, computer applications, digital databases, and the like are used in the accomplishment of logistics and maintenance support functions. Software may be evaluated in terms of language levels or complexity, number of programs, program length on the basis of the number of source code lines, cost per maintenance subroutine, or something of a comparable nature.

As with equipment, reliability, maintainability, and quality are significant considerations in the development of software. Although software does not degrade in the same way as equipment, the reliability of software is still important and must be measured. There has been a great deal of discussion on how software reliability is to be measured, and there still is not complete agreement as to the specific measures or levels of acceptable performance; however, one definition of *software reliability* is "the probability of failure-free operation of a software component or system in a specified environment for a specified time." A *failure* is defined as an unacceptable departure of program operation from program requirements.

In any event, errors occur in the initial development of software, and software reliability is a function of the number of inherent errors contained within the software that have not been eliminated. Such errors may be classified as faulty or omitted logic, addressability errors, missing commentary, regression or integration problems, counting or calculation problems, or general documentation problems. Usually, the overall measure is in terms of the number of errors per 1,000 source code lines. Higher-level languages will probably contain fewer errors, since fewer lines of code are required (as compared with assembly languages). Conversely, language complexity may be introduced in the process.

In the specification and development of system software, applicable measurement factors must be identified and the system must be evaluated to include not only coverage of equipment and personnel factors but software factors as well. The logistics support resource requirements for the overall system include consideration of equipment, personnel, facilities, data, consumables, and software.

4.10 Technical Data and Information System Factors

In recent years, the logistics and maintenance support infrastructure has evolved significantly with the advent of computers and associated software, the availability of information networks (LANs and WANs), the introduction of electronic data interchange (EDI) and electronic commerce (EC) methods, and related technologies. The objectives have included (1) simplifying the task of generating and processing technical data through better packaging, by eliminating redundancies, by reducing processing times, and by making the information more accessible to all organizations in need; (2) providing an expeditious means for the introduction of design changes and for better implementation of configuration management requirements; (3) providing a mechanism for greater asset visibility relative to the traceability of components in transit and the location of items in inventories; and (4) enabling faster, timely, accurate, and more reliable communications among multiple locations on a current basis.

While it has been an objective to improve the overall communications processes through a wide spectrum and flow in data distribution, care must be exercised to ensure that such information is protected from those individuals who should not have the information/data. In the defense sector, in particular, there has been much emphasis on *net-eccentricity*, which can be defined as "the ability to provide a framework for full human and technical interoperability that (1) allows all DOD users and mission partners to share the information they need, when they need it, in a form they can understand and act on with confidence, and (2) protects information from those who should not have it."

In any event, the *information age* has had a major impact on logistics and some of the measures that may be applicable include the following:

1. *Logistics response time:* the time that is consumed from the point when a system support requirement is first identified until that requirement has been satisfied (this may include the time required for the provisioning and procurement of a new item, the time to ship an item from inventory to the location of need, the time required to acquire the necessary personnel or test equipment for maintenance, and so on).
2. *Data access time:* the time to locate and gain access to the data/information element needed.
3. *Item location time:* the time required to locate a given asset whether in use, transit, or in some inventory (asset visibility).
4. *Information processing time:* the time required for the processing of messages.
5. *Change implementation time:* the time to process and implement a design change.
6. *Cost:* the cost of transmission per bit of data, the cost per data access incident, and so on.

Recently, there has been an extensive amount of effort directed toward the capability of being able to rapidly identify, locate, and track (in the event of movement) certain system elements and components within and throughout the overall logistics and maintenance support infrastructure. This has resulted in recent advances in automatic information technology (AIT) and the development of new bar coding methods, active and passive radio frequency identification (RFID) tags, and global positioning systems (GPS) in particular. The ultimate objective is to provide the necessary information on item status in a reliable and timely manner, and to further simplify the logistics and support processes, increasing both effectiveness and efficiency in facilitating system life-cycle support. The application of RFID technologies, in particular, has been growing almost exponentially during the past 5 years, or more.

5 LOGISTICS AND MAINTENANCE SUPPORT IN THE SYSTEM LIFE CYCLE

The logistics and maintenance support infrastructure includes those elements and associated activities described in Section 1 and illustrated in Figures A.1, 1, and 2. This includes all supply chain logistics and the sustaining maintenance support of the system throughout its life cycle. These functions should be addressed as an integrated entity, and the infrastructure should be considered in the same context as one of the many subsystems that make up the total system configuration.

The basic infrastructure requirements stem from the overall system-level requirements and must be developed during the conceptual design phase and be allocated downward early in preliminary system design. Inherent within are the characteristics (attributes) that must be responsive to these system requirements and incorporated into the design as it applies, not only to the prime mission-related elements of the system but to each of the elements of the logistics and maintenance support infrastructure as well. In other words, the requirements for the *design for supportability* must be applicable to both (1) the prime elements of the system to ensure that they can be effectively and efficiently supported, and (2) the various aspects and the supporting infrastructure to ensure that they can fulfill the necessary life-cycle requirements for the system.

Referring to Figure 10, requirements pertaining to the *design for supportability* must be considered from the beginning, along with the requirements for reliability, maintainability, human factors and safety, producibility, sustainability, disposability, and so on. Trade-off analyses may be conducted among these various design-to requirements, with the results being allocated as appropriate to the various elements of the support infrastructure. Given these basic design-to requirements, the ultimate system configuration then evolves through the phases of preliminary system design and detail design and development. As the design configuration takes shape, the iterative process of *supportability analysis* (SA) is implemented with the following objectives: (1) to identify the specific logistics and maintenance support resource requirement for the system as it is configured at the time; and (2) to assess the overall design relative to the degree of "supportability" attributes incorporated. As the design process evolves, the day-to-day design participation and the conductance of periodic design reviews occur throughout. Finally, the supportability requirements for the system are verified and validated through the overall system test and evaluation effort.

5.1 System Requirements

System-level requirements are initially defined through the development of operational requirements and the maintenance concept. From these basic requirements, the identification and prioritization of technical performance measures (TPMs) occurs. Specific quantitative design-to factors are described along with their relative degrees of importance in the design. Included within are those factors that are particularly relevant to the logistics and maintenance support infrastructure, otherwise known as *performance-based logistics* (PBL) factors. These PBL factors represent the design-to requirements necessary for implementing the *design for supportability* objectives.[12]

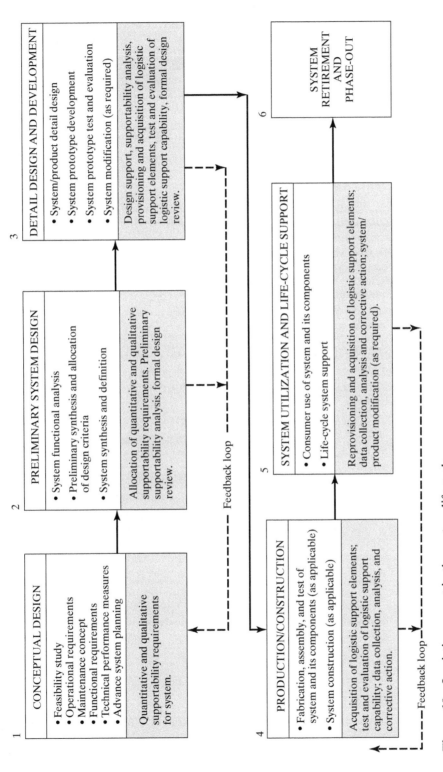

Figure 10 Logistic support in the system life cycle.

Figure 11 illustrates a hierarchical relationship, where the requirements at the system level lead into the requirements for the support infrastructure, which, in turn, lead into the requirements for the various elements of support. At the system level (for example), such requirements may include the specification of some measure of *cost effectiveness, system effectiveness, operational availability (Ao), maintenance downtime* (MDT), or equivalent. In order to comply with the overall system requirements, there must be some requirements for the design of the support infrastructure if the system is to accomplish its overall mission. Such requirements should be specified covering both the technical and economic sides of the spectrum.

5.2 Requirements Allocation

The purpose of allocation is to translate requirements at the system level into lower-level design criteria. Referring to Figure 11, for example, *If there is a specified quantitative effectiveness requirement for the logistics and maintenance support infrastructure (e.g., logistics responsiveness), what should be specified for the various elements of support (i.e., spares/repair parts, test and support equipment, maintenance personnel, facilities, etc.) in terms of quantitative design-to factors?* There should be some specific quantitative design goals which, when applied, would help to ensure that the overall system-level requirements will be met. Although these lower-level requirements may vary significantly from one instance to the next, a few examples are noted here:

1. Test equipment utilization in the intermediate maintenance shop shall be at least 80%, and test equipment reliability shall be at least 90%.
2. Self-test thoroughness for the system (using the built-in test capability) shall be 95% or better.
3. Personnel skill levels at the organizational level of maintenance shall be equivalent to grade x or below.
4. The maintenance facility at the intermediate level shall be designed for a minimum of 75% utilization.
5. The transportation time between the location where organizational maintenance is accomplished and the intermediate maintenance shop shall not exceed 4 hours.
6. The turnaround time in the intermediate maintenance shop shall be 2 days (or less), and in the depot/supplier maintenance facility, it shall be 10 days (or less).
7. The probability of spares availability at the organizational level of maintenance shall be at least 95%.

In Figure 11, many of the measures pertaining to the elements of support, described in Section 4, are presented in a summary form. Depending on the type of system, its makeup, and its mission, different factors may be applicable. The factors shown here are included for the purposes of illustration.

---

[12]The PBL concept is currently being emphasized in the defense sector, with the objective of promoting the principles pertaining to *design for supportability*. Refer to (1) MIL-HDBK-502, *Department of Defense Handbook—Acquisition Logistics* (Washington, DC: Department of Defense, 1997), or (2) B. S. Blanchard, *Logistics Engineering and Management*, 6th ed. (Upper Saddle River, NJ: Pearson Prentice Hall, 2004), Section 1.5.

Design for Logistics and Supportability

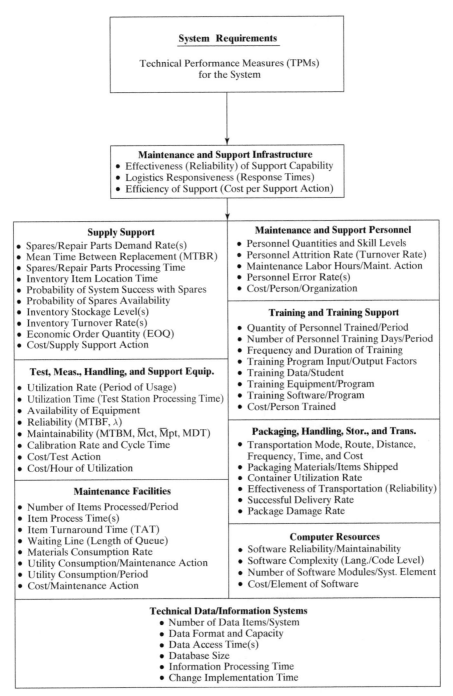

Figure 11 Selected technical performance measures for the logistics and maintenance support infrastructure. *Source*: B. S. Blanchard, *Logistics Engineering and Management*, 6th ed. (Upper Saddle River, NJ: Pearson Prentice Hall, 2004), Figure 1.11.

5.3 Design Review and Evaluation

Conducting a review of system design for the incorporation of supportability characteristics is accomplished as an inherent part of the process. Such design reviews cover both the prime mission-related elements of the system to ensure that they are designed to be supportable in an effective and efficient manner and the design of the logistics and maintenance support infrastructure to ensure that it adequately responds in fulfilling all system requirements. If the requirements appear to have been met, the design is approved and the program enters into the next phase. If not, the appropriate changes are initiated for corrective action.

In accomplishing a supportability review, a checklist may be developed to facilitate the review process. An abbreviated representative list, not to be considered as being all-inclusive, follows. It should be noted that the answer to those questions that are applicable should be *yes*:

1. Have the major logistics and maintenance support functions for the entire system life cycle been adequately defined? This includes an illustrated approach such as presented in Figure 1.

2. Have the basic supply chain functions been properly identified and defined? This includes an illustrated approach similar to that presented in Figure 2.

3. Has the system maintenance concept been adequately defined? This includes a definition of the levels of maintenance and the basic functions to be accomplished at each level. This includes an illustrated approach similar to that presented in Figure A.1.

4. Have the appropriate technical performance measures (TPMs) been defined at the system level that will lead directly to the establishment of PBL factors for the logistics and maintenance support infrastructure and its various elements? The objective here is to provide the appropriate quantitative design-to goals for both the logistics functions (in Figure 2) and the sustaining maintenance support functions (in Figure A.1).

5. Has a supportability analysis (SA) been accomplished throughout the system design and development process, commencing with the initial establishment of system support requirements during the conceptual design phase? Refer to Section 6 for coverage of the SA.

6. Does the supportability analysis (SA) evolve directly from the maintenance concept? This refers to the "traceability" from and the expansion of the requirements illustrated by the concept presented in Figure A.1.

7. Does the supportability analysis (SA) data package (i.e., logistics management information (LMI)) justify the system *design for supportability*? This includes the results from design trade-off studies where decisions pertaining to supportability were implemented.

8. Does the supportability analysis (SA) adequately lead to the identification and definition of all logistics and maintenance support resource requirements for the system life cycle (i.e., maintenance personnel, spares/repair parts and associated inventories, test and support equipment, maintenance facilities, training and training support, packaging and transportation, computer resources and maintenance software, and technical data and logistics information)?

9. Does the supportability analysis (SA) include the proper integration of the different models and the resulting analyses that are applied in completing SA objectives

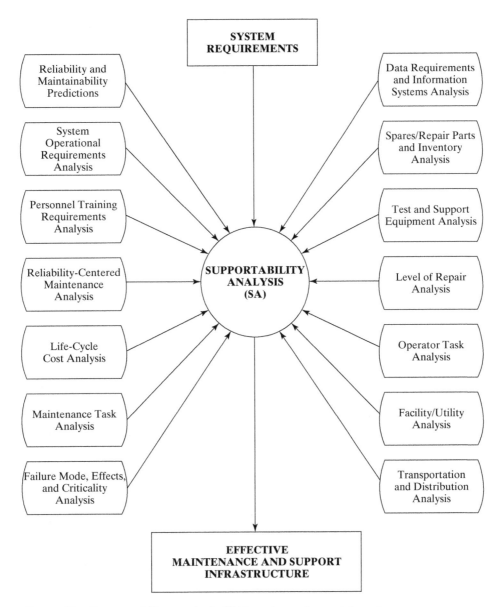

Figure 12 Supportability analysis (SA) and supplemental analyses.

(i.e., FMECA, RCM, OTA, MTA, LORA, and LCCA)? The objective is to ensure the proper integration of those activities shown in Figure 12 (as applicable).

10. Have the specific requirements, in terms of type and quantity, been defined for each level of logistics and maintenance support for
(a) Maintenance personnel?
(b) Spares/repair parts and associated inventories?

(c) Test and support equipment?

(d) Maintenance facilities?

(e) Training and training support?

(f) Packaging, handling, and transportation?

(g) Computer resources and maintenance software?

(h) Technical data and logistics information?

11. Have the appropriate design-to requirements (i.e., criteria) been specified for the logistics and maintenance support infrastructure and its various elements? This pertains to establishing the proper design-to requirements (e.g., PBL factors) for test and support equipment, maintenance facilities, packaging and transportation, and so on.

The questions here are representative of general areas of interest with regard to the *design for supportability*.

6 SUPPORTABILITY ANALYSIS (SA)[13]

The supportability analysis (SA) constitutes the integration and application of different analytical techniques/methods to solve a wide variety of problems. The SA, in its application, is the process employed on an iterative basis throughout system design and development that addresses the issue of *supportability* both from an inherent system design perspective and in terms of the logistics and maintenance support resources required for the sustaining support of the system through its life cycle. As such, the SA is an inherent part of the system engineering process.

The objective of the supportability analysis process, which follows the general approach for system analysis, is to (1) initially influence the design of a given system, and (2) aid in the identification of the logistics and maintenance support resources based on the assumed design configuration at the time. Specifically, the SA objectives are to[14]

1. Aid in the initial establishment of supportability requirements during conceptual design. In the selection of alternative technological approaches, accomplished as part of the feasibility analysis, supportability-related factors must be addressed in the decision-making process. Alternatives must be evaluated from a total life-cycle perspective.

2. Aid in the early establishment of supportability design criteria (as an *input* to the design process) through the development of system operational requirements, the logistics and maintenance support concept, identification and prioritization of technical performance measures (TPM and PBL factors), and in accomplishment of functional analysis

[13]The scope of activity defined in this section is sometimes identified by the term *logistic support analysis* (LSA), *maintenance analysis* (MA), *maintenance engineering analysis* (MEA), *maintenance level analysis* (MLA), or something equivalent. However, the objectives are the same in each case.

[14]For defense systems, the SA is found in MIL-HDBK-502, *DOD Handbook—Acquisition Logistics* (Washington, DC: Department of Defense May 1997). The data requirements, evolving from the SA (or "Logistics Management Information"), are specified in MIL-PRF-49506, *Performance Specification for Logistics Management Information* (Washington, DC: Department of Defense, November 1996). Emphasis here is on the SA process.

and allocation. Such criteria may include the specification of both quantitative (applicable TPM and PBL factors) and qualitative (nonquantitative) requirements.

3. Aid in the process of synthesis, analysis, and design optimization through accomplishment of trade-off studies and the evaluation of various design alternatives. Specific applications may include the evaluation of alternative materials being considered for selection, alternative packaging schemes, alternative repair policies, the incorporation of manual versus automatic provisions, alternative diagnostic and test approaches, alternative packaging and transportation methods, and so on.

4. Aid in the evaluation of a given design configuration relative to determining specific logistics and maintenance support resource requirements. Once system requirements are known and design data are available (during the preliminary system design and detail design and development phases), it is possible to determine the requirements for logistics and maintenance personnel quantities and skill levels, personnel training, spares/repair parts and related inventories, test and support equipment, maintenance facilities, packaging and transportation, computer resources and maintenance software, and technical data and information. Through the SA process, alternative candidates for system support may be evaluated, leading to the development of an optimum logistics and maintenance support infrastructure configuration. The results are generally integrated into a *logistics management information* (LMI) data package in some form.

5. Aid in the measurement and evaluation, or *assessment*, of an operating system to determine its overall effectiveness and the degree to which the system is supportable when performing in the user's environment. *Given a fully operational capability, can the system be effectively and efficiently supported throughout its planned life cycle?* Field data are collected, evaluated, and the results are analyzed and compared against the initially specified requirements for the system. The objective is not only to verify that the system is successfully accomplishing its mission but also to identify the high-cost/high-risk areas and incorporate any necessary modifications as part of a *continuous product/process improvement* (CPPI) effort.

The successful completion of the supportability analysis (SA) depends on the application of many different analytical tools in the conductance of design trade-offs and in the development of an effective LMI data package for the system in question. Figure 12 illustrates some of the input requirements, although the extent and depth of application must be properly tailored to the system requirement. For example, the results of reliability analyses (reliability predictions, FMECA, and FTA), maintainability analyses (maintainability predictions, RCM, LORA, and MTA), and human factors analyses (OTA, OSD, safety/hazard, and error), along with economic and life-cycle cost analyses (LCCA), are required inputs to the SA. Again, it is critical that such applications be properly tailored and accomplished in a timely manner and to the depth needed.

The supportability analysis process flow and the results are conveyed in Figure 13. The figure emphasizes the output in terms of the resource requirements pertaining to the logistics and maintenance support infrastructure, with the results described in an LMI database and distributed to many different locations. Regarding data formatting and packaging, the requirements may vary depending on the program, its organizational structure, the needs for specific information (type and amount of information, number of locations where the data are to be distributed, specific format, frequency, timeliness, etc.), and program reporting requirements. The objective, of course, is to provide the right information, to the

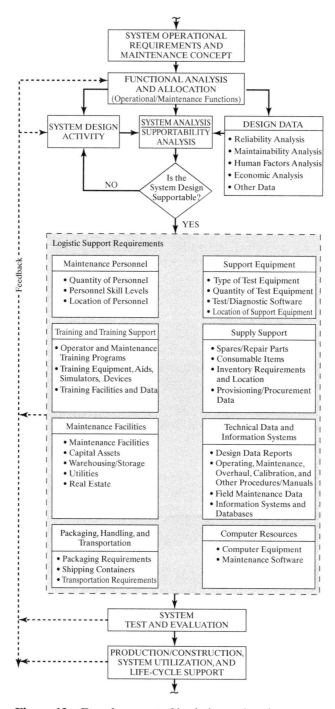

Figure 13 Development of logistics and maintenance support requirements through the supportability analysis. *Source*: B. S. Blanchard, *Logistics Engineering and Management*, 6th ed.(Upper Saddle River, NJ: Pearson Prentice Hall, 2004), Figure 5.6.

right location, and at the right time. This, in turn, requires the development of a database structure that is easily accessible and through which the appropriate data are made available to all members of the design team concurrently.

In the past, the development and processing of logistics-related data have been a major problem, with too much data being generated that are not timely or accessible and that have been costly. As a result, the concept of *continuous acquisition and life-cycle support* (CALS) was implemented in the 1980s. CALS pertains to the application of computerized technology in the development and processing of data, primarily in a digital format, with the objectives of reducing preparation and processing times, eliminating redundancies, shortening the system acquisition process, and reducing the overall program costs. Specific applications have included the automation of technical publications, the preparation of a data package for the provisioning and procurement of spares/repair parts, and the development of design data-defining products in a digital format. An objective is to develop some form of an integrated systems database structure that can (1) serve as a repository for all logistics and related data evolving from the supportability analysis, and (2) provide the necessary information at the right time and in the proper format in response to the reporting requirements for specific programs.

7 SUPPORTABILITY TEST AND EVALUATION

With the system and its logistics and maintenance support infrastructure designed, the next step is *validation*; that is, the verification that the resulting configuration will meet the initially specified requirements. This validation process is accomplished within the context of the system test and evaluation capability. With reference to Figure A.2, the objective is to gain some degree of confidence, as early as possible, that the requirements have been met. This validation can be accomplished to a certain extent through the selective application of analytical methods (e.g., simulation), Type 1 testing, Type 2 testing, and so on. On the other hand, a major part of the validation effort (as it pertains to logistics support) can be accomplished only through Types 2–4 testing.

As part of Type 2 testing, with the availability of preproduction prototype equipment, software, and other major elements of the system, the following tests can be conducted to validate the selected system elements:

1. *Reliability qualification testing.* Various elements of the system are subjected to an environment characteristic of what will be experienced by the ultimate user later on and are "operated" through a series of mission scenarios by personnel in the quantities and with the skills that have been recommended for system operation during the utilization phase. As the testing progresses, failures are noted, analyzed, and recorded. After the appropriate repair action(s) is completed, the testing continues until enough time has accumulated for the system element and the required MTBF value has been verified. From the results of this test, one should be able to verify frequency of unscheduled maintenance, prime failure modes and effects/causes, and the supporting resources required for the various repair actions.

2. *Maintainability demonstration.* The objective here is to test the selected system elements in a true operational environment (as if the system were installed and operating in a realistic user setting, sheltered or outside, mobile or fixed, in the mountains or in the desert, with the appropriate weather conditions, recommended personnel quantities and skills, operational procedures and maintenance data, etc.). During the test, different logistics and maintenance tasks (and task sequences) are simulated utilizing approved maintenance procedures,

maintenance times are recorded, and the resources required in accomplishing the tasks are noted; that is, personnel quantities and skills, tools and test equipment, spares/repair parts, packaging and transportation requirements, and so on. The simulation of unscheduled maintenance tasks commences by inducing a failure into the system in the beginning (with the operator/maintainer out of the area); then operating the system until such time as a symptom of failure occurs and subsequently performing the necessary corrective actions in accordance with the recommended procedures. In simulating a scheduled (preventive) maintenance task, testing continues, using a predetermined sample size of the tasks to be demonstrated, until the required values for $\overline{M}ct$, $\overline{M}pt$, $\overline{M}$, MLH/OH, and supporting resource requirements have been verified. From the results of this test, one should be able not only to assess the degree to which maintainability characteristics have been incorporated into the design but also to assess the adequacy of the logistics resources required in support of system maintenance activities.

3. *Personnel test and evaluation.* The objective here is to verify that the personnel requirements recommended for system operation and maintenance support are adequate; that is, the appropriate quantities, skill levels, and classifications. Personnel requirements are initially recommended through the accomplishment of an operator task analysis (OTA) for "operators" and the maintenance task analysis (MTA) for "maintainers." Selected personnel for the test, considered to be representative of those in the user's organizational structure, are identified, are given the appropriate recommended training, and are then scheduled to accomplish certain designated operational and/or maintenance tasks using approved procedures. Task sequences are verified, areas of difficulty are noted along with personnel-induced faults, and the procedures used are validated. Although some of these requirements can be verified through reliability testing and maintainability demonstration, there may be some additional tasks (particularly those that are highly complex) that should be evaluated. From the results of this test, one should be able to verify personnel and training requirements; that is, the quantities and skill levels of personnel and the adequacy of training requirements.

4. *Test and support equipment compatibility.* It may be necessary to ensure that significant items of test and support equipment will effectively accomplish the function(s) intended. In response, special tests may be conducted to verify compatibility between the test equipment and the element(s) of the system that it is intended to support. This may be particularly relevant for electronic-intensive systems and large test stations that are designed to support corrective and/or preventive maintenance activities in the shop. From the results of such testing, one should be able to verify test equipment performance requirements, accuracies, reliability, and required interfaces (if any). Although some of these requirements may be verified through reliability testing and maintainability demonstration, there may be some additional tasks that need to be evaluated, particularly for test procedures that are rather complex.

5. *Logistics validation.* Within the spectrum of Types 2 and 3 testing, one can certainly evaluate selected activities (and task sequences) pertaining to item purchasing and material flow, packaging and transportation, and warehousing and inventory control functions. Although a true evaluation of the overall logistics (supply chain) and support infrastructure cannot be accomplished at this stage, it is possible to acquire a good start in evaluating the purchasing process; the various means for packaging, handling, and transportation; the processing of selected items for warehousing (storing, order picking); the effectiveness of the data/information capability; and so on.

System evaluation continues through Type 3 and Type 4 testing. Obviously, as one continues, the validation process becomes more meaningful. For instance, the first opportunity for a good evaluation of the overall logistics supply chain and support infrastructure, as an integrated entity, does not occur until the initiation of the Type 3 testing activity, and then only a limited number of logistics activities may be accomplished. On the other hand, a complete validation of the entire logistics and support infrastructure may be completed only as part of Type 4 testing when contractor/customer/supplier activities are integrated and involve a number of different user operational sites (refer to Figure A.2). Finally, as an integral part of the system test and evaluation process, there is the ongoing activity of data collection, analysis, and the subsequent recommendation(s) for corrective action in the event of not meeting a specific requirement and/or for the purposes of product and/or process improvement.

8 SUMMARY AND EXTENSIONS

This chapter covers the subject of logistics and supportability in the broader context of the system life cycle; thus, the phrase *logistics and supportability engineering*. The assumption is that, to be successful in meeting the systems engineering objectives described throughout this text, one has to include both (1) the support-related functions associated with bringing a system into being, and (2) the sustaining support functions after the system is operational and over its life cycle. This pertains to both the *forward* flow of activities noted in Figure 1, to include the necessary logistics and supply chain activities, and the *reverse* flow of activities associated with the sustaining maintenance and support of the system during and throughout its period of utilization.

All functional requirements must be properly addressed through the system engineering process, and the *design for supportability* reflects that effort necessary to ensure that the proper characteristics (or attributes) are inherent within the ultimate system design configuration. The objective is to provide a logistics and maintenance support infrastructure that can effectively and efficiently support the system throughout its life cycle.

The past is replete with instances where the various logistics and maintenance support functions, reflected in Figures A.1 and 2, have been addressed somewhat on an individual-by-individual basis. The initial purchasing and supplier-related activities have not been well integrated with the follow-on material flow functions within a production/manufacturing capability; transportation and warehousing functions have in many cases been treated separately; major supply chain activities have been viewed independently from the subsequent activities associated with the sustaining life-cycle maintenance and support of a given system; and many of these very important functions have been addressed "after the fact" and have not been considered in the system design process.

Accordingly, these numerous and varied activities have not been well integrated from the beginning, have not been considered from a total *system* perspective, and have not been addressed early in design when major decisions are made that ultimately have a great impact on the logistics and maintenance support infrastructure later. Relative to the future, it is recommended that (1) the functions and activities covered herein be addressed from a total integrated overall system perspective; (2) the logistics and maintenance support infrastructure be considered as a significant element of the system being developed; and (3) this infrastructure be addressed within the system engineering design process from the beginning (i.e., from conceptual design on).

Finally, it should be noted that there is a great deal of activity and continuing growth in logistics and system life-cycle support, particularly in view of the increased emphasis on globalization, outsourcing, and international competition. The "complexity" of many of the systems being introduced today is increasing, particularly for those systems contained within a given system-of-systems (SOS) structure. Interoperability and networking requirements are becoming much greater than in the past. Thus, it becomes even more important that principles and concepts described throughout this chapter be thoroughly understood and properly implemented.

For current topical treatment, it would be worthwhile to become familiar with and periodically visit some of the relevant websites, such as:

1. American Production Inventory Control Society (APICS) — *http://www.apics.org.*
2. American Society of Transportation and Logistics (AST&L) — *http://www.astl.org.*
3. Council of Logistics Engineering Professionals (CLEP) — *www.logisticsengineering.org.*
4. Council of Supply Chain Management Professionals (CSCMP), Lombard, IL — *http://www.cscmp.org.*
5. Institute for Supply Management (ISM) — *http://www.ism.ws.*
6. International Society of Logistics (SOLE) — *http://www.sole.org.*
7. Material Handling Industry of America (MHIA) — *http://www.mhia.org.*

QUESTIONS AND PROBLEMS

1. How would you define *logistics*? What are the basic elements of logistics?

2. What are the basic differences between logistics as it is practiced in the business-oriented commercial sector and logistics as practiced in the defense sector? Identify some functions/activities that are common to both sectors. Identify some of the differences.

3. Define what is meant by a *supply chain* (SC) and *supply chain management* (SCM). What functions and activities are included?

4. How would you relate logistics to a *system*?

5. How does logistics fit into the system life cycle?

6. The text refers to the *logistics and maintenance support infrastructure*. What is it and what is included?

7. What is meant by the *design for supportability*? When should it be considered in the system life cycle? Why is it important?

8. Explain how *design for supportability* relates to the following:

(a) *design for reliability*

(b) *design for maintainability*

(c) *design for human factors*

9. How do *logistics* and the *design for supportability* relate to the system engineering process (if at all)?

10. Identify and describe at least three measures (metrics) that can be applied to each of the following:

    (a) Supply chain
    (b) Purchasing and material flow
    (c) Transportation and packaging
    (d) Warehousing and distribution
    (e) Maintenance organization
    (f) Training and training support
    (g) Spares, repair parts, and related inventories
    (h) Test and support equipment
    (i) Maintenance facility
    (j) Computer resources and maintenance software
    (k) Technical data and logistics information

11. What factors should be considered in determining supply support requirements, test and support equipment requirements, personnel requirements, training requirements, transportation and handling requirements, facility requirements, and technical data requirements?

12. Assuming that a single component with a reliability of 0.85 is used in a unique application in the system and that there is one backup spare component, determine the probability of system success having a spare available in time, t, when required.

13. Assuming that the component in Problem 12 is supported with two backup spares, determine the probability of system success.

14. Determine the probability of system success (having a spare available) for a configuration consisting of two operating components backed by two spares (assume that the components are interchangeable with a reliability of 0.875).

15. There are 10 systems located at a site scheduled to perform a 20-hour mission. The system has a predicted MTBF of 100 hours. What is the probability that at least eight of these systems will operate for the duration of the mission without failure?

16. An equipment unit contains 30 parts of the same type. The part has a predicted failure frequency of 10,000 hours. The equipment operates 24 hours a day and spares are procured at 90-day intervals. How many spares should be carried in the inventory to ensure a 95% probability of success?

17. Determine the economic order quantity (EOQ) for spares inventory replenishment when the cost per unit is $100, the cost of preparing for a shipment and sending a truck to the warehouse is $25, the estimated cost of managing an item in inventory is 25% of the inventory value, and the annual demand is 200 units.

18. What is the total cost per period given that the demand is 4 units, the cost per unit is $12.00, the procurement cost is $16.00, and the inventory management cost is $2.00?

19. What are the major considerations in determining an economic order quantity (EOQ)? Should the EOQ principle be applied in all instances when spares/repair parts are required (explain)?

20. Some spare parts have a higher priority (in importance) than others. What factor(s) determines this priority? What considerations should be addressed in the procurement of such items?

21. Assume that you have an intermediate-level shop full of test equipment. What steps would you take to ensure the adequacy of this equipment over time?

22. How could the maintenance requirements for test and support equipment affect the availability of the prime elements of the system?

23. Assume that you have been requested to set up a maintenance organization for system support. What factors would you consider? Describe the steps involved. How would you later evaluate this organization in terms of effectiveness (identify the metrics that you would use)?

24. What considerations would you address in selecting a specific mode of transportation (or combination of modes)?

25. How can *packaging* design impact the reliability of a system (if at all)?

26. Assume that you are responsible for the design and development of a new system and that this new system is to be included within a system-of-systems (SOS) structure. Describe some of the design requirements that should be imposed on the new system design.

27. What are some of the "challenges" that you may face and need to address in the design of a new system as part of a system-of-systems (SOS)?

28. How can the logistics and maintenance support infrastructure be validated?

29. Identify several new technologies and describe how the application of these will enhance logistics and system support. Provide some specific examples.

Appendix: Figures

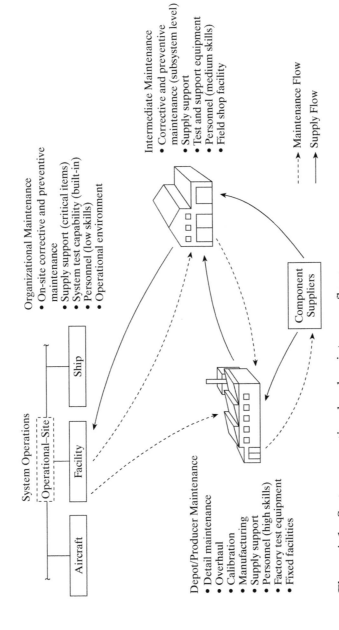

Figure A.1 System operational and maintenance flow.

609

Design for Logistics and Supportability

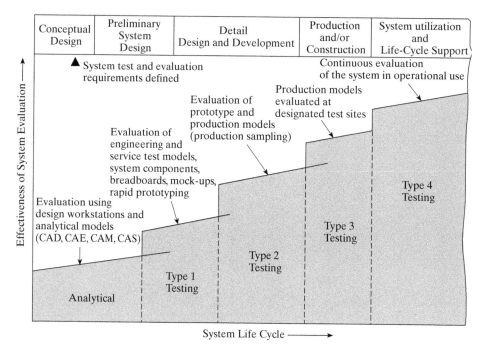

Figure A.2 Stages of system test and evaluation during the life cycle.

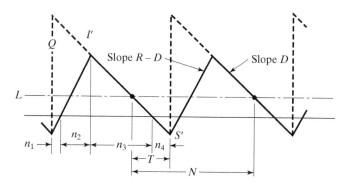

Figure A.3 General deterministic inventory system geometry.

Design for Producibility, Disposability, and Sustainability

Producibility and disposability are design-dependent parameters related to each other in subtle ways. The relationship is inherent in *bringing into being* and *ceasing to be*. Existing across and between these states is the concern for sustainability of system function and form over its life cycle and, simultaneously, for sustainability of the environment within which the system operates.

To bring a system into being is to achieve a high degree of organization and order; to cease being is to return to a state of disorganization and disorder, as in the concept of entropy. The degree of order exhibited by an engineered system is a remarkable manifestation of system design—design that is largely responsible for the realized outcomes of producibility, disposability, and sustainability.

Producers normally give priority to producibility over disposability and sustainability because producibility is more visibly internal to the firm and its profitability. But disposability and sustainability (as externalities) are becoming system design characteristics of increasing interest to customers and regulators alike. System and product disposal, along with the treatment of waste from production or construction processes, now requires attention during design of both the product and the process.

In this chapter, producibility, disposability, and sustainability are presented as important design considerations capable of making a significant contribution to stakeholder satisfaction. They are presented as closely related design-dependent parameters with the realization that applicable measures and metrics are different, not fully developed, and not yet readily available. Accordingly, these design considerations are presented in this chapter in a descriptive rather than a normative manner.

Upon completion of this chapter, the reader will have obtained insight regarding opportunities for addressing producibility, disposability, and sustainability as important

From Chapter 16 of *Systems Engineering and Analysis,* Fifth Edition, Benjamin S. Blanchard, Wolter J. Fabrycky. Copyright © 2011 by Pearson Education, Inc. Published by Pearson Prentice Hall. All rights reserved.

aspects of system life-cycle engineering. The topics presented are intended to impart general appreciation and understanding about the following:

- Technological and ecological factors promoting "green engineering";
- Producing for environmental quality and the importance of technological and ecological services in achieving "green engineering" and sustainability;
- The role of environmentally conscious design and manufacturing (ECDM) in advancing environmental sustainability;
- Differences in impact when producibility, disposability, and sustainability are considered as internalities, externalities, or both;
- Manufacturing, remanufacturing, and demanufacturing issues as they arise during the system life cycle, with emphasis on the retirement, recycling, and disposal of subsystems, components, and material elements;
- Some quantitative measures for production and production progress and the factors to be considered in design for producibility;
- Anticipated benefits to be realized from integrating sustainability into production, utilization, phaseout, and disposal; and
- Life-cycle cost and benefits as they relate to total system value, along with discussion of the often hidden sources of these costs and benefits.

The last section of this chapter offers a goal-oriented information flow schematic to guide further study and in extension of the concepts and methods presented. Also included are suggested links to the literature on industrial ecology and sustainability, as well as website references.

1 INTRODUCING PRODUCIBILITY, DISPOSABILITY, AND SUSTAINABILITY

Market opportunities are expanding for the efficient production (or construction) of "environmentally friendly" or "green" products. The key words are *efficient* and *green*, invoking the necessity for giving attention to producibility, disposability, and sustainability as product design characteristics. Production is internal to the producer, and not directly visible to the consumer, but disposability and sustainability are directly visible externalities of keen interest in society.

Figure 1 illustrates the pervasive interactions that encompass bringing into being, utilizing, and ceasing to exist. At the same time, the need for sustainability of the system that produces, supports, maintains, and disposes of the product (and the system itself) is increasing. Accordingly, sustainability is shown conceptually as a higher-level consideration, subsuming production, utilization, and disposal processes insofar as environmental impacts are concerned. Although environmental issues are externalities to producers, they are increasingly raising concerns among stakeholders and societies worldwide.

1.1 Technological and Ecological Services

Technical systems are the source of technological services. These services generally involve the substitution of energy for human effort. They include delivering diverse foods independent of season or local climate, supplying potable water, modifying the climate in buildings, providing communications capability, removing wastes, and so on. Civilization, especially in advanced nations, has developed an extensive infrastructure for the delivery of technological

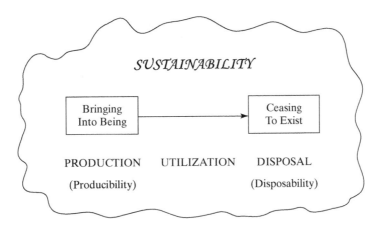

Figure 1 The overarching prominence of sustainability in the system life cycle.

services, including power generation and transmission, communication networks, highway systems and airports, homes and public buildings, waste management, and many others.

Humans also depend on ecosystem services. Ecosystem services are those functions of ecosystems that people desire, including the maintenance of atmospheric balance, carbon storage, flood control, production of food and fiber, and maintenance of air and water quality. Although these services are often taken for granted, people are beginning to realize that many functions of natural systems are not immutable and unchanging. They are affected by human actions and through applications of potent technologies.

The definition of an ecosystem service is a matter of individual and societal perception; it depends on subjective evaluation. However, people most familiar with natural systems are not usually in decision-making positions and are not usually consulted by those who are. Balancing human-made and natural systems is difficult to achieve, but balance will be facilitated if the operational characteristics of the natural systems become as well understood as the physical aspects of technological systems.

Responding to societal needs, regulatory agencies such as the U.S. Environmental Protection Agency, the Food and Drug Administration, and the Occupational Safety and Health Agency were formed by the government to assess real or anticipated environmental quality issues. Because regulatory decisions are a force for incorporating externalities into designs for products and services, they have far-reaching economic and social impacts and should be made carefully and wisely.

1.2 Factors Promoting Green Engineering

It is through production or construction that technology is put to use by humans. But, "green" is not a goal that can be achieved easily. Nevertheless, it should be an objective of producers to continuously reduce the environmental impact of products, production operations, utilization, and disposal practices. Effective and sustainable environmental improvement requires a system life-cycle approach to guide design decisions and operational policies. Sustainability of the system and its product, the production process, and disposal of the system and product itself are factors upon which ultimate net value depends.

The reuse value of recoverable materials and components from most products at the time of disposal is usually small in relation to initial production cost. Accordingly, it is difficult to persuade producers and the first buyer to make changes to the product that would improve the eventual ease of material recovery. This is especially true if those changes incur cost, or have even a small actual or imagined effect on initial product performance, functionality, or affordability.

But this situation is changing. Leading firms have increasingly taken initiatives to improve the environmental performance of their products and processes to reduce undesirable impacts. What are the changes in environmental consciousness that favor movement away from "end-of-pipe" cleanup solutions toward environmentally oriented product design and production? The following factors are now recognized as primary drivers encouraging this proactive (preventative) mode of thinking:

1. *Competitive differentiation.* Traditionally, marketplace decisions have been based on a product's performance, quality, and price. In worldwide competition, some nonprice factors are being included in the firm's competitive strategy. Indications are that environmental impacts, one of the nonprice factors, became an issue of considerable importance worldwide beginning in the 1990s.

2. *Customer consciousness.* Most people believe that a healthy natural environment not only enhances their quality of life but also provides assurance that the quality of life will be sustained. Increasingly, customers are becoming concerned about the environmental quality of the products they use and the quality of the environment in which they live. Consumer awareness of potential damage inflicted on the environment by abuses of technology is beginning to create pressure to develop green products and clean manufacturing operations. Regulatory bodies respond to this attitude change by developing legislation for producers that neglect environmental issues.

3. *Regulatory pressures.* Existing and pending environmental legislation will continue to impose environment-related restrictions on industrial and consumer product manufacturers. To address environmental problems, regulations inspired by the "polluter pays principle" are emerging. Motivated by environmental laws and customers' environmental consciousness, manufacturers are forced to consider how to produce greener products and to remain competitive at the same time.

4. *Profitability improvement.* An eco-design approach to human-made systems can have a significant impact on profitability through savings in manufacturing and other operating costs, by waste elimination–related strategies, and through increased market share. And a growing trend toward safer product design and production addresses the responsibility to provide a healthy and safe working environment.

5. *International standards.* Coordinated by the International Standards Organization (ISO), many manufacturers are participating in a worldwide effort to establish standards for environmental stewardship over the entire life cycle of the system and its product(s).[1]

1.3 Ecology-Based Manufacturing

One of the important issues in design for manufacturability is the harmonization of manufacturing activities with the global ecology. This is a relatively new concept pertaining to an

---

[1]Refer to ISO 14001, "Environmental Management System" (EMS). This standard was developed recently and is being introduced in many companies and implemented in conjunction with ISO 9000–9004 *Quality Standards* (ANSI/ASQC Q90–Q94).

ecologically conscious manufacturing system, namely, an ecology-based production system (or *eco-factory*). The basic requirements for an eco-factory are low-energy consumption, limited use of scarce natural resources, recycling, and reuse.

An eco-factory can be justified by the reduction in its propensity for damaging the environment, by maintaining productivity, by economy of the manufacturing process, and by producing high value-added products. In the eco-factory, the system approach consists of a product design technology that aims to design and develop products with a low ecological load, along with a production technology that also strives for low ecological load. Embracing an eco-factory approach could lead to complete implementation of the concept of integrated design (i.e., the design of products that are conceived not only to be assembled and used but also to be totally and efficiently disassembled and recycled). This implies that a closed-loop eco-factory should be evaluated both by environmental (ecological) criteria and by conventional criteria, such as economy, productivity, performance, and marketability.

Environmental loads have seldom been considered in conventional manufacturing activities. A closed eco-manufacturing system that aims at reducing consumption of energy and resources and incorporates recycling processes will be increasingly desirable and necessary in the future.

2 PRODUCIBILITY, DISPOSABILITY, AND SUSTAINABILITY IN THE LIFE CYCLE

Utilization, reuse, renewal, and disposal constitute major areas of consideration in the design of a system/product and in the implementation of the systems engineering process. The system that brings the product into being is the same system that is responsible for product producibility and disposability, as well as for sustainability of the whole. Accordingly, life-cycle concepts discussed in this chapter should be applied in a manner similar to those for reliability, maintainability, human factors, and supportability.

2.1 Concurrent Life-Cycle Relationships

There are four concurrent life cycles that make up a system about which design should be concerned. The initial emphasis is on designing the *product* for producibility and for disposability, applied as the top series of activities (i.e., the product life cycle). Then there are major interfaces involving the design of the production or construction process (the second life cycle), in design of the maintenance and support processes (the third life cycle), and in the design of the renewal/reuse/disposal capability (the fourth life cycle). These life cycles of the system should be addressed concurrently from the top down, in an integrated manner, and with appropriate feedback to facilitate continuous improvement.[2]

Life-cycle relationships as they apply specifically to the recycling and disposal of system components and material are shown in Figure 2. Components and material can be subjected to a classification and decision procedure that will focus attention on design characteristics supporting disposability (a design-dependent parameter). After the design is implemented, this classification can aid in the implementation of environment-friendly recycling and disposal procedures.

---

[2]The principles of *concurrent engineering* must be implemented herein if the overall benefits of design for producibility, sustainability, and disposability are to be realized.

Design for Producibility, Disposability, and Sustainability

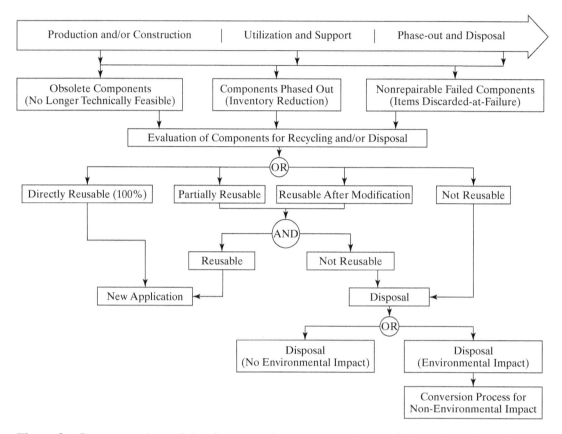

Figure 2 Component/material retirement, phaseout, recycling, and disposal relationships.

Specific design-to requirements should be established and specified as early as the conceptual design phase of the life cycle, functional analysis and requirements allocation are accomplished in the preliminary design phase, trade-offs and design optimization then occur, and so on. The objective is to influence the design in such a way that system elements and components can be produced effectively and efficiently and, when retired, can be disposed of efficiently and without causing undue detrimental impact(s) to the environment. Assimilative capacity of the environment is a threshold that should not be exceeded.

Within the context of the life-cycle activities in Figure 2, design synthesis, analysis, and evaluation are required throughout the product design and development cycle. During the design process, producers must be able to predict and compare the environmental impacts of various product and process alternatives. If environmental requirements for the product/system are specified well, design alternatives can be evaluated against them simultaneously with other requirements. In addition to design for performance and functionality, designers are constantly faced with choosing from among components, materials, and processes, as well as among end-of-life characteristics such as recyclability and disposability.

2.2 Environmentally Conscious Design and Manufacturing[3]

In the activity of life-cycle design, there is an evolutionary design paradigm that starts with the consideration of environmental impacts caused by products and product-related processes during the product and process design stage. This is known as *environmentally conscious design and manufacturing*. ECDM has been increasingly characterized in the last decade as being within the broader domain of Industrial Ecology.[4] The word *sustainability* is now appearing with greater frequency to enlarge upon the intent of ECDM.

The ultimate aim of ECDM during the life cycle is to reduce environmental impact or to improve the environmental friendliness of products, processes, and associated activities. There is a significant difference between ECDM and waste management, or pollution prevention. It is that the ECDM approach is proactive. Its objective is to reduce environmental impact during the early stages of system and product design. Product designers and manufacturing engineers (as well as managers) have a shared responsibility for planning and implementing environmentally safe and environment-friendly manufacturing operations. Much can be gained by careful analysis of products, their design, the materials used, and the manufacturing processes and practices employed.

ECDM can be generally divided into two categories: design for environment (DFE) and environmental management (EM). The DFE approach is the proactive activity that aims at prevention of environmental impacts, whereas EM is remedial in nature. Accordingly, it is ECDM and DFE together that is most effective as a system design strategy when implemented over the life cycle.

3 MEASURES OF PRODUCIBILITY AND PRODUCTION PROGRESS

Producibility is not synonymous with manufacturability. Producibility is a more comprehensive term, which includes not only ease of manufacture as a goal but also the ability of the designed entity to be manufactured, packaged, and shipped (the physical distribution dimension). Producibility, like manufacturability, is a characteristic of design. It can be improved by design decisions that incorporate life-cycle considerations.

3.1 Measures of Producibility

It is difficult to express producibility generically, or in quantitative terms using mathematical measures. However, it is possible to partition the concept into the measurable areas of manufacturing on the one hand and marketing on the other. In both areas, the issue of sustainability arises.

Manufacturability Measures. The more manufacturable a product is, the more quickly (and inexpensively) it may be produced. Therefore, manufacturability is often measured in terms of traditional industrial engineering methods. An example of manufacturability may be *manufacturing lead time* (MLT) or the time needed for a product to be in the manufacturing process. For example, a product must be processed on n_m machines and the average operational time, T_O, is required for each machine. Setup requirements for

---

[3]Some of the material in this section and in Section 7 were adapted from research presented in an unpublished doctoral dissertation by M. J. Goan, "An Integrated Approach to Environmentally Conscious Design and Manufacturing," Virginia Tech, Blacksburg, VA, June 1996.

[4]T. E. Graedel and B.R. Allenby, *Industrial Ecology* (Upper Saddle River, NJ: Prentice Hall, Inc., 2003).

these machines are T_{SU} and nonoperational time for the product is T_{NO}. If the number of units produced per batch is Q, then the MLT can be expressed as

$$MLT = \Sigma\,(T_{SUi}/Q_{TOi} + T_{Oi} + T_{NOi}), \quad i = 1 \text{ to } n_m \tag{1}$$

For a product-oriented measurement, a common metric is the average production time per unit T_P, which is simply the batch time per machine divided by the number of units per batch. With the preceding measurements already defined and a zero defect rate, T_p can be expressed as

$$T_P = [\text{batchtime/ machine}]/Q = T_{SU}/Q + T_O \tag{2}$$

The processes and methods for manufacturing will determine how quickly the product is manufactured. However, the costs of the processes are not a factor in these measures; these costs will affect the overhead rates and must be included in the life-cycle analysis (LCA) for process design trade-off studies. A process may be twice as fast as another but cost somewhat more per product unit, or per hour. But the benefit of quick processing time may be found in improved sustainability from the reduced emission of residuals, such as waste heat. Accordingly, process time should be considered within a broader environmental context.

These considerations embody a systems approach to design, wherein the product and its manufacturing processes in the producer's facilities are part of the calculation process. In the development stage of the system life cycle, these values are purely theoretical. Once production begins, these values become significant in terms of cost and profit. Continuous improvement in the manufacturing process must be considered to improve the manufacturability of the product. This type of program can be planned and may be similar to the manner in which "Reliability Growth Plans" account for such improvements in the life cycle. Thus, consideration of "design for manufacturability" (DFM) is desirable before overall process design is completed.

Subjective measures can be used to determine manufacturability. Industrial science methods use surveys and direct observation for determining the ease of manufacture. Sometimes, a lack of manufacturability can result in operator injuries over time. A common example of this occurrence is the appearance of "carpal tunnel syndrome" in poultry processing and in data entry operations.

Market Measures. Marketing measures, as defined herein, refer primarily to the flow of finished goods (i.e., the product) from the manufacturing site to the ultimate consumer. This aspect of product distribution, often considered and linked closely with design for producibility, is also included as a major element of logistic support (i.e., the *packaging*, *handling*, *storage*, and *transportation* elements).

The primary producibility measures that are of a "market" nature include the following:

1. The time that it takes to move a product from the source of manufacture to the ultimate customer (i.e., user). This includes the elements of time associated with packaging and handling, in-process warehousing and storage, and transportation. Given a demand for a product, how long does it take for delivery and installation? And, is the delivery and installation environmentally sustainable?

2. The cost of processing an item from the source of manufacture to the customer. This includes the cost of packaging, handling, storage, energy, and transportation. Given a demand for a product, how much does it cost for delivery and installation? What are the sustainability issues and costs for alternative methods?

In addition to the design of a product for producibility, the product must be designed so that it can be economically packaged (using standard and conventional materials), easily handled (using standard handling methods), and transportable (using commercial transportation provisions) without causing degrading effects on either the product itself or the environment. Accordingly, the requirements for producibility must be closely coordinated with the requirements for logistics and supportability.

3.2 Measuring Manufacturing Progress

Production and related operations require a coordinated and integrated set of activities that are often repeated over time. This repetition makes possible improvements in the production process such as a reduction in the time to produce a unit, an increase in the rate at which selected activities are performed with a corresponding increase in the number of units produced, a reduction in overall time in process, and a reduction in the cost per unit of output.

Learning takes place within an individual or within an organization as a function of the number of times a task is repeated. It is commonly accepted that the amount of time required to complete a given task (or unit of product) will be less each time the task is undertaken. The unit time will decrease at a decreasing rate, and this time reduction will follow a predictable pattern.

The empirical evidence supporting the concept of a *learning curve* was first noted in the aircraft industry.[5] The reduction in direct labor hours required to build an aircraft was observed and found to be predictable. Since then, the learning curve has found applications in other industries as a means for adjusting costs for items produced beyond the first one and for determining throughput rates.

Most learning curves are based upon the assumption that the direct labor hours needed to complete a unit of product will decrease by a constant percentage each time the production quantity is doubled. A typical rate of improvement in the aircraft industry is 20% between doubled quantities. This establishes an 80% learning function and means that the direct labor hours needed to build the second aircraft will be 80% of the hours required to build the first. The fourth aircraft will require 80% of the hours that the second required, the eighth aircraft will require 80% of the fourth, and so on. This relationship is given in Table 1.

An analytical expression for the learning curve may be developed from the preceding assumptions. Let

x = the unit number;

Y_x = the number of direct labor hours required to produce the xth unit;

K = the number of direct labor hours required to produce the first unit;

ϕ = the slope parameter of the learning curve.

---

[5]T. P. Wright, "Factors Affecting the Cost of Airplanes," *Journal of Aeronautical Sciences*, Vol. 3, No. 4, February, 1936.

TABLE 1 Unit Cumulative and Cumulative Average Direct Labor Hours
for an 80% Improvement Function with Unit 1 Set at 100 Hours

Unit Number	Unit Direct Labor Hours	Cumulative Direct Labor Hours	Cumulative Average Direct Labor Hours
1	100.00	100.00	100.00
2	80.00	180.00	90.00
4	64.00	314.21	78.55
8	51.20	534.59	66.82
16	40.96	892.01	55.75
32	32.77	1,467.86	45.87
64	26.21	2,392.45	37.38

From the assumption of a constant percentage reduction in direct labor hours for doubled production units,

$$Y_x = K\phi^0 \qquad \text{where } x = 2^0 = 1$$
$$Y_x = K\phi^1 \qquad \text{where } x = 2^1 = 2$$
$$Y_x = K\phi^2 \qquad \text{where } x = 2^2 = 4$$
$$Y_x = K\phi^3 \qquad \text{where } x = 2^3 = 8$$

Therefore,

$$Y_x = K\phi^d \qquad \text{where } x = 2^d$$

Taking the common logarithm gives

$$\log Y_x = \log K + d \log \phi$$

where

$$\log x = d \log 2$$

Solving for d gives

$$d = \frac{\log Y_x - \log K}{\log \phi} \quad \text{and} \quad d = \frac{\log x}{\log 2}$$

from which

$$\frac{\log Y_x - \log K}{\log \phi} = \frac{\log x}{\log 2}$$

$$\log Y_x - \log K = \frac{\log x (\log \phi)}{\log 2}$$

Let

$$n = \frac{\log \phi}{\log 2}$$

Therefore,

$$\log Y_x - \log K = n \log x$$

Taking the antilog of both sides gives

$$\frac{Y_x}{K} = x^n$$

$$Y_x = Kx^n \tag{3}$$

Application of Equation 3 can be illustrated with reference to the example of an 80% progress function with unit 1 at 100 direct labor hours. Solving for Y_8, we see that the number of direct labor hours required to complete the eighth unit is

$$Y_8 = 100(8)^{\log 0.8/ \log 2}$$

$$= 100(8)^{-0.322}$$

$$= \frac{100}{1.9535} = 51.2$$

Progress functions for several improvement rates are shown in Figure 3. The curves are for doubled production quantities with a base of 1 (the 90% curve) and for doubled production quantities with a base of 10 (the 90%, 80%, and 70% curves). In the cases where the base is 10, batch sizes of 10 are involved. Equation 3 may be applied to batches of any size provided the constant percentage reduction assumption holds for the batch as a whole.

Information from the learning curve can be extended to cost estimates for labor by multiplying by the labor rate that applies. In doing this, the analyst must realize that the

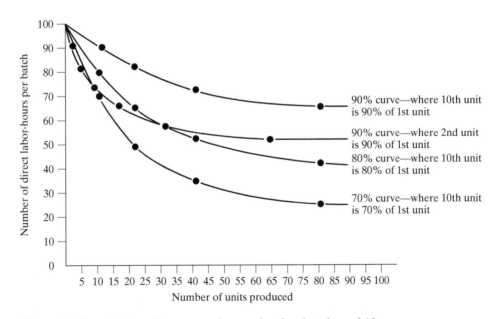

Figure 3 Sample learning curves for production batches of 10.

subsequent units may be completed months or years after the initial units. Adjustments for labor-rate increases might have to be made along with the adjustment for learning that is inherent in the application of the learning curve.

Given the preceding relationships, the application of learning curves has found wide usage in recent years in a variety of industries where multiple quantities of like items are being produced or where repetitive tasks such as maintenance are accomplished. Although the 80% factor in the aircraft industry may be assumed as being characteristic for this particular industry, other percentage factors may be more realistic for other industries depending on system complexities, the manufacturing processes used, and so on. For instance, a 90% learning curve may be more realistic in the production of sophisticated electronic equipment, whereas a 70% learning curve may be appropriate in the production of mechanical items. Thus, learning curves have been developed and adapted to fit the situation, and these curves have subsequently been used in the prediction of production rates, energy consumption, and unit cost.

4 DESIGN FOR PRODUCIBILITY

Producibility is a more comprehensive term than *manufacturability*. It includes not only ease of manufacture, as implied by the word *manufacturability*, but also the ability of the manufactured entity to be packaged, transported, and delivered to the point of use in a timely manner. Producibility, like manufacturability, is a characteristic of design that includes the entity itself as well as aspects of the supply chain. It can be enhanced by design decisions that incorporate life-cycle considerations.

4.1 A Classification of Manufacturing Processes

Manufacturing processes are generally classified into five categories with each having its own inherent producibility characteristics:

1. *Forming processes.* Processes in which an original shape is created from a molten or gaseous state, or from solid particles of an undefined shape.
2. *Deforming processes.* Processes that convert the original shape of solid to another shape without changing its mass or material composition.
3. *Removing processes.* Processes in which material removal occurs during the process itself.
4. *Joining processes.* Processes that unite individual workpieces to make subassemblies or final products.
5. *Material properties modification processes.* Processes that purposely change the material properties of a workpiece to achieve characteristics without changing its shape.

These categories are applied to a variety of engineering materials, which can be generally classified into metals, ceramics, plastics (polymers), and composites. The selection of materials depends on the product design and should incorporate environmental and sustainability considerations. Sustainable and recyclable materials should always be considered first in green production design. Different processes are applied to different engineering material as summarized in Table 2.

The selection of which process to apply to a particular material is influenced by several factors (e.g., the lot size of the parts to be made and the physical properties of the

TABLE 2 Matching Materials and Processes (Overview)

Materials	Processes				
	Forming	Deforming	Removing	Joining	Modifying
Metals	Widely used	Widely used	Widely used	Widely used	Widely used
Ceramics	Widely used	Not used	Seldom used	Not used	Not used
Polymers	Widely used	Seldom used	Seldom used	Seldom used	Not used
Composites	Widely used	Not used	Seldom used	Seldom used	Not used

materials) that affect process attributes (e.g., cost, production rate, flexibility, and part quality). But selection should be especially influenced by the environmental loads caused by manufacturing processes. Selection of an appropriate process often depends on a trade-off (balance) of all process attributes (or criteria) including the potential environmental impacts. Table 3 gives an example based on engineering requirements for making a desired product in an efficient, ecological, and profitable manner.

4.2 Manufacturability Principles

There are a few high-level design-for-manufacturability principles of general importance with sustainability implications. These are the following:

1. *Use gravity.* It is easier to work with lighter components, with up and down movements, with snap fit in vertical slot openings, and so on. Gravity can, in some instances, save human and natural energy.

2. *Use fewer parts.* An increase in the number of parts means an increase in design and manufacturing cost. Nonexistent components cost nothing to purchase, assemble, and test and consume no resources.

TABLE 3 Some Characteristics of Manufacturing Processes

Attribute	Processes				
	Forming	Deforming	Removing	Joining	Modifying
Costs	High tooling/ low labor cost	High tooling/ low labor cost	Medium tooling/ high labor cost	Low capacity/ high labor cost	Medium to high capital/low labor cost
Environmental costs	Expert judgment	Expert judgment	Expert judgment	Expert judgment	Expert judgment
Production rate	High	High	Medium (milling) to low (grinding)	Medium (welding) to low (adhesive)	Low
Part quality	Medium to low	Medium to low	Medium to high	Medium to high	High
Flexibility	Low	Low	High	High	Medium to high
Environmental factors	Inputs/outputs	Inputs/outputs	Inputs/outputs	Inputs/outputs	Inputs/outputs

3. *Design for ease of fabrication.* Design parts so that (a) tolerances are compatible with the assembly method employed and (b) fabrication costs are compatible with targeted product costs. This eliminates the waste of part rejections or tolerance failures during assembly.

4. *Reduce nonstandard parts.* The use of common and standard parts eliminates the development costs associated with designing and manufacturing, reducing the environmental load.

5. *Add more functionality per part.* The goal is to accomplish the functions required with fewer parts, or to allocate more functions per part. This is an objective of the functional analysis and packaging requirement. It can result in doing more with less.

In assembly, a few high-level guidelines are of general importance with some relationship to sustainability. These are the following:

1. *Employ automatic inserters.* Specify parts that can be automatically sequenced and inserted using DIP inserters, a variable center device (VCD), axial part inserters, or selective compliance assembly robot arm (SCARA) insertion robots. Minimizing setups and reorientation saves energy and time.

2. *Employ "preoriented" parts.* Parts that cannot be supplied preoriented in reels, tubes, or matrix arrays for easy insertion should be avoided due to the waste of time and energy.

3. *Minimize sudden and frequent changes in assembly direction.* Following a unidirectional assembly sequence, such as design for top-down assembly, is usually the most desirable from an effort standpoint.

4. *Maximize process compliance.* Process compliance consists of designing with standard parts, standard processes, ease of assembly, and so on. Use processes that are easy to install, maintain, and sustain.

5. *Maximize accessibility.* Designers should provide adequate clearances for accessing the part for future repair or replacement and for ease of disassembly to facilitate recycling.

6. *Minimize handling.* There are two aspects to minimize handling and its environmental impact: design of parts for ease of feeding (insertion) and design of parts so that they are easy to grasp, manipulate, or orient.

Informal guidelines have been developed for design with consideration to assembly. Mechanical design, predating the industrial revolution, probably has the most universally understood guidelines. Some of these are as follows:

1. *Assemble to a foundation.* This method allows for automated assembly by gripping to a foundation. The foundation must be designed for accurate machine positioning, since a large tolerance on the foundation location will be added to the assembled components.

2. *Assemble from as few positions as possible.* Repetitive machinery is more reliable with fewer components. Reliability of the production equipment will be reduced with an increase in components. This practice (monodirectional access) also improves maintainability.

3. *Make parts independently replaceable.* Subcomponents of an assembly should not require removal of other components to reach the faulty ones. This assembly practice also improves maintainability and saves energy.

4. *Order assembly so that most reliable goes in first, with the least reliable last.* This guideline concerns the testing of a product before delivery. If a particular component or subassembly requires a significant portion of the final test, production time devoted to troubleshooting is minimized.

5. *Assure commonality in design.* Commonality in design attempts to reduce the types of subcomponents in a system. The more standardized a product is, the less overhead is associated with supporting the variety of the parts before assembly. If the variety of tools used to assemble these types of parts can be reduced, a contribution will be made to sustainability.

4.3 Manufacturing and Demanufacturing Issues[6]

In a production system with demanufacturing capability, the basic structure of demanufacturing consists of two major divisions as shown in Figure 4: a production division and a disposal division. In the production division, a producer purchases raw materials or components from outside suppliers, and then produces, offers, and distributes various types of products to consumers. The production division is classified into three sectors: (1) recyclable products for producing the same type of products inside the factory, (2) recyclable products for producing different kinds of products inside the factory, and (3) wastes to be discarded. Accordingly, a production system with demanufacturing capacity actually contains the closed-loop flow (for division 1) or the open-loop flow (for division 2), or both, within the factory.

In a typical product development process, the product configuration is established during design. Product attributes are then imparted through several manufacturing operations. Returned (or recycled) products are processed by demanufacturing operations. Recycling and demanufacturing, which are the product's major end-of-life activities, should be integrated into the design and life-cycle processes. ECDM results will affect downstream process planning for product end-of-life disposition.

It is now recognized that demanufacturing plays an important role in the green product concept. Manufacturing firms respond to social and regulatory pressures for environmental

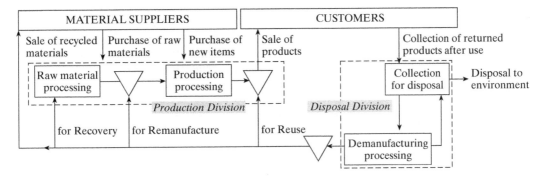

Figure 4   A production system with demanufacturing and disposal.

[6]Demanufacturing is used in this chapter instead of the term *disassembly*. Demanufacturing includes consideration of the planning and processing related to recycling as a whole. Disassembly, viewed as a recycling activity as part of demanufacturing, is employed to reuse, refurbish, remanufacture, and recycle. Further, it should be noted that many of the principles related to *design for disassembly* are similar to those for design for maintainability.

attention in several different ways. These responses are directed at decreasing disposal costs by demanufacturing the recycled products received from consumers for reuse, remanufacture, and recycling. Also, these responses have an impact on the firm's strategy for green product design and development. Accordingly, the collection, demanufacture, processing, recycling, and disposal of used products recycled after use present major design and manufacturing challenges to traditional manufacturing operations.

Demanufacturing operations are essential processes for the purposes of reuse, demanufacturing, recovery, and related sustainability enhancing practices. These are described below:

1. *Reuse* is the highest form of waste reduction and has the potential to increase the product's end-of-life value. Reuse may be most easily justified in the case of components with high manufacturing costs, long innovation cycles or lifetimes, and high ecological impact during production.

2. *Remanufacturing* is the refurbishing or partial rebuilding of a product returned from the customer (by collection) at the end of its life with the objective of giving it the functionality equivalent of a new product. Even though remanufacturing requires disassembly efforts that incur cost, a payoff may occur due to a lower overall ecological impact.

3. *Recovery* from products to obtain raw materials or reusable components is an important means of reducing disposal volume and cost while contributing to environmental sustainability.

The efforts listed previously may be important ways of reducing product life-cycle cost even though the activities contribute to demanufacturing costs. The significant demanufacturing costs for recycling and reuse include costs for retrieval, separation, disassembly, sorting, storage, transportation, identification, testing, reprocessing, and remarking. The value of the recycled or remanufactured product is a function of the intrinsic value of its components/parts and materials, as compared to the cost of recycling and disposal.

Costs of recycling and disposal can be significantly affected by product design decisions. Design for demanufacturing, design for disassembly, design for serviceability and/or supportability, and design for recycling are fundamental to effective and sustainable demanufacturing and recycling strategies and programs. Both manufacturing and demanufacturing processes are affected by design decisions, especially when the environmental impacts caused by the processes are considered. And, the product realization process is complicated by the nature of economic, logistic, and environmental considerations. To illustrate the relationship between value and cost throughout a product's life cycle, a cost/value flow diagram is presented in Section 6.

5 DESIGN FOR DISPOSABILITY

Examples of "green" products, "clean" processes, and the "eco-factory" are beginning to appear across a wide spectrum of industries, driven by the so-called green revolution. The green revolution derives from Industrial Ecology, where environmentally conscious design and manufacturing plays a central role in green product realization. Because the nature of product realization is cross-functional, interdisciplinary teams are needed to pursue the activities of design, manufacturing, utilization, support, and disposal/recycling.

5.1 Disposability, Sustainability, and Industrial Ecology

Waste and emissions caused by the supply chain result in serious impacts to sustainment of the local and global environment, such as global warming and acid rain. Accordingly, disposability must be recognized as an important design-dependent parameter in product design and development under the ECDM paradigm. Decisions about product design, production planning, process selection, logistics, inventory disposition, and recycling will likely increase in importance because of customer preferences and governmental regulations.

The relationship between the supply and environmental chains is the domain of industrial ecology in which manufacturing enterprises adopt environmental strategies to discharge their sustainment responsibility. Consequently, there is a need to implement new design and manufacturing ideas to deal quickly with opportunities for reducing environmental impacts caused by the supply chain. The evolutionary ECDM paradigm is an applicable ecological strategy for sustainability.

A green product may be defined as one that, at the ultimate end of its useful life, passes through disassembly and other reclamation processes to enable reuse of nonhazardous and renewable materials. ECDM strategies deliberately attempt to reduce the ecological impacts of industrial activity without sacrificing quality, cost, reliability, performance, or energy-use efficiency. However, efforts leading to environmental sustainment result in complicated problems. The general ECDM paradigm requires new design and specified analysis method and methodology to shift from reactive end-of-pipe treatment to integrated, multidisciplinary, proactive design approaches. Although the explicit consideration of environmental objectives and constraints in the product development cycle can lead to a green technology, the question as to what is green depends on how environmental problems and system boundaries are defined.

The concepts of design for environment are also employed in design of the manufacturing processes in the eco-factory. An eco-factory's initiatives focus primarily on various manufacturing strategies and technologies. The major emphasis for manufacturing technology is on production systems, restoration systems, and control and assessment. Eco-factory technology can enable sustainment of the environment and the effective use of natural resources by industry. In a sense, the eco-factory is viewed as a physical system developed on the basis of an ECDM approach to implement the concept of industrial ecology. Various technologies within each of these groupings are shown in Figure 5.

5.2 Manufacturing with Recycling Applications

Recycling has been recently recognized as one of the most effective means for solving environmental problems. Therefore, in the design of future manufacturing systems, the recycling of materials should be considered as a design objective.

Recycling of products to obtain raw materials or reusable components is an important means of reducing disposal costs and increasing total product value. In demanufacturing processes, demanufacturing involves refurbishing or partially rebuilding a returned product. The objective is to give it the functionality of a new product. This strategy has been identified by several firms as a major source of cost savings and competitive advantage. Effective recycling and demanufacturing require ease of product recovery through disassembly or separation, performance and market acceptance of the recycled materials and components, and the environmentally responsible disposal of nonrecoverables remaining after recycling (see Figure 2).

Recycling activities could occur during the production stage. One of the features in a plastic extrusion production process is that waste materials are created, reground, and

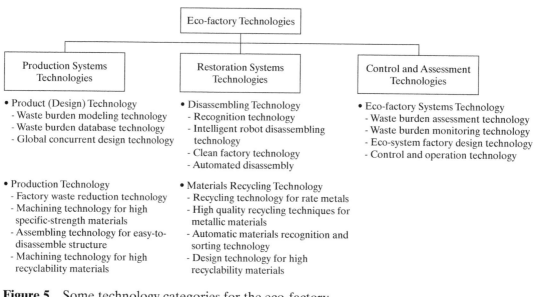

Design for Producibility, Disposability, and Sustainability

Eco-factory Technologies

Production Systems Technologies

Restoration Systems Technologies

Control and Assessment Technologies

- Product (Design) Technology
  - Waste burden modeling technology
  - Waste burden database technology
  - Global concurrent design technology

- Production Technology
  - Factory waste reduction technology
  - Machining technology for high specific-strength materials
  - Assembling technology for easy-to-disassemble structure
  - Machining technology for high recyclability materials

- Disassembling Technology
  - Recognition technology
  - Intelligent robot disassembling technology
  - Clean factory technology
  - Automated disassembly

- Materials Recycling Technology
  - Recycling technology for rate metals
  - High quality recycling techniques for metallic materials
  - Automatic materials recognition and sorting technology
  - Design technology for high recyclability materials

- Eco-factory Systems Technology
  - Waste burden assessment technology
  - Waste burden monitoring technology
  - Eco-system factory design technology
  - Control and operation technology

Figure 5 Some technology categories for the eco-factory.

mixed with virgin plastic raw material for use in subsequent (extrusion) production cycles. The recycling of these waste materials helps to reduce the need to use the more expensive virgin raw materials. Another increasingly popular recycling opportunity is the reuse of recaptured waste heat.

Reuse in demanufacturing is the highest form of waste reduction, and it has the potential to reduce the end-of-life cost of products. Reuse may be most easily justified in the case of components with high manufacturing costs, long innovation cycles, and long lifetimes.

6 DESIGN FOR SUSTAINABILITY

According to the U.S. Department of Commerce, sustainable production is the "creation of manufactured products that use processes that are non-polluting, conserve energy and natural resources, and are economically sound and safe for employees, communities, and consumers."[7] This definition is limited in scope to production but, by implication, can be nominally extended to operations and disposal.

Consider the vision expressed by Svante Arrhenius, Director of the Nobel Institute, who (in 1926) urged engineers "to design more efficient internal combustion engines capable of running on alternative fuels such as alcohol, and new research into battery power . . . Wind motors and solar engines hold great promise and would reduce the level of CO_2 emissions . . . Lighting with petroleum products should be replaced with more efficient electric lamps. Although not designated sustainability until recently, design for sustainability is not limited in scope nor is it of recent origin."[8]

---

[7]U.S. Department of Commerce, 1401 Constitution Ave., NW, Washington, DC, 20230
[8]S. Arrhenius, *Chemistry in Modern Life*, Van Nostrand Company, New York, NY, 1926.

6.1 Internal and Environmental Sustainability

With the rapid growth in population and widespread resource use on a finite planet, the generation of wastes associated with the production/construction, utilization, and disposal of goods and services is of great concern. These byproducts of market transactions impact the environment. The losses of environmental quality resulting from pollution of the air, water, and soil by effluents from manufacturing are one example of externalities. Noise pollution from factories, aircraft, trucks, and automobiles is another. Externalities have no direct markets; clean air, rivers, or low decibel noise cannot be bought or sold in a marketplace.

The loosely regulated market has poorly internalized environmental quality concerns. There are two reasons for this deficiency. First, environmental quality loss is often not seen or understood at the time the transaction occurs. Second, losses are usually insignificant for a single market transaction. Such losses arise from the aggregate, cumulative effects of manufacture, use, and disposal of products. Degradation over the long term of common property resources, such as air and water, is difficult to assess in the present.

Designers have always had the opportunity to consider sustainability; sustainability of the system/product in service (internal sustainability) as well as sustainability of the environment (external sustainability) when impacted by the system and product during and after use. Such concerns are reflected in the market, where many products are made to last, to be recycled, are biodegradable, and are expected to be free from toxic substances.

In the past, the existence of environmental effects or constraints imposed on the producer has not been a major issue for designers. This may be because the waste and pollution generated from manufacturing processes are most often considered to be byproducts of manufacturing and, hence, are assumed not to affect the product's environmental impact and quality. Consequently, the costs incurred become part of the rigid fixed cost incurred by the producer and are not linked to the product realization process. As a result, too little effort has been made to engineer for system and product sustainability through "green" products and associated "environment-friendly" manufacturing processes.

6.2 Metrics for Sustainability

Metrics for sustainability are generally unavailable, but would be of great value in measuring the concern and advancing the practice of green engineering. Producers need clear and consistent criteria to respond to requirements for sustainability. And there are increasingly restrictive international environmental regulations to be accommodated. Also, simple metrics would be invaluable in promoting product design changes leading to reduced resource depletion, environmental degradation, and negative impact on humans over the life cycle.

Most metrics in use today are voluntary, firm specific, or governmentally mandated. They exhibit little consistency. Metrics in use vary with market sector and country of origin. But the Organization for Economic Cooperation and Development (OECD) is now researching, developing, and publishing sustainability metrics. The framework being used includes the following:[9]

1. *Resources.* Includes depletion, degradation, and utilization efficiency.
2. *Product.* Includes design, durability, useful life, quality, and packaging.
3. *Employment.* Includes health, safety, security, and worker satisfaction.
4. *Economic.* Includes value added by investment and production ethics.

[9]OCED, *Eco-efficiency.* Paris, 1998.

5. *Society.* Includes community development and social impacts.
6. *Environment.* Includes waste production, emissions, and acoustics.
7. *Infrastructure.* Includes transport ease and communications.

Metrics will help in designing for sustainability in general and in designing to sustainability requirements in specific situations. Some areas of interest are (1) energy consumption reduction, (2) use of recycled materials, (3) design life extension, (4) quality and durability, (5) renewability, and (6) design for recycling or reuse. If sustainability is to join other design-dependent parameters on a coequal basis, metrics will be the means by which it will attain a "design to" status. Advancing sustainability metrics will have a positive effect on the ability of producers to benchmark and share best practices. They will also be invaluable in bringing more objectivity to the emerging marketplace for environmental credits.

7 LIFE-CYCLE VALUE–COST DIAGRAM

The system-oriented approach for achieving a balanced industrial eco-system requires that designers consider eco-product design and eco-manufacturing processes simultaneously at the early design stage. Life-cycle analysis is a guide to evaluate the environmental consequences of a product or process over its entire life, from raw material acquisition to final disposal. Life-cycle assessment has been classified as both a conceptual and a technical framework that may be used to evaluate the environmental performance of a product, process, or activity over the entire product life cycle, from the first stage of procuring raw materials to the final stage of waste management. Recycling analysis within a closed-loop system requires information from other activities, such as material selection analysis, manufacturing processes analysis, and product utilization analysis.

Integration approaches are required in constructing a cost–benefit analysis model, as shown in Figure 6. A cost/value flow diagram provides basic concepts of costs and values that vary throughout a product's life cycle. It also depicts a life-cycle cost/value, economic flow component that explains the complexity of economic, logistic, and environmental considerations for manufacturing systems with demanufacturing/recycling and sustainment capacity.

Under traditional manufacturing engineering and life-cycle costing, production/manufacturing cost–benefit analysis is employed in Phases I and II of Figure 6. Cost is committed during manufacturing of the product and related items, including environmental cost (Phase I). Value is added to the products and items as they are fashioned from raw materials. The extent of the added value is determined by the production/manufacturing costs arising from capital equipment, design engineering, purchases, labor, processing operations, and so on. During Phase II, a benefit is obtained only when the finished products or items are sold. The value decreases because the products and items depreciate physically or functionally during their life as a result of use, environmental stress, and by competition from newer products. For capital products, such as vehicles, a net salvage value may be realized. But, for most small items, the net value is zero or negative reflecting removal or disposal costs. Thus, traditional product life cycle and manufacturing cost–benefit analysis does not consider the cost because of environmental effects and the potential revenue from recycling.

A more complete economic cycle includes the lower half of the flow diagram in Figure 6. Products and items can be recycled in the salvaged condition. Collection, sorting, disassembly, contaminant removal, and material recovery are necessary during the recycling and demanufacturing process. These activities add to costs, while extending the actual net value of the product and the items recycled from consumers and industry (Phase III). In Phase IV,

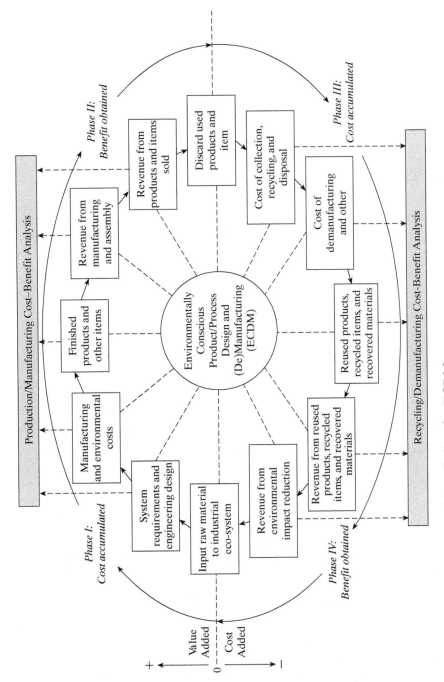

Figure 6 Life-cycle cost and benefit flow for ECDM.

631

recycled items and recovered materials are sold to the original or other product manufacturers. Other savings in the overall four-phase life cycle result from the reuse of recycled items and recovered materials, as well as from the reduction of environmental impacts, including waste-reduction energy saving.

Environmental costs are difficult to specify accurately, but show up as environmental loading costs, landfill fees, disposal costs, waste handling charges, and general environmental liability. Nevertheless, recycling makes good economic sense in many instances in addition to its environmental benefit. In such a complete economic cycle, the opportunities and feasibility of item recycling and materials recovery should be considered throughout product and process design. These considerations should be subjected to a decision evaluation model employed to select from among design alternatives for the product and process.

8 SUMMARY AND EXTENSIONS

Figure 7 summarizes the ECDM-related problems that were addressed in this chapter. Total product value is generally the goal. System design evaluation or design optimization is an essential procedure in eco-product development to maximize the total product value. In the context of ECDM, the product's total life-cycle value is created from production activities and demanufacturing efforts. Production, demanufacturing, and sustainment are significantly affected by product design and the production system by which they are realized. All may contribute to an increase or decrease in the value of a product.

Much attention is now being paid to the integration of design and manufacturing. But, integration of sustainment, with demanufacturing, and recycling are becoming equally important. The impact of environmental issues on product development ranges from production planning and product production to product disassembly and recycling, both economically and ecologically. However, none of the preceding considers the life-cycle design problem holistically throughout the product/material life cycle. In many ways, the previous types of problems and solutions are simply parts (or subproblems) of the whole DFE problem. Systematic integration of environmental considerations into life-cycle engineering and design is needed.

Ecology-oriented product development is not limited to production and marketing. The entire product life cycle must be considered. Beyond green product design, producers have to consider the environmental impact of the product-related processes and logistic support. Therefore, an integrative approach is desired that considers three elements (i.e., products, processes, and logistic support) during the whole life cycle, including the renewal and recycling phase.

Within the context of the eco-design of products and processes, the ECDM approach seeks to discover product innovations that will result in reducing harmful environmental impacts at any or all stages of the life cycle, while satisfying cost and performance as well as quality objectives. ECDM should be practiced in balance against other design considerations to get the "best" design. For ECDM to be implemented and integrated effectively into the eco-product development process, the following key elements are required throughout the life-cycle stages: life-cycle synthesis, life-cycle analysis, and life-cycle evaluation, to determine the best alternative with balance among the various design considerations.

From a product design perspective, total product value is determined by performance functionality, quality, environmental compatibility, and economic viability. From a manufacturing process perspective, the coordination of disassembly, demanufacturing, and recycling operations, with each other and with other manufacturing facilities, is complex task. There appears to be a need for analytical models that will allow designers and engineers to gain

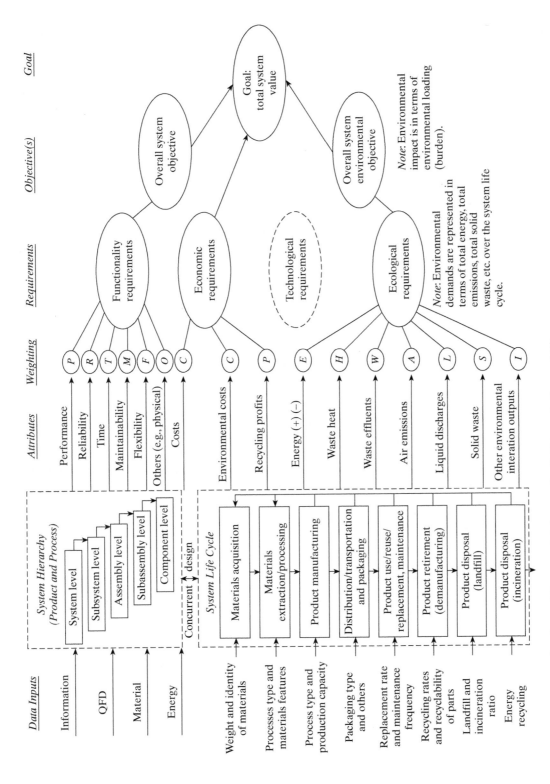

Figure 7 A goal-oriented flow model for producibility and disposability.

insight into product-design and manufacturing process issues that relate to the environment. In addition, life-cycle assessment is developed and employed as a tool to evaluate the environmental burden of manufacturing processes and of products. The results of LCA are an important factor in the evaluation of the environmental impacts of a product and its manufacturing process.

Life-cycle cost analysis (LCCA) is used for an economic assessment of the environmental issues. The life-cycle approach is one of the significant features of the ECDM framework. Another feature involves product/materials-oriented and system-oriented approaches. These approaches led to the integration of DFE, LCA, and LCCA into the ECDM concept.

In addition to the dissertation research acknowledged in footnote 3, two books on industrial ecology and sustainability have had a significant influence on the general thinking and specific topics presented in this chapter. These books are recommended as seminal works to those seeking a deeper understanding of this important emerging field. The books are T. E. Graedel and B. R. Allenby, *Industrial Ecology*, 2nd ed. (Upper Saddle River, NJ: Prentice Hall, 2003) and P. Stasinopoulos, M. H. Smith, K. Hargroves, and C. Desha, *Whole System Design: An Integrated Approach to Sustainable Engineering* (London, UK and Sterling, VA, USA: Earthscan, 2009).

The field of industrial ecology gained initial recognition in the late 1990s through the creation of the *Journal of Industrial Ecology*. This journal is published quarterly by the MIT Press and fosters understanding and practice in this emerging field. Interested individuals may obtain more information by visiting *http://mitpress.mit.edu*.

An international society for the field of industrial ecology now exists. In January 2000, a group of leaders from diverse fields who share an interest in promoting industrial ecology gathered at the New York Academy of Sciences and decided that the time had come to create the International Society for Industrial Ecology. The website for more information is *http://www.is4ie.org*.

More recently, there has now been established an International Society of Sustainability Professionals (ISSP), with a committment to make sustainability standard practice. ISSP is headquartered in Portland, Oregon, and the website for more information is *http://sustainabilityprofessionals.org*

A strongly recommended extension and expansion of many of the topics in this chapter is available from National Geographic. This is a 16-page summary containing beautiful pictures, appropriate graphics, technical statements, and brief discussions. It appeared in April 2009 as a supplement to *National Geographic* magazine and is entitled *Solutions for a Better World*.

QUESTIONS AND PROBLEMS

1. Write a brief explanation of the phrases "bringing into being" and "ceasing to be" as related to the concept of entropy.

2. Elaborate upon the interconnections between producibility, disposability, and sustainability as illustrated in Figure 1.

3. Humanity is sustained by both technological services and ecological services. Explain.

4. List some of the factors and incentives promoting "green engineering."

5. Define environmentally conscious design and manufacturing and contrast it with environmental management as in Figure 2.

6. Define *producibility*. Why is it important? When in the system life cycle should it be addressed?

7. Identify and describe some of the measures of producibility.

8. Define *disposability*. Why is it important? When in the system life cycle should it be addressed? How does it relate to *producibility*?

9. Identify and describe some of the measures of disposability.

10. Explain in your own words why producibility is an internality and disposability is normally an externality.

11. What incentives are there within the firm to make disposability an internality?

11. When internal incentives are insufficient, what policies might be adopted by society through government to make disposability an internality?

12. Describe what is meant by *environmental quality*.

13. Describe in your own words what is meant by *environmentally conscious design and manufacturing*.

14. What impact might the results of the functional analysis have on producibility?

15. What is meant by *demanufacture*? Provide some examples.

16. If you were tasked with the objective of reducing *waste* in your manufacturing process, what goals would you establish for the design of the product to be produced?

17. How does producibility relate to *mobility* and *transportability*?

18. Identify and describe some of the manufacturing processes that need to be addressed to enhance producibility (pick at least five).

19. Describe what is meant by a *learning curve*. Provide a simple illustration of a 70% and an 80% learning curve. Under what conditions can learning curves be applied?

20. Describe what is meant by the *eco-factory*. What elements should be included?

21. Define what is meant by *green engineering*.

22. Assume that you have been assigned by your organization to develop a retirement and material disposal plan. Prepare a topical outline of the plan (i.e., what is to be included). Develop a flow process identifying the activities (and their interfaces) that you would specify in the implementation of such a plan.

23. Go to the websites of ISIE and ISSP and list some similarities and differences in the goals and objectives of these associations.

Design for Affordability (Life-Cycle Costing)

From Chapter 17 of *Systems Engineering and Analysis,* Fifth Edition, Benjamin S. Blanchard, Wolter J. Fabrycky. Copyright © 2011 by Pearson Education, Inc. Published by Pearson Prentice Hall. All rights reserved.

Design for Affordability (Life-Cycle Costing)

Many systems are planned, designed, produced, deployed, and operated with too little concern for *affordability* and the total cost of the system over its intended life cycle; that is, *life-cycle cost* (LCC). The technical side is usually considered first, with the economic side deferred until later. Economic and cost factors arising from activities such as research, design, testing, production or construction, utilization and support, and phaseout and disposal are often treated independently and sporadically over the life cycle.

Accordingly, this chapter addresses economic and cost factors emphasizing the general theme of design for affordability. An introduction to life-cycle costing is followed by the identification of costs that occur over the system life cycle. Thereafter, a generic life-cycle costing process incorporating 12 steps introduces a preferred methodology. This methodology is applied first to a comprehensive life-cycle costing example based on money flow modeling. A second life-cycle costing example is then presented to demonstrate application of the methodology based on economic optimization modeling.

This chapter provides a deeper appreciation of the economic aspects of systems engineering than is obtainable only from models for economic evaluation. The topics presented are intended to impart insight into and an understanding of the following:

- The technical and economic environment giving rise to both the planned and the unanticipated (or hidden) costs generated over the system life cycle;
- Major cost categories and their subordinate elements that should be considered during the process of designing for the life cycle;
- The origin and life-cycle impact of costs arising from the activities of design, production, operation, support, maintenance, and disposal;
- Life-cycle cost and benefits as they relate to the determination of cost-benefit and/or cost-effectiveness of a system and/or its product;
- How to utilize the 12 steps in a life-cycle costing process, based on a money flow model.

- How to implement the economic optimization model, guided by the 12-step life-cycle costing process;
- The advantages and disadvantages associated with the money flow modeling paradigm and with the economic optimization paradigm; and
- Benefits to be expected from the application of life-cycle costing to the process of bringing systems and their products into being.

This chapter summarizes the pervasive applications and benefits to be derived from life-cycle cost and affordability analysis. The chapter ends with a few classical references and selected website addresses.

1 INTRODUCTION TO LIFE-CYCLE COSTING

In general, the complexity of systems is increasing and many of those systems in use are not fully meeting the needs of the customer (i.e., user) in terms of performance, effectiveness, and overall cost. New technologies are being introduced on a continuing basis, while the duty cycles for many systems in use are being extended. The length of time that it takes to develop and deploy a new system needs to be reduced, the industrial base is changing rapidly, and available resources are dwindling. These trends, combined with past practices in system design and development, have often led to an imbalance between the *economic* and the *technical* aspects of total system value, as illustrated in Figure 1. This imbalance during development may lead to the deployment of a system that is not as cost-effective as it could be.

Referring to the economic side of the balance in Figure 1, experience has indicated that not only have the acquisition costs associated with a new system been rising but also the costs of operating and maintaining systems already in use have been increasing substantially. This is due primarily to a combination of inflation and cost growth from several causes:[1]

1. Cost growth resulting from engineering changes occurring during the design and development of a system or product (for the purposes of improving performance, adding capability, etc.)
2. Cost growth resulting from changing suppliers in the procurement of system components
3. Cost growth resulting from system production or construction changes
4. Cost growth resulting from changes in the logistic support capability
5. Cost growth resulting from initial estimating inaccuracies and from changes in estimating procedures
6. Cost growth resulting from unforeseen problems

At a time when considerable cost growth is being experienced, budget allocations for many categories of systems are decreasing from year to year. The net result is that less money is available for acquiring and operating new systems and for maintaining and supporting systems that are already in being. The available funds for projects (i.e., buying power), when inflation and cost growth are considered, are declining.

---

[1]It has been noted that cost growth resulting from these various causes over the past several decades has increased well beyond the rate of inflation.

Design for Affordability (Life-Cycle Costing)

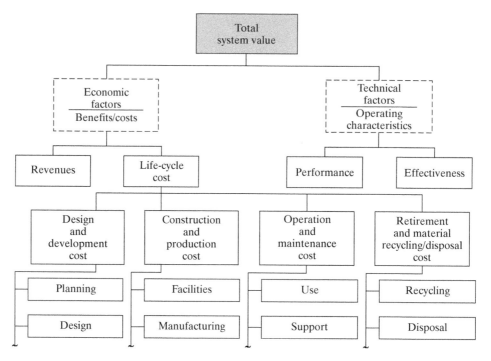

Figure 1 Economic and technical factors comprising total system value.

The current economic situation is further complicated by some additional problems related to the actual determination of system or product cost, such as follows:

1. Total system cost is often not visible, particularly those costs associated with system operation and support. The cost visibility problem can be called the "iceberg effect" illustrated in Figure 2. One must not only address system acquisition cost but other costs as well.
2. Economic factors are often improperly applied in estimating cost. For example, costs are identified and often included in the wrong category; variable costs are treated as fixed costs (and vice versa); indirect costs are treated as direct costs; and so on.
3. Accounting procedures do not always permit a realistic and timely assessment of total cost. In addition, it is often difficult (if not impossible) to determine costs on a functional basis.
4. Budgeting practices are often inflexible regarding the shift in funds from one category to another, or from year to year, shifts that could facilitate improvements in system acquisition and utilization.

The current trends of inflation and cost growth, combined with these additional problems, have caused inefficiencies in the utilization of scarce resources. Systems and products have been developed that are not as cost-effective as they might be. Further, it is anticipated that these conditions will worsen unless an increased degree of cost consciousness is applied in day-to-day activities.

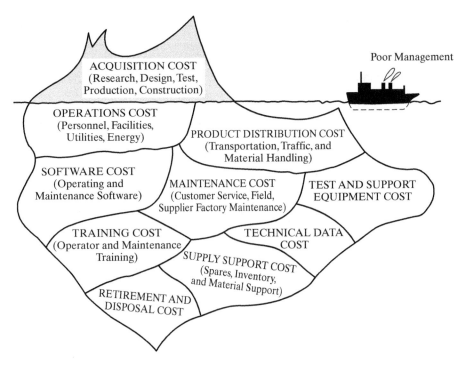

Figure 2 Visibility of the elements of total life-cycle cost.

Life-cycle cost refers to all costs associated with the system as applied to the defined life cycle. The life cycle, tailored to the specific system being addressed, forms the basis for life-cycle costing. In general, life-cycle cost includes the following:[2]

1. *Research and development cost.* Initial planning; market analysis; feasibility studies; product research; engineering design; design documentation; software; test and evaluation of engineering models; and associated management functions

2. *Production and construction cost.* Manufacturing engineering and operations analysis; manufacturing (fabrication, assembly, and test); facility construction; process development; production operations; quality control; and *initial* logistic support requirements (e.g., initial consumer support, the manufacture of spare parts, the production of test and support equipment, etc.)

3. *Operation and support cost.* Customer or user operations of the system in the field; product distribution (marketing and sales, transportation, and traffic management); and sustaining logistic support throughout the system or product life cycle (e.g., customer service, maintenance activities, supply support, test and support equipment, transportation and handling, technical data, facilities, system modifications, etc.)

[2]Describing the system/product life cycle may appear to be rather elementary, but experience indicates that many interpretations exist about what constitutes the life cycle. Because the description establishes the major reference point for life-cycle costing, it is essential that a *common understanding* be established as to what is meant by the life cycle and what is included (or excluded).

4. *Retirement and disposal cost.* Disposal of nonrepairable items throughout the life cycle; system/product retirement; material recycling; and applicable logistic support requirements

Life-cycle cost is determined by identifying the applicable functions in each phase of the life cycle, costing these functions, applying the appropriate costs by function on a year-to-year schedule, and then accumulating the costs for the entire life cycle. Life-cycle cost includes all producer, supplier, customer (user), maintainer, and related costs.[3]

2 COST CONSIDERATIONS OVER THE SYSTEM LIFE CYCLE

Experience indicates that a large percentage of the total cost for many systems is the direct result of the downstream activities associated with system operation and support, whereas the commitment of these costs is based on engineering and management decisions made in the early (conceptual and preliminary) design stages of the life cycle. The costs associated with the different phases of the life cycle are interrelated, and a decision made in any one phase can have an impact on other phases. Further, the costs associated with customer, producer, and supplier activities are all interrelated. Thus, in addressing the economic issues, one must look at total cost in the context of the overall life cycle, and particularly during the early stages of advance planning and conceptual design.

As is the case pertaining to the other characteristics of design discussed in the prior chapters, a cost target can be (1) initially specified for the system (as a "design-to-cost" (DTC) metric), (2) allocated to a lower-level function or element of the system, and (3) measured for the purpose of system evaluation as the life cycle unfolds. The emphasis should be on *total life-cycle cost* and not just individual cost elements. This is essential if the risks inherent in design and operating decisions are to be properly assessed on a continuous basis.[4]

Figure 3 shows a characteristic life-cycle cost commitment curve as caused by actions occurring during the various phases of the life cycle. As illustrated, more than half of the projected life-cycle cost is *committed* by the end of the system planning and conceptual design phase. However, actual project expenditures are relatively minimal at that point in time, but increase at an increasing rate. Reduction of the *gap* between cost committed and actual cost incurred is desired.

The initial step in life-cycle costing is to establish cost targets or goals; that is, one or more economic figures-of-merit to which the system and product should be designed, produced (or constructed), and supported for a designated interval of time. Second, these cost targets are then allocated to specific subsystems and system elements as design constraints or criteria. During the progression of design, feasible alternative configurations are evaluated in terms of compliance with the allocated targets, and a preferred approach is selected. As the system/product continues to evolve through various stages of development, life-cycle cost estimates are made and the results are compared against initially specified targets. Areas of noncompliance are noted and corrective action is initiated where appropriate. Cost emphasis throughout the system/product life cycle is highlighted in Figure 4 and discussed in the following sections.

[3]It should be realized that *all* life-cycle costs may be difficult (if not impossible) to predict or estimate. For instance, some indirect costs caused by the interaction effects of one system on another may be impossible to quantify. Thus, the emphasis should relate primarily to those costs that can be *directly* attributed to a given system or product. Also, an improvement is needed in the traceability of costs back to actual *causes*.

[4]Inherent within the conduct of a systems engineering program is the preparation and implementation of a *risk management plan*. The initial identification and assessment of risks, along with the subsequent actions required for risk abatement, are key factors to be addressed.

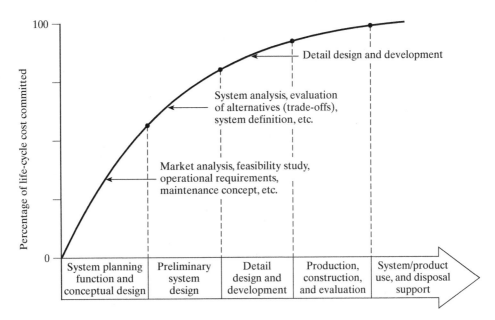

Figure 3 Activities affecting life-cycle cost.

2.1 Conceptual Design (Figure 4, Block 1)

It is well known that a major portion of the projected life-cycle cost for a given system is a consequence of decisions made during early planning and architecting as part of system conceptual design. These decisions deal with system operational requirements, performance and effectiveness factors, the maintenance concept, the system configuration (system packaging schemes and levels of diagnostics), production volumes, customer utilization factors, logistic support policies, disposal practices, and so on. Decisions made as a result of a market analysis or design evaluation serve to guide subsequent design and production activities, distribution functions, various aspects of sustaining system support, and phaseout. Accordingly, if ultimate life-cycle costs are to be minimized to achieve affordability, it is essential that a high degree of cost emphasis be applied in the early stages of system development.

In the early stages of planning and conceptual design, quantitative cost figures-of-merit should be established as requirements to which the system or product is to be designed, tested, produced (or constructed), operated, supported, and phased out. A *design-to-cost* TPM may be established as a system design requirement along with performance, effectiveness, reliability, maintainability, supportability, producibility, disposability, and so on. Cost must be addressed on a *proactive* basis rather than a *resultant* basis during the design and development process.

Design-to-cost figures-of-merit can be specified in terms of life-cycle cost at the system level. However, DTC requirements are sometimes established at lower levels to facilitate improved cost visibility and control throughout the life cycle. For example,

1. *Design to unit acquisition cost.* A factor that includes only research and development cost and production or construction cost.
2. *Design to unit operation and support cost.* A factor that includes only operation and maintenance support cost.

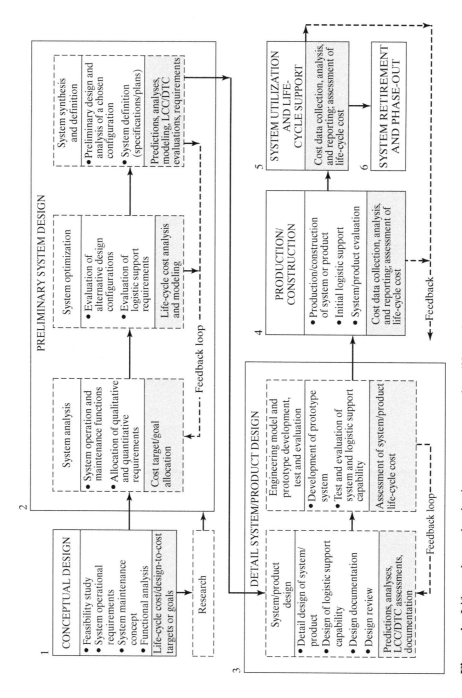

Figure 4 Life-cycle costing in the system or product life-cycle process.

When considering only a single segment of life-cycle cost, one must be certain that decisions are not based on that one segment alone, ignoring overall effects on total life-cycle cost. For example, one can propose a given design configuration on the basis of a low unit acquisition cost, but the projected operation and support cost (and resulting life-cycle cost) for that configuration may be considerably higher than necessary. Ideally, acquisition cost should not be addressed without considering operation and support cost (and vice versa). Both segments of cost must be viewed in terms of the life cycle.

2.2 Preliminary System Design (Figure 4, Block 2)

With the cost requirements established quantitatively, the next step involves an iterative process of synthesis, optimization, and system/product definition. The criteria defined in Block 1 are initially allocated, or apportioned, to various segments of the system to establish guidelines for the design or procurement of the needed element(s).

Allocation is accomplished from the system level down to the level necessary to provide adequate technical and cost control. The factors projected reflect the target cost per individual unit (i.e., a single system or one product in a population) and are based on system operational requirements, the system maintenance concept, and so on.

As the design process evolves, various approaches are considered in arriving at a preferred system configuration. Life-cycle cost analyses are accomplished in evaluating each feasible candidate with the objective of (1) ensuring that the candidate selected is compatible with the established cost targets, and (2) determining which of the various candidates being considered is preferred from an overall cost-effectiveness standpoint. Numerous trade-off studies are accomplished using life-cycle cost analysis as an evaluation criterion, until a preferred design configuration is chosen. Areas of compliance are justified, and noncompliant approaches are discarded. This is an iterative process, inherent within the overall morphology, with the necessary feedback and corrective action illustrated by Block 2 of Figure 4.

2.3 Detail System Design (Figure 4, Block 3)

As the system or product design is further refined, better design data become available. The life-cycle cost analysis is based on evaluation of the specific design characteristics (as reflected by design documentation and insight into engineering or prototype models), the prediction of cost-generating factors, the estimation of costs, and the projection of life-cycle cost as a cost profile. The results are compared with the initial requirement and corrective action is taken as necessary.

As before, this is an iterative process. But it applies at a lower level than that accomplished during preliminary system design. Here, it is the individual components comprising the system that are of interest.

2.4 Production/Construction, Utilization, Support, Retirement, and Disposal (Figure 4, Blocks 4–6)

Cost concerns in the latter stages of the system/product life cycle require a data collection, analysis, and assessment function. Through proper assessment, the high-cost contributors are to be identified and given attention. It is hoped that valuable information is gained and used for the purposes of product improvement as well as for the development of valid historical records useable for future application in similar situations.

As the life cycle unfolds, cost is employed as a major parameter in the evaluation of alternative design configurations and in the selection of a preferred approach. Subsequently, cost data are generated based on established design and production characteristics and used in the development of life-cycle cost projections. These projections are then compared with the initial requirements to determine the degree of compliance as well as the necessity for corrective action. In practice, life-cycle costing evolves from a series of rough estimates to a relatively refined process over time.

3 A GENERIC LIFE-CYCLE COSTING PROCESS

Inherent within the activities identified in Figure 4 are different and varied design and management decisions that, when made, and can have a significant impact on life-cycle cost. In particular, early system-level decisions made during the conceptual and preliminary design phases will have a significant influence on the activities and associated costs of system operations, maintenance and support, and retirement and material disposal (refer to Figure 3). Thus, it is essential that designers, support personnel, managers, and others engaged in decision making consider and evaluate the anticipated impact of their day-to-day actions on total life-cycle cost.

In conducting a life-cycle cost analysis, there are certain steps that the analyst should perform to obtain the desired overall outcome. For the purpose of guidance, the 12 steps outlined in Figure 5 will serve as a basis for discussion in this chapter.[5]

3.1 Define System Requirements and TPMs (Figure 5, Step 1)

Independent of the nature of the problem, an initial or first step is to define a baseline in terms of system expectations, operational requirements, the maintenance concept, the basic functions that the system must perform, and the applicable TPMs. This provides the overall framework within which the problem can be defined. The need for defining operational requirements and the maintenance concept is particularly important and a necessary input in determining the downstream costs associated with activities in the system utilization and sustaining support phase of the life cycle.

The maintenance concept identifies the functions that are anticipated for each level of maintenance, the effectiveness requirements in terms of maintenance frequency and times, and the major elements of logistic support to include personnel skill levels, test and support equipment, supply support requirements, and facilities. This information is not only required as an input to the system design process but also serves as the basis for determining operation and support costs.

3.2 Describe the System Life Cycle and Identify Activities by Phase (Figure 5, Step 2)

Given the operational requirements as a baseline as defined in Step 1, the next step is to describe the system life cycle and to identify the major activities in each phase. In general, all systems experience the major activities and cost categories of *research and development, production and construction, operation and support, and phaseout and disposal.* These costs and the associated activities are described more fully in Section 1.[6] Because these activities

[5]The specific steps in accomplishing a life-cycle cost analysis (and the depth of that analysis) may vary somewhat from one application to the next and must be tailored to the problem at hand. The approach presented in Figure 5 is offered to provide a basic understanding of the overall process and to identify items that need to be considered in performing the analysis.

[6]It is not uncommon for changes to occur in the projected life cycle, distribution rates, inventory profiles, and so on. However, the analyst needs to make some initial assumptions at this point and then account for possible variations by performing a sensitivity analysis.

Design for Affordability (Life-Cycle Costing)

1. ***Define system requirements and TPMs.*** Define operational requirements and the maintenance concept. Identify applicable technical performance measures (TPMs) and describe the system in functional terms, using a functional analysis at the system level.

2. ***Specify the system life cycle and identify activities by phase.*** Establish a baseline for the development of a cost breakdown structure (CBS) and for the estimation of cost for each year of the projected life cycle. Be sure all life-cycle activities are included.

3. ***Develop a cost breakdown structure.*** Provide a top-down/bottom-up cost structure. Include all categories for the initial allocation of costs (top-down) and the subsequent collection and summary of costs (bottom-up).

4. ***Identify input data requirements.*** Identify all input data requirements and all possible sources of input data. The type and amount of data will depend on the nature of the problem, the phase of the life cycle, and the depth of analysis.

5. ***Establish costs for each category in the CBS.*** Develop the appropriate cost-estimating relationships and estimate the costs for each category in the CBS on a year-by-year basis over the life cycle. Be sure all costs are included.

6. ***Select a cost model for analysis and evaluation.*** Select (or develop) a mathematical or computer-based model to facilitate the life-cycle costing process. The model must be valid for and sensitive to the specific system being evaluated.

7. ***Develop a cost profile and summary.*** Construct a cost profile showing the flow of costs over the life cycle. Provide a summary identifying the cost for each category in the CBS and calculate the percentage contribution in terms of the total.

8. ***Identify high-cost contributors and establish cause-and-effect relationships.*** Highlight those functions, system elements, or segments of processes that should be investigated for possible opportunities for design improvement and/or cost reduction.

9. ***Conduct a sensitivity analysis.*** Evaluate the model input–output data relationships and the results of the baseline analysis to ensure that the overall LCC analysis process is valid and that the model itself is well constructed and sensitive.

10. ***Identify priorities for problem resolution.*** Construct a Pareto diagram and conduct a Pareto analysis to identify priorities for problem resolution (i.e., those problems that are most important to remove in terms of their impact on value).

11. ***Identify additional alternatives.*** After developing an approach for the LCC evaluation of a given baseline configuration, it is then appropriate to extend the LCC analysis to the evaluation of multiple design alternatives.

12. ***Evaluate feasible alternatives and select a preferred approach.*** Develop a cost profile for each feasible design alternative, compare the alternatives equivalently, perform a break-even analysis, and select a preferred design approach.

Figure 5 Twelve basic steps in the life-cycle cost analysis process.

and costs are composed of many subordinate activities and costs, there exists a hierarchy of cost elements, known as a cost breakdown structure (CBS), that constitutes Step 3.

3.3 Develop a Cost Breakdown Structure (Figure 5, Step 3)[7]

Referring to Figure 1, the cost side of the balance should be broken down into specific cost categories. Figure 6 shows a generic CBS, which may be used as a frame of reference.

The cost breakdown structure links objectives and activities with resources and constitutes a logical subdivision of cost by functional activity area, major element of a system, and one or more discrete classes of common or like items. The cost breakdown structure, which is usually adapted or tailored to meet the needs of each individual program, should exhibit the following characteristics:

1. All life-cycle costs should be considered and identified in the CBS. This includes research and development cost, production and construction cost, operation and system support cost, and retirement and disposal cost.[8]

2. Cost categories are generally identified with a significant function, level of activity, or major item of material. Cost categories in the CBS must be well defined, and managers, engineers, accountants, and others must have the same understanding of what is included in a given cost category and what is not included.[9]

3. Costs must be broken down to the level necessary to provide management with the visibility required in evaluating various facets of system design and development, production, operational use, and support. Management must be able to identify high-cost areas and cause-and-effect relationships.

4. The CBS and the categories defined should be coded in a manner to facilitate the analysis of specific areas of interest while virtually ignoring other areas. For example, the analyst may wish to investigate supply support costs as a function of engineering design or distribution costs as a function of manufacturing, independent of other aspects of the system.

5. The CBS and the categories defined should be coded in such a manner as to enable the separation of producer costs, supplier costs, and consumer costs in an expeditious manner.

6. When related to a particular program, the cost structure should be directly compatible (through cross-indexing, coding, etc.) with planning documentation, the work breakdown structure (WBS), work packages, the organization structure, PERT and PERT-COST scheduling networks, Gantt charts, and so on. Costs that are reported through various management information systems must be compatible and consistent with those comparable cost factors in the CBS.

---

[7]In some instances, a *summary* work breakdown structure may be equated to a cost breakdown structure as long as *all* life-cycle activities, material requirements, and so on are included. On programs where a *contract* WBS (CWBS) is negotiated, only those work units that apply to the contract are included, and the other areas of activity are not addressed. In these instances, although there may be some similarities between the CWBS and the CBS, the CBS prevails because it addresses the total spectrum of cost.

[8]This does not imply that all cost categories are relevant to all analyses. The objective is to include all life-cycle costs and then identify those categories that are considered significant relative to the problem under consideration.

[9]On completion of an LCC analysis, all costs within the CBS should be *traceable* back to the applicable functional blocks identified through the functional analysis. The objective is to determine the costs associated with the accomplishment of a given *function*.

Design for Affordability (Life-Cycle Costing)

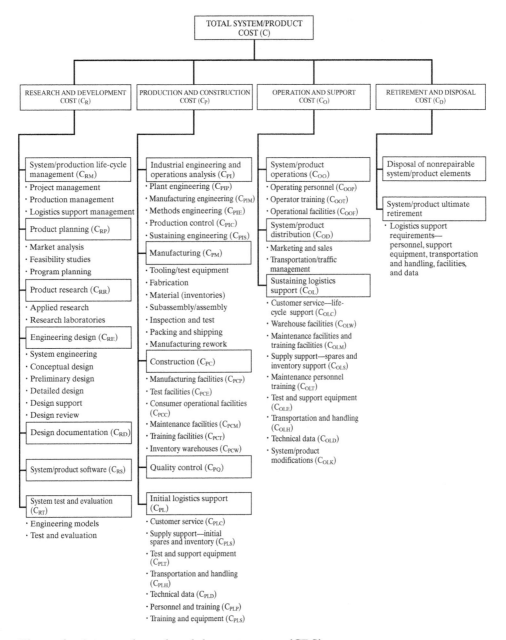

Figure 6   A general cost breakdown structure (CBS).

A CBS constitutes a *functional* breakdown of costs over the life cycle. It involves *all costs* related to customer, contractor, supplier, and user activities over the entire life cycle. Variable and fixed costs, direct and indirect costs, recurring and nonrecurring costs, inflationary and other cost-growth factors, and so on must be included. It provides the necessary *visibility* to the depth required for engineering and management decisions. If such visibility is not apparent, the applicable categories in the CBS may be extended downward as required (or summarized upward if such visibility is not required). Finally, the CBS may be utilized initially to facilitate a top-down allocation of costs stemming from a given design-to-cost requirement and later as a mechanism for a bottom-up summarization of cost for the purposes of evaluation (refer to Section 2).

Referring to Figure 6, the cost categories identified are too broad to ensure any degree of traceability, accountability, or control. The analyst cannot readily determine what is and what is not included, nor can he or she validate that the proper parameter relationships have been used in determining the specific cost factors that are inputs into the cost breakdown structure. Thus, the analyst requires a much more in-depth description of each cost category, the methods for calculating the costs in each category, and the assumptions upon which the costs are based. Accordingly, Figure 7 presents an example of a description of three of the categories identified in the CBS.[10]

3.4 Identify Data Input Requirements (Figure 5, Step 4)

While the initial presumption is that the completion of a life-cycle cost analysis requires considerable input data, the actual requirements depend on the phase in which the analysis is accomplished and the depth of the analysis performed. An LCC analysis can be accomplished at the system (or subsystem) level during conceptual design with very little actual input data. The analyst, using past experience and intuition, can accomplish his or her objective by making rough estimates with enough accuracy to adequately support the top-level design decisions being made at this point. A thorough understanding of the LCC analysis process is needed (i.e., the steps involved and what to look for), some knowledge pertaining to how the system will be operated and maintained by the customer in the field, a feeling for some of the major interrelationships between activities and costs, and a knowledge of applicable cost-estimating relationships (CERs). Accordingly, an analyst who has had some past experience in completing LCC analyses and who has a good understanding of system requirements should be able to complete the necessary task successfully in a timely manner.[11]

As one progresses through the life cycle and the system configuration becomes better defined, an LCC analysis may be accomplished to a greater depth (Figure 4, Blocks 2 and 3). Referring to the CBS in Figure 6, the analyst may proceed with an estimation of costs for the various categories shown, pushing down to the depth required to provide the desired visibility. In areas where the costs appear to be high, a greater depth may be required to identify the high-cost "drivers." For example, if the system in question is *transportation intensive*, then it may be necessary to expand Category C_{OD} to a lower level. This often involves an iterative process of analysis, feedback, more in-depth analysis, and so on. This will ultimately dictate the amount of input data required.

As the life-cycle cost analysis effort becomes more complex, the data-input requirements expand. For example, in conducting an evaluation of an existing system configuration currently in use (where the objective is to identify the highest contributors, determine

---

[10]An example of a CBS, with the cost categories described, along with the methods for calculating the costs in each category, is presented in W. J. Fabrycky and B. S. Blanchard, *Life-Cycle Cost and Economic Analysis* (Upper Saddle River, NJ: Prentice Hall, Inc., 1991), Appendix B.

[11]In conducting an LCC analysis, there is a tendency to begin by requesting data from various places in the organization (e.g., design data, reliability and maintainability data, production data, logistics data, etc.) This can result in too much data. The results can be costly. Further, the responsible design engineer/manager who requested that the LCC analysis be accomplished in the first place needs feedback as soon as possible. Thus, it is important to first understand the problem, know what input data are required, and complete the analysis in a timely manner.

Cost Category (Reference Figure 17.10)	Method of Determination (Quantitative Expression)	Cost Category Description and Justification
Spare/repair parts cost (C_{OLS})	$C_{OLS} = [C_{SO} + C_{SI} + C_{SD} + C_{SS} + C_{SC}]$ C_{SO} = Cost of organizational spare/repair parts C_{SI} = Cost of intermediate spare/repair parts C_{SD} = Cost of depot spare/repair parts C_{SS} = Cost of supplier spare/repair parts C_{SC} = Cost of consumables $C_{SO} = \sum_{N_{MS}} \left[(C_A)(Q_A) + \sum_{i=1} (C_{Mi})(Q_{Mi}) \right. $ $\left. + \sum_{i=1} (C_{Hi})(Q_{Hi}) \right]$ C_A = Average cost of material purchase order ($/order) Q_A = Quantity of purchase orders C_M = Cost of spare item i Q_M = Quantity of i items required or demand C_H = Cost of maintaining spare item i in the inventory ($/$ value of the inventory) Q_H = Quantity of i items in the inventory N_{MS} = Number of maintenance sites C_{SI}, C_{SD}, and C_{SS} are determined in a similar manner.	Initial spare/repair part costs are covered in C_{PL}. This category includes all replenishment spare/repair parts and consumable materials (e.g., oil, lubricants, fuel, etc.) that are required to support maintenance activities associated with prime equipment, operational support and handling equipment, test and support equipment, and training equipment at each level (organizational, intermediate, depot, supplier). This category covers the cost of purchasing; the actual cost of the material itself; and the cost of holding or maintaining items in the inventory. Costs are assigned to the applicable level of maintenance. Specific quantitative requirements for spares (Q_M) are derived from the Supportability Analysis (SA) discussed in Chapter 15. These requirements are based on the criteria described in Chapter 3. The optimum quantity of purchase orders (Q_A) is based on the EOQ criteria described in Section 15.3. Support equipment spares are based on the same criteria used in determining spare-part requirements for prime equipment.
Maintenance facilities cost C_{OLM}	$C_{OLM} = [(C_{PPM} + C_U) \times (\% \text{ allocation})(N_{MS})]$ C_{PPM} = Cost of maintenance facility support ($/site) C_U = Cost of utilities ($/site) N_{MS} = Number of maintenance sites *Alternate approach* $C_{OMF} = [(C_{PPO})(N_{MS})(S_O)]$ C_{PPO} = Cost of maintenance facility space ($/square foot/site). Utility cost allocation is included. S_O = Facility space requirements (square feet) Determine C_{OMF} for each appropriate echelon of maintenance.	Initial acquisition (construction) cost for maintenance facilities is included in C_{PCM}. This category covers the annual recurring costs associated with the occupancy and support (repair, modification, paint, etc.) of maintenance shops at all level throughout the system life cycle. On some occasions, a given maintenance shop will support more than one system, and in such cases, associated costs are allocated proportionately to each system concerned.
Engineering design (C_{RE})	$C_{RE} = \sum_{i=1}^{N} C_{RE_i}$ C_{RE_i} = Cost of specific design activity i N = Number of design activities	Includes all initial design effort associated with system/equipment definition and development. Specific areas include system engineering; design engineering (electrical, mechanical, drafting); reliability, maintainability, and human factors (Chapters 12, 13, and 14); functional analysis and allocation (Chapter 4); supportability analysis (Chapter 15); components; producibility; standardization; safety; etc. Design modifications are covered in C_{OLK}.

Figure 7 Sample breakout of cost categories and estimating relationships.

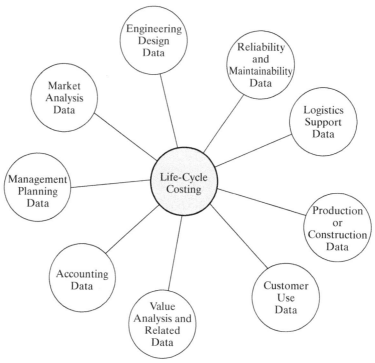

Figure 8 Possible sources of data for a life-cycle cost analysis.

cause-and-effect-relationships, and provide specific recommendations for design improvement), it may be appropriate to solicit data from diverse sources such as those in Figure 8, accomplish an in-depth analysis, study the interrelationships between system elements/activities and costs, and develop recommendations through a *continuous product/process improvement effort.*

3.5 Establish the Costs for Each Category in the CBS (Figure 5, Step 5)

In conducting an LCC analysis in the early phases of a project (where the greatest impact on life-cycle cost can be realized), available input data may be limited because of the lack of detail design definition. Thus, the cost analyst must rely primarily on the use of various cost-estimating techniques in the development of cost data, as is shown in Figure 9.

As the system design progresses, more complete design information becomes available and the analyst is able to develop cost estimates by comparing the characteristics of the new system with similar systems where historical data are recorded. The generation of cost data is based on *analogous* estimating methods. Finally, as the system design configuration becomes firm, design data (to include drawings and layouts, parts and material lists, specifications, predictions, etc., as conveyed in Figure 8) are produced that will enable the development of good engineering and manufacturing cost estimates. Further, the results from available reliability, maintainability, supportability, and disposability analyses can be used to aid in the prediction of operation, support, retirement, and material disposal/recycling costs.

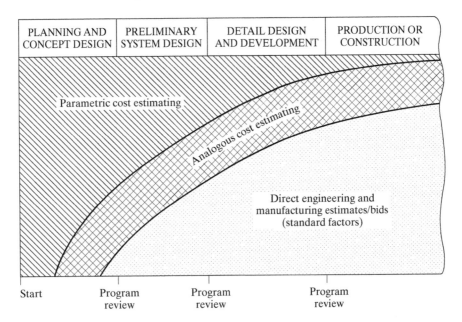

Design for Affordability (Life-Cycle Costing)

| PLANNING AND CONCEPT DESIGN | PRELIMINARY SYSTEM DESIGN | DETAIL DESIGN AND DEVELOPMENT | PRODUCTION OR CONSTRUCTION |

Parametric cost estimating

Analogous cost estimating

Direct engineering and manufacturing estimates/bids (standard factors)

Start   Program review   Program review   Program review

Figure 9   Cost estimation by program phase.

Parameter-Based Costing.   Referring to Figure 9, the most general approach for cost estimating during early program phases, where good input data are scarce, is by the use of *parametric* methods. Parametric cost-estimating relationships are basically "rules of thumb," developed from past experience, which relate various cost factors to explanatory variables of one form or another. These explanatory variables usually reflect characteristics of the system such as performance, effectiveness factors, physical features, or even other cost elements.

For example, from past experience a cost factor may be specified in terms of a unit of speed for a vehicle, a mile of range for a radar system, a unit of weight per mile for a transportation system, a unit of space for a facility, a unit of volume for materials or liquids, or some cost-to-cost factor. Cost-estimating relationships may take different forms (i.e., continuous or discontinuous, mathematical or nonmathematical, linear or nonlinear, etc.) and should relate to the specified TPMs for a given system.

Activity-Based Costing (ABC).   To be effective in total cost management (and in the accomplishment of cost-effectiveness analyses) requires complete cost visibility allowing for the traceability of all costs back to the activities, processes, or products that generate them. In traditional accounting structures employed in most organizations, a large percentage of the total cost cannot be traced back to "causes." For example, "overhead" or "indirect" costs that often constitute greater than 50% of the total include management costs, supporting organization costs, and other costs that are difficult to trace and assign to specific objects. With these costs being allocated across the board, it is impossible to identify the actual "causes" and to pinpoint the *true* high-cost contributors. As a result, the concept of *activity-based costing* has been introduced in recent years.[12]

[12] Activity-based costing is covered further in J. R. Canada, W. G. Sullivan, D.J. Kulonda, and J. A. White, *Capital Investment Analysis for Engineering and Management*, 3rd ed. (Upper Saddle River, NJ: Prentice Hall, Inc., 2005); and P. T. Kidd, *Agile Manufacturing: Forging New Frontiers* (Reading, MA: Addison-Wesley, 1994).

Activity-based costing is a methodology directed toward the detailing and assignment of costs to the activities that cause them to occur. The objective is to enable the "traceability" of *all* applicable costs to the process or product that generates them. The ABC approach allows for the initial allocation and later assessment of costs by function, and was developed to deal with the shortcomings of the traditional management accounting structure where large overhead factors are assigned to all elements of the enterprise across the board without concern for whether they directly apply or not. More specifically, the principles of ABC are as follows:

1. Costs are directly traceable to the applicable cost-generating process, product, or related object. Cause-and-effect relationships are established between a cost factor and a specific process or activity.

2. There is no distinction between direct and indirect (or overhead) costs. Although 80% to 90% of all costs are traceable, those nontraceable costs are not allocated across the board, but are allocated directly to the organizational unit(s) involved in the project.

3. Costs can be easily allocated on a *functional* basis. It is relatively easy to develop cost-estimating relationships in terms of the cost of activities per some activity measure (i.e., the cost per unit output).

4. The emphasis in ABC is on "resource consumption" (versus "spending"). Processes and products consume activities, and activities consume resources. With resource consumption being the objective, the ABC approach facilitates the evaluation of day-to-day decisions in terms of their impact on resource consumption downstream.

5. The ABC approach fosters the establishment of "cause-and-effect" relationships and, as such, enables the identification of the "high-cost contributors." Areas of risk can be identified with some specific activity and the decisions that are being made within.

6. The ABC approach tends to eliminate some of the cost doubling (or double counting) that occurs when attempting to differentiate on what should be included as a direct cost or as an indirect cost. By not having the necessary visibility, there is the potential of including the same costs in both categories.

Developing Cost Data. To develop cost data for a life-cycle cost analysis, the analyst should initially investigate all possible data sources to determine what is available for direct application in support of analysis objectives. If the required data are not available, the use of parametric cost-estimating techniques may be appropriate. However, one should first determine what can be derived from existing data banks, initial system planning data, supplier documentation, reliability and maintainability predictions, supportability analyses, test data, field data, and so on. Some of these data sources are discussed here:

1. *Existing data banks.* Actual historical information on existing systems, similar in configuration and function to the item(s) being developed, may be used when applicable. Often it is feasible to employ such data and apply adjustment factors as necessary to compensate for any differences in technology, configuration, projected operational environment, and time frame. Included in this category of existing data are standard cost factors that have been derived from historical experience that can be applied to specific functions or activities. Standard cost factors may cover such areas as the following:
   (a) The cost of engineering labor—dollars per labor hour for the principal engineer, senior engineer, technician, and so on

(b) The cost of manufacturing labor by classification—dollars per labor hour per classification
(c) Overhead rate—dollars per direct labor cost (or percentage)
(d) Training cost—dollars per student week
(e) Shipping cost—dollars per pound per mile
(f) The cost of fuel—dollars per gallon
(g) The cost of maintaining inventory—percent of the inventory value per year
(h) The cost of facilities—dollars per cubic foot of occupancy
(i) The cost of material x—dollars per pound or per foot.

These and comparable factors, where actual quantitative values can be directly applied, are usually established from known rates and costs in the marketplace and are a direct input to the analysis. However, care must be exercised to ensure that the necessary inflationary and deflationary adjustments are incorporated on a year-to-year basis.

2. *Advance system/product planning data.* Advanced planning data for the system or product being evaluated usually include market analysis data, definition of system operational requirements and the maintenance concept, the results of technical feasibility studies, and program management data. The cost analyst needs information pertaining to the proposed physical configuration and major performance features of the system, the anticipated mission to be performed and associated utilization factors, system effectiveness parameters, the geographical location and environmental aspects of the system, the maintenance concept and logistic support philosophy, and so on. This information serves as the baseline from which all subsequent program activities evolve. If the basic information is not available, the analyst must make some assumptions and proceed accordingly. These assumptions must then be thoroughly documented.

3. *Individual cost estimates, predictions, and analyses.* Throughout the early phases of a program, cost estimates are usually generated on a somewhat continuing basis. These estimates may cover research and development activities, production or construction activities, or system operating and support activities. Research and development activities, which are basically nonrecurring in nature, are usually covered by initial engineering cost estimates or by cost-to-complete projections. Such projections primarily reflect labor costs and include inflationary factors, cost growth resulting from design changes, and so on.

Production cost estimates are often presented in terms of both nonrecurring costs and recurring costs. Nonrecurring costs are handled in a manner similar to research and development costs. Conversely, recurring costs are frequently based on individual manufacturing cost standards, value engineering data, industrial engineering standards, and so on. Quite often, the individual standard cost factors that are used in estimating recurring manufacturing costs are documented separately and are revised periodically to reflect labor and material inflationary effects, supplier price changes, effects of learning, and so on.

System operating and support costs are based on the projected activities throughout the operational use and support phase of the life cycle and are usually the most difficult to estimate. Operating costs are a function of system or product mission requirements and utilization factors. Support costs are basically a function of the inherent reliability and maintainability characteristics in the system design and the logistics requirements necessary to support all scheduled and unscheduled maintenance actions throughout the programmed life cycle. Logistic support requirements include maintenance personnel and training, supply support (spares, repair parts, and inventories), test and support equipment, transportation and handling, facilities, and certain facets of technical data. Thus, individual operation and support cost estimates are based on the predicted frequency of maintenance or the mean of the maintenance (MTBM) factor and on the logistic support resources required when maintenance actions occur. These costs are derived from reliability and maintainability prediction data, supportability analysis (SA) data, and other supporting information, all of which are based on system/product engineering design data.

4. *Supplier documentation.* Proposals, catalogs, design data, and reports covering special studies conducted by suppliers (or potential suppliers) may be used as a data source when appropriate. Quite often, major elements of a system are either procured off the shelf or developed through a subcontracting arrangement of some type. Various potential suppliers will submit proposals for consideration, and these proposals may include not only acquisition cost factors but (in some instances) also life-cycle cost projections. If supplier cost data are used, the cost analyst must become completely knowledgeable as to what is and is not included. Omissions or the double counting of costs must not occur.

5. *Engineering test and field data.* During the latter phases of system development and production and when the system or product is being tested or is in operational use, the experience gained represents the best source of data for actual analysis and assessment purposes. Such data are collected and used as an input to the life-cycle cost analysis. Also, field data are used to the extent possible in assessing the life-cycle cost impact that may result from any proposed modifications on prime equipment, software, or the elements of logistic support.

These five main sources of data identified above for life-cycle costing purposes are presented in a summary manner to provide an overview as to what the cost analyst should look for. In pursuing the data requirements further, the analyst will find that a great deal of experience has been gained in determining research and development and production/construction costs. However, few historical cost data are currently available in the operations and support area. Accounting for operation and support costs has been lacking in the past, but this situation should ultimately rectify itself as the emphasis on life-cycle costing continues to increase.[13]

3.6 Select a Life-Cycle Cost Modeling Paradigm (Figure 5, Step 6)

After the establishment of the cost breakdown structure, it is necessary to select and adapt a model of some type to facilitate the life-cycle cost evaluation process. The model may be a set of computer subroutines or spreadsheets or a series of mathematical expressions, depending on the phase of the system life cycle and the nature of the problem at hand. But at the generic level, the model will be based on money flow modeling or on economic optimization modeling.

---

[13]The subject of cost estimation is a significant area that should be investigated further. Refer to P. F. Ostwald, *Engineering Cost Estimating*, 3rd ed. (Upper Saddle River, NJ: Prentice Hall, Inc., 1992); R. D. Stewart and R. M. Wyskida, *Cost Estimator's Reference Manual*, 2nd ed. (New York: John Wiley & Sons, Inc., 1995); and G. J. Thuesen and W. J. Fabrycky, *Engineering Economy*, 9th ed. (Upper Saddle River, NJ: Prentice Hall, Inc., 2001).

Regardless of the costing model and paradigm chosen, life-cycle costing itself includes a compilation of a variety of cost factors, reflecting the many different types of activities as indicated by the CBS. The objective in using a model is to evaluate a system in terms of total life-cycle cost, as well as the various individual segments of cost. Computer-based subroutines that reflect segments of cost may be structured differently depending on the system element covered. For instance, supply support costs may be compiled through the application of spare- and repair-part demand factors and inventory techniques. This, in turn, may require factors derived through the use of a reliability model. On the other hand, engineering design costs may be extracted directly from a proposal or a set of engineering cost projections.

3.7 Develop a Cost Profile and Summary (Figure 5, Step 7)

In developing a cost profile, there are different approaches that may be followed. However, the following steps are suggested as a place to begin:

1. Identify all activities throughout the life cycle that will generate costs in one form or another. This includes functions associated with planning, research and development, test and evaluation, production or construction, distribution, systems use and support, and retirement and material disposal (refer to Figure 4).

2. Relate each activity identified in item 1 to a specific function (refer to functional analysis), and to a specific cost category in the cost breakdown structure (refer to Figures 1 and 6). All functions and program activities should fall in one or more categories in the CBS, and there should be a traceability of requirements from the functional analysis to the CBS.

3. Develop a format for the recording of costs for each activity in the CBS and for each year in the life cycle. Figure 10 shows a basic format that is often used in the presentation of costs at the top level.

4. Through application of the cost-estimating methods discussed earlier, compute the costs for each CBS category and for each year in the life cycle. Considerations pertaining to the effects of learning curves, cost growth because of changes in suppliers and the development of new contractual agreements, changes in procurement price levels, changes in reliability and maintainability factors that influence system operating and maintenance support costs, and so on should be addressed as appropriate. The analyst may commence with the prediction of costs for each year in terms of *constant* dollars (i.e., today's dollars) because it will allow for the direct comparison of various activity levels from year to year. Also, using constant dollars tends to ensure consistency in accomplishing comparative studies.

5. Within each category in the CBS, the individual cost elements are next projected to include the appropriate inflationary factors. The modified values constitute a new cost stream and reflect realistic costs as they are anticipated for each year in the future (i.e., expected 2012 costs in 2012, 2013 costs in 2013, etc.). This cost stream represents a *budgetary* profile, which may be used for management decision making.

6. Construct a third profile considering the "time value of money" or the presentation of costs in terms of equivalency such as the *present equivalent* (PE) amount. When evaluating two or more alternative system configurations, each will include different levels of activity conducted in different years, different design approaches, and different maintenance and support requirements. Although all of the options being considered may be in full compliance with the specified system requirements, no two alternatives

PROGRAM ACTIVITY	COST CATEGORY DESIGNATION	COST BY PROGRAM YEAR ($)													TOTAL COST (CONSTANT $)	TOTAL COST (ACTUAL $)	TOTAL COST (PE $)	% CONTRI-BUTION
		1	2	3	4	5	6	7	8	9	10	11	12	13				
Alternative A																		
1. Research and development	C_R																	
a. Life-cycle management	C_{RM}																	
b. Product planning	C_{RP}																	
(1) Feasibility studies																		
(2) Program planning																		
2.																		
3.																		
Others																		
Total cost (constant)																		
Total cost (actual)																		
Total cost (PE)																		
Alternative B																		
1. Research and development	C_R																		
2.																		

Figure 10 Cost collection worksheet.

will be identical in terms of component makeup, use of materials, and so on. When individual profiles for each alternative are to be compared, the comparison must be on an equivalent basis by incorporating the time value of money.

Determining Production Costs. Production costs depend on system operational requirements as they pertain to the buildup of items in the inventory; that is, the demand for the system and its elements by the customer. The production rate, as well as the number of items needed, will have an impact on unit manufacturing cost. Fundamentals of producibility and their cost implications are of concern during life-cycle costing.

A significant aspect of production and other repetitive activities, such as maintenance actions, is the effect of learning. Human and organizational learning serves to reduce the unit cost of producing subsequent items or of performing subsequent tasks. The place for the cost results in the CBS depends upon the phase of the life cycle where the activity is actually carried out.

Considering Inflation. When developing time-phased cost profiles, the reality of inflation should be considered for each future year in the life cycle. During the past several decades, inflation has been a significant factor in the rising costs of products and services and in the reduction of the purchasing power of the dollar. Inflation is a broad term covering the general increase(s) in the unit cost of an item or activity and is related primarily to labor and material costs, as follows:

1. Inflation factors applied to labor costs are due to salary and wage increases, cost of living increases, and increases in overhead rates due to the rising costs of personnel fringe benefits, retirement benefits, insurance, and so on. Inflation factors should be determined for different categories of labor (i.e., engineering labor, technician labor, manufacturing labor, construction labor, customer service personnel labor, management labor, etc.) and should be estimated for each year in the life cycle.

2. Inflation factors applied to material costs are due to material availability (or unavailability), supply and demand characteristics, the increased costs of material processing, and increases in material handling and transportation costs. Inflation factors will often vary with each type of material and should be estimated for each year in the life cycle.

Increasing costs of an inflationary nature often occur as a result of new contract provisions with suppliers, new labor agreements and union contracts, revisions in procurement policies, shifts in sources of supply, the introduction of engineering changes, program schedule shifts, changes in productivity levels, changes in item quantities, and for other comparable reasons. Also, inflation factors are influenced to some extent by geographical location and competition. When reviewing the various causes of inflation, one must be extremely careful to avoid overestimating and double counting for the effects of inflation. For instance, a supplier's proposal may include provisions for inflation, and unless this fact is noted, there is a chance that an additional factor for inflation will be included for the same reason.

Inflation factors should be estimated on a year-to-year basis if at all possible. Because inflation estimates may change considerably with general economic conditions at the national level, cost estimates far out in the future (i.e., 5 years and more) should be reviewed at least annually and adjusted as required. Inflation factors may be established by using price indices or by the application of a uniform escalation rate.

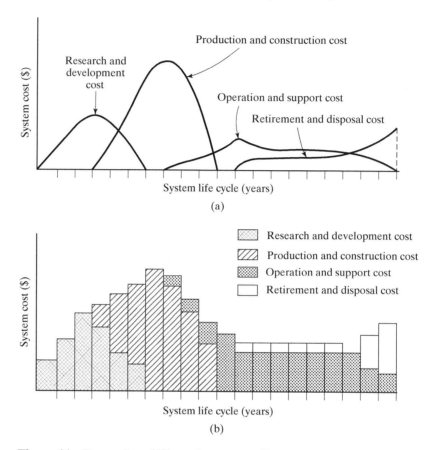

Design for Affordability (Life-Cycle Costing)

(a)

(b)

Figure 11 Examples of life-cycle cost profiles.

Displaying Life-Cycle Costs. Figure 11 shows the results from collecting and summarizing costs determined from guidance given in Figure 10. Cost streams may be plotted based on the actual inflated (or budgetary) costs projected through the life cycle. The important issue is to identify any areas (for any given category of the CBS) where there may be significant changes from 1 year to the next and to investigate the possible causes. Conversely, if the purpose is to compare alternative profiles, then equivalence calculations must be applied to the cost streams to determine the present equivalent amount to support a choice.

3.8 Identify High-Cost Contributors and Establish Cause-and-Effect Relationships (Figure 5, Step 8)

It is not unusual for certain categories of cost to stand out as *high-cost contributors*. The objective at this point is to be able to determine some of the causes for these high costs. For example, if 18% of the total projected life-cycle cost is in the category of "manufacturing–recurring cost," the analyst should refer back to the CBS and review the underlying assumptions. The question is one of determining if there are any assumptions that are questionable, or assumptions that may be considered invalid.

Assuming that the proper level of traceability is apparent, the analyst should then relate the applicable high-cost factors in the CBS back to the specific function(s) that is being performed, which causes this high cost. This may, in turn, lead to a particular element of the system that fails frequently and consumes considerable resources to maintain, a process that requires people with high skills to complete, or the identification of a method of transportation that is costly to operate. The objective is to identify specific design-related characteristics that tend to "drive" the costs of the system from a life-cycle perspective, whether it be the prime mission-related elements of the system or an element of the support infrastructure.

Having identified the causes, the question is: *Are there any alternative system design approaches that can be implemented that will allow for at least the same level of system effectiveness but at less overall cost?* Feasible candidate solutions should be evaluated in terms of life-cycle cost, and major areas of improvement may be realized through the modification process. This process of evaluation, identifying the high-cost contributors, determining the cause-and-effect relationships, and modifying the system for improvement can be accomplished on a continuing basis.

3.9 Conduct a Sensitivity Analysis (Figure 5, Step 9)

In the performance of a life-cycle cost analysis, there may be a few areas where the results might be "suspect" because of inadequate data input, poor prediction methods, invalid assumptions at the beginning, and so on. Identifying the high-cost areas provides a good place to start. There are some questions that need to be addressed: *How sensitive are the results of the analysis to possible variations of these uncertain input factors?* To what extent can certain input parameters be varied without causing the configuration being evaluated to be discarded for no longer being economically feasible?

To answer these questions, the analyst should select the high contributors (those that contribute more than 10% of the total cost), identify the critical input factors that directly impact cost (from Step 8), change some of these factors at the input stage, and determine the changes in cost at the output. Such input variations can be accomplished over a designated range, considering appropriate distributions, and so on. For example, key input parameters in the analysis of the communications system include the system *operating time* and the *MTBM*. Using the life-cycle cost model (from Step 6), the analyst may apply a multiple factor to the operating time and MTBM values and determine the delta cost associated with each variation. From this, trend curves are projected as an aid in generating what is hoped to be a better design alternative.

Through evaluation of the information presented, it might be evident that a *small* variation in operating time and MTBM may cause a significant delta increase in cost at the output. This provides an indication that there is a high degree of *risk* associated with making decisions based on the results.

On identifying the high-risk areas (as measured by the magnitude of the delta cost at the output stage), the analyst should make every effort to reduce this risk by improving the input data to the greatest extent possible. This may, in turn, lead to a revised MTBM prediction, a more in-depth description of system operational requirements, a more in-depth maintenance task analysis (MTA), and so on. The most significant challenge here is that the analyst needs to be familiar with the various parameter interrelationships, their impact on

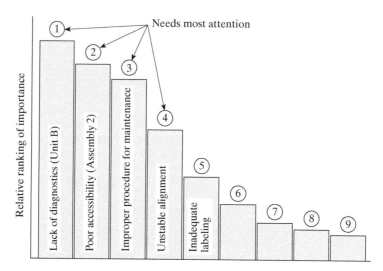

Figure 12   A Pareto ranking of major problem areas.

cost, the sensitivities of the model in the identification of cause-and-effect relationships, and the potential areas of risk associated with the LCC analysis results.

3.10 Construct a Pareto Diagram and Identify Priorities for Problem Resolution (Figure 5, Step 10)

The problem areas identified in Steps 8 and 9 should be evaluated and prioritized in terms of degree of importance. High-cost/high-risk areas should receive the most management attention and, as these are addressed, others may rise to the top. A Pareto diagram, illustrated in Figure 12, may be constructed to show the relative importance of different problem areas that (either directly or indirectly) cause high costs. "Importance" factors are based not only on high-cost areas but also on criticality as it pertains to the system and its ability to accomplish its designated mission.[14]

3.11 Identify Feasible Alternatives for Design Evaluation (Figure 5, Step 11)

Having acquired the necessary visibility relative to the most critical problems and their causes, the next step is to investigate possible alternative ways by which the applicable functions can be accomplished. Design trade-offs are conducted with the objective of selecting a few feasible candidates for further evaluation. The goal is to improve the balance illustrated in Figure 1 by continuing to meet the effectiveness requirements while, at the same time, reducing the overall life-cycle cost. Figure 13 shows a projected life-cycle cost profile for each of three candidate system design alternatives under consideration.

---

[14]The FMECA and the Ishikawa diagram can be used to help identify "cause-and-effect" relationships, and the results conveyed in Figure 12 should consider the results from the Pareto analysis. Although the examples in Figure 12 are primarily maintenance related, the results from such an analysis should identify areas of concern (e.g., a process-related problem, a system operational problem, an organizational problem, and so on).

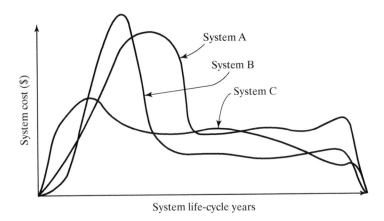

Design for Affordability (Life-Cycle Costing)

Figure 13 Alternative life-cycle cost profiles being considered for evaluation.

3.12 Evaluate Feasible Alternatives and Select Preferred Approach (Figure 5, Step 12)

The first 10 steps in Figure 5 address the life-cycle cost analysis approach as it applies to a single system configuration. Step 11 introduces the requirement to address alternatives and to evaluate each on an equivalent basis. It is now appropriate to extend the LCC analysis to the situation where multiple alternatives are being considered.

For each specific problem where there are possible alternative solutions and a decision is required to select a preferred approach, there is an overall analysis process usually followed, either intuitively or on a formal basis. This process is comparable with the generic analysis approach. Inherent within this process is the identification of the appropriate analytical techniques and tools and the selection of a modeling paradigm that is responsive to the need.

4 LIFE-CYCLE COSTING BY MONEY FLOW MODELING

The life-cycle costing process outlined in Section 3 will now be applied to a system example utilizing money flow modeling as the underlying evaluation paradigm. This hypothetical example is guided by and implements the 12 steps given in Figure 5.

4.1 Introduction to the Example[15]

Suppose that a metropolitan area authority plans to upgrade its capability to locate and communicate with mobile emergency and safety units in real time. The authority has decided

---

[15]A version of this example appeared in W. J. Fabrycky and B. S. Blanchard, *Life-Cycle Cost and Economic Analysis* (Upper Saddle River, NJ: Prentice Hall, 1991).

to procure a new generation of active equipment packages that will accomplish the functions of location and communications simultaneously. Deployment of the location/communications system (LCSys) will be done by the installation of packages in patrol aircraft, helicopters, ground vehicles, designated facilities, and a central control facility. In the product and system classifications, LCSys is a multiple entity population system.

LCSys is also a network centric system with associated maintenance and support facilities shown in Figure 14. There is a requirement for 80 package installations within the network. Two alternative LCSys configurations are being considered with the help of a communications consultant. The objective is to define system operational requirements, the

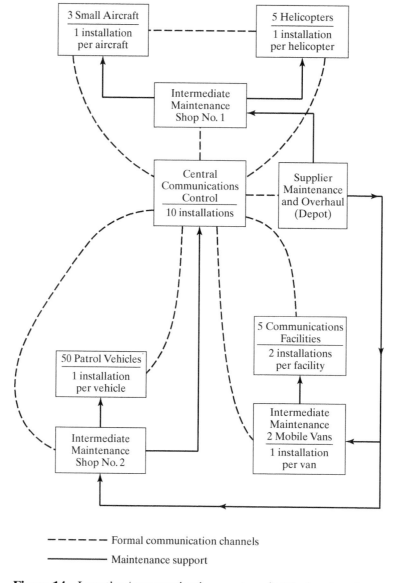

Figure 14 Location/communications system (LCSys) network.

maintenance concept, and program planning information so that the LCSys configuration with the lowest equivalent life-cycle cost can be specified for possible procurement.

LCSys Operational Objectives. A planning activity with the consultant has identified six operational objectives to be met by the LCSys. Although these objectives pertain largely to the equipment packages, the network system is composed of the packages collectively, the wireless links, and a maintenance support capability. The six objectives are as follows:

1. The equipment is to be installed in three low-flying light aircraft (one package per aircraft) to enable contact with loitering helicopters operating within a 200-mile radius and with the central control facility. It is anticipated that each aircraft will fly 15 times per month with an average duration of 3 hours. The equipment utilization requirement is 1.1 hours of operation for every hour of aircraft operation (air time plus ground time). The equipment packages must exhibit a MTBM of at least 480 hours and have a $\overline{M}ct$ that does not exceed 15 minutes.

2. The equipment is to be installed in each of five helicopters (one package per helicopter) to enable contact with patrol aircraft within a 200-mile range, with other helicopters within a 50-mile radius, and with the central control facility. It is anticipated that each helicopter will fly 25 times per month, with an average flight duration of 2 hours. The utilization requirement is 0.9 hour of equipment operation for every flight hour of helicopter operation. The equipment must meet a 500-hour MTBM requirement and have a $\overline{M}ct$ of 15 minutes or less.

3. The equipment is to be installed in each of 50 patrol vehicles (one package per vehicle) to enable contact with other vehicles within a range of 25 miles and with the central communications control facility from a range of 50 miles or less. Each vehicle will be in operation on the average of 5 hours per day, 5 days per week, and will be utilized 100% during that time. The required MTBM is 400 hours, and the $\overline{M}ct$ should be less than 30 minutes.

4. Two packages are to be installed in each of five fixed communication facilities strategically located throughout the metropolitan area. This will enable contact with patrol vehicles within a range of 25 miles and the central control facility from a range of up to 50 miles. Equipment utilization requirements are 120 hours per month, the required MTBM is 200 hours, and the $\overline{M}ct$ should not exceed 60 minutes.

5. Ten packages are to be installed in the central control facility to enable contact with patrol aircraft and loitering helicopters within a range of 400 miles, with the five fixed communication facilities within a range of 50 miles, and with patrol vehicles within a 50-mile radius. In addition, each package will be able to communicate with the intermediate maintenance facilities. The average equipment utilization requirement is 3 hours per day for 360 days per year. The MTBM requirement is 200 hours and the $\overline{M}ct$ should not exceed 45 minutes.

6. Two mobile vans will be used to support intermediate-level maintenance and operator assistance at the five fixed ground facilities. Each van will incorporate one package that will be used an average of 2 hours per day for a 360-day year. The required MTBM is 400 hours and the $\overline{M}ct$ is not to exceed 15 minutes.

There is no known system that will completely satisfy the need. However, there are two new candidate design configurations being offered by contract design firms that should suffice, provided that all design objectives can be met. The goal is to evaluate each configuration in terms of its performance and life-cycle cost and then to recommend a preferred LCSys configuration.

LCSys Performance Requirements. To achieve the stated operational objectives, there is a need to acquire LCSys packages that will meet certain performance and effectiveness requirements (e.g., location accuracy, voice transmission range, clarity of message, data transmission speed, etc.). Preliminary study of the aggregate performance metric determines that the second configuration (Alternative B) will likely exceed Alternative A when placed on a scale of 1–10. The minimum acceptable score for performance is set at 0.8. Further, budget limitations require that the cost of development and production (over the first 5 program years) not exceed $5,000,000 in inflated dollars.

Advanced program planning indicates that a full complement of packages must be in operation 5 years after the start of the program, and that this capability must be maintained through the 11th year. Significant program milestones and the projected number of packages in operational use are shown in Figure 15. This provides the basis for defining the program life cycle, the major life-cycle functions, and the life-cycle cost.

Prior to the identification of an applicable cost breakdown structure and the development of cost elements, the baseline configuration for the LCSys and its maintenance concept should be described in detail. To begin, an assumed packaging scheme is developed

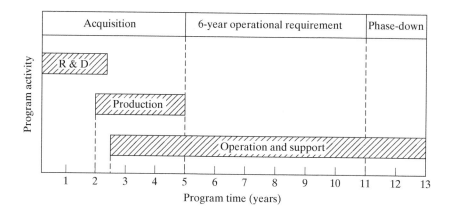

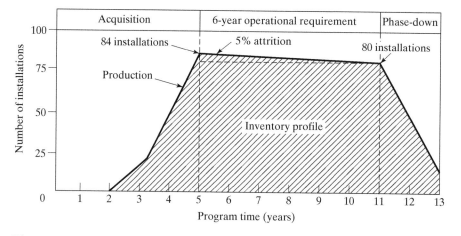

Figure 15 LCSys program plan and life-cycle profile.

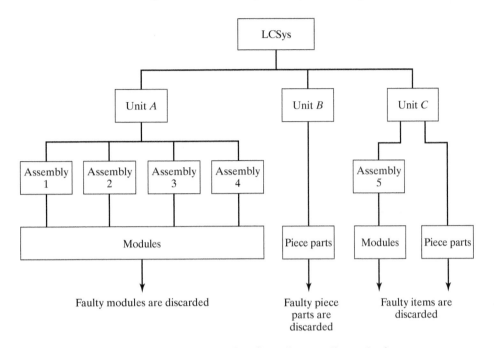

Figure 16 LCSys equipment packaging (baseline configuration).

as shown in Figure 16. This configuration (including units, assemblies, and modules) is developed from conceptual design considerations representative of each candidate being considered.[16]

LCSys Maintenance Concept. The maintenance concept for the system can be defined by statements and/or illustrations that include criteria addressing maintenance levels, support policies, effectiveness factors (e.g., maintenance time constraints, turnaround times, and transportation times), and basic logistic support requirements. The maintenance concept is an input to the system and package design. A detailed maintenance plan reflects the results of design and is used for the acquisition of logistics elements required for sustaining life-cycle support of the LCSys during its utilization phase. The maintenance concept is illustrated in Figure 17.[17]

There are three levels of maintenance under consideration. *Organizational maintenance* is performed in the aircraft or helicopters, in the patrol vehicles, or in the ground facilities by the user or operator. *Intermediate maintenance*, or the second level of maintenance, is accomplished in a remote technical facility by trained personnel having the skills necessary to perform the maintenance function. *Depot maintenance*, the highest level of maintenance,

---

[16]The configuration shown in Figure 16 does not represent a *final* design, but is close enough for life-cycle planning purposes (particularly for analyses accomplished in the early stages of the program). Further design definition will occur as the program progresses.

[17]The quantitative support factors presented in Figure 17 are minimum design requirements. For each alternative configuration being evaluated, the support factors may vary loosely based on specific design characteristics. However, the minimum requirements must be met.

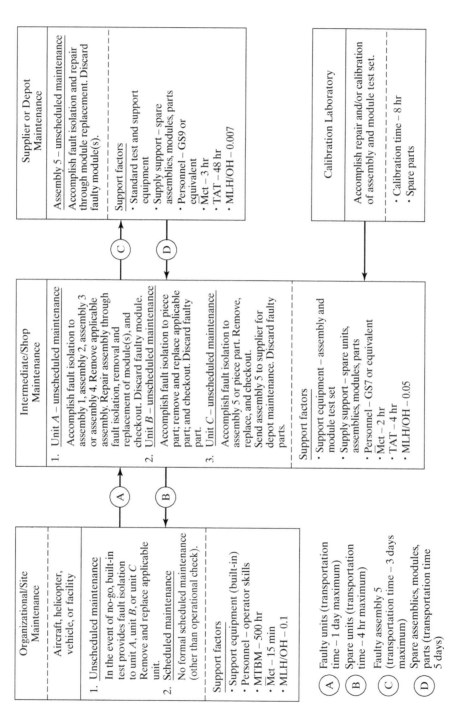

Figure 17 LCSys maintenance concept and repair policy.

involves the specialized repair of complex components at the supplier's facility by highly skilled supplier personnel.

The specific functions scheduled to be accomplished on the equipment at each level of maintenance are noted. In the event of a malfunction, fault isolation is performed to the applicable unit by using the built-in test capability (i.e., Unit A, B, or C). Units are removed and replaced at the organizational level and sent to the intermediate maintenance facility for corrective maintenance. At this facility, units are repaired through assembly and/or part replacement and assemblies are repaired through module replacement. In case of fault isolation to assembly 5 in Unit C, the faulty assembly is sent to the depot, where repair is accomplished through module replacement.

The maintenance concept identifies the functions that are anticipated for each level of maintenance, the effectiveness requirements in terms of maintenance frequency and times, and the major elements of logistics support that include personnel skill levels, test and support equipment needs, supply support requirements, and applicable facilities. This information is not only required as an input to the system design process but also serves as the basis for determining operation and support costs.

4.2 Life-Cycle Cost Analysis

With the problem (or need) defined and with a description of the LCSys operational requirements and maintenance concept completed, it is appropriate to proceed with the specific steps involved in the life-cycle cost analysis of each proposed configuration. These steps lead to the generation of cost data and are presented next.

Developing the Cost Breakdown Structure. The CBS assumed for this evaluation is the generic version presented in Figure 6. Although not all of the cost categories may be relevant or significant, the generic CBS serves as a good starting point. Initially, all costs must be considered, with the objective of concentrating on those categories that are high-cost contributors. The evaluation of two alternative LCSys package configurations and the selection of a preferred approach are required. There will be activities involving planning, management, engineering design, test and evaluation, production, distribution, operation, maintenance and logistic support, and equipment disposal. In an attempt to be more specific for the life-cycle cost analysis, the analysis should consider the following steps:

1. Identify all anticipated program activities that will generate costs over the life cycle for each alternative under consideration.
2. Relate each activity to a specific cost category in the CBS. Each activity should fall into one or more of the categories. If it does not, the CBS should be expanded or revised to include the activity.
3. Develop a matrix-type worksheet as in Figure 10 or a computer-based spreadsheet for the purpose of recording costs by year for each applicable category in the life cycle.[18]
4. Generate input cost data for each applicable activity listed in the matrix and record the results in the worksheet or spreadsheet.

For the LCSys system, the major cost categories are C_R, C_P, and C_O in the generic CBS of Figure 6. It is assumed in this example that C_D will be negligible. The following paragraphs give an overview of the major activities supporting these cost categories.

[18]The desired matrix could be a direct output from the life-cycle cost model or a spreadsheet. Choice must consider the data output requirements in terms of both content and format.

Research and Development Cost(C_R).   Research and development costs include those early life-cycle costs that will be incurred by the metropolitan agency responsible for the procurement of the LCSys (i.e., the customer) and those costs incurred by the supplier in the development of the system (i.e., the contractor). There will be some common costs to the customer associated with both alternatives relating to initial program planning, the accomplishment of feasibility studies, the development of operational requirements and the maintenance concept, the preparation of top-level system specifications, and general management activities. There will also be supplier costs that are peculiar to each alternative and that are included in the supplier's proposal.

Although the cost analyst may ultimately choose to evaluate only delta or incremental costs (i.e., those costs peculiar to one alternative or the other), the approach used here is to address total life-cycle cost. In determining these costs, the analyst may use a combination of customer cost projections and the proposals submitted by each potential supplier as source data for the life-cycle cost analysis.

Table 1 presents research and development costs. These costs are primarily nonrecurring, made up of management and engineering time with inflationary factors included. Labor costs are developed from personnel projections indicating the class of labor (i.e., manager, supervisor, senior engineer, engineer, technician, etc.) and the labor hours of effort required by class per month for each functional activity. The labor hours per month are then converted to dollars by applying standard cost factors. Both indirect and direct

TABLE 1   Research and Development Cost

Program Activity[a]	Cost Category Designator	Cost by Program Year (Inflated Dollars)			Total Actual Cost (Dollars)
		Year 1	Year 2	Year 3	
A. Customer costs: Alternative A					
1. System/product management	C_{RM}	115,256	121,019	137,069	373,344
2. Product planning	C_{RP}	45,704	0	0	45,704
B. Supplier costs: Alternative A					
1. System/product management	C_{RM}	101,957	107,055	112,407	321,419
2. Product planning	C_{RP}	28,814	30,255	0	59,069
3. Engineering design	C_{RE}	186,692	221,887	239,396	647,975
4. Design documentation	C_{RD}	28,951	49,517	52,707	131,175
5. System test and evaluation	C_{RT}	0	0	176,509	176,509
Total research and development cost for Alternative A	C_R	507,374	529,733	718,088	1,755,195
A. Customer costs: Alternative B					
1. System/product management	C_{RM}	115,256	121,019	137,069	373,344
2. Product planning	C_{RP}	45,704	0	0	45,704
B. Supplier costs: Alternative B					
1. System/product management	C_{RM}	93,091	97,746	102,634	293,471
2. Product planning	C_{RP}	0	0	0	0
3. Engineering design	C_{RE}	145,267	152,530	160,156	457,953
4. Design documentation	C_{RD}	34,976	36,724	39,921	111,621
5. System test and evaluation	C_{RT}	0	0	157,939	157,939
Total research and development cost for Alternative B	C_R	434,294	408,019	597,719	1,440,032

[a]Only applicable cost categories are listed. There are no costs associated with product research (C_{RR}) and system software (C_{RS}).

costs are included, along with a 5% per year inflation factor. Material costs are included in test and evaluation (category C_{RT}), since prototype models are to be produced to verify performance, effectiveness, and supportability characteristics.

The supplier costs for Alternatives A and B in Table 1 are derived from the individual supplier proposals. Direct costs, indirect costs, inflation factors for both labor and material, general and administrative expenses, and projected supplier profits are included.

Production and Construction Cost(C_P). This category includes the recurring and nonrecurring costs associated with the production of the required 80 operational packages plus 4 additional packages intended to compensate for possible attrition. Referring to Figure 15, the leading edge of the inventory profile represents production quantity requirements. The analyst must convert the projected population growth to a specific production profile to be used as the basis for determining production costs. If the option is permitted, each potential supplier may propose an entirely different production pattern while still meeting the desired inventory buildup. This could create a significant variation in the input planning factors and cost. But, in this example, the same production plan is assumed for both alternatives.[19]

Production and construction cost considers the appropriate factors indicated in the generic CBS of Figure 6, and the results are given in Table 2.[20] A description of these costs follows:

1. *System or product management cost* (C_{RM}) includes the ongoing management activity required throughout the production phase. Costs in this category are a continuation of the system or product management cost stream reflected in the R&D cost summary of Table 1.

2. *Manufacturing engineering and operations analysis cost* (C_{PI}) includes the functions of production planning, manufacturing engineering, methods engineering, and so on.

3. *Manufacturing recurring cost* (C_{PMR}) includes those activities related directly to the fabrication, assembly, inspection, and test of the 84 packages being produced. Each of the two potential suppliers submitted a proposal covering functions compatible with and in direct support of the production buildup illustrated in Figure 15. Since the majority of these activities are repetitive in nature, each supplier estimated the cost of the first package and then projected a learning curve to reflect the cost of subsequent packages. Accordingly, learning is incorporated for each alternative. It is assumed that each supplier will experience the same rate of learning during production of the LCSys packages.

4. *Manufacturing nonrecurring cost* (C_{PMN}) includes all costs associated with the acquisition and installation of special tooling, fixtures and jigs, and test equipment. These costs, from the supplier's proposal, are basically a one-time expenditure during year 2 in anticipation of production beginning in year 3.

5. *Quality control cost* (C_{PO}) includes the category of quality assurance. A sustaining level of activity is required to ensure that good overall product quality exists throughout the production process as well as the category of qualification testing, which constitutes the testing of a representative sample to verify that the level of quality in-

---

[19]It is this investment cost that should be the budget constraint. If the final production profile differs significantly, the budget constraint may be violated.

[20]The results in Tables 1, 2, and 4 include inflationary effects. Costs given in the written text are not adjusted for inflation.

TABLE 2 Production/Construction Costs

Program Activity	Cost Category Designator	Cost by Program Year (Inflated Dollars)				Total Actual Cost (Dollars)
		Year 2	Year 3	Year 4	Year 5	
A. Customer costs: Alternative A						
1. System/product management	C_{RM}	0	0	146,824	154,080	300,904
B. Supplier costs: Alternative A						
1. Industrial engineering and operations analysis	C_{PI}	30,082	51,679	54,816	57,525	194,102
2. Manufacturing						
a. Recurring cost	C_{PMR}	0	499,567	620,790	613,421	1,733,778
b. Nonrecurring cost	C_{PMN}	147,100	0	0	0	147,100
3. Quality control	C_{PQ}	0	72,925	76,578	74,782	224,285
4. Initial logistics support						
a. Supply support	C_{PLS}	0	33,350	70,041	73,503	176,894
b. Test and support equipment	C_{PLT}	16,545	225,810	0	0	242,355
c. Technical data	C_{PLD}	16,876	0	0	0	16,876
d. Personnel training	C_{PLP}	54,268	52,110	54,720	0	161,098
Total production and construction cost for Alternative A	C_P	264,871	935,441	1,023,769	973,311	3,197,392
A. Customer costs: Alternative B						
1. System/product management	C_{RM}	0	0	146,824	154,080	300,904
B. Supplier costs: Alternative B						
1. Industrial engineering and operations analysis	C_{PI}	45,123	57,421	60,298	63,277	226,119
2. Manufacturing						
a. Recurring cost	C_{PMR}	0	479,469	645,458	666,762	1,791,689
b. Nonrecurring cost	C_{PMN}	165,751	0	0	0	165,751
3. Quality control	C_{PQ}	0	56,847	60,572	81,293	198,712
4. Initial logistics support						
a. Supply support	C_{PLS}	0	63,712	70,554	74,039	208,305
b. Test and support equipment	C_{PLT}	16,545	225,810	0	0	242,355
c. Technical data	C_{PLD}	17,703	0	0	0	17,703
d. Personnel training	C_{PLP}	54,268	52,110	54,720	0	161,098
Total production and construction cost for Alternative B	C_P	299,390	935,369	1,038,426	1,039,451	3,312,636

herent in the items being produced is adequate. The costs from each supplier's proposal are included.

6. **Initial spares and inventory cost** (C_{PLS}) includes the acquisition of major units to support organizational maintenance and a few assemblies, plus parts to provide support at the intermediate level of maintenance. These items represent the inventory safety stock and are located at the intermediate maintenance shops and the supplier facility identified in Figure 14. Replenishment spares for the sustaining support of the system in operational use are covered in category (C_{OLS}). This category includes an initial limited procurement, whereas the replenishment spares are based on realistic

consumption and demand factors. The assumptions used in determining costs for each alternative are as follows:

(a) *Alternative A.* One complete set of units (Units *A–C*) are required as a backup for shop 1, shop 2, each of the two mobile vans, and for the supplier facility. The acquisition price for Unit *A* is $11,400, Unit *B* is $6,450, and Unit *C* is $7,950. Five sets are equivalent to $129,000, with $25,800 in year 3, $51,600 in year 4, and $51,600 in year 5. The cost of assemblies and parts is $15,000 ($3,000 in year 3, $6,000 in year 4, and $6,000 in year 5).

(b) *Alternative B.* Six sets of units (Units *A–C*) are required. The additional unit (above and beyond the requirements for Alternative A) is to be located at shop 2 to handle an anticipated increase in the number of maintenance actions. The acquisition price for Unit *A* is $10,590, Unit *B* is $6,180, and Unit *C* is $7,740. Six sets are equivalent to $147,060, with $49,020 in each of the years 3, 4, and 5. The cost of assemblies and parts is $24,000 ($6,000 in year 3, $9,000 in year 4, and $9,000 in year 5).

7. *Test and support equipment acquisition cost* (C_{PLT}) includes the assembly and module test set located in each intermediate maintenance facility and several items of commercial and standard equipment located at the supplier facility to support depot maintenance. (Refer to Figures 14 and 17.) The design cost associated with the test set is $15,000 expended in year 2. The production acquisition price of each test set is $45,000 and there are four test sets required. According to Table 3, there are packages being introduced into operational use in year 3 at all locations; thus, the four test sets must be available in year 3. The commercial and standard equipment needed for depot maintenance requires no additional design effort and can be acquired for $15,000. Thus, the costs for year 3 are $195,000. The test and support equipment requirements for each alternative are considered to be comparable. The sustaining annual maintenance and logistics requirements for test equipment are included in cost category (C_{OLE}).

8. *Technical data cost* (C_{PLD}) includes the preparation and publication of installation and test instructions, operating procedures, and maintenance procedures. These data are required to operate and maintain the system in the field throughout its programmed life cycle. The acquisition cost of this data is $15,300 for Alternative A and $16,050 for Alternative B. These costs are applicable in year 2.

9. *Personnel training cost* (C_{PLP}) includes the initial cost of training operators and maintenance technicians. For operator training, it is assumed that 20 operators are trained in year 2, 30 in year 3, and 30 in year 4. The cost of training is $1,500 per technician (for each alternative configuration). In the maintenance area, formal training will be given to two technicians assigned to each of the four intermediate maintenance facilities. These individuals will accomplish on-the-job training for the additional personnel in the shops and vans. The cost of maintenance training is $2,400 per student week, or $19,200 in year 2.

Operation and Support Cost (C_O).   This category includes the cost of operating and supporting the system throughout its anticipated life cycle. These are primarily user costs and are based on the program planning information illustrated in Figure 15. The inventory profile shown is expanded in Table 3 to indicate the number of packages in use and the total operating time (in hours) for all devices in each applicable year of the life cycle.

TABLE 3 LCSys Packages in Use and Operating Times

Category	Program Year												
	1	2	3	4	5	6	7	8	9	10	11	12	13
LCSys packages in use	—	—	20	50	80	80	80	80	80	80	80	29	10
1. Aircraft application (3)													
a. Number of packages	—	—	1	3	3	3	3	3	3	3	3	3	1
b. Operating time (hour)	—	—	594	1,782	1,782	1,782	1,782	1,782	1,782	1,782	1,782	1,782	594
2. Helicopter application (3)													
a. Number of packages	—	—	2	5	5	5	5	5	5	5	5	5	3
b. Operating time (hour)	—	—	1,080	2,700	2,700	2,700	2,700	2,700	2,700	2,700	2,700	2,700	1,620
3. Patrol vehicles (50)													
a. Number of packages	—	—	7	22	50	50	50	50	50	50	50	10	2
b. Operating time (hour)	—	—	9,100	28,600	65,000	65,000	65,000	65,000	65,000	65,000	65,000	13,000	2,600
4. Communication facilities (5)													
a. Number of packages	—	—	5	10	10	10	10	10	10	10	10	5	—
b. Operating time (hour)	—	—	7,200	14,400	14,400	14,400	14,400	14,400	14,400	14,400	14,400	7,200	—
5. Central communications (1)													
a. Number of packages	—	—	4	8	10	10	10	10	10	10	10	4	3
b. Operating time (hour)	—	—	4,320	8,640	10,800	10,800	10,800	10,800	10,800	10,800	10,800	5,400	3,240
6. Mobile vans (2)													
a. Number of packages	—	—	1	2	2	2	2	2	2	2	2	2	1
b. Operating time (hour)	—	—	720	1,440	1,440	1,440	1,440	1,440	1,440	1,440	1,440	1,440	720

LCSys use is based on the individual operating times stated in the objectives described as part of the problem definition and then determined from Equations 1 through 6:

$$\text{Operating time for aircraft application (hours)} \qquad (1)$$
$$= \text{(number of units in aircraft)(15flights/month)(12)}$$
$$\times \text{(3hoursflight)(1.1)}$$

$$\text{Operating time for helicopter application (hours)} \qquad (2)$$
$$= \text{(number of units in helicopters)(25flights/month)}$$
$$\times \text{(12 months)(2hours/flight)(0.9)}$$

$$\text{Operating time for patrol vehicle application (hours)} \qquad (3)$$
$$= \text{(number of units in patrol vehicles)(5hours/day)}$$
$$\times \text{(5days/week)(52weeks/year)(1.0)}$$

$$\text{Operating time for communication facility application (hours)} \qquad (4)$$
$$= \text{(number of units in facilities)(120hours/month)(12)}$$

$$\text{Operating time for central communication control application (hours)} \qquad (5)$$
$$= \text{(number of units)(3hours/day)(360days/year)}$$

$$\text{Operating time for mobile van application (hours)} \qquad (6)$$
$$= \text{(number of units in vans)(2hours/day)(360days/year)}$$

Although the actual utilization of the equipment will vary from operator to operator, from organization to organization, from one geographical area to the next, and so on, the factors included in Table 3 are average values and are employed in the baseline example. Also, it is assumed that each alternative configuration being evaluated will be operated in the same manner.

Operation and support costs include the individual costs associated with operations, distribution, and sustaining logistic support. The significant costs that are applicable to the LCSys are as follows:

1. *Operating personnel cost* (C_{OOP}) covers the total costs of operating the LCSys for the various applications. Since the operator is charged with a number of different duties, only that allocated portion of time is counted that is associated with the direct operation of the communications system. Operating personnel cost is determined using the data in Table 3 as a base:

$$C_{OOP} = \text{(number of units)(hours of system operation)} \qquad (7)$$
$$\text{(\% allocation)(labor cost)}$$

In determining operator costs, different hourly rates are applied for the various applications (e.g., \$43.50 per hour for the aircraft and helicopter application, \$37.50 per hour for the facility application, etc.) and varied allocation factors are applied because of a personnel workload different from one application to the next (e.g., 1%, 2%). The resulting costs, adjusted for inflation, are given in Table 4.

TABLE 4 Operation and Support Cost

Program Activity	Cost Category Designation	1	2	3	4	5	6	7	8	9	10	11	12	13	Total Cost Actual (Dollars)
							Cost by Program Year (Dollars)								
A. Alternative A															
1. Operating personnel	C_{OOP}	—	—	2,990	9,423	9,894	10,388	10,908	11,453	12,026	12,627	13,258	13,291	4,871	111,759
2. Transportation	C_{ODT}	—	—	27,783	72,930	122,523	128,649	135,082	141,817	148,927	156,374	164,192	36,000	6,364	1,140,641
3. Unscheduled maintenance	C_{OLA}	—	—	12,902	32,403	56,939	59,786	62,775	65,914	69,209	72,671	76,303	26,377	8,689	543,968
4. Maintenance facilities	C_{OLM}	—	—	483	1,218	2,140	2,247	2,360	2,478	2,602	2,732	2,868	991	317	20,436
5. Supply support	C_{OLS}	—	—	23,170	77,890	136,919	143,765	150,954	158,501	166,426	174,748	183,485	63,358	20,365	1,299,581
6. Maintenance personnel training	C_{OLT}	—	—	0	0	4,978	5,226	5,488	5,762	6,050	6,353	6,670	0	0	40,527
7. Test and support equipment	C_{OLE}	—	—	11,291	11,856	12,443	13,066	13,719	14,405	15,125	15,882	16,676	17,510	18,194	160,167
8. Transportation/handling	C_{OLH}	—	—	584	1,378	2,412	2,533	2,659	2,792	2,932	3,079	3,233	1,131	870	23,603
Total operation and support cost for Alternative A	C_O	—	—	79,203	207,098	348,248	365,660	383,945	403,122	423,297	444,466	466,685	159,288	59,670	3,340,682
B. Alternative B															
1. Operating personnel	C_{OOP}	—	—	2,990	9,423	9,894	10,388	10,908	11,453	12,026	12,627	13,258	13,921	4,871	111,759
2. Transportation	C_{ODT}	—	—	27,783	72,930	122,523	128,649	135,082	141,817	148,927	156,374	164,192	36,000	6,364	1,140,641
3. Unscheduled maintenance	C_{OLA}	—	—	16,355	39,678	70,152	73,660	77,343	81,210	85,270	89,534	94,010	32,740	10,251	670,203
4. Maintenance facilities	C_{OLM}	—	—	587	1,491	2,642	2,774	2,913	3,058	3,211	3,372	3,540	1,234	385	25,207
5. Supply support	C_{OLS}	—	—	37,507	95,393	169,081	177,536	186,413	195,733	205,520	215,796	226,586	78,874	24,438	1,612,877
6. Maintenance personnel training	C_{OLT}	—	—	0	0	4,978	5,226	5,488	5,762	6,050	6,353	6,670	0	0	40,527
7. Test and support equipment	C_{OLE}	—	—	16,936	17,784	18,662	19,598	20,577	21,601	22,683	23,824	25,009	26,267	27,290	240,231
8. Transportation/handling	C_{OLH}	—	—	730	1,684	2,895	3,039	3,192	3,351	3,519	3,695	3,879	1,358	475	27,817
Total operation and support cost for Alternative B	C_O	—	—	102,888	238,383	400,827	420,870	441,916	463,985	487,206	511,575	537,144	190,394	74,074	3,869,262

2. *Distribution and transportation cost* (C_{ODT}) covers the initial transportation and installation cost (i.e., the packing and shipping of packages from the supplier's manufacturing facility to the point of installation for operational use). The product of the reliabilities for the individual subsystems is used to determine total cost in this category where transportation and packing costs are based on dollars per hundredweight (cwt) (i.e., $90 per cwt for transportation and $120 per cwt for packing), and installation costs are a function of labor cost in dollars per labor-hour and the number of labor hours:

$$C_{ODT} = \text{(cost of packing)} + \text{(cost of transportation)} \qquad (8)$$
$$+ \text{(cost of system installation)}$$

Costs in this category are based on the number of packages indicated in Table 3, and on the appropriate transportation rate structures. The analyst should review the latest Interstate Commerce Commission (ICC) documentation on rates to determine the proper transportation costs. The figures used in this cost analysis are presented in Table 4.

3. *Unscheduled maintenance cost* (C_{OLA}) covers the personnel activity costs associated with the accomplishment of *unscheduled* or corrective maintenance on the LCSys. Specifically, this includes the direct and indirect costs in the performance of maintenance actions (a function of maintenance labor-hours and the cost per labor-hour), the material handling cost associated with given maintenance actions, and the cost of documentation for each maintenance action. These costs, for the two alternative system configurations being evaluated, are summarized in Table 5. Determining unscheduled maintenance cost depends on predicting the number of maintenance actions that are likely to occur throughout the life cycle (i.e., the expected frequency of unscheduled maintenance or the reciprocal of the MTBM). Since there is no *scheduled* maintenance permitted in this example, the MTBM factor assumed is directly equated with unscheduled maintenance actions. The frequency of maintenance is usually based on the failure rates for individual components of the system and is derived from reliability prediction data, maintainability prediction data, logistic support analysis data, or a combination thereof.

A review of the objectives in the problem definition indicates that the need relative to MTBM requirements differs from one application to the next. Since it is the goal to design a single system configuration for use in all applications (to the maximum extent practicable), the most stringent conditions must be met. Thus, the new system is required to exhibit an MTBM of 500 hours or greater. Response to this requirement by the two suppliers is illustrated in Figure 18. Note that the predicted MTBM for Alternative A is 650 hours and for Alternative B is 525 hours. These values are further broken down to unit levels compatible with the equipment-packaging scheme in Figure 16 and illustrated maintenance concept in Figure 17. Although failures are randomly distributed, these values are used to determine an average factor for the frequency of maintenance.

Figure 18 is the final result of an iterative design process that trades off the life-cycle cost of the system with its MTBM. The MTBMs shown are optimal values. Using the information in Figure 8, the next step is to calculate the average number of maintenance actions for each package configuration by year through the life cycle. This is accomplished by dividing the operating time in Table 3 by the MTBM. The results (rounded off to the nearest dollar) are presented in Table 5.

Personnel labor cost covering the accomplishment of unscheduled maintenance is a function of maintenance labor-hours and the cost per labor-hour. Maintenance

TABLE 5 Unscheduled Maintenance Costs

Program Activity		1	2	3	4	5	6	7	8	9	10	11	12	13
A. Alternative A														
1. Organizational maintenance	Total labor-hours	—	—	9.25	22.25	37.35	37.35	37.35	37.35	37.35	37.35	37.35	12.25	3.75
	Personnel cost	—	—	306	735	1,230	1,230	1,230	1,230	1,230	1,230	1,230	405	123
	Material handling cost	—	—	1,110	2,670	4,470	4,470	4,470	4,470	4,470	4,470	4,470	1,470	450
	Documentation cost	—	—	1,110	2,670	4,470	4,470	4,470	4,470	4,470	4,470	4,470	1,470	450
2. Intermediate maintenance	Total labor hours	—	—	111	267	447	447	447	447	447	447	447	147	45
	Personnel cost	—	—	4,329	11,223	17,433	17,433	17,433	17,433	17,433	17,433	17,433	5,733	1,755
	Material handling cost	—	—	1,110	2,670	4,470	4,470	4,470	4,470	4,470	4,470	4,470	1,470	450
	Documentation cost	—	—	2,220	5,340	8,940	8,940	8,940	8,940	8,940	8,940	8,940	2,940	900
3. Depot maintenance	Total labor hours	—	—	16	36	60	60	60	60	60	60	60	20	8
	Personnel cost	—	—	720	1,620	2,700	2,700	2,700	2,700	2,700	2,700	2,700	900	360
	Material handling cost	—	—	120	270	450	450	450	450	450	450	450	150	60
	Documentation cost	—	—	120	270	450	450	450	450	450	450	450	150	60
Total unscheduled maintenance cost		—	—	11,145	27,468	44,613	44,613	44,613	44,613	44,613	44,613	44,613	14,688	4,608
Total cost (adjusted for inflation)		—	—	12,906	33,401	56,926	59,781	62,770	65,893	69,195	72,675	76,288	26,380	8,599
B. Alternative B														
1. Organizational maintenance	Total labor hours	—	—	11.25	27.25	46	46	46	46	46	46	46	15.25	4.5
	Personnel cost	—	—	372	900	1,518	1,518	1,518	1,518	1,518	1,518	1,518	504	150
	Material handling cost	—	—	1,350	3,270	5,520	5,520	5,520	5,520	5,520	5,520	5,520	1,830	540
	Documentation cost	—	—	1,350	3,270	5,520	5,520	5,520	5,520	5,520	5,520	5,520	1,830	540
2. Intermediate maintenance	Total labor hours	—	—	135	327	552	552	552	552	552	552	552	183	54
	Personnel cost	—	—	5,265	12,753	21,528	21,528	21,528	21,528	21,528	21,528	21,528	7,137	2,106
	Material handling cost	—	—	1,350	3,270	5,520	5,520	5,520	5,520	5,520	5,520	5,520	1,830	540
	Documentation cost	—	—	2,700	6,540	11,040	11,040	11,040	11,040	11,040	11,040	11,040	3,660	1,080
3. Depot maintenance	Total labor hours	—	—	20	44	72	72	72	72	72	72	72	24	8
	Personnel cost	—	—	900	1,980	3,240	3,240	3,240	3,240	3,240	3,240	3,240	1,080	360
	Material handling cost	—	—	150	330	540	540	540	540	540	540	540	180	60
	Documentation cost	—	—	150	330	540	540	540	540	540	540	540	180	60
Total unscheduled maintenance cost		—	—	13,587	32,643	54,966	54,966	54,966	54,966	54,966	54,966	54,966	18,231	5,436
Total cost (adjusted for inflation)		—	—	15,734	39,694	70,137	73,654	77,337	81,185	85,252	89,540	93,992	32,743	10,144

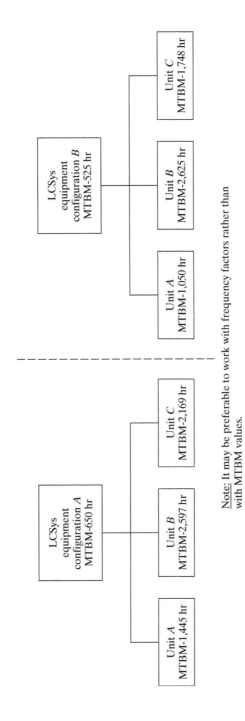

Figure 18 LCSys equipment maintenance factors.

Note: It may be preferable to work with frequency factors rather than with MTBM values.

labor-hour and labor-cost factors in this example are based on the number of technicians (with specific skill levels) assigned to a given maintenance action and the length of time that the technicians are assigned. The assumed factors are as follows:

Organizational maintenance: 0.25 MLH per maintenance action at a cost of $33.00 per MLH.

Intermediate maintenance: 3 MLH per maintenance action at a cost of $39.00 per MLH.

Supplier (or depot) maintenance: 4 MLH per maintenance action at a cost of $45.00 per MLH.

The labor-hour factors used are related to personnel requirements for direct maintenance where the elapsed times are indicated by the $\overline{\text{Mct}}$ values. The hourly rates include direct dollars plus a burden or overhead rate. Labor-hour calculations, personnel cost, material handling cost (i.e., $30 per maintenance action), and documentation cost (i.e., $30 per maintenance action at the depot level, $60 at the intermediate level, and $30 at the organizational level) are noted in Table 5 for each configuration evaluated. The total cost factors are also included in Table 4.

4. *Maintenance facilities cost* (C_{OLM}) is based on the occupancy, utilities, and facility maintenance costs as prorated to the communications system. Facilities cost in this instance is primarily related to the intermediate level of maintenance.

5. *Supply support cost* (C_{OLS}) includes the cost of spare parts needed as a result of package failures; spare parts required to fill the logistics pipeline to compensate for delays due to active repair times, turnaround times, and supplier lead times; spare parts required to replace repairable items that are condemned or phased out for one reason or another (e.g., those items that are damaged to the extent beyond which it is not economically feasible to do repairs); and the cost of maintaining the inventory throughout the designated period of support.

Referring to the illustrated maintenance concept in Figure 17, spare units are required in the intermediate maintenance shops to support unit replacements at the organizational level. Spare assemblies, modules, and certain designated piece parts are required to support intermediate maintenance actions, and some assemblies and parts are required to support supplier repair activities or depot maintenance. Actually, the maintenance concept indicates the type of spares required at each level of maintenance, and the system network in Figure 14 (with specific geographical locations defined) conveys the quantity of maintenance facilities providing overall system support. A logistic support analysis is accomplished to provide additional maintenance data as required.

The next step is to determine the number of spare parts required at each location. Too many spares, or a large inventory, could be quite costly in terms of investment and inventory maintenance. On the other hand, not enough spares could result in a stock-out condition, which will cause systems to be inoperative; thus, the defined objectives of the communications network will not be met. This condition may also be quite costly. The goal is to analyze the inventory requirements and obtain a balance between the cost of acquiring spares and the cost of maintaining inventory. The critical factor constitutes consumption of demand and the probability of having the right types of spare parts available when required. Relative to this analysis, demand rates are a function of unit, assembly, module, or reliability.

Another consideration is the turnaround time (TAT) and the transportation time between facilities for repairable items (i.e., the total time from the point of failure until the item is repaired and returned to inventory and ready for use, or the time that it takes to acquire an item from the source of supply). These time factors, identified in Figure 17, have a significant impact on spare-part requirements.

The process for determining the costs associated with supply support is fairly comprehensive, and the detailed steps used in this analysis are not included here. However, the concepts used are discussed. Actually, inventory requirements are covered in two categories. An initial procurement of spare units and assemblies, basically representing safety stock, is reflected under initial logistics support cost (i.e., category C_{PLS}). The spares required for sustaining support, to include both repairable and nonrepairable items, are covered in this category. These items are directly related to consumption and the number of unscheduled maintenance actions identified in Table 6. The associated costs include both material costs and the cost of maintaining inventory. Annual inventory holding cost is assumed to be 20% of the inventory value.

6. *Maintenance personnel training cost* (C_{OLT}) is covered in two categories. When the system is first introduced, there is a requirement to train operators and maintenance technicians. The cost for this initial training is included in category C_{PLP}. Subsequently, the cost of training relates to personnel attrition and the addition of new operators and/or maintenance technicians. A cost of $2,600 per year is assumed for formal sustaining training until that point in time when system phaseout commences.

7. *Test and support equipment cost* (C_{OLE}) is presented in two categories. Category P_{PLT} includes the design and acquisition of the test and support equipment required for the intermediate and depot levels of maintenance. This category includes the sustaining support of these items on a year-to-year basis (i.e., the unscheduled and scheduled maintenance actions associated with the test equipment). Unscheduled maintenance is a function of the use of the test equipment, which relates directly to the unscheduled maintenance actions noted in Table 6. In addition, the reliability and maintainability characteristics of the test equipment itself will significantly influence the cost of supporting that test equipment. Scheduled maintenance constitutes the periodic 180-day calibration of certain elements of the test equipment in the calibration laboratory (refer to Figure 4). Calibration is required to maintain the proper test traceability to primary and secondary standards. The costs associated with test equipment maintenance and logistic support can be derived through an in-depth logistic support analysis. However, the magnitude of the test equipment required in this case is relatively small when compared to other systems. Based on past experience with comparables, a factor of 5% of the acquisition cost ($9,750) is considered to be appropriate for the annual maintenance and logistics cost for the test equipment associated with Alternative A. A factor of 7.5% ($14,625) is assumed for Alternative B, since the test equipment use requirements will be greater.

8. *Transportation and handling costs* (C_{OLH}) include the annual costs associated with the movement of materials among the organizational, intermediate, and depot levels of maintenance. This is in addition to the costs of initial distribution and system installation covered in category C_{ODT}. For the LCSys, the movement of materials between the organizational and intermediate levels of maintenance is considered insignificant in terms of relative cost. However, the shipment of materials between the intermediate maintenance facilities and the supplier or depot maintenance is considered significant, and can be determined from Equation 9:

$$C_{OLH} = (\text{cost of packing}) + (\text{cost of transportation}) \qquad (9)$$
$$\times (\text{number of one-way shipments})$$

TABLE 6 Number of Unscheduled Maintenance Actions

Alternative	Application	Equipment or Unit	1	2	3	4	5	6	7	8	9	10	11	12	13
Alternative A	Aircraft application	System	—	—	1	3	3	3	3	3	3	3	3	3	1
		Unit A	—	—	1	1	1	1	1	1	1	1	1	1	1
		Unit B	—	—	—	1	1	1	1	1	1	1	1	1	—
		Unit C	—	—	1	1	1	1	1	1	1	1	1	1	1
	Helicopter application	System	—	—	2	4	4	4	4	4	4	4	4	4	3
		Unit A	—	—	1	2	2	2	2	2	2	2	2	2	1
		Unit B	—	—	—	1	1	1	1	1	1	1	1	1	—
		Unit C	—	—	1	1	1	1	1	1	1	1	1	1	1
	Patrol vehicle application	System	—	—	14	44	100	100	100	100	100	100	100	20	4
		Unit A	—	—	6	20	45	45	45	45	45	45	45	9	2
		Unit B	—	—	4	11	25	25	25	25	25	25	25	5	1
		Unit C	—	—	4	13	30	30	30	30	30	30	30	6	1
	Communications facilities	System	—	—	11	22	22	22	22	22	22	22	22	11	—
		Unit A	—	—	5	10	10	10	10	10	10	10	10	5	—
		Unit B	—	—	2	5	5	5	5	5	5	5	5	2	—
		Unit C	—	—	4	7	7	7	7	7	7	7	7	4	—
	Central communications control	System	—	—	7	13	17	17	17	17	17	17	17	8	5
		Unit A	—	—	3	6	8	8	8	8	8	8	8	4	2
		Unit B	—	—	2	3	4	4	4	4	4	4	4	2	1
		Unit C	—	—	2	4	5	5	5	5	5	5	5	2	2
	Mobile vans	System	—	—	2	3	3	3	3	3	3	3	3	3	2
		Unit A	—	—	—	1	1	1	1	1	1	1	1	1	1
		Unit B	—	—	—	—	—	—	—	—	—	—	—	—	—
		Unit C	—	—	1	1	1	1	1	1	1	1	1	1	1
Alternative B	Aircraft application	System	—	—	2	4	4	4	4	4	4	4	4	4	2
		Unit A	—	—	1	2	2	2	2	2	2	2	2	2	1
		Unit B	—	—	—	1	1	1	1	1	1	1	1	1	—
		Unit C	—	—	1	1	1	1	1	1	1	1	1	1	1
	Helicopter application	System	—	—	2	5	5	5	5	5	5	5	5	5	3
		Unit A	—	—	1	2	2	2	2	2	2	2	2	2	1
		Unit B	—	—	—	1	1	1	1	1	1	1	1	1	—
		Unit C	—	—	1	1	1	1	1	1	1	1	1	1	1
	Patrol vehicle application	System	—	—	17	54	124	124	124	124	124	124	124	25	5
		Unit A	—	—	9	27	62	62	62	62	62	62	62	12	5
		Unit B	—	—	3	16	25	25	25	25	25	25	25	5	1
		Unit C	—	—	5	27	37	37	37	37	37	37	37	8	2
	Communications facilities	System	—	—	14	27	27	27	27	27	27	27	27	14	—
		Unit A	—	—	7	14	14	14	14	14	14	14	14	7	—
		Unit B	—	—	3	5	5	5	5	5	5	5	5	3	—
		Unit C	—	—	4	8	8	8	8	8	8	8	8	4	—
	Central communication control	System	—	—	8	16	21	21	21	21	21	21	21	10	6
		Unit A	—	—	4	8	11	11	11	11	11	11	11	5	3
		Unit B	—	—	2	3	4	4	4	4	4	4	4	2	1
		Unit C	—	—	2	5	6	6	6	6	6	6	6	3	2
	Mobile vans	System	—	—	2	3	3	3	3	3	3	3	3	3	2
		Unit A	—	—	2	1	1	1	1	1	1	1	1	1	1
		Unit B	—	—	—	—	—	—	—	—	—	—	—	—	—
		Unit C	—	—	1	1	1	1	1	1	1	1	1	1	1

TABLE 7 Trips Between Maintenance Facilities and Supplier

	Alternative A	Alternative B
Year 3	8	10
Year 4	18	22
Years 5–11	30 per year	36 per year
Year 12	10	12
Year 13	4	4

where transportation and packing costs are \$90 per cwt and \$120 per cwt, respectively. The material being moved between the intermediate maintenance facilities and the supplier includes assembly 5 of Unit *C*, which is supported at the depot level of maintenance (refer to Figure 14). The estimated number of one-way trips is given in Table 7.

4.3 Evaluation of the Alternatives

Evaluation requires selection of the preferred alternative on the basis of present equivalent life-cycle cost, considering the imposed constraints. Because this is a multiple-criteria decision situation, it cannot be resolved solely on the basis of a life-cycle cost comparison of Alternatives A and B, as summarized in Table 8. Costs are listed for those major categories of the CBS that are relevant to this analysis, with high-cost contributors identified by their percentage contribution to the total.

When evaluating two or more alternatives on a relative basis, the future cost estimations for each alternative must be reduced to their present equivalent amounts. The present equivalent costs for Alternatives A and B are shown in Table 9, and the cost profiles are illustrated in Figure 19 based on an interest rate of 10%. The present equivalent cost for Alternative A is

$$\mathrm{PE} = \$507{,}734\overset{P/F,\,10,\,1}{(0.9091)} + \$794{,}604\overset{P/F,\,10,\,2}{(0.8265)}$$
$$+ \cdots + \$59{,}670\overset{P/F,\,10,\,13}{(0.2897)} = \$5{,}255{,}036$$

For Alternative B, the present equivalent cost is

$$\mathrm{PE} = \$434{,}294\overset{P/F,\,10,\,1}{(0.9091)} + \$707{,}409\overset{P/F,\,10,\,2}{(0.8265)}$$
$$+ \cdots + \$74{,}074\overset{P/F,\,10,\,13}{(0.2897)} = \$5{,}329{,}999$$

Presentation of the decision options and relevant data is best done with the aid of a decision evaluation display. The challenge is to determine whether the alternatives meet the specified requirements (i.e., performance threshold of 0.8, budget constraint of \$5 million, MTBM of 500 hours, etc.). The relevant decision evaluation display is developed from the cost and other criteria and is shown in Figure 20. In this example, both alternatives meet or more than satisfy all criteria. Accordingly, Alternative A with the lowest present equivalent life-cycle cost should be chosen unless the higher LCC for Alternative B can be justified. It could be tempting to commit to the higher life-cycle cost of

TABLE 8 Life-Cycle Cost Breakdown

Cost Category	Alternative A		Alternative B	
	Cost ($)	Percent of Total	Cost ($)	Percent of Total
1. Research and development cost (C_R)				
(a) System/product management (C_{RM})	573,392	11.0	550,296	10.4
(b) Product planning (C_{RP})	92,748	1.8	41,549	0.8
(c) Engineering design (C_{RE})	532,959	10.1	378,446	7.1
(d) Design data (C_{RD})	106,841	2.0	92,140	1.7
(e) System test and evaluation (C_{RT})	132,614	2.5	118,662	2.2
Subtotal	1,438,555	27.4	1,181,094	22.2
2. Production and construction cost (C_P)				
(a) System/product management (C_{RM})	195,954	3.7	195,954	3.7
(b) Industrial engineering and operations analysis (C_{PI})	136,847	2.6	160,907	3.0
(c) Manufacturing—recurring (C_{PMR})	1,180,226	22.5	1,215,095	22.8
(d) Manufacturing—nonrecurring (C_{PMN})	121,570	2.3	136,984	2.6
(e) Quality control (C_{PQ})	153,527	2.9	134,558	2.5
(f) Initial logistics support (C_{PLS})				
(1) Supply support—initial (C_{PLS})	118,535	2.3	142,030	2.7
(2) Test and support equipment (C_{PLT})	183,328	3.5	183,328	3.4
(3) Technical data (C_{PLD})	13,947	0.3	14,631	0.3
(4) Personnel training (C_{PLP})	121,375	2.3	121,375	2.3
Subtotal	2,225,310	42.3	2,304,862	43.2
3. Operation and support cost (C_O)				
(a) Operating personnel (C_{OOP})	52,092	1.0	52,092	1.0
(b) Distribution—transportation (C_{ODT})	549,170	10.5	549,170	10.3
(c) Unscheduled maintenance (C_{OLA})	258,924	4.9	319,134	6.0
(d) Maintenance facilities (C_{OLM})	9,729	0.2	11,994	0.2
(e) Supply support (C_{OLS})	616,532	11.7	767,492	14.4
(f) Maintenance personnel training (C_{OLT})	18,898	0.4	18,898	0.4
(g) Test and support equipment (C_{OLE})	74,674	1.4	112,002	2.1
(h) Transportation and handling (C_{OLH})	11,150	0.2	13,260	0.2
Subtotal	1,591,170	30.3	1,844,042	34.6
Grand total	5,255,036	100.0	5,329,999	100.0

Alternative B to capture a performance score of 0.92 versus the 0.84 anticipated for Alternative A. Only the customer (the municipality) can make this trade-off.

The results of this analysis support Alternative A as the preferred configuration on the basis of present equivalent life-cycle cost, because all other criteria are met. It is noted that the research and development (R&D) cost is somewhat higher for Alternative A even though the overall life-cycle cost is lower, owing to a significantly lower operation and support (O&S) cost. This tends to indicate that the equipment reliability design pertaining to Alternative A is somewhat better. Although this increased reliability results in higher R&D cost, the anticipated number of maintenance actions is less, resulting in lower O&S costs. This thinking might be included in the presentation to the customer.

TABLE 9 Cost Allocation by Program Year

Program Activity	Cost Category Designation	Cost by Program Year (Dollars)													Total Actual Cost (Dollars)
		1	2	3	4	5	6	7	8	9	10	11	12	13	
A. Alternative A															
1. Research and development	C_R	507,374	529,733	718,088											1,755,195
2. Production and construction	C_P		264,871	935,441	1,023,769	973,311									3,197,392
3. Operation and support	C_O			79,203	207,098	348,248	365,660	383,945	403,122	423,297	444,466	483,355	159,288	59,670	3,340,682
Total actual cost	C	507,374	794,604	1,732,732	1,230,867	1,321,559	365,660	383,945	403,122	423,297	444,466	466,685	159,288	59,670	8,293,269
Total present cost	C (10%)	461,254	656,740	1,301,802	840,682	820,556	206,415	197,041	188,056	179,520	171,386	163,573	50,749	17,286	5,255,036
Cumulative PC		461,254	1,117,994	2,419,796	3,260,478	4,081,034	4,287,449	4,484,490	4,672,546	4,852,066	5,023,452	5,187,025	5,237,774	5,255,036	5,255,036
B. Alternative B															
1. Research and development	C_R	434,294	408,019	597,719											1,440,032
2. Production and construction	C_P		299,390	935,369	1,038,426	1,039,451									3,312,636
3. Operation and support	C_O			102,888	238,383	400,827	420,870	441,916	463,985	487,206	511,575	537,144	190,394	74,074	3,869,262
Total actual cost	C	434,294	707,409	1,635,976	1,276,809	1,440,278	420,870	441,916	463,985	487,206	511,575	537,144	190,394	74,074	8,621,930
Total present cost	C (10%)	394,817	584,674	1,229,109	872,061	894,269	237,581	226,791	216,449	206,624	197,263	188,269	60,660	21,459	5,329,999
Cumulative PC		394,817	979,491	2,208,600	3,080,661	3,974,930	4,212,511	4,439,302	4,655,751	4,862,375	5,059,638	5,247,907	5,308,567	5,329,999	5,329,999

685

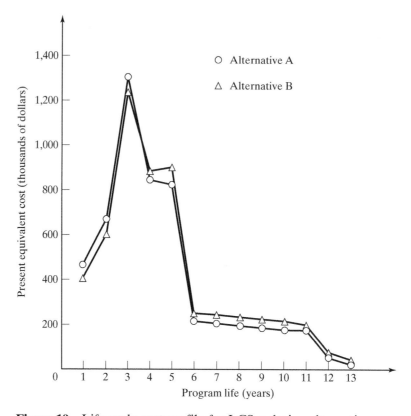

Figure 19 Life-cycle cost profile for LCSys design alternatives.

4.4 Break-Even and Sensitivity Analysis

Prior to a final decision about which alternative to select, the analyst should accomplish a break-even analysis to determine the point in time when Alternative A becomes more economical than Alternative B. Figure 21 indicates that the break-even point, or the point in time when Alternative A becomes less costly, is approximately 4 years and 3 months after the program starts. This point is early enough in the life cycle to support the decision. If this crossover point were much further out in time, the decision might be questioned.

It is noted in considering the analysis results that the delta present equivalent cost between the two alternatives is $74,963 (refer to Table 9), and the cost profiles are relatively close to each other (refer to Figure 19). These factors do not support a clear decision in favor of Alternative A without introducing some risk. In view of the possible inaccuracies associated with the input data, the analyst may wish to perform a sensitivity analysis to determine the effects of input variations on the present equivalent life-cycle cost. The analyst should determine how much variation could be tolerated before the decision shifts in favor of Alternative B.

Referring to Table 8, the analyst should select the *high-cost contributors* (those that contribute more than 10% of the total cost), determine the cause-and-effect relationships, and identify the various input data factors that directly affect cost. In instances where such factors are based on questionable prediction techniques, the analyst should vary these

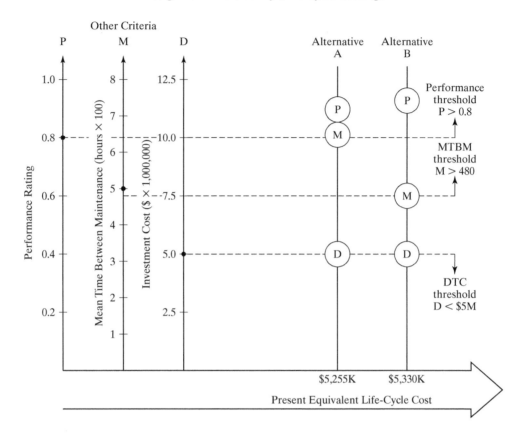

Figure 20 Design decision evaluation display for the LCSys.

factors over a probable range of values and assess the results. For instance, key input para-meters in this analysis include the system operating time (in hours) and the MTBM factor. Using the life-cycle cost model, the analyst will apply a multiple factor to the operating time values and the MTBM and determine the delta cost associated with each variation. From this, trend curves are projected.

A small variation in operating time and/or MTBM may cause the decision to shift in favor of Alternative B. This provides an indication that there is a high degree of risk associated with making a wrong decision. Thus, the analyst should make every effort to reduce this risk by improving the input data to the greatest extent possible. Also, when the results are particularly close, the magnitude of risk associated with the decision must be determined.

Another area of concern in decision making is the variation in the design-independent parameters such as the interest rate, inflation rate, labor rate, and so on. For example, any variation in interest rate will affect both configurations under study, but the magnitude of the effect will be different, depending on the nature of the money flows. Figure 22 shows the change in the life-cycle cost of Alternatives A and B. There is an interest rate for which there would be no LCC difference in the alternatives being evaluated. It is 16.9%.

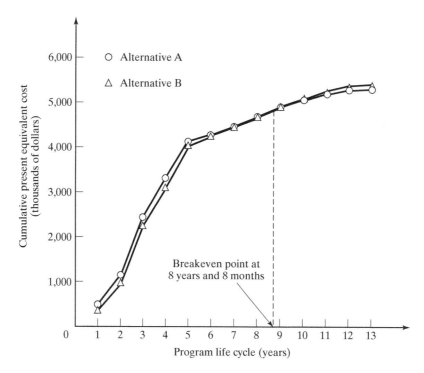

Design for Affordability (Life-Cycle Costing)

Figure 21   Break-even analysis for LCSys design alternatives.

5 LIFE-CYCLE COSTING BY ECONOMIC OPTIMIZATION

The life-cycle costing process outlined in Section 3 will now guide a generic example utilizing *economic optimization* as the underlying evaluation paradigm. This example pertains to a finite population of repairable equipment deployed to meet a demand (REPS).

The 12 steps exhibited in Figure 5 do not map as completely on this example as they did in the example of Section 4.

5.1 Introduction to the Example[21]

Suppose that a finite population of repairable equipment is to be procured and maintained in operation to meet a demand or need. As repairable equipment units fail or become

[21]A version of this example first appeared in W. J. Fabrycky and J. T. Hart, "Economic Optimization of a Finite Population System Deployed to Meet a Demand," *Proceedings of the Joint Conference*, American Association of Cost Engineers/American Institute of Industrial Engineers, Houston, TX, 1975. Also see Chapter 13 in W. J. Fabrycky and B. S. Blanchard, *Life-Cycle Cost and Economic Analysis* (Upper Saddle River, NJ: Prentice Hall, 1991).

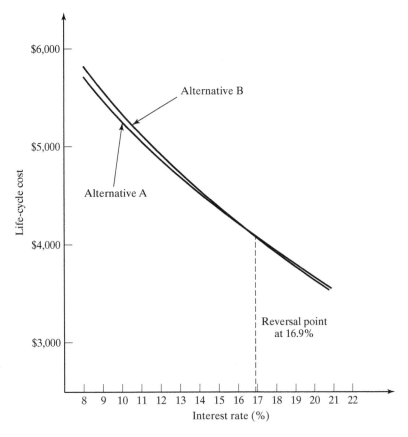

Figure 22 Sensitivity analysis for LCSys interest rate change.

unserviceable, they will be repaired and returned to service. As they age, the older units will be removed from the system and replaced with new units. The system design problem is to determine the population size, the replacement age of units, the number of repair channels, the design mean time between failures (MTBF), and the design mean time to repair (MTTR), so that design requirements will be met at a minimum life-cycle cost.

Both the airlines and the military acquire, operate, and maintain aircraft with finite population characteristics. In ground transit, vehicles such as taxicabs, rental automobiles, and trucks constitute repairable equipment populations. Production equipment types such as machine tools, weaving looms, and autoclaves are populations of equipment known as producer goods. In housing, populations of dwelling units come into being after a construction process. But, the repairable entity may exist as part of an inventory of components for a prime equipment population. For example, aircraft hydraulic actuators, water pumps, automobile starters and alternators, and automation controllers are all repairable components that must be acquired to meet a higher level system need.

Two problem versions are treated in this example. The first is to determine the population size, the replacement age of units, and the number of repair channels so that the sum of all life-cycle costs associated with the system will be minimized. This is a problem in optimizing operations and will be referred to as REPS Problem I. The second problem is to evaluate candidate system designs by predicting both the unit mean time between failures and the unit mean time to repair as a function of unit cost, as well as the population size, the replacement age of units, and the number of repair channels. This is a design situation referred to as REPS Problem II.

5.2 The Operational System

The repairable equipment population system, illustrated in Figure 23, is designed and deployed to meet a demand, D. Units within the system can be separated into two groups: those in operation and available to meet demand, and those out of operation and hence unavailable to meet demand. It is assumed that units are not discarded upon failure, but are repaired and returned to service.

As units age, they become less reliable and their maintenance costs increase. Accordingly, it is important to determine the optimum replacement age. It is assumed that the number of new units procured each year is constant and that the number of units in each age group is equal to the ratio of the total number required in the population and the desired number of age groups. Although the analysis deals with the life cycle of the units, the objective is to optimize the total system of which the units are a part. This is in keeping with the system/product relationship that exists for finite population systems.

In REPS Problem I, the decision process consists of specifying a population of units, a number of maintenance channels, and a replacement schedule for bringing new units into the system. For REPS Problem II, the decision process is extended to include establishing the unit's reliability and maintainability characteristics. In either case, the system is to be designed to meet the demand for equipment optimally.

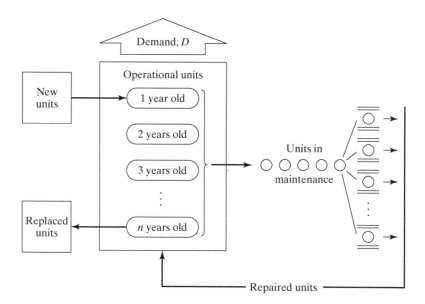

Figure 23 The repairable equipment population system (REPS).

Scope and Assumptions. Repairable equipment population systems normally come into being over a nonsteady-state buildup phase. They then operate over a steady-state interval of years, after which a phaseout period is entered. Only the steady-state mode of operation will be considered here.

The following assumptions are adopted in the development of the mathematical model and algorithm for REPS:

1. The interarrival times are exponentially distributed.
2. The repair times are exponentially distributed.
3. The number of units in the population is small such that finite population queuing formulations must be used.
4. The interarrival times are statistically independent of the repair times.
5. The repair channels are parallel and each is capable of similar performance.
6. The population size will always be larger than or at least equal to the number of service channels.
7. Each channel performs service on one unit at a time.
8. MTBF and MTTR values vary for each age group and represent the expected value for these variables for that age group.
9. Units completing repair return to operation with the same operational characteristics as their age group.

System Design Evaluation. Evaluation of REPS Problem I is based on the decision evaluation function. Here the objective is to find optimal values for controllable design variables in the face of uncontrollable system parameters. In REPS Problem II, the decision evaluation function is used to seek the best candidate system. This is accomplished by establishing values for controllable design-dependent parameters in the face of uncontrollable design-independent parameters and optimal values for design variables.

Three design variables are identified in the repairable equipment population system. These controllables are the number of units to deploy, the replacement age of units, and the number of repair channels. Optimal values are sought for these variables so that the sum of all costs associated with the repairable equipment population system will be minimized.

In Design Problem I, the focus is entirely on optimizing design variables as the only controllable factors. This situation arises when the system is in existence and the objective is to optimize its operation in the face of uncontrollable system parameters. The focus shifts to seeking the best candidate system in Design Problem II. In this activity, optimal values for design variables are secondary. They are needed as a means for comparing candidate systems equivalently. Then, when the best system design is identified, their specific values are implemented to assure its optimal operation during utilization.

Demand is the primary stimulus on the repairable equipment population system and the justification for its existence. This uncontrollable system parameter is assumed to be constant over time. Other uncontrollable system parameters are economic in nature. They include the shortage penalty cost that arises when there are insufficient units operational to meet demand, the cost of providing repair capability, and the time value of money on invested capital.

Some system parameters are uncontrollable in REPS Problem I, but controllable in REPS Problem II. These are the design MTBF and MTTR, the energy efficiency of equipment units, the design life of units, and the first cost and salvage value of these units. It is through these design-dependent system parameters that the best candidate system may be identified.

5.3 Evaluation Function Formulation

A mathematical model for system design evaluation can be formulated. The model uses annual equivalent life-cycle cost as the evaluation measure expressed as

$$AELCC = PC + OC + RC + SC \tag{10}$$

where

$$
\begin{aligned}
AELCC &= \text{annual equivalent life-cycle cost} \\
PC &= \text{annual equivalent population cost} \\
OC &= \text{annual operating cost} \\
RC &= \text{annual repair facility cost} \\
SC &= \text{annual shortage penalty cost}
\end{aligned}
$$

Annual Equivalent Population Cost. The annual equivalent cost of a deployed population of N equipment units is

$$PC = C_i N$$

with

$$C_i = P(^{A/P,i,n}) - B(^{A/F,i,n}) \tag{11}$$

Alternatively, $(P - B)(^{A/P,i,n}) + Bi$ could be used as given.

Book value, B, in Equation 11, is used to represent the original value of a unit minus its accumulated depreciation at any point in time. The depreciation of a unit over its lifetime by the straight-line method gives an expression for book value as

$$B = P - n\frac{P - F}{L} \tag{12}$$

where

$$
\begin{aligned}
C_i &= \text{annual equivalent cost per unit} \\
P &= \text{first or acquisition cost of a unit} \\
F &= \text{estimated salvage value of a unit} \\
B &= \text{book value of a unit at the end of year } n \\
L &= \text{estimated design life of the unit} \\
N &= \text{number of units in the population} \\
n &= \text{retirement age of units } n > 1 \\
i &= \text{annual interest rate}
\end{aligned}
$$

Annual Operating Cost. The annual cost of operating a population of N deployed equipment units is

$$OC = (EC + LC + PMC + \text{other})N \qquad (13)$$

where

EC = annual cost of energy consumed
LC = annual cost of operating labor
PMC = annual cost of preventive maintenance

Other annual operating costs may be incurred. These include all recurring annual costs of keeping the population of equipment units in service, such as storage cost, insurance premiums, and taxes.

Annual Repair Facility Cost. The annual cost of providing a repair facility to repair failed equipment units is

$$RC = C_r M \qquad (14)$$

where

C_r = annual fixed and variable repair cost per repair channel
M = number of repair channels

If there are a number of repair channel components with different estimated lives, then C_r is the sum of their annual costs. Some of the repair facility cost items that could be included are the cost of the building, maintenance supplies, test equipment, and so on expressed on a per channel basis. The administrative and maintenance manpower and other overhead costs would also be computed on a yearly basis and on a per channel basis.

Annual Shortage Penalty Cost. When failed equipment units cause the number in an operational state to fall below the demand, an out-of-operation or shortage cost is incurred. The annual shortage cost is the product of the shortage cost per unit short per year and the expected number of units short expressed as

$$SC = C_s[E(S)] \qquad (15)$$

The expected number of units short can be found from the probability distribution of n units short, P_n,

Define the quantity $N - D$ as the number of extra units to be held in the population. For $n = 0, 1, 2, \ldots, N - D$ there is no shortage of units. However, when

$$n = N - D + 1, \text{ a shortage of 1 unit exists}$$

$$n = N - D + 2, \text{ a shortage of 2 units exists}$$

$$\vdots$$

$$n = N, \text{ a shortage of } D \text{ units exists}$$

Design for Affordability (Life-Cycle Costing)

TABLE 10 System Parameters for REPS Optimization Problem

Parameter	Value	
Unit acquisition cost	$52,000	
Unit design life	6 years	
Unit salvage value at end of design life	$7,000	
Unit operating cost		
Energy and fuel	$500	
Operating labor	450	
Preventive maintenance	400	
Other operating costs	400	
Annual repair channel cost	$45,000	
Annual shortage cost	$73,000	
Annual interest rate	10%	
Age cohorts	MTBF	MTTR
0–1	0.20	0.03
1–2	0.24	0.04
2–3	0.29	0.05
3–4	0.29	0.05
4–5	0.26	0.06
5–6	0.22	0.07

The expected number of units short, $E(S)$, can be found by multiplying the number of units short by the probability of that occurrence as

$$E(S) = \sum_{j=1}^{D} j P_{(N-D+j)} \tag{16}$$

5.4 REPS Optimization Problem

In REPS Problem I, the decision maker has no control over system parameters but can only choose the number of equipment units to procure and deploy, the age at which units should be replaced, and the number of channels in the repair facility.

Assume that the demand D is for 15 identical equipment units. Table 10 lists system parameters for this design example.

Table 11 exhibits a set of design variables for the example under consideration. The example computations that follow are based on these variables and the system parameters in Table 10.

Annual Equivalent Population Cost. First compute the book value B of the units at retirement age after 4 years using Equation 12 as

$$\$52,000 - 4\left(\frac{\$52,000 - \$7,000}{6}\right) = \$22,000$$

TABLE 11 Design Variables for REPS Problem 2

Population, N	Repair Channels, M	Retirement Age, n
19	3	4

694

The annual equivalent unit cost per unit from Equation 11 is

$$C_i = \$52,000 \left[\frac{0.10(1.10)^4}{(1.10)^4 - 1} \right] - \$22,000 \left[\frac{0.10}{(1.10)^4 - 1} \right]$$
$$= \$16,404 - \$4,740 = \$11,664$$

from which the annual equivalent population cost is

$$PC = \$11,664(19) = \$221,616$$

Annual Operating Cost.   Annual operating cost for the deployed population is found from Equation 13 to be

$$OC = (\$500 + \$450 + \$400 + \$400)(19) = \$33,250$$

Annual Repair Facility Cost.   The annual equivalent repair channel cost for three channels from Equation 14 is

$$RC = \$45,000(3) = \$135,000$$

Annual Shortage Cost.   Calculation of the shortage cost is based on the MTBF and MTTR values from Table 10 for years 1–4. From these values, the average MTBF and MTTR for the population can be computed as[22]

$$MTBF = (1/4)(0.20 + 0.24 + 0.29 + 0.29) = 0.2550$$
$$MTTR = (1/4)(0.03 + 0.04 + 0.05 + 0.05) = 0.0425$$

The failure rate of an item and the repair rate at a repair channel are given by

$$\lambda = (1/MTBF) = (1/0.2550) = 3.9215$$
$$\mu = (1/MTTR) = (1/0.0425) = 23.5294$$

from which $\left(\dfrac{\lambda}{M} \right) = 1/6$. Next compute C_n for $n = 0, 1, \ldots, 3$ as

$$C_0 = \frac{19!(1/6)^0}{19!0!} = 1$$
$$C_1 = \frac{19!(1/6)^1}{18!1!} = 3.1665$$
$$C_2 = \frac{19!(1/6)^2}{17!2!} = 4.7496$$
$$C_3 = \frac{19!(1/6)^3}{16!3!} = 4.4856$$
$$\vdots$$

[22]Because the units are homogeneous, the aggregate MTBF and MTTR for the population can be found as

$$MTBF = \frac{1}{n} \sum_{j=1}^{n} MTBF_j$$
$$MTTR = \frac{1}{n} \sum_{j=1}^{n} MTTR_j$$

where the subscript j represents age groups and n is the number of these age groups. This follows from the superposition of Poisson processes.

Computing C_n for $n = 4, \ldots, 19$,

$$C_4 = \frac{19!(1/6)^4}{15!3!3^1} = 3.9813$$

$$C_5 = \frac{19!(1/6)^5}{14!3!3^2} = 3.3224$$

$$C_6 = \frac{19!(1/6)^6}{13!3!3^3} = 2.5840$$

$$\vdots$$

$$C_{18} = \frac{19!(1/6)^{18}}{1!3!5^{15}} = 0.0000$$

$$C_{19} = \frac{19!(1/6)^{19}}{0!3!3^{16}} = 0.0000$$

Now,

$$\sum_{n=0}^{19} C_n = 27.9390$$

and,

$$P_0 = \frac{1}{\sum\limits_{n=0}^{19} C_n} = \frac{1}{27.9390} = 0.0358$$

P_n for $n = 1, 2, \ldots, N$ can now be computed from $P_n = P_o C_n = (0.0358)C_n$ as follows:

$$P_0 = 0.0358 \times 1 = 0.0358$$
$$P_1 = 0.0358 \times 3.1665 = 0.1134$$
$$\vdots$$
$$P_5 = 0.0358 \times 3.3224 = 0.1189$$
$$P_6 = 0.0358 \times 2.5840 = 0.0925$$
$$\vdots$$
$$P_{18} = 0.0358 \times 0.0000 = 0.0000$$
$$P_{19} = 0.0358 \times 0.0000 = 0.0000$$

Now the expected number of units short can be calculated from Equation 16 as

$$E(S) = \sum_{j=1}^{D} j P_{(N-D+j)} = \sum_{j=1}^{15} j P_{(4+j)}$$

$$= 1(0.1189) + 2(0.0925) + \cdots + 15(0.0000)$$

$$= 1.00663$$

TABLE 12 Points in the Optimum Region

Retirement Age, n	Number of Units, N	Number of Repair Channels, M		
		2	3	4
3	19	$598,395	$465,985	$469,130
	18	592,920	464,770	465,755
4	19	600,720	463,350*	464,295
	20	610,775	466,610	468,755
5	19	643,050	480,375	467,735

*Optimum.

from which the annual shortage cost is

$$SC = C_S[E(S)]$$
$$= \$73,000(1.00663) = \$73,484$$

The total system annual equivalent cost may now be summarized as

$$TC = PC + OC + RC + SC$$
$$= \$221,616 + \$33,250 + \$135,000 + \$73,484 = \$463,350$$

As can be seen from Table 12, this solution is actually the optimum with neighboring points given for N, M, and n.

The shortage distribution can be calculated from the P_n values and plotted as a histogram of $Pr(S = s) = N - D + s$. In this example, $Pr(S = s) = P_{4+s}$

$$Pr(S = 0) = 0.622$$
$$Pr(S = 1) = 0.119$$
$$Pr(S = 2) = 0.093$$
$$Pr(S = 3) = 0.067$$
$$Pr(S = 4) = 0.046$$
$$Pr(S = 5) = 0.027$$
$$Pr(S = 6) = 0.015$$
$$Pr(S = 7) = 0.008$$
$$Pr(S = 8) = 0.003$$

The shortage probability histogram for this example is exhibited in Figure 24.

In Section 5.5, the adequacy of this "baseline" design will be considered in the face of design requirements. In addition, challengers in the form of alternative candidate systems will be presented and evaluated.

5.5 REPS Design Problem

In REPS Problem II, the decision maker has control over a set of design-dependent parameters in addition to the set of design variables. Accordingly, the decision evaluation function

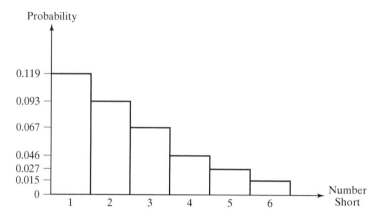

Figure 24   Shortage probability histogram.

applies. In this section, design-dependent parameters will be altered through design iteration to seek a REPS candidate more acceptable than the baseline design.

Considering Design Requirements.   In the absence of design requirements, the baseline design might be acceptable. However, assume that it is not acceptable due to the existence of the following design requirements:

1. The first cost of the deployed population must not exceed $900,000.
2. The probability of no units short of demand must be at least 0.70.
3. The mean MTBF for the unit over its life must be at least 0.20 year.
4. A unit must not be kept in service more than 4 years.

Note that the baseline design does not meet requirements 1 and 2, but it does meet requirements 3 and 4. Also, its annual equivalent life-cycle cost is noted to be $463,350. Figure 25 shows a design evaluation display for this situation with the baseline design indicated. However, because it does not meet all design requirements, one or more alternative candidates must be designed that will meet all requirements at an acceptable life-cycle cost.

Generating Candidate Systems.   Suppose that two candidate system designs are generated in the face of the demand for 15 equipment units and the design-independent parameters of Table 10. Design-dependent parameters for these candidates are given in Table 13.

Optimization over the design variables for each instance of the set of design-dependent parameters gives the results summarized in Table 14 for the candidate systems. Also included are the optimized results for the baseline design. The following observations are made:

1. Only Candidate System 2 meets all design requirements.
2. Candidate System 2 has the lowest cost per unit and, therefore, the lowest investment cost for the deployed population.
3. Candidate System 2 has the highest probability of meeting the demand for 15 service-able equipment units.

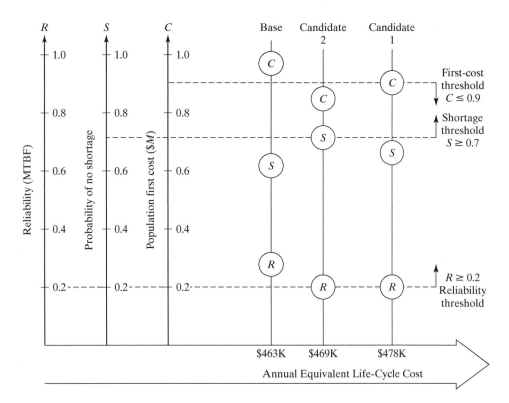

Figure 25 Design decision evaluation display for REPS.

TABLE 13 Design-Dependent Parameters for Candidate Systems

Parameter	Candidate System 1		Candidate System 2	
Unit acquisition cost	$45,000		$43,000	
Unit design life	6 years		6 years	
Unit salvage value at end of design life	$6,000		$5,000	
Unit operating cost				
Energy and fuel	$600		$800	
Operating labor	500		700	
Preventive maintenance	400		400	
Other operating costs	400		400	
Age cohorts	MTBF	MTTR	MTBF	MTTR
0–1	0.16	0.04	0.18	0.04
1–2	0.21	0.04	0.21	0.04
2–3	0.26	0.05	0.25	0.05
3–4	0.26	0.06	0.25	0.05
4–5	0.26	0.06	0.23	0.06
5–6	0.24	0.06	0.20	0.06

TABLE 14 Evaluation Summary for REPS Candidates

	Baseline Design	Candidate System 1	Candidate System 2
Number deployed, N	19	20	20
Repair channels, M	3	4	4
Retirement age, n	4	3	4
Unit cost, P	$52,000	$45,000	$43,000
Mean MTBF	0.26	0.21	0.22
Probability of no units short	0.622	0.71	0.730
Annual equivalent life-cycle cost	$463,350	$477,635	$468,825

4. Candidate System 2 consumes the most energy and requires the highest expenditure for operating labor.

5. Candidate System 2 has an annual equivalent life-cycle cost penalty of $468,825 less $463,350 or $5,475 over the baseline design, but has a lower investment cost than Candidate System 1.

There may be other candidate system designs that meet all design requirements and have life-cycle costs equal to or less than the LCC for Candidate System 2. The objective of the REPS design problem is to formalize the evaluation of candidates, with each candidate being characterized by its design-dependent parameter values; comparison of the candidates is made only after the optimized (minimized) values for life-cycle cost are found.

6 APPLICATIONS AND BENEFITS OF LIFE-CYCLE COSTING

The life-cycle cost analysis process in Figure 5 can be used as an aid in addressing a wide variety of design evaluation situations. However, the process reflected by the steps must be properly tailored to the problem at hand. One can accomplish an LCC analysis at the macrolevel, or one can go to great depths in terms of data gathering and analysis. The results of such an analysis must be to the proper depth, must be timely, and must be responsive to designer/manager involvement in the decision-making process. The important issue here is to *think economic analysis* and to *think life cycle* with the former being within the context of the latter.

Referring to Figure 4, there are numerous possible applications where life-cycle cost analysis methods can be applied. More specifically, LCC analysis may be used in the evaluation of the following:

1. Alternative technology approaches in accomplishing a feasibility analysis.
2. Alternative system operational profiles or mission scenarios.
3. Alternative maintenance and logistic support concepts.
4. Alternative approaches in considering various design methods in response to the functional analysis (i.e., trade-offs pertaining to the accomplishment of functions by the human or through automation, the selection of COTS items versus the design of new items, the appropriate mix of hardware versus software, and so on).

5. Alternative system design configurations relative to packaging schemes, levels of repair, diagnostic routines, built-in versus external tests, component selection and standardization, reliability versus maintainability, and so on.

6. Alternative sources of supply and procurement schemes.

7. Alternative production approaches (i.e., continuous versus discontinuous production, internal production versus outsourcing, quantity of production lines, number of inventory points and levels of inventory, inspection and test alternatives, etc.).

8. Alternative product distribution channels, packaging methods, transportation and handling approaches, warehouse and storage locations, and so on.

9. Alternative maintenance and logistic support policies (i.e., levels of maintenance performed, in-house versus contracted or third-party maintenance, alternative warranty and guarantee policies, etc.).

10. Alternative product disposal or recycling policies (i.e., various degrees of disposability/recyclability, alternative logistic support configurations for disposal or recycling functions, etc.).

Although there is nothing novel about addressing the above problems in the development of new systems (and the reengineering of existing systems), the emphasis here is to accomplish the necessary analysis and evaluation on the basis of economic considerations over the life cycle, and not just the short-term costs associated with design and procurement activities. At the same time, many benefits can be gained through the application of life-cycle costing methods in the evaluation and subsequent improvement of systems that are already in operational use. The identification of high-cost contributors and the follow-on system modification for improvement (realizing a reduction in projected life-cycle cost), implemented on an iterative and continuing basis, can result in many benefits in this time where resources are limited and there is a great deal of international competition. Success in this area heavily depends on the availability of good historical data and having the visibility needed for the implementation of a *continuous product/process improvement* capability.

Although the benefits are numerous, there are also some major impediments. Our current "thought processes," accounting practices, budgetary cycles, organizational objectives, and politically driven activities are more oriented to the short term. Further, on many occasions the visibility from an LCC analysis is not desired due to the fear of exposure for one reason or another.

To be successful in this area requires that the proper *organizational environment* be established that will allow it to happen. There must be a commitment to "life-cycle thinking" from the top down; the right type of data must be collected and available; and the analyst must have direct access to all applicable areas of activity. Given this, it is important to get involved by understanding the process, applying it to a known entity and evaluating the results, and establishing some cost-estimating relationships that can be applied to future LCC analyses.

Life-cycle costing, when coupled with noneconomic criteria, provides the basis for the consideration of affordability by the customer. Furthermore, life-cycle cost analysis

1. Forces long-range planning versus the more traditional short-term thinking and decision making. As a result, decisions can be based on more complete information leading to less risk.

2. Forces total cost visibility and the identification of the high-cost system elements, equipment, processes, and so on. This aids in pinpointing the specific functional areas where resource consumption is high and modifications for improvement are possible.

3. Enables a better understanding of the interrelationships between different system elements and categories of cost. The interaction effects become more visible through a life-cycle cost with sensitivity analysis.

4. Aids in the early identification of potential high-risk areas, their quantification, and the subsequent elimination of the possible causes of risk.

5. Allows for better overall resource management because of the long-term visibility that is provided.

7 SUMMARY AND EXTENSIONS

This chapter consolidates and advances the concepts, models, and methods of life-cycle costing. After discussion of the importance and place of design for affordability in systems engineering and analysis, this chapter sets forth a generic life-cycle costing process. This process is based on 12 steps that may be tailored to most life-cycle costing situations. Important aspects of a life-cycle cost analysis will not be overlooked if the 12-step process is invoked completely as appropriate. Accordingly, this generic process is the core of this chapter. It provides valuable guidance for doing life-cycle costing properly.

Two comprehensive examples of design for affordability based on life-cycle cost follow the generic 12-step process established in Figure 5. These examples are similar. They both pertain to populations of repairable entities. The first example involves a population of repairable packages making up a location/communications system called LCSys. The second example is for a general (unspecified) repairable equipment population system called REPS.

The first example (in Section 4) implements the 12-step process quite completely, utilizing money flow modeling as the underlying analysis paradigm. This requires a cost breakdown structure and the recording of cost estimates into a hierarchy of worksheets or spreadsheets. The money flow approach permits (1) phase-in and phase-out of the repairable entities, (2) decisions regarding levels of repair, and (3) cost inputs at a low level of detail. It does not permit (4) nonsteady-state failure and maintenance rates, (5) consideration of shortage penalty cost, and (6) determination of the optimum size of the population, the economic life of deployed entities, and the optimum capacity of the repair facility. The second example (in Section 5) is only guided by the 12-step process. It relies principally on the economic optimization paradigm of analysis. This requires the formulation of a series of mathematical costing functions followed by their aggregation into an evaluation function for optimization. It does not permit consideration of (1) through (3) above, but does permit (4) through (6). Accordingly, each evaluation paradigm is shown to have its own advantages and disadvantages.

Each example is very elaborate in its required input and in the computations required. Accordingly, a computer-based model is available for each one, along with input data files for the cases presented in Sections 4 and 5. These models are available from *http://www.a2i2.com.* Go to Products and then to

1. Life-Cycle Cost Calculator (LCCC). The database files for the alternatives presented in Section 4 are found as Lccc12-a.dbf for Alternative A and Lccc12-b.dbf for Alternative B.
2. Repairable Equipment Population System (REPS). The database files for the example presented in Section 5 are found as BASE.IND for design-independent parameters, and BASE.DEP, ALT_1.DEP, and ALT_2.DEP for design-dependent parameters pertaining to Alternatives A–C.

It is suggested that these models be downloaded, studied, and exercised in accordance with the examples presented in Sections 4 and 5.

Two websites are good sources of current professional events and recent information on cost analysis and life-cycle costing:

1. The Association for the Advancement of Cost Engineering (AACE) at *http://www. aacei.org.*
2. The International Society of Parametric Analysis (ISPA) at http://www.ispa-cost.org.

There are two classic books recommended to those who desire a greater depth of understanding about system cost and costing, including cost–benefit and effectiveness analysis:

1. Hendricks, H. H., and G. M. Taylor, *Program Budgeting and Benefit-Cost Analysis*, Goodyear Publishing Co., Inc., Pacific Palisades, CA, 1969.
2. Fisher, G. H., *Cost Considerations in Systems Analysis*, American Elsevier Publishing Co., Inc., New York, NY, 1971.

QUESTIONS AND PROBLEMS

1. Define *life-cycle cost* in your own words. What is included (excluded)?
2. What is meant by DTC? When in the system life cycle should it be applied? How does it relate to the TPMs (if at all)?
3. Why is life-cycle costing important?
4. What is meant by *functional costing?* How does it relate to the functional analysis?
5. Describe the steps involved in accomplishing a life-cycle cost analysis.

6. What is the cost *breakdown structure*? What are its purposes? What is included (or excluded)? How does it relate to a *work breakdown structure*?

7. How does the CBS relate to the functional analysis (if at all)?

8. Describe some of the more commonly used cost-estimating methods. Under what conditions should they be applied? Provide some examples.

9. What is *activity-based costing*? Why is it important (if at all)? What are some of the differences between this method of costing and some of the more conventional approaches?

10. Refer to Figure 9. Under what conditions should *revenues* be treated as an element of cost?

11. Refer to Figure 10. Would you consider the CBS in the figure to be adequate for all applications where an LCC analysis is required? If so, why? If not, why not?

12. Refer to Figure 13. What is meant by *parametric cost estimating?* How would you develop cost-estimating relationships?

13. Describe what is meant by a *learning curve*. How are learning curves developed? How can they be applied? What are some of the cautions that need to be addressed in applying a learning curve?

14. In this chapter, three different cost profiles are identified. What are they? Under what conditions can each be applied?

15. Refer to Table 1. Why is it sometimes beneficial to present the results of an LCC analysis in this type of a format?

16. What is meant by a *sensitivity* analysis? What type of information may be derived from the accomplishment of such? Why is the accomplishment of a sensitivity analysis important?

17. What is a *Pareto analysis?* What benefits can it provide?

18. Calculate the anticipated life-cycle cost for your personal automobile.

19. Referring to the following table, which configuration would you select? Why?

Cost Category	Configuration A		Configuration B	
	Present Cost	Percent of Total	Present Cost	Percent of Total
Research and development	$70,219	7.8	$53,246	4.2
Management	9,374	1.1	9,252	0.8
Engineering design	45,552	5.0	28,731	2.3
Test and evaluation	12,176	1.4	12,153	0.9
Technical data	3,117	0.3	3,110	0.2
Production	407,814	45.3	330,885	26.1
Construction	45,553	5.1	43,227	3.4
Manufacturing	362,261	40.2	287,658	22.7
Operations and maintenance	422,217	46.7	883,629	69.4
Operations	37,811	4.2	39,301	3.1
Maintenance	382,106	42.5	841,108	66.3
Maintenance personnel	210,659	23.4	407,219	32.2
Spares/repair parts	103,520	11.5	228,926	18.1
Test equipment	47,713	5.3	131,747	10.4
Transportation	14,404	1.6	51,838	4.1
Maintenance training	1,808	0.2	2,125	0.1
Facilities	900	0.1	1,021	Neg
Field data	3,102	0.4	18,232	1.4
Phaseout and disposal	2,300	0.2	3,220	0.3
Grand total	$900,250	100%	$1,267,760	100%

Design for Affordability (Life-Cycle Costing)

20. Refer to the table in Problem 19. What would be the likely impact on LCC if

    (a) The system MTBF is decreased?

    (b) The $\overline{\text{M}}$ct is increased?

    (c) The MLH/OH is increased?

    (d) System use is increased?

    (e) The fault-isolation capability in the system was inadequate?

What steps would you take to reduce these costs?

21. The BAF Corporation is considering the possibility of introducing Product Y into the market. A market analysis indicates that the corporation could sell all of the products that it can produce at a price of $200 each for at least 10 years into the future. The product is not repairable and is discarded at failure.

To manufacture the product, the corporation needs to invest in some capital equipment. Based on a survey of potential sources, there are three alternatives considered feasible to meet the need:

    (a) *Configuration A*   includes a machine that is semiautomatic (requiring a part-time machine operator), will produce 395 products per year, and can be purchased at a price of $15,000. The expected reliability (MTBF) of this machine is 165, and the anticipated cost per corrective maintenance action is $200. The average machine operating cost is $3.00 per hour, and the estimated salvage value after 10 years of operations is $1,000.

    (b) *Configuration B*   includes a machine that is automatic, will produce 525 products per year, and can be purchased at a price of $28,000. The expected reliability (MTBF) of this machine is 270, and the anticipated cost per corrective maintenance action is $300. Preventive maintenance is required every 6 months, and the cost per maintenance action is $250. The average machine operating cost is $0.60 per hour, and the estimated salvage value after 10 years of operations is $2,500.

    (c) *Configuration C*   includes a machine that is automatic, will produce 500 products per year, and can be purchased at a price of $23,000. The expected reliability (MTBF) of this machine is 195, and the anticipated cost per corrective maintenance action is $225. Preventive maintenance is required every 90 days, and the cost per maintenance action is $200. The average machine operating cost is $0.50 per hour, and the estimated salvage value after 10 years of operation is $2,200.

The anticipated machine use will be 8 hours per day and 270 days per year. Machine output (when considering a normal buildup rate) is expected to be 0.5 during the first year of operation and at full capacity during year 2 and on. The manufacturing cost (materials and labor associated with material procurement, material handling, quality control, inspection and test, and packaging) for Product Y is $50 using Machine A, $30 using Machine B, and $40 using Machine C. The expected allocated distribution cost (transportation, warehousing, etc.) is $10 per item. Assume that the interest rate is 10%.

    Which configuration would you select? Why?

22. A need has been identified that will require the addition of a new communications capability to a ground-based transportation system. This new communications capability, referred to as System *XYZ*, must be developed, and there are two different supplier configurations being considered for procurement. Based on the information provided, (a) compute the life-cycle cost for each of the two configurations; (b) plot the applicable cost streams (discounted and undiscounted); (c) do a break-even analysis; and (d) select a preferred approach. In computing present equivalent costs, assume a 15% discount factor. In solving this problem, *be sure to state all assumptions in a clear and concise manner.*

System *XYZ* is to be installed in a transportation vehicle, and the total number of systems in operational use for each year of the projected life cycle is noted. It is assumed that the vehicles will be distributed, with the first 12 systems in Area A, the next 12 systems in Area B, and so on.

Year Number									
1	2	3	4	5	6	7	8	9	10
0	0	10	20	40	60	60	60	35	25

It is assumed that each System *XYZ* will be used an average of 4 hours per day, 365 days per year. The vehicle operator is assigned to operate several different systems throughout the accomplishment of a mission, and it is assumed that 1% of his/her time is allocated to System *XYZ*.

System *XYZ* is a newly designed entity, and each of the two candidate configurations is packaged in Units *A–C*, as illustrated.

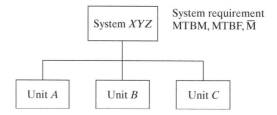

The predicted reliability and maintainability factors associated with each of the two candidate configurations are noted in the following table. Regarding the maintenance concept, two levels of maintenance are assumed (with an intermediate-level shop in each of the five geographical areas). Each of the configurations incorporates a built-in self-test capability that enables rapid system checkout and fault isolation to the unit level. No external support equipment is required for organizational maintenance in the vehicle. In the event of a "no-go" condition, fault isolation is accomplished to the unit and the applicable unit is removed, replaced with a spare, and the faulty unit is sent to the intermediate-level maintenance shop for corrective maintenance. Unit repair is accomplished through module replacement, with the modules being discarded on failure (i.e., the modules are assumed to be "nonrepairable"). Scheduled (preventive) maintenance is accomplished for Configuration *A* (Unit *A*) and Configuration *B* (Unit *B*), as given in the table, in the intermediate-level shop every 6 months. No supplier-level (or depot) maintenance is required; however, the supplier does provide backup supply and support functions when required.

Design for Affordability (Life-Cycle Costing)

Parameter*	Configuration *A*	Configuration *B*
System Level (organizational maintenance)		
MTBM	195 hours	249 hours
MTBM_u (or MTBF)	267 hours	377 hours
$\overline{\text{M}}$	30 minutes	30 minutes
Unit Level (intermediate maintenance)		
Unit *A*		
MTBM	382 hours	800 hours
MTBM_u	800 hours	800 hours
MTBM_s	730 hours	—
$\overline{\text{M}}\text{ct}$	5 hours	5 hours
$\overline{\text{M}}\text{ct}$	16 hours	—
Unit *B*		
MTBM	500 hours	422 hours
MTBM_u	500 hours	1,000 hours
MTBM_s		730 hours
$\overline{\text{M}}\text{ct}$	4 hours	5 hours
$\overline{\text{M}}\text{ct}$	—	12 hours
Unit *C*		
MTBM	2,000 hours	2,500 hours
$\overline{\text{M}}\text{ct}$	2 hours	3 hours

*Assume that MTBM_u= MTBF. When there is no scheduled maintenance, MTBM_u= MTBM.

The requirements for System *XYZ* dictate the program profile in the following figure. Assume that life-cycle costs are broken down into the three categories represented by the blocks in the program profile (i.e., design and development, production, and operations and maintenance).

Design for Affordability (Life-Cycle Costing)

In an attempt to simplify the problem, the following additional factors are assumed:

a. Design and development costs for System XYZ (to include labor and material):

> Configuration A: $80,000 ($50,000/year 1 and $30,000/year 2)
> Configuration B: $100,000 ($70,000/year 1 and $30,000/year 2)

b. Design and development costs for special support equipment at the intermediate level of maintenance:

> Configuration A: $30,000 ($20,000/year 1 and $10,000/year 2)
> Configuration B: $23,000 ($17,000/year 1 and $6,000/year 2)

c. System XYZ models for operational use are produced and delivered in the year before the identified need (i.e., 10 models are produced and delivered in year 2, etc.). The production costs for each System XYZ (Units A–C) are

> Configuration A: $21,000
> Configuration B: $23,000

d. Special support equipment is required at each intermediate maintenance shop (for the corrective maintenance of units) at the start of the year when System XYZ operational models are distributed (i.e., Area A at the beginning of year 3). In addition, a backup set of special support equipment is required at the supplier location when the first intermediate shop becomes operational. Special support equipment is produced and delivered at a cost of

> Configuration A support equipment: $13,000
> Configuration B support equipment: $12,000

e. Spare units are required at each intermediate-level maintenance shop at the time of activation. Assume that one Unit A, one Unit B, and one Unit C constitute a set of spares and that the cost of a set is equivalent to the cost of a production system (i.e., $21,000 for Configuration A and $23,000 for Configuration B). Also, assume that a set of spares is stocked at the supplier's facility at the time when the first intermediate shop is activated.

Additional spares constitute components (i.e., assemblies, modules, parts, etc.). Assume that the material costs are $250 per corrective maintenance actions, and $100 per preventive maintenance action. The cost factors include amortized inventory maintenance costs.

f. Maintenance facilities, as defined here, include the supporting resources required for System XYZ (i.e., at the intermediate shop), above and beyond spares/inventories, personnel, and data. A burden rate of $1 per direct maintenance man-hour associated with the prime equipment is assumed.

g. Maintenance data include the preparation and distribution of maintenance reports, failure reports, and related data associated with each maintenance action. Maintenance data costs are assumed to be $25 per maintenance action.

h. For each maintenance action at the system level, one low-skilled technician at $20 per direct maintenance man-hour is required on a full-time basis. It is assumed that this is an average value, applied throughout the life cycle, and it includes direct, indirect, and inflationary factors. The $\overline{M}$ is 30 minutes for each of the two configurations.

    i. For each corrective maintenance action involving Unit A, B, or C, two technicians are required on a full-time basis (i.e., the duration of the $\overline{M}ct$ value). One low-skilled technician at $20 per hour and one high-skilled technician at $30 per hour are required. Direct, indirect, and inflationary factors are considered in these average values.

    j. For each preventive maintenance action involving units (Configurations A and B), one high-skilled technician at $30 per hour is required on a full-time basis (i.e., the duration of the $\overline{M}pt$ value).

    k. For the operation of System XYZ, the allocated cost for the operator is $40 per hour.

23. Referring to the figure, there are four feasible configurations identified in the trade-off area. Which one would you select? Why?

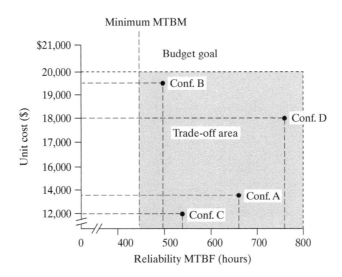

24. The figure that follows results from a break-even analysis, influenced by a sensitivity analysis. What steps would you take to reduce the identified risk?

25. Select a simple system of your choice and accomplish a life-cycle cost analysis.

26. Describe some of the benefits derived from accomplishing a life-cycle cost analysis. What are some of the impediments? How would you go about removing some of these impediments?

27. Perform the calculations to verify that the annual equivalent life-cycle cost for Candidate System 2 in Table 14 is truly $468,825.

28. Show that the probability of one or more units short for Candidate System 2 in Table 14 is truly 0.27.

29. Download the Life-Cycle Cost Calculator and use it to present an exhibit of the available graphics for each alternative in Section 4.

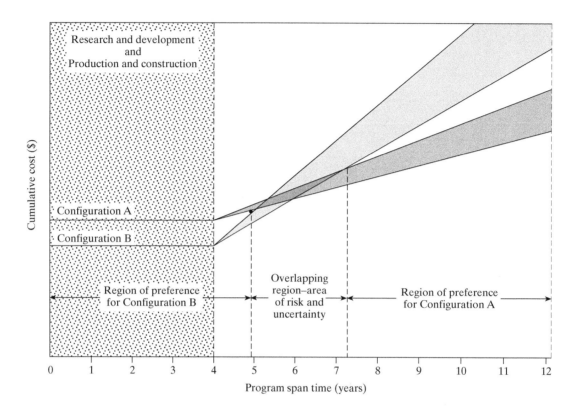

30. Download and use the Life-Cycle Cost Calculator to perform a sensitivity analysis on a design-independent variable different from the interest rate chosen for the example of Section 4.

31. Download REPS and use it to generate and evaluate another design alternative for the example of Section 5 by plotting the new alternative on the design evaluation display of Figure 25.

Systems Engineering Planning and Organization

From Chapter 18 of *Systems Engineering and Analysis,* Fifth Edition, Benjamin S. Blanchard, Wolter J. Fabrycky. Copyright © 2011 by Pearson Education, Inc. Published by Pearson Prentice Hall. All rights reserved.

Systems Engineering Planning and Organization

The successful implementation of the systems engineering process and the associated requirements critically depends not just on the "technology" alone but on the "organization and management" approach as well. The combined iteration and integration on both sides of the spectrum, as shown in Figure 1, is required.

An initial step in managing the systems engineering process is to develop an early implementation plan and an organizational structure that will be responsive to program requirements. Systems engineering planning starts with the identification of a customer need and the definition of requirements for a program (project) to design, develop, produce, and deliver a system that will be responsive and affordable. Although every program is somewhat different, its overall planning is usually promulgated through a *program management plan* (PMP) or equivalent. From this top-level plan, a *systems engineering management plan* (SEMP), or *systems engineering plan* (SEP), is derived to guide implementation of the technical activities.

A SEMP, which should be prepared during the conceptual design phase, provides the necessary guidance for the many design and management plans required for a given program. Included within the SEMP is the identification of systems engineering program tasks, a program work breakdown structure (WBS), task schedule and cost requirements, and the needed organizational structure for program implementation. In developing an organizational approach, it is essential that an environment be established that will allow for the effective and efficient coordination and integration of the various engineering and supporting disciplines that contribute to the overall system design process. Appropriate leadership must also be in place to promote good communications across organizational lines and to foster a truly interdisciplinary approach to system design and development.

The primary objective of systems engineering management is to facilitate the timely integration of numerous design considerations into a functioning system that will be of high value to the customer. Accordingly, the purpose in this chapter is to provide the reader with an

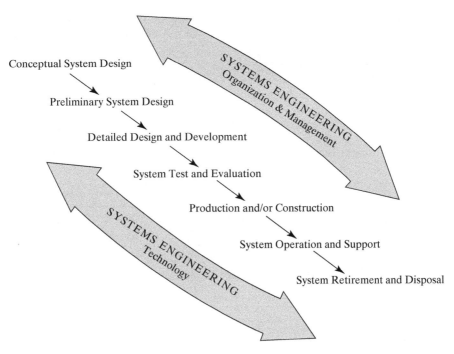

Conceptual System Design

Preliminary System Design

Detailed Design and Development

System Test and Evaluation

Production and/or Construction

System Operation and Support

System Retirement and Disposal

Figure 1 Management and technology applied to the systems engineering process.

understanding of the basic systems engineering planning and organizational requirements for a typical program through consideration of the following factors:

- Systems engineering program planning needs;
- The development of a systems engineering management plan (SEMP)—statement of work (SOW), systems engineering program tasks, work breakdown structure (WBS), scheduling of tasks, projecting costs for program tasks, interfacing with other planning activities; and
- Organization for systems engineering—developing the organizational structure, customer/producer/supplier relationships, customer organization and functions, producer organization and functions, supplier organization and functions, and staffing the organization (human resource requirements).

There are a number of resources available to the reader to augment and extend the topics in this chapter. Selected websites are suggested in the summary and extensions section at the end of the chapter.

1 SYSTEMS ENGINEERING PROGRAM PLANNING

As the need for a new (or reengineered) system is identified, early planning commences with the preparation of a *program management plan* (PMP), which, in turn, leads to the development of a *systems engineering management plan (SEMP)*. This early advance system planning results in a description of the tasks that need to be accomplished for bringing the system into being along with applicable schedules, program resource requirements, and organizational approach. This early planning commences with the identification of management-related requirements for the conceptual design phase, followed by the applicable requirements for the preliminary system design phase, and so on. These requirements must not only include those associated with the new system being developed but also those interface requirements associated with other systems that may be contained within the same system-of-systems (SOS) network as applicable.

As the program evolves through the conceptual design phase, system technical requirements are established through the definition of operational requirements, maintenance and support concept, identification and prioritization of technical performance measures (TPMs), functional analysis, and the allocation of design criteria from the system level and down to major subsystems and other elements of the system. These system-level requirements are then included in a *system specification* (Type *A*). Top-level specification forms the basis for all lower-tier specifications (i.e., Types *B, C, D,* and *E* specifications).

When preparing both plans and specifications, it is important to establish the proper relationships between the two and to ensure that the planning documentation references the proper specifications (and vice versa). In other words, the two areas of activity (documentation) must "talk to each other" and must be mutually supportive. Figure 2 illustrates these relationships, with emphasis on the planning side of the spectrum.

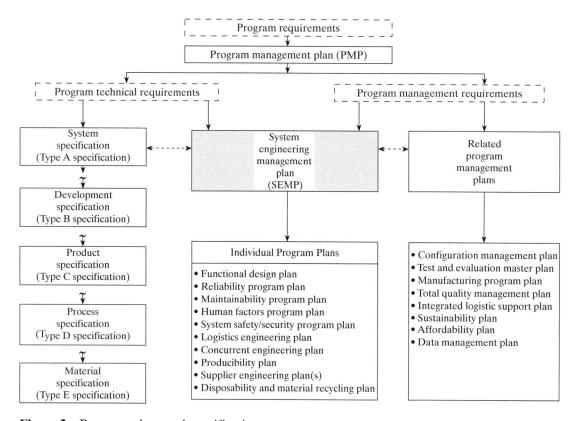

Figure 2 Program plans and specifications.

2 SYSTEMS ENGINEERING MANAGEMENT PLAN

The SEMP is the key management document covering the activities, milestones, organization, and resource requirements necessary to accomplish the functions/tasks discussed throughout this text. The objectives of the SEMP, which is usually developed during the conceptual design phase, are to provide the structure, policies, and procedures to foster the integration of the engineering and support activities needed for system design and development. Referring to Figure 2, the SEMP facilitates the integration of all design-oriented plans and provides the necessary communication links with other key planning activities (e.g., the configuration management plan, total quality management plan, sustainability plan, and data management plan).

In developing a SEMP, the format adopted may vary somewhat depending on the type and nature of the system being acquired, the organizational approach preferred, and so on. In any event, the plan must be properly tailored for the magnitude of the system level-of-effort anticipated. Two examples of SEMP format outlines are presented in Figures 3 and 4, respectively.

Referring to Figure 4 (which is preferred by the authors), Part I includes a description of program requirements, SOW, WBS, description of systems engineering tasks, associated schedules and cost projections, organizational structure, and all of the management functions necessary in carrying out a successful program in response to systems engineering objectives. Part II covers the systems engineering process. The purpose is to describe the process that is to be planned and managed, or the basis for all program planning and management requirements. Part IV covers the integration of the appropriate key engineering disciplines that are required for implementation of the process described in Part II. This includes the integration of the planning activities associated with reliability engineering, maintainability engineering, human factors and safety engineering, logistics and supportability, and other engineering disciplines, as applicable. Thus, the proposed SEMP includes a description of the management requirements for a given program, the activities to be managed, the interfaces to be maintained, and the resources required for program implementation.

2.1 Statement of Work

The SOW is a narrative description of the work required for a given project. Regarding the SEMP, it must be developed from the overall project SOW described in the PMP and it should include the following:

1. A summary statement of the tasks to be accomplished.
2. An identification of the input requirements from other tasks. These may include the results from other tasks accomplished within the project, tasks completed by the customer, or tasks accomplished by a supplier.
3. References to applicable specifications (to include the System Type *A* Specification), standards, procedures, and related documentation as necessary for the completion of the defined scope of work. These references should be identified as key requirements in the documentation tree.
4. A description of the specific results to be achieved. This may include deliverable equipment, software, design data, reports, or related documentation, along with the proposed schedule of delivery.

SYSTEMS ENGINEERING MANAGEMENT PLAN
(Outline: INCOSE, "Systems Engineering Handbook")

1. Title Page, Table of Contents, Scope, Applicable Documents
2. Systems Engineering Process

   2.1 Systems Engineering Process Planning—decision database (deliverables), process inputs, technical objectives, work breakdown structure, training, standards and procedures, resource allocation, constraints, work authorization, verification planning.

   2.2 Requirements Analysis—reliability and availability; maintainability, supportability, and integrated logistics support (ILS); survivability; electromagnetic compatibility; human engineering and human systems integration; safety, health hazards, and environmental impact; system security; producibility; test and evaluation; testability and integrated diagnostics; computer resources; transportability; infrastructure support; other engineering specialties.

   2.3 Functional Analysis—scope, approach, methods, procedures, tools (system-level functional block diagram).

   2.4 Synthesis—approach, methods to transform the functional architecture into a physical architecture, to define alternative system concepts, to define physical interfaces, and to select preferred product and process solutions.

   2.5 Systems Analysis and Control—trade studies, system/cost effectiveness analyses, risk management, configuration management, interface management, data management, systems engineering master schedule (SEMS), technical performance measurement (TPM), technical reviews (design reviews), supplier control, requirements traceability.

3. Transitioning Critical Technologies—activities, risks, criteria for selecting technologies and for transitioning these technologies.

4. Integration of the Systems Engineering Effort—team organization, technology verifications, process proofing, manufacturing of engineering test articles, development test and evaluation, implementation of software designs for system end items, sustaining engineering and problem solution support, other systems engineering implementation tasks.

5. Additional Systems Engineering Activities—long-lead items, engineering tools, design to cost/cost as an independent variable, value engineering, system integration plan, compatibility with supporting activities, other plans and controls.

6. Systems Engineering Scheduling—systems engineering master schedule (SEMS), systems engineering detailed schedule (SEDS).

7. Systems Engineering Process Metrics—cost and schedule performance measurement, other process control techniques (control charts).

8. Role and Function of Reviews and Audits.
9. Notes and Appendices.

Figure 3 Systems engineering management plan (SEMP) outline.
Source: INCOSE-TP-2003-016-02, *Systems Engineering Handbook* (San Diego, CA: International Council on Systems Engineering, June 2004).

In preparing a SOW, the following general guidelines are considered to be appropriate:

1. The SOW should be relatively short and to the point (not to exceed two pages) and must be written in a clear and precise manner.
2. Every effort must be made to avoid ambiguity and the possibility of misinterpretation by the reader.

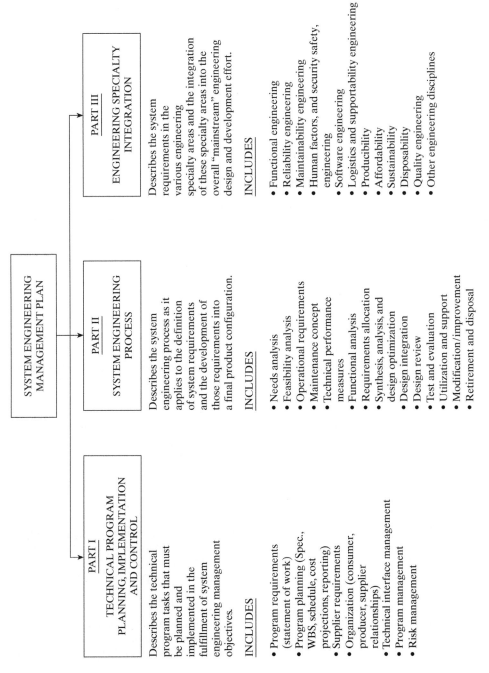

SYSTEM ENGINEERING MANAGEMENT PLAN

PART I
TECHNICAL PROGRAM PLANNING, IMPLEMENTATION AND CONTROL

Describes the technical program tasks that must be planned and implemented in the fulfillment of system engineering management objectives.

INCLUDES

- Program requirements (statement of work)
- Program planning (Spec., WBS, schedule, cost projections, reporting)
- Supplier requirements
- Organization (consumer, producer, supplier relationships)
- Technical interface management
- Program management
- Risk management

PART II
SYSTEM ENGINEERING PROCESS

Describes the system engineering process as it applies to the definition of system requirements and the development of those requirements into a final product configuration.

INCLUDES

- Needs analysis
- Feasibility analysis
- Operational requirements
- Maintenance concept
- Technical performance measures
- Functional analysis
- Requirements allocation
- Synthesis, analysis, and design optimization
- Design integration
- Design review
- Test and evaluation
- Utilization and support
- Modification/improvement
- Retirement and disposal

PART III
ENGINEERING SPECIALTY INTEGRATION

Describes the system requirements in the various engineering specialty areas and the integration of these specialty areas into the overall "mainstream" engineering design and development effort.

INCLUDES

- Functional engineering
- Reliability engineering
- Maintainability engineering
- Human factors, and security safety, engineering
- Software engineering
- Logistics and supportability engineering
- Producibility
- Affordability
- Sustainability
- Disposability
- Quality engineering
- Other engineering disciplines

Figure 4 Systems engineering management plan (SEMP) — a generic approach. *Source:* Modification of an outline that was included in the *Systems Engineering Management Guide*, Systems Engineering Management College, 1990, and prior editions.

717

3. Describe the requirements in sufficient detail to ensure clarity, considering both practical applications and possible legal interpretations. Do not *underspecify* or *overspecify*.

4. Avoid unnecessary repetition and the incorporation of extraneous material and requirements. This could result in unnecessary cost.

5. Do not repeat detailed specifications and requirements that are already covered in the applicable referenced documentation.

The SOW will be read by many different individuals with a variety of backgrounds (e.g., engineers, accountants, contract managers, schedulers, and lawyers), and there must be no unanswered questions as to the scope of work desired. It forms a basis for the definition and costing of detailed tasks, for the establishment of subcontractor and supplier requirements, and so on.

2.2 Systems Engineering Program Tasks

Systems engineering, as defined throughout this text, covers a broad spectrum of activity. It may even appear that the "systems engineer," or the "systems engineering organization," does everything. Although this is not intended, nor practical, the fulfillment of systems engineering objectives does require some involvement, either directly or indirectly, in almost every facet of program activity. The challenge is to identify those functions (or tasks) that deal with the overall *system* as an entity and, when successfully completed, will have a positive impact on the many related and subordinate tasks that must be accomplished. Although there are variations from one program to the next, the following tasks have been identified where strong *leadership* from the systems engineering organization (or a designated alternate activity) is required:

1. *Perform a needs analysis and conduct feasibility studies.* These activities should be the responsibility of the systems engineering organization because they deal with the system as an entity and are fundamental in the initial interpretation and subsequent definition of system requirements. A strong interface and good communications with the customer/user are required.

2. *Define system operational requirements and the maintenance concept, and identify and prioritize the technical performance measures.* The results of these activities are included in the overall defininition of system-level requirements and are the basis for top-down system design. Trade-off studies are accomplished, requiring an excellent understanding of system interfaces and the objectives of the customer/user.

3. *Accomplish a functional analysis at the system level and allocate requirements to the next lower level.* It is essential that a good *functional* baseline be defined from which specific resource requirements will be identified. This baseline serves as a common frame of reference used as an input source for many of the engineering and support activities accomplished later on. Although the systems engineering organization may not accomplish the functional analysis in total, a strong leadership role is essential in providing a necessary foundation at the system level. Good *baseline management* is essential.

4. *Prepare system specification, Type A.* This represents the top "technical" document for design and serves as the basis for the development of all subordinate specifications. The systems engineering organization should be responsible for the preparation of this top-level document and should ensure that all subordinate specifications "track" and are mutually supportive.

5. *Prepare the test and evaluation master plan* (TEMP). Because this document reflects the approach for system *validation*, the systems engineering organization should assume a leadership role to ensure that there is connectivity between the prioritized TPMs, the levels of design complexity, the risks associated with pursuing given design approaches, and so on. As specific quantitative requirements are initially defined, a viable plan must be established for the validation of these requirements. The systems engineering organization needs to ensure an integrated test and evaluation effort and that the system specification, the SEMP, and other design plans "talk" to each other.

6. *Prepare the systems engineering management plan* (SEMP). Because this plan constitutes the top-level "integrating" engineering document that ensures systems engineering objectives are fulfilled, it must be prepared, revised as necessary, and implemented under the leadership of the systems engineering organization (refer to Figures 2 and 4).

7. *Accomplish synthesis, analysis, and evaluation.* Although this area of activity is continuous and widespread across the design community, the systems engineering organization must provide an "oversight" function to ensure that the major day-to-day design decisions are in compliance with the system specification, the results of trade-off studies are properly documented, and the design risks have been properly identified and addressed.

8. *Plan, coordinate, and conduct formal design review meetings.* The conductance of formal design reviews is necessary to ensure that the systems engineering process is being properly implemented, the appropriate *functional* baselines are well defined, good configuration management is being implemented, and all members of the design team understand the basis for past decisions and are "tracking" the same design database. The systems engineering organization, or designated representative, should "Chair" the formal design review meetings.

9. *Monitor and review system test and evaluation activities.* This constitutes a systems engineering organization "oversight" function to review the results of test and evaluation (and thus "validation") to ensure that systems requirements are being met. If not, the problems need to be identified rapidly, followed by implementation of the appropriate corrective action.

10. *Coordinate and review all formal design changes and modifications for improvement.* The systems engineering organization, or designated representative, should "Chair" the Change Control Board (CCB) to ensure that (1) system requirements (i.e., all TPMs) will still be met should a change be approved; (2) all proposed changes have been evaluated and selected considering impacts on mission criticality, system effectiveness, life-cycle cost (LCC), and environmental factors; (3) a comprehensive modification and rework plan has been developed; and (4) the approved changes are incorporated efficiently and in a timely manner. The application of good configuration management is required here.

11. *Initiate and establish the necessary ongoing liaison activities throughout the production/ construction, utilization and sustaining support, and retirement and material disposal phases.* The systems engineering organization must maintain surveillance of production/construction activities to ensure that the system is being produced/constructed as designed. Further, an ongoing level of surveillance of consumer/user operations in the field is essential in order to provide the feedback necessary for the purpose of system "validation."

These 11 basic program tasks constitute an example of what might be appropriate for a typical large-scale program, although the specific requirements may vary from one instance to the next. The goal is to identify tasks that are oriented to the *system* and are *critical*. The overall objectives are to ensure that (1) the requirements for the system are initially well defined from the beginning, (2) the appropriate characteristics and attributes are *designed-in*, and (3) the system has been validated in terms of the initially specified requirements.

2.3 Work Breakdown Structure

One of the first steps in the program planning process after the generation of the SOW is the development of a WBS. The WBS is a product-oriented family tree that leads to the identification of the functions, activities, tasks, subtasks, work packages, and so on, that must be performed for the completion of a given program. It displays and defines the system (or product) to be developed, produced, operated and supported, and portrays all of the elements of work to be accomplished. The WBS is *not* an organizational chart in terms of project personnel assignments and responsibilities, but does represent an organization of work packages prepared for the purposes of program planning, budgeting, contracting, and reporting.

Figure 5 illustrates an approach to the development of the WBS. During the early stages of planning, a *summary work breakdown structure* (SWBS) should be prepared to include all elements of activity through the projected system life cycle, working from the top down. While the SWBS often includes only system *acquisition* activities, the approach recommended herein is to include all activities (design and development, production, utilization, and support) because the implementation of systems engineering principles and concepts requires that one consider the entire life cycle and all of the functions that need to be performed therein. Referring to the figure, the SWBS generally includes the following three levels of activity:

1. *Level 1.* Identifies the total anticipated scope of work related to the design and development, production, distribution, operation, support, and retirement of a system.
2. *Level 2.* Identifies the various projects, or categories of activity, that must be completed in response to program requirements. It may also include major elements of a system or significant project activities (e.g., subsystems, equipment, software, facilities, data, elements of support, program management, etc.). Program budgets are usually prepared at this level.
3. *Level 3.* Identifies the functions, activities, major tasks, or components of the system that are directly subordinate to Level 2 items. Program schedules are prepared at this level.

As program planning progresses and individual negotiations are consummated, the SWBS may be developed further and adapted to a particular contract, or procurement action, resulting in a *contract work breakdown structure* (CWBS). Referring to Figure 5, the SWBS can be broken down as shown; that is, a CWBS for the elements of work in the preliminary system design phase, another CWBS to reflect the elements of work in the detail design and development phase, and so on.

Figure 6 presents a sample SWBS reflecting all anticipated levels of activity throughout the life cycle of System *XYZ*. System engineering activities are reflected under Category 3B1100, an element of "Design and Development (2B1000)." Although such activities are accomplished throughout all phases of the life cycle and systems engineering activities cut across many of the other categories, this category serves as a focal point for the

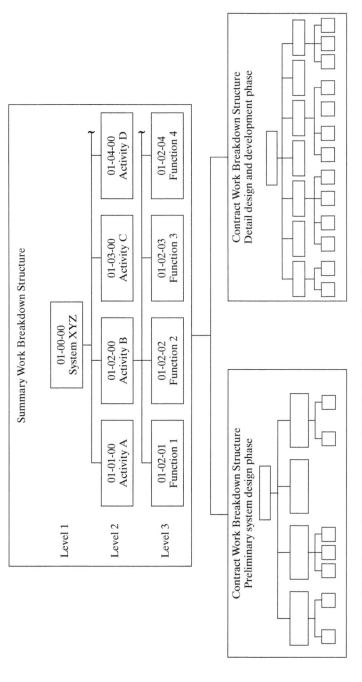

Figure 5 Work breakdown structure (WBS) development (partial).

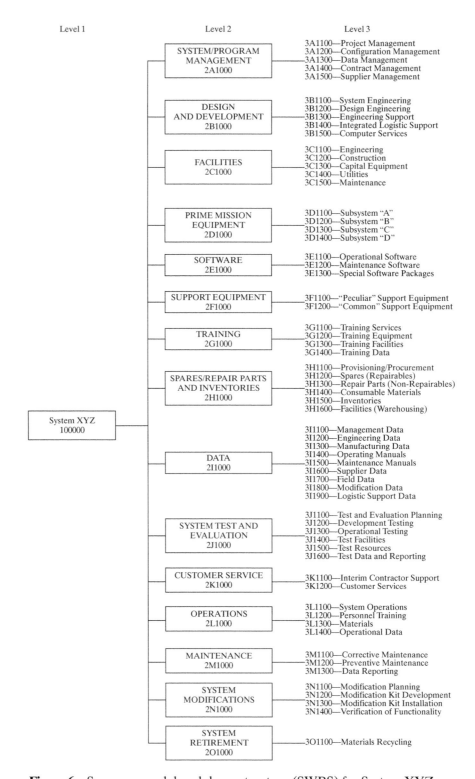

Level 1   Level 2   Level 3

3A1100—Project Management
SYSTEM/PROGRAM   3A1200—Configuration Management
MANAGEMENT   3A1300—Data Management
2A1000   3A1400—Contract Management
3A1500—Supplier Management

3B1100—System Engineering
DESIGN   3B1200—Design Engineering
AND DEVELOPMENT   3B1300—Engineering Support
2B1000   3B1400—Integrated Logistic Support
3B1500—Computer Services

3C1100—Engineering
FACILITIES   3C1200—Construction
2C1000   3C1300—Capital Equipment
3C1400—Utilities
3C1500—Maintenance

PRIME MISSION   3D1100—Subsystem "A"
EQUIPMENT   3D1200—Subsystem "B"
2D1000   3D1300—Subsystem "C"
3D1400—Subsystem "D"

SOFTWARE   3E1100—Operational Software
2E1000   3E1200—Maintenance Software
3E1300—Special Software Packages

SUPPORT EQUIPMENT   3F1100—"Peculiar" Support Equipment
2F1000   3F1200—"Common" Support Equipment

TRAINING   3G1100—Training Services
2G1000   3G1200—Training Equipment
3G1300—Training Facilities
3G1400—Training Data

3H1100—Provisioning/Procurement
SPARES/REPAIR PARTS   3H1200—Spares (Repairables)
AND INVENTORIES   3H1300—Repair Parts (Non-Repairables)
2H1000   3H1400—Consumable Materials
3H1500—Inventories
3H1600—Facilities (Warehousing)

3I1100—Management Data
3I1200—Engineering Data
3I1300—Manufacturing Data
DATA   3I1400—Operating Manuals
System XYZ   2I1000   3I1500—Maintenance Manuals
100000   3I1600—Supplier Data
3I1700—Field Data
3I1800—Modification Data
3I1900—Logistic Support Data

3J1100—Test and Evaluation Planning
SYSTEM TEST AND   3J1200—Development Testing
EVALUATION   3J1300—Operational Testing
2J1000   3J1400—Test Facilities
3J1500—Test Resources
3J1600—Test Data and Reporting

CUSTOMER SERVICE   3K1100—Interim Contractor Support
2K1000   3K1200—Customer Services

OPERATIONS   3L1100—System Operations
2L1000   3L1200—Personnel Training
3L1300—Materials
3L1400—Operational Data

MAINTENANCE   3M1100—Corrective Maintenance
2M1000   3M1200—Preventive Maintenance
3M1300—Data Reporting

SYSTEM   3N1100—Modification Planning
MODIFICATIONS   3N1200—Modification Kit Development
2N1000   3N1300—Modification Kit Installation
3N1400—Verification of Functionality

SYSTEM
RETIREMENT   3O1100—Materials Recycling
2O1000

Figure 6 Summary work breakdown structure (SWBS) for System XYZ.

722

establishment of initial budgeting requirements for systems engineering tasks and later for collecting costs resulting from task completion. As an initial step, the 11 systems engineering tasks discussed in the previous section can be included within Category 3B1100 of the SWBS and then can be amplified or adjusted to meet specific program requirements through the development of a CWBS, such as presented in Figure 7.

In summary, the WBS provides a mechanism for the purposes of initial program budgeting and subsequently for cost collection and the reporting of progress against individual work packages. It constitutes an excellent program management tool and is utilized in many program management situations.

2.4 The Scheduling of Tasks

Available scheduling methods include the use of *bar charts*, *milestone charts*, *Gantt charts*, *program networks*, or various combinations thereof. The selection of a specific technique is dependent both on the method(s) used in the overall PMP for a given program and on the nature of activities and the specific program phase (conceptual design, detail design and development, and production). As a start, Figure 8 identifies some of the major systems engineering activities, presented in the form of a combined bar/milestone chart.

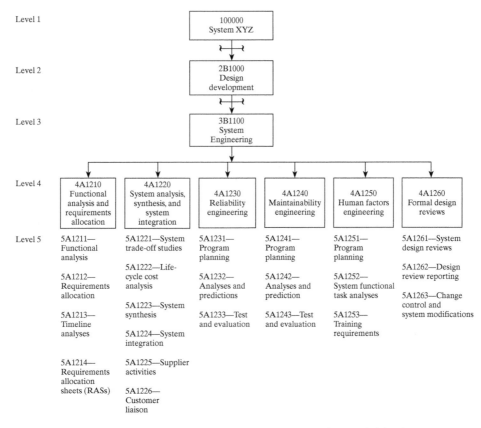

Figure 7 CWBS expansion showing systems engineering activities in the preliminary design phase (partial).

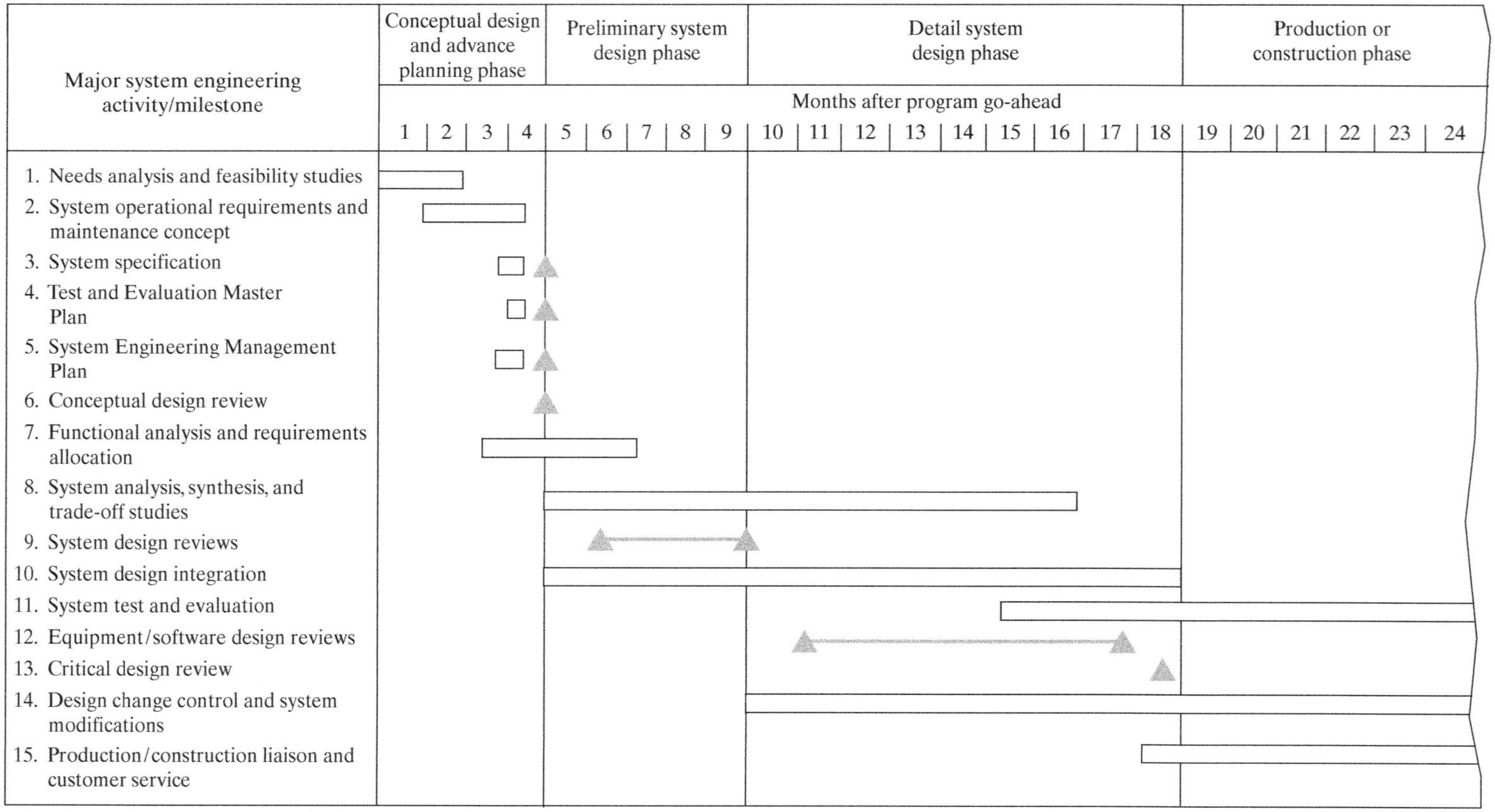

Figure 8 Major systems engineering activities and milestones (example).

Although application of the traditional bar or milestone chart is popular, it is often difficult to ensure that the proper interfaces are being addressed as the program manager "tracks" progress against each of the 15 tasks shown. In other words, one may attempt to assess progress for Task 10, "System Design Integration," in terms of the scheduled timeline. However, in accomplishing systems engineering activities, there may be numerous inputs from various sources that are required to complete this task in a satisfactory manner. Thus, if this scheduling approach is used, care must be taken to ensure that all of the inputs and outputs are properly defined and can be properly assessed.

A preferred method of scheduling is through the use of program networks such as the program evaluation and review technique (PERT), the critical path method (CPM), or various combinations of these. PERT and CPM are ideally suited for early planning where there are many interfaces (i.e., suppliers providing input from various parts of the world), precise task time data are not readily available, and the aspects of probability are introduced to aid in defining the risks associated with the many day-to-day design decisions.

In applying the PERT/CPM scheduling approach to a project, one must initially identify all interdependent *activities* and *events* for each applicable phase of the project. Activities refering to continuous levels of effort and events are related to program milestone dates based on management objectives. Managers and programmers work with engineering organizations to define these objectives and identify specific tasks and subtasks. When this is accomplished to the necessary level of detail, networks are developed which start with a summary network covering the key system engineering activities identified in the WBS. An example is illustrated in Figure 9. Figure 10 lists the activities that are reflected by the lines in the network, and Figure 11 provides an example of program network calculations.

When constructing networks, one starts with the end objective (e.g., the system is delivered to the customer) and works backward until the beginning or starting event is identified. Each event is labeled, coded, and checked in terms of program timeframe. Activities are then identified and checked to ensure that they are properly sequenced. Activity times are estimated and these times are stated in terms of their probability of occurrence. Some activities can be performed on a concurrent basis, but others must be accomplished in series. For each completed network, there is one beginning event and one ending event, with all activities leading to the ending event. Finally, a summary network can be expanded and broken down into lower-level networks (i.e., a network for a reliability engineering program, another network for a maintainability program, etc.).

The application of network scheduling is appropriate for both small- and large-scale projects and is of particular value for a one-of-a-kind system development effort where there are numerous interdependencies or for those programs where repetitive tasks are not predominant. Networking is readily adaptable to advance planning and forces the precise definition of tasks, task sequences, and task interrelationships. The technique enables management and engineering to predict with some degree of certainty the probable time that it will take to achieve an objective. It also enables the rapid assessment of progress and the detection of problems and delays, and is particularly adaptable to computer methods. The application of PERT/CPM scheduling is particularly appropriate for systems engineering where there are many different and varied activities that must be integrated in a timely manner, and where the early identification of potential areas of risk is critical.

2.5 Projecting Costs for Program Tasks

The WBS serves as the basis for the initial identification of functions, activities, tasks, and the grouping of tasks into work packages. Further, the WBS provides the framework against which cost projections are made and budgets are developed for projects. The subsequent

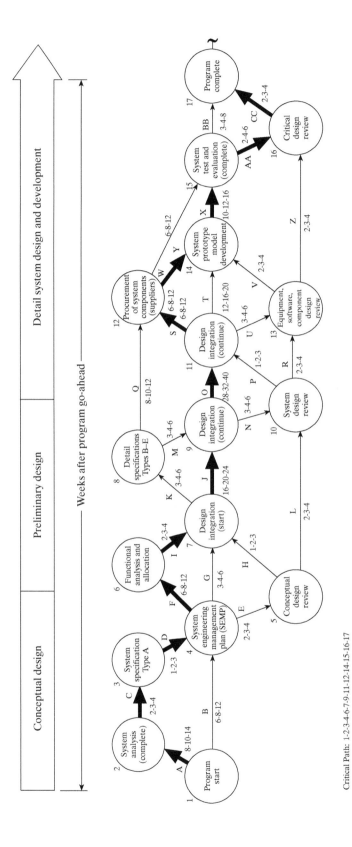

| Conceptual design | Preliminary design | Detail system design and development |

— Weeks after program go-ahead —

Critical Path: 1-2-3-4-6-7-9-11-12-14-15-16-17

Figure 9 Systems engineering program summary network schedule (example).

Activity	Description of Program Activity	Activity	Description of Program Activity
A	Perform needs analysis, conduct feasibility studies, and accomplish systems analysis (operational requirements, maintenance concept, and functional definition of the system).	Q	Identify the appropriate system component suppliers, impose the necessary specification requirements through contracts, and monitor supplier activities.
B	Conduct advance planning, perform initial management functions, and complete the Systems Engineering Management Plan (SEMP).	R	Conduct the necessary planning and prepare for the Equipment/Software/Component Design Reviews (there may be a series of individual design reviews covering different system components).
C	Prepare the System Specification (Type A).	S	Provide detail design data (as necessary) to support supplier operations.
D	Develop system-level technical requirements for inclusion in the Systems Engineering Management Plan (SEMP).	T	Develop a prototype model, with associated support, in preparation for system test and evaluation.
E	Prepare system-level design data and supporting materials for the Conceptual Design Review.	U	Prepare design data and supporting materials (as a result of detail design) for the Equipment/Software/Component Design Reviews.
F	Accomplish functional analysis and the allocation of overall system requirements to the subsystem level and below (as required).	V	Translate the results from the Equipment/Software/Component Design Reviews for incorporation into the prototype model(s) as applicable. The prototype model that is to be utilized in test and evaluation must reflect the latest design configuration.
G	Develop the necessary organizational and related infrastructure in preparation for the accomplishment of the required program design integration tasks.	W	Provide supplier components, with supporting data, for the development of the system prototype to be utilized in test and evaluation activities
H	Translate the results from the Conceptual Design Review to the appropriate design activities (i.e., approved design data, recommendations for improvement/corrective action).	X	Prepare for and conduct System Test and Evaluation (implement the requirements of the Test and Evaluation Master Plan).
I	Translate the results from the functional analysis and allocation activity into specific design criteria required as an input for the design integration process.	Y	Provide test data and logistic support, from the various suppliers, throughout the system test and evaluation phase. Test data are required to cover individual tests conducted at supplier facilities, and logistic support (i.e., spare/repair parts, test equipment, etc.) is necessary to support system testing activities.
J	Accomplish preliminary design and related design integration activities.	Z	Conduct the necessary planning and prepare for the Critical Design Review.
K	Translate the results from system level design into specific requirements at the subsystem level and below. Prepare Development, Process, Product, or Material Specifications as required.	AA	Test results, in the form of either design verification or recommendations for improvement/corrective action, are provided as an input into the Critical Design Review.
L	Conduct the necessary planning and prepare for the System Design Review.	BB	Prepare system test and evaluation report.
M	Translate the requirements contained within the various applicable specifications into specific design criteria required as an input for the design integration process.	CC	Translate the results from the Critical Design Review for incorporation into the final system configuration prior to entering the Production or Construction Phase of the Program.
N	Prepare design data and supporting materials (as a result of preliminary design) for the System Design Review.		
O	Accomplish detail design and related design integration activities.		
P	Translate the results from the System Design Review to the appropriate design activities (i.e., approved design data, recommendations for improvement/corrective action).		

Figure 10 List of activities for program network schedule.

1	2	3	4	5	6	7	8	9	10	11	12
Event number	Previous number	t_a	t_b	t_c	t_e	s^2	TE	TL	TS	TC	Probability (%)
17	16	2	3	4	3.0	0.111	115.2	115.2	0	110	6.4
	15	3	4	8	4.5	0.694	112.1	115.2	3.1	115	47.9
16	15	2	4	6	4.0	0.444	112.1	112.2	0	120	91.9
	13	2	3	4	3.0	0.111	86.5	112.2	25.7		
15	14	10	12	16	12.3	1.000	108.2	108.2	0		
	12	6	8	12	8.3	1.000	95.9	108.2	12.3		
14	13	2	3	4	3.0	0.111	86.5	95.9	9.4		
	12	6	8	12	8.3	1.000	95.9	95.9	0		
	11	12	16	20	16.0	1.778	95.3	95.9	0.6		
13	11	3	4	6	4.2	0.250	83.5		13.6		
	10	2	3	4	3.0	0.111	53.8		42.1		
12	11	6	8	12	8.3	1.000	87.6	87.6	0		
	8	8	10	12	10.0	0.444	60.8	87.6	26.8		
11	10	1	2	3	2.0	0.111	52.8	79.3	26.5		
	9	28	32	40	32.7	4.000	79.3	79.3	0		
10	9	3	4	6	4.2	0.250	50.8		30.7		
	5	2	3	4	3.0	0.111	21.3		59.0		
9	8	3	4	6	4.2	0.250	35.0	46.6	11.6		
	7	16	20	24	20.0	1.778	46.6	46.6	0		
8	7	3	4	6	4.2	0.250	30.8		15.8		
7	6	2	3	4	3.0	0.111	26.6	26.6	0		
	5	1	2	3	2.0	0.111	20.3	26.6	6.3		
	4	3	4	6	4.2	0.250	19.5	26.6	7.1		
6	4	6	8	12	8.3	1.000	23.6	23.6	0		
5	4	2	3	4	3.0	0.111	18.3		9.3		
4	3	1	2	3	2.0	0.111	15.3	15.3	0		
	1	6	8	12	8.3	1.000	8.3	15.3	7.0		
3	2	2	3	4	3.0	0.111	13.3	13.0	0		
2	1	8	10	14	10.3	1.000	10.3	10.3	0		

Figure 11   Example of program network schedule calculations.

728

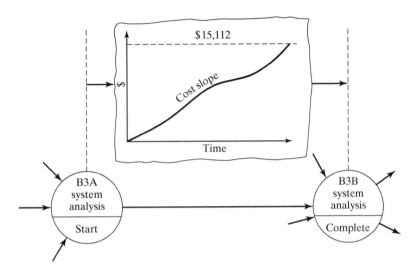

Figure 12   Activity-cost function.

identification of the required human and material resources that are needed for task completion is then accomplished and is based on project scheduling requirements.

Referring to Figure 9, while the emphasis indicated is on the element of *time*, a cost network can be superimposed upon the PERT/CPM network by estimating the total cost and cost function for each activity line. Figure 12 shows a sample activity cost function. Such functions can be generated for the entire network.

When using this method, there is a *time-cost* option that enables management to evaluate alternatives relative to the allocation of resources for activity accomplishment. In many instances, time can be saved by applying more resources, or, conversely, costs may be reduced by extending the time to complete an activity. Time and cost alternatives are evaluated with the objective of selecting the lowest-cost approach within the maximum allowable time limit.

2.6 Interfacing with Other Planning Activities

To be successful, systems engineering requires a closely integrated effort, both within and with many external program planning activities. In Figure 2, there are numerous individual design plans that must directly support the accomplishment of systems engineering activities (e.g., the electrical engineering design plan, reliability program plan, maintainability program plan, and concurrent engineering plan), and there are many other higher-level plans that need to be supportive as well (e.g., the configuration management plan, manufacturing program plan, total quality management plan, integrated logistic support plan, and sustainability plan). The SEMP must be developed showing the proper information links (i.e., "communication" route) to these various external plans, and these other plans must reflect and support the objectives specified within the SEMP. In the preparation of these external plans, it is essential that all applicable tasks, task schedules, and cost information be compatible with what is specified within the SEMP.

3 ORGANIZATION FOR SYSTEMS ENGINEERING

The initial planning for systems engineering begins during the early phases of conceptual design and evolves through the development of the SEMP described in Section 2. To implement this plan successfully requires an organizational structure that will promote, support, and generally enhance the application of systems engineering principles and concepts on the program in question. The proper organizational *environment* must be created that will (1) allow for the successful accomplishment of systems engineering requirements, and (2) facilitate the implementation of these requirements in an effective and efficient manner.

Organization is the combining of human resources in such a manner as to fulfill a need. Organizations constitute groups of individuals of varying levels of expertise combined into a social structure of some form to accomplish one or more functions. Organizational structures will vary with the functions to be performed, and the results will depend on established goals and objectives, the resources available, the communications and working relationships among the individual participants, the motivation of the personnel, and many other factors. The ultimate objective is to achieve the most effective and efficient use of human, material, and monetary resources through the establishment of decision-making and communication processes designed to accomplish specific objectives.

The fulfillment of systems engineering objectives highly depends on the proper mix of resources, the establishment of good communications, and on the development of good interpersonal skills by the participants. The uniqueness of tasks and the many different interfaces that exist require not only good communication skills but also an understanding of the system as an entity and the many design disciplines that may contribute to its development.

Of particular significance (and perhaps a major challenge in some instances) is the accomplishment of systems engineering organizational objectives when dealing with the interfaces associated with other systems within a given system-of-systems (SOS) structure. These "other" systems may be similar in nature and makeup, and the overall system operational and logistics support objectives may be somewhat comparable. In this situation, the systems engineering manager may have a relatively easy time in establishing the appropriate communication links and initiating the desired working relationships necessary to assure success in meeting the desired results. On the other hand, the mission and objectives of one, or more, of these other systems (within the same SOS network) may be of an "international" nature, located and managed through an international organization existing within a completely separate and unique environment with different language requirements, day-to-day operating procedures, methods of contracting, supplier requirements, and so on. In such cases, the systems engineering manager must be cognizant of these differences, knowledgeable of the applicable international working environment and associated cultural requirements, and familiar with the business and related practices followed in accomplishing project activities. Depending on the specific situation, the role of the systems engineering manager (and the organization of which he/she controls) may be particularly challenging, and the results will highly depend on his/her knowledge of major system interfaces, the other organizations affected, and on his/her communication skills and the interpersonal relationships developed in managing the system in question.

3.1 Developing the Organizational Structure

An initial step in the development of any type of an organizational structure is to determine the goals and objectives for the overall company/agency/institution involved, along with the functions and tasks that must be accomplished. Depending on the complexity and size of programs, the structure may assume a pure *functional* model, a *project* orientation, a *matrix* approach, or a combination thereof. Further, the structure may change in context as the system development effort evolves from the conceptual design phase through preliminary system design, detail design and development, production, and so on.

Regarding systems engineering, a prime objective during the early stages of conceptual design is to ensure the proper development of system-level requirements (i.e., the system architecture and the first six tasks described in Section 2.2). These activities are highly *customer/consumer* focused and are directed toward the *system* as an entity, and the accomplishment of such does not require a large organization per se. However, the selection of a few *key* personnel with the appropriate skills, background, and experience levels is essential.

As the program evolves into the preliminary and detail design and development phases, the number of personnel assigned may increase as the design requirements at the subsystem level (and below) may dictate the necessity for including expertise from many different design disciplines (e.g., reliability, maintainability, human factors, and logistics). In this context, the organizational structure may transition from a pure *project* configuration to a mixed *functional-project* or *matrix* approach. As the system and its elements enter the production/construction phase, the organizational structure may shift once again.

In addressing the overall organizational issue, the emphasis herein is intended to stress the accomplishment of the many and varied tasks described in Section 2.2, independent of which organizational element (department, section, or group of personnel) actually accomplishes the work. Experience has indicated that there are organizational departments or groups located within industrial firms or government agencies that have been designated as "systems engineering" and given the appropriate responsibilities, but who are not actually performing the tasks required. Conversely, there are organizational elements with different identities that are, in actuality, performing the desired functions effectively and efficiently. Further, for small projects where a single individual must assume different roles, the systems engineering responsibilities may be accomplished by the chief engineer, an electrical engineer, a mechanical engineer, or someone equivalent. On the one hand, the project manager may serve as the "systems engineer," or there may be a designated group of people performing the required tasks. The objective is to accomplish the appropriate level of effort utilizing the resources from different organizational entities across the board.

3.2 Consumer, Producer, and Supplier Relationships

In addressing the subject of "organization for systems engineering" properly, one needs to understand the environment in which systems engineering functions are performed. Although this may vary somewhat depending on the size of the project and the stage of design and development, this discussion is primarily directed to a large project operation, characteristic in the acquisition of many large-scale systems. By addressing large projects, it is anticipated that a better understanding of the role of systems engineering in a somewhat complex environment is provided. The reader must, of course, adapt and structure an approach for his/her program.

For a relatively large project, the systems engineering function may appear at several levels as shown in Figure 13. The requirements for systems engineering and the responsibility for implementing the 11 tasks described in Section 2.2 lie with the customer (or user) since it is at this level that the *system* is first addressed as an entity. The customer may establish a systems engineering organization to accomplish the required tasks, or these tasks may be relegated (in part or in total) to the producer through some form of contractual arrangement. In any event, the responsibility, along with the authority, for accomplishing systems engineering functions must be clearly defined.

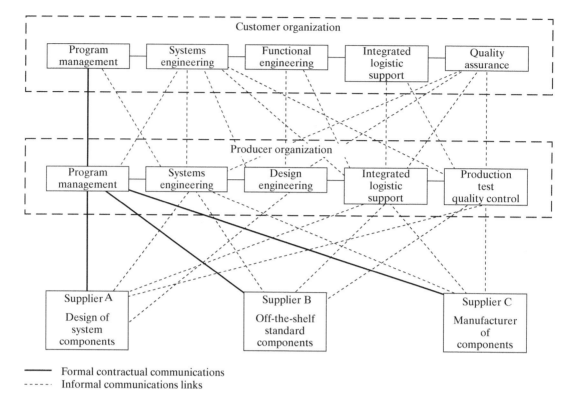

Figure 13 Customer, producer, and supplier organizational relationships.

In some instances, the customer may assume full responsibility for the overall design and development, production, integration, and installation of the system for operational use. The needs analysis, accomplishment of feasibility studies, definition of operational requirements and the maintenance concept, identification and prioritization of TPMs, preparation of the system specification (Type *A*), preparation of the SEMP, and so on, are accomplished within the customer's organization. Top-level functions are defined and specific program requirements (i.e., tasks) are allocated to individual producers, subcontractors, and component suppliers.

In other cases, although the customer provides the overall guidance in terms of issuing a general statement of work (SOW), or a contractual document of an equivalent nature, the producer (or *prime contractor*) is held responsible for the entire system design and development effort and for completing the 11 tasks described in Section 2.2. Although both the customer and producer have established systems engineering organizations (as shown in Figure 13), the basic responsibility for completing the tasks and fulfilling the objectives described throughout this text lies with the producer's organization, with supporting tasks being accomplished by individual suppliers as required. To accomplish this, the customer must not only delegate the appropriate level of *responsibility* for completing the functions specified but the necessary *authority* as well. Further, the customer must make accessible all of the necessary input data for the producer to complete successfully the conceptual design functions noted earlier.

Referring to Figure 13, note that there is an extensive amount of communication required not only within each of the customer and producer organizations, but between the various customer, producer, and supplier organizations as well. Although the solid lines pertain primarily to the more formal program management direction and that of a contractual nature, there are many informal channels of communication that must exist to ensure that a proper dialogue is established between the numerous and varied entities involved in the system development effort. Establishing these communication channels is particularly important when the program involves major suppliers located internationally and when there are suppliers which are providing a variety of services as part of a system-of-systems (SOS) network. In such instances, the communication process can become somewhat complex and increased management emphasis is needed to ensure success. The successful implementation of a *teaming* or *partnership* approach, along with the fostering of the concurrent engineering principles discussed earlier, heavily depends on good communications (both downward and upward) from the beginning.

3.3 Producer Organization and Functions

A primary building block for most organizational patterns is the functional approach, which involves the grouping of functional specialties or disciplines into separately identifiable entities. The intent is to perform similar work within one organizational component. In the pure functional structure, all engineering work is the responsibility of one executive, all manufacturing effort is the responsibility of another executive, and so on. In this case, the same organizational group will accomplish the same type of work for *all* ongoing projects on a concurrent basis.

A partial functional organization, including the identification of major work packages, is illustrated in Figure 14. The systems engineering function comes within the overall engineering organization and includes the tasks described in Section 2.2. The organization is responsible for accomplishing these tasks as they apply to all projects. This approach is often desirable for small firms or agencies since it is easier to manage a homogeneous group of similar functions and personnel with comparable backgrounds. Also, the duplication of effort is minimized. On the other hand, for large operations this is not necessarily preferable, owing to impractical centralization of responsibility. Thus, for large multiproduct firms, the pure functional approach may be modified somewhat and organized in terms of individual projects, with some cross-disciplinary activity introduced as appropriate.

Figure 15 illustrates a pure project organization that is solely responsible for the planning, design and development, production, operational use, and support of a unique or single system. It is time limited, and the commitment of the varied skills and resources required is purely for the purpose of accomplishing tasks associated with the particular system being developed. Each project organization will include its own engineering functions, testing functions, and support functions. Thus, there is a systems engineering group for Project A, another for Project B, and so on. Each group performs the same basic tasks described in Section 2.2. While the pure project organization does provide some benefits in terms of responsiveness to a particular customer, there may be some redundancies within a given company/agency due to the duplication of resources across multiple project lines.

Figure 16 illustrates a matrix configuration with a mix of the pure functional organization in Figure 14 and the project organization in Figure 15. For large single-system acquisitions, where the scope of work and funding are adequate, the project configuration may be the preferred approach. However, if there are many in-house projects where each individually cannot justify a separate project organization, the matrix configuration may be preferable. In this instance, it may be appropriate to accomplish some functions

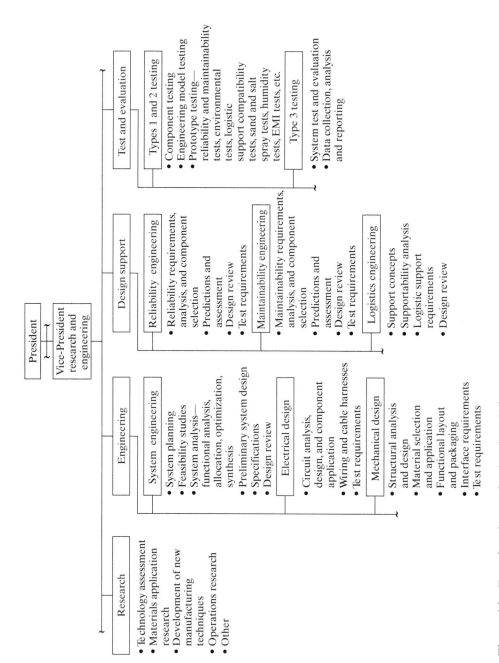

Figure 14 Functional organization (partial).

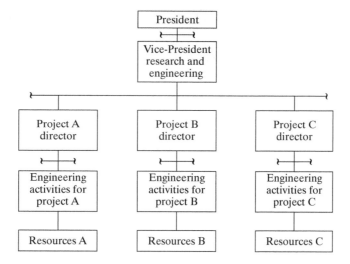

Figure 15 Simplified pure project organization.

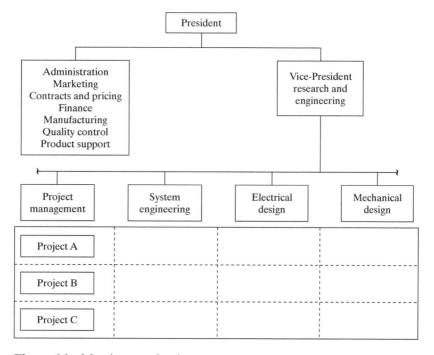

Figure 16 Matrix organization.

through a project structure, whereas others are accomplished by specialized staff groups. In this example, systems engineering tasks are accomplished through a staff function supporting different projects on an as-required basis. Although the resources are combined within the functional structure in this case, the direct involvement in the day-to-day project activities may be inhibited since the systems engineering activity is not actually a part of the project organization.

Figure 17 illustrates a slightly different version of a typical project-staff organizational configuration. In this example, each project includes a systems engineering group while many of the engineering support activities are provided on a task-by-task basis through various staff functions. Because most tasks described in Section 2.2 must be accomplished as an integral part of the system design and development process, combined with the fact that the systems engineering activity should assume a leadership role in a "technical" sense, it is preferable that such a function be an inherent part of the project organization. Further, the project type of organization facilitates the communications process between the customer and the producer. The systems engineering activity must provide a leadership role in fostering this communication.

As conveyed earlier, the interfaces between the systems engineering group and other segments of a producer's organization are numerous, particularly when involved in a large system acquisition. Referring to Figure 18, the systems engineering group for Project Y is challenged by the fact that there are many other activities to deal with on an almost day-to-day basis. There is a need to establish a good communication link with the many internal project activities noted; that is, design engineering, software engineering, the business unit, other project organizations, the logistic support function, production operations, and so on. An example of the required communication network is illustrated in Figure 19 (which explains the dotted lines in Figure 18).

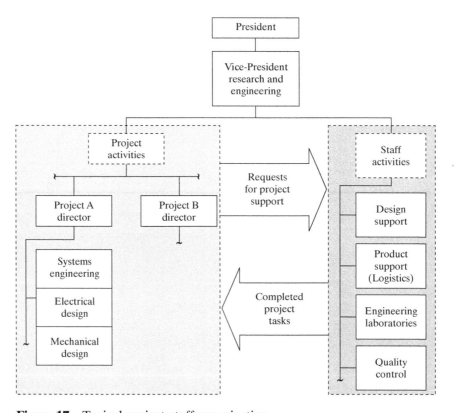

Figure 17 Typical project-staff organization.

736

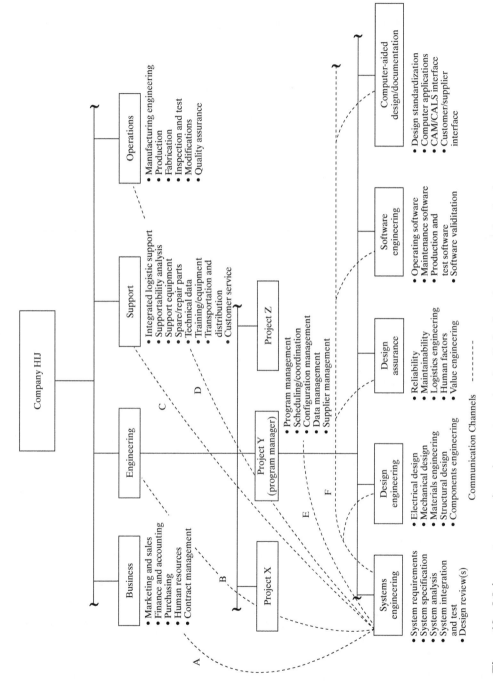

Figure 18 Major systems engineering communication links (producer organization).

Communication Channel (Figure 18.18)	Supporting Organization (Interface Requirements)
A	1. *Marketing and sales*—to acquire and sustain the necessary communications with the customer. Supplemental information pertaining to customer requirements, system operational and maintenance support requirements, changes in requirements, outside competition, etc., is needed. This is above and beyond the formal "contractual" channel of communications. 2. *Accounting*—to acquire both budgetary and cost data in support of economic analysis efforts (e.g., life-cycle cost analysis). 3. *Purchasing*—to assist in the identification, evaluation, and selection of component suppliers with regard to technical, quality, and life-cycle cost implications. 4. *Human resources*—to solicit assistance in the initial recruiting and hiring of qualified project personnel for system engineering, and in the subsequent training and maintenance of personnel skills. To conduct training programs for all project personnel across the board relative to system engineering concepts, objectives, and the implementation of program requirements. 5. *Contract management*—to keep abreast of contract requirements (of a technical nature) between the customer and the contractor. To ensure that the appropriate relationships are established and maintained with suppliers as they pertain to meeting the technical needs for system design and development.
B	To establish and maintain on-going liaison and close communications with other projects with the objective of transferring knowledge that can be applied for the benefit of Project Y. To solicit assistance from other company-wide functionally-oriented engineering laboratories and departments relative to the application of new technologies in support of system design and development.
C	To provide an input relative to project requirements for system support, and to solicit assistance in terms of the functional aspects associated with the design, development, test and evaluation, production, and sustaining maintenance of a support capability through the planned system life cycle.
D	To provide an input relative to project requirements for production (i.e., manufacturing, fabrication, assembly, inspection and test, and quality assurance), and to solicit assistance relative to the design for producibility and the implementation of quality engineering requirements in support of system design and development.
E	To establish and maintain close relationships and the necessary on-going communications with such project activities as scheduling (the monitoring of critical program activities through a network scheduling approach); configuration management (the definition of various configuration baselines and the monitoring and control of changes/modifications); data management (the monitoring, review, and evaluation of various data packages to ensure compatibility and the elimination of unnecessary redundancies); and supplier management (to monitor progress and ensure the appropriate integration of supplier activities).
F	To provide an input relative to *system-level* design requirements, and to monitor, review, evaluate, and ensure the appropriate integration of system design activities. This includes providing a *technical* lead in the definition of system requirements, the accomplishment of functional analysis, the conductance of system-level trade-off studies, and the other project tasks.

Figure 19   Description of major project interface requirements.

The ultimate objective in the design and development of any system is to establish a *team* approach, with the appropriate communications, enabling the application of concurrent engineering methods throughout. However, there are often problems when dealing with large programs involving many different functional organizational units. Barriers are developed which tend to inhibit the necessary day-to-day close working relationships, the timely transfer of essential information, and the communications discussed earlier. As there are many companies organized strictly along functional lines (versus the project-staff configuration shown in Figure 18), there is a need to ensure that the proper level of communication is maintained between all applicable organizational units, regardless of where they are permanently located within the overall company structure, or external to this structure.

With this objective in mind, the Department of Defense (DOD) initiated the concept of *integrated product and process development* (IPPD) in the early 1990s. IPPD can be defined as a "management technique that simultaneously integrates all essential acquisition activities through the use of multidisciplinary teams to optimize the design, manufacturing, and supportability processes." This concept promotes the communications and integration of the key functional areas, as they apply to the various phases of program activity. Although the specific nature of the activities involved and the degree of emphasis exerted will change somewhat as the system design and development effort evolves, the structure conveyed in Figure 20 is maintained throughout to foster the necessary communications across the more traditional functional lines of authority. In this regard, the concept of IPPD is directly in line with systems engineering objectives.[1]

Inherent within the IPPD concept is the establishment of *integrated product teams* (IPTs), with the objective of addressing certain designated and well-defined issues. An IPT, constituting a selected team of individuals from the appropriate disciplines, may be established to investigate a specific segment of design, a solution for some outstanding problem, design activities that have a large impact on a high-priority TPM, and so on. The objective is to create a *team* of qualified individuals that can effectively work together to solve some problem in response to a given requirement. Further, there may be several different teams established to address issues at different levels in the overall system hierarchical structure (i.e., issues at the system, subsystem, or component level). Referring to Figure 20, an IPT may be established to concentrate on those activities that significantly impact *performance*, *cost-of-ownership*, *integrated data*, and *configuration management* requirements. There may be another IPT assigned to "track" a specific supplier design requirement. The objective is to provide the necessary emphasis in critical areas and to reap the benefits of a *team* approach in arriving at the best solution possible.[2]

IPTs are often established by the program manager, or by some designated high-level authority in the organization. The representative team members must be well qualified in their respective areas of expertise, empowered to make on-the-spot decisions when necessary, proactive relative to team participation, success oriented, and resolved to solve the problem assigned. The program manager must clearly define the objectives and the expectations for the team in terms of results, and the team members must maintain a continuous "up-the-line" communications channel. The longevity of the IPT will depend on the nature of the problem and the effectiveness of the team in progressing toward meeting its objective.

---

[1]DOD Regulation 5000.2, "Mandatory Procedures for Major Defense Acquisition Programs (MDAPs) and Major Automated Information System (MAIS) Acquisition Programs" (Washington, DC: Department of Defense, 1996).

[2]The term IPT is also used as a designator for "integrated process team."

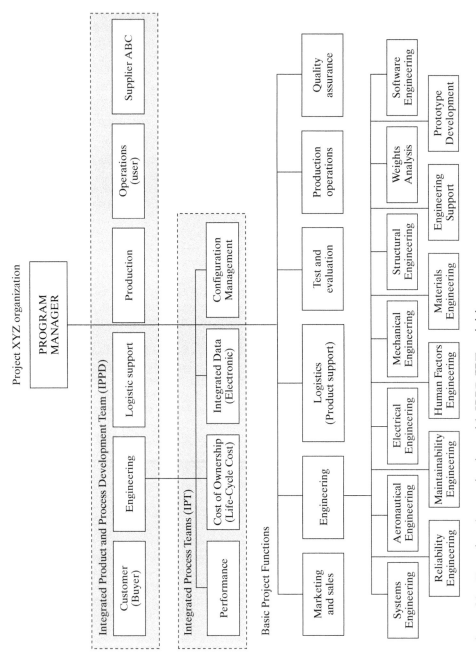

Figure 20 Functional organization with IPPD/IPT activities.

Care must be taken to avoid the establishment of too many teams, as the communication processes and interfaces become too complex when there are many teams in place. In addition, there are often conflicts when it comes to issues of importance and which issue is "traded-off" as a result. Further, when a team ceases to be effective in accomplishing its objectives, it should be disbanded. An established team that has "outlived its usefulness" can be counterproductive.

3.4 Staffing the Systems Engineering Organization

As a start, it is important to understand and address the issue of *culture* through the design and development of an organizational capability and establishing human resource requirements, leadership characteristics, staffing, motivational factors, the development of personnel, and so on. Culture pertains to an organization's personality, environment, atmosphere, and the like. The culture of an organization defines appropriate behavior and ethics; it can motivate individuals, can govern the way it processes information and deals with internal and external relations and values, and can be either positive or negative.

With regard to systems engineering, the successful accomplishment of the objectives and specific tasks described earlier highly depends on the culture, the overall capabilities, and the environment that has been created within the organization. There have been numerous instances when organizations have been established and designated as the "Systems Engineering Department" or the "Systems Engineering Group," in which the basic requirements are not being met. On the other hand, there have been many other instances when such organizations have been very successful. The key is to establish a structure and its processes and to create an *environment* which is positive. This requires promoting effective communications throughout a given program/project structure, gaining the necessary respect in terms of technical capability and assuming a leadership role for a program, and being able to influence the system design and development process on a continuing basis. This may be particularly challenging since the systems engineering organization must accomplish much without having direct control of all of the resources required to complete the job. In accomplishing its function successfully, it must be able to work effectively and efficiently with many different and varied external organizations often located throughout the world.

Given this, the nature of the systems engineering organization requires consideration of the following characteristics in developing an appropriate structure:

1. The personnel selected for the systems engineering group must, in general, be highly professional senior-level individuals with varied backgrounds and a wide breadth of knowledge—for example, an understanding of research, design and development, manufacturing, system operations and support, and material recycling and disposal applications. The emphasis is on overall system-level design and technology applications, with knowledge of user operations and sustaining life-cycle maintenance and support in mind.

2. The personnel selected must also be relatively conversant with some of the different design-related technologies and their specific applications in and throughout the system design and development process; for example, a general understanding of some of the principles and concepts of reliability, maintainability, human factors, safety and security, logistics and supportability, quality, value engineering, and so on as applicable throughout the system life cycle.

3. The systems engineering group must have *vision* and be *creative* in the selection of technologies for design, manufacturing, and support applications. Group personnel must constantly search for new opportunities and must be innovative, and applied research is often required in order to solve specific technical problems.

4. A *teamwork* approach must be initiated within the systems engineering group. The personnel assigned must be committed to the objectives of the organization; a certain degree of interdependence is required, and there must be mutual trust and respect.

5. A high degree of communication must prevail, both within the systems engineering group and with the many other related functions associated with a given project (refer to Figure 18). Communication is a two-way process and may be accomplished via written, verbal, and/or nonverbal means. Good communications must first exist within the systems engineering organization. With that established, it is then necessary to develop two-way communications externally, using both vertical and horizontal channels as required.

The term "environment" refers to the (1) the working environment within the systems engineering group itself, (2) the working environment external to the systems engineering organization but within and throughout the producer's overall organizational structure, and (3) the working environment associated with external organizations located in various remote locations. The creation of a favorable environment throughout an organization must start from the *top*. The president, or general manager, must first "believe in" and subsequently "support" the concepts and objectives of systems engineering. In addition, the appropriate level of responsibility and authority, and required resources, must be delegated and allocated from the top and on down to the Systems Engineering Group Manager.

To further facilitate the objectives described herein, the characteristics and managerial style of the Systems Engineering Group Manager should favor a "democratic" and "participative" style of leadership versus a more "autocratic" and "dictatorial" approach. An environment must be created to allow for individual initiative, creativity, flexibility, personal growth, and so on. While the manager must maintain authority and provide the necessary direction and control to effectively and efficiently accomplish the organization's goals and objectives, he/she can introduce some practices that directly support the democratic style of leadership and the team approach on a day-to-day basis; for example, through the solicitation of individual contributions leading to overall organizational growth, the introduction of personnel recognition and reward programs, the implementation of a good personnel promotional practices, and so on.

While developing the right mix of human resources and creating a good working environment from the beginning is essential, the challenge is to be able to maintain this condition over time as the organization continues to grow and be responsive to different program requirements. It is often relatively easy for an organization (and its management) to become too complacent and settle into somewhat of a "stagnant" situation. However, to continue to be effective and efficient in the fulfillment of systems engineering objectives requires an organization that is highly dynamic by nature. Personnel assigned to accomplish such functions must be strongly motivated by improving their respective knowledge through the ongoing pursuit of continuing education (or equivalent), improving their individual working skills and communication processes, and so on. To experience success in the field (in the long term) requires not only the initial staffing and development of the proper organizational structure but also the maintaining of such as well.

4 SUMMARY AND EXTENSIONS

This chapter introduces essential planning and organizational approaches within the overall spectrum of *systems engineering management*. The objective is to address systems engineering from a management perspective. An enterprise may develop the best technical model possible, but its application will not be effective or efficient in practice unless the proper management approach is implemented concurrently.

Included within this chapter is the development of a systems engineering management plan (SEMP) to include the identification of systems engineering tasks, a sample work breakdown structure (WBS), several approaches to task scheduling, a program cost projection, and major interfaces with other critical program activities. Following the basic planning activity is the development of a systems engineering organization. A variety of organizational structures are shown to include systems engineering within the context of a formal *functional* structure, a pure *project* structure, a *matrix* structure, and a combined *project-functional* structure. The major facets of an IPPD/IPT approach are also discussed. Finally, there is some discussion on the specific requirements and staffing of a typical systems engineering organization (department or group).

It should be emphasized here that systems engineering is more of a philosophy, a "thought process," and a life-cycle-oriented approach recommended for the design and development of systems. Systems engineering should not be viewed as just another engineering discipline or technical domain as is the case for electrical engineering, mechanical engineering, reliability engineering, quality engineering, and the like. Good systems engineering is a disciplined approach, albeit an engineering interdiscipline, applied through the proper and timely integration of different design (and subsequent validation and support) activities. Its purpose is to provide an effective and efficient product output in response to some identified customer need. It is highly interdisciplinary by nature, cutting across many different organizational lines.

To achieve systems engineering objectives does not require a large organizational structure per se, but just a few highly qualified senior personnel with experience and the proper background and level of expertise. Further, the principles and concepts of systems engineering can be initiated from within any type of organizational structure, whether functional, pure project, matrix, or a combination thereof. It just requires good direction from the top down (from an organizational perspective), the right managerial style, and good communications throughout the technical organization.

While the practice of systems engineering dates back a half-century or more, these principles and concepts are not being universally implemented today. One of the reasons for this relates to the lack of general understanding and acceptance in many organizations. Although the benefits derived can be numerous, there is a strong reluctance to think in terms of *systems*, to think *interdisciplinary*, and to think of the system in the context of its entire *life cycle*. This stems from our basic educational foundation where the emphasis is on the short term, thinking about individual products (and not systems), and being highly specialized in some particular discipline or engineering domain. The challenge for the future is to extend one's thinking to a broader approach through formal education and training.

The subjects of systems management, program management, project management, and various derivatives of these, constitute a rather broad field of study. Only the bare essentials, as they pertain to systems engineering, are included in this chapter. The objective is to provide insight for consideration in applying the principles and concepts of system engineering to realistic situations. The following websites provide current information and material in the field:

1. American Management Association (AMA) — *http://www.amanet.org.*
2. Project Management Institute (PMI) — *http://www.pmi.org.*
3. National Contract Management Association (NCMA) — *http://www.ncmahq.org*

QUESTIONS AND PROBLEMS

1. When in the life cycle should systems engineering planning be initiated? What is included in it? Why is it important?

2. What is the purpose of the SEMP? When in the life cycle should it be developed? How does it relate to each of the following: (a) the PMP, (b) the reliability program plan, (c) the integrated logistic support plan (ILSP), (d) the configuration management plan, and (e) the test and evaluation master plan (TEMP) and the supplier engineering plan?

3. How does the SEMP relate to Type A system specification (if at all)?

4. Select a system of your choice and develop a detailed outline of a SEMP for your system.

5. Identify and describe the tasks that should be included in the implementation of a systems engineering program.

6. What is the purpose of the WBS? What is the difference between an SWBS and a CWBS? If you were outsourcing and responsible for specifying the work requirements for a supplier, which would you use?

7. If you were assigned to manage a systems engineering program and to develop a master schedule for the work and tasks to be completed, what scheduling method would you use for determining the status? Why would you select this method?

8. Assume that you have been assigned to develop a SEMP for a system of your choice. Identify and describe the functions/tasks that need to be accomplished. Then construct a network (PERT/CPM) for the program and identify the critical path.

9. For the network developed in Problem 8, develop a cost projection, relating the costs to each scheduled activity.

10. Assume that in the evaluation of program status in Problem 8, you discover that an activity along the critical path is behind schedule. What steps would you take to correct the situation?

11. Identify some of the key objectives in organizing for systems engineering.

---

[3]For a more comprehensive coverage of systems engineering planning, organization, and management, refer to B. S. Blanchard, *System Engineering Management*, 4th ed. (Hoboken, NJ: John Wiley & Sons, Inc., 2008). It is also recommended that reader should review literature pertaining to organizational structures, organizational development and dynamics, etc.

12. With systems engineering in mind, describe some of the advantages and disadvantages of a pure *functional* organization structure, a pure *project* organization structure, and a *matrix* organization structure.

13. Assume that you have just been assigned to develop a new systems engineering capability in your company. Construct a hypothetical organizational chart for the company, identify the organizational elements responsible for the accomplishment of systems engineering tasks, and describe the interface relationships with other critical organizational groups.

14. If you were charged with the staffing of a systems engineering organization, what type of an individual would you hire? Describe background and experience expectations, personal characteristics, and specific desired skills.

15. As a systems engineering manager, what type of an organizational *environment* would you try to create to accomplish your objectives effectively?

16. What is the objective of an IPPD concept? What is an IPT and what are its objectives?

17. Refer to Figure 18. Assume that you have been appointed as the manager of the systems engineering organization and that the communication processes represented by the dotted lines were not functioning properly. What steps would you take to strengthen these communication links?

18. Assume that you are the manager of a newly established systems engineering organization and that, in accomplishing your required functions, you will be dealing with a large number of external organizations located both nationally and internationally. Describe some of the requirements that you would implement to ensure the successful accomplishment of your objectives.

Program Management, Control, and Evaluation

From Chapter 19 of *Systems Engineering and Analysis*, Fifth Edition, Benjamin S. Blanchard, Wolter J. Fabrycky. Copyright © 2011 by Pearson Education, Inc. Published by Pearson Prentice Hall. All rights reserved.

Program Management, Control, and Evaluation

Given a comprehensive systems engineering plan and an organizational structure, the challenge becomes one of *implementation*. The past provides many examples where relevant and timely planning was accomplished from the beginning, only to find that the subsequent implementation process did not follow the plan. As a result, the plan became impotent by neglect. For systems engineering objectives to be accomplished successfully, a two-step management approach is recommended. First, the planning and organization for systems engineering should be completed. Then, the follow-on program management, implementation and control, and evaluation activities described in this chapter should be activated.

The systems engineering management plan (SEMP) defines the specific requirements for the implementation of a systems engineering program (or project) with the objective of designing, developing, and bringing a new (or reengineered) system into being. Program goals and objectives, required systems engineering tasks, a work breakdown structure (WBS), task schedules, and cost projections should be included in the SEMP. In addition, an organizational approach is described, along with structure and organizational interface requirements. The SEMP, in conjunction with the *system specification* (Type *A*), basically addresses the *what* requirements from an overall program perspective.

With the *what* requirements as input, the essential next step is to implement the plan. Accordingly, the objectives of this chapter are to respond to the SEMP requirements by describing the *hows* as they pertain to plan implementation. The reader should become knowledgeable about the planning and organizational requirements for systems engineering, and also about the day-to-day program management, control, and evaluation requirements described in this chapter. Remaining topics that should be reviewed and assimilated are the following:

- Establishing specific organizational goals and objectives;
- Outsourcing requirements and the identification of suppliers;
- Providing day-to-day program leadership and direction;

- Implementing a program evaluation and feedback capability; and
- Conducting a risk analysis and management function.

Of special interest in this chapter is an approach for conducting a *performance evaluation* of a systems engineering organization, including the determination of a *level of maturity* for the organization. This kind of organizational assessment is in keeping with the increasing necessity for accountability that now exists in most areas of human endeavor.

1 ORGANIZATIONAL GOALS AND OBJECTIVES

The purpose of *organization*, considered in the systems engineering context, is to fulfill the requirements described in the SEMP and in the *system specification* (Type *A*). The *goals* of the organization are directed toward this end and pertain to two levels of activity: (1) the goals specified by the technical performance measures (TPMs) which quantify the design-dependent parameters (DDPs) making specific the characteristics that must be incorporated into the design of a given system; and (2) the goals of the organization relative to accomplishing the necessary activities (tasks) to ensure that the first objective is attained. A basic question is, *How effective and efficient is the organization functioning in the accomplishment of the first goal?*

In this instance, the question can be addressed to the applicable systems engineering organization relative to (1) its impact on the ultimate system/product design configuration itself; and (2) its effectiveness and efficiency in performing the 11 tasks described in the Appendix. It is not uncommon to address only the second goal, believing that the organization is doing well in accomplishing the identified tasks when, at the same time, the organization (and its personnel) has not impacted the design at all. Thus, when assessing the organizational capabilities later, the second goal must be evaluated in terms of the first.

Regarding the issue of organizational structure, the objective is to develop a complete systems engineering *capability* that will enable the accomplishment of the 11 program tasks (or functions of an equivalent nature) in an effective manner. This, in turn, requires (1) the establishment of a desired set of *metrics* against which each of the applicable tasks can be assessed; (2) the development of the necessary processes for the performance of these tasks; and (3) implementation of a data collection and information capability that will enable management to "track" performance and determine just how well the organization is functioning overall. The organizational objective is to establish a disciplined, well-defined approach for the performance of these tasks, establish measurable goals that can be quantitatively expressed and controlled, and initiate a provision for continuous process improvement.

Having identified the systems engineering tasks to be accomplished, along with the associated "metrics," management needs to assess the current status of the organization. At the same time, it may be appropriate to compare the results with other comparable entities. A *benchmarking* approach can be applied to aid in the development of future goals. Benchmarking can be defined as "an ongoing activity of comparing one's own process, product, or service against the best known similar activity, so that challenging but attainable goals can be set and a realistic course of action implemented to efficiently become and remain the best of the best in a reasonable period of time." The ultimate questions are, *Where are we today? How do we compare with others relative to the product, process, and/or organization? Where would we like to be in the future?*

The organizational goals and objectives must first be responsive to the established system operational requirements and the TPMs for the system being developed. Then, additional goals can be established for the systems engineering organization itself. Through the use of benchmarking, future goals can be identified and an organizational *growth plan* can be developed, which, when implemented, can aid in realizing the desired objectives.

2 OUTSOURCING AND THE IDENTIFICATION OF SUPPLIERS

As a result of recent trends pertaining to increased system/product demands, in less time and at a lower cost, and in a highly competitive international marketplace environment, there has been a great deal of emphasis on the practice of *outsourcing*. The term *outsourcing* refers to identifying, selecting, and contracting with one or more outside suppliers for the procurement and/or acquisition of materials and services for a given system. The term *supplier* refers to a broad class of external organizations that provide products, components, materials, and/or services to the producer. This may range from the design and delivery of a major subsystem or configuration item down to a small component. Specifically, suppliers may provide services to include (1) the design, development, and manufacture of a major element of a system; (2) the production and distribution of products already designed, providing a manufacturing source; (3) the distribution of commercial and standard component parts from an established inventory, serving as a warehouse and providing parts from various sources of supply; and/or (4) providing a service through the implementation of a process in response to some functional requirement.

For many contemporary systems, suppliers provide a large number of their elements (e.g., more than 75% of the major components in some instances), as well as items that are required for the sustaining maintenance and support of the system. Given the trends toward increased globalization and greater international competition, the suppliers associated with any relatively large-scale program are likely to be geographically located throughout the world, thus creating a worldwide "working environment," as conveyed in Figure 1. Further, when major suppliers are selected, particularly for the design and development of large system elements, there are likely to be a number of different suppliers selected for the production and delivery of some of the smaller components that make up the various subsystems and items of an equivalent level and complexity. At the same time, many of the suppliers involved may be supporting any number of different systems with different and unique mission objectives (requirements) on a concurrent basis. Thus, one may be dealing with a *layering* of suppliers, each with different operational objectives, as illustrated in Figure 2.

Referring to Section 1, while the SEMP and the system specification state the "requirements" for the implementation of a systems engineering program (i.e., the *whats*), it is the responsibility of the producer to determine the way in which these requirements will be fulfilled (i.e., the *hows*). Through the functional analysis, specific resource requirements are identified (i.e., equipment, software, facilities, data, etc.), with the next step being the identification of possible sources of supply. *Should the design and/or manufacture of an item of equipment, the development of a software package, or the completion of a process be accomplished internally within the producer's organization and facility or should an external source of supply be selected?* The objective is to establish the *where* requirements in determining the "source of supply" in responding to the identified resource needs.

In many industrial organizations, a "make-or-buy" committee, or an equivalent activity within the producer's organization, is established, with representation from program management, engineering, manufacturing, logistics, quality assurance, purchasing, and other supporting organizational activities as needed. Engineering participation should include the systems engineering group and the appropriate design disciplines. Decisions are based on an evaluation of a combination of factors such as criticality and time of need, item

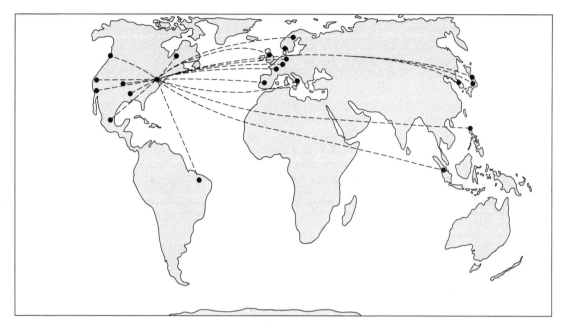

Figure 1 Potential suppliers for System XYZ.

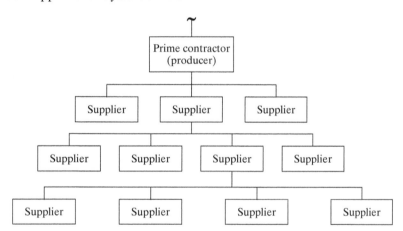

Figure 2 A typical structure involving the "layering" of suppliers.

complexity, availability of internal capabilities and required resources, related social and political factors, cultural issues, and cost. When considering international sources, in particular, such issues as language and cultural concerns, work practices and local holiday provisions, packaging and transportation requirements, customs regulations, and so on must be addressed in the overall evaluation and selection process.

From a systems engineering perspective, items that are relatively complex, that involve the application of new technologies, and that are critical to the overall system development effort should be handled internally if possible. These activities will, in all likelihood, require frequent monitoring and the application of relatively tight controls (both management and technical), which may be difficult to accomplish should a remotely located supplier be selected for the task.

The results from the deliberations of the "make-or-buy" committee will lead to specific recommendations as to the potential sources of supply for the fulfillment of various functional requirements in a given system development effort. Potential external "candidate" suppliers are identified; in turn, this will lead to the preparation of an applicable specification (Types *B*, *C*, *D*, and/or *E*) and the development of a formal Request for Proposal (RFP), Invitation for Bid (IFB), or the equivalent. A typical producer--supplier mix of requirements may evolve, as reflected in Figure 3. Interested and qualified suppliers will proceed with the preparation and submittal of a management and technical proposal in

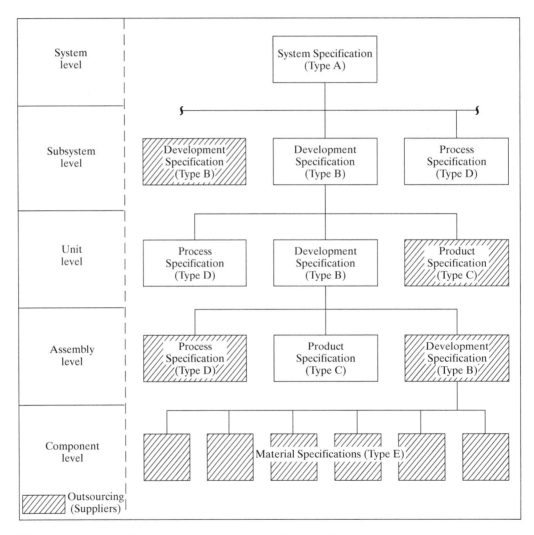

Figure 3 A "specification tree" showing supplier requirements.

response to the specification and RFP/IFB requirements, the producer will review and evaluate all proposals and make a selection based on the results, and the appropriate contractual provisions will subsequently be negotiated and implemented accordingly.

In the evaluation of supplier proposals, the various topics identified in Figure 4 should be addressed as applicable. Of particular interest are (1) the supplier's response to the producer's requirements in general, (2) the specification of the appropriate TPMs and traceability of these requirements from the top system-level requirements, (3) the supplier's recognition of the need for and importance of systems engineering, and (4) the specific tasks that

---

SUPPLIER EVALUATION CHECKLIST

D.1 GENERAL CRITERIA

D.2 PRODUCT DESIGN CHARACTERISTICS

    D.2.1 TECHNICAL PERFORMANCE MEASURES
    D.2.2 TECHNOLOGY APPLICATIONS
    D.2.3 PHYSICAL CHARACTERISTICS
    D.2.4 EFFECTIVENESS FACTORS
        1. RELIABILITY
        2. MAINTAINABILITY
        3. HUMAN FACTORS
        4. SAFETY FACTORS
        5. SUPPORTABILITY/SERVICEABILITY
        6. QUALITY FACTORS
    D.2.5 PRODUCIBILITY FACTORS
    D.2.6 DISPOSABILITY FACTORS
    D.2.7 SUSTAINABILITY FACTORS
    D.2.8 ECONOMIC FACTORS

D.3 PRODUCT MAINTENANCE AND SUPPORT INFRASTRUCTURE

    D.3.1 MAINTENANCE AND SUPPORT REQUIREMENTS
    D.3.2 DATA/DOCUMENTATION
    D.3.3 WARRANTY/GUARANTEE PROVISIONS
    D.3.4 CUSTOMER SERVICE
    D.3.5 ECONOMIC FACTORS

D.4 SUPPLIER QUALIFICATIONS

    D.4.1 PLANNING/PROCEDURES
    D.4.2 ORGANIZATIONAL FACTORS
    D.4.3 AVAILABLE PERSONNEL AND RESOURCES
    D.4.4 DESIGN APPROACH
    D.4.5 MANUFACTURING CAPABILITY
    D.4.6 TEST AND EVALUATION APPROACH
    D.4.7 MANAGEMENT CONTROLS
    D.4.8 EXPERIENCE FACTORS
    D.4.9 PAST PERFORMANCE
    D.4.10 MATURITY
    D.4.11 ECONOMIC FACTORS
    D.4.12 SOCIAL/CULTURAL FACTORS

Figure 4 Supplier evaluation checklist (example).

the supplier has proposed, which will, in turn, directly support the accomplishment of the appropriate systems engineering activities for the overall system development effort. The supplier's motivation toward initiating and maintaining good communications throughout should be readily visible.

3 PROGRAM LEADERSHIP AND DIRECTION

Referring to the Appendix, the successful accomplishment of the 11 tasks listed requires that the systems engineering organization assume a strong leadership role for the program/project. The systems engineering manager must take the initiative leading to the initial definition of system requirements, preparation of the system specification (Type A) and the SEMP, accomplishing the functional analysis at the system level, preparation of the TEMP, conducting formal design reviews, monitoring and controlling design changes and system modifications, and so on.

The manager may not actually accomplish all of these tasks by himself/herself, but should assume a leadership role in guiding and directing these and related activities. Further, he/she must demonstrate a leadership capability in first defining and subsequently monitoring supplier activities as they pertain to systems engineering activities. As a goal, the systems engineering manager should exhibit the following leadership characteristics, as a minimum:

1. *Acceptance:* earns respect and has the confidence of others;
2. *Administration skills:* organizes his/her own work and that of his/her subordinates; delegates responsibility and authority; measures, evaluates, and controls position activities;
3. *Attitude:* enthusiastic; optimistic; loyal to the superior, associates, and organization;
4. *Communication:* promotes communication within and between all related organizational elements;
5. *Creativeness:* has inquiring mind; develops original ideas; and initiates new approaches for problem solving;
6. *Decisiveness:* makes prompt decisions when necessary;
7. *Flexibility:* is adaptable; quickly adjusts to changing conditions; and copes with the unexpected;
8. *Human relations:* is sensitive to and understands personal interactions; has a "feel" for individuals and quick to recognize their problems; considerate of others; ability to motivate and get people to work together;
9. *Initiative:* self-starting; prompt to take hold; seeks and acts on new opportunities; exhibits high degree of energy in work; not easily discouraged; and possesses basic urge to get things done;
10. *Knowledge:* possesses knowledge (breadth and depth) of functional skills needed to fulfill position requirements; uses information and concepts from other related fields of knowledge; and generally understands the "big picture";
11. *Objectivity:* has an open mind and makes decisions without the influence of personal or emotional interests;
12. *Self-motivation:* has well-planned goals; willingly assumes greater responsibilities; realistically ambitious; and generally eager for self-improvement;
13. *Sociability:* makes friends easily; works well with others; and has a sincere interest in people;

14. *Verbal ability:* articulate; communicative; and is generally understood by persons at all organizational levels;

15. *Vision:* possesses foresight; sees new trends and opportunities; anticipates future events; and is not bound by tradition or custom.

Of particular relevance, the systems engineering manager must exhibit the appropriate leadership skills from the beginning in completing those initial tasks associated with the definition of system requirements and architecture (needs and feasibility analyses, operational requirements, maintenance concept, identification and prioritization of TPM factors, and functional analysis and allocation at the system level). It is the establishment of this early "foundation" that is critical and provides the basis for all follow-on activity. Key subsequent tasks, where the systems engineering manager should assume a leadership role, include the implementation of both program reviews and formal design reviews, the follow-on reporting (feedback) and iniation of corrective action as required, and related activities that will allow for an on-going assessment of the performance of the systems engineering organization in accomplishing its overall program-related objectives.

4 PROGRAM EVALUATION AND FEEDBACK

For the sake of discussion, program review and evaluation activities are broken down into two categories. First, there are the review and evaluation functions that are directly related to the systems engineering tasks accomplished in response to specific program/project requirements that is, the design and development of a new, or reengineered, system. Second, there are the review and evaluation requirements as they pertain to the systems engineering organization that is responsible for the tasks that are to be completed. These two areas of activity are discussed in the following sections.

4.1 Program Review and Evaluation

Periodic program management reviews are usually conducted throughout the system design and development process. The nature and frequency of these will vary with each program and will be a function of design complexity, the number and location of organizations participating in the program, the number and geographical location of suppliers, the number of problems being encountered, and so on. Systems being developed, incorporating new technologies and where the risks are greater, will require more frequent reviews than for other programs.

The objective of these management-oriented reviews is to determine the status of selected program tasks, assess the degree of progress with respect to fulfilling contractual requirements, review program schedule and cost data, review supplier activities, and coordinate the results of the technical design reviews. These reviews are directed toward dealing with issues of a "program" nature, and are not intended to duplicate the formal design review process which deals with design-related problems of a "technical" nature. However, of mutual interest is the "tracking" of the critical TPMs.

The program management data capability that includes the information needed for these program reviews must be designed in such a way as to provide the right type of information, to selected locations and in the proper format, in a timely manner and at the desired frequency, with the right degree of reliability, and at the right cost. The nature and the content of the data will be based on a number of factors, both of a technical nature (e.g., TPMs) and of a program nature (e.g., high-volume suppliers). It is not uncommon for one to go ahead and design and develop, or procure, a large computer-based management information system (MIS) capability with good intent, but which turns out to be too complex and nonresponsive by providing the wrong type of information, and very costly to operate. Such a capability must be tailored to the situation and sensitive to the needs of the particular project.

Regarding the more conventional type of reporting requirements, Figure 5 presents the results of a PERT/CPM program network, and Figure 6 shows a projected life-cycle cost stream, evaluated at different program milestones. While these represent only two examples, the data from a management information system (MIS) should readily point out existing problem areas. Also, potential areas where problems are likely to occur, if program operations continue as originally planned, should become visible. To deal with these situations, a corrective-action procedure should be established that will include the following:

1. Problems are identified and ranked in order of importance.
2. Each problem is evaluated on the basis of the ranking, addressing the most critical ones first. Alternative possibilities for corrective action are considered in terms of (a) effects on program schedule and cost, (b) effects on system performance and effectiveness, and (c) the risks associated with the decision as to whether to take corrective action.
3. When the decision to take corrective action is reached, planning is initiated to take steps to resolve the problem. This may be in the form of a change in management policy, a contractual change, an organizational change, or a system/equipment configuration change.
4. After corrective action has been implemented, some follow-up activity is required to (a) ensure that the incorporated change(s) actually has resolved the problem, and (b) assess other aspects of the program to ensure that additional problems have not been created as a result of the change.

Figure 7 identifies three areas that need attention, as detected at program review M1.

4.2 Evaluation of the Systems Engineering Organization

Having established certain program/project goals, the next step is to determine the extent to which the systems engineering organization has progressed toward meeting these goals, that is, the measure of the organization's capability to meet the desired level of performance. Given the objectives of systems engineering and the recommended tasks that must be performed, there are some questions that should be addressed: *To what extent is the organization completing these tasks effectively and efficiently? Does management understand the principles and concepts of systems engineering? Is there a commitment from the top down toward the implementation of the systems engineering process? If so, what policies are currently being implemented to support this? Have standards, measurable goals, and the appropriate processes been established for the successful accomplishment of systems engineering objectives? Has the organization developed a plan for continuous improvement?*

Although there are many questions of this nature that can be asked, the objective is to determine the organization's *level of maturity*, where it may "fit" in the hierarchical structure as compared with other organizations functioning in a similar area of activity (through *benchmarking*), and where there are weaknesses that need to be addressed. In essence, there is a need to develop a model to assist in the evaluation of an organization's current capability.

Network/Cost Status Report

	Project: System XYZ					Contract Number: 6BSB-1002						Report Date: 8/15/05		
Item/Identification						Time Status						Cost Status		
WBS No.	Cost Account	Beginning Event	Ending Event	Exp. Elap. Time (te) (Weeks)		Earliest Completion Date (D_E)	Latest Completion Date (D_L)	Slack D_L-D_E (Weeks)	Actual Date Completed	Cost. Est. ($)	Actual Cost to Date ($)	Latest Revised Est. ($)	Overrun (Underrun) ($)
4A1210	3310	8	9	4.2		3/4/05	4/11/05	11.6	4/4/05	2500	2250	2250	(250)
4A1230	3762	R100	R102	3.0		5/15/05	4/28/05	−3.3		4500	4650	5000	500
5A1224	3521	7	9	20.0		6/20/05	8/3/05	0		6750	5150	6750	0

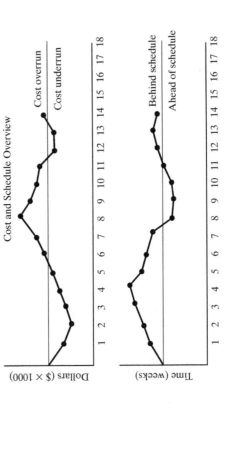

Figure 5   Status report relating cost to a PERT/CPM network and a CWBS (example).

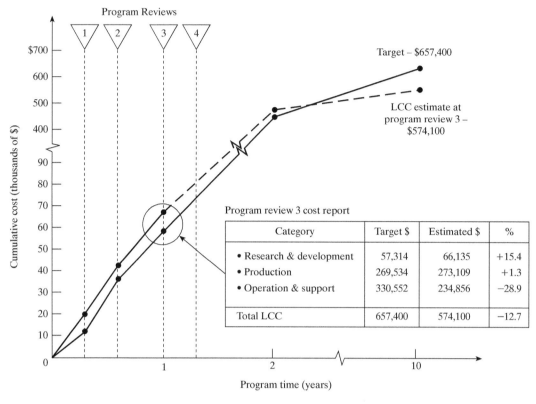

Figure 6 Life-cycle cost projection at a major review point.

In response, there has been a concerted and continuing effort since the late 1980s to develop a model that will address the organizational assessment issue. Although there have been numerous models used to varying degrees through the years, a series of recently developed specific projects/models is noteworthy. Through the early efforts of the Software Engineering Institute (SEI) at Carnegie Mellon University, a process improvement model oriented to software development, Software Capability Maturity Model (SW-CMM), was first introduced in 1989. As a result of experience and continuous upgrading, Version 1.1 of SW-CMM was released in 1993. Based on this experience, and through the combined efforts of many in industry, government, and academia, the System Engineering Capability Maturity Model (SE-CMM) was developed and released for use in 1994.[1] At the same time, and with the coordination and support of the International Council on Systems Engineering (INCOSE), the Systems Engineering Capability Assessment Model (SECAM) was released in 1994.[2] This model was updated to Version 1.50a and released in

---

[1] Software Engineering Institute (SEI), *A Systems Engineering Capability Maturity Model* (SE-CMM), Version 1.1, SECMM-95-01 (Pittsburgh, PA: Carnegie Mellon University, 1995).

[2] A good reference that provides a historical basis for the SECAM and its applications is B. A. Andrews and E. R. Widmann, "A Synopsis of Metrics and Observations from Systems Engineering Process Assessments Conducted Using the INCOSE SECAM," in *Proceedings of the Sixth Annual International Symposium of the INCOSE*, Vol. 1 (San Diego, CA: INCOSE, 1996), p. 1071. Additional references are included in the *Proceedings* from earlier INCOSE symposia.

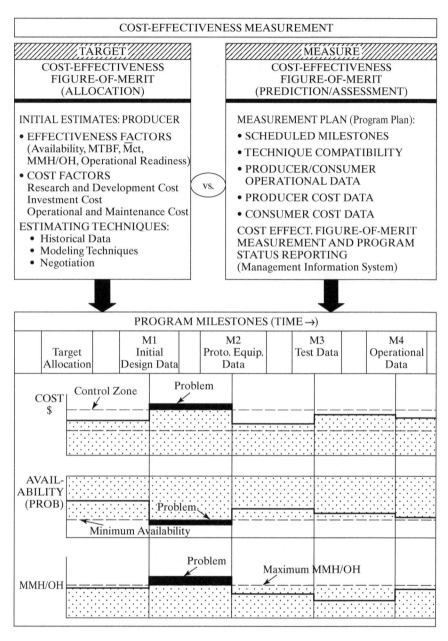

Figure 7 Technical performance measurement and evaluation.

June 1996. These two models were then successfully merged into EIA/IS-731 in 1998 as a result of a collaborative effort involving EIA (Electronic Industries Alliance), EPIC (Enterprise Process Improvement Collaboration), and INCOSE.[3]

[3]GEIA (Government Electronics and Information Technology Association), EIA/IS 731: *Systems Engineering Capability Model* (SECM), Washington, DC, 2001 (Website: *http://www.geia.org/sstc/G47/page6 .htm,* October 2001).

Subsequent to the initial release of EIA/IS-731, there have been a number of individual efforts to develop comparable models for different purposes. In addition to the models covering software development and systems engineering, an effort was initiated to develop a model for *Integrated Product and Process Development* (IPPD). Further, there have been efforts to address other critical areas where a measure of organizational maturity is desired. Given the trend relative to developing a series of different models for individual purposes, an effort was initiated in 1998 to study the feasibility of developing one comprehensive model that would represent an integrated approach and combine the capabilities of the SW-CMM, SE-CMM, SECAM, and the IPPD model. The result of this effort has produced a Capability Maturity Model Integration (CMMI), the latest configuration of which is Version 1.2, August 2006. The overall objective is to eliminate the "stovepipe" models and to adapt CMMI as the ultimate measurement tool for the various areas of concern.[4]

To get some idea of the detailed approach for implementation, given that the emphasis throughout this text is on systems engineering, it would be appropriate at this point to consider the Systems Engineering Capability Model (SECM), discussed in EIA/IS-731. One of the first steps in its development was, of course, to define the goals and objectives of a systems engineering organization. Having accomplished this, essential systems engineering and management tasks that an organization must perform to ensure a successful effort were identified and included in three basic focus-area categories; that is, a *technical category focus area*, a *management category focus area*, and an *environment category focus area*. Establishment of the focus-area categories then led to the identification of specific *focus areas*, which led to *themes*, which in turn led to a description of specific *practices*. The results of this progression are summarized in a presentation of the major topics shown in Figure 8.

Given a description of the desired practices, the next step was to identify different *capability levels* (levels of "maturity") or the degree(s) of capability an organization should strive to meet, evolving from current capability to a future level, indicating growth potential.

FOCUS-AREA CATEGORIES		
TECHNICAL	MANAGEMENT	ENVIRONMENTAL
• Define Stakeholder and System-Level Requirements • Define the Technical Problem • Define the Solution • Assess and Select • Integrate System • Verify System • Validate System	• Plan and Organize • Monitor and Control • Integrate Design Disciplines • Coordinate with Suppliers • Manage Risk • Manage Data • Manage Configurations • Ensure Quality	• Define and Improve the Systems Engineering Process • Manage Competency • Manage Technology • Manage Systems Engineering Support Environment

Figure 8 SECM focus areas and categories (EIA/IS-731).

[4]A good reference covering the history and background leading to the development of the CMMI is *Systems Engineering*, The Journal of the International Council on Systems Engineering, Vol. 5, No. 1 (2002) is published by John Wiley & Sons, Inc., NY. There are a series of articles in this Journal issue that deal with CMMI, the status of EIA/IS-731, and related topics. Two additional references are (1) M. B. Chrissis, M. Konrad, and S. Shrum, 2nd ed., *CMMI: Guidelines for Process Integration and Product Improvement* (Addison—Wesley, Boston, MA, 2006); and (2) M. K. Kulpa and K. A. Johnson, *Interpretingthe CMMI: A Process Improvement Approach* (Auerbach Publications, 2003).

Six capability levels were established and related to individual focus areas. A focus area includes a list of practices describing the activities that an organization must successfully perform. In Figure 9, the levels of capability (from "Level 0" to "Level 5") are identified as *initial, performed, managed, defined, measured,* and *optimized.* These levels are supported by a description of specific practices that are desired in order to meet the requirements for a given level. The objective, of course, is to progress to Level 5.

In applying this model in the appraisal (or assessment) of an organization's capability, there are different phases: *preassessment, on-site assessment,* and *postassessment.* During the "preassessment" phase, it is necessary to solicit management support of the organization to be evaluated and to develop the process for evaluation. Included in this phase is the development of a rather extensive questionnaire (which contains many different questions for the EIA/IS-731 requirement, or a minimum of 40 questions pertaining to Level 1, 91 questions for Level 2, 156 questions for Level 3, 56 questions for Level 4, and 83 questions for Level 5).[5] The "on-site assessment" phase includes the following steps: administering the questionnaire, analyzing the results, developing some additional exploratory questions, conducting interviews with focus groups, analyzing exploratory data, summarizing the results and coordinating with management, and preparing the final evaluation report. This phase is usually conducted during a 1-week period, by a team of three to five people working with a group of department managers, project leaders, and workforce practitioners, and results in rapid feedback and minimizing any impact on internal projects and the day-to-day scheduled work. The "postassessment" phase involves management briefings and the preparation of a plan for future action as required.

The results of the assessment, utilizing the SECM, should include a summary chart/graphic showing the different focus areas and the degrees to which each has achieved a given "level of capability." In Figure 10, it can be seen that Focus Area 1 in the *management* category has achieved "capability" at Level 3, and that Focus Area 2 (in the same category) is only at Level 1. Given these results, the final assessment report (and plan for future action) should include some specific recommendations for improvement, particularly in regard to Focus Area 2, and the action(s) that need to be initiated in order to progress to the next higher level. The objective is, of course, to make progress in all of the focus areas with the proper balance being achieved across the board.

The preceding description provides only a rough idea as to the objectives and content of the SECM. For more in-depth coverage, a detailed review of EIA/IS-731 is recommended. Relative to the future, although this model will, in all probability, continue to be applied in selected areas and oriented to the assessment of a systems engineering organization as an entity, acquiring a good understanding of the CMMI is also recommended, as this is a more comprehensive model and is gaining in popularity.

In comparing the SECM with the CMMI, it is clear that the basic architectures are quite similar.[6] The SECM includes focus areas and categories; the CMMI takes the same basic approach, although the specific topics and nomenclature are different. In Figure 11, there are four Process Area Categories: *process management, project management, engineering,* and *support.* Within each of these categories, there are a number of specific

---

[5]S. Alessi, "A Simple Statistic for Use with Capability Maturity Models," *Systems Engineering,* The Journal of the International Council on Systems Engineering, Vol. 5, No. 3 (2002): 242–252 (published by John Wiley & Sons, Inc. Hoboken, NJ).

[6]I. Minnich, "EIA/IS-731 Compared to CMMI-SE/SW," *Systems Engineering,* The Journal of the International Council on Systems Engineering, Vol. 5, No. 1 (2002): 62–72 (published by John Wiley & Sons, Inc., Hoboken, NJ).

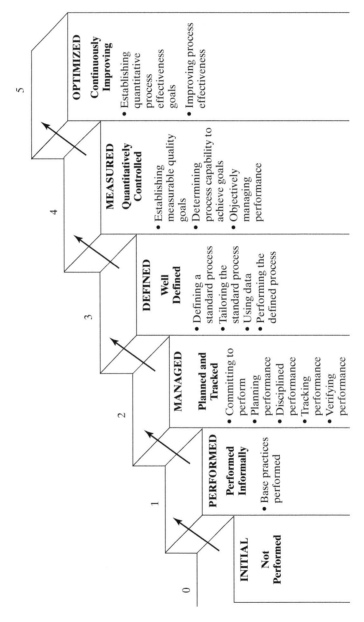

Figure 9 Improvement path for systems engineering process capability.

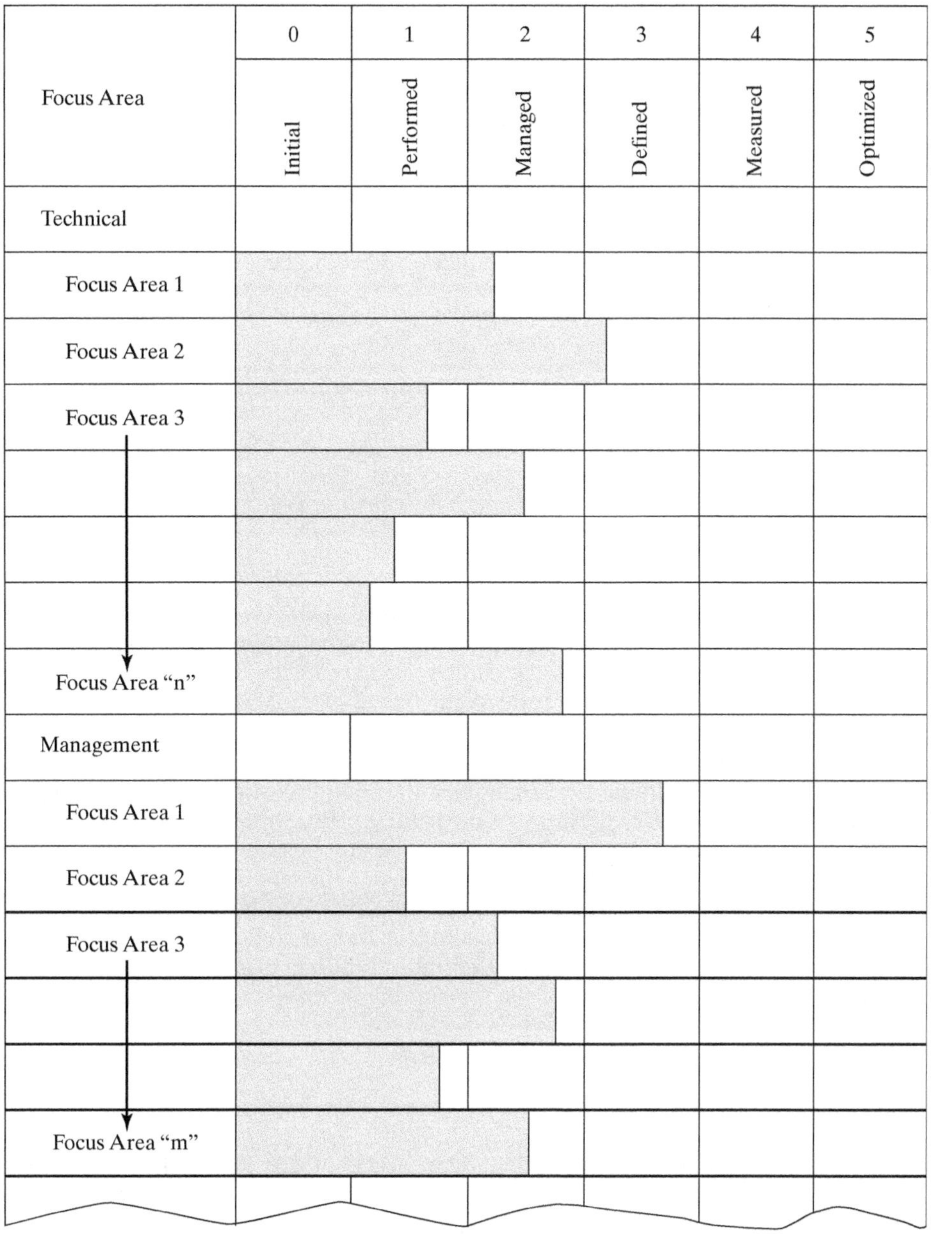

Focus Area	0	1	2	3	4	5
	Initial	Performed	Managed	Defined	Measured	Optimized
Technical						
Focus Area 1						
Focus Area 2						
Focus Area 3						
Focus Area "n"						
Management						
Focus Area 1						
Focus Area 2						
Focus Area 3						
Focus Area "m"						

Figure 10   Focus-area capability assessment.

Process Areas, for which detailed questions have been prepared for purposes of assessment. Note that the activities in CMMI are much broader in scope than those in SECM. In regard to "levels of capability," the CMMI also has established six levels (i.e., "Levels 0–5"), including *incomplete*, *performed*, *managed*, *defined*, *quantitatively managed*, and *optimized*.

PROCESS AREA CATEGORIES			
PROCESS MANAGEMENT	PROJECT MANAGEMENT	ENGINEERING	SUPPORT
• Organizational Process Focus • Organizational Process Definition • Organizational Innovation and Deployment	• Project Planning • Project Monitoring and Control • Supplier Management • Integrated Project Management • Risk Management • Quantitative Project Management	• Requirements Management • Requirements Development • Technical Solution • Product Integration • Verification • Validation	• Configuration Management • Process and Product Quality Assurance • Measurement and Analysis • Decision Analysis and Resolution • Causal Analysis and Resolution

Figure 11 CMMI process areas and categories. *Source:* I. Minnich, "EIA/IS-731 Compared to CMMI SE/SW," *Systems Engineering*, The Journal of the International Council on Systems Engineering, Vol. 5, No. 1 (2002), Table II. (Published by John Wiley & Sons, Inc.).

For purposes of assessment, the Standard CMMI Assessment Method for Process Improvement (SCAMPI) is accomplished through the application of questionnaires, local visits and interviews, and the like. Specific scoring rules for each capability level are used, and the highest resulting score reflects the "level of capability" attained for the process area being evaluated. In any event, the overall approach here is similar to that described earlier for SECM.

In summary, it should be emphasized that the discussion presented throughout this section (i.e., Section 4.2) pertains to only one approach leading to the evaluation of a systems engineering organization. There may be any number of other and more desirable approaches developed and applied in accomplishing the same overall purpose. The objective herein is to introduce the requirement, to describe what has been accomplished in this area to date, and to recommend that the implementation of such be considered as an essenial element in accomplishing effective program planning, organization, management, and control. The proper evaluation of the systems engineering organization, through whatever approach implemented, is essential.

5 RISK MANAGEMENT

Inherent in any formal program activity is the aspect of *risk.* Risk is the potential that something will go wrong as a result of one or a series of events. A Risk Management Plan (describing the planning for, assessment, analysis, and the handling of risk) should be included as a major section in the SEMP for any given program. There are four basic categories of risk, as noted in Figure 12:

1. *Technical risk:* the possibility that a technical requirement of the system will not be achieved. Technical risk exists if, in the system design and development process, it appears that the system will not meet a specific performance objective, such as one or more of the appropriate TPMs.

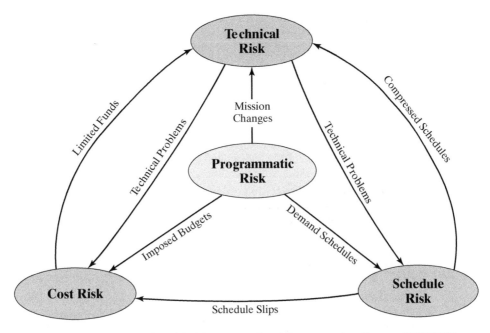

Figure 12 Typical relationship(s) among the risk categories. *Source*: INCOSE-TP-2003-016-02, *Systems Engineering Handbook* (San Diego, CA: International Council on Systems Engineering (INCOSE), 2004), Figure 6-2, p. 61.

2. *Cost risk:* the possibility that the specified allocated budget will be exceeded. While the emphasis is generally directed to the initial program/project acquisition budget only (i.e., initial project resources for the design and development of a system identified in the applicable CWBS), the concern for budgeted costs over the entire system life cycle is also relevant. Thus, category of "cost risk" should be addressed to both the initial project budget and funding levels and the subsequent projected life-cycle cost for the system.

3. *Schedule risk:* the possibility that the project will fail to meet the scheduled milestones. Schedule risk can be incurred if the overall program schedule is not being met, or if any lower-level task is behind schedule. Of particular importance are the respective schedules for the tasks along the critical path.

4. *Programmatic risk:* the occurrence of events, imposed on the program/project, which are the result of external influences. Decisions made at a higher level in the organization, delayed funding, external environmental impacts, and so on can have a significant impact on the overall program/project in terms of technical, cost, and schedule objectives.

As illustrated in Figure 12, all of these individual categories of risk interact with each other. The potential of risk becomes increasingly greater as complexities and new technologies are introduced in the design of systems. In addition , the possibility of introducing risk becomes greater as the number of program suppliers increases through outsourcing. Thus, it is important that a comprehensive risk management effort be included and inherent within the scope of a systems engineering management program.

Risk management constitutes an iterative process involving the following five steps:

1. *Risk planning* includes the development of a risk management plan for a given program. The plan (or an abbreviated summary referencing such) should be included in the SEMP.

2. *Risk identification* includes the screening of all requirements (technical, cost, schedule, and programmatic) and to identify those that are likely not to be met. Figure 7 shows an example of TPM requirements that will, in all probability, not be met unless corrective action is initiated immediately.

3. *Risk assessment* pertains to determining the probability of failure to meet a specified requirement and the possible outcomes and the consequences of not meeting the requirement. This pertains to both the probability of occurrence and the estimated magnitude of the risk.

4. *Risk analysis* is accomplished to determine the way(s) in which the risk can be eliminated or minimized if not eliminated altogether. Possible solutions are determined through the accomplishment of design trade-offs and resulting in a recommended engineering change proposal (ECP).

5. *Risk handling* includes the activities associated with the incorporation of changes and system modifications (of products and/or processes) recommended as a solution to the identified problem(s). The purpose here is to ensure closure to the process of reducing overall program risk as an end objective.

The implementation of a formalized risk management program is critical to the successful accomplishment of systems engineering objectives. Whether accomplished within the systems engineering organization or not, the requirements for such are very important—thus, the systems engineering organization must be a key participant in such activities.[7]

6 SUMMARY AND EXTENSIONS

Given initial systems engineering planning and organization requirements for a typical program, this chapter commences with a summary of organizational goals and objectives, outsourcing requirements and the identification of program suppliers, program leadership and direction, evaluation of both program/project activities and performance of the systems engineering organization, and the importance of implementing a risk management effort within the context of systems engineering management.

Discussed within this chapter is an approach for conducting a *performance evaluation* of a typical systems engineering organization, utilizing the SECM and CMMI models to

---

[7]Risk planning, identification, assessment, analysis, handling, and management are covered rather extensively in INCOSE-TP-2003-016-02, *Systems Engineering Handbook* (San Diego, CA: International Council on Systems Engineering /INCOSE, 2004). Also, the reader should check the website for the Society for Risk Analysis (SRA); refer to Appendix H.

facilitate the evaluation process. The process includes the identification of activity categories, focus areas and themes, and an assessment of current organizational capability and "level of maturity." At this point, it should be noted that the development of CMMI and related models constitutes an ongoing and iterative process. Recent activities have led to the development of four different CMMI models, devoted primarily to (1) software engineering; (2) systems engineering and software engineering; (3) systems engineering, software engineering, and integrated product and process development (IPPD); and (4) systems engineering, software engineering, IPPD, and supplier sourcing. The development of these and related models is expected to continue as new disciplines are considered for incorporation. A good source for the latest information on CMMI models is *http://www.sei.cmu.edu/cmmi/models/*.

It should be emphasized that the process described for the evaluation of a systems engineering organization utilizing the SECM and CMMI models represents only one approach. There may be many other and alternative ways of accomplishing such an evaluation, and the presentation of the SECM and CMMI models is included to provide some guidelines for consideration in the event of need. The objective is to encourage both the evaluation of program activities and the organization responsible for accomplishing the specific tasks involved.

A second, and most critical, area that needs emphasis is the development and implementation of a *risk management* capability. Inherent within the systems engineering process are the many and varied decisions made during the early phases of conceptual and preliminary system design and the results of these decisions could have a significant impact on the program and ultimate system configuration later. In addition, throughout a given program, there are numerous challenges associated with fulfilling the requirements pertaining to the development of a system within the context of a system-of-systems (SOS) network, implementing the many different possible supply chain alternatives in an international environment, and so on. Thus, a critical activity in the management of a systems engineering program is the establishment of a risk management capability and the early development of a formal risk management plan. The purpose is to establish a vehicle, early in the life cycle, for identifying potential problem areas (both of an administrative and a technical nature) so that any required corrective action can be initiated during the early stages of system design and development. A good source describing requirements and activities in this area is the Society for Risk Analysis (SRA), McLean, VA, with more information available from the website *http://www.sra.org*.

In summary, the prime objectives in the implementation of systems engineering include timely initial planning, effective organizational leadership, and a highly integrated approach involving the different design and related disciplines. Good communications, upward and downward and across organizational lines, is essential throughout. Given the current trends toward more outsourcing through the utilization of external suppliers, increased globalization, and greater international competition, the challenges associated with the implementation of systems engineering appear to be greater than ever before. The development of a good SEMP and systems specification (Type *A*), the identification and selection of the right suppliers to accomplish the required functions/tasks, the establishment of an effective "team" approach with both producer and supplier commitment and activity integration, and the proper ongoing day-to-day management and control activities are critical, particularly when the various system design and development functions are likely to be geographically dispersed and/or spread throughout the world. Accordingly, it is important that these and related challenges pertaining to systems engineering management be recognized early and from the top down, that is, starting with the customer.

QUESTIONS AND PROBLEMS

1. Assume that you have just been assigned to develop a systems engineering organization. What steps would you follow in developing the goals and objectives for this organization? What would you include? Support your answer with examples.

2. Describe *benchmarking.* For your organization, describe the steps that you would follow in establishing the appropriate benchmarks. Provide an example for the same.

3. Assume that in the establishment of benchmarks for your current organization, it is noted that your organization is deficient in certain critical areas. What steps would you take to correct the situation?

4. As a manager of a newly established systems engineering organization, you need to develop a *management information system* (MIS) to provide you with the visibility needed to properly manage the organization on a day-to-day basis. Describe the steps that you would take to accomplish this objective. What information would you include (type, format, frequency, etc.)? What reports should be provided, at what frequency, and to whom?

5. What factors (criteria) should be considered in the identification of *outsourcing* requirements? Provide an example (describe a specific situation).

6. Assume that a decision has been made in favor of outsourcing and that you need to develop a request for proposal (RFP). What information should you include in the RFP? (Prepare a detailed outline.)

7. Why is the development of a *specification tree* important?

8. Refer to Figure 3. How would you determine the specific TPMs that should be included in each of the specifications noted?

9. Assume that, in response to a RFP, there are a number of different suppliers, each having submitted a proposal. What factors (criteria) should be considered in the selection of a specific supplier? Provide a list of factors (in order of importance).

10. Refer to Figure 4. Develop a supplier checklist tailored to your own specific needs. (Extend the checklist to include a second level—D.4.2.1, as an example).

11. Assume that you are responsible for the activities of many different suppliers located throughout the world. What steps would you take to ensure that the proper level of effort (and integration) is being accomplished throughout and on a continuing basis?

12. Refer to Figure 5. Assume that one of the activities along the critical path reflects a supplier task that has been reported as being behind schedule. What steps would you take to correct the situation?

13. Refer to the cost projection in Figure 6. Although the overall estimate looks favorable in terms of the projected life-cycle cost, the results from Program Review 3 are not favorable. What would you do (if anything) to ensure that the ultimate projection is a realistic one and is being maintained?

14. Refer to Figure 7. There are three problem areas noted. What steps would you take in correcting (i.e., eliminating) each of these problems?

15. Assume that you are the manager of a systems engineering department. What would you do to determine just how well your organization is functioning, that is, in assessing organizational capability?

16. Develop an *assessment model* tailored to your particular organization. Identify the appropriate activity categories and the focus areas within each category. Develop at least three questions that might be addressed in the evaluation for each focus area.

17. Refer to Figure 10. Assume that the results of an assessment of your organization turned out to be similar to what is presented in the figure. As manager of the organization, you want to show growth (from "managed" to "defined" in Figure 9) within a year. What steps would you take to initiate such? What steps would you take to ensure continual growth within the organization through the years to come (for the long term)?

18. Describe what is meant by *risk*. How can it be measured? What aspects of risk would be of particular interest to a systems engineering manager? Provide some examples.

19. Develop a detailed outline for a risk management plan. How would this plan relate to the SEMP?

20. Why is the development of a risk management capability important?

Appendix

Systems Engineering Program Tasks

Systems engineering, as defined throughout this text, covers a broad spectrum of activity. It may even appear that the "systems engineer," or the "systems engineering organization," does everything. Although this is not intended, nor practical, the fulfillment of systems engineering objectives does require some involvement, either directly or indirectly, in almost every facet of program activity. The challenge is to identify those functions (or tasks) that deal with the overall *system* as an entity and, when successfully completed, will have a positive impact on the many related and subordinate tasks that must be accomplished. Although there are variations from one program to the next, the following tasks have been identified where strong *leadership* from the systems engineering organization (or a designated alternate activity) is required:

1. *Perform a needs analysis and conduct feasibility studies.* These activities should be the responsibility of the systems engineering organization because they deal with the system as an entity and are fundamental in the initial interpretation and subsequent definition of system requirements. A strong interface and good communications with the customer/user are required.

2. *Define system operational requirements and the maintenance concept, and identify and prioritize the technical performance measures.* The results of these activities are included in the overall defininition of system-level requirements and are the basis for top-down system design. Trade-off studies are accomplished, requiring an excellent understanding of system interfaces and the objectives of the customer/user.

3. *Accomplish a functional analysis at the system level and allocate requirements to the next lower level.* It is essential that a good *functional* baseline be defined from which specific resource requirements will be identified. This baseline serves as a common frame of reference used as an input source for many of the engineering and support activities accomplished later on. Although the systems engineering organization may not accomplish the functional analysis in total, a strong leadership role is essential in providing a necessary foundation at the system level. Good *baseline management* is essential.

4. *Prepare system specification, Type A.* This represents the top "technical" document for design and serves as the basis for the development of all subordinate specifications. The systems engineering organization should be responsible for the preparation of this top-level document and should ensure that all subordinate specifications "track" and are mutually supportive.

5. *Prepare the test and evaluation master plan.* Because this document reflects the approach for system *validation*, the systems engineering organization should assume a leadership role to ensure that there is connectivity between the prioritized TPMs, the levels of design complexity, the risks associated with pursuing given design approaches, and so on. As specific quantitative requirements are initially defined, a viable plan must be established for the validation of these requirements. The systems engineering organization needs to ensure an integrated test and evaluation effort and that the system specification, the SEMP, and other design plans "talk" to each other.

6. *Prepare the systems engineering management plan* (SEMP). Because this plan constitutes the top-level "integrating" engineering document that ensures systems engineering objectives are fulfilled, it must be prepared, revised as necessary, and implemented under the leadership of the systems engineering organization.

7. *Accomplish synthesis, analysis, and evaluation.* Although this area of activity is continuous and widespread across the design community, the systems engineering organization must provide an "oversight" function to ensure that the major day-to-day design decisions are in compliance with the system specification, the results of trade-off studies are properly documented, and the design risks have been properly identified and addressed.

8. *Plan, coordinate, and conduct formal design review meetings.* The conductance of formal design reviews is necessary to ensure that the systems engineering process is being properly implemented, the appropriate *functional* baselines are well defined, good configuration management is being implemented, and all members of the design team understand the basis for past decisions and are "tracking" the same design database. The systems engineering organization, or designated representative, should "Chair" the formal design review meetings.

9. *Monitor and review system test and evaluation activities.* This constitutes a systems engineering organization "oversight" function to review the results of test and evaluation (and thus "validation") to ensure that systems requirements are being met. If not, the problems need to be identified rapidly, followed by implementation of the appropriate corrective action.

10. *Coordinate and review all formal design changes and modifications for improvement.* The systems engineering organization, or designated representative, should "Chair" the Change Control Board (CCB) to ensure that (1) system requirements (i.e., all TPMs) will still be met should a change be approved; (2) all proposed changes have been evaluated and selected considering impacts on mission criticality, system effectiveness, life-cycle cost (LCC), and environmental factors; (3) a comprehensive modification and rework plan has been developed; and (4) the approved changes are incorporated efficiently and in a timely manner. The application of good configuration management is required here.

11. *Initiate and establish the necessary ongoing liaison activities throughout the production/ construction, utilization and sustaining support, and retirement and material disposal phases.* The systems engineering organization must maintain surveillance of production/construction activities to ensure that the system is being produced/constructed as designed. Further, an ongoing level of surveillance of consumer/user operations in the field is essential in order to provide the feedback necessary for the purpose of system "validation."

These 11 basic program tasks constitute an example of what might be appropriate for a typical large-scale program, although the specific requirements may vary from one instance to the next. The goal is to identify tasks that are oriented to the *system* and are *critical*. The overall objectives are to ensure that (1) the requirements for the system are initially well defined from the beginning, (2) the appropriate characteristics and attributes are *designed-in*, and (3) the system has been validated in terms of the initially specified requirements.

Appendix:

Functional Analysis

A critical step in implementing the systems engineering process is the accomplishment of the functional analysis and the definition of the system in "functional" terms. Functions are initially identified as part of defining the need and the basic requirements for the system. System operational requirements and the maintenance concept are defined, and the functional analysis is expanded to establish a functional baseline, from which the resource requirements for the system are identified; that is, equipment, software, people, facilities, data, the various elements of maintenance and support, and so on. The functional analysis is initiated during the conceptual design phase. As design and development continues, the functional analysis is accomplished to a greater depth, to the subsystem level and below, during the preliminary system design phase. This appendix provides guidance as to the detailed steps involved in accomplishing a functional analysis and in the development of functional flow block diagrams (FFBDs).

Functional analysis includes the process of translating top-level system requirements into specific qualitative and quantitative design-to requirements. Given an identified need for a system, supported by the definition of system operational requirements and the maintenance concept, it is necessary to translate this information into meaningful design criteria. This translation task constitutes an iterative process of breaking down system-level requirements into successive levels of detail; a convenient mechanism for communicating this information is through the various levels of FFBDs.

1 FUNCTIONAL FLOW BLOCK DIAGRAMS

Functional flow block diagrams (FFBDs) are developed to describe the system and its elements in functional terms. These diagrams reflect both operational and support activities as they occur throughout the system life cycle, and they are structured in a manner that illustrates the hierarchical aspects of the system. Some of the key features of the overall functional flow process are noted as follows:

1. The functional block diagram approach should include coverage of all activities throughout the system life cycle, and the method of presentation should reflect proper activity sequences and interface interrelationships.
2. The information included within the functional blocks should be concerned with *what* is required before looking at *how* it should be accomplished.
3. The process should be flexible to allow for expansion if additional definition is required or reduction if too much detail is presented. The objective is to progressively

and systematically work down to the level where resources can be identified with how a task should be accomplished.

In the development of functional flow diagrams, some degree of standardization is necessary (for communication) in defining the system. Thus, certain basic practices and symbols should be used, whenever possible, in the physical layout of functional diagrams. The paragraphs below provide some guidance in this direction.

1. *Function block.* Each separate function in a functional diagram should be presented in a single box enclosed by a solid line. Blocks used for reference to other flows should be indicated as partially enclosed boxes labeled "Ref." Each function may be as gross or detailed as required by the level of functional diagram on which it appears, but it should stand for a definite, finite, discrete action to be accomplished by equipment, personnel, facilities, software, or any combination thereof. Questionable or tentative functions should be enclosed in dotted blocks.

2. *Function numbering.* Functions identified in the functional flow diagrams at each level should be numbered in a manner which preserves the continuity of functions and provides information with respect to function origin throughout the system. Functions on the top-level functional diagram should be numbered 1.0, 2.0, 3.0, and so on. Functions which further indenture these top functions should contain the same parent identifier and should be coded at the next decimal level for each indenture. For example, the first indenture of function 3.0 would be 3.1, the second 3.1.1, the third 3.1.1.1, and so on. For expansion of a higher-level function within a particular level of indenture, a numerical sequence should be used to preserve the continuity of the function. For example, if more than one function is required to amplify function 3.0 at the first level of indenture, the sequence should be 3.1, 3.2, 3.3, ..., 3.*n*. For expansion of function 3.3 at the second level, the numbering shall be 3.3.1, 3.3.2, ..., 3.3.*n*. Where several levels of indentures appear in a single functional diagram, the same pattern should be maintained. While the basic ground rule should be to maintain a minimum level of indentures in any one particular flow, it may become necessary to include several levels to preserve the continuity of functions and to minimize the number of flows required to functionally depict the system.

3. *Functional reference.* Each functional diagram should contain a reference to its next higher functional diagram through the use of a reference block. For example, function 4.3 should be shown as a reference block in the case where the functions 4.3.1, 4.3.2, . . . , 4.3.*n*, and so on, are being used to expand function 4.3. Reference blocks shall also be used to indicate interfacing functions as appropriate.

4. *Flow connection.* Lines connecting functions should indicate only the functional flow and should not represent either a lapse in time or any intermediate activity. Vertical and horizontal lines between blocks should indicate that all functions so interrelated must be performed in either a parallel or a series sequence. Diagonal lines may be used to indicate alternative sequences (cases where alternative paths lead to the next function in the sequence).

5. *Flow directions.* Functional diagrams should be laid out so that the functional flow is generally from left to right and the reverse flow, in the case of a feedback functional loop, from right to left. Primary input lines should enter the function block from the left side; the primary output, or go line, should exit from the right, and the no-go line should exit from the bottom of the box.

6. *Summing gates.* A circle should be used to depict a summing gate. As in the case of functional blocks, lines should enter or exit the summing gate as appropriate. The summing gate is used to indicate the convergence, divergence, parallel, or alternative functional paths and is annotated with the term AND or OR. The term AND is used to indicate that parallel functions leading into the gate must be accomplished before proceeding to the next function, or that paths emerging from the AND gate must be accomplished after the preceding functions. The term OR is used to indicate that any of the several alternative paths (alternative functions) converge to, or diverge from, the OR gate. The OR gate thus indicates that alternative paths may lead or follow a particular function.

7. *Go and no-go paths.* The symbols G and $\overline{G}$ are used to indicate go and no-go paths, respectively. The symbols are entered adjacent to the lines leaving a particular function to indicate alternative functional paths.

8. *Numbering procedure for changes.* Additions of functions to existing data should be accomplished by locating the new function in its correct position without regard to sequence of numbering. The new function should be numbered using the first unused number at the level of indenture appropriate for the new function.

2 SOME EXAMPLES OF APPLICATION

With the objective of illustrating how some of these general guidelines are employed, Figures 1–7 are included to present a few simple applications.

1. Figure 1 provides an example of the basic format used in the development of functional flow block diagrams in general.
2. Figure 2 shows a manufacturing capability (top level or Blocks 1.0–7.0), an expansion of the design function (Block 2.0), and an expansion of the operating functions of the manufacturing plant (Blocks 5.1.1–5.1.16).
3. Figure 3 shows two levels of operational flow diagrams for a space system.
4. Figure 4 shows a maintenance functional flow diagram for the space system that evolves from the operation flow in Figure 3.
5. Figure 5 shows two levels of operational flow diagrams and two levels of maintenance flow diagrams for a radar system.
6. Figure 6 shows two levels of operational flow diagrams and a maintenance functional flow diagram for an automotive system.
7. Figure 7 shows two levels of operational flow diagrams and two levels of maintenance flow diagrams for a lawn mowing system.

Although these sample block diagrams do not cover the selected systems entirely, it is hoped that the material is presented in enough detail to provide an appropriate level of guidance for the development of functional block diagrams.

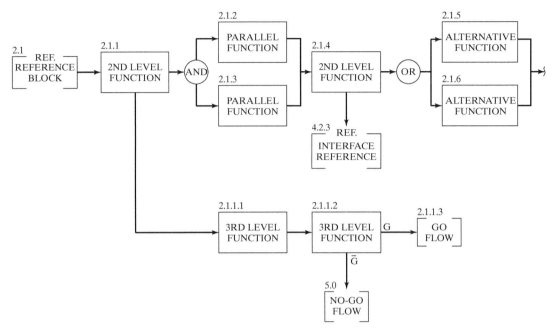

Figure 1   General format for the development of functional flow block diagrams.

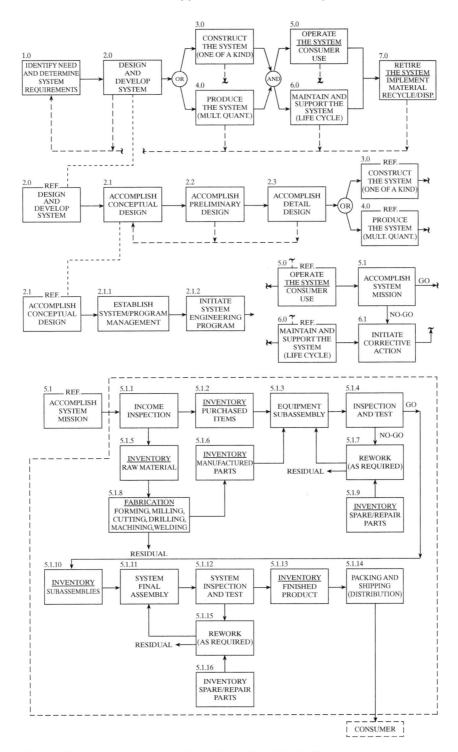

Figure 2 A manufacturing functional flow block diagram.

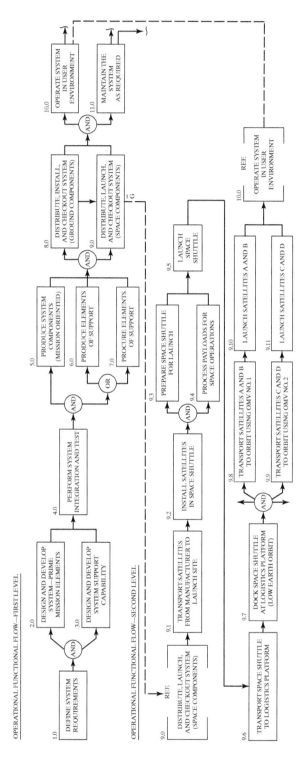

Figure 3 Operational functional flow diagram for a space system.

MAINTENANCE FUNCTIONAL FLOW

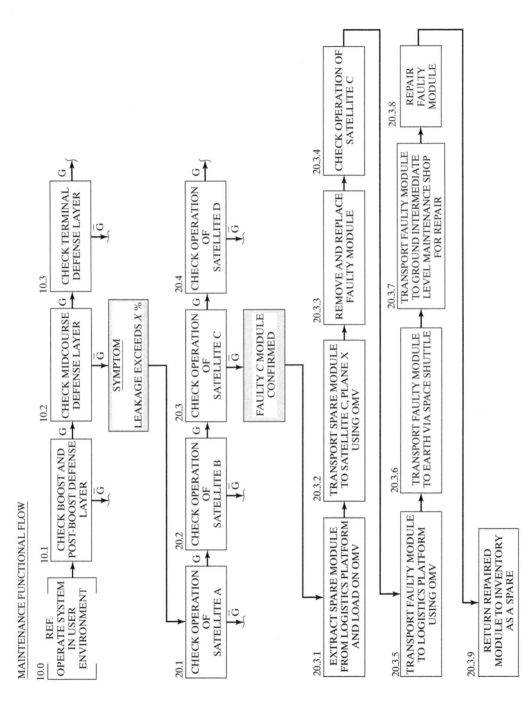

Figure 4 Space system maintenance functional flow diagram (example).

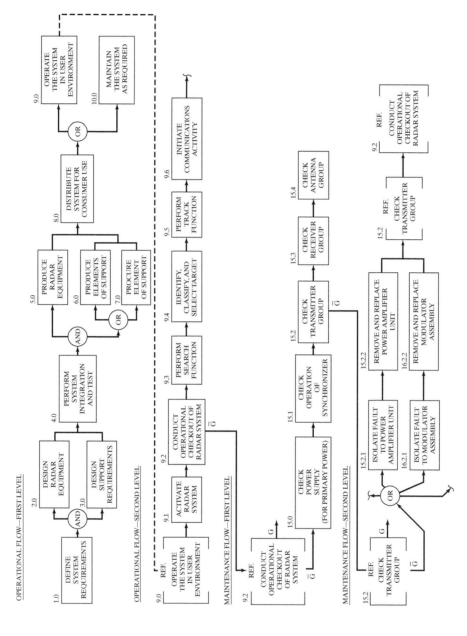

Figure 5 Radar system functional flow diagram (example).

Appendix: Functional Analysis

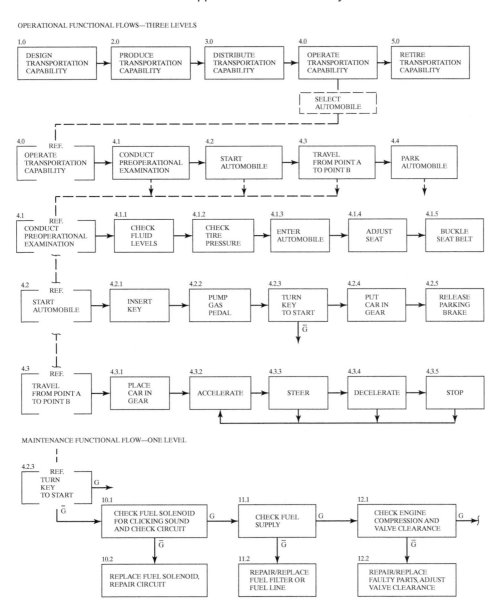

Figure 6 Operational and maintenance functional flow diagram for an automobile.

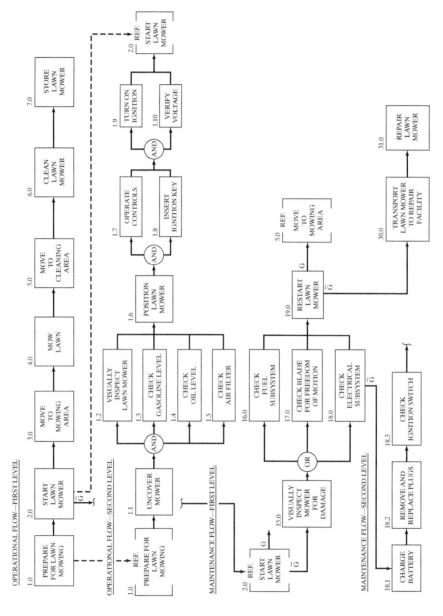

Figure 7 Operational and maintenance functional flow diagram for a lawn-mowing system.

Appendix:

Design and Management Checklists

Throughout the system design and development process, there are numerous occasions when a design review and evaluation effort (in some form) is appropriate. Such design reviews may be accomplished either "informally" or "formally" at discrete times in the life cycle. During the day-to-day design participation and integration process, there may be a series of informal mini-reviews covering critical areas in design. In addition, formal reviews are conducted at specific times as the level of design definition evolves from the system level to the subsystem level, and down to the detailed component level. The formal design reviews include the conceptual, system, equipment/software, and critical design reviews. The purpose of these reviews is to ensure that the design configuration (as envisioned at the time of review) is in compliance with the initially specified requirements for the system.

In conducting a design review, whether on an informal or a formal basis, the development and utilization of a checklist often serves as a useful aid in meeting the intended objective. The checklist may include any number of items, topics, and/or questions reflecting the ultimate requirements and focusing on the design configuration being reviewed. Checklists may be developed to cover the "technical" characteristics of design, the "management" requirements for a given program, or a combination of these. Each item on a checklist represents a specific requirement, and the review serves as a medium for determining whether or not that requirement is being met.

The process that leads to the development of checklists is illustrated in Figure 1. Although including a comprehensive technical design and management checklist is not feasible for the space allocated, an example of each (i.e., a design review checklist and a management review checklist) is discussed in the sections to follow.

1 DESIGN REVIEW CHECKLIST

To facilitate the design review and evaluation process from a "technical" perspective, a design review checklist may be developed. Referring to the figure, each of the topics listed represents an area of interest, and each topic is covered by a series of questions that are presented with the objective of promoting greater visibility in the area and forcing the reviewer to evaluate whether or not a certain design requirement is being met. Each of the questions on the checklist should be supported with specific qualitative and quantitative design criteria, with such criteria having evolved from the definition of systems requirements (i.e., operational requirements, maintenance and support concept, and TPMs). For

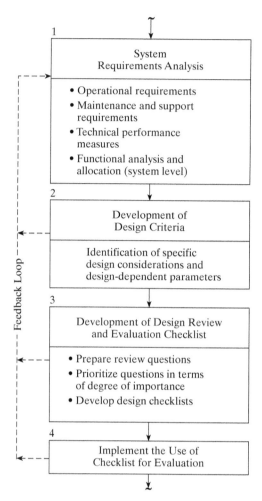

Figure 1 Development of design review checklists.

example, the following questions under "accessibility" might be applicable: Are all access requirements compatible with the frequency of need, where the greater frequency requires more access? Are the access openings for maintenance at least 7.9 inches by 7.2 inches? Can access doors be opened without requiring the use of external tools? For "fasteners," one may ask the following questions: Are the fasteners used throughout the system of the "standard" type, versus being unique and peculiar? Are the number of different types of fasteners held to a minimum? Can the fasteners be opened using common and standard tools?

The questions in Figure 2 cover only three of the 34 different areas of interest. For each of the questions presented, the response should be YES to indicate successful compliance with a given requirement. Questions may be specified with quantitative design-to (TPM) factors included, or may be of a qualitative nature. In any event, the development of a design review checklist must be "tailored" to the system and its elements, and should include only those questions that are relevant to the system being evaluated.

2 MANAGEMENT REVIEW CHECKLIST

A management review checklist may be developed in a similar manner. Key topic areas are identified, questions are developed to provide the emphasis needed, an evaluation is accomplished, and the results are assessed in terms of compliance with program requirements and/or benchmark objectives.

The total spectrum of management covers a broad area to include customer, producer/contractor, and supplier functions and activities. Included is an example of one of these areas; that is, a supplier evaluation checklist. Given that there are likely to be many suppliers involved in a typical program for the acquisition of a large-scale system, and given that such suppliers are generally remotely located from the producer/prime contractor, the importance of developing a good review and evaluation tool for supplier performance cannot be overemphasized. To complete the process, a series of questions are presented in Figure 3, as an example. Again, the checklist questions must be tailored to the particular program of interest.

12.0 **Packaging and Mounting**

12.1 Is the packaging design attractive from the standpoint of consumer appeal (color, shape, size)?

12.2 Is functional packaging incorporated to the maximum extent possible? Interaction effects between packages should be minimized. It should be possible to limit maintenance to the removal of one module (the one containing the fail part) when a failure occurs and not require the removal of two, three, or four modules in order to resolve the problem.

12.3 Is the packaging design compatible with the level of repair analysis decisions? Repairable items are designed to include maintenance provisions such as test points, accessibility, and plug-in components. Items to be discarded upon failure should be encapsulated and relatively low in cost. Maintenance provisions within the disposable module are not required.

12.4 Are disposable modules incorporated to the maximum extent practical? It is highly desirable to reduce overall support through a no-maintenance-design concept as long as the items being discarded are relatively high in reliability and low in cost.

12.5 Are plug-in modules and components used to the maximum extent possible (unless the use of plug-in components significantly degrades the equipment reliability)?

12.6 Are accesses between modules adequate to allow for hand grasping?

12.7 Are modules and components mounted such that the removal of any single item for maintenance will not require the removal of other items? Component stacking should be avoided where possible.

12.8 In areas where module stacking is necessary because of limited space, are the modules mounted in such a way that access priority has been assigned in accordance with the predicted removal and replacement frequency? Items that require frequent maintenance should be more accessible.

12.9 Are modules and components, not of a plug-in variety, mounted with four fasteners or less? Modules should be securely mounted, but the number of fasteners should be held to a minimum.

12.10 Are shock-mounting provisions incorporated where shock and vibration requirements are excessive?

12.11 Are provisions incorporated to preclude installation of the wrong module?

12.12 Are plug-in modules and components removable without the use of tools? If tools are required, they should be of the standard variety.

12.13 Are guides (slides or pins) provided to facilitate module installation?

12.14 Are modules and components labeled?

12.15 Are module and component labels located on top or immediately adjacent to the item and in plain sight?

12.16 Are the labels permanently affixed so that they will not come off during a maintenance action or as a result of environment? Is the information on the label adequate? Disposable modules should be so labeled. In equipment racks, are the heavier items mounted at the bottom of the rack? Unit weight should decrease with the increase in installation height.

12.17 Are operator panels optimally positioned? For personnel in the standing position, panels should be located between 40 and 70 inches above the floor. Critical or precise controls should be between 48 and 64 inches above the floor.

Figure 2 Typical design review questions (example).

19.0 Software

19.1 Have all system software requirements for operating and maintenance functions been identified? Have these requirements been developed through the system-level functional analysis to provide traceability?

19.2 Is the software complete in terms of scope and depth of coverage?

19.3 Is the software compatible relative to the equipment with which it interfaces? Is operating software compatible with maintenance software? With other elements of the system?

19.4 Are the language requirements for operating software and maintenance software compatible?

19.5 Is all software adequately covered through good documentation (logic functional flows, coded programs, and so on)?

19.6 Has the software been adequately tested and verified for accuracy (performance), reliability, and maintainability?

22.0 Testability

22.1 Have self-test provisions been incorporated where appropriate?

22.2 Is reliability degradation due to the incorporation of built-in tests minimized? The BIT capability should not significantly impact the reliability of the overall system.

22.3 Is the extent or depth of self-testing compatible with the level of repair analysis?

22.4 Are self-test provisions automatic?

22.5 Have direct fault indicators been provided (a fault light, an audio signal, or a means of determining that a malfunction positively exists)? Are continuous condition monitoring provisions incorporated where appropriate?

22.6 Are test points provided to enable checkout and fault isolation beyond the level of self-test? Test points for fault isolation within an assembly should not be incorporated if the assembly is to be discarded at failure. Test point provisions must be compatible with the level of repair analysis.

22.7 Are test points accessible? Accessibility should be compatible with the extent of maintenance performed. Test points on the operator's front panel are not required for a depot maintenance action.

22.8 Are test points functionally and conveniently grouped to allow for sequential testing (following a signal flow), testing of similar functions, or frequency of use when access is limited?

22.9 Are test points provided for a direct test of all replaceable items?

22.10 Are test points adequately labeled? Each test point should be identified with a unique number, and the proper signal or expected measured output should be specified on a label located adjacent to the test point.

22.11 Are test points adequately illuminated to allow the technician to see the test point number and labeled signal value?

22.12 Can the component malfunctions that could possibly occur be detected through a no-go indication at the system level? Are false alarm rates minimized? This is a measure of test thoroughness.

22.13 Will the prescribed maintenance software provide adequate diagnostic information?

Figure 2 Typical design review questions (*Continued*).

Supplier Review Checklist
1.   Is the supplier's proposal responsive to the needs specified in the producer's RFP?
2.   Is the supplier's proposal directly supportive of the requirements in the System *Specification* (*Type A*) and the *System's Engineering Management Plan* (*SEMP*)?
3.   Have the performance characteristics and effectiveness factors been adequately specified for the product(s) proposed? Are they meaningful, measurable, and traceable from system-level requirements?
4.   In the event that new design is required, has the design process within the supplier's organization been adequately defined? Does the process incorporate use of the appropriate computer-aided tools (e.g., CAD, CAM, CAS)? Does the process directly support (i.e., communicate with) the system developer's design process?
5.   Is the supplier's design adequately defined through good data/documentation (i.e., drawings, material and parts lists, trade-off analysis reports, databases, etc.)?
6.   Has the supplier proposed a requirement for test, evaluation, and validation? Is this compatible with the requirements for system test, evaluation, and validation, and has this been properly integrated into the *Test and Evaluation Master Plan* (*TEMP*)?
7.   Has the supplier defined the life-cycle support requirements for the product(s) proposed; i.e., the resource requirements for the sustaining maintenance support, retirement, and material recycling and disposal phases of the product's life cycle? Have these requirements been minimized to the extent practicable through good design?
8.   In the event of the need to incorporate design changes, does the supplier have a good policy/procedure in place for "change control"? Does the supplier have a good *Configuration Management Plan* (*CMP*) in place and being implemented?
9.   Has the supplier developed a comprehensive production/construction plan? Are key manufacturing processes identified, along with their characteristics?
10.   Does the supplier have a good quality assurance program in place? Have the appropriate "control" methods been identified; e.g., statistical process control? Does the supplier have a good "rework plan" in place to handle rejected items as necessary?
11.   Does the supplier's proposal include a good comprehensive management plan? Does the plan cover program tasks, a WBS, task schedules, organizational structure and responsibilities, cost projection(s), program monitoring and control procedures, etc.? Has the responsibility for applicable systems engineering tasks been defined?
12.   Does the supplier's proposal address all aspects of *total cost*; i.e., acquisition cost, operation and maintenance support cost, material recycling/disposal cost, life-cycle cost?
13.   Does the supplier have previous experience in the design, development, production, and life-cycle support of product(s) comparable to the product being proposed? Was the experience favorable in terms of delivering and supporting high-quality products in a timely manner and within cost?
14.   Has the supplier identified all relevant *technical and management risks* for the proposed program?

Figure 3   Supplier management review checklist (example).

Appendix:

Probability Theory
and Analysis

Some models will give satisfactory results if variation is not incorporated. But, these models usually apply to physical phenomena where certainty is generally observed. Models formulated to analyze and evaluate human-made systems must incorporate probabilistic elements to be useful in the system engineering process. Accordingly, this appendix presents probability concepts and theory, probability distribution models, and an introduction to Monte Carlo analysis.

1 PROBABILITY CONCEPTS AND THEORY

If one tosses a coin, the outcome will not be known with certainty until either a head or a tail is observed. Prior to the toss, one can assign a probability to the outcome from knowledge of the physical characteristics of the coin. One may know that the diameter of an acorn ranges between 0.80 and 3.20 cm, but the diameter of a specific acorn to be selected from an oak tree will not be known until the acorn is measured. Experiments such as tossing a coin and selecting an acorn provide outcomes called *random events*. Most events in the decision environment are random and probability theory provides a means for quantifying these events.

1.1 The Universe and the Sample

The terms *universe* and *population* are used interchangeably. A *universe* consists of all possible objects, stages, and events within an arbitrarily defined boundary. A universe may be finite or it may be infinite. If it is finite, the universe may be very large or very small. If the universe is large, it may sometimes be assumed to be infinite for computational purposes. A universe need not always be large; it may be defined as a dozen events or as only one object. The relative usefulness of the universe as an entity will be paramount in its definition.

A *sample* is a part or portion of a universe. It may range in size from 1 to 0 less than the size of the universe. A sample is drawn from the population, and observations are made. This is done either because the universe is infinite in size or scope or because the population is large and/or inaccessible as a whole. The sample is used because it is smaller, more accessible, and more economical, and because it suggests certain characteristics of the population.

It is usually assumed that the sample is typical of the population in regard to the characteristics under consideration. The sample is then assessed, and inferences are made in regard to the population as a whole. To the extent that the sample is representative of the population, these inferences may be correct. The problem of selecting a representative sample from a population is an area in statistics to which an entire chapter might be devoted.

Subsequent discussion assumes that the sample is a *random sample*, that is, one in which each object or state or event that constitutes the population has an equally likely chance or probability of being selected and represented in the sample. It is rather simple to state this definition; it may be much more difficult to implement it in practice.

1.2 The Probability of an Event

A measure of the relative certainty of an event, before the occurrence of the event, is its probability. The usual representation of a probability is a number $P(A)$ assigned to the outcome A. This number has the following property: $0 \leq P(A) \leq 1$, with $P(A) = 0$ if the event is certain not to occur and $P(A) = 1$ if the event is certain to occur.

Because probability is only a measure of the certainty (or uncertainty) associated with an event, its definition is rather tenuous. The concept of relative frequency is sometimes employed to establish the number $P(A)$. Sometimes probabilities are established a priori. Other times they are simply a subjective estimate. Consider the example of tossing a fair coin. In a lengthy series of tosses, the coin may have come up heads as often as tails. Then the limiting value of the relative frequency of a head will be 0.5 and will be stated as $P(H) = 0.5$.

Two definitions pertaining to events are needed in the development of probability theorems:

1. Events A and B are said to be *mutually exclusive* if both cannot occur at the same time.
2. Event A is said to be *independent* of event B if the probability of the occurrence of A is the same regardless of whether or not B has occurred.

The probability of the occurrence of either one or another of a series of mutually exclusive events is the sum of probabilities of their separate occurrences. If a fair coin is tossed and success is defined as the occurrence of either a head or a tail, then the probability of a head or a tail is

$$P(H + T) = P(H) + P(T)$$
$$= 0.5 + 0.5 = 1.0 \tag{1}$$

The key to use of the addition theorem is the proper definition of mutually exclusive events. Such events must be distinct from one another. If one event occurs, it must be impossible for the second to occur at the same time. For example, assume that the probability of having a flat tire during a given period on each of four tires on an automobile is 0.3. Then the probability of having a flat tire on any of the four tires during this time period is not given by the addition of these four probabilities. If $P(T_1) = P(T_2) = P(T_3) = P(T_4) = 0.3$ are the respective probabilities of failure for each of the four tires, then

$$P(T_1 + T_2 + T_3 + T_4) \neq P(T_1) + P(T_2) + P(T_3) + P(T_4)$$

$$\neq 0.3 + 0.3 + 0.3 + 0.3 = 1.2$$

This cannot be true because the failure of tires is not mutually exclusive. During the time period established, two or more tires may fail, whereas in the example of coin tossing, it is not possible to obtain a head and a tail on the same toss.

Appendix: Probability Theory and Analysis

1.3   The Multiplication Theorem

The probability of occurrence of independent events is the product of the probabilities of their separate events. Implicit in this theorem is the successful occurrence of two events simultaneously or in succession. Thus, the probability of the occurrence of two heads in two tosses of a coin is

$$P(H \cdot H) = P(H)P(H)$$
$$D = (0.5)(0.5) = 0.25 \qquad (2)$$

The tire-failure problem can now be resolved by considering the probabilities of each tire not failing. The probability of each tire not failing is given by $P(\overline{T_i}) = 0.7$. The probability of no tire failing is then given by

$$P[(\overline{T_1})(\overline{T_2})(\overline{T_3})(\overline{T_4})] = P(\overline{T_1})\,P(\overline{T_2})\,P(\overline{T_3})\,P(\overline{T_4})$$
$$= (0.7)(0.7)(0.7)(0.7) = 0.2401$$

Thus, the probability of a tire failing, or of one or more tires failing, is

$$P(T_1 + T_2 + T_3 + T_4) = 1 - 0.2401 = 0.7599$$

This approach is valid, since the probability of one tire not failing is independent of the success or failure of the other three tires.

1.4   The Conditional Theorem

The probability of the occurrence of two dependent events is the probability of the first event times the probability of the second event, given that the first has occurred. This may be expressed as

$$P(W_1 \cdot W_2) = P(W_1)P(W_2|W_1) \qquad (3)$$

This theorem is similar to the multiplication theorem, except that consideration is given to the lack of independence between events.

As an example, consider the probability of selecting two successive white balls from an urn containing three white and two black balls. This problem reduces to a calculation of the product of the probability of selecting a white ball times the probability of selecting a second white ball, given that the first attempt has been successful, or

$$P(W_1 \cdot W_2) = \left(\frac{3}{5}\right)\left(\frac{2}{4}\right) = \frac{3}{10}$$

The conditional theorem makes allowances for a change in probabilities between two successive events. This theorem will be helpful in constructing finite discrete probability distributions.

1.5 The Central Limit Theorem

Although many real-world variables are normally distributed, this assumption cannot be universally applied. However, the distribution of the means of samples or the sums of random variables approximates the normal distribution provided certain assumptions hold. The Central Limit Theorem states: If x has a distribution for which the moment-generating function exists, then the variable $\bar{x}$ has a distribution that approaches normality as the size of the sample tends toward infinity. The sample size required for any desired degree of convergence is a function of the shape of the parent distribution. Fairly good results have been demonstrated with a sample of $n = 4$ for both the rectangular and triangular distributions.

2 PROBABILITY DISTRIBUTION MODELS

The pattern of the distribution of probabilities over all possible outcomes is called a probability distribution. *Probability distribution models* provide a means for assigning the likelihood of occurrences of all possible values. Variables described in terms of a probability distribution are conveniently called *random variables*. The specific value of a random variable is determined by the distribution.

A probability distribution is completely defined when the probability associated with every possible outcome is defined. In most instances, the outcomes themselves are represented by numbers or different values of a variable, such as the diameter of an acorn. When the pattern of the probability distribution is expressed as a function of this variable, the resulting function is called a *probability distribution function*.

An example empirical probability distribution function may be developed as follows. A maintenance mechanic attends four machines and his services are needed only when a machine fails. He would like to estimate how many machines will fail each shift. From previous experience, and using the relative frequency concept of probability, the mechanic knows that 40% of the time only one machine will fail at least once during the shift. Further, 30% of the time two machines will fail, three machines will fail 20% of the time, and all four will fail 10% of the time.

The probability distribution of the number of failed machines may be expressed as $P(1) = 0.4$, $P(2) = 0.3$, $P(3) = 0.2$, and $P(4) = 0.1$. This probability distribution is exhibited in Figure 1.

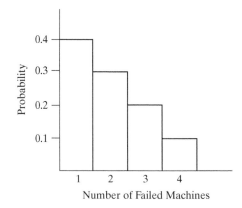

Figure 1 A probability distribution of the number of failed machines.

Appendix: Probability Theory and Analysis

The probability distribution function for this case may be defined as

$$P(x) = \frac{5-x}{10} \qquad \text{if } x = 1, 2, 3, 4$$

$$P(x) = 0 \qquad \text{otherwise}$$

Although the function $P(x) = (5-x)/10$ uniquely represents the probability distribution pattern for the number of failed machines, the function itself belongs to a wider class of functions of the type $P(x) = (a-x)/b$. All functions of this type indicate similar patterns, yet each pair of numbers (a, b) uniquely defines a specific probability distribution. These numbers (a, b) are called *parameters*.

In the sense that they serve to define the probability distribution function, it is possible to look upon parameters as properties of the distribution function. The choice of representation of parameters is not unique, and the most desirable representation would reflect a measure of the properties of the universe under study. Two most commonly sought measures are the *mean*, an indication of central tendency, and the *variance*, a measure of dispersion.

The probability distribution just presented is discrete in that it assigns probabilities to an event that can only take on integer values. Continuous probability distributions are used to define the probability of the occurrence of an event that may take on values over a continuum. Under certain conditions, it may be desirable to use a continuous probability distribution to approximate a discrete probability distribution. By so doing, tedious summations may be replaced by integrals. In other instances, it may be desirable to make a continuous distribution discrete as when calculations are to be performed on a digital computer. Several discrete and continuous probability distribution models are presented subsequently.

2.1 The Binomial Distribution

The binomial distribution is a basic discrete sampling distribution. It is applicable where the probability is sought of exactly x occurrences in n trials of an event that has a constant probability of occurrence p. The requirement of a constant probability of occurrence is satisfied when the population being sampled is infinite in size, or where replacement of the sampled unit takes place.

The probability of exactly x occurrences in n trials of an event that has a constant probability of occurrence p is given as

$$P(x) = \frac{n!}{x!(n-x)!} p^x q^{n-x} \qquad 0 \le x \le n \tag{4}$$

where $q = 1 - p$. The mean and variance of this distribution are given by np and npq, respectively.

As an example of the application of the binomial distribution, assume that a fair coin is to be tossed five times. The probability of obtaining exactly two heads is

$$P(2) = \frac{5!}{2!(5-2)!} (0.5)^2 (1-0.5)^3$$

$$= 10(0.03125) = 0.3125$$

A probability distribution may be constructed by solving for the probability of exactly zero, one, two, three, four, and five heads in five tosses. If $p = 0.5$, as in this example, the resulting

distribution is symmetrical. If the distribution is skewed to the right; if the distribution is skewed to the left.

2.2 The Uniform Distribution

The uniform or rectangular probability distribution may be either discrete or continuous. The continuous form of this simple distribution is

$$f(x) = \frac{1}{a} \qquad 0 \le x \le a \tag{5}$$

The discrete form divides the interval 0 to a into $n + 1$ cells over the range 0 to n, with $1/(n + 1)$ as the unit probabilities. The mean and variance of the rectangular probability distribution are given as $a/2$ and $a^2/12$ for the continuous case, and as $n/2$ and $n^2/12 + n/6$ for the discrete case.

The general form of the rectangular probability distribution is shown in Figure 2. The probability that a value of x will fall between the limits 0 and a is equal to unity. One may determine the probability associated with a specific value of x, or a range of x, by integration for the continuous case. The probability associated with a specific value of x for the discrete distributions of the previous section was found from the functions given. Determination of the probability associated with a range of x required a summation of individual probabilities. This is a fundamental difference in dealing with discrete and continuous probability distributions.

Values are drawn at random from the rectangular distribution with x allowed to take on values ranging from 0 through 9. These random rectangular variates may be used to randomize a sample or to develop values drawn at random from other probability distributions as is illustrated in the last section of this appendix.

2.3 The Poisson Distribution

The Poisson is a discrete distribution useful in its own right and as an approximation to the binomial. It is applicable when the opportunity for the occurrence of an event is large, but when the actual occurrence is unlikely. The probability of exactly x occurrences of an event of probability p in a sample n is

$$P(x) = \frac{(\mu)^x e^{-\mu}}{x!} \qquad 0 \le x \le \infty \tag{6}$$

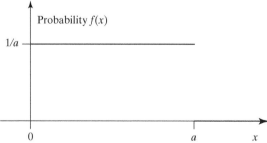

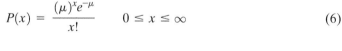

Figure 2 The general form of the rectangular distribution.

The mean and variance of this distribution are equal and given by μ, where $\mu = np$.

As an example of the application of the Poisson distribution, assume that a sample of 100 items is selected from a population of items which are 1% defective. The probability of obtaining exactly three defectives in the sample is found from Equation 6 as

$$P(3) = \frac{(1)^3(2.72)^{-1}}{3!} = 0.061$$

The Poisson distribution may be used as an approximation to the binomial distribution. Such an approximation is good when n is relatively large, p is relatively small, and in general, $pn < 5$. These conditions were satisfied in the previous example.

2.4 The Exponential Distribution

The exponential probability distribution is given by

$$f(x) = \frac{1}{a}e^{-x/a} \qquad 0 \leq x \leq \infty \tag{7}$$

The mean and variance of this distribution are given by a and a^2, respectively. Its form is illustrated in Figure 3.

As an example of the application of the exponential probability distribution, consider the selection of a light bulb from a population of light bulbs whose life is known to be exponentially distributed with a mean $\mu = 1,000$ hours. The probability of the life of this sample bulb not exceeding 1,000 hours would be expressed as $P(x \leq 1,000)$. This would be the proportional area under the exponential function over the range $x = 0$ to $x = 1,000$, or

$$P(x \leq 1,000) = \int_0^{1,000} f(x)dx$$

$$D = \int_0^{1,000} \frac{1}{1,000}e^{-x/1,000}dx$$

$$= -e^{-x/1,000}\Big|_0^{1,000}$$

$$= 1 - e^{-1} = 0.632$$

Note that 0.632 is that proportion of the area of an exponential distribution to the left of the mean. This illustrates that the probability of the occurrence of an event exceeding the mean value is only $1 - 0.632 = 0.368$.

2.5 The Normal Distribution

The normal or Gaussian probability distribution is one of the most important of all distributions. It is defined by

$$f(x) = \frac{1}{\sigma\sqrt{2\pi}}e^{[-(x-\mu)^2/2\sigma^2]} \qquad -\infty \leq x \leq +\infty \tag{8}$$

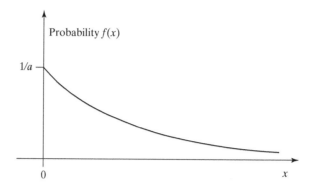

Figure 3   The general form of the exponential distribution.

The mean and variance are μ and σ^2, respectively. Variation is inherent in nature, and much of this variation appears to follow the normal distribution, the form of which is given in Figure 4.

The normal distribution is symmetrical about the mean and possesses some interesting and useful properties regarding its shape. Where distances from the mean are expressed in terms of standard deviations, σ, the relative areas defined between two such distances will be constant from one distribution to another. In effect, all normal distributions, when defined in terms of a common value of μ and σ, will be identical in form, and corresponding probabilities may be tabulated. Normally, cumulative probabilities are given from $-\infty$ to any value expressed as standard deviation units. This table gives the probability from $-\infty$ to Z, where Z is a standard normal variate defined as

$$Z = \frac{x - \mu}{\sigma} \tag{9}$$

This is shown as the shaded area in Figure 4.

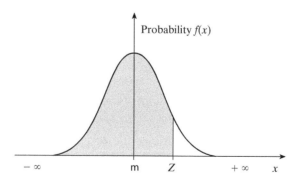

Figure 4   The normal probability distribution.

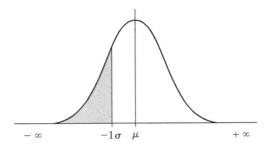

Figure 5   The area from $-\infty$ to -1σ
under the normal distribution.

The area from $-\infty$ to -1σ is indicated as the shaded area in Figure 5. The probability of x falling in this range is 0.1587. Likewise, the area from $-\infty$ to $+2\sigma$ is 0.9773. If the probability of a value falling in the interval -1σ to $+2\sigma$ is required, the following computations are made.

$$P(\text{area } -\infty \text{ to } +2\sigma) \;=\; 0.9773$$

$$-P(\text{area } -\infty \text{ to } -1\sigma) \;=\; 0.1587$$

$$P(\text{area } -1\sigma \text{ to } +2\sigma) \;=\; 0.8186$$

This situation is shown in Figure 6.

2.6   The Lognormal Distribution

The lognormal probability distribution is related to the normal distribution. If a random variable $Y = \ln X$ is normally distributed with mean μ and variance σ^2, then the random variable X follows the lognormal distribution. Accordingly, the probability distribution function of the random variable X is defined as

$$f(x) = \frac{1}{x\sigma\sqrt{2\pi}}\, e^{[-(\ln x - \mu)^2/2\sigma^2]}, \qquad x > 0 \tag{10}$$

where $\mu \in (-\infty, \infty)$ is called the scale parameter and $\sigma > 0$ is called the shape parameter. The mean and variance of the lognormal distribution are $e^{\mu + \sigma^2/2}$ and $e^{2\mu + \sigma^2}(e^{\sigma^2} - 1)$ respectively.

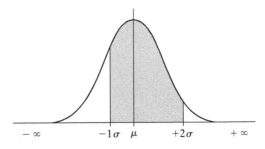

Figure 6   The area from -1σ to $+2\sigma$
under the normal distribution.

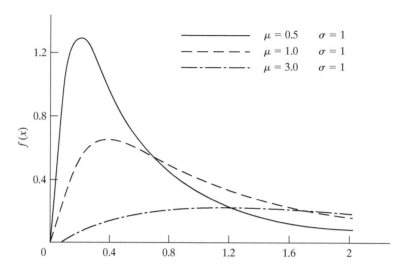

Figure 7 Lognormal distribution with constant shape parameter.

Figure 7 shows a lognormal distribution with a constant shape parameter and various scale parameters. Figure 8 shows a lognormal distribution with a constant scale parameter and various shape parameters.

If $X \sim \ln(\mu, \sigma^2)$ then $\ln X \sim N(\mu, \sigma^2)$. This implies that if n data points $x_1, x_2, \ldots, x_n$ are lognormal, then the logarithms of these data points, $\ln x_1, \ln x_2, \ldots, \ln x_n$ will be normally distributed and may be used for parameter estimation, goodness-of-fit testing, and hypothesis testing.

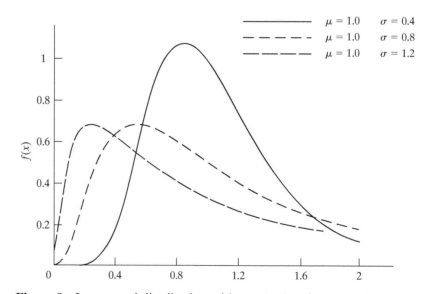

Figure 8 Lognormal distribution with constant scale parameter.

2.7 The Weibull Distribution

The probability distribution function of a random variable X, which follows the Weibull distribution, is given by

$$f(x) = \frac{\alpha}{\beta^\alpha} x^{\alpha-1} e^{-x/\beta)^\alpha}, \qquad x \geq 0 \tag{11}$$

where $\alpha > 0, \beta > 0$ are the shape and scale parameters respectively, and defined on $(0, \infty)$. A Weibull variate X has mean $\dfrac{\beta}{\alpha}\Gamma\left(\dfrac{1}{\alpha}\right)$ and variance $\dfrac{\beta^2}{\alpha}\left\{2\Gamma\left(\dfrac{2}{\alpha}\right) - \dfrac{1}{\alpha}\left[\Gamma\left(\dfrac{1}{\alpha}\right)\right]^2\right\}$, where $\Gamma(\,)$ is the gamma function

$$\Gamma(z) = \int_0^\infty t^{z-1} e^{-t} dt$$

Weibull distributions with different shape and scale parameters are illustrated in Figures 9 and 10, respectively. The Weibull distribution has some interesting characteristics:

1. For $\alpha = 1$, the Weibull distribution is the same as the exponential distribution with parameter β.
2. For $\alpha = 3.4$, the Weibull distribution approximates the normal distribution.
3. If $X \sim$ Weibull (α, β) then $X^a \sim$ exponential $(\beta\alpha)$.

The Weibull distribution is frequently used as a time-to-failure model and in reliability analysis, especially in instances where failure data cannot be fitted by the exponential distribution.

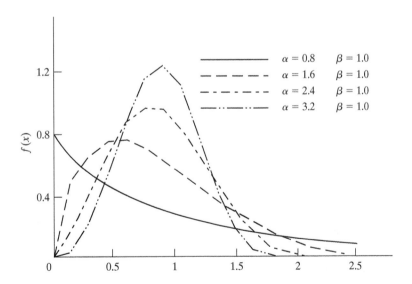

Figure 9 Weibull distribution with various shape parameters.

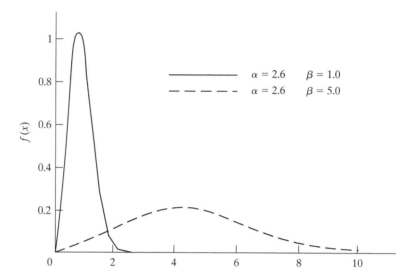

Figure 10 Weibull distribution with various scale parameters.

3 MONTE CARLO ANALYSIS

The decision environment is made up of many random variables. Thus, models used to explain operational systems must often incorporate probabilistic elements. In some cases, formal mathematical solutions are difficult or impossible to obtain from these models. Under such conditions it may be necessary to use a method known as *Monte Carlo analysis*. When applied to an operational system, Monte Carlo analysis provides a powerful means of simulation.

3.1 A Simple Monte Carlo Example

As an introduction to the idea of Monte Carlo analysis, consider its application to the determination of the area of a circle with a diameter of 1 inch. Proceed as follows:

1. Enclose the circle of a 1-inch square as shown in Figure 11.
2. Divide two adjoining sides of the square into tenths, or hundredths, or thousandths, and so on, depending on the accuracy desired.
3. Secure a sequence of pairs of random rectangular variates.
4. Use each pair of rectangular variates to determine a point within the square and possibly within the circle. This process is illustrated in Table 1 for 100 trials.
5. Compute a ratio of the number of times a point falls within the circle to the total number of trials. The value of this ratio is an approximate area for the circle expressed as a fraction of the 1 inch^2 represented by the square. It is 79/100, or 0.79, in this example.

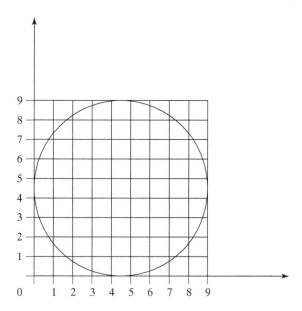

Figure 11 Area of circle by Monte Carlo analysis.

The example just presented has a well-known mathematical solution as follows:

$$A = \pi r^2 = 3.1416(0.50)^2 = 0.7854.$$

TABLE 1 Determining the area of a circle

Trial	Random number	In	Out
1	73	x	
2	26	x	
3	19		x
4	84	x	
5	81	x	
6	47	x	
7	18		x
8	44	x	
$\vdots$	$\vdots$	$\vdots$	$\vdots$
100	35	x	
Total		79	21

3.2 Steps in the Monte Carlo Procedure

Monte Carlo analysis may be implemented in accordance with a step-by-step procedure that is usually independent of the application. The steps are:

1. **Formalize the system logic**—The system chosen for study usually operates in accordance with a certain logical pattern. Therefore, before beginning the actual process of Monte Carlo analysis, it is necessary to formalize the operational procedure by the

construction of a model. This may require the development of a step-by-step flow diagram outlining the logic. If the actual simulation process is to be performed on a digital computer, it is mandatory to prepare an accurate logic diagram. From this, the computer can be programmed to pattern the process under study.

2. **Determine the probability distributions**—Each random variable in the operation refers to an event in the system being studied. Therefore, an important step in Monte Carlo analysis is determining the behavior of these random variables. This involves the development of empirical frequency distributions to describe the relevant variables by the collection of historical data. Once this is done, the frequency distribution for each variable may be studied statistically to ascertain whether it conforms to a known theoretical distribution.

3. **Develop the cumulative probability distributions**—Convert each probability distribution to its cumulative equivalent, with the cumulative probability exhibited on the ordinate ready to receive random number inputs. It is the cumulative probability distributions that serve to convert random rectangular variates to values drawn at random from the underlying probability distribution.

4. **Perform the Monte Carlo process**—The Monte Carlo process proceeds by exercising the system logic and saving the results of each trial. Analysis of a sample of trials gives insight into the estimated behavior of the system under study. The validity of the estimate depends upon the fidelity of the logic and the number of trials in the sample.

Appendix:

Probability and Statistical Tables

TABLE 1 RANDOM RECTANGULAR VARIATES

Random variates from the rectangular distribution, $f(x) = \frac{1}{10}$, are presented. (These tables are reproduced with permission from the RAND Corporation, *A Million Random Digits with 100,000 Normal Deviates* [New York: The Free Press, 1955], pp. 130–131.)

TABLE 2 CUMULATIVE POISSON PROBABILITIES

Cumulative probabilities $\times$ 1,000 for the Poisson distribution are given for μ up to 24. The tabular values were computed from $\Sigma(\mu^x e^{-\mu}/x!)$. (These tables are reproduced from W. J. Fabrycky, P. M. Ghare, and P. E. Torgersen, *Applied Operations Research and Management Science* [Upper Saddle River, N.J.: Prentice Hall, Inc., 1984].)

TABLE 3 CUMULATIVE NORMAL PROBABILITIES

Cumulative probabilities are given from $-\infty$ to $Z = (x - \mu)/\sigma$ for the standard normal distribution. (Tabular values are adapted with permission from E. L. Grant and R. S. Leavenworth, *Statistical Quality Control*, 4th ed. [New York: McGraw-Hill Book Co., 1972].)

TABLE 1 Random Rectangular Variates

14541	36678	54343	94932	25238	84928	30668	34992	69955	06633
88626	98899	01337	48085	83315	33563	78656	99440	55584	54178
31466	87268	62975	19310	28192	06654	06720	64938	67111	55091
52738	52893	51373	43430	95885	93795	20129	54847	68674	21040
17444	35560	35348	75467	26026	89118	51810	06389	02391	96061
62596	56854	76099	38469	26285	86175	65468	32354	02675	24070
38338	83917	50232	29164	07461	25385	84838	07405	38303	55635
29163	61006	98106	47538	99122	36242	90365	15581	89597	03327
59049	95306	31227	75288	10122	92687	99971	97105	37597	91673
67447	52922	58657	67601	96148	97263	39110	95111	04682	64873
57082	55108	26992	19196	08044	57300	75095	84330	92314	11370
00179	04358	95645	91751	56618	73782	38575	17401	38686	98435
65420	87257	44374	54312	94692	81776	24422	99198	51432	63943
52450	75445	40002	69727	29775	32572	79980	67902	97260	21050
82767	26273	02192	88536	08191	91750	46993	02245	38659	28026
17066	64286	35972	32550	82167	53177	32396	34014	20993	03031
86168	32643	23668	92038	03096	51029	09693	45454	89854	70103
33632	69631	70537	06464	83543	48297	67693	63137	62675	56572
77915	56481	43065	24231	43011	40505	90386	13870	84603	73101
90000	92887	92668	93521	44072	01785	27003	01851	40232	25842
55809	70237	10368	58664	39521	11137	20461	53081	07150	11832
50948	64026	03350	03153	75913	72651	28651	94299	67706	92507
27138	59012	27872	90522	69791	85482	80337	12252	83388	48909
03534	58643	75913	63557	25527	47131	72295	55801	44847	48019
48895	34733	58057	00195	79496	93453	07813	66038	55245	43168
57585	23710	77321	70662	82884	80132	42281	17032	96737	93284
95913	24669	42050	92757	68677	75567	99777	49246	93049	79863
12981	37145	95773	92475	43700	85253	33214	87656	13295	09721
62349	64163	57369	65773	86217	00135	33762	72398	16343	02263
68193	37564	56257	50030	53951	84887	34590	22038	40629	29562
56203	82226	83294	60361	29924	09353	87021	08149	11167	81744
31945	23224	08211	02562	20299	85836	94714	50278	99818	62489
68726	52274	59535	80873	35423	05166	06911	25916	90728	20431
79557	25747	55585	93461	44360	18359	20493	54287	43693	88568
05764	29803	01819	51972	91641	03524	18381	65427	11394	37447
30187	66931	01972	48438	90716	21847	35114	91839	26913	68893
30858	43646	96984	80412	91973	81339	05548	49812	40775	14263
85117	38268	18921	29519	33359	80642	95362	22133	40322	37826
59422	12752	56798	31954	19859	32451	04433	62116	14899	38825
73479	91833	91122	45524	73871	77931	67822	95602	23325	37718
83648	66882	15327	89748	76685	76282	98624	71547	49089	33105
19454	91265	09051	94410	06418	34484	37929	61070	62346	79970
49327	97807	61390	08005	71795	49290	52285	82119	59348	55986
54482	51025	12382	35719	66721	84890	38106	44136	95164	92935
30487	19459	25693	09427	10967	36164	33893	07087	16141	12734
42998	68627	66295	59360	44041	76909	56321	12978	31304	97444
03668	61096	26292	79688	05625	52198	74844	69815	76591	35398
45074	91457	28311	56499	60403	13658	81838	54729	12365	24082
58444	99255	14960	02275	37925	03852	81235	91628	72136	53070
82912	91185	89612	02362	93360	20158	24796	38284	55328	96041

TABLE 1 Random Rectangular Variates (*Continued*)

44553	29642	20317	69470	57789	27631	68040	73201	51302	66497
01914	36106	71351	69176	53353	57353	42430	68050	47862	61922
00768	37958	69915	17709	31629	49587	07136	42959	56207	03625
29742	67676	62608	54215	97167	07008	77130	15806	53081	14297
07721	20143	56131	56112	23451	48773	38121	74419	11696	42614
99158	07133	04325	43936	83619	77182	55459	28808	38034	01054
97168	13859	78155	55361	04871	78433	58538	78437	14058	79510
07508	63835	83056	74942	70117	91928	10383	93793	31015	60839
68400	66460	67212	28690	66913	90798	71714	07698	31581	31086
88512	62908	65455	64015	00821	23970	58118	93174	02201	16771
94549	31145	62897	91582	94064	14687	47570	83714	45928	32685
02307	86181	44897	60884	68072	77693	83413	61680	55872	12111
28922	89390	66771	39185	04266	55216	91537	36500	48154	04517
73898	85742	97914	74170	10383	16366	37404	73282	20524	85004
66220	81596	18533	84825	43509	16009	00830	13177	54961	31140
64452	91627	21897	31830	62051	00760	43702	22305	79009	15065
26748	19441	87908	06086	62879	99865	50739	98540	54002	98337
61328	52330	17850	53204	29955	48425	84694	11280	70661	27303
89134	85791	73207	93578	62563	37205	97667	61453	01067	31982
91365	23327	81658	56441	01480	09677	86053	11505	30898	82143
54576	02572	60501	98257	40475	81401	31624	27951	60172	21382
39870	60476	02934	39857	06430	59325	84345	62302	98616	13452
82288	29758	35692	21268	35101	77554	35201	22795	84532	29927
57404	93848	87288	30246	34990	50575	49485	60474	17377	46550
22043	17104	49653	79082	45099	24889	04829	49097	58065	23492
61981	00340	43594	22386	41782	94104	08867	68590	61716	36120
96056	16227	74598	28155	23304	66923	07918	15303	44988	79076
64013	74715	31525	62676	75435	93055	37086	52737	89455	83016
59515	37354	55422	79471	23150	79170	74043	49340	61320	50390
38534	33169	40448	21683	82153	23411	53057	26069	86906	49708
41422	50502	40570	59748	59499	70322	62416	71408	06429	70123
38633	80107	10241	30880	13914	09228	68929	06438	17749	81149
48214	75994	31689	25257	28641	14854	72571	78189	35508	26381
54799	37862	06714	55885	07481	16966	04797	57846	69080	49631
25848	27142	63477	33416	60961	19781	65457	23981	90348	24499
27576	47298	47163	69614	29372	24859	62090	81667	50635	08295
52970	93916	81350	81057	16962	56039	27739	59574	79617	45698
69516	87573	13313	69388	32020	66294	99126	50474	04258	03084
94504	41733	55936	77595	55959	90727	61367	83645	80997	62103
67935	14568	27992	09784	81917	79303	08616	83509	64932	34764
63345	09500	40232	51061	09455	36491	04810	06040	78959	41435
87119	21605	86917	97715	91250	79587	80967	39872	52512	78444
02612	97319	10487	68923	58607	38261	67119	36351	48521	69965
69860	16526	41420	01514	46902	03399	12286	52467	80387	10561
27669	67730	53932	38578	25746	00025	98917	18790	51091	24920
59705	91472	01302	33123	35274	88433	55491	27609	02824	05245
36508	74042	44014	36243	12724	06092	23742	90436	33419	12301
13612	24554	73326	61445	77198	43360	62006	31038	54756	88137
82893	11961	19656	71181	63201	44946	14169	72755	47883	24119
97914	61228	42903	71187	54964	14945	20809	33937	13257	66387

TABLE 2 Cumulative Poisson Probabilities × 1,000

μ \ x	0	1	2	3	4	5	6	7	8	9	10	11	12	13	14
0.1	905	995	1,000												
0.2	819	982	999	1,000											
0.3	741	963	996	1,000											
0.4	670	938	992	999	1,000										
0.5	607	910	986	998	1,000										
0.6	549	878	977	997	1,000										
0.7	497	844	966	994	999	1,000									
0.8	449	809	953	991	999	1,000									
0.9	407	772	937	987	998	1,000									
1.0	368	736	920	981	996	999	1,000								
1.1	333	699	900	974	995	999	1,000								
1.2	301	663	879	966	992	998	1,000								
1.3	273	627	857	957	989	998	1,000								
1.4	247	592	833	946	986	997	999	1,000							
1.5	223	558	809	934	981	996	999	1,000							
1.6	202	525	783	921	976	994	999	1,000							
1.7	183	493	757	907	970	992	998	1,000							
1.8	165	463	731	891	964	990	997	999	1,000						
1.9	150	434	704	875	956	987	997	999	1,000						
2.0	135	406	677	857	947	983	995	999	1,000						
2.2	111	355	623	819	928	975	993	998	1,000						
2.4	091	308	570	779	904	964	988	997	999	1,000					
2.6	074	267	518	736	877	951	983	995	999	1,000					
2.8	061	231	469	692	848	935	976	992	998	999	1,000				
3.0	050	199	423	647	815	916	966	988	996	999	1,000				
3.2	041	171	380	603	781	895	955	983	994	998	999	1,000			
3.4	033	147	340	558	744	871	942	977	992	997	999	1,000			
3.6	027	126	303	515	706	844	927	969	988	996	998	999	1,000		
3.8	022	107	269	473	668	816	909	960	984	994	998	999	1,000		
4.0	018	092	238	433	629	785	889	949	979	992	997	999	1,000		
4.2	015	078	210	395	590	753	867	936	972	989	996	999	1,000		
4.4	012	066	185	359	551	720	844	921	964	985	994	998	999	1,000	
4.6	010	056	163	326	513	686	818	905	955	980	992	997	999	1,000	
4.8	008	048	143	294	476	651	791	887	944	975	990	996	999	1,000	
5	007	040	125	265	440	616	762	867	932	968	986	995	998	999	1,000
6	002	017	062	151	285	446	606	744	847	916	957	980	991	996	1,000

TABLE 2 Cumulative Poisson Probabilities × 1,000 (Continued)

x \ μ	0	1	2	3	4	5	6	7	8	9	10	11	12	13	14
8	000	003	014	042	100	191	313	453	593	717	816	888	936	966	983
9	000	001	006	021	055	116	207	324	456	587	706	803	876	926	959
10		000	003	010	029	067	130	220	333	458	583	697	792	864	917
11		000	001	005	015	038	079	143	232	341	460	579	689	781	854
12		000	001	002	008	020	046	090	155	242	347	462	576	682	772
13			000	001	004	011	026	054	100	166	252	353	463	573	675
14				000	002	006	014	032	062	109	176	260	358	464	570
15				000	001	003	008	018	037	070	118	185	268	363	466

x \ μ	15	16	17	18	19	20	21	22	23	24	25	26	27	28	29
7	998	999	1,000												
8	992	996	998	999	1,000										
9	978	989	995	998	999	1,000									
10	951	973	986	993	997	998	999	1,000							
11	907	944	968	982	991	995	998	999	1,000						
12	844	899	937	963	979	988	994	997	999	999	1,000				
13	764	835	890	930	957	975	986	992	996	998	999	1,000			
14	669	756	827	883	923	952	971	983	991	995	997	999	999	1,000	
15	568	664	749	819	875	917	947	967	981	989	994	997	998	999	1,000

x \ μ	0	1	2	3	4	5	6	7	8	9	10	11	12	13	14
16					000	001	004	010	022	043	077	127	193	275	368
17					000	001	002	005	013	026	049	085	135	201	281
18						000	001	003	007	015	030	055	092	143	208
19						000	001	002	004	009	018	035	061	098	150
20							000	001	002	005	011	021	039	066	105
21								000	001	003	006	013	025	043	072
22								000	001	002	004	008	015	028	048
23									000	001	002	004	009	017	031
24									000	000	001	003	005	011	020

TABLE 2 Cumulative Poisson Probabilities × 1,000 (Continued)

μ＼x	15	16	17	18	19	20	21	22	23	24	25	26	27	28	29
16	467	566	659	742	812	868	911	942	963	978	987	993	996	998	999
17	371	468	564	655	736	805	861	905	937	959	975	985	991	995	997
18	287	375	469	562	651	731	799	855	899	932	955	972	983	990	994
19	215	292	378	469	561	647	725	793	849	893	927	951	969	980	988
20	157	221	297	381	470	559	644	721	787	843	888	922	948	966	978
21	111	163	227	302	384	471	558	640	716	782	838	883	917	944	963
22	077	117	169	232	306	387	472	556	637	712	777	832	877	913	940
23	052	82	123	175	238	310	389	472	555	635	708	772	827	873	908
24	034	056	087	128	180	243	314	392	473	554	632	704	768	823	868

μ＼x	30	31	32	33	34	35	36	37	38	39	40	41	42	43	44
16	999	1,000													
17	999	999	1,000												
18	997	998	999	1,000											
19	993	996	998	999	999	1,000									
20	987	992	995	997	999	999	1,000								
21	976	985	991	994	997	998	999	999	1,000						
22	959	973	983	989	994	996	998	999	999	1,000					
23	936	956	971	981	988	993	996	997	999	999	1,000				
24	904	932	953	969	979	987	992	995	997	998	999	999	1,000		

TABLE 3 Cumulative Normal Probabilities

Z	0.09	0.08	0.07	0.06	0.05	0.04	0.03	0.02	0.01	0.00
−3.5	0.00017	0.00017	0.00018	0.00019	0.00019	0.00020	0.00021	0.00022	0.00022	0.00023
−3.4	0.00024	0.00025	0.00026	0.00027	0.00028	0.00029	0.00030	0.00031	0.00033	0.00034
−3.3	0.00035	0.00036	0.00038	0.00039	0.00040	0.00042	0.00043	0.00045	0.00047	0.00048
−3.2	0.00050	0.00052	0.00054	0.00056	0.00058	0.00060	0.00062	0.00064	0.00066	0.00069
−3.1	0.00071	0.00074	0.00076	0.00079	0.00082	0.00085	0.00087	0.00090	0.00094	0.00097
−3.0	0.00100	0.00104	0.00107	0.00111	0.00114	0.00118	0.00122	0.00126	0.00131	0.00135
−2.9	0.0014	0.0014	0.0015	0.0015	0.0016	0.0016	0.0017	0.0017	0.0018	0.0019
−2.8	0.0019	0.0020	0.0021	0.0021	0.0022	0.0023	0.0023	0.0024	0.0025	0.0026
−2.7	0.0026	0.0027	0.0028	0.0029	0.0030	0.0031	0.0032	0.0033	0.0034	0.0035
−2.6	0.0036	0.0037	0.0038	0.0039	0.0040	0.0041	0.0043	0.0044	0.0045	0.0047
−2.5	0.0048	0.0049	0.0051	0.0052	0.0054	0.0055	0.0057	0.0059	0.0060	0.0062
−2.4	0.0064	0.0066	0.0068	0.0069	0.0071	0.0073	0.0075	0.0078	0.0080	0.0082
−2.3	0.0084	0.0087	0.0089	0.0091	0.0094	0.0096	0.0099	0.0102	0.0104	0.0107
−2.2	0.0110	0.0113	0.0116	0.0119	0.0122	0.0125	0.0129	0.0132	0.0136	0.0139
−2.1	0.0143	0.0146	0.0150	0.0154	0.0158	0.0162	0.0166	0.0170	0.0174	0.0179
−2.0	0.0183	0.0188	0.0192	0.0197	0.0202	0.0207	0.0212	0.0217	0.0222	0.0228
−1.9	0.0233	0.0239	0.0244	0.0250	0.0256	0.0262	0.0268	0.0274	0.0281	0.0287
−1.8	0.0294	0.0301	0.0307	0.0314	0.0322	0.0329	0.0336	0.0344	0.0351	0.0359
−1.7	0.0367	0.0375	0.0384	0.0392	0.0401	0.0409	0.0418	0.0427	0.0436	0.0446
−1.6	0.0455	0.0465	0.0475	0.0485	0.0495	0.0505	0.0516	0.0526	0.0537	0.0548
−1.5	0.0559	0.0571	0.0582	0.0594	0.0606	0.0618	0.0630	0.0643	0.0655	0.0668
−1.4	0.0681	0.0694	0.0708	0.0721	0.0735	0.0749	0.0764	0.0778	0.0793	0.0808
−1.3	0.0823	0.0838	0.0853	0.0869	0.0885	0.0901	0.0918	0.0934	0.0951	0.0968
−1.2	0.0985	0.1003	0.1020	0.1038	0.1057	0.1075	0.1093	0.1112	0.1131	0.1151
−1.1	0.1170	0.1190	0.1210	0.1230	0.1251	0.1271	0.1292	0.1314	0.1335	0.1357
−1.0	0.1379	0.1401	0.1423	0.1446	0.1469	0.1492	0.1515	0.1539	0.1562	0.1587
−0.9	0.1611	0.1635	0.1660	0.1685	0.1711	0.1736	0.1762	0.1788	0.1814	0.1841
−0.8	0.1867	0.1894	0.1922	0.1949	0.1977	0.2005	0.2033	0.2061	0.2090	0.2119
−0.7	0.2148	0.2177	0.2207	0.2236	0.2266	0.2297	0.2327	0.2358	0.2389	0.2420
−0.6	0.2451	0.2483	0.2514	0.2546	0.2578	0.2611	0.2643	0.2676	0.2709	0.2743
−0.5	0.2776	0.2810	0.2843	0.2877	0.2912	0.2946	0.2981	0.3015	0.3050	0.3085
−0.4	0.3121	0.3156	0.3192	0.3228	0.3264	0.3300	0.3336	0.3372	0.3409	0.3446
−0.3	0.3483	0.3520	0.3557	0.3594	0.3632	0.3669	0.3707	0.3745	0.3783	0.3821
−0.2	0.3859	0.3897	0.3936	0.3974	0.4013	0.4052	0.4090	0.4129	0.4168	0.4207
−0.1	0.4247	0.4286	0.4325	0.4364	0.4404	0.4443	0.4483	0.4522	0.4562	0.4602
−0.0	0.4641	0.4681	0.4721	0.4761	0.4801	0.4840	0.4880	0.4920	0.4960	0.5000

TABLE 3 Cumulative Normal Probabilities   (*Continued*)

Z	0.00	0.01	0.02	0.03	0.04	0.05	0.06	0.07	0.08	0.09
+0.0	0.5000	0.5040	0.5080	0.5120	0.5160	0.5199	0.5239	0.5279	0.5319	0.5359
+0.1	0.5398	0.5438	0.5478	0.5517	0.5557	0.5596	0.5636	0.5675	0.5714	0.5753
+0.2	0.5793	0.5832	0.5871	0.5910	0.5948	0.5987	0.6026	0.6064	0.6103	0.6141
+0.3	0.6179	0.6217	0.6255	0.6293	0.6331	0.6368	0.6406	0.6443	0.6480	0.6517
+0.4	0.6554	0.6591	0.6628	0.6664	0.6700	0.6736	0.6772	0.6808	0.6844	0.6879
+0.5	0.6915	0.6950	0.6985	0.7019	0.7054	0.7088	0.7123	0.7157	0.7190	0.7224
+0.6	0.7257	0.7291	0.7324	0.7357	0.7389	0.7422	0.7454	0.7486	0.7517	0.7549
+0.7	0.7580	0.7611	0.7642	0.7673	0.7704	0.7734	0.7764	0.7794	0.7823	0.7852
+0.8	0.7881	0.7910	0.7939	0.7967	0.7995	0.8023	0.8051	0.8079	0.8106	0.8133
+0.9	0.8159	0.8186	0.8212	0.8238	0.8264	0.8289	0.8315	0.8340	0.8365	0.8389
+1.0	0.8413	0.8438	0.8461	0.8485	0.8508	0.8531	0.8554	0.8577	0.8599	0.8621
+1.1	0.8643	0.8665	0.8686	0.8708	0.8729	0.8749	0.8770	0.8790	0.8810	0.8830
+1.2	0.8849	0.8869	0.8888	0.8907	0.8925	0.8944	0.8962	0.8980	0.8997	0.9015
+1.3	0.9032	0.9049	0.9066	0.9082	0.9099	0.9115	0.9131	0.9147	0.9162	0.9177
+1.4	0.9192	0.9207	0.9222	0.9236	0.9251	0.9265	0.9279	0.9292	0.9306	0.9319
+1.5	0.9332	0.9345	0.9357	0.9370	0.9382	0.9394	0.9406	0.9418	0.9429	0.9441
+1.6	0.9452	0.9463	0.9474	0.9484	0.9495	0.9505	0.9515	0.9525	0.9535	0.9545
+1.7	0.9554	0.9564	0.9573	0.9582	0.9591	0.9599	0.9608	0.9616	0.9625	0.9633
+1.8	0.9641	0.9649	0.9656	0.9664	0.9671	0.9678	0.9686	0.9693	0.9699	0.9706
+1.9	0.9713	0.9719	0.9726	0.9732	0.9738	0.9744	0.9750	0.9756	0.9761	0.9767
+2.0	0.9773	0.9778	0.9783	0.9788	0.9793	0.9798	0.9803	0.9808	0.9812	0.9817
+2.1	0.9821	0.9826	0.9830	0.9834	0.9838	0.9842	0.9846	0.9850	0.9854	0.9857
+2.2	0.9861	0.9864	0.9868	0.9871	0.9875	0.9878	0.9881	0.9884	0.9887	0.9890
+2.3	0.9893	0.9896	0.9898	0.9901	0.9904	0.9906	0.9909	0.9911	0.9913	0.9916
+2.4	0.9918	0.9920	0.9922	0.9925	0.9927	0.9929	0.9931	0.9932	0.9934	0.9936
+2.5	0.9938	0.9940	0.9941	0.9943	0.9945	0.9946	0.9948	0.9949	0.9951	0.9952
+2.6	0.9953	0.9955	0.9956	0.9957	0.9959	0.9960	0.9961	0.9962	0.9963	0.9964
+2.7	0.9965	0.9966	0.9967	0.9968	0.9969	0.9970	0.9971	0.9972	0.9973	0.9974
+2.8	0.9974	0.9975	0.9976	0.9977	0.9977	0.9978	0.9979	0.9979	0.9980	0.9981
+2.9	0.9981	0.9982	0.9983	0.9983	0.9984	0.9984	0.9985	0.9985	0.9986	0.9986
+3.0	0.99865	0.99869	0.99874	0.99878	0.99882	0.99886	0.99889	0.99893	0.99896	0.99900
+3.1	0.99903	0.99906	0.99910	0.99913	0.99915	0.99918	0.99921	0.99924	0.99926	0.99929
+3.2	0.99931	0.99934	0.99936	0.99938	0.99940	0.99942	0.99944	0.99946	0.99948	0.99950
+3.3	0.99952	0.99953	0.99955	0.99957	0.99958	0.99960	0.99961	0.99962	0.99964	0.99965
+3.4	0.99966	0.99967	0.99969	0.99970	0.99971	0.99972	0.99973	0.99974	0.99975	0.99976
+3.5	0.99977	0.99978	0.99978	0.99979	0.99980	0.99981	0.99981	0.99982	0.99983	0.99983

Appendix

Interest Factor Tables

TABLES 1–8 INTEREST FACTORS FOR ANNUAL COMPOUNDING

Values for six interest formulas are given for interest rates from 8% to 20%. (These tables are reproduced from W. J. Fabrycky, G. J. Thuesen, and D. Verma, *Economic Decision Analysis* [Upper Saddle River, N.J.: Prentice Hall, Inc., 1998].)

From Appendix E of *Systems Engineering and Analysis,* Fifth Edition, Benjamin S. Blanchard, Wolter J. Fabrycky. Copyright © 2011 by Pearson Education, Inc. Published by Pearson Prentice Hall. All rights reserved.

TABLE 1 6% Interest Factors for Annual Compounding

	Single Payment		Equal-Payment Series			
	Compound-amount factor	Present-worth factor	Compound-amount factor	Sinking-fund factor	Present-worth factor	Capital-recovery factor
n	To find F given P $F/P, i, n$	To find P given F $P/F, i, n$	To find F given A $F/A, i, n$	To find A given F $A/F, i, n$	To find P given A $P/A, i, n$	To find A given P $A/P, i, n$
1	1.060	0.9434	1.000	1.0000	0.9434	1.0600
2	1.124	0.8900	2.060	0.4854	1.8334	0.5454
3	1.191	0.8396	3.184	0.3141	2.6730	0.3741
4	1.262	0.7921	4.375	0.2286	3.4651	0.2886
5	1.338	0.7473	5.637	0.1774	4.2124	0.2374
6	1.419	0.7050	6.975	0.1434	4.9173	0.2034
7	1.504	0.6651	8.394	0.1191	5.5824	0.1791
8	1.594	0.6274	9.897	0.1010	6.2098	0.1610
9	1.689	0.5919	11.491	0.0870	6.8017	0.1470
10	1.791	0.5584	13.181	0.0759	7.3601	0.1359
11	1.898	0.5268	14.972	0.0668	7.8869	0.1268
12	2.012	0.4970	16.870	0.0593	8.3839	0.1193
13	2.133	0.4688	18.882	0.0530	8.8527	0.1130
14	2.261	0.4423	21.015	0.0476	9.2950	0.1076
15	2.397	0.4173	23.276	0.0430	9.7123	0.1030
16	2.540	0.3937	25.673	0.0390	10.1059	0.0990
17	2.693	0.3714	28.213	0.0355	10.4773	0.0955
18	2.854	0.3504	30.906	0.0324	10.8276	0.0924
19	3.026	0.3305	33.760	0.0296	11.1581	0.0896
20	3.207	0.3118	36.786	0.0272	11.4699	0.0872
21	3.400	0.2942	39.993	0.0250	11.7641	0.0850
22	3.604	0.2775	43.392	0.0231	12.0416	0.0831
23	3.820	0.2618	46.996	0.0213	12.3034	0.0813
24	4.049	0.2470	50.816	0.0197	12.5504	0.0797
25	4.292	0.2330	54.865	0.0182	12.7834	0.0782
26	4.549	0.2198	59.156	0.0169	13.0032	0.0769
27	4.822	0.2074	63.706	0.0157	13.2105	0.0757
28	5.112	0.1956	68.528	0.0146	13.4062	0.0746
29	5.418	0.1846	73.640	0.0136	13.5907	0.0736
30	5.744	0.1741	79.058	0.0127	13.7648	0.0727
31	6.088	0.1643	84.802	0.0118	13.9291	0.0718
32	6.453	0.1550	90.890	0.0110	14.0841	0.0710
33	6.841	0.1462	97.343	0.0103	14.2302	0.0703
34	7.251	0.1379	104.184	0.0096	14.3682	0.0696
35	7.686	0.1301	111.435	0.0090	14.4983	0.0690

TABLE 2 7% Interest Factors for Annual Compounding

	Single Payment		Equal-Payment Series			
	Compound-amount factor	Present-worth factor	Compound-amount factor	Sinking-fund factor	Present-worth factor	Capital-recovery factor
n	To find F given P $F/P, i, n$	To find P given F $P/F, i, n$	To find F given A $F/A, i, n$	To find A given F $A/F, i, n$	To find P given A $P/A, i, n$	To find A given P $A/P, i, n$
1	1.070	0.9346	1.000	1.0000	0.9346	1.0700
2	1.145	0.8734	2.070	0.4831	1.8080	0.5531
3	1.225	0.8163	3.215	0.3111	2.6243	0.3811
4	1.311	0.7629	4.440	0.2252	3.3872	0.2952
5	1.403	0.7130	5.751	0.1739	4.1002	0.2439
6	1.501	0.6664	7.153	0.1398	4.7665	0.2098
7	1.606	0.6228	8.654	0.1156	5.3893	0.1856
8	1.718	0.5820	10.260	0.0975	5.9713	0.1675
9	1.838	0.5439	11.978	0.0835	6.5152	0.1535
10	1.967	0.5084	13.816	0.0724	7.0236	0.1424
11	2.105	0.4751	15.784	0.0634	7.4987	0.1334
12	2.252	0.4440	17.888	0.0559	7.9427	0.1259
13	2.410	0.4150	20.141	0.0497	8.3577	0.1197
14	2.579	0.3878	22.550	0.0444	8.7455	0.1144
15	2.759	0.3625	25.129	0.0398	9.1079	0.1098
16	2.952	0.3387	27.888	0.0359	9.4467	0.1059
17	3.159	0.3166	30.840	0.0324	9.7632	0.1024
18	3.380	0.2959	33.999	0.0294	10.0591	0.0994
19	3.617	0.2765	37.379	0.0268	10.3356	0.0968
20	3.870	0.2584	40.996	0.0244	10.5940	0.0944
21	4.141	0.2415	44.865	0.0223	10.8355	0.0923
22	4.430	0.2257	49.006	0.0204	11.0613	0.0904
23	4.741	0.2110	53.436	0.0187	11.2722	0.0887
24	5.072	0.1972	58.177	0.0172	11.4693	0.0872
25	5.427	0.1843	63.249	0.0158	11.6536	0.0858
26	5.807	0.1722	68.676	0.0146	11.8258	0.0846
27	6.214	0.1609	74.484	0.0134	11.9867	0.0834
28	6.649	0.1504	80.698	0.0124	12.1371	0.0824
29	7.114	0.1406	87.347	0.0115	12.2777	0.0815
30	7.612	0.1314	94.461	0.0106	12.4091	0.0806
31	8.145	0.1228	102.073	0.0098	12.5318	0.0798
32	8.715	0.1148	110.218	0.0091	12.6466	0.0791
33	9.325	0.1072	118.933	0.0084	12.7538	0.0784
34	9.978	0.1002	128.259	0.0078	12.8540	0.0778
35	10.677	0.0937	138.237	0.0072	12.9477	0.0772

TABLE 3 8% Interest Factors for Annual Compounding

	Single Payment		Equal-Payment Series			
	Compound-amount factor	Present-worth factor	Compound-amount factor	Sinking-fund factor	Present-worth factor	Capital-recovery factor
n	To find F given P $F/P, i, n$	To find P given F $P/F, i, n$	To find F given A $F/A, i, n$	To find A given F $A/F, i, n$	To find P given A $P/A, i, n$	To find A given P $A/P, i, n$
1	1.080	0.9259	1.000	1.0000	0.9259	1.0800
2	1.166	0.8573	2.080	0.4808	1.7833	0.5608
3	1.260	0.7938	3.246	0.3080	2.5771	0.3880
4	1.360	0.7350	4.506	0.2219	3.3121	0.3019
5	1.469	0.6806	5.867	0.1705	3.9927	0.2505
6	1.587	0.6302	7.336	0.1363	4.6229	0.2163
7	1.714	0.5835	8.923	0.1121	5.2064	0.1921
8	1.851	0.5403	10.637	0.0940	5.7466	0.1740
9	1.999	0.5003	12.488	0.0801	6.2469	0.1601
10	2.159	0.4632	14.487	0.0690	6.7101	0.1490
11	2.332	0.4289	16.645	0.0601	7.1390	0.1401
12	2.518	0.3971	18.977	0.0527	7.5361	0.1327
13	2.720	0.3677	21.495	0.0465	7.9038	0.1265
14	2.937	0.3405	24.215	0.0413	8.2442	0.1213
15	3.172	0.3153	27.152	0.0368	8.5595	0.1168
16	3.426	0.2919	30.324	0.0330	8.8514	0.1130
17	3.700	0.2703	33.750	0.0296	9.1216	0.1096
18	3.996	0.2503	37.450	0.0267	9.3719	0.1067
19	4.316	0.2317	41.446	0.0241	9.6036	0.1041
20	4.661	0.2146	45.762	0.0219	9.8182	0.1019
21	5.034	0.1987	50.423	0.0198	10.0168	0.0998
22	5.437	0.1840	55.457	0.0180	10.2008	0.0980
23	5.871	0.1703	60.893	0.0164	10.3711	0.0964
24	6.341	0.1577	66.765	0.0150	10.5288	0.0950
25	6.848	0.1460	73.106	0.0137	10.6748	0.0937
26	7.396	0.1352	79.954	0.0125	10.8100	0.0925
27	7.988	0.1252	87.351	0.0115	10.9352	0.0915
28	8.627	0.1159	95.339	0.0105	11.0511	0.0905
29	9.317	0.1073	103.966	0.0096	11.1584	0.0896
30	10.063	0.0994	113.283	0.0088	11.2578	0.0888
31	10.868	0.0920	123.346	0.0081	11.3498	0.0881
32	11.737	0.0852	134.214	0.0075	11.4350	0.0875
33	12.676	0.0789	145.951	0.0069	11.5139	0.0869
34	13.690	0.0731	158.627	0.0063	11.5869	0.0863
35	14.785	0.0676	172.317	0.0058	11.6546	0.0858

TABLE 4 9% Interest Factors for Annual Compounding

	Single Payment		Equal-Payment Series			
	Compound-amount factor	Present-worth factor	Compound-amount factor	Sinking-fund factor	Present-worth factor	Capital-recovery factor
n	To find F given P $F/P, i, n$	To find P given F $P/F, i, n$	To find F given A $F/A, i, n$	To find A given F $A/F, i, n$	To find P given A $P/A, i, n$	To find A given P $A/P, i, n$
1	1.090	0.9174	1.000	1.0000	0.9174	1.0900
2	1.188	0.8417	2.090	0.4785	1.7591	0.5685
3	1.295	0.7722	3.278	0.3051	2.5313	0.3951
4	1.412	0.7084	4.573	0.2187	3.2397	0.3087
5	1.539	0.6499	5.985	0.1671	3.8897	0.2571
6	1.677	0.5963	7.523	0.1329	4.4859	0.2229
7	1.828	0.5470	9.200	0.1087	5.0330	0.1987
8	1.993	0.5019	11.028	0.0907	5.5348	0.1807
9	2.172	0.4604	13.021	0.0768	5.9953	0.1668
10	2.367	0.4224	15.193	0.0658	6.4177	0.1558
11	2.580	0.3875	17.560	0.0570	6.8052	0.1470
12	2.813	0.3555	20.141	0.0497	7.1607	0.1397
13	3.066	0.3262	22.953	0.0436	7.4869	0.1336
14	3.342	0.2993	26.019	0.0384	7.7862	0.1284
15	3.642	0.2745	29.361	0.0341	8.0607	0.1241
16	3.970	0.2519	33.003	0.0303	8.3126	0.1203
17	4.328	0.2311	36.974	0.0271	8.5436	0.1171
18	4.717	0.2120	41.301	0.0242	8.7556	0.1142
19	5.142	0.1945	46.018	0.0217	8.9501	0.1117
20	5.604	0.1784	51.160	0.0196	9.1286	0.1096
21	6.109	0.1637	56.765	0.0176	9.2923	0.1076
22	6.659	0.1502	62.873	0.0159	9.4424	0.1059
23	7.258	0.1378	69.532	0.0144	9.5802	0.1044
24	7.11	0.1264	76.790	0.0130	9.7066	0.1030
25	8.623	0.1160	84.701	0.0118	9.8226	0.1018
26	9.399	0.1064	93.324	0.0107	9.9290	0.1007
27	10.245	0.0976	102.723	0.0097	10.0266	0.0997
28	11.167	0.0896	112.968	0.0089	10.1161	0.0989
29	12.172	0.0822	124.135	0.0081	10.1983	0.0981
30	13.268	0.0754	136.308	0.0073	10.2737	0.0973
31	14.462	0.0692	149.575	0.0067	10.3428	0.0967
32	15.763	0.0634	164.037	0.0061	10.4063	0.0961
33	17.182	0.0582	179.800	0.0056	10.4645	0.0956
34	18.728	0.0534	196.982	0.0051	10.5178	0.0951
35	20.414	0.0490	215.711	0.0046	10.5668	0.0946

TABLE 5 10% Interest Factors for Annual Compounding

	Single Payment		Equal-Payment Series			
n	Compound-amount factor	Present-worth factor	Compound-amount factor	Sinking-fund factor	Present-worth factor	Capital-recovery factor
	To find F given P $F/P, i, n$	*To find P given F* $P/F, i, n$	*To find F given A* $F/A, i, n$	*To find A given F* $A/F, i, n$	*To find P given A* $P/A, i, n$	*To find A given P* $A/P, i, n$
1	1.100	0.9091	1.000	1.0000	0.9091	1.1000
2	1.210	0.8265	2.100	0.4762	1.7355	0.5762
3	1.331	0.7513	3.310	0.3021	2.4869	0.4021
4	1.464	0.6830	4.641	0.2155	3.1699	0.3155
5	1.611	0.6209	6.105	0.1638	3.7908	0.2638
6	1.772	0.5645	7.716	0.1296	4.3553	0.2296
7	1.949	0.5132	9.487	0.1054	4.8684	0.2054
8	2.144	0.4665	11.436	0.0875	5.3349	0.1875
9	2.358	0.4241	13.579	0.0737	5.7950	0.1737
10	2.594	0.3856	15.937	0.0628	6.1446	0.1628
11	2.853	0.3505	18.531	0.0540	6.4951	0.1540
12	3.138	0.3186	21.384	0.0468	6.8137	0.1468
13	3.452	0.2897	24.523	0.0408	7.1034	0.1408
14	3.798	0.2633	27.975	0.0358	7.3667	0.1358
15	4.177	0.2394	31.772	0.0315	7.6061	0.1315
16	4.595	0.2176	35.950	0.0278	7.8237	0.1278
17	5.054	0.1979	40.545	0.0247	8.0216	0.1247
18	5.560	0.1799	45.599	0.0219	8.2014	0.1219
19	6.116	0.1635	51.159	0.0196	8.3649	0.1196
20	6.728	0.1487	57.275	0.0175	8.5136	0.1175
21	7.400	0.1351	64.003	0.0156	8.6487	0.1156
22	8.140	0.1229	71.403	0.0140	8.7716	0.1140
23	8.953	0.1117	79.543	0.0126	8.8832	0.1126
24	9.850	0.1015	88.497	0.0113	8.9848	0.1113
25	10.835	0.0923	98.347	0.0102	9.0771	0.1102
26	11.918	0.0839	109.182	0.0092	9.1610	0.1092
27	13.110	0.0763	121.100	0.0083	9.2372	0.1083
28	14.421	0.0694	134.210	0.0075	9.3066	0.1075
29	15.863	0.0630	148.631	0.0067	9.3696	0.1067
30	17.449	0.0573	164.494	0.0061	9.4269	0.1061
31	19.194	0.0521	181.943	0.0055	9.4790	0.1055
32	21.114	0.0474	201.138	0.0050	9.5264	0.1050
33	23.225	0.0431	222.252	0.0045	9.5694	0.1045
34	25.548	0.0392	245.477	0.0041	9.6086	0.1041
35	28.102	0.0356	271.024	0.0037	9.6442	0.1037

Appendix: Interest Factor Tables

TABLE 6 12% Interest Factors for Annual Compounding

	Single Payment		Equal-Payment Series			
	Compound-amount factor	Present-worth factor	Compound-amount factor	Sinking-fund factor	Present-worth factor	Capital-recovery factor
n	To find F given P $F/P, i, n$	To find P given F $P/F, i, n$	To find F given A $F/A, i, n$	To find A given F $A/F, i, n$	To find P given A $P/A, i, n$	To find A given P $A/P, i, n$
1	1.120	0.8929	1.000	1.0000	0.8929	1.1200
2	1.254	0.7972	2.120	0.4717	1.6901	0.5917
3	1.405	0.7118	3.374	0.2964	2.4018	0.4164
4	1.574	0.6355	4.779	0.2092	3.0374	0.3292
5	1.762	0.5674	6.353	0.1574	3.6048	0.2774
6	1.974	0.5066	8.115	0.1232	4.1114	0.2432
7	2.211	0.4524	10.089	0.0991	4.5638	0.2191
8	2.476	0.4039	12.300	0.0813	4.9676	0.2013
9	2.773	0.3606	14.776	0.0677	5.3283	0.1877
10	3.106	0.3220	17.549	0.0570	5.6502	0.1770
11	3.479	0.2875	20.655	0.0484	5.9377	0.1684
12	3.896	0.2567	24.133	0.0414	6.1944	0.1614
13	4.364	0.2292	28.029	0.0357	6.4236	0.1557
14	4.887	0.2046	32.393	0.0309	6.6282	0.1509
15	5.474	0.1827	37.280	0.0268	6.8109	0.1468
16	6.130	0.1631	42.753	0.0234	6.9740	0.1434
17	6.866	0.1457	48.884	0.0205	7.1196	0.1405
18	7.690	0.1300	55.750	0.0179	7.2497	0.1379
19	8.613	0.1161	63.440	0.0158	7.3658	0.1358
20	9.646	0.1037	72.052	0.0139	7.4695	0.1339
21	10.804	0.0926	81.699	0.0123	7.5620	0.1323
22	12.100	0.0827	92.503	0.0108	7.6447	0.1308
23	13.552	0.0738	104.603	0.0096	7.7184	0.1296
24	15.179	0.0659	118.155	0.0085	7.7843	0.1285
25	17.000	0.0588	133.334	0.0075	7.8431	0.1275
26	19.040	0.0525	150.334	0.0067	7.8957	0.1267
27	21.325	0.0469	169.374	0.0059	7.9426	0.1259
28	23.884	0.0419	190.699	0.0053	7.9844	0.1253
29	26.750	0.0374	214.583	0.0047	8.0218	0.1247
30	29.960	0.0334	241.333	0.0042	8.0552	0.1242
31	33.555	0.0298	271.293	0.0037	8.0850	0.1237
32	37.582	0.0266	304.848	0.0033	8.1116	0.1233
33	42.092	0.0238	342.429	0.0029	8.1354	0.1229
34	47.143	0.0212	384.521	0.0026	8.1566	0.1226
35	52.800	0.0189	431.664	0.0023	8.1755	0.1223

TABLE 7 15% Interest Factors for Annual Compounding

	Single Payment		Equal-Payment Series			
	Compound-amount factor	Present-worth factor	Compound-amount factor	Sinking-fund factor	Present-worth factor	Capital-recovery factor
n	To find F given P $F/P, i, n$	To find P given F $P/F, i, n$	To find F given A $F/A, i, n$	To find A given F $A/F, i, n$	To find P given A $P/A, i, n$	To find A given P $A/P, i, n$
1	1.150	0.8696	1.000	1.0000	0.8696	1.1500
2	1.323	0.7562	2.150	0.4651	1.6257	0.6151
3	1.521	0.6575	3.473	0.2880	2.2832	0.4380
4	1.749	0.5718	4.993	0.2003	2.8550	0.3503
5	2.011	0.4972	6.742	0.1483	3.3522	0.2983
6	2.313	0.4323	8.754	0.1142	3.7845	0.2642
7	2.660	0.3759	11.067	0.0904	4.1604	0.2404
8	3.059	0.3269	13.727	0.0729	4.4873	0.2229
9	3.518	0.2843	16.786	0.0596	4.7716	0.2096
10	4.046	0.2472	20.304	0.0493	5.0188	0.1993
11	4.652	0.2150	24.349	0.0411	5.2337	0.1911
12	5.350	0.1869	29.002	0.0345	5.4206	0.1845
13	6.153	0.1625	34.352	0.0291	5.5832	0.1791
14	7.076	0.1413	40.505	0.0247	5.7245	0.1747
15	8.137	0.1229	47.580	0.0210	5.8474	0.1710
16	9.358	0.1069	55.717	0.0180	5.9542	0.1680
17	10.761	0.0929	65.075	0.0154	6.0472	0.1654
18	12.375	0.0808	75.836	0.0132	6.1280	0.1632
19	14.232	0.0703	88.212	0.0113	6.1982	0.1613
20	16.367	0.0611	102.444	0.0098	6.2593	0.1598
21	18.822	0.0531	118.810	0.0084	6.3125	0.1584
22	21.645	0.0462	137.632	0.0073	6.3587	0.1573
23	24.891	0.0402	159.276	0.0063	6.3988	0.1563
24	28.625	0.0349	184.168	0.0054	6.4338	0.1554
25	32.919	0.0304	212.793	0.0047	6.4642	0.1547
26	37.857	0.0264	245.712	0.0041	6.4906	0.1541
27	43.535	0.0230	283.569	0.0035	6.5135	0.1535
28	50.066	0.0200	327.104	0.0031	6.5335	0.1531
29	57.575	0.0174	377.170	0.0027	6.5509	0.1527
30	66.212	0.0151	434.745	0.0023	6.5660	0.1523
31	76.144	0.0131	500.957	0.0020	6.5791	0.1520
32	87.565	0.0114	577.100	0.0017	6.5905	0.1517
33	100.700	0.0099	664.666	0.0015	6.6005	0.1515
34	115.805	0.0086	765.365	0.0013	6.6091	0.1513
35	133.176	0.0075	881.170	0.0011	6.6166	0.1511

TABLE 8 20% Interest Factors for Annual Compounding

	Single Payment		Equal-Payment Series			
	Compound-amount factor	Present-worth factor	Compound-amount factor	Sinking-fund factor	Present-worth factor	Capital-recovery factor
n	To find F given P $F/P, i, n$	To find P given F $P/F, i, n$	To find F given A $F/A, i, n$	To find A given F $A/F, i, n$	To find P given A $P/A, i, n$	To find A given P $A/P, i, n$
1	1.200	0.8333	1.000	1.0000	0.8333	1.2000
2	1.440	0.6945	2.200	0.4546	1.5278	0.6546
3	1.728	0.5787	3.640	0.2747	2.1065	0.4747
4	2.074	0.4823	5.368	0.1863	2.5887	0.3863
5	2.488	0.4019	7.442	0.1344	2.9906	0.3344
6	2.986	0.3349	9.930	0.1007	3.3255	0.3007
7	3.583	0.2791	12.916	0.0774	3.6046	0.2774
8	4.300	0.2326	16.499	0.0606	3.8372	0.2606
9	5.160	0.1938	20.799	0.0481	4.0310	0.2481
10	6.192	0.1615	25.959	0.0385	4.1925	0.2385
11	7.430	0.1346	32.150	0.0311	4.3271	0.2311
12	8.916	0.1122	39.581	0.0253	4.4392	0.2253
13	10.699	0.0935	48.497	0.0206	4.5327	0.2206
14	12.839	0.0779	59.196	0.0169	4.6106	0.2169
15	15.407	0.0649	72.035	0.0139	4.6755	0.2139
16	18.488	0.0541	87.442	0.0114	4.7296	0.2114
17	22.186	0.0451	105.931	0.0095	4.7746	0.2095
18	26.623	0.0376	128.117	0.0078	4.8122	0.2078
19	31.948	0.0313	154.740	0.0065	4.8435	0.2065
20	38.338	0.0261	186.688	0.0054	4.8696	0.2054
21	46.005	0.0217	225.026	0.0045	4.8913	0.2045
22	55.206	0.0181	271.031	0.0037	4.9094	0.2037
23	66.247	0.0151	326.237	0.0031	4.9245	0.2031
24	79.497	0.0126	392.484	0.0026	4.9371	0.2026
25	95.396	0.0105	471.981	0.0021	4.9476	0.2021
26	114.475	0.0087	567.377	0.0018	4.9563	0.2018
27	137.371	0.0073	681.853	0.0015	4.9636	0.2015
28	164.845	0.0061	819.223	0.0012	4.9697	0.2012
29	197.814	0.0051	984.068	0.0010	4.9747	0.2010
30	237.376	0.0042	1181.882	0.0009	4.9789	0.2009
31	284.852	0.0035	1419.258	0.0007	4.9825	0.2007
32	341.822	0.0029	1704.109	0.0006	4.9854	0.2006
33	410.186	0.0024	2045.931	0.0005	4.9878	0.2005
34	492.224	0.0020	2456.118	0.0004	4.9899	0.2004
35	590.668	0.0017	2948.341	0.0003	4.9915	0.2003

Appendix

Finite Queuing Tables

From Appendix F of *Systems Engineering and Analysis,* Fifth Edition, Benjamin S. Blanchard, Wolter J. Fabrycky. Copyright © 2011 by Pearson Education, Inc. Published by Pearson Prentice Hall.

Appendix

Finite Queuing Tables

TABLES 1–3 FINITE QUEUING FACTORS

The probability of a delay, D, and the efficiency factor, F, are given for populations of 10, 20, and 30 units. Each set of values is keyed to the service factor, X, and the number of channels, M. (These tabular values are adapted with permission from L. G. Peck and R. N. Hazelwood, *Finite Queuing Tables* [New York: John Wiley & Sons, Inc., 1958].)

TABLE 1 Finite Queuing Factors—Population 10

X	M	D	F	X	M	D	F	X	M	D	F
0.008	1	0.072	0.999		2	0.177	0.990		3	0.182	0.986
0.013	1	0.117	0.998	0.085	1	0.660	0.899		2	0.528	0.921
0.016	1	0.144	0.997		3	0.037	0.999		1	0.954	0.610
0.019	1	0.170 *	0.996		2	0.196	0.988	0.165	4	0.049	0.997
0.021	1	0.188	0.995		1	0.692	0.883		3	0.195	0.984
0.023	1	0.206	0.994	0.090	3	0.043	0.998		2	0.550	0.914
0.025	1	0.224	0.993		2	0.216	0.986		1	0.961	0.594
0.026	1	0.232	0.992		1	0.722	0.867	0.170	4	0.054	0.997
0.028	1	0.250	0.991	0.095	3	0.049	0.998		3	0.209	0.982
0.030	1	0.268	0.990		2	0.237	0.984		2	0.571	0.906
0.032	2	0.033	0.999		1	0.750	0.850		1	0.966	0.579
	1	0.285	0.988	0.100	3	0.056	0.998	0.180	5	0.013	0.999
0.034	2	0.037	0.999		2	0.258	0.981		4	0.066	0.996
	1	0.302	0.986		1	0.776	0.832		3	0.238	0.978
0.036	2	0.041	0.999	0.105	3	0.064	0.997		2	0.614	0.890
	1	0.320	0.984		2	0.279	0.978		1	0.975	0.549
0.038	2	0.046	0.999		1	0.800	0.814	0.190	5	0.016	0.999
	1	0.337	0.982	0.110	3	0.072	0.997		4	0.078	0.995
0.040	2	0.050	0.999		2	0.301	0.974		3	0.269	0.973
	1	0.354	0.980		1	0.822	0.795		2	0.654	0.873
0.042	2	0.055	0.999	0.115	3	0.081	0.996		1	0.982	0.522
	1	0.371	0.978		2	0.324	0.971	0.200	5	0.020	0.999
0.044	2	0.060	0.998		1	0.843	0.776		4	0.092	0.994
	1	0.388	0.975	0.120	4	0.016	0.999		3	0.300	0.968
0.046	2	0.065	0.998		3	0.090	0.995		2	0.692	0.854
	1	0.404	0.973		2	0.346	0.967	0.210	1	0.987	0.497
0.048	2	0.071	0.998		1	0.861	0.756		5	0.025	0.999
	1	0.421	0.970	0.125	4	0.019	0.999		4	0.108	0.992
0.050	2	0.076	0.998		3	0.100	0.994		3	0.333	0.961
	1	0.437	0.967		2	0.369	0.962		2	0.728	0.835
0.052	2	0.082	0.997		1	0.878	0.737		1	0.990	0.474
	1	0.454	0.963	0.130	4	0.022	0.999	0.220	5	0.030	0.998
0.054	2	0.088	0.997		3	0.110	0.994		4	0.124	0.990
	1	0.470	0.960		2	0.392	0.958		3	0.366	0.954
0.056	2	0.094	0.997		1	0.893	0.718		2	0.761	0.815
	1	0.486	0.956	0.135	4	0.025	0.999		1	0.993	0.453
0.058	2	0.100	0.996		3	0.121	0.993	0.230	5	0.037	0.998
	1	0.501	0.953		2	0.415	0.952		4	0.142	0.988
0.060	2	0.106	0.996		1	0.907	0.699		3	0.400	0.947
	1	0.517	0.949	0.140	4	0.028	0.999		2	0.791	0.794
0.062	2	0.113	0.996		3	0.132	0.991		1	0.995	0.434
	1	0.532	0.945		2	0.437	0.947	0.240	5	0.044	0.997
0.064	2	0.119	0.995		1	0.919	0.680		4	0.162	0.986
	1	0.547	0.940	0.145	4	0.032	0.999		3	0.434	0.938
0.066	2	0.126	0.995		3	0.144	0.990		2	0.819	0.774
	1	0.562	0.936		2	0.460	0.941		1	0.996	0.416
0.068	3	0.020	0.999		1	0.929	0.662	0.250	6	0.010	0.999
	2	0.133	0.994	0.150	4	0.036	0.998		5	0.052	0.997
	1	0.577	0.931		3	0.156	0.989		4	0.183	0.983
0.070	3	0.022	0.999		2	0.483	0.935		3	0.469	0.929
	2	0.140	0.994		1	0.939	0.644		2	0.844	0.753
	1	0.591	0.926	0.155	4	0.040	0.998		1	0.997	0.400
0.075	3	0.026	0.999		3	0.169	0.987	0.260	6	0.013	0.999
	2	0.158	0.992		2	0.505	0.928		5	0.060	0.996
	1	0.627	0.913		1	0.947	0.627		4	0.205	0.980
0.080	3	0.031	0.999	0.160	4	0.044	0.998		3	0.503	0.919
	2	0.866	0.732		4	0.533	0.906		7	0.171	0.982

TABLE 1 Finite Queuing Factors—Population 10 (*Continued*)

X	Ṁ	D	F	X	M	D	F	X	M	D	F
	1	0.998	0.384		3	0.840	0.758		6	0.413	0.939
0.270	6	0.015	0.999		2	0.986	0.525		5	0.707	0.848
	5	0.070	0.995	0.400	7	0.026	0.998		4	0.917	0.706
	4	0.228	0.976		6	0.105	0.991		3	0.991	0.535
	3	0.537	0.908		5	0.292	0.963	0.580	8	0.057	0.995
	2	0.886	0.712		4	0.591	0.887		7	0.204	0.977
	1	0.999	0.370		3	0.875	0.728		6	0.465	0.927
0.280	6	0.018	0.999		2	0.991	0.499		5	0.753	0.829
	5	0.081	0.994	0.420	7	0.034	0.998		4	0.937	0.684
	4	0.252	0.972		6	0.130	0.987		3	0.994	0.517
	3	0.571	0.896		5	0.341	0.954	0.600	9	0.010	0.999
	2	0.903	0.692		4	0.646	0.866		8	0.072	0.994
	1	0.999	0.357		3	0.905	0.700		7	0.242	0.972
0.290	6	0.022	0.999		2	0.994	0.476		6	0.518	0.915
	5	0.093	0.993	0.440	7	0.045	0.997		5	0.795	0.809
	4	0.278	0.968		6	0.160	0.984		4	0.953	0.663
	3	0.603	0.884		5	0.392	0.943		3	0.996	0.500
	2	0.918	0.672		4	0.698	0.845	0.650	9	0.021	0.999
	1	0.999	0.345		3	0.928	0.672		8	0.123	0.988
0.300	6	0.026	0.998		2	0.996	0.454		7	0.353	0.954
	5	0.106	0.991	0.460	8	0.011	0.999		6	0.651	0.878
	4	0.304	0.963		7	0.058	0.995		5	0.882	0.759
	3	0.635	0.872		6	0.193	0.979		4	0.980	0.614
	2	0.932	0.653		5	0.445	0.930		3	0.999	0.461
	1	0.999	0.333		4	0.747	0.822	0.700	9	0.040	0.997
0.310	6	0.031	0.998		3	0.947	0.646		8	0.200	0.979
	5	0.120	0.990		2	0.998	0.435		7	0.484	0.929
	4	0.331	0.957	0.480	8	0.015	0.999		6	0.772	0.836
	3	0.666	0.858		7	0.074	0.994		5	0.940	0.711
	2	0.943	0.635		6	0.230	0.973		4	0.992	0.571
0.320	6	0.036	0.998		5	0.499	0.916	0.750	9	0.075	0.994
	5	0.135	0.988		4	0.791	0.799		8	0.307	0.965
	4	0.359	0.952		3	0.961	0.621		7	0.626	0.897
	3	0.695	0.845		2	0.998	0.417		6	0.870	0.792
	2	0.952	0.617	0.500	8	0.020	0.999		5	0.975	0.666
0.330	6	0.042	0.997		7	0.093	0.992		4	0.998	0.533
	5	0.151	0.986		6	0.271	0.966	0.800	9	0.134	0.988
	4	0.387	0.945		5	0.553	0.901		8	0.446	0.944
	3	0.723	0.831		4	0.830	0.775		7	0.763	0.859
	2	0.961	0.600		3	0.972	0.598		6	0.939	0.747
0.340	7	0.010	0.999		2	0.999	0.400		5	0.991	0.625
	6	0.049	0.997	0.520	8	0.026	0.998		4	0.999	0.500
	5	0.168	0.983		7	0.115	0.989	0.850	9	0.232	0.979
	4	0.416	0.938		6	0.316	0.958		8	0.611	0.916
	3	0.750	0.816		5	0.606	0.884		7	0.879	0.818
	2	0.968	0.584		4	0.864	0.752		6	0.978	0.705
0.360	7	0.014	0.999		3	0.980	0.575		5	0.998	0.588
	6	0.064	0.995		2	0.999	0.385	0.900	9	0.387	0.963
	5	0.205	0.978	0.540	8	0.034	0.997		8	0.785	0.881
	4	0.474	0.923		7	0.141	0.986		7	0.957	0.777
	3	0.798	0.787		6	0.363	0.949		6	0.995	0.667
	2	0.978	0.553		5	0.658	0.867	0.950	9	0.630	0.938
0.380	7	0.019	0.999		4	0.893	0.729		8	0.934	0.841
	6	0.083	0.993		3	0.986	0.555		7	0.994	0.737
	5	0.247	0.971	0.560	8	0.044	0.996				

TABLE 2 Finite Queuing Factors—Population 20   (*Continued*)

X	M	D	F	X	M	D	F	X	M	D	F
0.005	1	0.095	0.999		1	0.837	0.866		3	0.326	0.980
0.009	1	0.171	0.998	0.052	3	0.080	0.998		2	0.733	0.896
0.011	1	0.208	0.997		2	0.312	0.986		1	0.998	0.526
0.013	1	0.246	0.996		1	0.858	0.851	0.100	5	0.038	0.999
0.014	1	0.265	0.995	0.054	3	0.088	0.998		4	0.131	0.995
0.015	1	0.283	0.994		2	0.332	0.984		3	0.363	0.975
0.016	1	0.302	0.993		1	0.876	0.835		2	0.773	0.878
0.017	1	0.321	0.992	0.056	3	0.097	0.997		1	0.999	0.500
0.018	2	0.048	0.999		2	0.352	0.982	0.110	5	0.055	0.998
	1	0.339	0.991		1	0.893	0.819		4	0.172	0.992
0.019	2	0.053	0.999	0.058	3	0.105	0.997		3	0.438	0.964
	1	0.358	0.990		2	0.372	0.980		2	0.842	0.837
0.020	2	0.058	0.999		1	0.908	0.802	0.120	6	0.022	0.999
	1	0.376	0.989	0.060	4	0.026	0.999		5	0.076	0.997
0.021	2	0.064	0.999		3	0.115	0.997		4	0.219	0.988
	1	0.394	0.987		2	0.392	0.978		3	0.514	0.950
0.022	2	0.070	0.999		1	0.922	0.785		2	0.895	0.793
	1	0.412	0.986	0.062	4	0.029	0.999	0.130	6	0.031	0.999
0.023	2	0.075	0.999		3	0.124	0.996		5	0.101	0.996
	1	0.431	0.984		2	0.413	0.975		4	0.271	0.983
0.024	2	0.082	0.999		1	0.934	0.768		3	0.589	0.933
	1	0.449	0.982	0.064	4	0.032	0.999		2	0.934	0.748
0.025	2	0.088	0.999		3	0.134	0.996	0.140	6	0.043	0.998
	1	0.466	0.980		2	0.433	0.972		5	0.131	0.994
0.026	2	0.094	0.998		1	0.944	0.751		4	0.328	0.976
	1	0.484	0.978	0.066	4	0.036	0.999		3	0.661	0.912
0.028	2	0.108	0.998		3	0.144	0.995		2	0.960	0.703
	1	0.519	0.973		2	0.454	0.969	0.150	7	0.017	0.999
0.030	2	0.122	0.998		1	0.953	0.733		6	0.059	0.998
	1	0.553	0.968	0.068	4	0.039	0.999		5	0.166	0.991
0.032	2	0.137	0.997		3	0.155	0.995		4	0.388	0.968
	1	0.587	0.962		2	0.474	0.966		3	0.728	0.887
0.034	2	0.152	0.996		1	0.961	0.716		2	0.976	0.661
	1	0.620	0.955	0.070	4	0.043	0.999	0.160	7	0.024	0.999
0.036	2	0.168	0.996		3	0.165	0.994		6	0.077	0.997
	1	0.651	0.947		2	0.495	0.962		5	0.205	0.988
0.038	3	0.036	0.999		1	0.967	0.699		4	0.450	0.957
	2	0.185	0.995	0.075	4	0.054	0.999		3	0.787	0.860
	1	0.682	0.938		3	0.194	0.992		2	0.987	0.622
0.040	3	0.041	0.999		2	0.545	0.953	0.180	7	0.044	0.998
	2	0.202	0.994		1	0.980	0.659		6	0.125	0.994
	1	0.712	0.929	0.080	4	0.066	0.998		5	0.295	0.978
0.042	3	0.047	0.999		3	0.225	0.990		4	0.575	0.930
	2	0.219	0.993		2	0.595	0.941		3	0.879	0.799
	1	0.740	0.918		1	0.988	0.621		2	0.996	0.555
0.044	3	0.053	0.999	0.085	4	0.080	0.997	0.200	8	0.025	0.999
	2	0.237	0.992		3	0.257	0.987		7	0.074	0.997
	1	0.767	0.906		2	0.643	0.928		6	0.187	0.988
0.046	3	0.059	0.999		1	0.993	0.586		5	0.397	0.963
	2	0.255	0.991	0.090	5	0.025	0.999		4	0.693	0.895
	1	0.792	0.894		4	0.095	0.997		3	0.938	0.736
0.048	3	0.066	0.999		3	0.291	0.984		2	0.999	0.500
	2	0.274	0.989		2	0.689	0.913	0.220	8	0.043	0.998
	1	0.815	0.881		1	0.996	0.554		7	0.115	0.994
0.050	3	0.073	0.998	0.095	5	0.031	0.999		6	0.263	0.980
	2	0.293	0.988		4	0.112	0.996		5	0.505	0.943

TABLE 2 Finite Queuing Factors—Population 20 (*Continued*)

X	M	D	F	X	M	D	F	X	M	D	F
	4	0.793	0.852		4	0.998	0.555	0.500	14	0.033	0.998
	3	0.971	0.677	0.380	12	0.024	0.999		13	0.088	0.995
0.240	9	0.024	0.999		11	0.067	0.996		12	0.194	0.985
	8	0.068	0.997		10	0.154	0.989		11	0.358	0.965
	7	0.168	0.989		9	0.305	0.973		10	0.563	0.929
	6	0.351	0.969		8	0.513	0.938		9	0.764	0.870
	5	0.613	0.917		7	0.739	0.874		8	0.908	0.791
	4	0.870	0.804		6	0.909	0.777		7	0.977	0.698
	3	0.988	0.623		5	0.984	0.656		6	0.997	0.600
0.260	9	0.039	0.998		4	0.999	0.526	0.540	15	0.023	0.999
	8	0.104	0.994	0.400	13	0.012	0.999		14	0.069	0.996
	7	0.233	0.983		12	0.037	0.998		13	0.161	0.988
	6	0.446	0.953		11	0.095	0.994		12	0.311	0.972
	5	0.712	0.884		10	0.205	0.984		11	0.509	0.941
	4	0.924	0.755		9	0.379	0.962		10	0.713	0.891
	3	0.995	0.576		8	0.598	0.918		9	0.873	0.821
0.280	10	0.021	0.999		7	0.807	0.845		8	0.961	0.738
	9	0.061	0.997		6	0.942	0.744		7	0.993	0.648
	8	0.149	0.990		5	0.992	0.624		6	0.999	0.556
	7	0.309	0.973	0.420	13	0.019	0.999	0.600	16	0.023	0.999
	6	0.544	0.932		12	0.055	0.997		15	0.072	0.996
	5	0.797	0.848		11	0.131	0.991		14	0.171	0.988
	4	0.958	0.708		10	0.265	0.977		13	0.331	0.970
	3	0.998	0.536		9	0.458	0.949		12	0.532	0.938
0.300	10	0.034	0.998		8	0.678	0.896		11	0.732	0.889
	9	0.091	0.995		7	0.863	0.815		10	0.882	0.824
	8	0.205	0.985		6	0.965	0.711		9	0.962	0.748
	7	0.394	0.961		5	0.996	0.595		8	0.992	0.666
	6	0.639	0.907	0.440	13	0.029	0.999		7	0.999	0.583
	5	0.865	0.808		12	0.078	0.995	0.700	17	0.047	0.998
	4	0.978	0.664		11	0.175	0.987		16	0.137	0.991
	3	0.999	0.500		10	0.333	0.969		15	0.295	0.976
0.320	11	0.018	0.999		9	0.540	0.933		14	0.503	0.948
	10	0.053	0.997		8	0.751	0.872		13	0.710	0.905
	9	0.130	0.992		7	0.907	0.785		12	0.866	0.849
	8	0.272	0.977		6	0.980	0.680		11	0.953	0.783
	7	0.483	0.944		5	0.998	0.568		10	0.988	0.714
	6	0.727	0.878	0.460	14	0.014	0.999		9	0.998	0.643
	5	0.915	0.768		13	0.043	0.998	0.800	19	0.014	0.999
	4	0.989	0.624		12	0.109	0.993		18	0.084	0.996
0.340	11	0.029	0.999		11	0.228	0.982		17	0.242	0.984
	10	0.079	0.996		10	0.407	0.958		16	0.470	0.959
	9	0.179	0.987		9	0.620	0.914		15	0.700	0.920
	8	0.347	0.967		8	0.815	0.846		14	0.867	0.869
	7	0.573	0.924		7	0.939	0.755		13	0.955	0.811
	6	0.802	0.846		6	0.989	0.651		12	0.989	0.750
	5	0.949	0.729		5	0.999	0.543		11	0.998	0.687
	4	0.995	0.588	0.480	14	0.022	0.999	0.900	19	0.135	0.994
0.360	12	0.015	0.999		13	0.063	0.996		18	0.425	0.972
	11	0.045	0.998		12	0.147	0.990		17	0.717	0.935
	10	0.112	0.993		11	0.289	0.974		16	0.898	0.886
	9	0.237	0.981		10	0.484	0.944		15	0.973	0.833
	8	0.429	0.954		9	0.695	0.893		14	0.995	0.778
	7	0.660	0.901		8	0.867	0.819		13	0.999	0.722
	6	0.863	0.812		7	0.962	0.726	0.950	19	0.377	0.981
	5	0.971	0.691		6	0.994	0.625		18	0.760	0.943

TABLE 3 Finite Queuing Factors—Population 30 *(Continued)*

X	M	D	F	X	M	D	F	X	M	D	F
0.004	1	0.116	0.999		1	0.963	0.772		3	0.426	0.976
0.007	1	0.203	0.998	0.044	4	0.040	0.999		2	0.847	0.873
0.009	1	0.260	0.997		3	0.154	0.996	0.075	5	0.069	0.998
0.010	1	0.289	0.996		2	0.474	0.977		4	0.201	0.993
0.011	1	0.317	0.995		1	0.974	0.744		3	0.486	0.969
0.012	1	0.346	0.994	0.046	4	0.046	0.999		2	0.893	0.840
0.013	1	0.374	0.993		3	0.171	0.996	0.080	6	0.027	0.999
0.014	2	0.067	0.999		2	0.506	0.972		5	0.088	0.998
	1	0.403	0.991		1	0.982	0.716		4	0.240	0.990
0.015	2	0.076	0.999	0.048	4	0.053	0.999		3	0.547	0.959
	1	0.431	0.989		3	0.189	0.995		2	0.929	0.805
0.016	2	0.085	0.999		2	0.539	0.968	0.085	6	0.036	0.999
	1	0.458	0.987		1	0.988	0.689		5	0.108	0.997
0.017	2	0.095	0.999	0.050	4	0.060	0.999		4	0.282	0.987
	1	0.486	0.985		3	0.208	0.994		3	0.607	0.948
0.018	2	0.105	0.999		2	0.571	0.963		2	0.955	0.768
	1	0.513	0.983		1	0.992	0.663	0.090	6	0.046	0.999
0.019	2	0.116	0.999	0.052	4	0.068	0.999		5	0.132	0.996
	1	0.541	0.980		3	0.227	0.993		4	0.326	0.984
0.020	2	0.127	0.998		2	0.603	0.957		3	0.665	0.934
	1	0.567	0.976		1	0.995	0.639		2	0.972	0.732
0.021	2	0.139	0.998	0.054	4	0.077	0.998	0.095	6	0.057	0.999
	1	0.594	0.973		3	0.247	0.992		5	0.158	0.994
0.022	2	0.151	0.998		2	0.634	0.951		4	0.372	0.979
	1	0.620	0.969		1	0.997	0.616		3	0.720	0.918
0.023	2	0.163	0.997	0.056	4	0.086	0.998		2	0.984	0.697
	1	0.645	0.965		3	0.267	0.991	0.100	6	0.071	0.998
0.024	2	0.175	0.997		2	0.665	0.944		5	0.187	0.993
	1	0.670	0.960		1	0.998	0.595		4	0.421	0.973
0.025	2	0.188	0.996	0.058	4	0.096	0.998		3	0.771	0.899
	1	0.694	0.954		3	0.288	0.989		2	0.991	0.664
0.026	2	0.201	0.996		2	0.695	0.936	0.110	7	0.038	0.999
	1	0.718	0.948		1	0.999	0.574		6	0.105	0.997
0.028	3	0.051	0.999	0.060	5	0.030	0.999		5	0.253	0.988
	2	0.229	0.995		4	0.106	0.997		4	0.520	0.959
	1	0.763	0.935		3	0.310	0.987		3	0.856	0.857
0.030	3	0.060	0.999		2	0.723	0.927		2	0.997	0.605
	2	0.257	0.994		1	0.999	0.555	0.120	7	0.057	0.998
	1	0.805	0.918	0.062	5	0.034	0.999		6	0.147	0.994
0.032	3	0.071	0.999		4	0.117	0.997		5	0.327	0.981
	2	0.286	0.992		3	0.332	0.986		4	0.619	0.939
	1	0.843	0.899		2	0.751	0.918		3	0.918	0.808
0.034	3	0.083	0.999	0.064	5	0.038	0.999		2	0.999	0.555
	2	0.316	0.990		4	0.128	0.997	0.130	8	0.030	0.999
	1	0.876	0.877		3	0.355	0.984		7	0.083	0.997
0.036	3	0.095	0.998		2	0.777	0.908		6	0.197	0.991
	2	0.347	0.988	0.066	5	0.043	0.999		5	0.409	0.972
	1	0.905	0.853		4	0.140	0.996		4	0.712	0.914
0.038	3	0.109	0.998		3	0.378	0.982		3	0.957	0.758
	2	0.378	0.986		2	0.802	0.897	0.140	8	0.045	0.999
	1	0.929	0.827	0.068	5	0.048	0.999		7	0.115	0.996
0.040	3	0.123	0.997		4	0.153	0.995		6	0.256	0.987
	2	0.410	0.983		3	0.402	0.979		5	0.494	0.960
	1	0.948	0.800		2	0.825	0.885		4	0.793	0.884
0.042	3	0.138	0.997	0.070	5	0.054	0.999		3	0.979	0.710
	2	0.442	0.980		4	0.166	0.995	0.150	9	0.024	0.999

TABLE 3 Finite Queuing Factors—Population 30 (*Continued*)

X	M	D	F	X	M	D	F	X	M	D	F
	8	0.065	0.998		7	0.585	0.938		7	0.901	0.818
	7	0.155	0.993		6	0.816	0.868		6	0.981	0.712
	6	0.322	0.980		5	0.961	0.751		5	0.999	0.595
	5	0.580	0.944		4	0.998	0.606	0.290	14	0.023	0.999
	4	0.860	0.849	0.230	12	0.023	0.999		13	0.055	0.998
	3	0.991	0.665		11	0.056	0.998		12	0.117	0.994
0.160	9	0.036	0.999		10	0.123	0.994		11	0.223	0.986
	8	0.090	0.997		9	0.242	0.985		10	0.382	0.969
	7	0.201	0.990		8	0.423	0.965		9	0.582	0.937
	6	0.394	0.972		7	0.652	0.923		8	0.785	0.880
	5	0.663	0.924		6	0.864	0.842		7	0.929	0.795
	4	0.910	0.811		5	0.976	0.721		6	0.988	0.688
	3	0.996	0.624		4	0.999	0.580		5	0.999	0.575
0.170	10	0.019	0.999	0.240	12	0.031	0.999	0.300	14	0.031	0.999
	9	0.051	0.998		11	0.074	0.997		13	0.071	0.997
	8	0.121	0.995		10	0.155	0.992		12	0.145	0.992
	7	0.254	0.986		9	0.291	0.981		11	0.266	0.982
	6	0.469	0.961		8	0.487	0.955		10	0.437	0.962
	5	0.739	0.901		7	0.715	0.905		9	0.641	0.924
	4	0.946	0.773		6	0.902	0.816		8	0.830	0.861
	3	0.998	0.588		5	0.986	0.693		7	0.950	0.771
0.180	10	0.028	0.999		4	0.999	0.556		6	0.993	0.666
	9	0.070	0.997	0.250	13	0.017	0.999	0.320	15	0.023	0.999
	8	0.158	0.993		12	0.042	0.998		14	0.054	0.998
	7	0.313	0.980		11	0.095	0.996		13	0.113	0.994
	6	0.546	0.948		10	0.192	0.989		12	0.213	0.987
	5	0.806	0.874		9	0.345	0.975		11	0.362	0.971
	4	0.969	0.735		8	0.552	0.944		10	0.552	0.943
	3	0.999	0.555		7	0.773	0.885		9	0.748	0.893
0.190	10	0.039	0.999		6	0.932	0.789		8	0.901	0.820
	9	0.094	0.996		5	0.992	0.666		7	0.977	0.727
	8	0.200	0.990	0.260	13	0.023	0.999		6	0.997	0.625
	7	0.378	0.973		12	0.056	0.998	0.340	16	0.016	0.999
	6	0.621	0.932		11	0.121	0.994		15	0.040	0.998
	5	0.862	0.845		10	0.233	0.986		14	0.086	0.996
	4	0.983	0.699		9	0.402	0.967		13	0.169	0.990
0.200	11	0.021	0.999		8	0.616	0.930		12	0.296	0.979
	10	0.054	0.998		7	0.823	0.864		11	0.468	0.957
	9	0.123	0.995		6	0.954	0.763		10	0.663	0.918
	8	0.249	0.985		5	0.995	0.641		9	0.836	0.858
	7	0.446	0.963	0.270	13	0.032	0.999		8	0.947	0.778
	6	0.693	0.913		12	0.073	0.997		7	0.990	0.685
	5	0.905	0.814		11	0.151	0.992		6	0.999	0.588
	4	0.991	0.665		10	0.279	0.981	0.360	16	0.029	0.999
0.210	11	0.030	0.999		9	0.462	0.959		15	0.065	0.997
	10	0.073	0.997		8	0.676	0.915		14	0.132	0.993
	9	0.157	0.992		7	0.866	0.841		13	0.240	0.984
	8	0.303	0.980		6	0.970	0.737		12	0.392	0.967
	7	0.515	0.952		5	0.997	0.617		11	0.578	0.937
	6	0.758	0.892	0.280	14	0.017	0.999		10	0.762	0.889
	5	0.938	0.782		13	0.042	0.998		9	0.902	0.821
	4	0.995	0.634		12	0.093	0.996		8	0.974	0.738
0.220	11	0.041	0.999		11	0.185	0.989		7	0.996	0.648
	10	0.095	0.996		10	0.329	0.976	0.380	17	0.020	0.999
	9	0.197	0.989		9	0.522	0.949		16	0.048	0.998
	8	0.361	0.974		8	0.733	0.898		15	0.101	0.995

TABLE 3 Finite Queuing Factors—Population 30 (*Continued*)

X	M	D	F	X	M	D	F	X	M	D	F
	14	0.191	0.988		16	0.310	0.977		22	0.038	0.998
	13	0.324	0.975		15	0.470	0.957		21	0.085	0.996
	12	0.496	0.952		14	0.643	0.926		20	0.167	0.990
	11	0.682	0.914		13	0.799	0.881		19	0.288	0.980
	10	0.843	0.857		12	0.910	0.826		18	0.443	0.963
	9	0.945	0.784		11	0.970	0.762		17	0.612	0.936
	8	0.988	0.701		10	0.993	0.694		16	0.766	0.899
	7	0.999	0.614		9	0.999	0.625		15	0.883	0.854
0.400	17	0.035	0.999	0.500	20	0.032	0.999		14	0.953	0.802
	16	0.076	0.996		19	0.072	0.997		13	0.985	0.746
	15	0.150	0.992		18	0.143	0.992		12	0.997	0.690
	14	0.264	0.982		17	0.252	0.983		11	0.999	0.632
	13	0.420	0.964		16	0.398	0.967	0.600	23	0.024	0.999
	12	0.601	0.933		15	0.568	0.941		22	0.059	0.997
	11	0.775	0.886		14	0.733	0.904		21	0.125	0.993
	10	0.903	0.823		13	0.865	0.854		20	0.230	0.986
	9	0.972	0.748		12	0.947	0.796		19	0.372	0.972
	8	0.995	0.666		11	0.985	0.732		18	0.538	0.949
0.420	18	0.024	0.999		10	0.997	0.667		17	0.702	0.918
	17	0.056	0.997	0.520	21	0.021	0.999		16	0.837	0.877
	16	0.116	0.994		20	0.051	0.998		15	0.927	0.829
	15	0.212	0.986		19	0.108	0.994		14	0.974	0.776
	14	0.350	0.972		18	0.200	0.988		13	0.993	0.722
	13	0.521	0.948		17	0.331	0.975		12	0.999	0.667
	12	0.700	0.910		16	0.493	0.954	0.700	25	0.039	0.998
	11	0.850	0.856		15	0.663	0.923		24	0.096	0.995
	10	0.945	0.789		14	0.811	0.880		23	0.196	0.989
	9	0.986	0.713		13	0.915	0.827		22	0.339	0.977
	8	0.998	0.635		12	0.971	0.767		21	0.511	0.958
0.440	19	0.017	0.999		11	0.993	0.705		20	0.681	0.930
	18	0.041	0.998		10	0.999	0.641		19	0.821	0.894
	17	0.087	0.996	0.540	21	0.035	0.999		18	0.916	0.853
	16	0.167	0.990		20	0.079	0.996		17	0.967	0.808
	15	0.288	0.979		19	0.155	0.991		16	0.990	0.762
	14	0.446	0.960		18	0.270	0.981		15	0.997	0.714
	13	0.623	0.929		17	0.421	0.965	0.800	27	0.053	0.998
	12	0.787	0.883		16	0.590	0.938		26	0.143	0.993
	11	0.906	0.824		15	0.750	0.901		25	0.292	0.984
	10	0.970	0.755		14	0.874	0.854		24	0.481	0.966
	9	0.994	0.681		13	0.949	0.799		23	0.670	0.941
	8	0.999	0.606		12	0.985	0.740		22	0.822	0.909
0.460	19	0.028	0.999		11	0.997	0.679		21	0.919	0.872
	18	0.064	0.997		10	0.999	0.617		20	0.970	0.832
	17	0.129	0.993	0.560	22	0.023	0.999		19	0.991	0.791
	16	0.232	0.985		21	0.056	0.997		18	0.998	0.750
	15	0.375	0.970		20	0.117	0.994	0.900	29	0.047	0.999
	14	0.545	0.944		19	0.215	0.986		28	0.200	0.992
	13	0.717	0.906		18	0.352	0.973		27	0.441	0.977
	12	0.857	0.855		17	0.516	0.952		26	0.683	0.953
	11	0.945	0.793		16	0.683	0.920		25	0.856	0.923
	10	0.985	0.724		15	0.824	0.878		24	0.947	0.888
	9	0.997	0.652		14	0.920	0.828		23	0.985	0.852
0.480	20	0.019	0.999		13	0.972	0.772		22	0.996	0.815
	19	0.046	0.998		12	0.993	0.714		21	0.999	0.778
	18	0.098	0.995		11	0.999	0.655	0.950	29	0.226	0.993
	17	0.184	0.989	0.580	23	0.014	0.999		28	0.574	0.973